Engineering Acoustics

Wiley Series in Acoustics, Noise and Vibration

Engineering Acoustics

Noise and Vibration Control

Malcolm J. Crocker
Auburn University
USA

Jorge P. Arenas
Institute of Acoustics, University Austral of Chile
Valdivia
Chile

Registered Offices
John Wiley & Sons, Inc., 111 River Street, Hoboken, NJ 07030, USA
John Wiley & Sons Ltd, The Atrium, Southern Gate, Chichester, West Sussex, PO19 8SQ, UK

Editorial Office
The Atrium, Southern Gate, Chichester, West Sussex, PO19 8SQ, UK

For details of our global editorial offices, customer services, and more information about Wiley products visit us at www.wiley.com.

Wiley also publishes its books in a variety of electronic formats and by print-on-demand. Some content that appears in standard print versions of this book may not be available in other formats.

Library of Congress Cataloging-in-Publication Data

Names: Crocker, Malcolm J., author. | Arenas, Jorge P., author.
Title: Engineering acoustics : noise and vibration control / Malcolm J
    Crocker, Auburn University, AL, US, Jorge P Arenas, Institute of
    Acoustics, Univ. Austral de Chile, Valdivia, Chile.
Description: First edition. | Hoboken, NJ : John Wiley & Sons, Inc., 2021.
    | Series: Wiley series in acoustics noise and vibration | Includes
    bibliographical references and index.
Identifiers: LCCN 2020010029 (print) | LCCN 2020010030 (ebook) | ISBN
    9781118496428 (hardback) | ISBN 9781118693896 (adobe pdf) | ISBN
    9781118693827 (epub)
Subjects: LCSH: Acoustical engineering. | Noise control. | Damping
    (Mechanics)
Classification: LCC TA365 .C76 2020 (print) | LCC TA365 (ebook) | DDC
    620.3/7–dc23
LC record available at https://lccn.loc.gov/2020010029
LC ebook record available at https://lccn.loc.gov/2020010030

Cover Design: Wiley
Cover Illustration: created by Malcolm Crocker

Set in 10/12.5pt STIXTwoText by SPi Global, Pondicherry, India
Printed and bound by CPI Group (UK) Ltd, Croydon, CR0 4YY

10   9   8   7   6   5   4   3   2   1

Dedicated to my wife Ruth

— Malcolm J. Crocker

Dedicated to my wife Ester

—Jorge P. Arenas

Contents

Series Preface

This book series will embrace a wide spectrum of acoustics, noise and vibration topics from theoretical foundations to real world applications. Individual volumes included will range from specialist works of science to advanced undergraduate and graduate student texts. Books in the series will review the scientific principles of acoustics, describe special research studies and discuss solutions for noise and vibration problems in communities, industry and transportation.

The first books in the series include those on biomedical ultrasound, effects of sound on people, engineering acoustics, noise and vibration control, environmental noise management, sound intensity and wind farm noise – books on a wide variety of related topics.

The books I have edited for Wiley, *Encyclopedia of Acoustics* (1997), *Handbook of Acoustics* (1998) and *Handbook of Noise and Vibration Control* (2007), included over 400 chapters written by different authors. Each author had to restrict the chapter length on their special topics to no more than about 10 pages. The books in the current series will allow authors to provide much more in-depth coverage of their topic.

The series will be of interest to senior undergraduate and graduate students, consultants, and researchers in acoustics, noise and vibration and, in particular, those involved in engineering and scientific fields, including aerospace, automotive, biomedical, civil/structural, electrical, environmental, industrial, materials, naval architecture, and mechanical systems. In addition, the books will be of interest to practitioners and researchers in fields such as audiology, architecture, the environment, physics, signal processing, and speech.

Malcolm J. Crocker
Series editor

Preface

Over the past decades, the authors of this book have been working together in several areas of acoustics and vibration, which has led to a number of joint research papers published in scholarly journals and congress proceedings, as well as book chapters. Our goal in writing this book was first to cover the fundamental theory relevant to engineering acoustics, noise, and vibration, and second to describe practical ways in which noise and vibration can be controlled and reduced. Each of the sixteen chapters has several worked examples designed to make the theoretical and empirical prediction methods accessible for readers. This book is aimed at senior undergraduates, graduate students, and practitioners in the noise and vibration fields. Although the use of SI units is emphasized in the book, English units are given in addition in some cases, in particular in Chapter 13, for the convenience of readers in the USA.

The book begins with fundamentals (Chapters 1–3) and continues with human aspects – hearing, speech, and the effects of noise and vibration on people (Chapters 4–6). At this point, two chapters are included on noise measurement (Chapters 7 and 8). Chapter 9 deals with principles of noise and vibration control. The remaining Chapters 10–16 deal with specific practical problems. These chapters include: acoustical design of reactive and passive mufflers and silencers (Chapter 10), control of noise and vibration of machines (Chapter 11), noise and vibration control in buildings (Chapter 12), noise and vibration of air-conditioning systems (Chapter 13), surface transportation noise (Chapter 14), aircraft and airport noise (Chapter 15), and community noise and vibration (Chapter 16).

The first author has five decades of experience in undergraduate and graduate teaching, research, and consulting in acoustics, noise, and vibration. The research was sponsored by companies and government agencies. The second author has over two decades of experience in undergraduate and graduate teaching, research, and consulting in acoustics, noise, and vibration and in performing research funded by government and private sources. He has also been a consultant with industry in noise and vibration.

Although our understanding of the acoustics, noise, and vibration fundamentals has remained largely unchanged, the last half century (1970–2020) has seen dramatic changes in our ability to make calculations, useful predictions, and measurements. Before the wide availability of electronic calculators (1975), most calculations were made using log tables and slide rules. Large, expensive computers did not appear until the 1960s and 1970s, and it was not until the early 1980s that personal computers and laptops became available. Now computers big and small are everywhere. Computational advances have revolutionized our ability to make acoustics, noise, and vibration calculations.

Because they cover important topics, Chapters 8, 10, 12, and 13 have received expanded treatment in this book. Contracts and grants received by the first author enabled him to conduct research on sound intensity in which he conceived the first use of sound intensity measurements in determining the transmission loss of partitions. During this period, he chaired the ANSI SI-12 committee which produced the first ever standard, S12-12-1992, Engineering Method for the Determination of the Sound Power Levels of Noise Sources Using Sound Intensity. Some of the research results are summarized in Chapter 8.

The first author received a nine-year contract from a company which produced over 80% of the new mufflers for automobiles manufactured in the USA in the 1960s and 1970s. During this research, he together with a graduate student, C.-I. J. Young, produced the first acoustical finite element model to predict muffler noise attenuation. This, together with research on measuring source impedance of engines and prediction of the transmission loss of concentric tube resonators with J.W. Sullivan, resulted in improved modeling of the acoustics of exhaust systems. These research results are included in Chapter 10. Intense noise was caused by the Saturn V moon rocket at launch. During the 1960s there was great concern about the vibration and potential for fatigue during the launch of the Saturn V vehicle. First during a 1966–1997 short course – Program for Advanced Study – and then through personal contacts and communications with Richard H. Lyon, the first author began studies on SEA. He and a colleague produced the first papers demonstrating the usefulness of SEA in predicting the transmission of sound through single and double partitions. These results and others are included in Section 12.5 of Chapter 12. The first author also received several research contracts from ASHRAE, AMCA, and ARI to study the acoustics of air-conditioning systems. These studies resulted in a new in-duct method of measuring the sound power of fans. During this period, the first author served as chair of the ASHRAE Technical Committee 2.6 Sound and Vibration. This experience and background information has been included in Chapter 13 in addition to some text and figures from an earlier chapter written for CRC Press by A.J. Price and M.J. Crocker.

Chapter 8 reviews the use of sound intensity measurements to determine the sound power of machinery, noise source identification, and the transmission loss of partitions. Accurate sound intensity measurements only became possible with the development of two-channel mini-computers, the fast-Fourier transform and the sound intensity algorithms. The use of sound intensity measurements allows the accurate determination of the sound power of machinery *in situ* even in the presence of extremely noisy ambient conditions. Sound intensity measurements can also be used to determine the transmission loss of walls without the need for a reception room. In addition, such measurements allow the transmission loss of different parts of a wall partition to be determined, which the conventional two-room method does not.

Chapter 10 deals with reactive and passive mufflers and silencers. Reactive mufflers are used on all automobiles and trucks, and proper acoustical design is of course most important. Although useful theory and measurements date back to the 1950s, it was not until finite element modeling was first used in 1971 on non-concentric reactive mufflers that the attenuation of such mufflers could be predicted accurately. This chapter also reviews and compares different attempts to predict the attenuation of passive mufflers used in industrial systems.

Chapter 12 discusses various aspects of sound and vibration in buildings. An important aspect of building acoustics is the unwanted transmission of sound from one room to an adjoining one. One measure of the effectiveness of a party wall to reduce this transmission is the sound reduction index, commonly called the transmission loss (TL). This chapter presents different theoretical models for TL and also measurements of TL for a wide variety of wall structures. A useful theoretical approach to determine the TL of single and multiple layer walls is statistical energy analysis (SEA), as described in Section 12.5 of Chapter 12.

Chapter 13 deals with the noise and vibration generated by air-conditioning (HVAC) systems in buildings. Although such systems are widely used they are often inadequately or incorrectly designed acoustically. Correct design at the beginning often costs a little more, but corrections made to such systems later, after they are installed, can be very expensive.

Malcolm J. Crocker
Jorge P. Arenas

30 September 2020

Acknowledgements

The authors are greatly indebted to many colleagues for their patience and considerable support with this project. The staff at John Wiley including Paul Petralia, Eric Willner, Anne Hunt, Lauren Poplawski, Becky Cowan and Karthika Sridharan have been most helpful. We were also supported by Margarita Maksotskaya at Auburn University, who provided splendid assistance in bringing this book project to a successful conclusion. Our colleagues are thanked for reading the book in manuscript. Pouria Oliazadeh read every chapter and provided helpful comments throughout. In addition, Gene Chung, Steven Hambric, Colin Hansen, David Herrin, Reginald Keith, Robin Langley, Florent Masson and Tomas Ulrich, kindly acted as reviewers of selected chapters with which they had expertise. The first author is grateful for the support of Richard H. Lyon, a brilliant acoustician and one of the originators and developers of SEA. The first author learned so much about the subject from him. Richard Lyon almost was an unofficial advisor for the first author's PhD dissertation and section 12.5 of Chapter 12 is dedicated to Dr. Lyon's memory. Last and not least, the authors thank their wives, Ruth Crocker and Ester Arteaga for their support, patience and understanding during the lengthy period of preparation of this book.

1

Introduction

1.1 Introduction

Real-world problems in the control of noise and vibration in aircraft, appliances, buildings, industry, and vehicles require the measurement of particular environmental parameters such as sound pressure, force, acceleration, velocity, displacement, etc. This process is often performed by using acoustical and vibration transducers. Vibration and acoustical sensors are transducers which convert a measured physical property (e.g. the vibration of a body or the propagation of a sound wave) into an electrical signal (voltage or charge). These electrical signals are often conditioned to provide signals suitable for the measurement devices. The signals are then amplified, attenuated, or transformed so that they can subsequently be analyzed and/or processed to provide the data of particular interest in the time domain and frequency domain. The information provided by these analyses is widely used to assess sources of noise and vibration, and design proper engineering control measures. For some cases, such as simple measurements of the A-weighted sound pressure level, only limited amounts of processing are needed. In other cases with more sophisticated measurements, special analysis and processing is required. Such examples include modal analysis, sound intensity, wavelet analysis, machinery condition monitoring, beamforming, and acoustical holography, with which quite complicated signal analysis and processing may be needed. Some years ago, almost all measurements were made with analog equipment. Many analog instruments are still in use around the world. However, by using analog-to-digital conversion, increasing use is now made of digital signal processing to extract the required data. This is done either in dedicated instruments or by transferring measurement results onto computers for later processing by software.

Real-time analysis in the frequency domain has many applications, including noise and vibration studies where the signal is nonstationary with time. Such applications include machinery vibration analysis, bearing noise, transient analysis, acoustic emission, speech analysis, music, and others. The goal of this chapter is to define the main types of signals used in noise, shock, and vibration control and also to serve as an introduction to signal analysis. The discussion in this chapter is kept mainly descriptive and those readers requiring a further mathematical discussion of signal analysis are referred to more detailed treatments available in several books [1–9].

1.2 Types of Noise and Vibration Signals

Depending on their time histories, noise and vibration signals can basically be divided into *stationary* and *nonstationary*. Examples of the various types of signals in the time and frequency domains are shown in Figure 1.1 [10].

Engineering Acoustics: Noise and Vibration Control, First Edition. Malcolm J. Crocker and Jorge P. Arenas.
© 2021 John Wiley & Sons Ltd. Published 2021 by John Wiley & Sons Ltd.

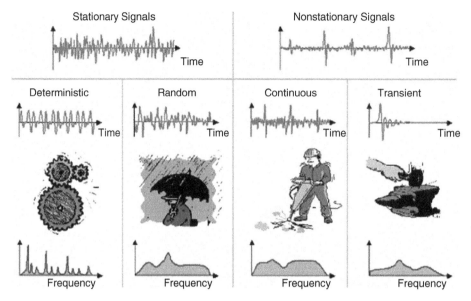

Figure 1.1 Examples of different types of signals and their spectral content [10].

1.2.1 Stationary Signals

Stationary signals can be divided into *deterministic* and *random* signals. Stationary deterministic signals can be described by a mathematical function. They are made of a combination of sinusoidal signals (pure tones) with different amplitudes and frequencies. The spectrum of a stationary deterministic signal is characterized by content (power) at discrete frequencies (a line spectrum). The measured displacement signal of a simple mass-spring system and the noise and vibration signals from machinery rotating at constant speeds are both examples of stationary deterministic signals.

Stationary random signals cannot be described by explicit mathematical functions but instead they must be described by their statistical properties (mean value, variance, standard deviation, crest factor, kurtosis, amplitude probability, etc.). In contrast to stationary deterministic signals, they have a continuous distribution of spectral content. If the random signal has constant statistical values which do not change with time, we refer to the random signal as stationary. The sound produced by rain or a waterfall, the noise produced by turbulent air coming out of a ventilation system, the noise produced by a passing vehicle, and the airborne noise of a circular saw during idle are examples of random signals. A random signal which has a flat (constant) spectral content over a wide frequency range is called *white noise*.

1.2.2 Nonstationary Signals

Nonstationary signals are divided into *transient* and *continuous* signals. Transient signals are signals which start and end at zero level and last a finite and relatively short amount of time. They are characterized by a certain amount of "energy" they contain in the same way that continuous signals are characterized by a "power" value. Examples of transient signals are the sound of a car door closing, a shock wave generated by an impact, the noise produced by a sheet metal stamping press, and the noise of an electric spark.

Nonstationary continuous signals are signals consisting of one or more of the following: sinusoidal components with variable amplitudes and/or frequencies, random signals with statistical properties which change with time, and transients which appear with varying intervals and with varying characteristics in time and frequency. Examples of nonstationary continuous signals are the acceleration on the chassis or frame of a truck driving on a rough road, the wind speed for wind-induced vibrations, the vibrations or chatter induced in machine tools during machining, the vibration produced by a jackhammer, and speech.

1.3 Frequency Analysis

In noise and vibration control, signal analysis means determining from a measurement or a set of measurements certain descriptive characteristics of the environment that will help in identifying the sources of the noise and vibration. Frequency analysis is probably the most widely used method for studying noise and vibration problems. The frequency content of a noise or vibration signal is usually related to a specific component of a given system, such as a machine, so that frequency analysis is often the key to obtain a better understanding of the causes or sources of the noise and vibration.

1.3.1 Fourier Series

Sometimes in acoustics and vibration we encounter signals which are pure tones (or very nearly so), e.g. the 120 Hz hum from an electric motor. In the case of a pure tone, the time history of the signal is simple harmonic and could be represented by the waveform $x(t) = A \sin(2\pi f_1 t)$ in Figure 1.2a. The pure tone can be represented in the frequency domain by a spike of height A at frequency f_1, Hz (see Figure 1.2a).

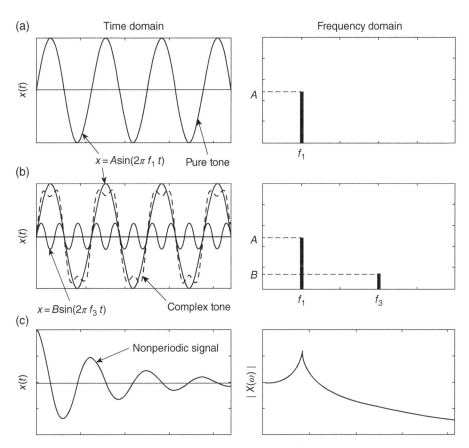

Figure 1.2 Time and frequency domain representations of (a) pure tone; (b) complex tone; and (c) nonperiodic deterministic signal.

More often, sound signals are encountered which are periodic, but not simple harmonic. These are known as complex tones. Such sound signals are produced by most musical instruments (both wind and string). They can also be produced mechanically or electronically (a square wave is an example of a periodic signal or complex tone). The broken line plotted in Figure 1.2b is an example of a complex tone which is made up by the superposition (addition) of two simple harmonic signals, $x(t) = A \sin (2\pi f_1 t) + B \sin (2\pi f_3 t)$. Note in this case we have chosen $f_3 = 3f_1$. The signal $A \sin (2\pi f_1 t)$ is known as the fundamental (or first harmonic) and $B \sin (2\pi f_3 t)$ is the third harmonic. In this particular case the second harmonic and the fourth and higher harmonics are completely absent from the complex tone $x(t)$. The frequency domain representation of the complex tone is also given in Figure 1.2b.

In fact, Fourier [11] showed in 1822 that any periodic signal may be analyzed as a combination of sinusoids:

$$x(t) = \frac{A_0}{2} + \sum_{n=1}^{\infty} (A_n \cos n\omega t + B_n \sin n\omega t), \tag{1.1}$$

or in complex notation:

$$x(t) = \sum_{n=-\infty}^{\infty} C_n e^{jn\omega t}, \tag{1.2}$$

where $\omega = 2\pi f$; f is the fundamental frequency; $T = 1/f = 2\pi/\omega$, is the period of the signal; $j = \sqrt{-1}$, and A_n and B_n are the Fourier coefficients calculated from [4, 6, 8]

$$A_n = \frac{2}{T} \int_0^T x(t) \cos (n\omega t) dt, \tag{1.3a}$$

$$B_n = \frac{2}{T} \int_0^T x(t) \sin (n\omega t) dt. \tag{1.3b}$$

The sine and cosine terms in Eq. (1.1) can have values of the subscript n equal to 1, 2, 3, ..., ∞. Hence, the signal $x(t)$ will be made up of a fundamental frequency ω and multiples, 2, 3, 4, ..., ∞ times greater. The $A_0/2$ term represents the D.C. (direct current) component (if present). The nth term of the Fourier series is called the nth harmonic of $x(t)$. The amplitude of the nth harmonic is

$$C_n = \sqrt{A_n^2 + B_n^2} \tag{1.4}$$

and its square, $C_n^2 = A_n^2 + B_n^2$, is sometimes called *energy* of the nth harmonic. Thus, the graph of the sequence C_n^2 is called the *energy spectrum* of $x(t)$ and shows the amplitudes of the harmonics.

Example 1.1 Buzz-saw noise is commonly generated by supersonic fans in modern turbofan aircraft engines. A buzzing sound can be represented by the periodic signal shown in Figure 1.3. Find the Fourier series and the energy spectrum for this signal.

Solution

We are required to represent $x(t) = At$ over the interval $0 \leq t \leq 1$, $T = 1$, and the fundamental frequency is $\omega = 2\pi/T = 2\pi$. Then, we determine the corresponding Fourier coefficients using Eqs. (1.3a) and (1.3b)

$$A_0 = \frac{2}{T} \int_0^T x(t) dt = 2 \int_0^1 At dt = A,$$

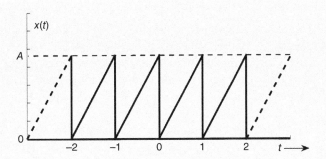

Figure 1.3 Periodic sound signal.

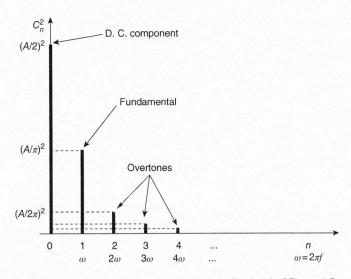

Figure 1.4 Frequency spectrum of the periodic signal of Figure 1.3.

$$A_n = \frac{2}{1}\int_0^1 At \cos{(2\pi nt)}dt = \frac{A}{2\pi^2 n^2}\left[\cos{(2\pi nt)} + 2\pi nt \sin{(2\pi nt)}\right]_{t=0}^{t=1} = 0, \quad \text{for } n \neq 0,$$

$$B_n = \frac{2}{1}\int_0^1 At \sin{(2\pi nt)}dt = \frac{A}{2\pi^2 n^2}\left[\sin{(2\pi nt)} - 2\pi nt \cos{(2\pi nt)}\right]_{t=0}^{t=1} = \frac{-A}{n\pi}.$$

Therefore, substituting for the Fourier coefficients in Eq. (1.1) we get

$$x(t) = \frac{A}{2} - \frac{A}{\pi}\sum_{n=1}^{\infty}\frac{\sin{(2\pi nt)}}{n}, \quad \text{for } 0 \leq t \leq 1.$$

The energy spectrum of $x(t)$ is shown in Figure 1.4. For this case the frequency spectrum is discrete, described by the Fourier coefficients (Eq. (1.4)). Note that for a Fourier series with only sine terms, as in Example 1.1, the amplitude of the nth harmonic is $C_n = |B_n|$. The energy spectrum has spikes at multiples of the fundamental frequency (harmonics) with a height equal to the value of B_n^2. Thus, the spectrum is a series of spikes at the frequencies $\omega = 2\pi f$, and overtones $2\omega = 2 \times 2\pi f$, $3\omega = 3 \times 2\pi f$, ... with amplitudes $(A/\pi)^2$, $(A/2\pi)^2$, $(A/3\pi)^2$, ... respectively.

1.3.2 Nonperiodic Functions and the Fourier Spectrum

Equation (1.1) is known as a Fourier series and can only be applied to periodic signals. Very often a sound signal is not a pure or a complex tone but is impulsive in time. Such a signal might be caused in practice by an impact, explosion, sonic boom, or the damped vibration of a mass-spring system (see Chapter 2 of this book) as shown in Figure 1.2c. Although we cannot find a Fourier series representation of the wave in Figure 1.2c because it is nonperiodic (it does not repeat itself), we can find a Fourier spectrum representation since it is a deterministic signal (i.e. it can be predicted in time). The mathematical arguments become more complicated [1, 7, 9] and will be omitted here. Briefly, the Fourier spectrum may be obtained by assuming that the period of the motion, T, becomes infinite. Then since $f = \omega/2\pi = 1/T$, the fundamental frequency approaches zero and Eq. (1.2) passes from a summation of harmonics to an integral:

$$x(t) = \frac{1}{2\pi} \int_{-\infty}^{\infty} X(\omega)e^{j\omega t}d\omega. \tag{1.5}$$

Just as C_n was complex in Eq. (1.2), $X(\omega)$ is complex in Eq. (1.5), having both a magnitude and a phase. The Fourier spectrum (magnitude), $|X(\omega)|$, of the wave in Figure 1.2c is plotted to the right of Figure 1.2c. $|X(\omega)|$ may be thought of as the amplitude of the time signal at each value of frequency ω.

1.3.3 Random Noise

So far we have discussed periodic and nonperiodic signals. In many practical cases the sound or vibration signal is not deterministic (i.e. it cannot be predicted) and it is random in time (see Figure 1.5). For a random signal, $x(t)$, mathematical descriptions become difficult since we have to use statistical theory [1, 7, 9]. Theoretically, for random signals the Fourier transform $X(\omega)$ does not exist unless we consider only a finite sample length of the random signal, for example, of duration τ in the range $0 < t < \tau$. Then the Fourier transform is

$$X(\omega, \tau) = \int_{0}^{\tau} x(t)e^{-j\omega t}dt, \tag{1.6}$$

where $X(\omega,\tau)$ is the finite Fourier transform of $x(t)$. Note that $X(\omega)$ is defined for both positive and negative frequencies. In the real world $x(t)$ must be a real function, which implies that the complex conjugate of X must satisfy $X(-\omega) = X^*(\omega)$; i.e. $X(\omega)$ exhibits conjugate symmetry. Finite Fourier transforms can easily be calculated with special analog-to-digital computers (see Section 1.5).

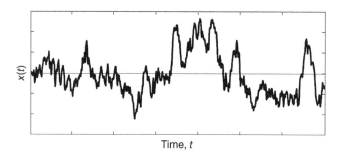

Figure 1.5 Random noise signal.

Example 1.2 An R–C (resistance–capacitance) series circuit is a classic first-order low-pass filter (see Section 1.4). Transient response describes how energy that is contained in a circuit will become dissipated when no input signal is applied. The transient response of an R–C series circuit (for $t > 0$) is given by

$$x(t) = \frac{1}{RC} e^{-t/RC}.$$

Find the Fourier spectrum representation of this transient response.

Solution

Substituting $x(t)$ into Eq. (1.6) we obtain

$$\boldsymbol{X}(\omega) = \frac{1}{RC} \int_0^\infty e^{-t/RC} e^{-j\omega t} dt = \frac{1}{RC} \int_0^\infty e^{-(1/RC + j\omega)t} dt = \frac{1}{RC} \left[\frac{e^{-(1/RC + j\omega)t}}{-(1/RC + j\omega)} \right]_{t=0}^{t=\infty}$$

$$= \frac{1/RC}{1/RC + j\omega}.$$

Therefore,

$$|\boldsymbol{X}(\omega)| = \frac{1/RC}{\sqrt{(1/RC)^2 + \omega^2}}.$$

The transient response and its Fourier spectrum are shown in Figure 1.6.

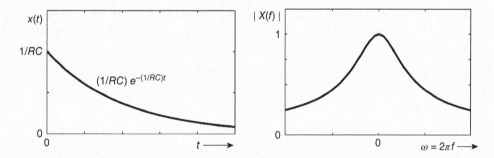

Figure 1.6 Time and frequency domain representations of the transient response of an R–C series circuit.

Example 1.3 The impulse response of a dynamic system is its output in response to a brief input pulse signal, called an impulse. The impulse response of the damped vibration of a one-degree-of-freedom mass-spring system of mass M, stiffness K, and coefficient of damping R (see Chapter 2 of this book) is given by

$$x(t) = Ae^{-\alpha t} \sin \lambda t,$$

where $A = (M\omega_d)^{-1}$, $\alpha = R/2M$ and $\lambda = \omega_d$ is known as the damped "natural" angular frequency. Find the Fourier spectrum representation of this impulse response.

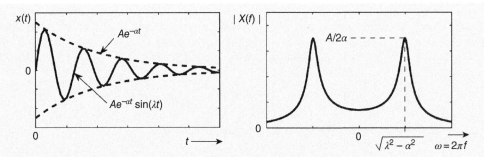

Figure 1.7 Time and frequency domain representations of the transient response of the impulse response of a damped vibration of a mass-spring system.

Solution

Using the mathematical property $e^{j\theta} = \cos\theta + j\sin\theta$, we can write $\sin\lambda t = \frac{e^{j\lambda t} - e^{-j\lambda t}}{2j}$. Then, Eq. (1.6) is

$$X(\omega) = \frac{A}{2j}\left[\int_0^\infty e^{-(\alpha - j\lambda + j\omega)t}dt - \int_0^\infty e^{-(\alpha + j\lambda + j\omega)t}dt\right]$$

$$= \frac{A}{2j}\left[\frac{1}{\alpha - j\lambda + j\omega} - \frac{1}{\alpha + j\lambda + j\omega}\right] = \frac{A\lambda}{[\alpha + j(\omega - \lambda)][\alpha + j(\omega + \lambda)]}$$

$$= \frac{A\lambda}{(\alpha + j\omega)^2 + \lambda^2}.$$

The impulse response and its Fourier spectrum are shown in Figure 1.7. We notice that replacing α and λ by the corresponding values in terms of the stiffness K, mass M, and damping constant R, of the damped mass-spring system, the Fourier spectrum becomes (compare with Eq. (2.18))

$$X(\omega) = \frac{1}{j\omega R + K - M\omega^2}.$$

1.3.4 Mean Square Values

In the case of the pure tone a useful quantity to determine is the mean square value, i.e. the time average of the signal squared $\langle x^2(t)\rangle_t$ [8]

$$\langle x^2(t)\rangle_t = \frac{1}{T}\int_0^T x^2(t)dt, \tag{1.7}$$

where $\langle\rangle_t$ denotes a time average.

For the pure tone in Figure 1.2a then we obtain

$$\langle x^2(t)\rangle_t = A^2/2, \tag{1.8}$$

where A is the signal amplitude.

The root mean square value is given by the square root of $\langle x^2(t) \rangle_t$ or

$$x_{rms} = A/\sqrt{2}. \tag{1.9}$$

For the general case of the complex pure tone in Eq. (1.1) or (1.2) we obtain:

$$\langle x^2(t) \rangle_t = \frac{A_0^2}{4} + \frac{1}{2} \sum_{n=1}^{\infty} (A_n^2 + B_n^2), \tag{1.10}$$

or

$$\langle x^2(t) \rangle_t = C_0^2 + \frac{1}{2} \sum_{n=1}^{\infty} |\mathbf{C}_n|^2, \tag{1.11}$$

since $|\mathbf{C}_n|^2 = (A_n^2 + B_n^2)$. The mean square value then is the sum of the squares of all the harmonic components of the wave weighted by a constant of 1/2.

Example 1.4   Determine the mean square and rms values of the signal in Figure 1.3.

Solution

We can use Eq. (1.7) to determine its mean square value,

$$\langle x^2(t) \rangle_t = \frac{1}{T} \int_0^T x^2(t) dt = \int_0^1 A^2 t^2 dt = A^2 \left[\frac{t^3}{3} \right]_{t=0}^{t=1} = \frac{A^2}{3}.$$

The same result is obtained from its Fourier series representation using Eq. (1.10):

$$\langle x^2(t) \rangle_t = \frac{A_0^2}{4} + \frac{1}{2} \sum_{n=1}^{\infty} (A_n^2 + B_n^2) = \frac{A^2}{4} + \frac{1}{2} \sum_{n=1}^{\infty} \left(0 + \frac{A^2}{n^2 \pi^2} \right) = \frac{A^2}{4} + \frac{A^2}{2\pi^2} \sum_{n=1}^{\infty} \frac{1}{n^2}.$$

$$\text{Since } \sum_{n=1}^{\infty} \frac{1}{n^2} = \frac{\pi^2}{6}, \text{ then } \langle x^2(t) \rangle_t = \frac{A^2}{4} + \frac{A^2}{12} = \frac{A^2}{3}.$$

Recalling that the root mean square value is given by the square root of the mean square value, the rms value of this saw tooth signal is $A/\sqrt{3}$. Note the difference between the rms value obtained in this example and that in Eq. (1.9) for a sine wave.

1.3.5   Energy and Power Spectral Densities

In the case of nonperiodic signals (see Section 1.3.2), a quantity called the *energy density function* or equivalently the *energy spectral density*, $S(f)$, is defined:

$$S(f) = |\mathbf{X}(f)|^2. \tag{1.12}$$

The energy spectral density $S(f)$ is the "energy" of the sound or vibration signal in a bandwidth of 1 Hz. Note that $S(\omega) = 2\pi S(f)$ where $S(\omega)$ is the "energy" in a 1 rad/s bandwidth. We use the term "energy" because if $x(t)$ were converted into a voltage signal, $S(\omega)$ would have the units of energy if the voltage were applied across a 1 Ω resistor. In the case of the pure tone, if $x(t)$ is assumed to be a voltage, then the mean square value in Eq. (1.8) represents the power in watts.

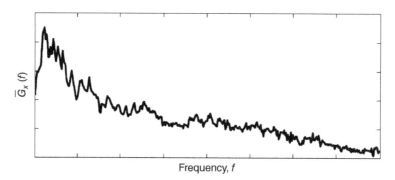

Figure 1.8 Power spectral density of random noise.

In the case of random sound or vibration signal we define a *power spectral density $G_x(f)$*. This may be derived through the filtering – squaring – averaging approach or the finite Fourier transform approach. We will consider both approaches in turn.

Suppose we filter the time signal through a filter of bandwidth Δf, then the mean square value

$$M_x^2(f, \Delta f) = \lim_{T \to \infty} \frac{1}{T} \int_0^T x^2(t, f, \Delta f) dt, \tag{1.13}$$

where $x(t,f,\Delta f)$ is the filtered frequency component of the signal after it is passed through a filter of bandwidth Δf centered on frequency f. In the practical case, the filter bandwidth, Δf, could be, for example, a one-third octave or smaller. The *power spectral density* is defined as:

$$G_x(f) = \lim_{\Delta f \to 0} \left[\frac{M_x^2(f, \Delta f)}{\Delta f} \right]. \tag{1.14}$$

The power spectral density may also be defined via the finite Fourier transform [12, 13]

$$G_x(f) = \lim_{T \to \infty} \frac{1}{T} |X(f)|^2. \tag{1.15}$$

Note the difference between Eqs. (1.15) and (1.12). We notice that G_x must be a *power* spectral density because of the division by time. The unit of power spectral density is U^2/Hz, where U is the unit of the measured signal. The square root of the power spectral density, often called the *rms spectral density*, has a unit of $U/\sqrt{Hz}$. For discussion in greater detail on analysis of random signals the reader is referred to References 1, 2, 7–9. Often in practice we define a new spectral density $\overline{G}_x(f) = 2G_x(f)$ which does not contain energy at negative frequencies but only exists in the region $0 < f < \infty$. The power spectral density of the random noise in Figure 1.5 is plotted in Figure 1.8.

1.4 Frequency Analysis Using Filters

Sound and vibration signals can be combined, but they can also be broken down into frequency components as shown by Fourier over 200 years ago. The ear seems to work as a frequency analyzer. We also can make instruments to analyze sound signals into frequency components.

In order to determine experimentally the contribution of the overall signal in some particular frequency band we *filter* the signal. Most of the important points concerning filters can be observed from a simple R–C passive circuit. By passive we mean that no electrical power is supplied. However, practical filters are usually made from more complicated R–C passive filters or active filters, involving R–C components, inductances, amplifiers, and a power supply. Modern digital signal processing can also be used to filter a signal. Digital filters are implemented through a series of digital operations (addition, multiplication, and time delay) on a digitized signal [3, 5, 14].

In a high-pass filter, only high-frequency signals are passed without attenuation. An ideal high-pass filter would pass no signal below the *cutoff* frequency and would have a vertical "skirt," which is impossible to achieve by use of a simple RC passive filter. Practical high-pass filters have more complicated circuits and normally incorporate inductances and/or active components. High-pass filters are often used when the displacement signal from a transducer is analyzed. This is because frequently the high-frequency displacement is of interest and large amplitude low-frequency signals must be filtered out to prevent the dynamic range of the instrumentation from being exceeded.

In a low-pass filter only low-frequency signals are passed without attenuation. In practice, the skirt of a simple RC filter is usually not steep enough and a more complicated circuit is also needed. In modern noise and vibration instrumentation, low-pass filters are often used as anti-aliasing filters at the input of an analog-to-digital converter (ADC) to suppress unwanted high-frequency signals. Aliasing errors occur when signals at high frequency (above the upper frequency limit) are interpreted as lower frequency signals below the upper frequency limit.

In a band-stop (or band-rejection) filter most frequencies are passed without attenuation; but those frequencies in a specific range are attenuated to very low levels. It is the opposite of a band-pass filter. Band-stop filters are commonly used as anti-hum filters and to remove specific interference frequencies in a complex signal.

By combining the simple high-pass and low-pass filters, a simple band-pass filter is produced. In a band-pass filter all simple harmonic frequency signals within a band are passed with little, if any, attenuation. All signals outside this band are attenuated. In practice it is usually desirable to make the skirts of such a filter steeper. This requires the use of additional electrical circuit elements and/or the use of active filters. Figure 1.9 illustrates the filtering process and shows only the magnitude of the signals in the frequency domain [10].

Band-pass filters are frequently used in acoustics and vibration. A frequency weighting network such as A-weighting is a special kind of band-pass filter. Frequency analysis using band-pass filters is commonly carried out using (i) constant frequency band filters and (ii) constant percentage filters. The following symbol notation is used in the characterization of a band-pass filter: f_L and f_U are the lower and upper cutoff frequencies, and f_C and Δf are the band center frequency and the frequency bandwidth, respectively. Thus $\Delta f = f_U - f_L$. See Figure 1.10.

In constant bandwidth filters, the bandwidth $\Delta f = f_U - f_L$ is kept constant regardless of the setting of the filter center frequency.

The constant percentage filter (usually one-octave or one-third-octave band types) most parallels the way the human auditory system analyzes sound and, although digital processing has mostly overtaken analog processing of signals, it is still frequently used. The human audible range is spanned by just a few octaves, so octave analysis produces a relatively coarse classification. Resolution can be improved by breaking each octave band into fractional-octave bands, which preserves the logarithmic band spacing. Thus, if the frequency scale is divided into contiguous frequency bands, the ratio f_U/f_L is the same for each band. The center frequency f_C for each band is defined as the geometric mean which is in the middle between f_L and f_U on a logarithmic frequency scale and is always less than the arithmetic average. Thus, the ratio of center frequencies of contiguous bands is the same as f_U/f_L for any one band. Octave and third-octave band filters are widely used, in particular for acoustical measurements [12].

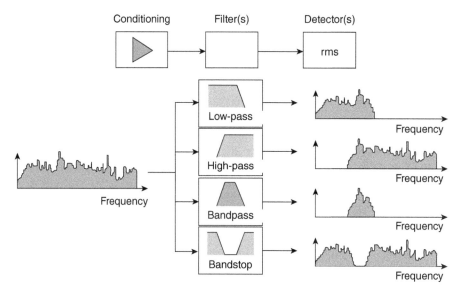

Figure 1.9 Different types of filter. Low-pass, high-pass, band-pass, and band-stop [10].

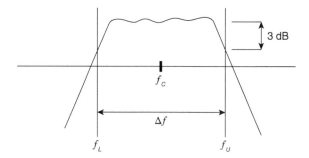

Figure 1.10 Typical frequency response of a filter of center frequency f_C and upper and lower cutoff frequencies, f_U and f_L.

a) One-Octave Bands

For one-octave bands, the cutoff frequencies f_L and f_U are defined as follows:

$$f_L = f_C/\sqrt{2},$$

$$f_U = \sqrt{2}f_C.$$

The center frequency (or geometric mean) is $f_C = \sqrt{f_L f_U}$. Thus $f_U/f_L = 2$. The bandwidth Δf is given by

$$\Delta f = f_U - f_L = f_C\left(\sqrt{2} - 1/\sqrt{2}\right) = f_C/\sqrt{2},$$

so $\Delta f \approx 70\% \; (f_C)$.

b) One-Third-Octave Bands

For one-third-octave bands, the cutoff frequencies, f_L and f_U, are defined as follows:

$$f_L = f_C/\sqrt[6]{2} = f_C/2^{1/6},$$

$$f_U = f_C 2^{1/6}.$$

The center frequency (geometric mean) is given by $f_C = \sqrt{f_L f_U}$. Thus $f_U/f_L = 2^{1/3}$. The bandwidth Δf is given by

$$\Delta f = f_U - f_L = f_C\left(2^{1/6} - 2^{-1/6}\right),$$

so $\Delta f \approx 23\%\ (f_C)$.

We thus see clearly why the filter bands we have just discussed are called constant percentage. This is because the bandwidth is a constant percentage of the filter center frequency f_C. Of course, the filter bandwidth does not have to be defined in terms of a fraction of an octave but can be defined simply in terms of the percentage of the center frequency. Since it is impracticable to make measurements at a large number of fixed frequencies, noise measurements are made at a selected number of standardized frequencies called *preferred center frequencies*. Standard one-octave and one-third-octave band specifications take advantage of the fact that $2^{10/3} \approx 10$. The ISO recommendation values for preferred center frequencies are given in Table 1.1 [15]. Specifications for octave-band and fractional-octave-band filters are defined by the IEC 1260:1995 and the ANSI S1.11:2004 standards [16, 17].

Note that:

1) The center frequencies of one-octave bands are related by 2, and 10 frequency bands are used to cover the human hearing range. They have center frequencies of 31.5, 63, 125, 250, 500, 1000, 2000, 4000, 8000, 16 000 Hz.

Table 1.1 Preferred center frequencies for noise measurements according to ISO R 266 [15].

Preferred frequencies, Hz	1/1oct.	1/3oct.	Preferred frequencies, Hz	1/1oct.	1/3oct.	Preferred frequencies, Hz	1/1oct.	1/3oct.
16	×	×	200		×	2500		×
20		×	250	×	×	3150		×
25		×	315		×	4000	×	×
31.5	×	×	400		×	5000		×
40		×	500	×	×	6300		×
50		×	630		×	8000	×	×
63	×	×	800		×	10 000		×
80		×	1000	×	×	12 500		×
100		×	1250		×	16 000	×	×
125	×	×	1600		×			
160		×	2000	×	×			

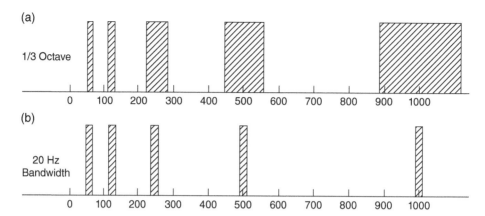

Figure 1.11 Comparison between bandwidths of (a) constant percentage and (b) constant bandwidth filters at the same frequency.

2) The center frequencies of one-third octave bands are related by $2^{1/3}$ and 10 cover a decade of frequency, and thus 30 frequency bands are used to cover the human hearing range: 20, 25, 31.5, 40, 50, 63, 80, 100, 125, 160, ..., 16 000 Hz.

Thus we note that at high frequencies a constant bandwidth analyzer will obviously have a narrower bandwidth than the constant percentage filter. Figure 1.11 shows a comparison of the bandwidths of constant percentage and constant band filters at the same frequencies.

Example 1.5 Determine the percentage bandwidth of the center frequency of a 1/12-octave band filter.

Solution

1/12-octave filters are obtained by dividing each one-octave band into 12 geometrically equal sub-sections, i.e. $f_U/f_L = 2^{1/12}$. By the same procedure as for octave and one-third-octave filters, we get the result that for 1/12-octave bands the cutoff frequencies, f_L and f_U, are $f_C \times 2^{-1/24}$ and $f_C \times 2^{1/24}$, respectively. Then, the bandwidth is given by $\Delta f = f_U - f_L = f_C(2^{1/24} - 2^{-1/24})$, so $\Delta f \approx 6\% \ (f_C)$.

There are two main types of constant percentage filters in common use: (i) those with a fixed center frequency and bandwidth which is a certain percentage of the center frequency, and (ii) those with a variable (or tunable) center frequency and a bandwidth which can be set to certain selected percentages of the center frequency. The first type of filter is perhaps in most common use. In practical instruments, many different parallel filters each with a different center frequency are assembled in one unit. The instrument is provided with a root mean square detector and a display.

On the other hand, instruments for constant bandwidth filter analysis are normally constructed so that the center frequency of a single filter can be tracked effectively throughout the frequency range of interest. Often different bandwidth settings are available on the same instrument (e.g. 1, 5, 10, 20 Hz). The narrower the bandwidth chosen, the slower the tracking rate should be to obtain reliable results. A rule which should be used in spectrum analysis is that the duration, T, of the noise sample length (or of the analysis time) must be at least as long as the reciprocal of the bandwidth Δf,

$$T\Delta f \geq 1. \tag{1.16}$$

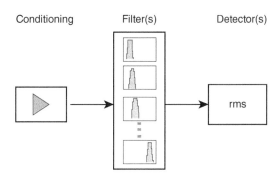

Figure 1.12 Simplified block diagram of a parallel-filter real-time analyzer [10].

This fundamental principle, also known as the *uncertainty principle*, puts a limit on the corresponding resolutions in the time and frequency domain, meaning that narrow resolution in one domain means wide resolution in the other domain [10].

If fine frequency resolution is not needed or if the signal is broadband in nature, then octave band readings are sufficient. One-third octave band readings (or narrower) should be used if the signal spectrum is not smooth or if pure tones are present. For diagnostic work on machinery, it may be necessary to use constant bandwidth filters (e.g. if the fan blade passing frequency and its higher harmonics must be separated).

When changes in level and frequency of a signal occur in a short period of time, *real-time frequency analysis* is required to observe rapid variations in the signal and showing the results on a continuously updated display. Real-time analysis can be performed using a frequency analyzer made up of a set of parallel filters and a detector (see Figure 1.12). The input signal is previously conditioned in terms of level (gain/attenuation) and high- and/or low-pass filtering. A digital analyzer will require an anti-aliasing filter at the input before the ADC. The conditioner is then connected to a large number of parallel band-pass filter channels (usually between 30 and 40 for a standard one-third octave band model). The detector detects the power in the transmitted signal in terms of its mean square or rms value. In Figure 1.12, only one detector is shown, and this is supposed to work as detector for all the filters in the situation of a parallel filter bank [10].

1.5 Fast Fourier Transform Analysis

Although equipment for analyzing noise and vibration signals with constant frequency bandwidth filters and constant percentage bandwidth (CPB) filters are still available, these instruments have largely been replaced by Fast Fourier Transform (FFT) Digital Fourier analyzers. These types of analyzers give similar results in a fraction of the time and at a lower cost. These Fourier analyzers are manufactured by several companies and make use of the FFT algorithm which was published by Cooley and Tukey in l965 [18]. This algorithm is much faster than conventional digital Fourier transform algorithms and has made the digital FFT analyzer a useful, efficient means of signal analysis [19].

In digital FFT analyzers, the analog signal needs to be converted to digital. The signal is not only sampled discretely in time but quantized as well with discrete amplitude values. Because of sampling, the digital system is limited in frequency, which may cause some problems, such as aliasing. During the sampling process, the amplitude of each sample is represented and stored as finite binary numbers in a computer. Rounding errors produce a signal-correlated noise, called *quantization noise*, which is another disadvantage of the digital system. So the sampling rate and resolution are two basic considerations of the ADC [20].

FFT analyzers operate on discrete blocks of data where each sample block is captured then analyzed while the next block is being captured, and so on. Unlike equipment which works on purely analog or hybrid analog/digital principles, the digital FFT analyzer makes all the analysis digitally. Since all the analysis results are in digital form, numerous calculations can be performed by such analyzers. The Fourier transform $X(f)$ of a time signal $x(t)$ may be calculated (see Eq. (1.6)). In addition, the auto-power spectral density $G_x(f)$ (Eq. (1.15)) may be calculated. A dual-channel FFT analyzer is able to sample two input signals simultaneously and compute several joint functions. This type of analyzer is widely used in modal testing, electroacoustics, and vibroacoustics applications. If two signals $x_1(t), x_2(t)$ are fed into the computer at once, then the cross-spectral density $G_{12}(f)$ can be calculated [12, 13],

$$\mathbf{G}_{12}(f) = \frac{1}{T}\left[\mathbf{X}_1(f)\mathbf{X}_2^*(f)\right], \tag{1.17}$$

where the asterisk denotes the complex conjugate.

In general $\mathbf{G}_{12}(f)$ is a complex quantity having both an amplitude and a phase. The phase is the relative phase between the two signals. The real and imaginary parts of the cross-spectrum are referred to as the co-spectrum and quad-spectrum, respectively. The auto-power spectral densities

$$G_{11}(f) = \frac{1}{T}\left[\mathbf{X}_1(f)\mathbf{X}_1^*(f)\right], \tag{1.18}$$

$$G_{22}(f) = \frac{1}{T}\left[\mathbf{X}_2(f)\mathbf{X}_2^*(f)\right], \tag{1.19}$$

are real quantities.

If $x_1(t)$ were an input (for example, a measured force) and $x_2(t)$ an output (for example, a measured displacement), then the *transfer function* $\mathbf{H}_{12}(f)$ can be computed [5, 13],

$$\mathbf{H}_{12}(f) = \frac{\mathbf{X}_2(f)}{\mathbf{X}_1(f)}. \tag{1.20}$$

The transfer function $\mathbf{H}_{12}(f)$ is a complex quantity because it will have amplitude and phase. In addition, the *coherence function* (also called coherency squared) between the input and output may be calculated, defined by [12, 13]

$$\gamma_{12}^2 = \frac{|\mathbf{G}_{12}(f)|^2}{G_{11}(f) \times G_{22}(f)}. \tag{1.21}$$

The coherence function varies between 0 and 1. If the coherence is 0 then the input and output are completely random with respect to each other. On the contrary, if the coherence is one, all the power of the output signal is due to the input signal, indicating a completely linearly dependence between the two signals. In cases where there are multiple inputs and a single output (e.g. several cylinders [inputs] on a diesel engine), but one microphone position (output), the situation becomes more complicated. However, in such cases the coherence function may be used to estimate the contribution to the output from each input. Main applications of the coherence function are in checking the validity of frequency response measurements and the calculation of the signal, S, to noise, N, ratio as a function of frequency [13]

$$S/N = \frac{\gamma^2(f)}{1 - \gamma^2(f)}. \tag{1.22}$$

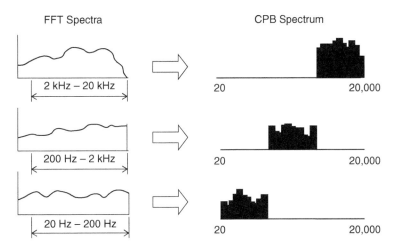

Figure 1.13 Conversion from FFT spectra to a constant percentage bandwidth (CPB) spectrum [21].

Measurement of the coherence function, transfer function, cross- and auto-power spectral densities have been successfully used to identify sources and predict noise levels in machinery such as diesel engines, punch presses, and other noise problems. The necessary theory is dealt with in detail in references [1, 2, 12] and in the application manuals supplied by the instrument manufacturers.

The FFT analyzer can also be used as a real-time analyzer up to a certain frequency limit (which depends on the accuracy desired). CPB (such as one-octave and one-third octave band) analysis can also be approximately performed from FFT spectra, as illustrated in Figure 1.13, where each decade is converted separately. The bandwidth of the individual lines in the original FFT spectra must be less than the percentage bandwidth being converted to at the lowest frequency in the FFT band. The conversion is achieved by calculating the lower and upper cutoff frequencies of each constant percentage band, and then integrating up the power in the FFT lines (and parts of lines) between the limits [21]. Care must be taken at low frequencies since just a few frequency samples may be available to calculate the band value.

With the development of and decreasing prices of computer technologies, traditional FFT-based instruments are becoming replaced by computer-based instruments because of the processing, interfacing, and networking power of computers. Thus, digital signal processors can be replaced by specific software running on a personal computer. Computation speed reduction is not a problem because normally, calculations are made faster than they can be displayed. For data acquisition, since computers have hard drives of huge data storage capabilities, computers have advantages over traditional data logging equipment [20]. In addition, users can now develop their own programs to design instruments for noise and vibration signal analysis.

References

1 Piersol, A.G. and Bendat, J.S. (2010). *Random Data: Analysis and Measurement Procedures*, 4e. Hoboken, NJ: Wiley.

2 Piersol, A.G. and Bendat, J.S. (1993). *Engineering Applications of Correlation and Spectral Analysis*, 2e. New York: Wiley.

3 Oppenheim, A.V., Willsky, A.S., and Hamid, S. (1997). *Signals and Systems*, 2e. Harlow: Pearson.

4 Bracewell, R.N. (1999). *The Fourier Transform and Its Applications*, 3e. New York: McGraw-Hill.

5 Tohyama, M. and Koike, T. (1998). *Fundamentals of Acoustic Signal Processing*. London: Academic Press.

6 Hansen, E.W. (2014). *Fourier Transforms: Principles and Applications*. New York: Wiley.

7 Newland, D.E. (2005). *An Introduction to Random Vibrations, Spectral & Wavelet Analysis*, 3e. Mineola, NY: Dover.

8 Lathi, B.P. and Ding, Z. (2009). *Modern Digital and Analog Communication Systems*, 4e. Oxford: Oxford University Press.

9 Magrab, E.B. and Blomquist, D.S. (1971). *Measurement of Time-Varying Phenomena: Fundamentals and Applications*. New York: Interscience.

10 Herlufsen, H., Gade, S., and Zaveri, H.K. (2007). Analyzers and signal generators. In: *Handbook of Noise and Vibration Control* (ed. M.J. Crocker), 470–485. New York: Wiley.

11 Fourier, J. (1822). *Théorie analytique de la chaleur*. Paris: Firmin Didot Père et Fils.

12 Randall, R.B. (1987). *Frequency Analysis*, 3e. Naerum, Denmark: Bruel & Kjaer.

13 Piersol, A.G. (2007). Signal Analysis. In: *Handbook of Noise and Vibration Control* (ed. M.J. Crocker), 493–500. New York: Wiley.

14 Oppenheim, A.V. and Schafer, R.W. (2009). *Discrete-Time Signal Processing*, 3e. Upper Saddle River, NJ: Prentice-Hall.

15 ISO R 266:1997 (1997) *Acoustics – Preferred Frequencies*. Geneva: International Standards Organization.

16 IEC 1260:1995-07 (1995) *Electroacoustics – Octave-band and Fractional-octave-band Filters, Class 1*. Geneva: International Electrotechnical Commission.

17 ANSI S1.11-2004 (2004) *Specification for Octave-band and Fractional-octave-band Analog and Digital Filters, Class 1*. New York: American National Standards Institute.

18 Cooley, J.W. and Tukey, J.W. (1965). An algorithm for the machine computation of the complex Fourier series. *Math. Comput.* 19 (90): 297–301.

19 Duhamel, P. and Vetterli, M. (1990). Fast Fourier Transforms: a tutorial review and a state of the art. *Signal Process.* 19: 259–299.

20 Li, Z. and Crocker, M.J. (2007). Equipment for data acquisition. In: *Handbook of Noise and Vibration Control* (ed. M.J. Crocker), 486–492. New York: Wiley.

21 Randall, R.B. (2007). Noise and vibration data analysis. In: *Handbook of Noise and Vibration Control* (ed. M.J. Crocker), 549–564. New York: Wiley.

2

Vibration of Simple and Continuous Systems

2.1 Introduction

The vibrations in machines and structures result in oscillatory motion that propagates in air and/or water and that is known as sound. The simplest type of oscillation in vibration and sound phenomena is known as simple harmonic motion, which can be shown to be sinusoidal in time. Simple harmonic motion is of academic interest because it is easy to treat and manipulate mathematically; but it is also of practical interest. Most musical instruments make tones that are approximately periodic and simple harmonic in nature. Some machines (such as electric motors, fans, gears, etc.) vibrate and make sounds that have pure tone components. Musical instruments and machines normally produce several pure tones simultaneously. The simplest vibration to analyze is that of a mass–spring–damper system. This elementary system is a useful model for the study of many simple vibration problems. In this chapter we will discuss some simple theory that is useful in the control of noise and vibration. For more extensive discussions on sound and vibration fundamentals, the reader is referred to more detailed treatments available in several books [1–7]. We start off by discussing simple harmonic motion. This is because very often oscillatory motion, whether it be the vibration of a body or the propagation of a sound wave, is like this idealized case. Next, we introduce the ideas of period, frequency, phase, displacement, velocity, and acceleration. Then we study free and forced vibration of a simple mass–spring system and the influence of damping forces on the system. In Section 2.4 we discuss the vibration of systems of several degrees of freedom and Section 2.5 describes the vibration of continuous systems. This chapter also serves as an introduction to some of the topics that follow in this book.

2.2 Simple Harmonic Motion

The motion of vibrating systems such as parts of machines, and the variation of sound pressure with time is often said to be simple harmonic. Let us examine what is meant by simple harmonic motion.

Suppose a point P is revolving around an origin O with a constant angular velocity ω, as shown in Figure 2.1.

If the vector OP is aligned in the direction OX when time $t = 0$, then after t seconds the angle between OP and OX is ωt. Suppose OP has a length A, then the projection on the X-axis is $A\cos(\omega t)$ and on the Y-axis, $A\sin(\omega t)$. The variation of the projected length on either the X-axis or the Y-axis with time is said to represent simple harmonic motion.

It is easy to generate a displacement vs. time plot with this model, as is shown in Figure 2.2. The projections on the X-axis and Y-axis are as before. If we move the circle to the right at a constant speed, then the point P

Engineering Acoustics: Noise and Vibration Control, First Edition. Malcolm J. Crocker and Jorge P. Arenas.
© 2021 John Wiley & Sons Ltd. Published 2021 by John Wiley & Sons Ltd.

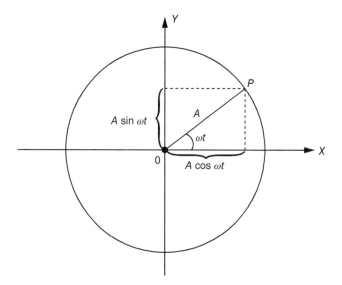

Figure 2.1 Representation of simple harmonic motion by projection of the rotating vector A on the *X*- or *Y*-axis.

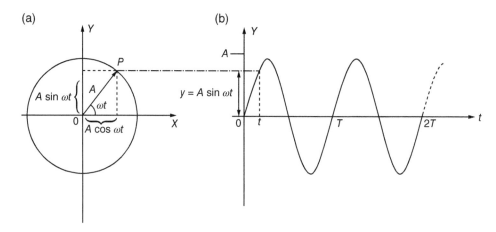

Figure 2.2 Simple harmonic motion.

traces out a curve $y = A \sin(\omega t)$, horizontally. If we move the circle vertically upwards at the same speed, then the point P would trace out a curve $x = A \cos(\omega t)$, vertically.

2.2.1 Period, Frequency, and Phase

The motion is seen to repeat itself every time the vector OP rotates once (in Figure 2.1) or after time T seconds (in Figures 2.2 and 2.3). When the motion has repeated itself, the displacement y is said to have gone through one *cycle*. The number of cycles that occur per second is called the *frequency f*. Frequency may be expressed in cycles per second or, equivalently in hertz, or as abbreviated, Hz. The use of hertz or Hz is preferable because this has become internationally agreed upon as the unit of frequency. (Note cycles per second = hertz.) Thus

$$f = 1/T \text{ hertz.} \tag{2.1}$$

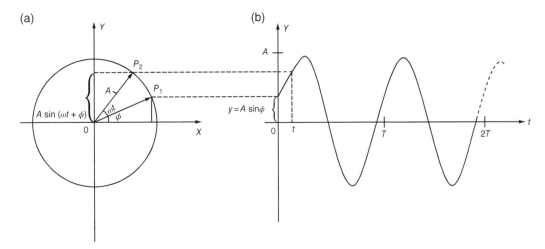

Figure 2.3 Simple harmonic motion with initial phase angle ϕ.

The time T is known as the *period* and is usually measured in seconds. From Figures 2.2 and 2.3, we see that the motion repeats itself every time ωt increases by 2π, since $\sin(0) = \sin(2\pi) = \sin(4\pi) = 0$, and so on. Thus $\omega T = 2\pi$ and from Eq. (2.1),

$$\omega = 2\pi f. \tag{2.2}$$

The angular frequency, ω, is expressed in radians per second (rad/s).

The motion described by the displacement y in Figure 2.2 or the projection OP on the X- or Y-axes in Figure 2.2 is said to be *simple harmonic*. We must now discuss something called the *initial phase angle*, which is sometimes just called *phase*. For the case we have chosen in Figure 2.2, the phase angle is zero. If, instead, we start counting time from when the vector points in the direction OP_1, as shown in Figure 2.3, and we let the angle $XOP_1 = \phi$, this is equivalent to moving the time origin t seconds to the right in Figure 2.2. Time is started when P is at P_1 and thus the initial displacement is $A\sin(\phi)$. The *initial phase angle* is ϕ. After time t, P_1 has moved to P_2 and the displacement

$$y = A \sin(\omega t + \phi). \tag{2.3}$$

If the initial phase angle $\phi = 0°$, then $y = A \sin(\omega t)$; if the phase angle $\phi = 90°$, then $y = A \sin(\omega t + \pi/2) = A \cos(\omega t)$. For mathematical convenience, complex exponential notation is often used. If the displacement is written as

$$y = A e^{j\omega t}, \tag{2.3a}$$

and we remember that $A e^{j\omega t} = A[\cos(\omega t) + j \sin(\omega t)]$, we see in Figure 2.1 that the real part of Eq. (2.3a) is represented by the projection of the point P onto the x-axis, $A \cos(\omega t)$, and of the point P onto the Y- (or imaginary axis), $A \sin(\omega t)$. Simple harmonic motion, then, is often written as the real part of $A e^{j\omega t}$, or in the more general form $y = A e^{j(\omega t + \phi)}$. If the constant A is made complex, then the displacement can be written as the real part of $y = A e^{j\omega t}$, where $A = A e^{j\phi}$.

2.2.2 Velocity and Acceleration

So far we have examined the displacement y of a point. Note that, when the displacement is in the OY direction, we say it is positive; when it is in the opposite direction to OY, we say it is negative. Displacement, velocity, and acceleration are really vector quantities in mathematics; that is, they have magnitude and direction.

The velocity v of a point is the rate of change of position with time of the point x in m/s. The acceleration a is the rate of change of velocity with time. Thus, using simple calculus:

$$v = \frac{dy}{dt} = \frac{d}{dt}[A \sin (\omega t + \phi)] = A\omega \cos (\omega t + \phi) \tag{2.4}$$

and

$$a = \frac{dv}{dt} = \frac{d}{dt}[A\omega \cos (\omega t + \phi)] = -A\omega^2 \sin (\omega t + \phi). \tag{2.5}$$

Equations (2.3)–(2.5) are plotted in Figure 2.4.

Note, by trigonometric manipulation we can rewrite Eqs. (2.4) and (2.5) as (2.6) and (2.7):

$$v = A\omega \cos (\omega t + \phi) = A\omega \sin \left(\omega t + \frac{\pi}{2} + \phi\right) \tag{2.6}$$

and

$$a = -A\omega^2 \sin (\omega t + \phi) = +A\omega^2 \sin (\omega t + \pi + \phi) \tag{2.7}$$

and from Eq. (2.3) we see that $a = -\omega^2 y$.

Equations (2.3), (2.6), and (2.7) tell us that for simple harmonic motion the *amplitude* of the velocity is ω or $2\pi f$ greater than the *amplitude of* the displacement, while the *amplitude* of the acceleration is ω^2 or $(2\pi f)^2$ greater. The *phase* of the velocity is $\pi/2$ or 90° ahead of the displacement, while the acceleration is π or 180° ahead of the displacement.

Note we could have come to the same conclusions and much more quickly if we had used the complex exponential notation. Writing

$$y = \boldsymbol{A}e^{j\omega t}$$

then

$$v = \boldsymbol{A}j\omega e^{j\omega t} = j\omega y$$

and

$$a = \boldsymbol{A}(j)^2\omega^2 e^{j\omega t} = -\boldsymbol{A}\omega^2 e^{j\omega t} = -\omega^2 y.$$

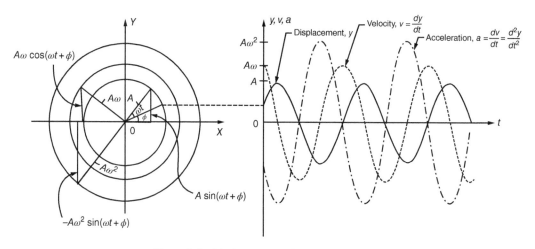

Figure 2.4 Displacement, velocity, and acceleration.

Example 2.1 In a simple harmonic motion of frequency 10 Hz, the displacement amplitude is 2 mm. Calculate the maximum velocity amplitude and maximum acceleration amplitude.

Solution

Since $\omega = 2\pi f = 2\pi(10) = 62.83$ rad/s. The velocity amplitude is calculated as
$v = \omega \times 2 = 62.83 \times 2 = 125.7$ mm/s and the acceleration amplitude is $a = \omega^2 \times 2 = (62.83)^2 \times 2 = 7896$ mm/s^2.

2.3 Vibrating Systems

2.3.1 Mass–Spring System

a) Free Vibration – Undamped

Suppose a mass of M kilogram is placed on a spring of stiffness K newton-metre (see Figure 2.5a), and the mass is allowed to sink down a distance d metres to its equilibrium position under its own weight Mg newtons, where g is the acceleration of gravity 9.81 m/s^2. Taking forces and deflections to be positive upward gives

$$-Mg = -Kd. \tag{2.8}$$

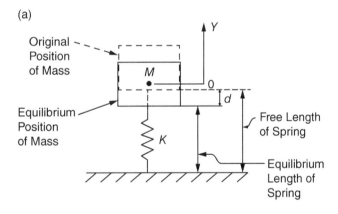

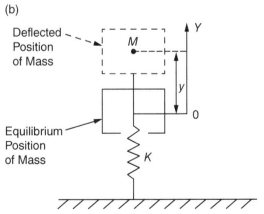

Figure 2.5 Movement of mass on a spring: (a) static deflection due to gravity and (b) oscillation due to initial displacement y_0.

Thus the static deflection d of the mass is

$$d = Mg/K. \tag{2.8a}$$

The distance d is normally called the *static deflection* of the mass; we define a new displacement coordinate system, where $Y = 0$ is the location of the mass after the gravity force is allowed to compress the spring.

Suppose now we displace the mass a distance y from its equilibrium position and release it; then it will oscillate about this position. We will measure the deflection from the equilibrium position of the mass (see Figure 2.5b). Newton's law states that force is equal to mass × acceleration. Forces and deflections are again assumed positive upward, and thus

$$-Ky = M\frac{d^2y}{dt^2}. \tag{2.9}$$

Let us assume a solution to Eq. (2.9) of the form $y = A\sin(\omega t + \phi)$. Then upon substitution into Eq. (2.9) we obtain

$$-KA\sin(\omega t + \phi) = M\left[-A\omega^2 \sin(\omega t + \phi)\right].$$

We see our solution satisfies Eq. (2.9) only if

$$\omega^2 = K/M.$$

The system vibrates with free vibration at an angular frequency ω rad/s. This frequency, ω, which is generally known as the *natural* angular frequency, depends only on the stiffness K and mass M. We normally signify this so-called natural frequency with the subscript n. And so

$$\omega_n = \sqrt{K/M};$$

and from Eq. (3.2)

$$f_n = \frac{1}{2\pi}\sqrt{\frac{K}{M}}\ \text{Hz}. \tag{2.10}$$

The frequency, f_n hertz, is known as the *natural frequency* of the mass on the spring. This result, Eq. (2.10), looks physically correct since if K increases (with M constant), f_n increases. If M increases with K constant, f_n decreases. These are the results we also find in practice.

Example 2.2   A machine of mass 600 kg is mounted on four springs of stiffness 2×10^5 N/m each. Determine the natural frequency of the system

Solution

We model the system as a hanging spring-mass system (see Figure 2.5). Equation (2.9) governs the displacement of the machine from its static-equilibrium position. Since we have four equal springs, the equivalent stiffness is $4 \times 2 \times 10^5 = 8 \times 10^5$ N/m. The natural frequency is then determined using Eq. (2.10) as

$$f_n = \frac{1}{2\pi}\sqrt{\frac{8 \times 10^5}{600}} = 5.81\ \text{Hz}.$$

We have seen that a solution to Eq. (2.9) is $y = A\sin(\omega t + \varphi)$ or the same as Eq. (2.3). Hence we know that *any system that has a restoring force that is proportional to the displacement* will have a displacement that is *simple harmonic*. This is an alternative definition to that given in Section 2.2 for *simple harmonic motion*.

b) Free Vibration – Damped

Many mechanical systems can be adequately described by the simple mass–spring system just discussed above. However, for some purposes it is necessary to include the effects of losses (sometimes called damping). This is normally done by including a viscous damper in the system (see Figure 2.6). See Refs. [8, 9] for further discussion on passive damping. With viscous or "coulomb" damping the friction or damping force F_d is assumed to be proportional to the velocity, dy/dt. If the constant of proportionality is R, then the damping force F_d on the mass is

$$F_d = -R\frac{dy}{dt},\tag{2.11}$$

and Eq. (2.9) becomes

$$-R\frac{dy}{dt} - Ky = M\frac{d^2y}{dt^2},\tag{2.12}$$

or equivalently

$$M\ddot{y} + R\dot{y} + Ky = 0,\tag{2.13}$$

where the dots represent single and double differentiation with respect to time.

The solution of Eq. (2.13) is most conveniently found by assuming a solution of the form: y is the real part of $Ae^{j\lambda t}$ where A is a complex number and λ is an arbitrary constant to be determined. By substituting $y = Ae^{j\lambda t}$ into Eq. (2.13) and assuming that the damping constant R is small, $R < (4MK)^{1/2}$ (which is true in most engineering applications), the solution is found that:

$$y = Ae^{-(R/2M)t}\sin(\omega_d t + \phi).\tag{2.14}$$

Here ω_d is known as the damped "natural" angular frequency:

$$\omega_d = \omega_n\sqrt{1 - \left(\frac{R}{2\sqrt{MK}}\right)^2},\tag{2.15}$$

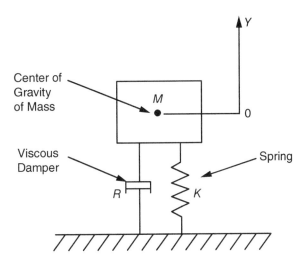

Figure 2.6 Movement of damped simple system.

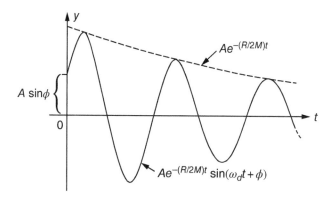

Figure 2.7 Motion of a damped mass–spring system, $R < (4MK)^{1/2}$.

where ω_n is the undamped natural frequency $\sqrt{K/M}$. The motion described by Eq. (2.14) is plotted in Figure 2.7.

The amplitude of the motion decreases with time unlike that for undamped motion (see Figure 2.3). If the damping is increased until R equals $(4MK)^{1/2}$, the damping is then called critical, $R_{\text{crit}} = (4MK)^{1/2}$. In this case, if the mass in Figure 2.6 is displaced, it gradually returns to its equilibrium position and the displacement never becomes negative. In other words, there is no oscillation or vibration. If $R > (4MK)^{1/2}$, the system is said to be overdamped.

The ratio of the damping constant R to the critical damping constant R_{crit} is called the damping ratio δ:

$$\delta = \frac{R}{R_{\text{crit}}} = \frac{R}{2M\omega_n}. \tag{2.16}$$

In most engineering cases, the damping ratio, δ, in a structure is hard to predict and is of the order of 0.01–0.1. There are, however, several ways to measure damping experimentally [8, 9].

Example 2.3   A 600-kg machine is mounted on springs such that its static deflection is 2 mm. Determine the damping constant of a viscous damper to be added to the system in parallel with the springs, such that the system is critically damped.

Solution

The static deflection is given by Eq. (2.8a) as $d = Mg/K$. Therefore $K = Mg/d = 600(9.8)/2 \times 10^{-3} = 294 \times 10^4$ N/m. The system is critically damped when the damped constant $R_{\text{crit}} = (4MK)^{1/2} = (4 \times 600 \times 294 \times 10^4)^{1/2} = 84\,000$ Ns/m.

(c) Forced Vibration – Damped

If a damped spring–mass system is excited by a simple harmonic force at some arbitrary angular forcing frequency ω (see Figure 2.8), we now obtain the equation of motion Eq. (2.17):

$$M\ddot{y} + R\dot{y} + Ky = \boldsymbol{F}e^{j\omega t} = |\boldsymbol{F}|\, e^{j(\omega t + \phi)}. \tag{2.17}$$

The force $\boldsymbol{F}$ is normally written in the complex form for mathematical convenience. The real force acting is, of course, the real part of $\boldsymbol{F}$ or $|\boldsymbol{F}|\cos(\omega t)$, where $|\boldsymbol{F}|$ is the force amplitude.

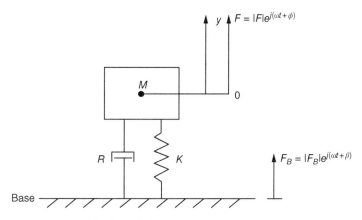

Figure 2.8 Forced vibration of damped simple system.

If we assume a solution of the form $y = Ae^{j\omega t}$ then we obtain from Eq. (2.17):

$$A = \frac{|F|}{j\omega R + K - M\omega^2}. \tag{2.18}$$

We can write $A = |A|\, e^{j\alpha}$, where α is the phase angle between force and displacement. The phase, α, is not normally of much interest, but the amplitude of motion $|A|$ of the mass is. The amplitude of the displacement is

$$|A| = \frac{|F|}{\sqrt{\omega^2 R^2 + (K - M\omega^2)^2}}. \tag{2.19}$$

This can be expressed in alternative form:

$$\frac{|A|}{|F|/K} = \frac{1}{\sqrt{4\delta^2(\omega/\omega_n)^2 + \left(1 - (\omega/\omega_n)^2\right)^2}}. \tag{2.20}$$

Equation (2.20) is plotted in Figure 2.9. It is observed that if the forcing frequency ω is equal to the natural frequency of the structure, ω_n, or equivalently $f = f_n$, a condition called resonance, then the amplitude of the motion is proportional to $1/(2\delta)$. The ratio $|A|/(|F|/K)$ is sometimes called the *dynamic magnification factor* (DMF). The number $|F|/K$ is the *static deflection* the mass would assume if exposed to a *constant* nonfluctuating force $|F|$. If the damping ratio, δ, is small, the displacement amplitude A of a structure excited at its *natural* or *resonance* frequency is very high. For example, if a simple system has a damping ratio, δ, of 0.01, then its dynamic displacement amplitude is 50 times (when exposed to an oscillating force of $|F|$ N) its static deflection (when exposed to a static force of amplitude $|F|$ N), that is, DMF = 50.

Situations such as this should be avoided in practice, wherever possible. For instance, if an oscillating force is present in some machine or structure, the frequency of the force should be moved away from the natural frequencies of the machine or structure, if possible, so that resonance is avoided. If the forcing frequency f is close to or coincides with a natural frequency f_n, large amplitude vibrations can occur with consequent vibration and noise problems and the potential of serious damage and machine malfunction.

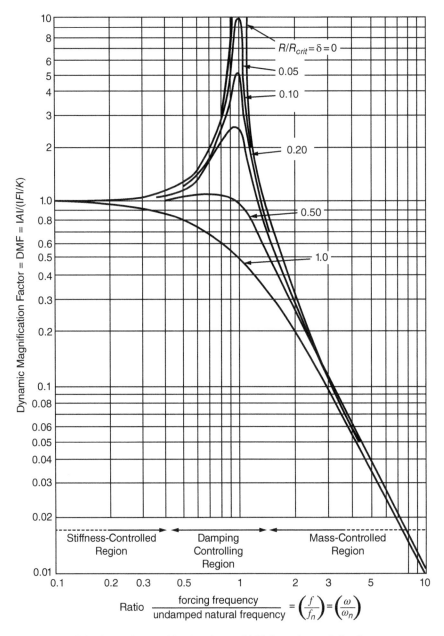

Figure 2.9 Dynamic magnification factor (DMF) for a damped simple system.

The force on the idealized damped simple system will create a force on the base $F_B = R\dot{y} + Ky$. Substituting this into Eq. (2.17) and rearranging and finally comparing the amplitudes of the imposed force $|F|$ with the force transmitted to the base $|F_B|$ gives

$$\frac{|F_B|}{|F|} = \left[\frac{1 + 4\delta^2(\omega/\omega_n)^2}{4\delta^2(\omega/\omega_n)^2 + \left(1 - (\omega/\omega_n)^2\right)^2} \right]^{1/2}. \tag{2.21}$$

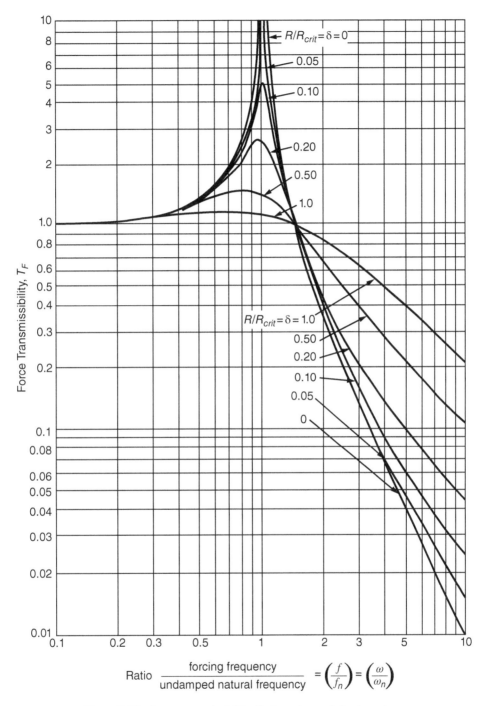

Figure 2.10 Force transmissibility, T_F, for a damped simple system.

Equation (2.21) is plotted in Figure 2.10. The ratio $|F_B|/|F|$ is sometimes called the force transmissibility T_F. The force amplitude transmitted to the machine support base, F_B, is seen to be much greater than 1, if the exciting frequency is at the system resonance frequency. The results in Eq. (2.21) and Figure 2.10 have important applications to machinery noise problems that will be discussed again in detail in Chapter 9 of this book.

Briefly, we can observe that these results can be utilized in designing vibration isolators for a machine. The natural frequency ω_n of a machine of mass M resting on its isolators of stiffness K and damping constant R must be made much less than the forcing frequency ω. Otherwise, large force amplitudes will be transmitted to the machine base. Transmitted forces will excite vibrations in machine supports and floors and walls of buildings, and the like, giving rise to additional noise radiation from these other areas.

Chapter 9 of this book gives a more complete discussion on vibration isolation.

Example 2.4   What is the maximum stiffness of an undamped isolator to provide 80% isolation for a 300-kg machine operating at 1000 rpm?

Solution

The excitation frequency is $f = 1000/60 = 16.7\,\text{Hz}$, or $\omega = 1000 \times (2\pi/60) = 104.7\,\text{rad/s}$. For 80% isolation the maximum force transmissibility is 0.2.

Using Eq. (2.21) with $\delta = 0$ and noting that isolation only occurs when $\omega/\omega_n > \sqrt{2}$ we get that $0.2 \geq [(\omega/\omega_n)^2 - 1]^{-1}$ which is solved giving $\omega/\omega_n \geq 2.45$. This result can be also obtained from Figure 2.10. Therefore, the system's maximum allowable natural frequency is $f_n = 6.8\,\text{Hz}$, or $\omega_n = \omega/2.45 = 104.7/2.45 = 42.7\,\text{rad/s}$. Consequently, the maximum isolator stiffness is $K = M\omega_n^2 = (300) \times (42.7)^2 = 5.47 \times 10^5\,\text{N/m}$.

2.4   Multi-Degree of Freedom Systems

The simple mass-spring-damper system excited by a harmonic force was discussed in the preceding sections assuming a single mass which could move in one axis only. This single-degree-of-freedom system idealization is reasonable when the mass is fairly rigid, the springs are lightweight and its motion can be described by means of one variable. For simple systems vibrating at low frequencies, it is also often possible to represent continuous systems with discrete or lumped parameter models. However, real systems have more than just one degree of freedom and, consequently, more than one natural frequency of vibration. For example, systems with more than one mass or systems in which a mass has considerable translation or rotation in more than one direction need to be modeled as multi-degree of freedom systems. In multi-degree of freedom systems, we have to consider the relationship between the motions of the various masses, i.e. their relative motion.

The general form of the equation that governs the forced vibration of an n-degree-of-freedom linear system with viscous damping can be written in matrix form as

$$[M]\ddot{\boldsymbol{q}} + [R]\dot{\boldsymbol{q}} + [K]\boldsymbol{q} = \boldsymbol{f}(t), \tag{2.22}$$

where $[M]$ is the $n \times n$ mass matrix, $[R]$ is the $n \times n$ damping matrix (that incorporates viscous damping terms in the matrix formulation), $[K]$ is the $n \times n$ stiffness matrix, $\boldsymbol{q}$ is the n-dimensional column vector of time-dependent displacements, and $\boldsymbol{f}(t)$ is the n-dimensional column vector of dynamic forces that act on the system. Therefore, the system governed by Eq. (2.22) exhibits motion which is governed by a set of n simultaneously second-order differential equations. These equations can be derived using either Newton's laws for free body diagrams or energy methods. In particular, it can be shown that the mass and stiffness matrix are symmetric. This fact is assured if energy methods are used to derive the differential equations. However, symmetric mass and stiffness matrices can also be obtained after algebraic manipulation of the equations. In general, damping matrices are not symmetric unless the system is proportionally damped, i.e. the damping matrix is a linear combination of the mass matrix and stiffness matrix.

The algebraic complexity of the solution grows exponentially with the number of degrees of freedom and the general solution of Eq. (2.22) can be difficult to obtain for systems with a large number of degrees of freedom. Therefore, approximate and numerical approaches are often required to obtain the vibration properties and system response of a multi-degree of freedom system.

2.4.1 Free Vibration – Undamped

By free vibration, we mean that the system is set into motion by some forces which then cease (at $t = 0$) and the system is then allowed to vibrate freely for $t > 0$ with no external forces applied. First we will consider a free undamped multi-degree of freedom system, i.e. $[R] = [0]$ and $\boldsymbol{f}(t) = \boldsymbol{0}$. Therefore, Eq. (2.22) now becomes

$$[M]\ddot{\boldsymbol{q}} + [K]\boldsymbol{q} = \boldsymbol{0}. \tag{2.23}$$

Similarly to the case of the single-degree-of-freedom system discussed in Section 2.3, we assume harmonic solutions in the form

$$\boldsymbol{q} = \boldsymbol{A}e^{j\omega t}, \tag{2.24}$$

where $\boldsymbol{A}$ is the vector of amplitudes. Substituting Eq. (2.24) into (2.23) yields

$$\left([K] - \omega^2[M]\right)\boldsymbol{A} = \boldsymbol{0}. \tag{2.25}$$

Equation (2.25) has a nontrivial solution if and only if the coefficient matrix $([K] - \omega^2[M])$ is singular, that is the determinant of this coefficient matrix is zero,

$$\det\left([K] - \omega^2[M]\right) = 0. \tag{2.26}$$

Equation (2.26) is called the *characteristic equation* (or characteristic polynomial) which leads to a polynomial of order n in ω^2. The roots of this polynomial, denoted as ω_i^2 (for $i = 1, 2, ..., n$), are called the *characteristic values* (or eigenvalues). The square root of these numbers, ω_i are called the *natural frequencies* of the undamped multi-degree of freedom system and they can be arranged in increasing order of magnitude by $\omega_1 \leq \omega_2 \leq ... \leq \omega_n$. The lowest frequency ω_1 is referred to as the fundamental frequency. The characteristic equation has only real roots due to the symmetry of the mass and stiffness matrices. In general, all the roots are distinct except in *degenerate* cases.

Note that ω_i^2 are the eigenvalues of the matrix $[M]^{-1}[K]$, where $[M]^{-1}$ is the inverse of $[M]$. Associated with each characteristic value ω_i^2, there is an n-dimensional column linearly independent vector called the *characteristic vector* (or eigenvector) $\boldsymbol{A}_i$ which is referred to as the i-th *natural mode* (normal mode, principal mode or mode shape) [10–13]. $\boldsymbol{A}_i$ is obtained from the homogeneous system of equations represented by Eq. (2.27) as

$$\left([K] - \omega_i^2[M]\right)\boldsymbol{A}_i = \boldsymbol{0}. \tag{2.27}$$

Since the system of equations represented by Eq. (2.27) is homogeneous, the mode shape is not unique. However, if ω_i^2 is not a repeated root of the characteristic equation then there is only one linearly independent nontrivial solution of Eq. (2.27). The eigenvector is unique only to an arbitrary multiplicative constant [13]. It can be shown that the mode shapes are orthogonal. This property is important and allows a set of n decoupled differential equations of motion of a multi-degree of freedom system to be obtained by using a modal transformation [10].

Solving Eq. (2.27) and replacing it into Eq. (2.24), we obtain a set of n linearly independent solutions $\boldsymbol{q}_i = \boldsymbol{A}_i \exp\{j\omega_i\, t\}$ of Eq. (2.23). Thus, the total solution can be expressed as a linear combination of them,

$$\boldsymbol{q} = \sum_{i=1}^{n} \beta_i \boldsymbol{A}_i e^{\,j\omega_i t}, \tag{2.28}$$

where β_i are arbitrary constants which can be determined from initial conditions [usually with initial displacements and velocities $\boldsymbol{q}(t=0)$ and $\dot{\boldsymbol{q}}(t=0)$]. Equation (2.28) represents the superposition of all modes of vibration of the multi-degree of freedom system.

Example 2.5 It is illustrative to consider an example of a two-degree-of-freedom system, as the one shown in Figure 2.11, because its analysis can easily be extrapolated to systems with many degrees of freedom.

Solution

The two-coordinates x_1 and x_2 uniquely define the position of the system illustrated in Figure 2.11 if it is constrained to move in the x-direction. The equations of motion of the system are:

$$m_1 \ddot{x}_1 = -k_1 x_1 - k_2(x_1 - x_2) \tag{2.29a}$$

and

$$m_2 \ddot{x}_2 = k_2(x_1 - x_2). \tag{2.29b}$$

We observe that the equations of motion are coupled, that is to say the motion x_1 is influenced by the motion x_2 and vice versa. Equation (2.29) can be written in matrix form as

$$[M]\ddot{\boldsymbol{q}} + [K]\boldsymbol{q} = \boldsymbol{0}, \tag{2.30}$$

where $\boldsymbol{q} = \begin{bmatrix} x_1 \\ x_2 \end{bmatrix}$, $[M] = \begin{bmatrix} m_1 & 0 \\ 0 & m_2 \end{bmatrix}$ and $[K] = \begin{bmatrix} k_1 + k_2 & -k_2 \\ -k_2 & k_2 \end{bmatrix}$.

Equation (2.26) gives the characteristic equation

$$m_1 m_2 \omega^4 - [k_2 m_1 + (k_1 + k_2)m_2]\omega^2 + k_1 k_2 = 0. \tag{2.31}$$

For simplicity, consider the situation where $m_1 = m_2 = m$ and $k_1 = k_2 = k$. Then, Eq. (2.31) becomes

$$m^2 \omega^4 - 3mk\omega^2 + k^2 = 0. \tag{2.32}$$

Solving Eq. (2.32) gives the natural frequencies of the system as

$$\omega_1 = 0.618\sqrt{k/m} \text{ and } \omega_2 = 1.618\sqrt{k/m}. \tag{2.33}$$

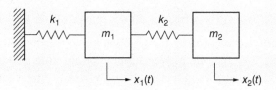

Figure 2.11 Two-degree-of-freedom system.

Note that Eq. (2.32) has four roots, the additional two being $-\omega_1$ and $-\omega_2$. However, since these negative frequencies have no physical meaning, they can be ignored. For each positive natural frequency there is an associated eigenvector that is obtained from Eq. (2.27). Substitution of Eq. (2.33) into Eq. (2.27) and solving for A_i, yields:

$$\begin{bmatrix} 1.618 & -1 \\ -1 & 0.618 \end{bmatrix} \begin{bmatrix} X_1 \\ X_2 \end{bmatrix} = \begin{bmatrix} 0 \\ 0 \end{bmatrix} \quad \text{for} \quad \omega_1 \tag{2.34a}$$

and

$$\begin{bmatrix} -0.618 & -1 \\ -1 & -1.618 \end{bmatrix} \begin{bmatrix} X_1 \\ X_2 \end{bmatrix} = \begin{bmatrix} 0 \\ 0 \end{bmatrix} \quad \text{for} \quad \omega_2, \tag{2.34b}$$

where X_1 and X_2 are the elements of vector A_i. Equations (2.34a) and (2.34b) are homogenous, so that no unique solution is possible. Indeed, a solution with all its components multiplied by the same constant is also a solution [11]. Choosing arbitrarily $X_1 = 1$ and solving Eq. (2.34) we get the eigenvectors

$$A_1 = \begin{bmatrix} 1 \\ 1.618 \end{bmatrix} \text{ and } A_2 = \begin{bmatrix} 1 \\ -0.618 \end{bmatrix}.$$

When used to describe the motion of a multi-degree of freedom system, the *mode shape* refers to the amplitude ratio. These ratios are possible to obtain because their absolute values are arbitrary [12]. Thus, we express the mode shapes as the ratio of the amplitudes X_1/X_2. Then, for ω_1, $X_1/X_2 = 0.618$ and for ω_2, $X_1/X_2 = -1.618$. These ratios can be represented in the mode plot of Figure 2.12. We note that when this simple two-degree of freedom system vibrates at the first (fundamental) natural frequency ω_1, the two masses vibrate in phase (Figure 2.12a). When the system vibrates at the second natural frequency ω_2, the two masses vibrate out of phase (Figure 2.12b).

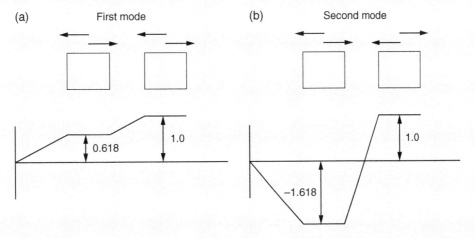

Figure 2.12 Mode shapes for the two-degree of freedom system shown in Figure 2.11; (a) first mode, (b) second mode.

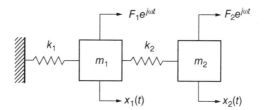

Figure 2.13 Harmonically forced two-degree-of-freedom system.

2.4.2 Forced Vibration – Undamped

By forced vibration, we mean that the system is vibrating under the influence of continuous (external) forces that do not cease. The total response of a multi-degree of freedom system due to a force excitation is the sum of a homogeneous solution and a particular solution. The homogenous solution depends upon the system properties while the particular solution is the response due to the particular form of excitation. The homogenous solution is often ignored for a system subjected to a periodic vibration for being of lesser practical importance than the particular solution. For a general form of excitation, a closed-form solution of a multi-degree of freedom system can be very difficult to obtain and numerical methods are often used.

The equations of motion of an n-degree-of-freedom undamped linear system excited by simple harmonic forces at some arbitrary angular forcing frequency ω (all excitation terms at the same phase) can be expressed in matrix form as

$$[M]\ddot{\boldsymbol{q}} + [K]\boldsymbol{q} = \boldsymbol{F}e^{j\omega t}, \tag{2.35}$$

where $\boldsymbol{F}$ is an n-dimensional complex column vector of dynamic amplitude forces. We assume harmonic solutions of the form

$$\boldsymbol{q} = \boldsymbol{A}e^{j\omega t}, \tag{2.36}$$

where $\boldsymbol{A}$ is a vector of undetermined amplitudes. Substituting Eq. (2.36) into (2.35) leads to

$$\left([K] - \omega^2[M]\right)\boldsymbol{A} = \boldsymbol{F}. \tag{2.37}$$

A unique solution of Eq. (2.37) exists unless

$$\det\left([K] - \omega^2[M]\right) = 0, \tag{2.38}$$

which has the same form as Eq. (2.26). Equation (2.38) is satisfied only when the forcing frequency coincides with one of the system's natural frequencies. In this condition, called resonance, the response of the system grows linearly with time and thus use of the solution Eq. (2.36) is unsuitable. When a solution of Eq. (2.37) exists, the amplitudes can be determined as [13]

$$\boldsymbol{A} = \left([K] - \omega^2[M]\right)^{-1}\boldsymbol{F}. \tag{2.39}$$

If we consider the two-degree of freedom system discussed in Example 2.5 but now harmonic force excitations of frequency ω and amplitude F_1 and F_2 are applied to the masses m_1 and m_2, respectively (see Figure 2.13), the equations of motion are

$$m_1\ddot{x}_1 + (k_1 + k_2)x_1 - k_2x_2 = F_1e^{j\omega t} \tag{2.40a}$$

and

$$m_2\ddot{x}_2 - k_2 x_1 + k_2 x_2 = F_2 e^{j\omega t}. \tag{2.40b}$$

The particular solution is given by Eq. (2.36) as

$$\begin{bmatrix} x_1(t) \\ x_2(t) \end{bmatrix} = \begin{bmatrix} A_1 \\ A_2 \end{bmatrix} e^{j\omega t}. \tag{2.41}$$

Therefore, Eq. (2.37) becomes

$$\begin{bmatrix} -\omega^2 m_1 + k_1 + k_2 & -k_2 \\ -k_2 & -\omega^2 m_2 + k_2 \end{bmatrix} \begin{bmatrix} A_1 \\ A_2 \end{bmatrix} = \begin{bmatrix} F_1 \\ F_2 \end{bmatrix}, \tag{2.42}$$

which has to be simultaneously solved to find the displacement amplitudes A_1 and A_2.

Example 2.6 Let consider the two-degree of freedom system of Example 2.5. Assume that a force $F_0 e^{j\omega t}$ is applied to mass m_1 and no force is applied to mass m_2. Then, Eq. (2.37) becomes

$$\begin{bmatrix} -\omega^2 m_1 + k_1 + k_2 & -k_2 \\ -k_2 & -\omega^2 m_2 + k_2 \end{bmatrix} \begin{bmatrix} A_1 \\ A_2 \end{bmatrix} = \begin{bmatrix} F_0 \\ 0 \end{bmatrix}. \tag{2.43}$$

Solution

Solving the system of Eq. (2.43) simultaneously, we obtain that

$$\frac{A_1}{F_0/k_1} = \frac{(1 - r_2^2)}{(1 - r_1^2 + k_2/k_1)(1 - r_2^2) - k_2/k_1}, \tag{2.44}$$

$$\frac{A_2}{F_0/k_1} = \frac{1}{(1 - r_1^2 + k_2/k_1)(1 - r_2^2) - k_2/k_1}, \tag{2.45}$$

and the ratio

$$\frac{A_2}{A_1} = \frac{1}{1 - r_2^2}, \tag{2.46}$$

where $r_1 = \omega\sqrt{m_1/k_1}$ and $r_2 = \omega\sqrt{m_2/k_2}$.

It is noted from Eq. (2.44) that the steady-state amplitude of the mass m_1 will become zero when $r_2 = 1$, i.e. when the excitation frequency is $\sqrt{k_2/m_2}$. Thus when the stiffness and mass of the secondary mass-spring system are chosen correctly, the main mass theoretically does not move. At this frequency the secondary mass is exactly 180° out-of-phase with the force applied to the primary mass and the mass has an amplitude $A_2 = -F_0/k_2$. This is the concept of the *dynamic vibration absorber* (also called neutralizer) used in machinery vibration control applications [11, 13]. The applied force is canceled by an equal and opposite force from the secondary spring. The dynamic vibration absorber was invented in 1909 by Hermann Frahm. This technique works when the excitation is at a fixed frequency at or close to resonance. Since the total system has two natural frequencies (one either side of the excitation frequency), a change in the frequency of the excitation force could excite the modified system at one of these frequencies, making the vibration absorber ineffective.

Example 2.7   Repeat the problem discussed in Example 2.6 but now assume that a force $F_0 e^{j\omega t}$ is applied to the mass m_2 and no force is applied to the mass m_1.

Solution

Equation (2.37) becomes now

$$\begin{bmatrix} -\omega^2 m_1 + k_1 + k_2 & -k_2 \\ -k_2 & -\omega^2 m_2 + k_2 \end{bmatrix} \begin{bmatrix} A_1 \\ A_2 \end{bmatrix} = \begin{bmatrix} 0 \\ F_0 \end{bmatrix}, \tag{2.47}$$

which leads to the following results

$$\frac{A_1}{F_0/k_2} = \frac{k_2/k_1}{(1 - r_1^2 + k_2/k_1)(1 - r_2^2) - k_2/k_1}, \tag{2.48}$$

$$\frac{A_2}{F_0/k_2} = \frac{1 - r_1^2 + k_2/k_1}{(1 - r_1^2 + k_2/k_1)(1 - r_2^2) - k_2/k_1}, \tag{2.49}$$

where $r_1 = \omega\sqrt{m_1/k_1}$ and $r_2 = \omega\sqrt{m_2/k_2}$.

2.4.3 Effect of Damping

If there is damping present (as there always is in real systems) the homogenous solution of a harmonically forced vibration system decays away with time. It has to be noted that when damping is included in the mathematical model, the eigenvalues and eigenvectors can be complex numbers, unlike in the undamped case. Although in practice the damping of a structural system is often small, its effect on the system response at or near resonance may be significant. If the damping matrix is a linear combination of the mass and the stiffness matrix (*proportional damping*), the system of differential Eq. (2.22) can be uncoupled using the modal matrix method [13]. This method is based on calculating the eigenvalues and eigenvectors of the system and the application of a modal transformation in a new set of coordinates called *modal coordinates*. This technique is not possible to apply if the damping matrix is arbitrary. In this case, a state-space representation is often used to uncouple the system [10]. This technique reduces the order of the differential equations at the expense of doubling the number of degrees of freedom.

For the case of an n-degree of freedom system with viscous damping and subject to a single-frequency harmonic excitation, we can assume harmonic solutions in the form of Eq. (2.36) and use the same arguments employed to obtain Eq. (2.39). Thus, the amplitudes of Eq. (2.36) are now expressed as [13]

$$\boldsymbol{A} = \left([K] + j\omega[R] - \omega^2 [M]\right)^{-1} \boldsymbol{F}. \tag{2.50}$$

Several examples are discussed in textbooks on vibration theory [10–13].

Equation (2.50) shows that if the forcing frequency is very low in comparison to the lowest natural frequency, the term $[K]$ is dominant and the vibration amplitudes are controlled mainly by the system's stiffness. If the system is excited significantly above their resonance frequency region, the term $-\omega^2[M]$ dominates and the system is mass-controlled. Damping only has an appreciable effect around the resonance frequencies. The effects of these frequency regions on the sound transmitted through a forced vibrating panel are discussed in Chapter 12.

Example 2.8 As an illustrative example consider the forced two-degree of freedom system of Example 2.6, where $k_1 = k_2 = k$ and $m_1 = m_2 = m$. In addition, two equal dampers of damping constant R are connected in parallel to the springs. The displacement amplitudes A_1 and A_2 can be determined from Eq. (2.50). Figure 2.14 shows the response of $|A_1|$ and $|A_2|$ with the forcing frequency. Note that both $|A_1|$ and $|A_2|$ reach maximum values at the same frequencies given by Eq. (2.33). It is also noted that the mass m_1 theoretically does not move when the excitation frequency is $\sqrt{k/m}$.

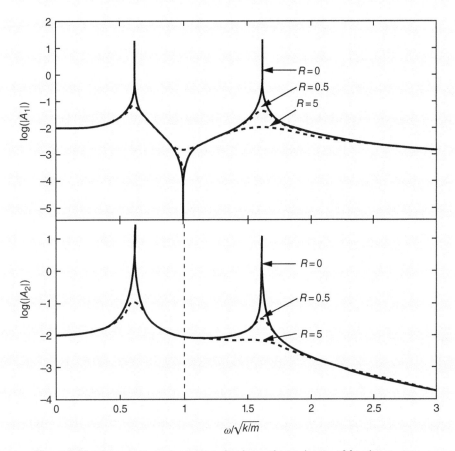

Figure 2.14 Forced response spectra of a damped two-degree of freedom system.

Example 2.9 A small electric motor is fixed on a rigid rectangular plate resting on springs. The total mass of the motor and the plate is 45.5 kg. The system is found to have a natural frequency of 15.9 Hz. It is proposed to suppress the vibration when the motor operates at 764 rpm by attaching an undamped vibration absorber underneath the motor, as shown in Figure 2.15. Determine the necessary stiffness of the absorber if $m_2 = 4.5$ kg.

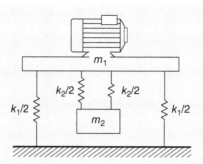

Figure 2.15 Undamped dynamic vibration absorber defined in Example 2.9.

Solution

The natural frequency of the original system is 15.9 Hz = 100 rad/s. Then, the stiffness $k_1 = m_1(\omega)^2 = 45.5$ $(100)^2 = 455\,000$ N/m. Now, the operating frequency of the motor is 764/60 = 12.7 Hz = 80 rad/s, so the absorber should have the natural frequency $\omega_2 = \sqrt{k_2/m_2} = 80$ rad/s. Then, the total stiffness of the absorber is

$$k_2 = m_2(\omega_2)^2 = 4.5(80)^2 = 28800 \text{ N/m}.$$

2.5 Continuous Systems

All structural systems such as beams, columns, and plates are continuous systems with an infinite number of degrees of freedom. Consequently, a continuous system has an infinite number of natural frequencies and corresponding mode shapes. Although easier, modeling a structure using a finite number of degrees of freedom provides just an approximation of the behavior of the system. The analysis of continuous systems requires the solution of partial differential equations. However, analytical solutions to partial differential equations are often difficult to obtain and numerical or approximate methods are usually employed to analyze continuous systems in particular at high frequencies. However, flexural vibration of some common structural elements can be analytically studied. Sound radiation can be produced by the vibration of these structural elements. Such is the case of the vibration of thin beams, thin plates and thin cylindrical shells that will be discussed in the following sections.

2.5.1 Vibration of Beams

If we ignore the effects of axial loads, rotary inertia, and shear deformation, the equation governing the free transverse vibrations $w(x,t)$ of a uniform beam is given by the Euler–Bernoulli beam theory as [10, 13]

$$EI\frac{\partial^4 w}{\partial x^4} + \rho S\frac{\partial^2 w}{\partial t^2} = 0, \tag{2.51}$$

where E is the Young's modulus, ρ is the mass density, I is the cross-sectional moment of inertia, and S is the cross-sectional area. Assuming harmonic vibrations in the form

$$w(x,t) = X(x)e^{j\omega t}, \tag{2.52}$$

and substituting $w(x,t)$ from Eq. (2.52) into Eq. (2.51) we get

$$\frac{d^4X}{dx^4} - \frac{\rho S}{EI}\omega^2 X = 0.$$ (2.53)

The solution of Eq. (2.53) is

$$X(x) = C_1 \cos \lambda x + C_2 \sin \lambda x + C_3 \cosh \lambda x + C_4 \sinh \lambda x,$$ (2.54)

where $\lambda = (\omega^2 \rho S/EI)^{1/4}$ and the C's are arbitrary constants that depend upon the boundary conditions (the deflections, slope, bending moment, and shear force constraints). Classical boundary conditions for a beam are

$$\text{a) Simply supported end} : w = 0 \text{ and } \frac{d^2 w}{dx^2} = 0,$$ (2.55)

$$\text{b) Clamped end} : w = 0 \text{ and } \frac{dw}{dx} = 0,$$ (2.56)

$$\text{c) Free end} : \frac{d^2 w}{dx^2} = 0 \text{ and } \frac{d^3 w}{dx^3} = 0.$$ (2.57)

A very important practical case is a *cantilever* beam (clamped-free beam) of length L. In this case, the deflection and slope are zero at the clamped end, while the bending moment and shear force are zero at the free end, i.e.

$$\text{1) at } x = 0, w = 0 \text{ and } \frac{dw}{dx} = 0,$$ (2.58)

$$\text{2) at } x = L, \frac{d^2 w}{dx^2} = 0 \text{ and } \frac{d^3 w}{dx^3} = 0.$$ (2.59)

Substituting the boundary conditions Eq. (2.58) and Eq. (2.59) into Eq. (2.54), we find that $C_2 = -C_4$, and we obtain the equation

$$\cos(\lambda L) \cosh(\lambda L) + 1 = 0.$$ (2.60)

The roots of the transcendental Eq. (2.60) can be obtained numerically. The first four roots are $\lambda_1 L = 1.875$, $\lambda_2 L = 4.694$, $\lambda_3 L = 7.855$, and $\lambda_4 L = 10.996$. For large values of n, the roots can be calculated using the equation

$$\lambda_n L \approx \left(n - \frac{1}{2}\right)\pi.$$ (2.61)

Noting that $\lambda_n = \left(\omega_n^2 \rho S/EI\right)^{1/4}$, we can solve for ω_n so that the first four natural frequencies of the cantilever beam are

$$\omega_1 = 3.5156\sqrt{\frac{EI}{\rho S L^4}}, \omega_2 = 22.0336\sqrt{\frac{EI}{\rho S L^4}}, \omega_3 = 61.7010\sqrt{\frac{EI}{\rho S L^4}}, \text{and } \omega_4 = 120.9120\sqrt{\frac{EI}{\rho S L^4}}.$$

The mode shapes are given by [10, 13]

$$X_n(x) = A_n\left[\sin \lambda_n x - \sinh \lambda_n x + \frac{\cos \lambda_n L + \cosh \lambda_n L}{\sin \lambda_n L - \sinh \lambda_n L}\left(\cos \lambda_n x - \cosh \lambda_n x\right)\right],$$ (2.62)

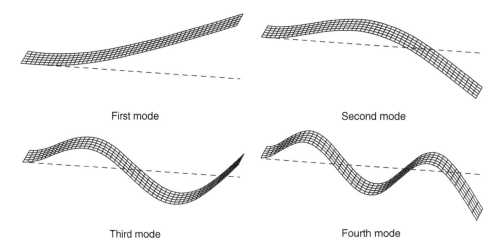

First mode

Second mode

Third mode

Fourth mode

Figure 2.16 First four mode shapes of a cantilever beam.

where A_n is an arbitrary constant. Thus, the total solution for the free transverse vibration of the cantilever beam is

$$w(x, t) = \sum_{n=1}^{\infty} X_n(x) \exp(j\omega_n t). \tag{2.63}$$

Figure 2.16 shows the first four mode shapes for a cantilever beam.

Example 2.10   Determine the natural frequencies of a uniform beam which is simply supported at both ends.

Solution

Applying the boundary conditions $w = 0$ and $\dfrac{d^2 w}{dx^2} = 0$ at $x = 0$ in Eq. (2.54) leads to $C_1 + C_3 = 0$ and $-\lambda^2 C_1 + \lambda^2 C_3 = 0$. These equations are satisfied if $C_1 = C_3 = 0$.

Applying the boundary conditions $w = 0$ and $\dfrac{d^2 w}{dx^2} = 0$ at $x = L$ in Eq. (2.54) yields

$C_2 \sin(\lambda L) + C_4 \sinh(\lambda L) = 0$ and $-\lambda^2 C_2 \sin(\lambda L) + \lambda^2 C_4 \sinh(\lambda L) = 0$. Therefore, nontrivial solutions are obtained when $C_4 = 0$ and $\sin(\lambda L) = 0$, so $\lambda = n\pi/L$ (for $n = 1,2,...$). Since $\lambda = (\omega^2 \rho S/EI)^{1/4}$, we find that the natural frequencies ω_n are given by

$$\omega_n = (n\pi)^2 \sqrt{\frac{EI}{\rho S L^4}}, \text{ for } n = 1, 2, ... \tag{2.64}$$

Example 2.11   Determine the lowest natural frequency for a cantilever steel beam of thickness $a = 6$ mm, width $b = 10$ mm, and length $L = 0.5$ m. Repeat the calculation when the beam is simply supported at both ends.

Solution

For steel we have $E = 210 \times 10^9 \, \text{N/m}^2$ and $\rho = 7800 \, \text{kg/m}^3$. The cross-sectional moment of inertia of a rectangular beam is $I = ab^3/12 = 0.006 \times (0.01)^3/12 = 5 \times 10^{-10} \, \text{m}^4$. The cross-sectional area of the beam is $S = 0.006 \times 0.01 = 6 \times 10^{-5} \, \text{m}^2$. Then,

$$\omega_1 = 3.5156\sqrt{\frac{(210 \times 10^9)(5 \times 10^{-10})}{(7800)(6 \times 10^{-5})(0.5)^4}} = 210.6 \, \text{rad/s which corresponds to 33.5 Hz.}$$

If the beam is now simply supported, we use Eq. (2.64) with $n = 1$,

$$\omega_n = (\pi)^2\sqrt{\frac{(210 \times 10^9)(5 \times 10^{-10})}{(7800)(6 \times 10^{-5})(0.5)^4}} = 591.3 \, \text{rad/s, which corresponds to 94.1 Hz.}$$

2.5.2 Vibration of Thin Plates

The present section is concerned with systems possessing two dimensions which are large compared with the third, e.g. plates whose lengths and widths are much greater than their thicknesses. There are numerous applications of vibrating plates in electroacoustical equipment such as loudspeakers, microphones, earphones, ultrasonic transducers, etc. In addition, plates can be found as constituting elements in several mechanical systems such as cars, trains, aircraft, and machinery. Plates, considered as plane systems, are a particular case of shells, whose surface may be of any shape. The general theory of vibrations of shells, which constitutes a large branch of mechanics, has been discussed by Leissa [14]. In 1828, Poisson and Cauchy established, for the first time, an approximate differential equation for flexural vibrations of a plate of infinite extent, and Poisson obtained an approximate solution for a particular case of the vibration of a circular plate. The development by Ritz in 1909 of the method for calculating the bending of plates on the basis of the energy balance was an important advance. This method made it possible to solve more complicated cases of the vibration of plates having different shapes [11]. Further advances were accomplished by the accurate calculation of the vibration distribution in rectangular plates with uniform and mixed boundary conditions [15].

a) Free Vibrations of a Rectangular Plate
For the requirements of acoustical engineering, it is sufficient to consider the flexural vibrations of thin plates ($\lambda >> h$, where h is the thickness of the plate) to which we can apply an approximate method analogous to that employed for bar [11].
The assumptions are:

1) The transverse cross-sections of the plate still remain plane in the presence of strains, and
2) Neither longitudinal nor transverse strains occur on the middle (neutral) plane.

These conditions are satisfied approximately when the flexural wavelength is at least equal to a six times the thickness h of the plate. Under this assumption, the wave equation for transverse vibration of the plate is [14]

$$B\left(\frac{\partial^4 W}{\partial x^4} + 2\frac{\partial^4 W}{\partial x^2 \partial y^2} + \frac{\partial^4 W}{\partial y^4}\right) + \rho h \frac{\partial^2 W}{\partial t^2} = f(x, y, t), \tag{2.65}$$

where $W(x,y,t)$ is the instantaneous transverse displacement, ρ is the density of the plate, $B = Eh^3/12(1-\nu^2)$ is the stiffness of the plate for bending, ν is the Poisson's ratio, E is the modulus of elasticity (Young's modulus) and $f(x,y,t)$ are the external forces referring to unit surface area of the plate. The sum of the first three terms of Eq. (2.65) is a double Laplacian of W, $\nabla^4 W$, expressed in rectangular coordinates. It is useful to define the solution for steady-state conditions as

$$W(x,y,t) = W_0(x,y)e^{j\omega t}, \tag{2.66}$$

where $W_0(x,y)$ is the amplitude of the transverse displacement.

A vibrating structure, such as a plate, at any instant contains some kinetic energy and some strain (or potential) energy. The kinetic energy is associated with the mass and the strain energy is associated with the stiffness. In addition, any structure also dissipates some energy as it deforms. This conversion of ordered mechanical energy into thermal energy is called damping. A simple way to describe the energy loss is given by the analysis of the one-dimensional system of Eq. (2.17). Thus, for sinusoidal vibration, use of a spring with an appropriately defined stiffness is completely equivalent to the use of an elastic spring and a dashpot. The internal losses in a plate arise not because of the motion of the plate as a whole body, but depend on the mutual displacements of the neighboring elements of the plate, and these are proportional to the changes in time of $\nabla^4 W$. Therefore, in order to account for energy dissipation in a plate, one may simply introduce a complex modulus of elasticity

$$\overline{E} = E(1 + j\eta), \tag{2.67}$$

where η is the internal loss factor of the plate. As an example, Figure 2.17 shows the computation of the velocity level response in decibels of the first resonance in a clamped-clamped rectangular panel and the effect of varying its internal loss factor η.

The free vibration of plates has two important characteristics:

1) The velocity of propagation of flexural waves in the plate depends on frequency, and
2) The second component of the deflection, arising as a result of the stiffness of the plate, brings about additional changes, compared to a beam, in the distribution of the vibration.

As in the case of beams, plates can be fixed at their edges in various manners. Solutions for vibrating plates subjected to different boundary conditions have been discussed in several books for the 27 possible combinations [14, 17]. In this section we will discuss a uniform simply supported plate of length a and width b, mainly because this problem is illustrative and it has a closed-form solution. In this case, the plate at its supports cannot perform motion perpendicular to its surface, but can rotate around the edge, and hence the displacement and bending moment are zero at the corresponding boundaries (see Eq. (2.55) for a beam)

$$W(0,y) = W(a,y) = W(x,0) = W(x,b) = 0 \text{ and } M_x(0,y) = M_x(a,y) = M_y(x,0) = M_y(x,b) = 0. \tag{2.68}$$

The solution for the amplitude of flexural vibrations of a simply-supported plate vibrating in mode (m,n), where $m-1$ and $n-1$ are the number of nodal lines in the x and y directions, respectively, can be obtained by separation of variables and is given by [14]

$$W_0(x,y) = A_{mn} \sin(k_m x) \sin(k_n y), \tag{2.69}$$

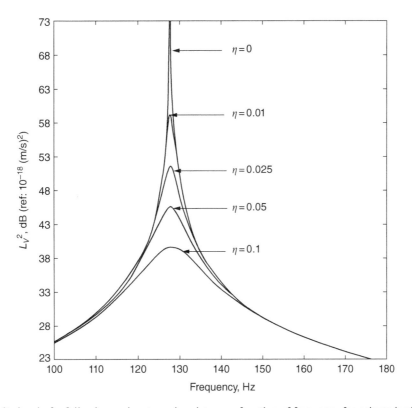

Figure 2.17 Velocity level of a fully-clamped rectangular plate as a function of frequency for selected values of η. The resonance of its first mode is shown. The plate is a plywood board with properties $E = 6 \times 10^9$ N/m^2, $\rho_s = 6.5$ kg/m^2, $S = a \times b = 0.7 \times 0.6$ m^2, and $\nu = 0.3$. (*Source:* from Ref. [16].)

where A_{mn} is the modal amplitude, $k_m = m\pi/a$, $k_n = n\pi/b$ and m and n are integers. Equation (2.69) gives the mode shapes of the simply-supported plate. Figure 2.18 shows the first six mode shapes of a rectangular plate. Natural frequencies of each plate's mode (m,n) are given by

$$\omega_{mn} = \pi^2 \left(\frac{m^2}{a^2} + \frac{n^2}{b^2} \right) \sqrt{\frac{B}{\rho_s}}, \tag{2.70a}$$

where ρ_s is the mass of the plate per unit surface area. Note that Eq. (2.70a) can be approximated by

$$f_{m,n} = 0.453 c_L h \left[\left(\frac{m}{a} \right)^2 + \left(\frac{n}{b} \right)^2 \right], \tag{2.70b}$$

where $f_{m,n}$ is the characteristic modal frequency (Hz) and c_L is the longitudinal wave speed in the plate material (m/s).

Thus, the total solution for the free vibration of a simply-supported plate is

$$W(x,y,t) = \sum_{m=1}^{\infty} \sum_{n=1}^{\infty} A_{mn} \sin \left(\frac{m\pi x}{a} \right) \sin \left(\frac{n\pi y}{b} \right) \exp\left(j\omega_{mn} t \right). \tag{2.71}$$

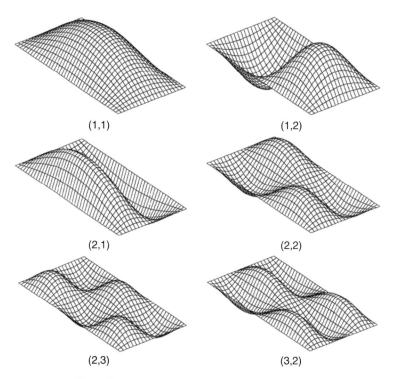

(1,1)   (1,2)

(2,1)   (2,2)

(2,3)   (3,2)

Figure 2.18   First six modes of a rectangular plate.

Note that for a square plate $(a = b)$ $\omega_{mn} = \omega_{nm}$. Therefore modes (m,n) and (n,m) have the same frequency and they are called *degenerate*. This fact also happens when the length or width are integer multiple of each other.

Example 2.12   Determine the natural frequency of the fundamental mode of a plywood board of thickness 8 mm, mass density 812.5 kg/m^3, and dimensions $a \times b = 0.7 \times 0.6$ m^2.

Solution

The plywood board has properties $E = 6 \times 10^9$ N/m^2, $S = a \times b = 0.7 \times 0.6$ m^2, and $\nu = 0.3$. The mass of the plate per unit surface area $\rho_s = \rho \times h = 812.5 \times 0.008 = 6.5$ kg/m^2. and the bending stiffness $B = Eh^3/12(1 - \nu^2)$ $= 6 \times 10^9(0.008)^3/12 \,(1 - [0.3]^2) = 281.3$ N/m. Then, the fundamental mode (the lowest natural frequency) is given for $m = n = 1$. Replacing the values in Eq. (2.70a) yields

$$\omega_{11} = \pi^2 \left(\frac{1}{(0.7)^2} + \frac{1}{(0.6)^2} \right) \sqrt{\frac{281.3}{6.5}} = 313 \text{ rad/s which corresponds to } 49.8 \text{ Hz}.$$

---

b) Forced Vibration of a Rectangular Plate

Let us assume that a distributed harmonic force acts on a rectangular plate (referred to unit plate area)

$$f = f_0(x,y)e^{j\omega t}. \tag{2.72}$$

Under the influence of this force, a plate deflection W is produced with the distribution

$$W = W_0(x,y)e^{j\omega t}. \tag{2.73}$$

The distribution of the deflection amplitude W_0 can be represented by the resultant of a series of sinusoidal vibrations. The frequency is equal to the frequency of free vibration of the plate, that is to say, the following equation must be satisfied [18]

$$\nabla^4 \Psi_{mn} - k_{mn}^4 \Psi_{mn} = 0 \tag{2.74}$$

where $k_{mn} = \omega_{mn}/c_b$ is the wavenumber of the (m,n) mode of free vibration, c_b is the velocity of bending waves in the plate and Ψ_{mn} are the dimensionless coefficients which determine the relative distribution in relation to the maximum amplitude of deflection.

Using the orthogonality of the function Ψ_{mn} and Fourier series, the resultant distribution of vibration on the rectangular plate is given by [16]

$$W_0(x,y) = \frac{1}{\rho_S S} \sum_{m=1}^{\infty} \sum_{n=1}^{\infty} \frac{\Psi_{mn}(x,y) \int_0^b \int_0^a f_0(x',y')\Psi_{mn}(x',y')dx'dy'}{\gamma(\omega_{mn}^2 - \omega^2)}, \tag{2.75}$$

where (x',y') denotes the coordinates of the position of the force f (referred to unit plate area), with respect to which the integration is carried out over the surface area of the plate. The ω_{mn} are real numbers for a plate without losses. For the frequency of forced vibration $\omega = \omega_{mn}$, the amplitude of the deflection grows theoretically to infinity. In a plate with losses, ω_{mn} are complex numbers, as discussed in Section 2.4.3.

For example, for a simply-supported rectangular plate, the distribution Ψ_{mn} (see Eq. (2.69)) is

$$\Psi_{mn} = \sin(k_m x)\sin(k_n y), \tag{2.76}$$

and $\gamma = 1/4$. Then, we can obtain that for a point force F concentrated at the point (x_0, y_0)

$$W_0(x,y) = \frac{4F}{\rho_S ab} \sum_{m=1}^{\infty} \sum_{n=1}^{\infty} \frac{\sin(k_m x)\sin(k_n y)\sin(k_m x_0)\sin(k_n y_0)}{(\omega_{mn}^2 - \omega^2)}. \tag{2.76}$$

The vibration velocity distribution on the rectangular plate can be obtained as

$$V_0(x,y) = j\omega W_0(x,y). \tag{2.77}$$

In practice, to calculate the plate response to a concentrated force, it is necessary to truncate the infinite summation. More complicated cases of boundary conditions have been studied in the literature [19].

All the theory presented above and the subsequent applications have been developed assuming light fluid loading, so that the plate response is not affected by the surrounding environment, which acts as added mass and radiation damping [20]. This criterion is not valid for submerged structures.

Example 2.13 The numerical results for the velocity level at the coordinate point $(x,y) = (0.04, 0.03)$ for a plywood board with properties $E = 6 \times 10^9$ N/m^2, $\rho_S = 6.5$ kg/m^2, $a \times b = 0.7 \times 0.6$ m^2, $\nu = 0.293$ and $\eta = 0.01$ are presented in Figure 2.19. The reference velocity is 10^{-18} (m/s)^2. It is observed that excitation close to a corner excites more modes in the plate. Excitation at the middle point of the plate excites mostly the modes (m,n)

where m and n are both even, while no contribution to the velocity level is included from modes having nodal lines passing through the center of the plate.

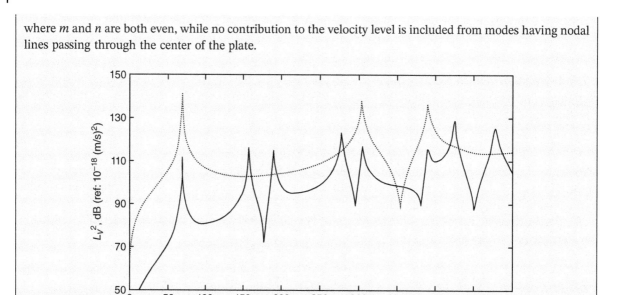

Figure 2.19 Computed velocity level of a simply-supported rectangular plate excited: ____ at a corner; ____ at the center. (*Source:* from Ref. [16].)

References

1 Bies, D.A. and Hansen, C.H. (2003). *Engineering Noise Control–Theory and Practice*, 3e. London: E & FN Spon.

2 Bell, L.H. (1982). *Industrial Noise Control – Fundamentals and Applications*. New York: Marcel Decker.

3 Hall, D.E. (1987). *Basic Acoustics*. New York: Wiley.

4 Kinsler, L.E., Frey, A.R., Coppens, A.B., and Sanders, J.V. (1999). *Fundamentals of Acoustics*, 4e. New York: Wiley.

5 Fahy, F.J. and Walker, J.G. (eds.) (1998). *Fundamentals of Noise and Vibration*. London: E & FN Spon.

6 Lighthill, M.J. (1978). *Waves in Fluids*. Cambridge: Cambridge University Press.

7 Pierce, A.D. (1981). *Acoustics: An Introduction to Its Physical Properties and Applications*. New York: McGraw-Hill (reprinted by the Acoustical Society of America, 1989).

8 Inman, D.J. (2007). Passive damping. In: *Handbook of Noise and Vibration Control* (ed. M.J. Crocker), 225–231. New York: Wiley.

9 Ungar, E.E. (2007). Damping of structures and use of damping materials. In: *Handbook of Noise and Vibration Control* (ed. M.J. Crocker), 734–744. New York: Wiley.

10 Shabana, A.A. (1997). *Vibration of Discrete and Continuous Systems*, 2e. New York: Springer-Verlag.

11 Meirovitch, L. (2001). *Fundamentals of Vibrations*. New York: McGraw-Hill.

12 Lalor, N. (1998). Fundamentals of vibration. In: *Fundamentals of Noise and Vibration* (eds. F.J. Fahy and J. Walker), 61–114. London: E & FN Spon.

13 Kelly, S.G. (2000). *Fundamentals of Mechanical Vibrations*, 2e. McGraw-Hill.

14 Leissa, A.W. (1993). *Vibration of Plates*. New York: Acoustical Society of America.

15 Berry, A., Guyader, J.L., and Nicolas, J. (1990). A general formulation for the sound radiation from rectangular, baffled plates with arbitrary boundary conditions. *J. Acoust. Soc. Am.* 88: 2792–2802.

16 Arenas, J.P. (2001) Analysis of the acoustic radiation resistance matrix and its applications to vibro-acoustic problems. PhD thesis. Auburn University.

17 Blevins, R.D. (2015). *Formulas for Dynamics, Acoustics and Vibration*. New York: Wiley.

18 Malecki, I. (1969). *Physical Foundations of Physical Acoustics*. Oxford: Pergamon Press.

19 Arenas, J.P. (2003). On the vibration analysis of rectangular clamped plates using the virtual work principle. *J. Sound Vib.* 266: 912–918.

20 Tao, J.S., Liu, G.R., and Lam, K.Y. (2001). Sound radiation of a thin infinite plate in light and heavy fluids subject to multi-point excitation. *Appl. Acoust.* 62: 573–587.

3

Sound Generation and Propagation

3.1 Introduction

The fluid mechanics equations, from which the acoustics equations and results may be derived, are quite complicated. However, because most acoustical phenomena involve very small perturbations from steady-state conditions, it is possible to make significant simplifications to these fluid equations and to linearize them. The results are the equations of linear acoustics. The most important equation, the wave equation, is presented in this chapter together with some of its solutions. Such solutions give the sound pressure explicitly as functions of time and space, and the general approach may be termed the wave acoustics approach. This chapter presents some of the useful results of this approach but also briefly discusses some of the other alternative approaches, sometimes termed ray acoustics and energy acoustics, that are used when the wave acoustics approach becomes too complicated. The main purpose of this chapter is to present some of the most important acoustics formulas and definitions, without derivation, which are used in the other chapters of this book.

3.2 Wave Motion

Some of the basic concepts of acoustics and sound wave propagation used throughout the rest of this book are discussed here. For further discussion of some of these basic concepts and/or a more advanced mathematical treatment of some of them, the reader is referred to the *Handbook of Acoustics* [1] and other texts [2–18] which are also useful for further discussion on fundamentals and applications of the theory of noise and vibration problems.

Wave motion is easily observed in the waves on stretched strings and as ripples on the surface of water. Waves on strings and surface water waves are very similar to sound waves in air (which we cannot see), but there are some differences that are useful to discuss. If we throw a stone into a calm lake, we observe that the water waves (*ripples*) travel out from the point where the stone enters the water. The ripples spread out circularly from the source at the wave speed, which is independent of the wave height.

Somewhat like the water ripples, sound waves in air travel at a constant speed, which is proportional to the square root of the absolute temperature and is almost independent of the sound wave strength. The wave speed is known as the speed of sound. Sound waves in air propagate by transferring momentum and energy between air particles. Sound wave motion in air is a disturbance that is imposed onto the random motion of the air molecules (known as Brownian motion). The mean speed of the molecular random motion and rate of molecular interaction increases with the absolute temperature of the gas. Since the momentum and sound

Engineering Acoustics: Noise and Vibration Control, First Edition. Malcolm J. Crocker and Jorge P. Arenas.
© 2021 John Wiley & Sons Ltd. Published 2021 by John Wiley & Sons Ltd.

energy transfer occurs through the molecular interaction, the sound wave speed is dependent solely upon the absolute temperature of the gas and *not* upon the strength of the sound wave disturbance. There is no net flow of air away from a source of sound, just as there is no net flow of water away from the source of water waves. Of course, unlike the waves on the surface of a lake, which are circular or two-dimensional, sound waves in air in general are spherical or three-dimensional.

As water waves move away from a source, their curvature decreases, and the *wavefronts* may be regarded almost as straight lines. Such waves are observed in practice as *breakers* on the seashore. A similar situation occurs with sound waves in the atmosphere. At large distances from a source of sound, the spherical wavefront curvature decreases, and the wavefronts may be regarded almost as plane surfaces.

Plane sound waves may be defined as waves that have the same acoustical properties at any position on a plane surface drawn perpendicular to the direction of propagation of the wave. Such plane sound waves can exist and propagate along a long straight tube or duct (such as an air-conditioning duct). In such a case, the waves propagate in a direction along the duct axis and the plane wave surfaces are perpendicular to this direction (and are represented by duct cross-sections). Such waves in a duct are one-dimensional, like the waves traveling along a long string or rope under tension (or like the ocean breakers described above).

Although there are many similarities between one-dimensional sound waves in air, waves on strings, and surface water waves, there are some differences. In a fluid such as air, the fluid particles vibrate back and forth in the same direction as the direction of wave propagation; such waves are known as longitudinal, compressional, or sound waves. On a stretched string, the particles vibrate at right angles to the direction of wave propagation; such waves are usually known as transverse waves. The surface water waves described are partly transverse and partly longitudinal, with the complication that the water particles move up and down and back and forth horizontally. (This movement describes elliptical paths in shallow water and circular paths in deep water. The vertical particle motion is much greater than the horizontal motion for shallow water, but the two motions are equal for deep water.) The water wave direction is, of course, horizontal.

Surface water waves are not compressional (like sound waves) and are normally termed surface gravity waves. Unlike sound waves, where the wave speed is independent of frequency, long-wavelength surface water waves travel faster than short-wavelength waves, and thus water wave motion is said to be dispersive. Bending waves on beams, plates, cylinders, and other engineering structures are also dispersive. There are several other types of waves that can be of interest in acoustics: shear waves, torsional waves, and boundary waves (see chapter 12 in the *Encyclopedia of Acoustics* [19]), but the discussion here will concentrate on sound wave propagation in fluids.

3.3 Plane Sound Waves

The propagation of sound may be illustrated by considering gas in a tube with rigid walls and having a rigid piston at one end. The tube is assumed to be infinitely long in the direction away from the piston. We shall assume that the piston is vibrating with simple harmonic motion at the left-hand side of the tube (see Figure 3.1) and that it has been oscillating back and forth for some time. We shall only consider the piston motion and the oscillatory motion it induces in the fluid from when we start our clock. Let us choose to start our clock when the piston is moving with its maximum velocity to the right through its normal equilibrium position at $x = 0$. See the top of Figure 3.1, at $t = 0$. As time increases from $t = 0$, the piston straight away starts slowing down with simple harmonic motion, so that it stops moving at $t = T/4$ at its maximum excursion to the right. The piston then starts moving to the left in its cycle of oscillation, and at $t = T/2$ it has reached its equilibrium position again and has a maximum velocity (the same as at $t = 0$) but now in the negative x direction. At $t = 3\,T/4$, the piston comes to rest again at its maximum excursion to the left. Finally at $t = T$ the piston reaches its equilibrium position at $x = 0$ with the same maximum velocity we imposed on it at $t = 0$. During the time T, the piston has undergone one complete cycle of oscillation. We

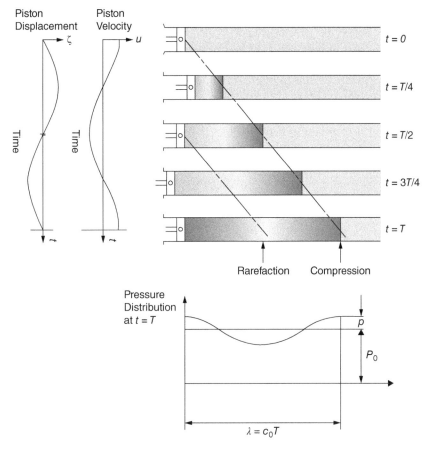

Figure 3.1 Schematic illustration of the sound pressure distribution created in a tube by a piston undergoing one complete simple harmonic cycle of operation in period T seconds.

assume that the piston continues vibrating and makes f oscillations each second, so that its frequency $f = 1/T$ (Hz).

As the piston moves backward and forward, the gas in front of the piston is set into motion. As we all know, the gas has mass and thus inertia and it is also compressible. If the gas is compressed into a smaller volume, its pressure increases. As the piston moves to the right, it compresses the gas in front of it, and as it moves to the left, the gas in front of it becomes rarified. When the gas is compressed, its pressure increases above atmospheric pressure, and, when it is rarified, its pressure decreases below atmospheric pressure. The pressure difference above or below the atmospheric pressure, p_0, is known as the sound pressure, p, in the gas. Thus the sound pressure $p = p_{tot} - p_0$, where p_{tot} is the total pressure in the gas. If these pressure changes occurred at constant temperature, the fluid pressure would be directly proportional to its density, ρ, and so $p/\rho = $ constant. This simple assumption was made by Sir Isaac Newton, who in 1660 was the first to try to predict the speed of sound. But we find that, in practice, regions of high and low pressure are sufficiently separated in space in the gas (see Figure 3.1) so that heat cannot easily flow from one region to the other and that the adiabatic law, $p/\rho^\gamma = $ constant, is more closely followed in nature.

As the piston moves to the right with maximum velocity at $t = 0$, the gas ahead receives maximum compression and maximum increase in density, and this simultaneously results in a maximum pressure increase. At the instant the piston is moving to the left with maximum negative velocity at $t = T/2$, the gas behind the piston, to the right, receives maximum rarefaction, which results in a maximum density and pressure decrease. These piston displacement and velocity perturbations are superimposed on the much greater random motion of the gas molecules (known as the Brownian motion). The mean speed of the molecular random motion in the gas depends on its absolute temperature. The disturbances induced in the gas are known as acoustic (or sound) disturbances. It is found that momentum and energy pulsations are transmitted from the piston throughout the whole region of the gas in the tube through molecular interactions (sometimes simply termed molecular collisions).

If a disturbance in a thin cross-sectional element of fluid in a duct is considered, a mathematical description of the motion may be obtained by assuming that (i) the amount of fluid in the element is conserved, (ii) the net longitudinal force is balanced by the inertia of the fluid in the element, (iii) the compressive process in the element is adiabatic (i.e. there is no flow of heat in or out of the element), and (iv) the undisturbed fluid is stationary (there is no fluid flow). Then the following equation of motion may be derived:

$$\frac{\partial^2 p}{\partial x^2} - \frac{1}{c^2}\frac{\partial^2 p}{\partial t^2} = 0, \tag{3.1}$$

where p is the sound pressure, x is the coordinate, and t is the time.

This equation is known as the one-dimensional equation of motion, or acoustic wave equation. Similar wave equations may be written if the sound pressure p in Eq. (3.1) is replaced with the particle displacement ξ, the particle velocity u, condensation s, fluctuating density ρ', or the fluctuating absolute temperature T'. The derivation of these equations is in general more complicated. However, the wave equation in terms of the sound pressure in Eq. (3.1) is perhaps most useful since the sound pressure is the easiest acoustical quantity to measure (using a microphone) and is the acoustical perturbation we sense with our ears. It is normal to write the wave equation in terms of sound pressure p, and to derive the other variables, ξ, u, s, ρ', and T' from their relations with the sound pressure p [16]. The sound pressure p is the acoustic pressure perturbation or fluctuation about the time-averaged, or undisturbed, pressure p_0.

The speed of sound waves c is given for a perfect gas by

$$c = (\gamma R T)^{1/2}. \tag{3.2}$$

The speed of sound is proportional to the square root of the absolute temperature T. The ratio of specific heats γ and the gas constant R are constants for any particular gas. Thus Eq. (3.2) may be written as

$$c = c_0 + 0.6T_c, \tag{3.3}$$

where, for air, $c_0 = 331.6$ m/s, the speed of sound at $0\,°C$, and T_c is the temperature in degrees Celsius. Note that Eq. (3.3) is an approximate formula valid for T_c near room temperature. The speed of sound in air is almost completely dependent on the air temperature and is almost independent of the atmospheric pressure. For a complete discussion of the speed of sound in fluids, see chapter 5 in the *Handbook of Acoustics* [1].

A solution to Eq. (3.1) is

$$p = f_1(ct - x) + f_2(ct + x), \tag{3.4}$$

where f_1 and f_2 are arbitrary functions such as sine, cosine, exponential, log, and so on. It is easy to show that Eq. (3.4) is a solution to the wave equation Eq. (3.1) by differentiation and substitution into Eq. (3.1). Varying x and t in Eq. (3.4) demonstrates that $f_1(ct - x)$ represents a wave traveling in the positive x-direction with wave speed c, while $f_2(ct + x)$ represents a wave traveling in the negative x-direction with wave speed c (see Figure 3.2).

The solution given in Eq. (3.4) is usually known as the *general solution* since, in principle, any type of sound waveform is possible. In practice, sound waves are usually classified as impulsive or steady in time. One particular case of a steady wave is of considerable importance. Waves created by sources vibrating sinusoidally in

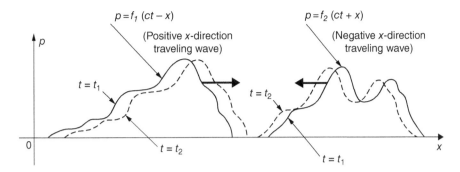

Figure 3.2 Plane waves of arbitrary waveform.

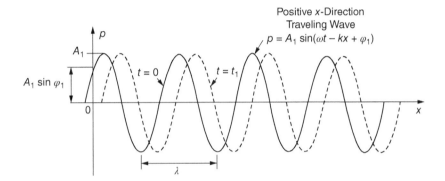

Figure 3.3 Simple harmonic plane waves.

time (e.g. a loudspeaker, a piston, or a more complicated structure vibrating with a discrete angular frequency ω) both in time t and space x in a sinusoidal manner (see Figure 3.3):

$$p = A_1 \sin\left(\omega t - kx + \varphi_1\right) + A_2 \sin\left(\omega t + kx + \varphi_2\right). \tag{3.5}$$

At any point in space, x, the sound pressure p is simple harmonic in time. The first expression on the right of Eq. (3.5) represents a wave of amplitude A_1 traveling in the positive x-direction with speed c, while the second expression represents a wave of amplitude A_2 traveling in the negative x-direction. The symbols φ_1 and φ_2 are phase angles, and k is the *acoustic wavenumber*. It is observed that the wavenumber $k = \omega/c$ by studying the ratio of x and t in Eqs. (3.4) and (3.5). At some instant t the sound pressure pattern is sinusoidal in space, and it repeats itself each time kx is increased by 2π. Such a repetition is called a wavelength λ. Hence, $k\lambda = 2\pi$ or $k = 2\pi/\lambda$. This gives $\omega/c = 2\pi/c = 2\pi/\lambda$, or

$$\lambda = \frac{c}{f}. \tag{3.6}$$

The wavelength of sound becomes smaller as the frequency is increased. In air, at 100 Hz, $\lambda \approx 3.5\,\text{m} \approx 10\,\text{ft}$. At 1000 Hz, $\lambda \approx 0.35\,\text{m} \approx 1\,\text{ft}$. At 10000 Hz, $\lambda \approx 0.035\,\text{m} \approx 0.1\,\text{ft}. \approx 1\,\text{in}$.

At some point x in space, the sound pressure is sinusoidal in time and goes through one complete cycle when ω increases by 2π. The time for a cycle is called the period T. Thus, $\omega T = 2\pi$, $T = 2\pi/\omega$, and

$$T = \frac{1}{f}. \tag{3.7}$$

Example 3.1   The human audible frequency range is from 20 Hz to 20 kHz. Calculate the extremes of wavelength for audible sounds at 20 °C.

Solution

From Eq. (3.3) the speed of sound at 20 °C is $c = 331.6 + 0.6(20) = 343.6$ m/s. Therefore, using Eq. (3.6) the wavelength of a sound of 20 Hz is $\lambda = 343.6/20 = 17.18$ m. The wavelength of a sound of 20 kHz is $\lambda = 343.6/20000 = 0.01718$ m $= 17.18$ mm.

3.3.1 Sound Pressure

With sound waves in a fluid such as air, the sound pressure at any point is the difference between the total pressure and normal atmospheric pressure. The sound pressure fluctuates with time and can be positive or negative with respect to the normal atmospheric pressure.

Sound varies in magnitude and frequency and it is normally convenient to give a single number measure of the sound by determining its time-averaged value. The time average of the sound pressure at any point in space, over a sufficiently long time, is zero and is of no interest or use. The time average of the square of the sound pressure, known as the mean square pressure, however, is not zero. If the sound pressure at any instant t is $p(t)$, then the mean square pressure, $\langle p^2(t)\rangle_t$, is the time average of the square of the sound pressure over the time interval T:

$$\langle p^2(t)\rangle_t = \frac{1}{T}\int_0^T p^2(t)dt, \tag{3.8}$$

where $\langle\rangle_t$ denotes a time average.

It is usually convenient to use the square root of the mean square pressure:

$$p_{rms} = \sqrt{\langle p^2(t)\rangle_t} = \sqrt{\frac{1}{T}\int_0^T p^2(t)dt,}$$

which is known as the root mean square (rms) sound pressure. This result is true for all cases of continuous sound time histories including noise and pure tones. For the special case of a pure tone sound, which is simple harmonic in time, given by $p = P\cos(\omega t)$, the rms sound pressure is

$$p_{rms} = P/\sqrt{2}, \tag{3.9}$$

where P is the sound pressure amplitude.

3.3.2 Particle Velocity

As the piston vibrates, the gas immediately next to the piston must have the same velocity as the piston. A small element of fluid is known as a *particle*, and its velocity, which can be positive or negative, is known as the *particle velocity*. For waves traveling away from the piston in the positive x-direction, it can be shown that the particle velocity, u, is given by

$$u = \frac{p}{\rho c}, \tag{3.10}$$

where $\rho =$ fluid density (kg/m^3) and $c =$ speed of sound (m/s).

If a wave is reflected by an obstacle, so that it is traveling in the negative x-direction, then

$$u = -\frac{p}{\rho c}. \tag{3.11}$$

The negative sign results from the fact that the sound pressure is a scalar quantity, while the particle velocity is a vector quantity. These results are true for any type of plane sound waves, not only for sinusoidal waves.

3.3.3 Impedance and Sound Intensity

We see that for the one-dimensional propagation considered, the sound wave disturbances travel with a constant wave speed c, although there is no net, time-averaged movement of the air particles. The air particles oscillate back and forth in the direction of wave propagation (x-axis) with velocity u. We may show that for any plane wave traveling in the positive x direction at any instant

$$\frac{p}{u} = \rho c, \tag{3.12}$$

and for any plane wave traveling in the negative x-direction

$$\frac{p}{u} = -\rho c. \tag{3.13}$$

The quantity ρc is known as the *characteristic impedance* of the fluid, and for air, $\rho c = 428$ kg s/m^2 at $0\,°$C and 415 kg s/m^2 at $20\,°$C.

The intensity of sound, I, is the time-averaged sound energy that passes through unit cross-sectional area in unit time. For a plane progressive wave, or far from any source of sound (in the absence of reflections):

$$I = \frac{p^2_{rms}}{\rho c}, \tag{3.14}$$

where $\rho =$ the fluid density (kg/m^3) and $c =$ speed of sound (m/s).

In the general case of sound propagation in a three-dimensional field, the sound intensity is the (net) flow of sound energy in unit time flowing through unit cross-sectional area. The intensity has magnitude and direction

$$I_r = pu_r = \langle p{\cdot}u_r \rangle_t = \frac{1}{T}\int_0^T p{\cdot}u_r dt, \tag{3.15}$$

where p is the total fluctuating sound pressure and u_r is the total fluctuating sound particle velocity in the r-direction at the measurement point. The total sound pressure p and particle velocity u_r include the effects of incident and reflected sound waves.

We note, in general, for sound propagation in three dimensions that the instantaneous sound intensity $\mathbf{I}$ is a vector quantity equal to the product of the scalar sound pressure and the instantaneous vector particle velocity u. Thus $\mathbf{I}$ has magnitude and direction. The vector intensity $\mathbf{I}$ may be resolved into components $\mathbf{I}_x$, $\mathbf{I}_y$, and $\mathbf{I}_z$. For a more complete discussion of sound intensity and its measurement see Chapter 8 in this book, chapters 45 and 156 in the *Handbook of Acoustics* [1] and the book by Fahy [13].

3.3.4 Energy Density

Consider the case again of the oscillating piston in Figure 3.1. We shall consider the sound energy that is produced by the oscillating piston, as it flows along the tube from the piston source. We observe that the wavefront and the sound energy travel along the tube with velocity c metres/second. Thus after one second, a column of fluid of length c m contains all of the sound energy provided by the piston during the previous second. The total amount of energy E in this column equals the time-averaged sound intensity multiplied by the cross-sectional area S, which is from Eq. (3.9):

$$E = SI = S\frac{p^2_{rms}}{\rho c}. \tag{3.16}$$

The sound energy per unit volume is known as the energy density ε,

$$\varepsilon = \left[Sp^2_{rms}/\rho c\right]/cS = \frac{p^2_{rms}}{\rho c^2}. \tag{3.17}$$

This result in Eq. (3.17) can also be shown to be true for other acoustic fields as well, as long as the total sound pressure is used in Eq. (3.17), and provided the location is not very close to a sound source.

3.3.5 Sound Power

Again in the case of the oscillating piston, we will consider the sound power radiated by the piston into the tube. The sound power radiated by the piston, W, is

$$W = SI. \tag{3.18}$$

But from Eqs. (3.10) and (3.14) the power is

$$W = S(p_{rms}u_{rms}), \tag{3.19}$$

and close to the piston, the rms particle velocity, u_{rms}, must be equal to the rms piston velocity. From Eq. (3.19), we can write

$$W = S\rho c_0 v^2_{rms} = 4\pi r^2 \rho c_0 v^2_{rms}, \tag{3.20}$$

where r is the piston and duct radius, and v_{rms} is the rms velocity of the piston.

3.4 Decibels and Levels

The range of sound pressure magnitudes and sound powers of sources experienced in practice is very large. Thus, logarithmic rather than linear measures are often used for sound pressure and sound power. The most common measure of sound is the *decibel*. Decibels are also used to measure vibration, which can have a similar large range of magnitudes. The decibel represents a relative measurement or ratio. Each quantity in decibels is expressed as a ratio relative to a *reference sound pressure*, *sound power*, or *sound intensity*, or in the case of vibration relative to a *reference displacement*, *velocity*, or *acceleration*. Whenever a quantity is expressed in decibels, the result is known as a *level*.

The decibel (dB) is the ratio R_1 given by

$$\log_{10}R_1 = 0.1 \text{ and thus } 10\log_{10}R_1 = 1 \text{ dB}. \tag{3.21}$$

Thus, $R_1 = 10^{0.1} = 1.26$. The decibel is seen to represent the ratio 1.26. A larger ratio, the *bel*, is sometimes used. The bel is the ratio R_2 given by $\log_{10}R_2 = 1$. Thus, $R_2 = 10^1 = 10$. The bel represents the ratio 10 and is thus much larger than a decibel. For simplicity, in the following sections of this book we denote $\log(\ ) = \log_{10}(\ )$.

3.4.1 Sound Pressure Level

The sound pressure level L_p is given by

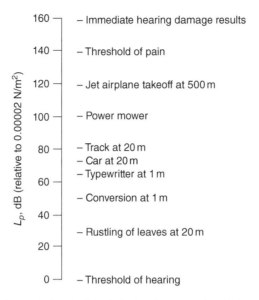

Figure 3.4 Some typical sound pressure levels, L_p.

$$L_p = 10 \log \left(\frac{\langle p^2 \rangle_t}{p^2_{ref}} \right) = 10 \log \left(\frac{p^2_{rms}}{p^2_{ref}} \right) = 20 \log \left(\frac{p_{rms}}{p_{ref}} \right) \text{dB,} \qquad (3.22)$$

where p_{ref} is the reference pressure, $p_{ref} = 20 \, \mu\text{Pa} = 0.00002 \, \text{N/m}^2 \, (= 0.0002 \, \mu\text{bar})$ for air. This reference pressure was originally chosen to correspond to the quietest sound (at 1000 Hz) that the average young person can hear. The sound pressure level is often abbreviated as SPL. Figure 3.4 shows some sound pressure levels of typical sounds.

3.4.2 Sound Power Level

The sound power level of a source, L_W, is given by

$$L_W = 10 \log \left(\frac{W}{W_{ref}} \right) \text{dB,} \qquad (3.23)$$

where W is the sound power of a source and $W_{ref} = 10^{-12} \, \text{W}$ is the reference sound power.

Some typical sound power levels are given in Figure 3.5.

3.4.3 Sound Intensity Level

The sound intensity level L_I is given by

$$L_I = 10 \log \left(\frac{I}{I_{ref}} \right) \text{dB,} \qquad (3.24)$$

where I is the component of the sound intensity in a given direction and $I_{ref} = 10^{-12} \, \text{W/m}^2$ is the reference sound intensity.

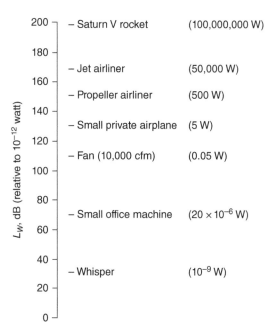

Figure 3.5 Some typical sound power levels, L_W.

3.4.4 Combination of Decibels

If the sound pressures p_1 and p_2 at a point produced by two independent sources are combined, the mean square pressure is

$$p^2{}_{rms} = \frac{1}{T}\int_0^T (p_1 + p_2)^2 dt = \left\langle p_1{}^2 + 2p_1p_2 + p_2{}^2 \right\rangle_t$$

$$= \left\langle p_1{}^2 \right\rangle_t + \left\langle p_2{}^2 \right\rangle_t + 2\langle p_1p_2 \rangle_t \equiv \overline{p_1{}^2} + \overline{p_2{}^2} + 2\overline{p_1p_2}, \tag{3.25}$$

where $\langle\rangle_t$ and the overbar indicate the time average $\frac{1}{T}\int()dt$.

Except for some special cases, such as two pure tones of the same frequency or the sounds from two correlated sound sources, the cross term $2\langle p_1p_2 \rangle_t$ disappears if $T \to \infty$. Then in such cases the mean square sound pressures $\overline{p_1{}^2}$ and $\overline{p_2{}^2}$ are additive, and the total mean square sound pressure at some point in space, if they are completely independent noise sources, may be determined using Eq. (3.26).

$$p^2{}_{rms} = \overline{p_1{}^2} + \overline{p_2{}^2}. \tag{3.26}$$

Let the two mean square pressure contributions to the total noise be $p^2{}_{rms1}$ and $p^2{}_{rms2}$ corresponding to sound pressure levels L_{p1} and L_{p2}, where $L_{p2} = L_{p1} - \Delta$. The total sound pressure level is given by the sum of the individual contributions in the case of uncorrelated sources, and the total sound pressure level is given by forming the total sound pressure level by taking logarithms of Eq. (3.26)

$$L_{pt} = 10\log_{10}\left(\frac{p^2{}_{rms1} + p^2{}_{rms2}}{p^2{}_{ref}}\right) = 10\log_{10}\left(10^{L_{p1}/10} + 10^{L_{p2}/10}\right)$$

$$= 10\log_{10}\left(10^{L_{p1}/10} + 10^{(L_{p1}-\Delta)/10}\right) = 10\log_{10}\left[10^{L_{p1}/10}\left(1 + 10^{-\Delta/10}\right)\right]$$

$$= L_{p1} + 10\log_{10}\left(1 + 10^{-\Delta/10}\right), \tag{3.27}$$

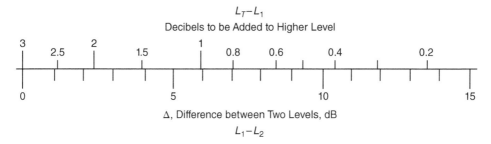

Figure 3.6 Diagram for combination of two sound pressure levels or two sound power levels of uncorrelated sources.

where L_{pt} is the combined sound pressure level due to both sources, L_{p1} is the greater of the two sound pressure level contributions, and Δ is the difference between the two contributions, all in dB. Equation (3.27) is presented in Figure 3.6.

Example 3.2   If two independent noise sources each create sound pressure levels operating on their own of 80 dB, at a certain point, what is the total sound pressure level?

Solution

The difference in levels is 0 dB; thus the total sound pressure level is $80 + 3 = 83$ dB.

Example 3.3   If two independent noise sources have sound power levels of 70 and 73 dB, what is the total level?

Solution

The difference in levels is 3 dB; thus the total sound power level is $73 + 1.8 = 74.8$ dB.

Figure 3.6 and these two examples do *not* apply to the case of two pure tones of the same frequency.

Note: For the special case of two pure tones of the same amplitude and frequency, if $p_1 = p_2$ (and the sound pressures are in phase at the point in space of the measurement):

$$L_{p_{total}} = 10 \log \left[\frac{1}{T} \int_0^T (p_1 + p_2)^2 dt \right] = L_{p1} + 10 \log 4 = L_{p2} + 6 \text{ dB}. \tag{3.28}$$

Example 3.4   If $p_1 = p_2 = 1$ Pa and the two sound pressures are of the same amplitude and frequency and in phase with each other, then the total sound pressure level

$$L_p(total) = 20 \log \left[\frac{2}{2 \times 10^{-5}} \right] = 100 \text{ dB}.$$

Example 3.5   If $p_1 = p_2 = 1$ Pa and the two sound pressures are of the same amplitude and frequency, but in opposite phase with each other, then the total sound pressure level

$$L_p(total) = 20 \log \left[\frac{0}{2 \times 10^{-5}} \right] = -\infty \text{ dB}.$$

For such a case as in Example 3.2 above, for pure tone sounds, instead of 83 dB, the total sound pressure level can range anywhere between 86 dB (for in-phase sound pressures) and $-\infty$ dB (for out-of-phase sound pressures). For the Example 3.3 above, the total sound power radiated by the two pure-tone sources depends on the phasing and separation distance.

Example 3.6   In a certain factory space the noise level with an electric motor in operation is 93 dB. When the motor is turned off the background noise is 85 dB. What is the sound pressure level due to the motor?

Solution

Since $10^{93/10} = 10^{85/10} + 10^{L_p/10}$, then $L_p = 10\log(10^{9.3} - 10^{8.5}) = 92.2$ dB.

3.5   Three-dimensional Wave Equation

In most sound fields, sound propagation occurs in two or three dimensions. The three-dimensional version of Eq. (3.1) in Cartesian coordinates is

$$\frac{\partial^2 p}{\partial x^2} + \frac{\partial^2 p}{\partial y^2} + \frac{\partial^2 p}{\partial z^2} - \frac{1}{c^2}\frac{\partial^2 p}{\partial t^2} = 0. \tag{3.29}$$

This equation is useful if sound wave propagation in rectangular spaces such as rooms is being considered. However, it is helpful to recast Eq. (3.29) in spherical coordinates if sound propagation from sources of sound in free space is being considered. It is a simple mathematical procedure to transform Eq. (3.29) into spherical coordinates, although the resulting equation is quite complicated. However, for propagation of sound waves from a spherically symmetric source (such as the idealized case of a pulsating spherical balloon known as an omnidirectional or monopole source) (Table 3.1), the equation becomes quite simple (since there is no angular dependence):

$$\frac{1}{r^2}\frac{\partial}{\partial r}\left(r^2\frac{\partial p}{\partial r}\right) - \frac{1}{c^2}\frac{\partial^2 p}{\partial t^2} = 0. \tag{3.30}$$

After some algebraic manipulation Eq. (3.30) can be written as

$$\frac{\partial^2(rp)}{\partial r^2} - \frac{1}{c^2}\frac{\partial^2(rp)}{\partial t^2} = 0. \tag{3.31}$$

Here, r is the distance from the origin and p is the sound pressure at that distance.

Equation (3.30) is identical in form to Eq. (3.1) with p replaced by rp and x by r. The general and simple harmonic solutions to Eq. (3.30) are thus the same as Eqs. (3.4) and (3.5) with p replaced by rp and x with r. The general solution is

$$rp = f_1(ct - r) + f_2(ct + r) \tag{3.32}$$

or

$$p = \frac{1}{r}f_1(ct - r) + \frac{1}{r}f_2(ct + r), \tag{3.33}$$

where f_1 and f_2 are arbitrary functions. The first term on the right of Eq. (3.33) represents a wave traveling outward from the origin; the sound pressure p is seen to be inversely proportional to the distance r. The

Table 3.1 Models of idealized spherical sources: Monopole, Dipole, and Quadrupole[a].

Monopole Distribution Representation	Velocity Distribution on Spherical Surface	Oscillating Sphere Representation	Oscillating Force Model

[a] For simple harmonic sources, after one half-period the velocity changes direction; positive sources become negative and vice versa, and forces reverse direction with dipole and quadrupole force models.

second term in Eq. (3.33) represents a sound wave traveling inward toward the origin, and in most practical cases such waves can be ignored (if reflecting surfaces are absent).

The simple harmonic (pure-tone) solution of Eq. (3.31) is

$$p = \frac{A_1}{r} \sin (\omega t - kr + \phi_1) + \frac{A_2}{r} \sin (\omega t + kr + \phi_2). \tag{3.33}$$

We may now write that the constants A_1 and A_2 may be written as $A_1 = \hat{p}_1 r$ and $A_2 = \hat{p}_2 r$, where $\hat{p}_1$ and $\hat{p}_2$ are the sound pressure amplitudes at unit distance (usually 1 m) from the origin.

3.6 Sources of Sound

The second term on the right of Eq. (3.33), as before, represents sound waves traveling inward to the origin and is of little practical interest. However, the first term represents simple harmonic waves of angular frequency ω traveling outward from the origin, and this may be rewritten as [4]

$$p = \frac{\rho c k Q}{4\pi r} \sin (\omega t - kr + \phi_1), \tag{3.34}$$

where Q is termed the *strength of an omnidirectional (monopole) source* situated at the origin, and $Q = 4\pi A_1 / \rho c k$. The mean-square sound pressure p^2_{rms} may be found [4] by time averaging the square of Eq. (3.34) over a period T:

$$p^2_{rms} = \frac{(\rho c k)^2 Q^2}{32\pi^2 r^2}. \tag{3.35}$$

From Eq. (3.35), the mean-square pressure is seen to vary with the inverse square of the distance r from the origin of the source, for such an idealized omnidirectional point sound source, everywhere in the sound field. Again, this is known as the *inverse square law*. If the distance r is doubled, the sound pressure level decreases by $20\log(2) = 20(0.301) = 6$ dB. If the source is idealized as a sphere of radius a pulsating with a simple harmonic velocity amplitude U, we may show that Q has units of volume flow rate (cubic metres per second). If the source radius is small in wavelengths so that $a \le \lambda$ or $ka \le 2\pi$, then we can show that the strength $Q = 4\pi a^2 U$.

Many sources of sound are not like the simple omnidirectional monopole source just described. For example, an unbaffled loudspeaker produces sound both from the back and front of the loudspeaker. The sound from the front and the back can be considered as two sources that are 180° out of phase with each other. This system can be modeled [13, 14] as two out-of-phase monopoles of source strength Q separated by a distance l. Provided $l \ll \lambda$, the sound pressure produced by such a dipole system is

$$p = \frac{\rho c k Q l \cos(\theta)}{4\pi r} \left[\frac{1}{r} \sin(\omega t - kr + \phi) + k \cos(\omega t - kr + \phi) \right], \tag{3.36}$$

where θ is the angle measured from the axis joining the two sources (the loudspeaker axis in the practical case). Unlike the monopole, the dipole field is not omnidirectional. The sound pressure field is directional. It is, however, symmetric and shaped like a figure-eight with its lobes on the dipole axis, as shown in Figure 3.11b.

The sound pressure of a dipole source has near-field and far-field regions that exhibit similar behaviors to the particle velocity near-field and far-field regions of a monopole.

Close to the source (the near field), for some fixed angle θ, the sound pressure falls off rapidly, $p \propto 1/r^2$, while far from the source (the far field $kr \ge 1$), the pressure falls off more slowly, $p \propto 1/r$. In the near field, the sound pressure level decreases by 12 dB for each doubling of distance r. In the far field the decrease in sound pressure level is only 6 dB for doubling of r (like a monopole). The phase of the sound pressure also changes with distance r, since close to the source the sine term dominates and far from the source the cosine term dominates. The particle velocity may be obtained from the sound pressure (Eq. (3.36)) and use of Euler's equation (see Eq. (3.37)). It has an even more complicated behavior with distance r than the sound pressure, having three distinct regions.

An oscillating force applied at a point in space gives rise to results identical to Eq. (3.36), and hence there are many real sources of sound that behave like the idealized dipole source described above, for example, pure-tone fan noise, vibrating beams, unbaffled loudspeakers, and even wires and branches (which sing in the wind due to alternate side vortex shedding).

The next higher order source is the quadrupole. It is thought that the sound produced by the mixing process in an air jet gives rise to stresses that are quadrupole in nature. Quadrupoles may be considered to consist of two opposing point forces (two opposing dipoles) or equivalently four monopoles (see Table 3.1). We note that some authors use slightly different but equivalent definitions for the source strength of monopoles, dipoles, and quadrupoles. The definitions used in Sections 3.6 and 3.7 of this chapter are the same as in Crocker and Price [4] and Fahy [13] and result in expressions for sound pressure, sound intensity, and sound power which, although equivalent, are different in form from those in Ref. [20], for example.

The expression for the sound pressure for a quadrupole is even more complicated than for a dipole. Close to the source, in the near field, the sound pressure $p \propto 1/r^3$. Farther from the sound source, $p \propto 1/r^2$; while in the far field, $p \propto 1/r$.

Sound sources experienced in practice are normally even more complicated than dipoles or quadrupoles. The sound radiation from a vibrating piston is described in Refs. [6, 16, 17, 21]. Chapters 9 and 11 in the *Handbook of Acoustics* [1] also describe radiation from dipoles and quadrupoles and the sound radiation from vibrating cylinders in chapter 9 of the same book [1].

The discussion in Ref. [21] considers steady-state radiation. However, there are many sources in nature and created by people that are transient. As shown in chapter 9 of the *Handbook of Acoustics*, [1] the harmonic analysis of these cases is often not suitable, and time-domain methods have given better results and understanding of the phenomena. These are the approaches adopted in chapter 9 of the *Handbook of Acoustics* [1].

3.6.1 Sound Intensity

The radial particle velocity in a nondirectional spherically spreading sound field is given by Euler's equation as

$$u = -\frac{1}{\rho}\int \frac{\partial p}{\partial r}dt \tag{3.37}$$

and substituting Eqs. (3.34) and (3.37) into (3.15) and then using Eq. (3.35) and time averaging gives the magnitude of the radial sound intensity in such a field as

$$\langle I \rangle_t = \frac{p^2_{rms}}{\rho c}, \tag{3.38}$$

the same result as for a plane wave. The sound intensity decreases with the inverse square of the distance r. Simple omnidirectional monopole sources radiate equally well in all directions. More complicated idealized sources such as dipoles, quadrupoles, and vibrating piston sources create sound fields that are directional. Of course, real sources such as machines produce even more complicated sound fields than these idealized sources. (For a more complete discussion of the sound fields created by idealized sources, see Ref. [21] and chapters 3 and 8 in the *Handbook of Acoustics* [1].) However, the same result as Eq. (3.38) is found to be true for any source of sound as long as the measurements are made sufficiently far from the source. The intensity is not given by the simple result of Eq. (3.38) close to idealized sources such as dipoles, quadrupoles, or more complicated real sources of sound such as vibrating structures.

Close to such sources Eq. (3.15) must be used for the instantaneous radial intensity, and

$$\langle I \rangle_t = \langle pu \rangle_t \tag{3.39}$$

for the time-averaged radial intensity.

The time-averaged radial sound intensity in the far field of a dipole is given by [4]

$$\langle I \rangle_t = \frac{\rho c k^4 (Ql)^2 \cos^2\theta}{32\pi^2 r^2}. \tag{3.40}$$

3.7 Sound Power of Sources

3.7.1 Sound Power of Idealized Sound Sources

The *sound power W* of a sound source is given by integrating the intensity over any imaginary closed surface S surrounding the source (see Figure 3.7):

$$W = \int_S \langle I_n \rangle_t dS. \tag{3.41}$$

The normal component of the intensity I_n must be measured in a direction perpendicular to the elemental area dS. If a spherical surface, whose center coincides with the source, is chosen, then the sound power of an omnidirectional (monopole) source is

$$W_m = \langle I_r \rangle_t 4\pi r^2, \tag{3.42}$$

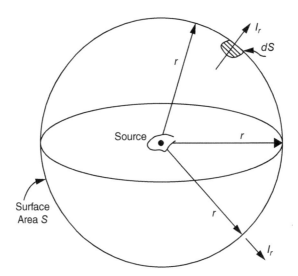

Figure 3.7 Imaginary surface area S for integration.

$$W_m = \frac{p^2_{rms}}{\rho c} 4\pi r^2,$$ (3.43)

and from Eq. (3.35) the sound power of a monopole is [4, 13]

$$W_m = \frac{\rho c k^2 Q^2}{8\pi}.$$ (3.44)

It is apparent from Eq. (3.44) that the sound power of an idealized (monopole) source is independent of the distance r from the origin, at which the power is calculated. This is the result required by conservation of energy and also to be expected for all sound sources.

Equation (3.43) shows that for an omnidirectional source (in the absence of reflections) the sound power can be determined from measurements of the mean square sound pressure made with a single microphone. Of course, for real sources, in environments where reflections occur, measurements should really be made very close to the source, where reflections are presumably less important.

The sound power of a dipole source is obtained by integrating the intensity given by Eq. (3.40) over a sphere around the source. The result for the sound power is

$$W_d = \frac{\rho c k^4 (Ql)^2}{24\pi}.$$ (3.45)

The dipole is obviously a much less efficient radiator than a monopole, particularly at low frequency.

Example 3.7 Two monopoles of equal sound power $W = 0.1$ watt at 150 Hz, but pulsating with a phase difference of 180° are spaced $\lambda/12$ apart. Determine the sound power of this dipole at 150 Hz.

Solution

We know that $l = \lambda/12$, $\lambda = 343/150 = 2.29$ m, and $W_m = 0.1$ watt. If we compare the sound power of a dipole W_d with that of a monopole W_m (Eqs. (3.44) and (3.45)) we find that

$$\frac{W_d}{W_m} = \frac{1}{3}(kl)^2. \text{ Since } k = 2\pi/\lambda, \text{ we obtain } W_d = \frac{1}{3}(kl)^2 W_m = \frac{1}{3}\left(\frac{2\pi\lambda}{12\lambda}\right)^2 0.1 = 9 \times 10^{-3} \text{ watt.}$$

Therefore, the sound power radiated by the dipole is 9 mW.

In practical situations with real directional sound sources and where background noise and reflections are important, use of Eq. (3.43) becomes difficult and less accurate, and then the sound power is more conveniently determined from Eq. (3.41) with a sound intensity measurement system. See Ref. [22] in this book and chapter 106 in the *Handbook of Acoustics* [1].

We note that since $p/u_r = \rho c$ (where ρ = mean air density kg/m^3 and c = speed of sound 343 m/s) for a plane wave or sufficiently far from any source, that

$$I_r = \frac{1}{T}\int_0^T \frac{p^2(t)}{\rho c} = \frac{p^2_{rms}}{\rho c}, \tag{3.46}$$

where Eq. (3.46) is true for random noise as well as for a single-frequency sound, known as a pure tone.

Note that for such cases we only need to measure the mean-square sound pressure with a simple sound level meter (or at least a simple measurement system) to obtain the sound intensity from Eq. (3.46) and then from that the sound power W watts from Eq. (3.41) is

$$W = \int_S \frac{p^2_{rms}}{\rho c}\, dS = 4\pi r^2 \frac{p^2_{rms}}{\rho c}, \tag{3.47}$$

for an omnidirectional source (monopole) with no reflections and no background noise. This result is true for noise signals and pure tones that are produced by omnidirectional sources and in the so-called far acoustic field.

For the special case of a pure-tone (single-frequency) source of sound pressure amplitude, $\hat{p}$, we note that $I_r = \hat{p}^2/2\rho c$ and $W = 2\pi r^2 \hat{p}^2/\rho c$ from Eq. (3.47).

For measurements on a hemisphere, $W = 2\pi r^2 p^2_{rms}/\rho c$ and for a pure-tone source $I_r = \hat{p}^2/2\rho c$, and $W = \pi r^2 \hat{p}^2/\rho c$, from Eq. (3.47).

Note that in the general case, the source is *not* omnidirectional, or more importantly, we must often measure quite close to the source so that we are in the *near* acoustic field, not the *far* acoustic field. However, if appreciable reflections or background noise (i.e. other sound sources) are present, then we *must* measure the intensity I_r in Eq. (3.41). Figure 3.8 shows two different enclosing surfaces that can be used to determine the sound power of a source. The sound intensity I_n must always be measured perpendicular (or normal) to the enclosing surfaces used. Measurements are normally made with a two-microphone probe (see Ref. [22]). The most common microphone arrangement is the face-to-face model (see Figure 3.9).

The microphone arrangement shown also indicates the microphone separation distance, Δr, needed for the intensity calculations [22]. In the face-to-face arrangement a solid cylindrical spacer is often put between the two microphones to improve the performance.

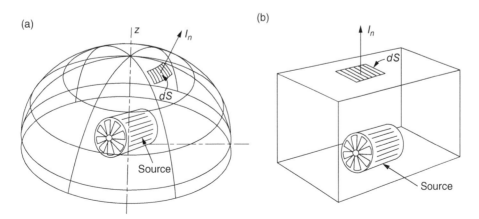

Figure 3.8 Sound intensity I_n, being measured on (a) segment dS of an imaginary hemispherical enclosure surface and (b) an elemental area dS of a rectangular enclosure surface surrounding a source having a sound power W.

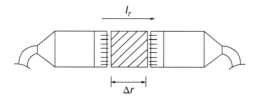

Figure 3.9 Sound intensity probe microphone arrangement commonly used.

Example 3.8 By making measurements around a source (an engine exhaust pipe) it is found that it is largely omnidirectional at low frequency (in the range of 50–200 Hz). If the measured sound pressure level on a spherical surface 10 m from the source is 60 dB at 100 Hz, which is equivalent to a mean-square sound pressure p^2_{rms} of $(20 \times 10^{-3})^2$ (Pa)^2, what is the sound power in watts at 100 Hz frequency?

Solution

Assuming that $\rho = 1.21$ kg/m^3 and $c = 343$ m/s, so $\rho c = 415 \approx 400$ rayls:

$$p^2_{rms} = \left(20 \times 10^{-3}\right)^2 = 400 \times 10^{-6}\ \text{Pa}^2,$$

then from Eq. (3.47):

$$W = 4\pi r^2 \left(400 \times 10^{-6}\right)/\rho c = 4\pi\left(100 \times 400 \times 10^{-6}\right)/400$$

$$\approx 4\pi \times 10^{-4} \approx 1.26 \times 10^{-3}\ \text{watts.}$$

Example 3.9 If the sound intensity level, measured using a sound intensity probe at the same frequency, as in Example 3.8, but at 1 m from the exhaust exit, is 80 dB (which is equivalent to 0.0001 W/m^2), what is the sound power of the exhaust source at this frequency?

Solution

From Eq. (3.41) $W = \int_S I_r dS = (0.0001) \times 4\pi(1)^2$ (for an omnidirectional source). Then $W = 1.26 \times 10^{-3}$ watts (the same result as Example 3.8).

Sound intensity measurements do and should give the same result as sound pressure measurements made in a free field.

Far away from omnidirectional sound sources, provided there is no background noise and reflections can be ignored:

$$p^2_{rms} = \rho c W / (4\pi r^2),$$ (3.48)

$$\frac{p^2_{rms}}{p^2_{ref}} = \frac{W}{W_{ref}} \times \frac{1}{r^2} \times \rho c \times \frac{W_{ref}}{p^2_{ref}4\pi(1)^2},$$ (3.49)

and by taking 10 log throughout this equation

$$L_p = L_W - 20 \log r - 11 \text{ dB},$$ (3.50)

where L_p is the sound pressure level, L_W is the source sound power level, and r is the distance, in metres, from the source center. (Note we have assumed here that $\rho c = 415 \cong 400$ rayls.) If $\rho c \cong 400$ rayls (kg/m^{2}s), then since $I = p^2_{rms}/\rho c$

$$I/I_{ref} = \left(p^2_{rms}/p^2_{ref}\right)\left(p^2_{ref}/I_{ref}\rho c\right).$$

So,

$$L_I = L_p + 10 \log \left[\frac{400 \times 10^{-12}}{1} \frac{1}{400 \times 10^{-12}}\right]$$ (3.51)

$$L_I = L_p + 0 \text{ dB}.$$

3.8 Sound Sources Above a Rigid Hard Surface

In practice many real engineering sources (such as machines and vehicles) are mounted or situated on hard reflecting ground and concrete surfaces. If we can assume that the source of sound power W radiates only to a half-space solid angle 2π, and no power is absorbed by the hard surface (Figure 3.10), then

$$I = W/2\pi r^2,$$

$$L_p \cong L_I = L_W - 20 \log r - 8 \text{ dB},$$ (3.52)

where L_W is the sound power level of the source and r is the distance in metres.

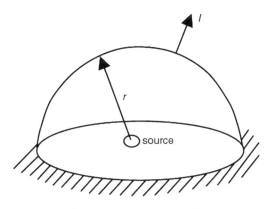

Figure 3.10   Source above a rigid surface.

In this discussion we have assumed that the sound source radiates the same sound intensity in all directions; that is, it is omnidirectional. If the source of sound power W becomes directional, the mean square sound pressure in Eqs. (3.48) and (3.46) will vary with direction, and the sound power W can only be obtained from Eqs. (3.41) and (3.47) by measuring either the mean-square pressure (p^2_{rms}) all over a surface enclosing the source (in the far acoustic field, the far field) and integrating Eq. (3.47) over the surface, or by measuring the intensity all over the surface in the *near* or *far* acoustic field and integrating over the surface (Eq. (3.41)). We shall discuss source directivity in Section 3.9.

Example 3.10   If the sound power level of a source is 120 dB (which is equivalent to 1 acoustical watt), what is the sound pressure level at 50 m (a) for sound radiation to whole space and (b) for radiation to half space?

Solution

a) For whole space: $I = 1/4\pi(50)^2 = 1/10^4\pi$ (W/m^2), then

$$L_I = 10\log\left(\frac{10^{-4}}{\pi 10^{-12}}\right) \left(\text{since } I_{ref} = 10^{-12}\,\text{W/m}^2\right)$$

$$= 10\log\left(10^8\right) - 10\log\pi$$

$$= 80 - 5 = 75\,\text{dB}.$$

Since we may assume $r = 50$ m is in the far acoustic field, $L_p \cong L_I = 75$ dB as well (we have also assumed $\rho c \cong$ 400 rayls).
For half space: $I = 1/2\pi(50)^2 = 2/10^4\pi$ (W/m^2), then

$$L_I = 10\log\left(\frac{2 \times 10^{-4}}{\pi 10^{-12}}\right) \left(\text{since } I_{ref} = 10^{-12}\,\text{W/m}^2\right)$$

$$= 10\log 2 + 10\log\left(10^8\right) - 10\log\pi$$

$$= 3 + 80 - 5 = 78\,\text{dB}$$

and $L_p \cong L_I = 78$ dB also.

It is important to note that the sound power radiated by a source can be significantly affected by its environment. For example, if a simple constant-volume velocity source (whose strength Q will be unaffected by the environment) is placed on a floor, its sound power will be doubled (and its sound power level increased by 3 dB). If it is placed at a floor–wall intersection, its sound power will be increased by four times (6 dB); and if it is placed in a room corner, its power is increased by eight times (9 dB). See Table 3.2. Many simple sources of sound (ideal sources, monopoles, and real small machine sources) produce more sound power when put near reflecting surfaces, provided their surface velocity remains constant. For example, if a monopole is placed touching a hard plane, an image source of equal strength may be assumed.

3.9 Directivity

The sound intensity radiated by a dipole is seen to depend on $\cos^2\theta$ (see Figure 3.11). Most real sources of sound become directional at high frequency, although some are almost omnidirectional at low frequency. This phenomenon depends on the source dimension, d, which must be small in size compared with a wavelength λ, so $d/\lambda \ll 1$ for them to behave almost omnidirectionally.

Table 3.2 Simple source near reflecting surfaces[a].

Intensity	Source	Condition	Number of Images	p^2_{rms}	Power	D	DI
I		Free field	None	p^2_{rms}	W	1	0 dB
$4I$		Reflecting plane	1	$4p^2_{\text{rms}}$	$2W$	4	6 dB
$16I$		Wall-floor intersection	3	$16p^2_{\text{rms}}$	$4W$	16	12 dB
$64I$		Room corner	7	$54p^2_{\text{rms}}$	$8W$	64	18 dB

[a] Q and DI are defined in Eqs. (3.53), (3.58), and (3.60).

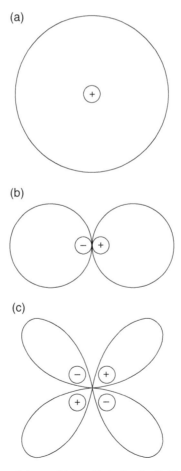

Figure 3.11 Polar directivity plots for the radial sound intensity in the far field of (a) monopole, (b) dipole, and (c) (lateral) quadrupole.

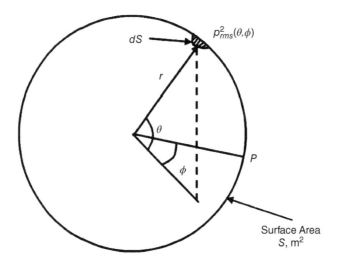

Figure 3.12 Geometry used in derivation of directivity factor.

3.9.1 Directivity Factor ($Q(\theta, \phi)$)

In general, a directivity factor $Q_{\theta,\phi}$ may be defined as the ratio of the radial intensity $\langle I_{\theta, \phi}\rangle_t$ (at angles θ and ϕ and distance r from the source) to the radial intensity $\langle I_s\rangle_t$ at the same distance r radiated from an omnidirectional source of the same total sound power (Figure 3.12). Thus

$$D_{\theta,\phi} = \frac{\langle I_{\theta,\phi}\rangle_t}{\langle I_s\rangle_t}. \tag{3.53}$$

For a directional source, the mean square sound pressure measured at distance r and angles θ and ϕ is $p^2_{rms}(\theta,\phi)$.

In the far field of this source ($r \gg \lambda$), then

$$W = \int_S \frac{p^2_{rms}(\theta,\phi)}{\rho c} dS. \tag{3.54}$$

But if the source were omnidirectional of the same power W, then

$$W = \int_S \frac{p^2_{rms}}{\rho c} dS, \tag{3.55}$$

where p^2_{rms} is a constant, independent of angles θ and ϕ.

We may therefore write:

$$W = \frac{1}{\rho c}\int_S p^2_{rms}(\theta, \phi)dS = \frac{p^2_{rms}}{\rho c}\int_S dS \tag{3.56}$$

and

$$p^2_{rms} = p^2_S = \frac{1}{S}\int_S p^2(\theta, \phi)dS, \tag{3.57}$$

where p^2_S is the space-averaged mean-square sound pressure.

We define the directivity factor Q as

$$Q(\theta, \phi) = \frac{p^2_{rms}}{p^2_S} \text{ or } Q(\theta, \phi) = \frac{p^2_{rms}(\theta, \phi)}{p^2_{rms}} \tag{3.58}$$

the ratio of the mean-square sound pressure at distance r to the space-averaged mean-square pressure at r, or equivalently the directivity Q may be defined as the ratio of the mean-square sound pressure at r divided by the mean-square sound pressure at r for an omnidirectional sound source of the same sound power W, watts.

3.9.2 Directivity Index

The directivity index DI is just a logarithmic version of the directivity factor Q. It is expressed in decibels.
A directivity index $DI_{\theta, \phi}$ may be defined, where

$$DI_{\theta, \phi} = 10 \log Q_{\theta, \phi}, \tag{3.59}$$

$$DI(\theta, \phi) = 10 \log Q(\theta, \phi). \tag{3.60}$$

Note if the source power remains the same when it is put on a hard rigid infinite surface $Q(\theta, \phi) = 2$ and $DI(\theta, \phi) = 3$ dB.

Example 3.11

a) If a constant-volume velocity source of sound power level 120 dB (which is equivalent to 1 acoustic watt) radiates to whole space and it has a directivity factor of 12 at 50 m, what is the sound pressure level in that direction?
b) If this constant-volume velocity source is put very near a hard reflecting floor, what will its sound pressure level be in the same direction?

Solution

a) We have that $I = 1/4\pi(50)^2 = 1/10^4\pi$ (W/m^2), then

$$\langle L_I \rangle_S = 10 \log \left(\frac{10^{-4}}{\pi 10^{-12}} \right) \left(\text{since } I_{ref} = 10^{-12} \text{ W/m}^2 \right)$$

$$= 10 \log \left(10^8 \right) - 10 \log \pi = 75 \text{ dB}.$$

But for the directional source $L_p(\theta, \phi) = \langle L_p \rangle_S + DI(\theta, \phi)$, then assuming $\rho c = 400$ rayls, $L_p(\theta, \phi) = 75 + 10 \log 12 = 75 + 10 + 10 \log 1.2 = 85.8$ dB.
b) If the direction is away from the floor, then

$$L_p(\theta, \phi) = 85.8 + 6 = 91.8 \text{ dB}.$$

3.10 Line Sources

Sometimes noise sources are distributed more like idealized *line sources*. Examples include the sound radiated from a long pipe containing fluid flow or the sound radiated by a stream of vehicles on a highway.

If sound sources are distributed continuously along a straight line and the sources are radiating sound independently, so that the sound power/unit length is W' watts/metre, then assuming cylindrical spreading (and we are located in the far acoustic field again and $\rho c = 400$ rayls):

$$I = \frac{W'}{2\pi r}, \tag{3.61}$$

so,

$$L_I = 10\log\left(\frac{I}{I_{ref}}\right) = 10\log\left(\frac{W'}{2\times 10^{-12}\pi r}\right),$$

then

$$L_p \cong L_I = 10\log\left(W'/r\right) + 112\,\text{dB}, \tag{3.62}$$

and for half-space radiation (such as a line source on a hard surface, such as a road)

$$L_p \cong L_I = 10\log\left(W'/r\right) + 115\,\text{dB}. \tag{3.63}$$

3.11 Reflection, Refraction, Scattering, and Diffraction

For a homogeneous plane sound wave at normal incidence on a fluid medium of different characteristic impedance ρc, both reflected and transmitted waves are formed (see Figure 3.13).

From energy considerations (provided no losses occur at the boundary) the sum of the reflected intensity I_r and transmitted intensity I_t equals the incident intensity I_i:

$$I_i = I_r + I_t, \tag{3.64}$$

and dividing throughout by I_i,

$$\frac{I_r}{I_i} + \frac{I_t}{I_i} = R + T = 1, \tag{3.65}$$

where R is the *energy reflection coefficient* and T is the *transmission coefficient*. For plane waves at normal incidence on a plane boundary between two fluids (see Figure 3.13):

$$R = \frac{(\rho_1 c_1 - \rho_2 c_2)^2}{(\rho_1 c_1 + \rho_2 c_2)^2}, \tag{3.66}$$

and

$$T = \frac{4\rho_1 c_1 \rho_2 c_2}{(\rho_1 c_1 + \rho_2 c_2)^2}. \tag{3.67}$$

Some interesting facts can be deduced from Eqs. (3.66) and (3.67). Both the reflection and transmission coefficients are independent of the direction of the wave since interchanging $\rho_1 c_1$ and $\rho_2 c_2$ does not affect the values of R and T. For example, for sound waves traveling from air to water or water to air, almost

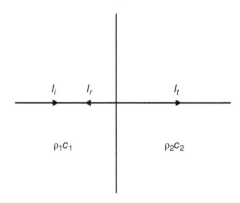

Figure 3.13 Incident intensity I_i, reflected intensity I_r, and transmitted intensity I_t in a homogeneous plane sound wave at normal incidence on a plane boundary between two fluid media of different characteristic impedances.

complete reflection occurs, independent of direction; the reflection coefficients are the same and the transmission coefficients are the same for the two different directions.

Example 3.12 A plane sound wave in air is incident normally on a boundary between air and water. If each medium can be assumed to be infinite in extent, compute the energy reflection and transmission coefficients.

Solution

At standard temperature and atmospheric pressure we have that in water $\rho = 998$ kg/m^3 and $c = 1480$ m/s. Then, the characteristic impedance of water is $\rho c = 1\,480\,000$ rayls.
   Then, $R = [(415 - 1\,480\,000)/(415 + 1\,480\,000)]^2 = 0.999$ and
   $T = 4(415)(1\,480\,000)/(415 + 1\,480\,000)^2 = 0.001$, or simply $T = 1 - R = 0.001$. Therefore, when a sound wave passes from air to water most of its energy is reflected back because of the impedance offered by the liquid medium.

As discussed before, when the characteristic impedance ρc of a fluid medium changes, incident sound waves are both reflected and transmitted. It can be shown that if a plane sound wave is incident at an oblique angle on a plane boundary between two fluids, then the wave transmitted into the changed medium changes direction. This effect is called *refraction*. Temperature changes and wind speed changes in the atmosphere are important causes of refraction.

Wind speed normally increases with altitude, and Figure 3.14 shows the refraction effects to be expected for an idealized wind speed profile. Atmospheric temperature changes alter the speed of sound c, and temperature gradients can also produce sound shadow and focusing effects, as seen in Figures 3.15 and 3.16.

When a sound wave meets an obstacle, some of the sound wave is deflected. The *scattered* wave is defined to be the difference between the resulting wave with the obstacle and the undisturbed wave without the presence of the obstacle. The scattered wave spreads out in all directions interfering with the undisturbed wave. If the obstacle is very small compared with the wavelength, no sharp-edged sound shadow is created behind the obstacle. If the obstacle is large compared with the wavelength, it is normal to say that the sound wave is *reflected* (in front) and *diffracted* (behind) the obstacle (rather than *scattered*).

In this case when the obstacle is large a strong sound shadow is caused in which the wave pressure amplitude is very small. In the zone between the sound shadow and the region fully "illuminated" by the source, the sound wave pressure amplitude oscillates. These oscillations are maximum near the shadow boundary and minimum well inside the shadow. These oscillations in amplitude are normally termed *diffraction bands*. One of the most common examples of diffraction caused by a body is the diffraction of sound over the sharp edge of a *barrier* or *screen*. For a plane homogeneous sound wave it is found that a strong shadow is caused by

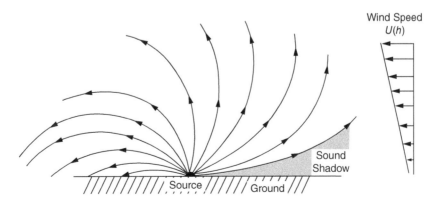

Figure 3.14 Refraction of sound in air with wind speed $U(h)$ increasing with altitude h.

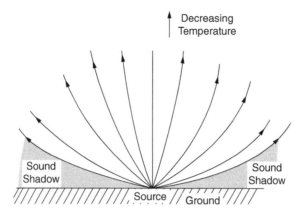

Figure 3.15 Refraction of sound in air with normal temperature lapse (temperature decreases with altitude).

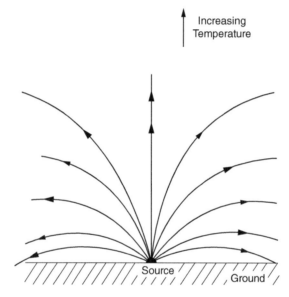

Figure 3.16 Refraction of sound in air with temperature inversion.

high-frequency waves, where $h/\lambda \geq 1$ and a weak shadow where $h/\lambda \leq 1$, where h is the barrier height and λ is the wavelength. For intermediate cases where $h/\lambda \approx 1$, a variety of interference and diffraction effects are caused by the barrier.

Scattering is caused not only by obstacles placed in the wave field but also by fluid regions where the properties of the medium such as its density or compressibility change their values from the rest of the medium. Scattering is also caused by turbulence (see chapters 5 and 28 in the *Handbook of Acoustics* [1]) and from rain or fog particles in the atmosphere and bubbles in water and by rough or absorbent areas on wall surfaces.

3.12 Ray Acoustics

There are three main modeling approaches in acoustics, which may be termed wave acoustics, ray acoustics, and energy acoustics. So far in this chapter we have mostly used the wave acoustics approach in which the acoustical quantities are completely defined as functions of space and time. This approach is practical in certain cases where the fluid medium is bounded and in cases where the fluid is unbounded as long as the fluid is homogenous. However, if the fluid properties vary in space due to variations in temperature or due to wind gradients, then the wave approach becomes more difficult and other simplified approaches such as the ray acoustics approach described here and in chapter 3 of the *Handbook of Acoustics* [1] are useful. This approach can also be extended to propagation in fluid-submerged elastic structures, as described in chapter 4 of the *Handbook of Acoustics* [1]. The energy approach is described in Section 3.13.

In the ray acoustics approach, rays are obtained that are solutions to the simplified eikonal equation (Eq. (3.68))

$$\left(\frac{\partial S}{\partial x}\right)^2 + \left(\frac{\partial S}{\partial y}\right)^2 + \left(\frac{\partial S}{\partial z}\right)^2 - \frac{1}{c^2} = 0. \tag{3.68}$$

The ray solutions can provide good approximations to more exact acoustical solutions. In certain cases they also satisfy the wave equation [14]. The eikonal $S(x, y, z)$ represents a surface of constant phase (or wavefront) that propagates at the speed of sound c. It can be shown that Eq. (3.68) is consistent with the wave equation only in the case when the frequency is very high [7]. However, in practice, it is useful, provided the changes in the speed of sound c are small when measured over distances comparable with the wavelength. In the case where the fluid is homogeneous (constant sound speed c and density ρ throughout), S is a constant and represents a plane surface given by $S = (\alpha x + \beta y + \gamma z)/c$, where α, β, and γ are the direction cosines of a straight line (a ray) that is perpendicular to the wavefront (surface S). If the fluid can no longer be assumed to be homogeneous and the speed of sound $c(x, y, z)$ varies with position, the approach becomes approximate only. In this case some paths bend and are no longer straight lines. In cases where the fluid has a mean flow, the rays are no longer quite parallel to the normal to the wavefront. This ray approach is described in more detail in several books [6, 12, 15, 16] and in chapter 3 of the *Handbook of Acoustics* [1] (where in this chapter the main example is from underwater acoustics).

The ray approach is also useful for the study of propagation in the atmosphere and is a method to obtain the results given in Figures 3.14–3.16. It is observed in these figures that the rays always bend in a direction toward the region where the sound speed is less. The effects of wind gradients are somewhat different since in that case the refraction of the sound rays depends on the relative directions of the sound rays and the wind in each fluid region.

3.13 Energy Acoustics

In enclosed spaces the wave acoustics approach is useful, particularly if the enclosed volume is small and simple in shape and the boundary conditions are well defined. In the case of rigid walls of simple geometry, the wave equation is used, and after the applicable boundary conditions are applied, the solutions for the natural (eigen) frequencies for the modes (standing waves) are found. See Refs. [23, 24], and chapter 6 in the *Handbook of Acoustics* [1] for more details. However, for large rooms with irregular shape and absorbing boundaries, the wave approach becomes impracticable and other approaches must be sought. The ray acoustics approach together with the multiple-image-source concept is useful in some room problems, particularly in auditorium design or in factory spaces where barriers are involved. However, in many cases a statistical approach where the energy in the sound field is considered is the most useful. See Refs. [25, 26] and also chapters 60–62 in the *Handbook of Acoustics* [1] for more detailed discussion of this approach. Some of the fundamental concepts are briefly described here.

For a plane wave progressing in one direction in a duct of unit cross-section area, all of the sound energy in a column of fluid c metres in length must pass through the cross-section in one second. Since the intensity $\langle I \rangle_t$ is given by $p^2_{\mathrm{rms}}/\rho c$, then the total sound energy in the fluid column c metres long must also be equal to $\langle I \rangle_t$. The energy per unit volume ε (joules per cubic metre) is thus

$$\varepsilon = \frac{\langle I \rangle_t}{c}, \tag{3.69}$$

or

$$\varepsilon = \frac{p^2_{rms}}{\rho c^2}.$$ (3.70)

The energy density ε may be derived by alternative means and is found to be the same as that given in Eq. (3.69) in most acoustic fields, except very close to sources of sound and in standing-wave fields. In a room with negligibly small absorption in the air or at the boundaries, the sound field created by a source producing broadband sound will become very reverberant (the sound waves will reach a point with equal probability from any direction). In addition, for such a case the sound energy may be said to be diffuse if the energy density is the same anywhere in the room. For these conditions, the time-averaged intensity incident on the walls (or on an imaginary surface from one side) is

$$\langle I \rangle_t = \frac{1}{4} \varepsilon c,$$ (3.71)

or

$$\langle I \rangle_t = \frac{p^2_{rms}}{4\rho c}.$$ (3.72)

In any real room, the walls will absorb some sound energy (and convert it into heat).

3.14 Near Field, Far Field, Direct Field, and Reverberant Field

Near to a source, we call the sound field, the *near acoustic field*. Far from the source, we call the field the *far acoustic field*. The extent of the near field depends on:

1) The type of source: (monopole, dipole, size of machine, type of machine, etc.)
2) Frequency of the sound.

In the *near field* of a source, the sound pressure and particle velocity tend to be very nearly *out of phase* ($\approx 90°$).

In the *far field*, the sound pressure and particle velocity are very nearly in phase. Note, far from any source, the sound wave fronts flatten out in curvature, and the waves appear to an observer to be like plane waves. In-plane progressive waves, the sound pressure and particle velocity are in phase (provided there are no reflected waves). Thus far from a source (or in a plane progressive wave) $p/u = \rho c$. Note ρc is a real number, so the sound pressure p and particle velocity u must be in phase.

Figure 3.17 shows the example of a finite monopole source with a normal simple harmonic velocity amplitude U. On the surface of the monopole, the surface velocity is equal to the particle velocity. The particle velocity decreases in inverse proportion to the distance from the source center O.

It is common to make the assumption that $kr = 2\pi f r/c = 10$ is the boundary between the near and far fields. Note this is only one criterion and that there is no sharp boundary, but only a gradual transition. First we should also think of the type and the dimensions of the source and assume, say that $r \gg d$, where d is a source dimension. We might say that $r > 10d$ should also be applied as a secondary criterion to determine when we are in the far field.

3.14.1 Reverberation

In a confined space there will be reflections, and far from the source the reflections will dominate. We call this reflection-dominated region the *reverberant field*. The region where reflections are unimportant and where a doubling of distance results in a sound pressure drop of 6 dB is called the *free* or *direct* field (see Figure 3.18).

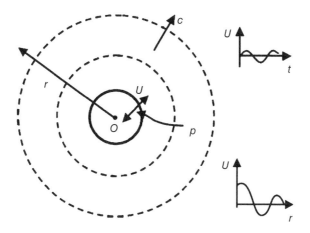

Figure 3.17 Example of monopole. On the monopole surface, velocity of surface U = particle velocity in the fluid.

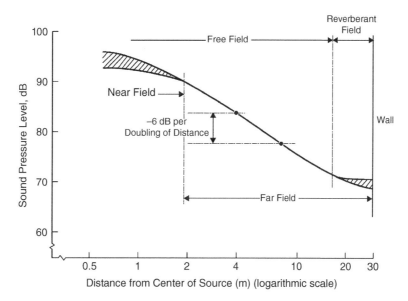

Figure 3.18 Sound pressure level in an interior sound field.

3.14.2 Sound Absorption

The sound absorption coefficient α of sound-absorbing materials (curtains, drapes, carpets, clothes, fiberglass, acoustical foams, etc.), is defined as

$$\alpha = \text{sound energy absorbed/sound energy incident,}$$

$$\alpha = \text{sound power absorbed/sound power incident,}$$

$$\alpha = \text{sound intensity absorbed/sound intensity incident} = \frac{I_a}{I_i}. \tag{3.73}$$

Note that α also depends on the angle of incidence. The absorption coefficient of materials depends on frequency as well. Thicker materials absorb more sound energy (particularly important at low frequency). See Figure 3.19.

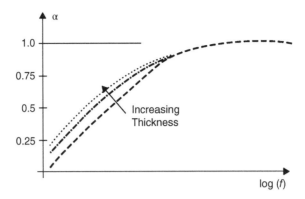

Figure 3.19 Sound absorption coefficient α of typical absorbing materials as a function of frequency.

If all the sound energy is absorbed, $\alpha = 1$ (none reflected). If no sound energy is absorbed, $\alpha = 0$:

$$0 \leq \alpha \leq 1.$$

If $\alpha = 1$, the sound absorption is perfect (e.g. an open window).

The behavior of sound-absorbing materials is described in more detail in Chapter 9 of this book.

3.14.3 Reverberation Time

In a reverberant space, the reverberation time T_R is normally defined to be the time for the sound pressure level to drop by 60 dB when the sound source is cut off (see Figure 3.20). Different reverberation times are desired for different types of spaces (see Figure 3.21). The Sabine formula is often used, $T_R = T_{60}$ (for 60 dB):

$$T_{60} = \frac{55.3V}{cS\overline{\alpha}} = \frac{0.161V}{S\overline{\alpha}},$$

where V is room volume (m³), c is the speed of sound (m/s), S is wall area (m²), and $\overline{\alpha}$ is the angle-averaged wall absorption coefficient, or

$$T_{60} = \frac{55.3V}{c \sum_{i=1}^{n} S_i \alpha_i}, \tag{3.74}$$

where S_i is ith wall area of absorption coefficient α_i.

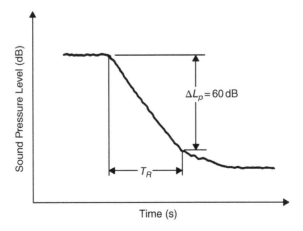

Figure 3.20 Measurement of reverberation time T_R.

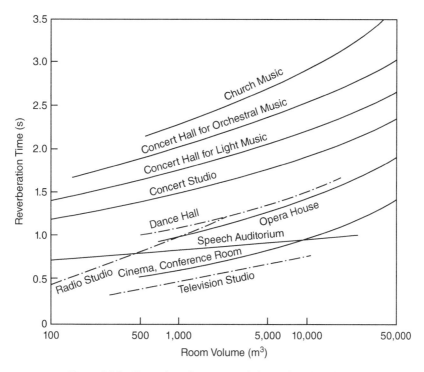

Figure 3.21 Examples of recommended reverberation times.

In practice, when the reverberation time is measured (see Figure 3.20), it is normal practice to ignore the first 5-dB drop in sound pressure level and find the time between the 5-dB and 35-dB drops and multiply this time by 2 to obtain the reverberation time T_R.

Example 3.13 A room has dimensions $5 \times 6 \times 10$ m^3. What is the reverberation time T_{60} if the floor (6×10 m) has absorbing material $\bar{a} = 0.5$ placed on it?

Solution

We will assume that $\bar{a} = 0$ on the other surfaces (that are made of hard painted concrete.)

$$T = \frac{55.3V}{cS\bar{a}} = \frac{55.3(5 \times 6 \times 10)}{343(6 \times 10)0.5} = 1.6 \text{ s}.$$

Notice that the Sabine reverberation time formula $T_{60} = 0.16\,V/S\bar{a}$ still predicts a reverberation time as $\bar{a} \rightarrow 1$, which does not agree with the physical world. This is approximately the case of an anechoic room (see Figure 3.22). Some improved formulas have been devised by Eyring and Millington-Sette that overcome this problem. Sabine's formula is acceptable, provided $\bar{a} \leq 0.5$.

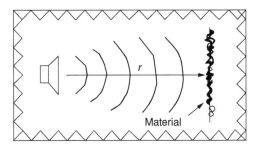

Figure 3.22 Sound source in anechoic room.

Example 3.14   A classroom has dimensions $4 \times 6 \times 10\,\mathrm{m}^3$ and a reverberation time of 1.5 seconds. (a) Determine the total sound absorption of the classroom; (b) if 35 students are in the classroom, and each is equivalent to 0.45 sabins (m^2) sound absorption, what is the new reverberation time of the classroom?

Solution

a) the volume of the classroom is $V = 240\,\mathrm{m}^3$. Therefore

$$S\bar{\alpha} = \frac{55.3(240)}{343(1.5)} = 25.8 \text{ sabins } (\mathrm{m}^2).$$

b) The total sound absorption is now $25.8 + 35(0.45) = 41.55$ sabins (m^2). Then

$$T = \frac{55.3(240)}{343(41.55)} = 0.93\,\mathrm{s}.$$

3.15 Room Equation

If we have a *diffuse* sound field (the same sound energy at any point in the room) and the field is also *reverberant* (the sound waves may come from any direction, with equal probability), then the sound intensity striking the wall of the room is found by integrating the plane wave intensity over all angles θ, $0 < \theta < 90°$. This involves a weighting of each wave by $\cos\theta$, and the average intensity for the wall in a reverberant field becomes

$$I_{rev} = \frac{p^2_{rms}}{4\rho c}. \tag{3.75}$$

Note the factor 1/4 compared with the plane wave case.

For a point in a room at distance r from a source of power W watts, we will have a direct field intensity contribution $W/4\pi r^2$ from an omnidirectional source to the mean square pressure and also a reverberant contribution.

We may define the reverberant field as the field created by waves after the first reflection of direct waves from the source. Thus the energy/second absorbed at the first reflection of waves from the source of sound power W is $W\bar{\alpha}$, where $\bar{\alpha}$ is the average absorption coefficient of the walls. The power thus supplied to the reverberant field is $W(1 - \bar{\alpha})$ (after the first reflection). Since the power lost by the reverberant field must equal the power supplied to it for steady-state conditions, then

$$S\bar{\alpha}\left(\frac{p^2_{rms}}{4\rho c}\right) = W(1-\bar{\alpha}),$$ (3.76)

where p^2_{rms} is the mean-square sound pressure contribution caused by the reverberant field.

There is also the direct field contribution to be accounted for. If the source is a broadband noise source, these two contributions: (i) the direct term $p^2_{d,rms} = \rho c W / 4\pi r^2$ and (ii) the reverberant contribution, $p^2_{rev,rms} = 4\rho c W (1-\bar{\alpha})/S\bar{\alpha}$. So,

$$p^2_{tot} = \rho c W \left[\frac{1}{4\pi r^2} + \frac{4(1-\bar{\alpha})}{S\bar{\alpha}}\right],$$ (3.77)

and after dividing by p^2_{ref}, and W_{ref} and taking 10 log, we obtain

$$L_p = L_W + 10\log\left[\frac{1}{4\pi r^2} + \frac{4}{R}\right] + 10\log\left(\frac{\rho c}{400}\right),$$ (3.78)

where R is the so-called room constant $S\bar{\alpha}/(1-\bar{\alpha})$.

3.15.1 Critical Distance

The critical distance r_c (or sometimes called the reverberation radius) is defined as the distance from the sound source where the direct field and reverberant field contributions to p^2_{rms} are equal:

$$\frac{1}{4\pi r^2} = \frac{4}{R}$$ (3.79)

thus,

$$r_c = \sqrt{\frac{R}{16\pi}}.$$ (3.80)

Figure 3.23 gives a plot of Eq. (3.78) (the so-called room equation).

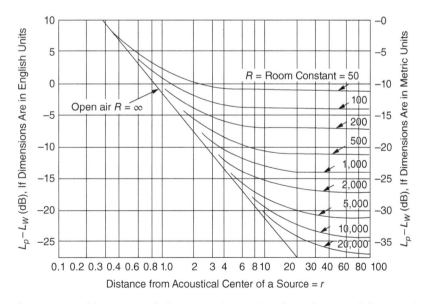

Figure 3.23 Sound pressure level in a room (relative to sound power level) as a function of distance r from sound source.

3.15.2 Noise Reduction

If we are situated in the reverberant field, we may show from Eq. (3.78) that the noise level reduction, ΔL, achieved by increasing the sound absorption is

$$\Delta L = L_{p1} - L_{p2} = 10 \log\left(\frac{4}{R_1}\right) - 10 \log\left(\frac{4}{R_2}\right), \tag{3.81}$$

$$\therefore \Delta L = 10 \log\left(\frac{R_2}{R_1}\right) \approx 10 \log\left(\frac{S_2 \bar{\alpha}_2}{S_1 \bar{\alpha}_1}\right). \tag{3.82}$$

Then $A = S\bar{\alpha}$ is sometimes known as the absorption area, m^2 (sabins). This may be assumed to be the area of perfect absorbing material, m^2 (like the area of a perfect open window that absorbs 100% of the sound energy falling on it). If we consider the sound field in a room with a uniform energy density ε created by a sound source that is suddenly stopped, then the sound pressure level in the room will decrease.

By considering the sound energy radiated into a room by a directional broadband noise source of sound power W, we may sum together the mean squares of the sound pressure contributions caused by the direct and reverberant fields and after taking logarithms obtain the sound pressure level in the room:

$$L_p = L_W + 10 \log\left(\frac{Q_{\theta,\phi}}{4\pi r^2} + \frac{4}{R}\right) + 10 \log\left(\frac{\rho c}{400}\right), \tag{3.83}$$

where $Q_{\theta,\phi}$ is the directivity factor of the source (see Section 3.9) and R *is* the so-called room constant:

$$R = \frac{S\bar{\alpha}}{1 - \bar{\alpha}}. \tag{3.84}$$

A plot of the sound pressure level against distance from the source is given for various room constants in Figure 3.23. It is seen that there are several different regions. The near and far fields depend on the type of source [21] and the free field and reverberant field. The free field is the region where the direct term $Q_{\theta,\phi}/4\pi r^2$ dominates, and the reverberant field is the region where the reverberant term $4/R$ in Eq. (3.83) dominates. The so-called critical distance $r_c = (Q_{\theta,\phi} R/16\pi)^{1/2}$ occurs where the two terms are equal.

3.16 Sound Radiation From Idealized Structures

The sound radiation from plates and cylinders in bending (flexural) vibration is discussed in Refs. [9, 27] and chapter 10 in the *Handbook of Acoustics* [1]. There are interesting phenomena observed with free-bending waves. Unlike sound waves, these are dispersive and travel faster at higher frequency. The bending-wave speed is $c_b = (\omega \kappa c_l)^{1/2}$, where κ is the radius of gyration $h/(12)^{1/2}$ for a rectangular cross-section, h is the thickness, and c_L is the longitudinal wave speed $\{E/[\rho(1 - \sigma^2)]\}^{1/2}$, where E is Young's modulus of elasticity, ρ is the material density, and σ is Poisson's ratio. When the bending-wave speed equals the speed of sound in air, the frequency is called the critical frequency (see Figure 3.24). The critical frequency is

$$f_c = \frac{c^2}{2\pi \kappa c_L}. \tag{3.85}$$

Above this frequency, f_c, the coincidence effect is observed because the bending wavelength λ_b is greater than the wavelength in air λ (Figure 3.25), and trace wave matching always occurs for the sound waves in air at some angle of incidence (see Figure 3.26). This has important consequences for the sound radiation from structures and also for the sound transmitted through the structures from one air space to the other (see Chapter 12 of this book).

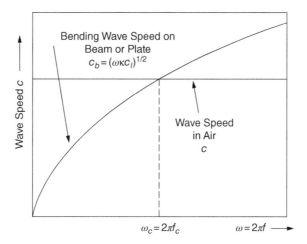

Figure 3.24 Dependence on frequency of bending-wave speed c_b on a beam or panel and wave speed in air c.

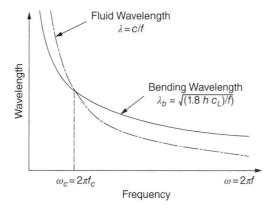

Figure 3.25 Variation with frequency of bending wavelength λ_b on a beam or panel and wavelength in air λ.

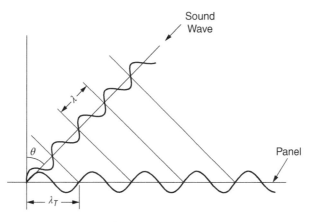

Figure 3.26 Diagram showing trace wave matching between waves in air of wavelength λ and waves in panel of trace wavelength λ_T.

For free-bending waves on infinite plates above the critical frequency, the plate radiates efficiently, while below this frequency (theoretically) the plate cannot radiate any sound energy at all [27]. For finite plates, reflection of the bending waves at the edges of the plates causes standing waves that allow radiation (although inefficient) from the plate corners or edges even below the critical frequency. In the plate center, radiation from adjacent quarter-wave areas cancels. But radiation from

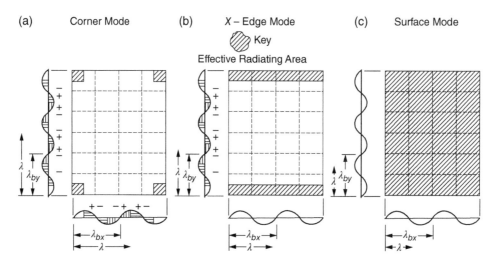

Figure 3.27 Wavelength relations and effective radiating areas for corner, edge, and surface modes. The acoustic wavelength is λ; while λ_{bx} and λ_{by} are the bending wavelengths in the *x*- and *y*-directions, respectively. (see also the *Handbook of Acoustics* [1], chapter 1.)

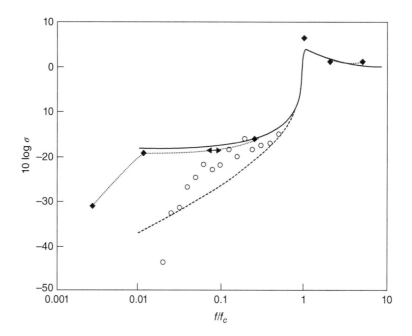

Figure 3.28 Comparison of theoretical and measured radiation ratios σ for a mechanically excited, simply-supported thin steel plate (300 × 300 × 1.22 mm). (—) Theory (simply-supported), (- - -) theory (clamped edges), (⋯⋯) theory [28], and (○) measured [29] (see [30]).

the plate corners and edges, which are normally separated sufficiently in acoustic wavelengths, does not cancel. At very low frequency, sound is radiated mostly by corner modes, then up to the critical frequency, mostly by edge modes. Above the critical frequency the radiation is caused by surface modes with which the whole plate radiates efficiently (see Figure 3.27). Radiation from bending waves in plates and cylinders is discussed in detail in Refs. [9, 27] and chapter 10 of the *Handbook of Acoustics* [1]. Figure 3.28 shows some comparisons between theory and experiment for the level of the radiation efficiencies for sound radiation for several practical cases of simply-supported and clamped panel structures with acoustical and point mechanical excitation.

Example 3.15   Determine the critical frequency for a 3 mm thick steel plate.

Solution

The longitudinal wave speed in aluminum is $c_L = 5100$ m/s. Now, replacing $\kappa = h/(12)^{1/2}$ in Eq. (3.85) yields

$$f_c = \frac{c^2}{1.8hc_L} = \frac{(343)^2}{1.8(0.003)(5100)} = 4272 \text{ Hz}.$$

Sound transmission through structures is discussed in Chapter 12 of this book and chapters 66, 76, and 77 of the *Handbook of Acoustics* [1].

Figure 3.29a and b show the logarithmic value of the radiation efficiency 10log σ plotted against frequency for stiffened and unstiffened plates. See Ref. [27] for further discussion on the radiation efficiency σ_{rad}, which is also known as radiation ratio.

3.17   Standing Waves

Standing-wave phenomena are observed in many situations in acoustics and the vibration of strings and elastic structures. Thus they are of interest with almost all musical instruments (both wind and stringed) (see Part XIV in the *Encyclopedia of Acoustics* [19]); in architectural spaces such as auditoria and reverberation rooms; in volumes such as automobile and aircraft cabins; and in numerous cases of vibrating structures, from tuning forks, xylophone bars, bells and cymbals to windows, wall panels, and innumerable other engineering systems including aircraft, vehicle, and ship structural members. With each standing wave is associated a mode shape (or shape of vibration) and an eigen (or natural) frequency. Some of these systems can be idealized as simple one-, two-, or three-dimensional systems. For example, with a simple wind instrument such as a flute, Eq. (3.1) together with the appropriate spatial boundary conditions can be used to predict the predominant frequency of the sound produced. Similarly, the vibration of a string on a violin can be predicted with an equation identical to Eq. (3.1) but with the variable p replaced by the lateral string displacement. With such a string, solutions can be obtained for the fundamental and higher natural frequencies (overtones) and the associated standing wave mode shapes (which are normally sine shapes). In such a case for a string with fixed ends, the so-called overtones are just integer multiples (2, 3, 4, 5, ...) of the fundamental frequency.

The standing wave with the flute and string can be considered mathematically to be composed of two waves of equal amplitude traveling in opposite directions. Consider the case of a lateral wave on a string under tension. If we create a wave at one end, it will travel forward to the other end. If this end is fixed, it will be reflected. The original (incident) and reflected waves interact (and if the reflection is equal in strength) a perfect standing wave will be created. In Figure 3.30 we show three different frequency waves that have interacted to cause standing waves of different frequencies on the string under tension. A similar situation can be conceived to exist for one-dimensional sound waves in a tube or duct. If the tube has two hard ends, we

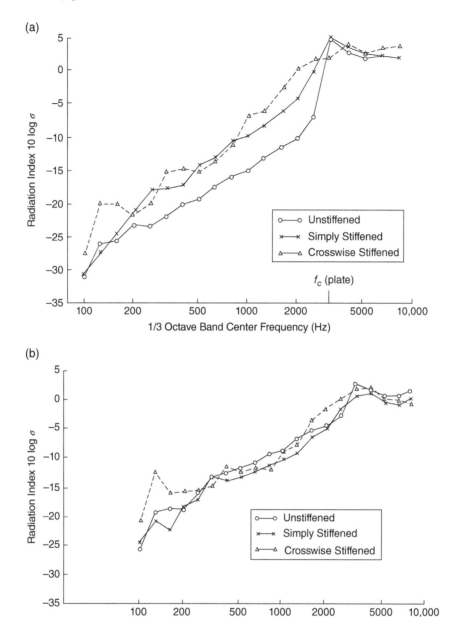

Figure 3.29 Measured radiation ratios of unstiffened and stiffened plates for (a) point mechanical excitation and (b) diffuse sound field excitation. (*Source:* Reproduced from Ref. [31] with permission. See [30].)

can create similar standing one-dimensional sound waves in the tube at different frequencies. In a tube, the regions of high sound pressure normally occur at the hard ends of the tube, as shown in Figure 3.31. See Refs. [14, 32, 33].

A similar situation occurs for bending waves on bars, but because the equation of motion is different (dispersive), the higher natural frequencies are not related by simple integers. However, for the case of a beam with simply supported ends, the higher natural frequencies are given by 2^2, 3^2, 4^2, 5^2, ..., or 4, 9, 16, 25 times the fundamental frequency, ..., and the mode shapes are sine shapes again.

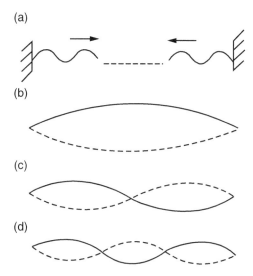

Figure 3.30 Waves on a string: (a) Two opposite and equal traveling waves on a string resulting in standing waves, (b) first mode, $n = 1$, (c) second mode, $n = 2$, and (d) third mode, $n = 3$.

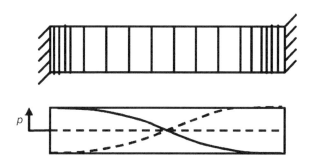

Figure 3.31 Sound waves in a tube. First mode standing wave for sound pressure in a tube. This mode is called the fundamental and occurs at the fundamental frequency.

The standing waves on two-dimensional systems (such as bending vibrations of plates) may be considered mathematically to be composed of four opposite traveling waves. For simply supported rectangular plates the mode shapes are sine shapes in each direction. For three-dimensional systems such as the air volumes of rectangular rooms, the standing waves may be considered to be made up of eight traveling waves. For a hard-walled room, the sound pressure has a cosine mode shape with the maximum pressure at the walls, and the particle velocity has a sine mode shape with zero normal particle velocity at the walls. See chapter 6 in the *Handbook of Acoustics* [1] for the natural frequencies and mode shapes for a large number of acoustical and structural systems.

For a three-dimensional room, normally there are standing waves in three directions with sound pressure maxima at the hard walls.

To understand the sound propagation in a room, it is best to use the three-dimensional wave equation in Cartesian coordinates:

$$\nabla^2 p - \frac{1}{c^2}\frac{\partial^2 p}{\partial t^2} = 0, \tag{3.86}$$

or

$$\frac{\partial^2 p}{\partial x^2} + \frac{\partial^2 p}{\partial y^2} + \frac{\partial^2 p}{\partial z^2} - \frac{1}{c^2}\frac{\partial^2 p}{\partial t^2} = 0. \tag{3.87}$$

This equation can have solutions that are "random" in time or are for the special case of a pure-tone, "simple harmonic."

The simple harmonic solution is of considerable interest to us because we find that in rooms there are solutions only at certain frequencies. It may be of some importance now to mention both the sinusoidal solution and the equivalent solution using complex notation that is very frequently used in acoustics and vibration theory.

For a one-dimensional wave, the simple harmonic solution to the wave equation is

$$p = \hat{p}_1 \cos\left[k(ct - x)\right] + \hat{p}_2 \cos\left[k(ct + x)\right], \tag{3.88}$$

where $k = \omega/c = 2\pi f/c$ (the wavenumber).

The first term in Eq. (3.88) represents a wave of amplitude $\hat{p}_1$ traveling in the $+x$-direction. The second term in Eq. (3.88) represents a wave of amplitude $\hat{p}_2$ traveling in the $-x$-direction.

The equivalent expression to Eq. (3.88) using complex notation is

$$p = \operatorname{Re}\left\{\tilde{p}_1 e^{j(\omega t - kx)}\right\} + \operatorname{Re}\left\{\tilde{p}_2 e^{j(\omega t + kx)}\right\}, \tag{3.89}$$

where $j = \sqrt{-1}$, Re{} means real part; and $\tilde{p}_1$ and $\tilde{p}_2$ are complex amplitudes of the sound pressure; remember $k = \omega/c$; $kc = 2\pi f$. Both Eqs. (3.88) and (3.89) are solutions to Eq. (3.86).

For the three-dimensional case (x, y, and z propagation), the sinusoidal (pure tone) solution to Eq. (3.87) is

$$p = \operatorname{Re}\left\{\tilde{p}\exp\left(j\left[\omega t \pm k_x x \pm k_y y \pm k_z z\right]\right)\right\}. \tag{3.90}$$

Note that there are 2^3 (eight) possible solutions given by Eq. (3.90). Substitution of Eq. (3.90) into Eq. (3.87) (the three-dimensional wave equation) gives (from any of the eight (2^3) equations):

$$\omega^2 = c^2\left(k_x^2 + k_y^2 + k_z^2\right), \tag{3.91}$$

from which the wavenumber k is

$$k = \frac{\omega}{c} = \sqrt{k_x^2 + k_y^2 + k_z^2}, \tag{3.92}$$

and the so-called direction cosines with the x, y and z directions are $\cos\theta_x = \pm k_x/k$, $\cos\theta_y = \pm k_y/k$, and $\cos\theta_z = \pm k_z/k$ (see Figure 3.32).

Equations (3.91) and (3.92) apply to the cases where the waves propagate in unbounded space (infinite space) or finite space (e.g. rectangular rooms).

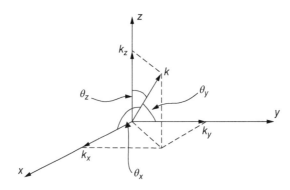

Figure 3.32 Direction cosines and vector *k*.

For the case of rectangular rooms with hard walls, we find that the sound (particle) velocity perpendicular to each wall must be zero. By using these boundary conditions in each of the eight solutions to Eq. (3.87), we find that $\omega^2 = (2\pi f)^2$ and k^2 in Eqs. (3.91) and (3.92) are restricted to only certain discrete values:

$$k^2 = \left(\frac{n_x \pi}{A}\right)^2 + \left(\frac{n_y \pi}{B}\right)^2 + \left(\frac{n_z \pi}{C}\right)^2,$$ (3.93)

or

$$\omega^2 = c^2 k^2.$$

Then the room natural frequencies are given by

$$f_{n_x n_y n_z} = \frac{c}{2}\sqrt{\left(\frac{n_x}{A}\right)^2 + \left(\frac{n_y}{B}\right)^2 + \left(\frac{n_z}{C}\right)^2},$$ (3.94)

where A, B, C are the room dimensions in the x, y, and z directions, and $n_x = 0, 1, 2, 3,...$; $n_y = 0, 1, 2, 3,...$ and $n_z = 0, 1, 2, 3, ...$ Note n_x, n_y, and n_z are the number of half waves in the x, y, and z directions. Note also for the room case, the eight propagating waves add together to give us a standing wave. The wave vectors for the eight waves are shown in Figure 3.33.

There are three types of standing waves resulting in three modes of sound wave vibration: axial, tangential, and oblique modes. Axial modes are a result of sound propagation in only one room direction. Tangential modes are caused by sound propagation in two directions in the room and none in the third direction. Oblique modes involve sound propagation in all three directions.

We have assumed there is no absorption of sound by the walls. The standing waves in the room can be excited by noise or pure tones. If they are excited by pure tones produced by a loudspeaker or a machine that creates sound waves exactly at the same frequency as the eigenfrequencies (natural frequencies) f_E of the room, the standing waves are very pronounced. Figures 3.34 and 3.35 show the distribution of particle velocity and sound pressure for the $n_x = 1$, $n_y = 1$, and $n_z = 1$ mode in a room with hard reflecting walls. See Refs. [23, 24] for further discussion of standing-wave behavior in rectangular rooms.

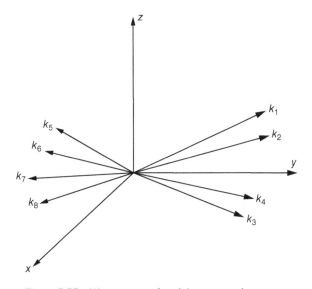

Figure 3.33 Wave vectors for eight propagating waves.

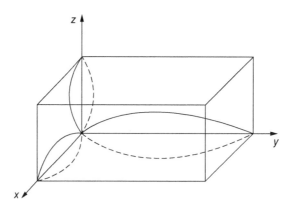

Figure 3.34 Standing wave for $n_x = 1$, $n_y = 1$, and $n_z = 1$ (particle velocity shown).

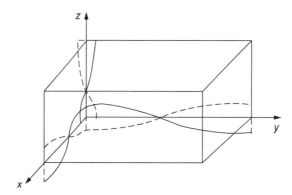

Figure 3.35 Standing wave for $n_x = 1$, $n_y = 1$, and $n_z = 1$ (sound pressure shown).

Example 3.16 Calculate all the possible natural frequencies for normal modes of vibration under 100 Hz within a rectangular room $3.1 \times 4.7 \times 6.2 \, \text{m}^3$.

Solution

Table 3.3 gives all the possible natural frequencies for modes under 100 Hz using Eq. (3.94) and $c = 343$ m/s.

One can see in Table 3.3 that the frequency spacing becomes smaller with increasing frequency and that there may be *degenerate modes* present (i.e. when two or more modes have the same characteristic frequency but different values of n_x, n_y, and n_z). Modes which are close to each other in frequency can easily "beat," while degenerate modes can greatly increase the response of the room at particular frequencies where degeneracy occurs. This can give rise to the "boomy" sensation (at low frequencies) which is often found in regular-shaped rooms of similar wall dimensions [4].

Table 3.3 Frequencies (less than 100 Hz) for a $3.1 \times 4.7 \times 6.2 \, \text{m}^3$ rectangular room, for $c = 343$ m/s.

n_x	n_y	n_z	f_E
0	0	1	27.7
0	1	0	36.5
0	1	1	45.8

Table 3.3 (Continued)

n_x	n_y	n_z	f_E
0	0	2	} 55.3
1	0	0	
1	0	1	61.9
0	1	2	} 66.3
1	1	0	
1	1	1	71.8
0	2	0	73.0
1	0	2	78.2
0	2	1	78.0
0	0	3	83.0
1	1	2	86.3
0	1	3	90.7
0	2	2	} 91.6
1	2	0	
1	2	1	95.7
1	0	3	99.7

The position of the sound source within the room is also an important parameter, since for many source positions certain types of modes may not be excited. For example, if the source is located in one of the corners of the room, then it is possible to excite every normal mode, while if the source is located at the center of a rectangular room then only the even modes (one eighth of the total number of possible modes) can be excited. Similarly if we keep the position of the source constant and measure the sound pressure throughout the room we see differences in level depending on where we are standing in the room relative to the normal modes. In this way the room superimposes its own acoustical response characteristics upon those of the source. Hence we cannot measure the true frequency response of a sound source (e.g. loudspeaker) in a reverberant room because of the effect of the modal response of the room. This interference can be removed by making all the wall surfaces highly sound-absorbent. Then all the modes are sufficiently damped so we are able to measure the true output of the source [4]. Such rooms are called anechoic (see Figure 3.22).

3.18 Waveguides

Waveguides can occur naturally where sound waves are channeled by reflections at boundaries and by refraction. Even the ocean can sometimes be considered to be an acoustic waveguide that is bounded above by the air–sea interface and below by the ocean bottom (see chapter 31 in the *Handbook of Acoustics* [1]). Similar channeling effects are also sometimes observed in the atmosphere [34]. Waveguides are also encountered in musical instruments and engineering applications. Wind instruments may be regarded as waveguides. In addition, waveguides comprised of pipes, tubes, and ducts are frequently used in engineering systems, for example, exhaust pipes, air-conditioning ducts and the ductwork in turbines and turbofan engines. The sound propagation in such waveguides is similar to the three-dimensional situation discussed in Section 3.17 but with some differences. Although rectangular ducts are used in air-conditioning systems, circular ducts are also frequently used, and theory for these must be considered as well. In real waveguides, airflow is often present and complications due to a mean fluid flow must be included in the theory.

For low-frequency excitation, only plane waves can propagate along the waveguide (in which the sound pressure is uniform across the duct cross-section). However, as the frequency is increased, the so-called first cut-on frequency is reached above which there is a standing wave across the duct cross-section caused by the first higher mode of propagation.

For excitation just above this cut-on frequency, besides the plane-wave propagation, propagation in higher order modes can also exist. The higher mode propagation in a rectangular duct can be considered to be composed of four traveling waves in each direction. Initially, these vectors (rays) are almost perpendicular to the duct walls and with a phase speed along the duct that is almost infinite. As the frequency is increased, these vectors point increasingly toward the duct axis, and the phase speed along the duct decreases until at very high frequency it is only just above the speed of sound c. However, for this mode, the sound pressure distribution across duct cross-section remains unchanged. As the frequency increases above the first cut-on frequency, the cut-on frequency for the second higher order mode is reached and so on. For rectangular ducts, the solution for the sound pressure distribution for the higher duct modes consists of cosine terms with a pressure maximum at the duct walls, while for circular ducts, the solution involves Bessel functions. Chapter 7 in the *Handbook of Acoustics* [1] explains how sound propagates in both rectangular and circular guides and includes discussion on the complications created by a mean flow, dissipation, discontinuities, and terminations. Chapter 161 in the *Encyclopedia of Acoustics* [19] discusses the propagation of sound in another class of waveguides, that is, acoustical horns.

3.19 Other Approaches

3.19.1 Acoustical Lumped Elements

When the wavelength of sound is large compared to physical dimensions of the acoustical system under consideration, then the lumped-element approach is useful. In this approach it is assumed that the fluid mass, stiffness, and dissipation distributions can be "lumped" together to act at a point, significantly simplifying the analysis of the problem. The most common example of this approach is its use with the well-known Helmholtz resonator (see Chapter 9 of this book) in which the mass of air in the neck of the resonator vibrates at its natural frequency against the stiffness of its volume.

A similar approach can be used in the design of loudspeaker enclosures and the concentric resonators in automobile mufflers in which the mass of the gas in the resonator louvers (orifices) vibrates against the stiffness of the resonator (which may not necessarily be regarded completely as a lumped element). Dissipation in the resonator louvers may also be taken into account. Chapter 21 in the *Handbook of Acoustics* [1] reviews the lumped-element approach in some detail.

3.19.2 Numerical Approaches: Finite Elements and Boundary Elements

In cases where the geometry of the acoustical space is complicated and where the lumped-element approach cannot be used, then it is necessary to use numerical approaches. In the late 1960s, with the advent of powerful computers, the acoustical finite element method (FEM) became feasible. In this approach, the fluid volume is divided into a number of small fluid elements (usually rectangular or triangular), and the equations of motion are solved for the elements, ensuring that the sound pressure and volume velocity are continuous at the node points where the elements are joined. The FEM has been widely used to study the acoustical performance of elements in automobile mufflers and cabins (See Chapter 10 of this book.).

The boundary element method (BEM) was developed a little later than the FEM. In the BEM approach the elements are described on the boundary surface only, which reduces the computational dimension of the problem by one. This correspondingly produces a smaller system of equations than the FEM. BEM involves the use of a surface mesh rather than a volume mesh. BEM, in general, produces a smaller set of equations that grows more slowly with frequency, and the resulting matrix is full; whereas the FEM matrix is sparse (with elements near and on the diagonal). Thus computations with FEM are generally less time-consuming than with BEM. For sound propagation problems involving the radiation of sound to infinity, the BEM is more suitable because the radiation condition at infinity can be easily satisfied with the BEM, unlike with the FEM. However, the FEM is better suited than the BEM for the determination of the natural frequencies and mode shapes of cavities (See Chapter 10 of this book.).

Recently, FEM and BEM commercial software has become widely available. The FEM and BEM are described in Refs. [35, 36] and in chapters 12 and 13 in the *Handbook of Acoustics* [1].

3.19.3 Acoustic Modeling Using Equivalent Circuits

Electrical analogies have often been found useful in the modeling of acoustical systems. There are two alternatives. The sound pressure can be represented by voltage and the volume velocity by current, or alternatively the sound pressure is replaced by current and the volume velocity by voltage. Use of electrical analogies is discussed in chapter 11 of the *Handbook of Acoustics* [1]. They have been widely used in loudspeaker design and are in fact perhaps most useful in the understanding and design of transducers such as microphones where acoustic, mechanical, and electrical systems are present together and where an overall equivalent electrical circuit can be formulated (see *Handbook of Acoustics* [1], chapters 111–113).

Beranek makes considerable use of electrical analogies in his books [10, 11]. In chapter 14 in the *Handbook of Acoustics* [1] their use in the design of automobile mufflers is described. Chapter 10 in this book also reviews the use of electrical analogies in muffler and silencer acoustical design.

References

1 Crocker, M.J. (ed.) (1998). *Handbook of Acoustics*. New York: Wiley.
2 Malecki, I. (1969). *Physical Foundations of Technical Acoustics*. Oxford: Pergamon Press.
3 Skudrzyk, E. (1971). *The Foundations of Acoustics*. New York: Springer (reprinted by the Acoustical Society of America in 2008).
4 Crocker, M.J. and Price, A.J. (1975). *Noise and Noise Control*, vol. I. Cleveland, OH: CRC Press.
5 Lighthill, M.J. (2001). *Waves in Fluids*, 2e. Cambridge: Cambridge University Press.
6 Pierce, A.D. (1981). *Acoustics: An Introduction to Its Physical Principles and Applications*. New York: McGraw-Hill (reprinted by the Acoustical Society of America, 1989).
7 Crocker, M.J. and Kessler, F.M. (1982). *Noise and Noise Control*, vol. II. Boca Raton, FL: CRC Press.
8 Morse, P.M. and Ingard, K.U. (1986). *Theoretical Acoustics*. Princeton, NJ: Princeton University Press.
9 Junger, M.J. and Feit, D. (1986). *Sound, Structures, and Their Interaction*. Cambridge, MA: MIT Press.
10 Beranek, L.L. (1986). *Acoustics*. New York: Acoustical Society of America (reprinted with changes).
11 Beranek, L.L. (1988). *Acoustical Measurements*, rev. ed. New York: Acoustical Society of America.
12 Crighton, D.G., Dowling, A.P., Ffowcs Williams, J.E. et al. (1992). *Modern Methods in Analytical Acoustics*. Berlin: Springer-Verlag.
13 Fahy, F.J. (1995). *Sound Intensity*, 2e. London: E&FN Spon, Chapman & Hall.
14 Fahy, F.J. and Walker, J.G. (eds.) (1998). *Fundamentals of Noise and Vibration*. London and New York: E/FN Spon.
15 Filippi, P., Habault, D., Lefebvre, J., and Bergassoli, A. (1999). *Acoustics: Basic Physics Theory & Methods*. San Diego, CA: Academic Press.

16 Kinsler, L.E., Frey, A.R., Coppens, A.B., and Sanders, J.V. (1999). *Fundamentals of Acoustics*, 4e. New York: Wiley.

17 Blackstock, D.T. (2000). *Fundamental of Physical Acoustics*. New York: Wiley.

18 Bruneau, M. and Scelo, T. (2006). *Fundamentals of Acoustics*. London: ISTE.

19 Crocker, M.J. (1997). *Encyclopedia of Acoustics*. New York: Wiley.

20 Fuller, C. (2007). Active vibration control. In: *Handbook of Noise and Vibration Control* (ed. M.J. Crocker), 770–784. New York: Wiley.

21 Nelson, P.A. (2007). Sound sources. In: *Handbook of Noise and Vibration Control* (ed. M.J. Crocker), 43–51. New York: Wiley.

22 Jacobsen, F. (2007). Sound intensity measurements. In: *Handbook of Noise and Vibration Control* (ed. M.J. Crocker), 534–548. New York: Wiley.

23 Kuttruff, K.H. (2007). Sound propagation in rooms. In: *Handbook of Noise and Vibration Control* (ed. M.J. Crocker), 52–68. New York: Wiley.

24 Hansen, C.H. (2007). Room acoustics. In: *Handbook of Noise and Vibration Control* (ed. M.J. Crocker), 1240–1246. New York: Wiley.

25 Manning, J.E. (2007). Statistical energy analysis. In: *Handbook of Noise and Vibration Control* (ed. M.J. Crocker), 241–254. New York: Wiley.

26 Hansen, C.H. (2007). Sound absorption in rooms. In: *Handbook of Noise and Vibration Control* (ed. M.J. Crocker), 1247–1256. New York: Wiley.

27 Guyader, J.-L. (2007). Sound radiation from structures and their response to sound. In: *Handbook of Noise and Vibration Control* (ed. M.J. Crocker), 79–100. New York: Wiley.

28 Ver, I.L. and Holmer, C.I. (1971). Interaction of sound waves with solid structures. In: *Noise and Vibration Control* (ed. L.L. Beranek), 270–361. New York: McGraw-Hill.

29. Pierri, R.A. (1977) Study of a dynamic absorber for reducing the vibration and noise radiation of plate-like structures. MSc thesis. University of Southampton.

30 Braun, S.G., Ewins, D.J., and Rao, S.S. (2001). *Encyclopedia of Vibration*. San Diego, CA: Academic.

31 Fahy, F.J. and Gardonio, P. (2007). *Sound and Structural Vibration – Radiation, Transmission and Response*, 2e. Oxford: Academic Press.

32 Bies, D.A. and Hansen, C.H. (2009). *Engineering Noise Control – Theory and Practice*, 4e. London and New York: Spon Press.

33 Fahy, F.J. (2001). *Foundations of Engineering Acoustics*. San Diego, CA: Academic Press.

34 Attenborough, K. (2007). Sound propagation in the atmosphere. In: *Handbook of Noise and Vibration Control* (ed. M.J. Crocker), 67–78. New York: Wiley.

35 Astley, R.J. (2007). Numerical acoustical modeling (finite element modeling). In: *Handbook of Noise and Vibration Control* (ed. M.J. Crocker), 101–115. New York: Wiley.

36 Herrin, D.W., Wu, T.W., and Seybert, A.F. (2007). Boundary element modeling. In: *Handbook of Noise and Vibration Control* (ed. M.J. Crocker), 116–127. New York: Wiley.

4

Human Hearing, Speech and Psychoacoustics

4.1 Introduction

The human ear is a marvelous and very sensitive biomechanical system for detecting sound. If it were only slightly more sensitive, we would be able to hear the Brownian (random) motion of the air molecules and we would have a perpetual buzz in our ears! The ear has a wide frequency response from about 15 or 20 Hz to about 20 kHz. Also, the ear has a large dynamic range; the ratio of the loudest sound pressure we can tolerate to the faintest we can hear is about 10 million (10^7). There are three essential reasons to consider the ear in this book. Sound pressure levels are now so high in industrialized societies that many individuals are exposed to intense noise and *permanent damage* results. Large numbers of other individuals are exposed to noise from aircraft, surface traffic, construction equipment or machines and appliances, and disturbance and *annoyance* results. Lastly there are subjective reasons. An understanding of people's subjective response to noise allows environmentalists and engineers to reduce noise in more effective ways. The human auditory response to sound concerns the science called psychoacoustics. For example, noise should be reduced in the frequency range in which the ear is most sensitive. Noise reduction should be by a magnitude which is subjectively significant. There are several other subjective parameters which are important in hearing.

4.2 Construction of Ear and Its Working

The ear can be divided into three main parts (Figure 4.1): the *outer*, *middle*, and *inner* ear. The outer ear consisting of the fleshy pinna and ear canal conducts the sound waves onto the eardrum. The middle ear converts the sound waves into mechanical motion of the auditory ossicles, and the inner ear converts the mechanical motion into neural impulses which travel along the auditory nerves to the brain. The anatomy and functioning of the ear are described more completely in various other references and textbooks [1–8] and will only be discussed briefly here.

4.2.1 Construction of the Ear

The fleshy appendage on the side of the head (the pinna) is not as well developed in humans as in some animals. Its function is to focus sound into the ear canal. It helps us to localize the source of sound, particularly in the vertical direction, and is more effective at higher frequencies. The ear canal is about 25 mm long and ends at the tympanic membrane (eardrum) which is under tension and has the thickness of at sheet of paper.

The eardrum is connected to the malleus, the first of the three small bones known as the auditory ossicles (see Figure 4.2). The middle ear air cavity is connected to the back of the mouth by the Eustachian tube. The

Engineering Acoustics: Noise and Vibration Control, First Edition. Malcolm J. Crocker and Jorge P. Arenas.
© 2021 John Wiley & Sons Ltd. Published 2021 by John Wiley & Sons Ltd.

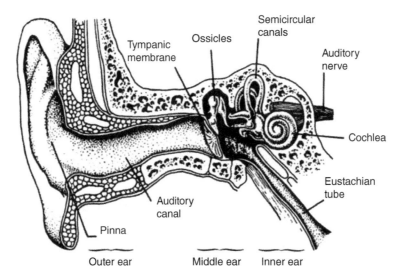

Figure 4.1   Simplified cross-section through the human ear.

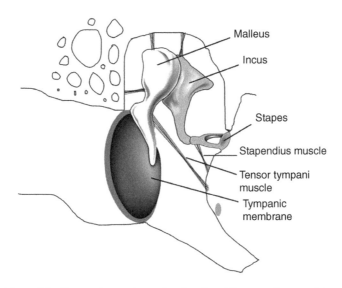

Figure 4.2   Tympanic membrane (eardrum) and three auditory ossicles.

smallest of the ossicles, the stapes, which is about half the size of a grain of rice and the smallest bone in the human body, is connected to a small oval window in the cochlea. The cochlea consists of spiral fluid-filled cavities inside the bone of the skull. The cochlea is comprised of a passageway which makes two and one half turns rather like a snail shell. Connected to the cochlea are the semicircular canals which are the balance mechanism and unrelated to hearing.

The passageway of the cochlea is separated into a lower and upper gallery (scala tympani and vestibuli, respectively) by a membranous duct (Figure 4.3). The upper and lower galleries are connected together only at the apex. Figure 4.4 is a schematic representation showing the cochlea "unrolled."

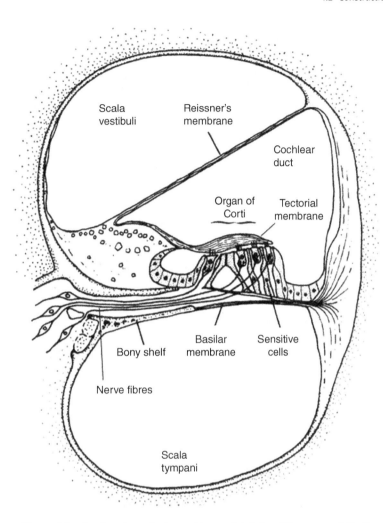

Figure 4.3 Section through the cochlea and details of the organ of Corti.

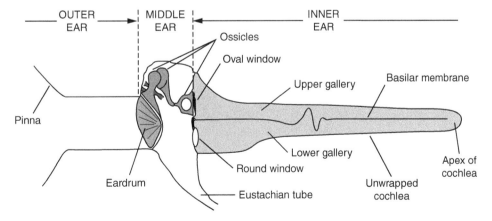

Figure 4.4 Cochlea "unwrapped" to show working of the ear schematically.

4.2.2 Working of the Ear Mechanism

When a sound wave reaches the ear, it travels down the auditory canal until it reaches the eardrum. It sets the eardrum in motion and this vibration is transmitted across the 2 mm gap to the oval window by the lever system comprised of the auditory ossicles. It is thought that this mechanical system is an impedance matching device. The characteristic impedance, ρc, of air is approximately one two-thousandth of the impedance of the cochlea fluid. The area of the tympanic membrane is 20 or 30 times larger than that of the oval window. Some believe that it is not the area of the oval window which is important, but rather the area of the footplate of the stapes. The tympanic membrane is about 20 times greater in area than the footplate area. However, not all of the tympanic membrane vibrates because it is firmly attached at its periphery. The ratio of the part of the tympanic membrane that moves to the footplate area is about 14 to 1.

Also, the pivot of the ossicle system may be assumed closer to the oval window than the eardrum, hence providing a mechanical advantage of two or three times. The net result is that low-pressure, high particle velocity amplitude air waves arriving at the eardrum are converted into high-pressure low particle velocity amplitude fluid waves in the cochlea, approximately matching the air to fluid impedances. We probably remember from electrical theory that in order to obtain maximum power transfer, impedances must be matched.

There is a "safety" device built into the inner ear mechanism. Attached to the malleus and stapes are two muscles: the tensor tympani and the stapedius. If continuous intense sounds are experienced, the muscles contract and rotate the ossicles so that the force passed onto the oval window does not increase correspondingly to the sound pressure.

This effect is called the acoustic reflex and many types of experiments indicate that the reflex attenuates low-frequency sound levels up to about 20 dB [9]. However, these muscles are rapidly fatigued by continuous narrow-band intense noise. In addition, the muscles are relatively slow in their contraction making the reflex ineffective in presence of impulse or impact sounds. There seems to be some evidence that the muscles also contract when we speak, to prevent us hearing so much of our own speech.

The Eustachian tube is used to equalize the pressure across the eardrum by swallowing. This explains why our ears "pop" in airplanes when we ascend and the atmospheric pressure changes. We may experience some pain in an airplane when landing again if we have a cold; mucus, blocking the Eustachian tube, can prevent us from equalizing the pressure by swallowing. Movement of the footplate of the stapes which is connected to the oval window causes pressure waves in the fluid of the upper gallery of the cochlea (Figures 4.3 and 4.4). The fluid in the lower gallery is separated from that in the upper gallery by the cochlear duct containing the organ of Corti. The organ has about 35 000 sensitive hair cells distributed along its length which are connected in a complicated way to about 18 000 nerve fibers which are combined into the auditory nerve which runs into the brain. The pressure waves cause the basilar membrane to deflect and a shearing motion occurs between the basilar and tectorial membranes. The hair cells sense the shearing motion and if the stimulus is great enough the neuron to which each hair cell is attached sends an impulse along the nerve fiber to the brain cortex [8]. Each neuron takes about 1/1000th of a second to recharge and so individual neurons are limited to "firing" no more than 1000 times/second. With the neurons, a triggering level must be reached before they "fire" and so they have an all-or-nothing response. The brain must interpret the neural impulses to give us the sensation of hearing and, as we can imagine, the way in which this is done is not well understood.

4.2.3 Theories of Hearing

Pythagoras in the sixth century BCE was perhaps the first to recognize that sound is an airborne vibration [10]. Hippocrates in the fourth century BCE recognized that the air vibrations are picked up by the eardrum but thought that the vibrations were transmitted directly to the brain by bones. In 175 CE, Galen of Pergamum, a Greek physician, realized that it was nerves that transmitted the sound sensations to the brain. Galen and most other early scientists and philosophers proposed, mistakenly, however, that somewhere deep in the head was a sealed pocket of implanted air which was the "seat" of hearing. This view was popularly held until 1760 when Domenico Cotugno declared that the inner ear (cochlea) was completely filled with fluid [10].

In 1543, Andreas Vesalius published his treatise on anatomy giving a description of the middle ear and in 1561 Gabriello Fallopio described the cochlea itself.

In 1605 Gaspard Bauhin put forward a resonance theory for the ear. In his model, different air cavities were excited by different frequencies. However, he knew little of the construction of the inner ear. Du Verney, in 1633, developed a more advanced theory by postulating that different parts of the ridge of bone which twists up the inside of the cochlea resonated at different frequencies which depended upon its width. Du Verney's theory was held until 1851 when Alfonso Corti, using a microscope, discovered that the thousands of hair cells on the basilar membrane were attached to the ridge of bone in the cochlea.

A few years later, Hermann von Helmholtz used Corti's findings to suggest a new theory of hearing. In Helmholtz's theory, as it became refined, different parts of the basilar membrane resonated at different frequencies. Later workers showed that Helmholtz was not exactly right (the basilar membrane is not under tension). However, in 1928 Georg von Békésy did show that waves do travel along the basilar membrane and different sections of the basilar membrane do respond more than others to a certain sound. The region of maximum response is frequency-dependent and as Helmholtz had predicted, von Békésy found that the high-frequency sound is detected nearer to the oval window and the low-frequency sound, nearer to the apex (Figures 4.3 and 4.4).

4.3 Subjective Response

So far we have traced the sound signal down the ear canal to the eardrum, through the auditory ossicles, through the oval window to the cochlear fluid to the basilar membrane and the hair cells, and finally to the neural impulses sent to the brain. How does the brain interpret these signals? Our study now enters the realm of psychology. While the physicist or engineer talks about sound pressure level and frequency, the psychologist talks about loudness and pitch, respectively. The human auditory response to sound is studied by psychoacoustics. In Section 4.3 we shall discuss the relationships between some of the engineering descriptions of sound and the psychological or subjective descriptions of psychoacoustics.

4.3.1 Hearing Envelope

Figure 4.5 presents the auditory field for an average, normal young person who has not suffered any hearing loss or damage. The lower curve represents the hearing threshold, that is, the quietest audible sound at any frequency. The upper curve represents the discomfort threshold, that is, the sound pressure level at any frequency at which there is a sensation of discomfort and even pain in the ears. Speech is mainly in the frequency range of about 250–6000 Hz and at sound pressure levels between about 30–80 dB at 1–2 m (depending upon frequency). Of course, the sound pressure level of speech can approach 90 dB at about 0.2–0.3 m from someone if they are shouting loudly. The sound of vowels is mostly in the low-frequency range from about 250 to 1000 Hz, while the sound of consonants is mainly in the higher frequency range of about 1000–6000 Hz. Music is spread over a somewhat greater frequency range and a greater dynamic range than speech. (The dynamic range represents the difference in levels between the lowest and highest sound pressure levels experienced.)

4.3.2 Loudness Measurement

The way in which the brain interprets the neural pulses is still a matter for research. However, various experiments have been conducted on groups of people to determine people's average sensation of loudness, etc. We should stress that no one's hearing is exactly the same as any other and hence we must find statistical responses.

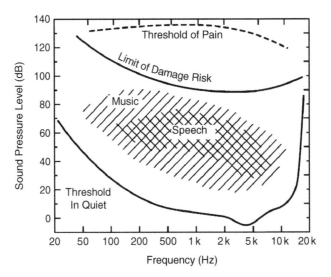

Figure 4.5 Human auditory field envelope.

Figure 4.6 shows equal loudness contours for pure tone sounds. Note that the lowest curve in Figure 4.6 is labeled MAF (minimum audible field). This is the hearing threshold, the quietest sounds, on average, at any frequency that average young people can hear.

We should note that there are two ways that we can measure the hearing threshold and the equal loudness contours. The first way is to present the listener with a free progressive wave field at a discrete frequency. This is the method which was used to obtain the results in Figure 4.6. Such measurements are normally made with the listener facing the source in an anechoic room where there are no reflections. The second way (which is

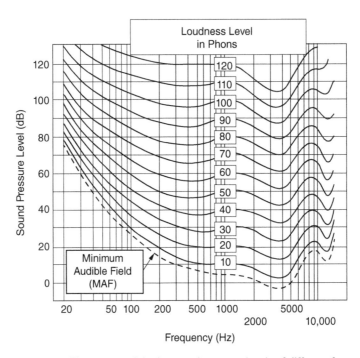

Figure 4.6 Equal loudness contours. The contours join the sound pressure levels of different frequency pure tones that are judged to be equally loud. The numbers on each contour are the loudness levels in phons.

used more frequently) is to present the listener with sounds played through earphones. There are some small differences in the results obtained by the two methods. The equal loudness contours are determined as follows. For the 60-phon curve, the listener is first presented with a pure tone at 1000 Hz at a sound pressure level of 60 dB. Then he is presented with a pure tone at, for example, 500 Hz. The pure tones are presented alternately to the listener every few seconds. The level of the 500 Hz tone is adjustable and the listener is asked to adjust the level of the 500 Hz tone until the two tones sound equally as loud. The procedure is repeated with pure tones at other frequencies which are continually adjusted to sound equal in loudness to the 1000 Hz pure tones at 60 dB. The curve drawn through these points (when the result is averaged for a large number of people) is called the 60-phon equal loudness contour. The 70-phon contour is obtained by finding the sound pressure level of pure tones which seem equally as loud as 1000 Hz pure tone at 70 dB and so on. Finally, the MAF is the contour joining all pure tones which are just audible. Figure 4.6 was determined for a group of young people with healthy ears in the age group 18–25, each person listening with both ears, facing the source, in a free progressive wave acoustic field.

We see from the equal loudness contours that the ear is most sensitive to sound at about 4000 Hz. This is mainly due to a quarter wavelength resonance in the ear canal at this frequency. The increase in sensitivity at 12 000 Hz is mainly caused by diffraction effects around the head.

The equal loudness contours are not flat as the frequency is changed. Notice at 1000 Hz we can just detect sounds of 4 dB, while at 100 Hz they must be 25 dB. Thus the intensity must be 21 dB higher at 100 Hz than 1000 Hz for us to detect a pure tone. This represents a 100-fold increase in intensity or a 10-fold increase in sound pressure amplitude. As the sound pressure level is increased, the equal loudness contours became flatter. A 40 dB tone at 1000 Hz appears equally as loud as a 51 dB tone at 100 Hz, an 11 dB increase. A 70 dB tone at 1000 Hz is as loud as a 75 dB tone at 100 Hz, only a 5 dB increase.

A further experiment has been performed on the loudness of pure tones. This experiment is to determine the increase in sound pressure level for a pure tone apparently to double in loudness. It has been found that for pure tones at 1000 Hz, when the sound pressure level is increased by about 10 dB, the sound appears twice as loud. Fletcher and Munson first commented on this in 1933 [11]. Alternatively we may say that since equal loudness contours join all pure tones with the same loudness levels in phons, then a doubling in loudness occurs when the loudness level increases by 10 phon. This relationship is shown in Figure 4.7 and in the equation

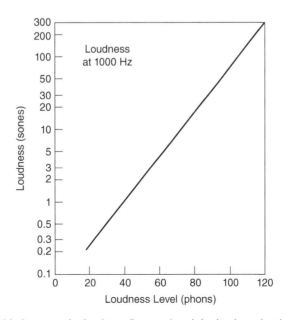

Figure 4.7 Relationship between the loudness (in sones) and the loudness level (in phons) of a sound.

$$S = 2^{(P-40)/10}, \tag{4.1}$$

where S is loudness (sone) and P is loudness level (phon).

Example 4.1 Equation (4.1) is quite useful since it converts the logarithmic loudness level, phon, into the more useful linear loudness measure, sone. We see from Eq. (4.1) or Figure 4.7 that a loudness level of 40 phons has a loudness of 1 sone, a loudness level of 50 phons has a loudness of 2 sones, and a loudness level of 60 phons has a loudness of 4 sones and so on.

Example 4.2 Given three pure tones with the following frequencies and sound pressure levels: 100 Hz at 60 dB, 200 Hz at 70 dB, and 1000 Hz at 80 dB:

a) Calculate the total loudness in sones of these three pure tones.
b) Find the sound pressure level of a single 2000 Hz pure tone which has the same loudness as all the three pure tones combined.

Solution

a) The loudness level in phons of each tone is found from Figure 4.6 and the corresponding loudness in sones is found from Eq. (4.1). Then we have that
100 Hz @ 60 dB has $P = 50$ phons and $S = 2$ sones
200 Hz @ 70 dB has $P = 70$ phons and $S = 8$ sones
1000 Hz @ 80 dB has $P = 80$ phons and $S = 16$ sones
Therefore, the total loudness of these three pure tones is $2 + 8 + 16 = 26$ sones.
b) The total loudness of the combined tones (26 sones) corresponds to a loudness level of $P = 10\log_2(26) + 40$. Since $\log_2(26) = \ln(26)/\ln(2) = 4.7$, then $P = 87$ phons.
   Now, from Figure 4.6 we observe that a pure tone of 2000 Hz and loudness level of 87 phons has a sound pressure level of 82 dB.

So far we have discussed the loudness of pure tones. However, many of the noises we experience, although they may contain pure tones, are predominantly broadband. Similar loudness rating schemes have been worked out for broadband noise. In 1958–1960, Zwicker [12], taking into account masking effects (see later discussion in Section 4.3.3), devised a graphical method to compute the loudness of a broadband noise. It should be noted that Zwicker's method can be used both for diffuse and free field conditions and for broadband noise even when pronounced pure tones are present. However, the method is somewhat time-consuming and involves measuring the area under a curve. It has been well described elsewhere [13].

Stevens [14] at about the same time (1957–1961) developed quite different procedures for calculating loudness. Stevens' method, which he named Mark VI, is simpler than Zwicker's method but is only designed to be used for diffuse sound fields and when the spectrum is relatively flat and does not contain any prominent pure tones.

The procedure used in the Stevens Mark VI method is to plot the noise spectrum in either octave or one-third-octave bands onto the loudness index contours. The loudness index (in sones) is determined for each octave (or one-third-octave) band and the total loudness S is then given by

$$S = S_{\text{max}} + 0.3\left(\sum S - S_{\text{max}}\right), \tag{4.2}$$

where S_{max} is the maximum loudness index and ΣS is the sum of all the loudness indices. The 0.3 constant (used for octave bands) in Eq. (4.2) is replaced by 0.15 for one-third-octave bands. The Zwicker method is based on the critical band concept and, although more complicated than the Stevens method, can be used either with diffuse or frontal sound fields. Complete details of the procedures are given in the ISO standard [15] and Kryter [16] has discussed the critical band concept in his book.

Example 4.3 The octave band sound pressure levels (see column 2 of Table 4.1) of the noise from machines in a factory are given for the center frequencies (see column 1 of Table 4.1). Calculate the loudness level.

Solution

From Figure 4.8 the loudness indices, S, given in column 3 of Table 4.1 are found.

We obtain that $S_{max} = 26.5$ and $\Sigma S = 134.2$. Thus from Eq. (4.2) the loudness is: $S = 26.5 + 0.3 (134.2 - 26.5) = 59$ sones (OD, Octave Diffuse) and from Eq. (4.1), or Figure 4.7, the loudness level is: $P = 99$ phons (OD).

Table 4.1 Octave band levels of factory noise and Stevens' loudness indices.

Octave band center frequency, Hz	Octave band level, dB	Band loudness index S
31.5	75	3.0
63	79	6.2
125	82	10.5
250	85	15.3
500	85	18.7
1000	87	26.5
2000	82	23.0
4000	75	17.5
8000	68	13.5

There are several other aspects of loudness which we have not had room to discuss in this book, e.g. impulsive noise, and monaural and binaural loudness. Readers will find these well covered in Kryter's book [16].

4.3.3 Masking

The masking phenomenon is well known to most people. A loud sound at one frequency can cause another quieter sound at the same frequency or a sound close in frequency to become inaudible. This effect is known as *masking*. Broadband sounds can have an even more complicated masking effect and can mask louder sounds over a much wider frequency range than narrow-band sounds. These effects are important in human assessment of product sound quality.

Noise that contains pure tones is generally annoying and unpleasant. In the case of automobile interior noise, engine and exhaust system noise consists mostly of fundamental pure tones together with integer multiplies and broadband engine combustion noise. Similarly cooling fan noise consists of pure tones and broadband aerodynamic fan noise. In the case of automobiles, exhaust and cooling fan noise have some strong pure tone components. Wind noise and tire noise are mostly broadband in nature and can mask some of the engine, exhaust, and cooling fan noise, particularly at high vehicle speed. Figure 4.9 shows the masking effect of broadband white noise. The solid lines represent the sound pressure level of a pure test tone (given in the ordinate) that can just be heard when the masking sound is white noise at the level shown on each curve. For example, it is seen that white noise at 40 dB masks pure tones at about 55 dB at 100 Hz, 60 dB at 1000 Hz, and 65 dB at 5000 Hz.

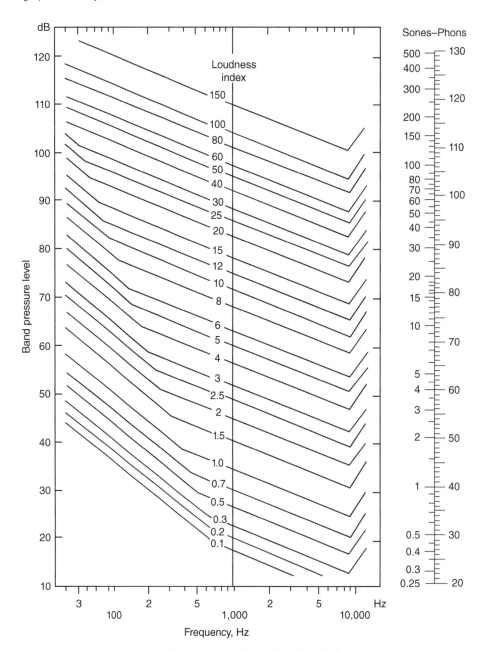

Figure 4.8 Contours of equal loudness index.

The masking of noise has been studied extensively for more than 70 years. The curves in Figure 4.9 [17, 18] are very similar to those of much earlier research work by Fletcher and Munson in 1950 but are shown here since they have been extended to lower and higher frequencies than these earlier published results [19]. One interesting fact is that the curves shown in Figure 4.9 are all almost parallel and are separated by about 10-dB intervals, which is also the interval between the masker levels. This suggests that the masking of noise is almost independent of level at any given frequency. This fact was also established by Fletcher and Munson in 1950 as is shown in Figure 4.10 [20]. Here the amount of masking is defined to be the upward shift in

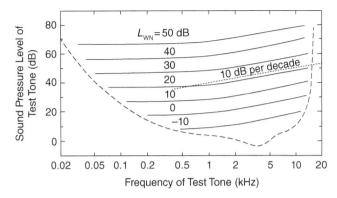

Figure 4.9 Contours joining sound pressure levels of pure tones at different frequencies that are masked by white noise at the spectral density level L_{WN} shown on each contour [17, 18].

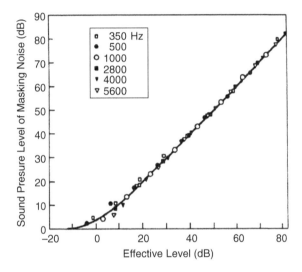

Figure 4.10 Masking of tones by noise at different frequencies and sound pressure levels [20]. Source: Reprinted with permission from [20], American Institute of Physics.

decibels in the baseline hearing threshold caused by the masking noise [19]. Figure 4.9 shows that once the masker has reached a sufficient level to become effective, the amount of masking attained is a linear function of the masker level. This means, for instance, that a 10-dB increase in the masker level causes a 10-dB increase in the masked threshold of the signal being masked. It was found that this effect is independent of the frequency of the tone being masked and applies both to the masking of pure tones and speech [19, 20].

Figure 4.11 shows the masking effect of a narrow-band noise of bandwidth 160 Hz centered at 1000 Hz. The curves in Figure 4.11 join the sound pressure levels of test tones that are just masked by the 1000-Hz narrow-band noise at the sound pressure level shown on each curve. It is observed that the curves are symmetric around 1000 Hz for low levels of the narrow-band noise, while for high levels, the curves become asymmetric. It is seen that high levels of the narrow band of noise mask high-frequency tones much better than low-frequency tones. Figures 4.9 and 4.11 show that using either white noise or narrow-band noise it is more difficult to mask low-frequency than high-frequency sounds. This fact is of some importance in evaluating the sound quality of a product since it is of no use to reduce or change the noise in one frequency range if it is masked by the noise at another frequency. There are other masking effects that are not normally as important

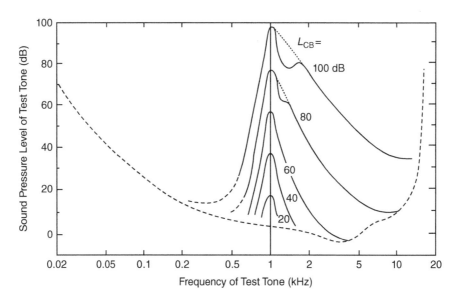

Figure 4.11 Masking effect of a narrow-band noise of bandwidth 160 Hz centered at 1000 Hz. The contours join sound pressure levels of pure tones that are just masked by the 1000-Hz narrow-band noise at the sound pressure level shown on each contour [17, 18].

for sound quality evaluations as the frequency effects discussed so far but still need to be described [16–19, 21–30].

When a masking noise stops, the human hearing system is unable to hear the primary sound signal immediately. This effect is known as *postmasking* (and is sometimes also known as forward masking.) The time it takes for the primary sound to be heard (normally called the delay time, t_d) depends both upon the sound pressure level of the masking noise and its duration. Figure 4.12 shows how the primary sound is affected by different sound pressure levels of the masking noise. Also shown in Figure 4.12 are dashed lines that correspond to an exponential decay in level with a time constant τ of 10 ms. It is observed that the human hearing mechanism decay is not exponential, but rather that it is nonlinear and that the decay process is complete after a decay time of about 200 ms. This fact is of practical importance since some vehicle and machinery noise is quite impulsive in character such as caused by diesel engines, automobile door closings, brake squeal, warning signals, or impacts that may mask the sounds of speech or other wanted sounds. Figure 4.13 shows how sounds are affected by different durations of

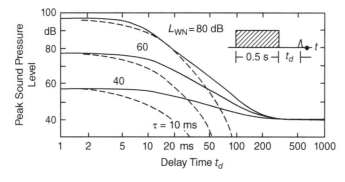

Figure 4.12 Postmasking at different masker sound pressure levels [17].

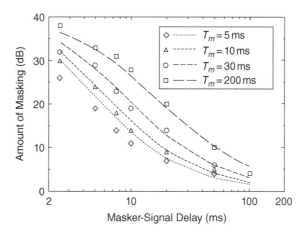

Figure 4.13 Postmasking of 5-ms, 2-kHz tones preceded by bursts of uniform masking noise are plotted as a function of the delay between masker and signal offsets. The parameter is masking duration T_m as indicated. The symbols are data from Zwicker [31]. Source: Reprinted with permission from [31], American Institute of Physics.

the masking noise. Figure 4.13 presents the level of a just-audible 2-kHz test tone as a function of delay time.

4.3.4 Pitch

Like loudness (Section 4.3.2), pitch is another subjective aspect of hearing. Just as people have invented scales to express loudness, others have invented scales for pitch. Stevens et al. [32] were the first to produce a scale in mels. A pure tone of 1000 Hz at a sound pressure level of 40 dB has a pitch of 1000 mels. As a result of subjective experiments, the pitch scale is found to be approximately linear with frequency below 1000 Hz, but approximately logarithmic above 1000 Hz. It has been suggested by some that noise measurements should be made in bands of equal mels (mel was named after the musical term, melody). However, this suggestion has not been adopted. A formula to convert frequency, f (in hertz) into mel, m, is [33]

$$m = 2595 \log \left(1 + \frac{f}{700} \right). \tag{4.3}$$

Masking noise may change the pitch of a tone. If the masking noise is of a higher frequency, the pitch of the masked tone is reduced slightly; if the masking noise is of a lower frequency the pitch is increased slightly. This can be explained [34, 35] by a signal/noise ratio argument. The locus of the position on the basilar membrane at which the tone is normally perceived could be changed by the masking noise [34].

Example 4.4 What is the equivalent frequency of 2595 mel?

Solution

We take the inverse of Eq. (4.3): $f = 700(10^{m/2595} - 1) = 700 \, (10 - 1) = 6300 \, \text{Hz}$.

4.3.5 Weighted Sound Pressure Levels

Figure 4.6 in this chapter shows that the ear is most sensitive to sounds in the mid-frequency range around 1000–4000 Hz. It has a particularly poor response to sound at low frequency. It became apparent to scientists in the 1930s that electrical filters could be designed and constructed with a frequency response approximately equal to the inverse of these equal loudness curves. Thus A-, B-, and C-weighting filters were constructed to approximate the inverse of the 40-, 70-, and 90-phon contours (i.e. for low-level, moderate, and intense sounds), respectively (see Figure 4.6). In principle, then, these filters, if placed between the microphone and the meter display of an instrument such as a sound level meter, should give some indication of the loudness of a sound (but for pure tones only).

The levels measured with the use of the filters shown in Figure 4.14 are commonly called the A-, B-, and C-weighted sound levels. The terminology A-, B-, and C-weighted sound pressure levels is preferred by ISO to reduce any confusion with sound power level and will be used wherever possible throughout this book. The A-weighting filter has been much more widely used than either the B- or C-weighting filter, and the A-weighted sound pressure level measured with it is still simply termed by ANSI as the sound level or noise level (unless the use of some other filter is specified). Several other weightings have also been proposed in the past [13]. However, because it is simple, giving a single number, and it can be measured with a low-cost sound level meter, the A-weighted sound pressure level has been used widely to give an estimate of the loudness of noise sources such as vehicles, even though these produce moderate to intense noise. Beranek and Ver have reviewed the use of the A-weighted sound pressure level as an approximate measure of loudness level [36].

The A-weighted sound pressure levels are often used to gain some approximate measure of the loudness levels of broadband sounds and even of the acceptability of the noise. Figure 4.15 shows that there is reasonable correlation between the subjective response of people to vehicle noise and the A-weighted sound pressure levels measured of the vehicle noise. The A-weighted sound pressure level forms the basis of many other descriptors for determining human response to noise described later in Chapter 6. The A-weighted sound pressure level descriptor is also used as a limit for new vehicles (Chapter 14) and noise levels in buildings (Chapter 12) in several countries. Although the A-weighting filter was originally intended for use with low-level sounds of about 40 dB, it is now commonly used to rate high-level noise such as in industry where

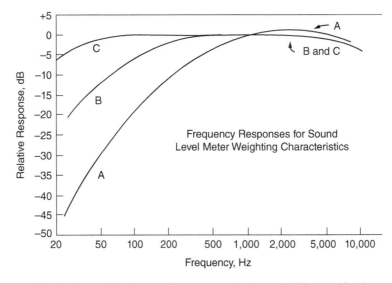

Figure 4.14 A-, B-, and C-weighting filter characteristics used with sound level meters.

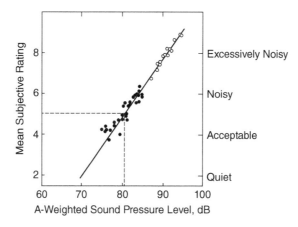

Figure 4.15 Relation between subjective response and A-weighted sound pressure level for diesel engine trucks undergoing an acceleration test: ●, values measured in 1960, ○, values measured in 1968. (Source: Adapted from Refs. [37–39].)

A-weighted sound pressure levels may exceed 90 dB. At such high levels the A-weighted sound pressure level and the loudness level are normally in disagreement.

Example 4.5 The factory noise spectrum (given in Table 4.1), was calculated to have a loudness level of 99 phon (see Example 4.2). Calculate the approximate A-weighted sound pressure level from the octave band levels given in Table 4.1.

Solution

Since we do not have the directly measured A-weighted sound level, we can calculate this approximately using Figure 4.14. The A-weighting corrections at the octave band center frequencies have been read off Figure 4.14 and entered in Column 3 of Table 4.2. The so-called A-weighted octave band sound pressure levels have been calculated in Column 4 and these values have been combined to give the A-weighted sound pressure level of 89.8, i.e. 90 dB. Note this A-weighted sound pressure level is dominated by the A-weighted band levels in the 500, 1000, and 2000 Hz octave bands. These three band levels combine to give 89.5 dB. Similar calculations of the C-weighted and linear (nonfiltered) sound pressure levels give 91.9 and 92.0 dB, respectively. Thus we see that the A-weighted sound pressure level is 9 dB below the

Table 4.2 Combination of octave-band sound pressure levels of factory noise to give the A-weighted sound pressure level.

Octave band center frequency, Hz	Octave band level, dB	A-weighting correction, dB	A-weighted octave-band levels, dB
31.5	75	−42	33
63	79	−28	51
125	82	−18	64
250	85	−9.0	76
500	85	−3.0	82
1000	87	0	87
2000	82	+1.5	83.5
4000	75	+0.5	75.5
8000	68	−2.0	66

level in phons and no closer than the linear unweighted sound pressure level of 92 dB. A-weighted levels should not be used to calculate the loudness level unless the noise is a pure tone – then a good loudness level estimate can be made using the A, B, or C filters (depending on the noise level).

Another problem with A-weighting is that it does not allow for the fact that loudness increases with the bandwidth of the noise and also with the duration of the noise event for very short impulsive-type sounds of duration less than about 200 ms. The concept of the critical band is of fundamental importance in psychoacoustics. It is of concern in studies of loudness, pitch, hearing thresholds, annoyance, speech intelligibility, masking, and fatigue caused by noise, phase perception, and even the pleasantness of music.

Figure 4.16 shows the loudness level of bands of filtered white noise centered at 1000 Hz as a function of bandwidth for the different constant sound pressure levels shown on each curve. The curves were obtained by a matching procedure in which listeners equated the loudness of a 1000-Hz pure tone with bands of noise of increasing bandwidth. The level at which the pure tone was judged to be equal in loudness to the band of noise is shown as the ordinate. Thus the curves do not represent equal loudness contours, but rather they show how the loudness of the band of noise centered at 1000 Hz changes as a function of bandwidth. The loudness of a sound does not change until its bandwidth exceeds the so-called critical bandwidth. The critical bandwidth at 1000 Hz is about 160 Hz. (Notice that, except for sounds of very low level of about 20 phons, for which loudness is almost independent of bandwidth, the critical bandwidth is almost independent of level and that the slopes of the loudness curves are very similar for sounds of different levels.) Critical bands are discussed further in Section 4.3.6 of this chapter.

The solid line in Figure 4.17 shows that sounds of very short duration are judged to be very quiet and to become louder as their duration is increased. However, once the duration has reached about 100–200 ms, then the loudness level reaches an asymptotic value. Also shown by broken lines in Figure 4.17 are A-weighted sound pressure levels recorded by a sound level meter using the "impulse," "fast," and "slow" settings. It is observed that the A-weighted sound pressure level measured by the fast setting on the sound level meter is closest of the three settings to the loudness level of the sounds.

Further methods of rating loudness, noisiness, and annoyance of noise are discussed in Chapter 6.

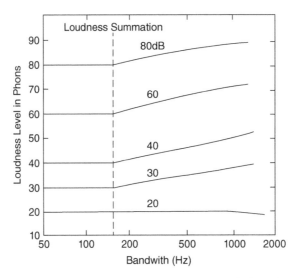

Figure 4.16 Loudness level in phons of a band of filtered white noise centered at 1000 Hz as a function of its bandwidth. The overall sound pressure level of each band of noise was held constant as its bandwidth was increased, and this level is shown on each curve. The dashed line indicates that the bandwidth at which the loudness starts to increase is about the same at all of the levels tested, except for the lowest level for which no increase in loudness occurs. (Source: From Ref. [40]; used with permission.)

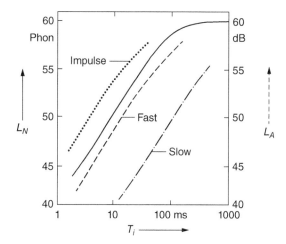

Figure 4.17 Dependence of loudness level L_N (left ordinate) on duration T_i of 1-kHz tone impulses of constant sound pressure level compared with measurements of A-weighted sound pressure level L_A (right ordinate) using the time constants "impulse," "fast," or "slow." [17].

4.3.6 Critical Bands

Another important factor is the way that the ear analyzes the frequency of sounds. The critical band concept already discussed in Section 4.3.5 is important here as well. It appears that the human hearing mechanism analyzes sound like a group of parallel frequency filters. Figure 4.18 shows the bandwidth of these filters as a function of their center frequency. These filters are often called *critical bands* and the bandwidth each possesses is known as its *critical bandwidth*. As a band of noise of a constant sound pressure level increases in bandwidth, its loudness does not increase until the critical bandwidth is exceeded, after which the loudness

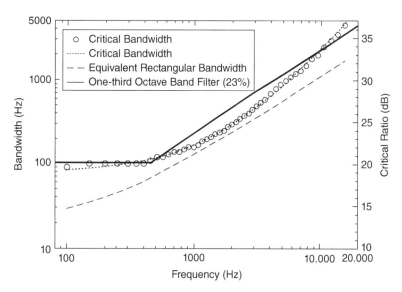

Figure 4.18 Critical bandwidth, critical ratio, and equivalent rectangular bandwidth as a function of frequency. (O) Critical bandwidth from Zwicker [41] is compared to the equivalent rectangular bandwidths according to a cochlear map function (- - -) (dashed line). (Source: Based in part on Ref. [42].)

continually increases. Thus the critical band may be considered to be that bandwidth at which subjective responses abruptly change [28]. It is observed that up to 500 Hz the critical bandwidth is about 100 Hz and is independent of frequency, while above that frequency it is about 21% of the center frequency and is thus almost the same as one-third octave band filters, which have a bandwidth of 23% of the center frequency shown by the solid line. This fact is also of practical importance in sound quality considerations since it explains some of the masking phenomena observed.

The *critical ratio* shown in Figure 4.18 originates from the early work of Fletcher and Munson in 1937 [28]. They conducted studies on the masking effects of wide-band noise on pure tones at different frequencies. They concluded that a pure tone is only masked by a narrow critical band of frequencies surrounding the tone, and that the power (mean-square sound pressure) in this band is equal to the power (mean-square sound pressure) in the tone [28]. The critical band can easily be calculated. From these assumptions, the critical bandwidth (in hertz) is defined to be the ratio of the sound pressure level of the tone to sound pressure level in a 1-Hz band (i.e. the spectral density) of the masking noise. This ratio is called the critical ratio to distinguish it from the directly measured critical band [28]. A good correspondence can be obtained between the critical band and the critical ratio by multiplying the critical ratio by a factor of 2.5. The critical ratio is given in decibels in Figure 4.18 and is shown by the broken line.

4.3.7 Frequency (Bark)

It is well known from music that humans do not hear the frequency of sound on a linear scale. A piano keyboard is a good example. For each doubling of frequency (known as an octave), the same distance is moved along the keyboard in a logarithmic fashion. If the critical bands are placed next to each other, the bark scale is produced. The unit *bark* was chosen to honor the German physicist Heinrich Barkhausen. Figure 4.19 illustrates the relationship between the bark (Z) as the ordinate and the frequency as the abscissa; on the left (Figure 4.19a), frequency is given using a linear scale, and on the right (Figure 4.19b) the frequency is given with a logarithmic scale. Also shown in Figure 4.19 are useful fits for calculating bark from frequency. At low frequency a linear fit is useful (Figure 4.19a), while at high frequency a logarithmic fit is more suitable (Figure 4.19b). One advantage of the bark scale is that, when the masking patterns of narrow-band noises are

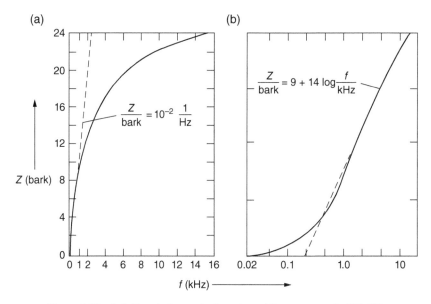

Figure 4.19 Relations between bark scale and frequency scale [17, 18].

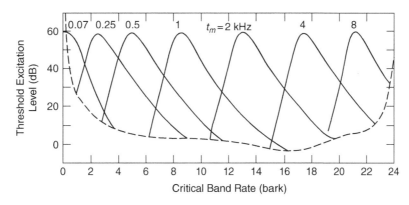

Figure 4.20 Masking patterns [17, 18] of narrow-band noises centered at different frequencies f_m.

plotted against the bark scale, their shapes are largely independent of the center frequency, except at very low frequency. (See Figure 4.20.)

An approximate analytical expression for auditory frequency in barks as a function of auditory frequency in hertz is given by [43]

$$Z = 28.32 - \frac{12098.7}{(1750 + f)^{0.81}}. \tag{4.4}$$

Care should be taken to note that the ear does not hear sounds at a fixed number of fixed center frequencies as might be suspected from Figure 4.20. Rather, at any frequency f_m considered, the average ear has a given bandwidth.

Example 4.6 Convert $f = 6$ kHz into its corresponding value in bark.

Solution

Using Eq. (4.4), $Z = 28.32 - 12098.7(6000 + 1750)^{-0.81} = 28.32 - 12098.7(7.07 \times 10^{-4}) = 28.32 - 8.56 = 19.76$. This result can also be obtained from Figure 4.19.

4.3.8 Zwicker Loudness

The loudness of sounds was discussed in Sections 4.3.2 and 4.3.5, where it was shown that A-weighted sound pressure level measurements underestimate the loudness of broadband noise. (See Figure 4.16.) Methods to evaluate the loudness of broadband noise based on multiband frequency analysis have been devised by Stevens [44], Kryter [16], and Zwicker [12]. The Stevens method was originally based on octave band analysis, but Kryter's and Zwicker's methods are based on one-third octave band analysis. Kryter's method has been standardized for aircraft certification noise measurements, while Zwicker's method has been standardized internationally and is most normally used to evaluate the loudness of many common sound sources including speech, music, machinery, and vehicles.

The procedure to evaluate loudness using Zwicker's method is shown in Figure 4.21. Figure 4.21a shows a narrow band centered at 1000 Hz, which corresponds to 8.5 bark. Figure 4.21b shows the narrow band of noise at 1000 Hz, including masking effects caused by spectral broadening in the cochlea due to inner ear mechanics. Figure 4.21c shows the specific loudness/critical band ratio pattern (sone/bark), known as the *Zwicker diagram*. The transition from the masking pattern in Figure 4.21b to the loudness pattern in Figure 4.21c can be considered to be obtained by simply taking the square root of the sound pressure or the fourth root of the sound

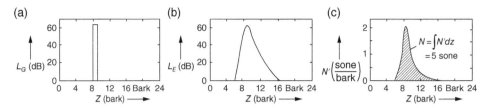

Figure 4.21 Schematic illustration of Zwicker's loudness model [17, 18].

intensity. The shaded area in Figure 4.21c is directly proportional to the perceived loudness. While Figure 4.21 illustrates the spectral process of obtaining the perceived loudness of a sound by the Zwicker method, Figure 4.22 shows the process including temporal effects. Figure 4.22a shows two impulses in the time domain, one with a broken line of 10-ms duration and the other with a solid line of 100-ms duration. As discussed earlier in connection with Figure 4.16, very short-duration pulses of noise are perceived to be quieter than longer ones up to a duration of about 100–200 ms.

Figure 4.22 is constructed by assuming that the hearing mechanism behaves like a parallel bank of 24 critical band filters. Figure 4.22b represents the processing of the loudness in each of the 24 channels of an empirical *loudness meter* used to model the hearing mechanism. Finally, Figure 4.22c shows the time dependence

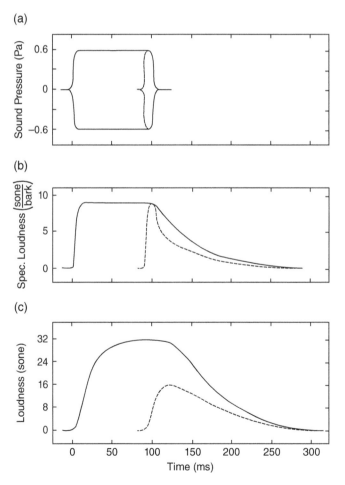

Figure 4.22 Illustration of temporal effects in loudness processing [17, 18].

of the total loudness summed up over all 24 channels of the empirical loudness meter. Figure 4.22b shows that the short 10-ms tone burst decays much more rapidly than the 100-ms tone burst. The results shown in Figures 4.21 and 4.22 are important in evaluating the sound quality of machinery that has impulsive noise components, such as diesel engines and machines in which impacts occur.

4.3.9 Loudness Adaptation

The term *loudness adaptation* refers to the apparent decrease in loudness that occurs when a subject is presented with a sound signal for a long period of time [45]. The effect has been studied extensively by presenting tones for an extended period of time to one ear and then allowing the subject to adjust the level of a second comparison tone of the same frequency to the other ear. Such experiments have demonstrated that loudness adaptation of as much as 30 dB can be observed for very quiet sounds below 30 dB and much less for louder sounds of the order of 70 dB. However, other research has shown that loudness adaptation is reduced or even absent when binaural interactions are minimized. Recent research has shown that loudness adaptation is quite complicated. It varies from person to person. More adaptation occurs for very quiet sounds of about 30 dB and for high-frequency tones than for low-frequency tones or noise [19]. This phenomenon is obviously of interest to those concerned with sounds such as interior vehicle noise, that people experience over an extended period.

4.3.10 Empirical Loudness Meter

Figure 4.23 shows a block diagram of an empirical Zwicker-type dynamic loudness meter (DLM) that includes the spectral and temporal loudness processing portrayed in Figures 4.21 and 4.22 [22]. First, the *spectral processing* (1 and 2) shown in Figure 4.22 using the critical band filter bank concept, *upward spread of masking* (7), and *spectral summation* (8) are illustrated. Second, the *temporal processing* discussed in relation to Figure 4.23 is observed as shown in the blocks marked *envelope extraction*, *postmasking*, and *temporal integration* (see steps 3, 6, and 9.) Lastly and of most importance is step 5 (labeled *loudness transformation*),

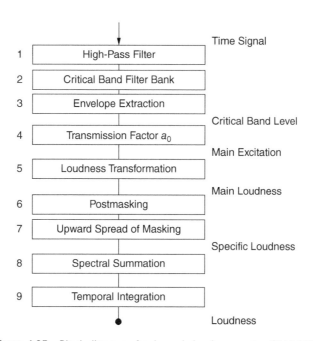

Figure 4.23 Block diagram of a dynamic loudness meter (DLM) [46].

which represents the fact that the loudness is assumed to be proportional to the square root of the sound pressure (or the fourth root of the sound intensity. See Figure 4.21c.).

One advantage of the DLM discussed and shown in Figure 4.23 is that, by suitable modification of the DLM loudness transformation (step 5), the loudness perception of both normal hearing and hearing impaired subjects can be simulated [22]. This fact is important since the aging population in industrialized countries has suffered varying degrees of hearing impairment that should be allowed for in sound quality simulations. Some fraction of the younger population may have suffered mild to moderate hearing loss because of exposure to intense noise during recreational or leisure time activities [22].

4.4 Hearing Loss and Diseases (Disorders)

Some are born with the severe handicap of deafness. Others suffer sudden hearing loss later in life or (much more commonly) gradually lose their hearing over a period of time. For all these persons, deafness is a crippling loss of one of the most important senses. Deafness can be very slight or complete. There are different legal and medical definitions of deafness. For our purpose we will define the onset of impaired hearing when a person's hearing level is 25 dB, averaged at 500, 1000, and 2000 Hz. We will define total deafness when a person's hearing level is 92 dB averaged at the same important speech frequencies [47] (see Section 5.7). Someone who is hard-of-hearing, then, may have a hearing level anywhere between these two extremes of 25 and 92 dB. The average hearing level of a nonimpaired person is 0 dB.

Deafness not only causes a lowering in the level at which sound is heard, but in many cases also causes a loss in hearing quality so that consonants cannot be distinguished and words are confused. In addition, some deaf people also suffer from tinnitus or a ringing in the ears which is disturbing and competes with other sounds.

Because of their affliction, deaf people tend to withdraw from society. This is sometimes reinforced by society itself. In ancient Rome, the deaf were classed with the mentally incompetent. In modern societies, education, advances in medicine, and use of hearing aids are now improving the plight of the deaf. Strict control of noise levels in industry will also reduce the number of people receiving needless hearing damage and consequent deafness. A review on noise-induced hearing loss is available in the literature [48].

Hearing disorders (deafness) can be divided into two main types: conduction deafness and sensory-neural deafness. Conduction deafness is related to disorders in the outer and middle ear, while sensory-neural deafness occurs in the inner ear in the organ of Corti, the auditory nerve, or the auditory cortex in the brain.

4.4.1 Conduction Hearing Loss

This normally manifests itself as a fairly uniform decrease in hearing over most frequencies. Background noise usually causes people to speak louder and those with conduction deafness can then often hear. This type of deafness can normally be overcome by a hearing aid with sufficient amplification. There are several causes of conduction deafness, some of which may be easily corrected. For example the ear canal may become blocked with wax. This can now be cleaned out, medically restoring hearing.

Until 65 years ago, middle ear infections were the largest cause of hearing loss. Such infections, often occurring in children, could be caused by inflammations of the tonsils or adenoids. Unchecked, these inflammations could spread to the Eustachian tube which would swell and become blocked. Fluids would form in the middle ear cavity causing pressure on the eardrum and pain. Frequently infection could spread to the mastoid bone surrounding the middle ear cavity. Death often resulted. Mastoiditis was treated by surgically removing some of the affected mastoid bone. With the advent of penicillin and other antibiotics, there is no need for middle ear infections to progress so far and now they rapidly can be cleared up with drugs in almost every case.

Another common cause of hearing loss occurs when bony material forms around the footplate of the stapes preventing it moving and transmitting vibrations through the oval window into the cochlea. This condition known as Otosclerosis normally causes a severe loss in hearing at most frequencies of 50 dB or more. Otosclerosis can often be helped quite successfully with a hearing aid. By providing appreciable amplification of the sound pressure at the eardrum, sufficient vibration amplitude can be transmitted to the oval window despite the calcification.

However, calcification can now be successfully treated by surgical methods. Toward the end of the nineteenth century, two German surgeons, first Johann Kessel, and later Karl Passow, tried to remove the calcified stapes and make an opening to the inner ear [47]. Although some operations were briefly successful, this technique had to be abandoned because of the crudity of the surgical instruments available and the danger of admitting infection into the cochlea. However, in the U.S. in 1938, Julius Lempert pioneered the surgical technique known as fenestration. In this operation the incus and malleus and some of the middle ear bone are removed and the ear canal is thus extended and put into contact with one of the semicircular canals. The middle ear is thus bypassed completely. This operation is normally quite successful and restores hearing, although at a lower level.

Other surgical techniques have recently become preferred and fenestration is seldom used now. In 1952, Samuel Rosen accidentally freed the stapes during ear surgery and this technique has been refined into an operation known as stapes mobilization. It is difficult to accomplish, however, because just the right amount of force is needed.

Too much force can damage the inner ear. Also, because otosclerosis is a progressive disease, the condition of calcification can recur. A surgical technique used for many years is *stapedectomy*: the complete removal of the stapes and its replacement with a plastic part. The operation must be conducted using a microscope and can be accomplished in less than 30 minutes. After the stapes is chipped away it is replaced with an artificial one which is attached both to the incus and the oval window and almost perfect hearing is restored. Another microsurgical option is *stapedotomy* which is thought by many surgeons to be safer and less prone to complications. In this procedure, a skilled surgeon drills a tiny hole in the stapes using a laser, in which to secure a prosthetic.

In the middle of 1970s the first *cochlear implants* were performed. A cochlear implant is an electronic device used in individuals with a severe or profound hearing loss. Basically the implant consists of a microphone, speech processor, transmitter and receiver, and a stimulator which converts the signals into electrical impulses to be sent by electrodes to the cochlea. The surgeon makes an incision in the skin behind the ear were the implant is placed. Then the surgeon drills into the mastoid to reach the inner ear where the electrode array is inserted directly into the cochlea. Although it is still an expensive procedure, many deaf people have recovered the sense of hearing after this surgery.

4.4.2 Sensory-Neural Hearing Loss

There are several causes of sensory-neural hearing loss and all are associated with disorders of the inner ear, the auditory nerve fibers, the auditory cortex in the brain, or combinations of all three. Unlike conductive deafness, sensory-neural deafness is often most severe at higher frequencies. Background noise can thus mask the consonants in speech and make it unintelligible. Although a hearing aid brings some relief, it is perhaps not so useful as with conductive deafness. Very profound sensory-neural hearing loss may be restored only by using a cochlear implant.

Congenital deafness is often a cause of sensory-neural deafness. Here genes producing imperfect hearing mechanisms are inherited. During the first three months of pregnancy, if a mother contracts certain viral diseases such as mumps, influenza, or German measles, these can impair the development of the hearing mechanism in the fetus. During birth itself, brain damage can affect the hearing mechanism. In early childhood, diseases such as meningitis can damage the inner ear or auditory nerve, producing deafness. Perforations in the eardrum can also be caused. However, the eardrum has a remarkable ability to heal itself, even if ruptured by intense sound.

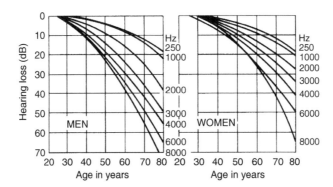

Figure 4.24 Shift in hearing threshold at different frequencies against age for men and women [52].

All these forms of sensory-neural hearing loss mentioned are presently medically untreatable. One exception is Ménière's disease, in which the inner ear becomes distended. Drugs can sometimes improve this condition.

Another reported cause of sensory-neural deafness is drug-induced hearing loss which is called *ototoxicity*. There are many approved drugs that can cause ototoxicity through direct effect on cochlear biology and they include anti-cancer chemotherapy drugs, some kinds of antibiotics, quinine, and loop diuretics used to treat hypertension [49].

These drugs can cause either temporary or permanent hearing loss and vertigo and tinnitus. Recent studies have suggested that certain industrial chemicals may also potentiate the effect of noise exposure [50, 51]. In this case, damage to the cochlea is the direct result of chemical injury to the fragile hair cells within the organ of Corti.

4.4.3 Presbycusis

Another form of sensory-neural hearing loss is presbycusis. This loss in hearing sensitivity occurs in all societies. Figure 4.24 shows the shift in hearing threshold at different frequencies against age [52]. As can be seen, presbycusis mainly affects the high frequencies, above 2000 or 3000 Hz. It affects men more than women. The curves shown in Figure 4.24 show the average hearing loss (with average hearing at the age of 25 assumed to be zero hearing loss). The group of people tested was subject to both presbycusis (the so-called natural hearing loss with age), and *sociocusis* (a term coined by Glorig [53] to include all the effects of noise damage in our everyday lives with the exception of excessive industrial noise).

Presbycusis is believed to be caused by central nervous system deterioration with aging, in addition to changes in the hearing mechanism in the inner ear. Hinchcliffe [54] believes that these changes explain the features normally found in presbycusis: loss of discrimination of normal or distorted speech, increase in susceptibility to masking particularly by low-frequency noise, and a general loss in sensitivity. Rosen [55] suggested that degenerative arterial disease is a major cause of presbycusis. Others have suggested that a diet low in saturated fat may, in addition to protecting a person against coronary heart disease, also protect him or her against sensory-neural hearing loss (presbycusis).

The sensory-neural hearing loss caused by intense noise is discussed in Chapter 5.

4.5 Speech Production

Since speech and hearing must be compatible, it is not surprising to find that the speech frequency range corresponds to the most sensitive region of the ear's response (Section 4.3.2) and generally extends from 100 to 10 000 Hz. The general mechanism for speech generation involves the contraction of the chest muscles

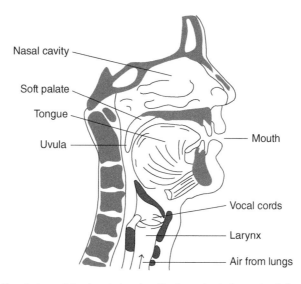

Figure 4.25 Sectional view of the head showing the important elements of the voice mechanism.

to force air out of the lungs and up through the vocal tract. This flow of air is modulated by various components of the vocal mechanism (Figure 4.25) to produce sounds which make up part of our speech pattern. The modulation effect first takes place at the larynx, across which are stretched the vocal cords. These are composed of two bands of membranes separated by a slit which can open and close to modulate the flow of air [56]. The modulation frequency depends upon the tension in the muscles attached to the vocal cords, and on the size of the slit in the membranes (about 24 cm for males and 15 cm for females). The sound emitted by the vocal cords has a buzz-type sound corresponding to a sawtooth waveform containing a large number of harmonically related components.

This sound/air flow is then further modified by its flow through the numerous cavities of the throat, nose, and mouth, many of which can be voluntarily changed at will by, for example, changing the position of the tongue or shape of the lips, to produce a large variety of voiced sounds. It is possible to produce some sounds without the use of the vocal chords and these are known as unvoiced or breathe sounds. These are usually caused by turbulent air flow through the upper parts of the vocal tract and especially by the lips, teeth, and tongue. It is in this way that the unvoiced fricative consonants *f* and *s* are formed. In some cases, part of the vocal tract can be blocked by constriction and then suddenly released to give the unvoiced consonants *p* and *g* [57].

Generally, vowels have a fairly definite frequency spectrum whereas many unvoiced consonants such as *s* and *f* tend to exhibit very broadband characteristics. Furthermore, when several vowels and/or consonants are joined together their individual spectra appear to change somewhat. The time duration of individual speech sounds also tends to vary widely over a range of 20–300 ms.

In the general context of speech, vowels, and consonants become woven together to produce not only linguistically organized words, but sounds which have a distinctive personal characteristic as well. The vowels usually have greater energy than consonants and give the voice its character. This is probably due to the fact that vowels have definite frequency spectra with superimposed periodic short-duration peaks. However, it is the consonants which give speech its intelligibility. It is therefore essential in the design of rooms for speech to preserve both the vowel and consonant sounds for all listeners. Consonants are generally transient, short-duration sounds of relatively low energy. Therefore, for speech, it is necessary to have a room with a short reverberation time to avoid blurring of consecutive consonants; we would expect therefore speech intelligibility to decrease with increasing reverberation time. At the same time we find that in order to produce a speech signal level well above the reverberant sound level (i.e. high signal-to-noise ratio), we require increased sound absorption in the room. This necessitates a lower reverberation time. Although this may lead us to think that an anechoic room would be most suitable for speech intelligibility, some sound reflections are required both to boost the level of the direct sound and to give the listener a feeling of volume. Therefore, an optimum reverberation time is established. This is usually under one second for rooms with volumes under $8500\,m^3$. If the speech power emitted by a male speaker is averaged over a relatively long period (i.e. five seconds), the overall sound power level is found to be 75 dB. This corresponds to an average sound pressure level of 65 dB at 1 m from the lips of the speaker and directly in front of him or her. Converting the sound power level to sound power shows that the long time averaged sound power for men is 30 μW. The average female voice is found to emit approximately 18 μW. However, if we average over a very short time (i.e. 1/8 second) we find that the power emitted in some vowel sounds can be 50 μW, while in other soft spoken consonants it is only 0.03 μW. Generally, the human voice has a dynamic range of approximately 30 dB throughout its frequency range [58]. At maximum vocal effort (loud shouting) the sound power from the male voice may reach 3000 μW.

Table 4.3 gives the long-term rms sound pressure levels at 1 m from the average male mouth for normal vocal effort as given by the American National Standards Institute [59] for both one-third-octave and one-octave bands. Although approximately 80% of the energy in speech lies below 600 Hz (including most vowels), it is in the higher frequencies that most consonants have most of their energy. These low-energy transient consonants contribute to the intelligibility perceived. For example, it has been found [60] that if speech is passed through a high-pass filter having a cutoff frequency of 1000 Hz then 90% of the spoken words can be understood. However, if the same speech is passed through a low-pass filter, then a cutoff frequency of 3000 Hz is required to produce the same percentage word intelligibility. Speech sounds below 200 Hz and above 6000 Hz do not significantly contribute to intelligibility but they do add to the natural qualities of the voice [57]. Calculation of the intelligibility of speech is discussed in Chapter 6.

Table 4.3 Male voice speech sound pressure levels +12 dB at 1 m from lips for both one-third- and one-octave bands. These levels represent the speech peaks that contribute to intelligibility. The voice peak sound power levels, $L_{W,pk}$, can be evaluated by adding 10.8 to the above values as shown for octave bands.

Center frequency, Hz	$L_{p,pk}$ (one-third-octave)	$L_{p,pk}$ (octave)	$L_{W,pk}$
200	67.0		
250	68.0	72.5	83.3
315	69.0		
400	70.0		
500	68.5	74.0	84.8
630	66.5		
800	65.0		
1000	64.0	68.0	78.8
1250	62.0		
1600	60.5		
2000	59.5	62.0	72.8
2500	58.0		
3150	56.0		
4000	53.0	57.0	67.8
5000	51.0		

Since speech is emitted from the mouth, it is not surprising to find that the acoustic radiation from this small aperture set in a larger object (the head) is subject to fairly strong directivity effects. These directivity effects become more marked at high frequencies. Figures 4.26 and 4.27 show the relative A-weighted sound pressure levels for the human voice in the horizontal and vertical planes, respectively. These experiments

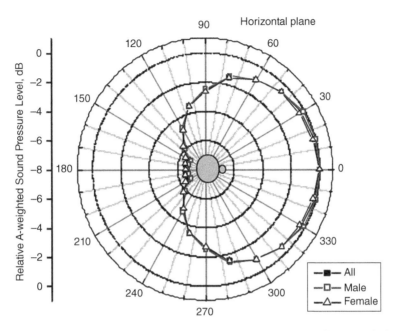

Figure 4.26 Directivity patterns for the human voice in a horizontal plane. (Source: From Ref. [61] with permission.)

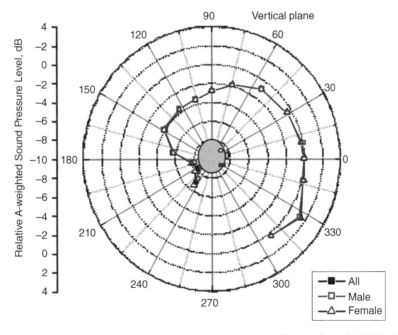

Figure 4.27 Directivity patterns for the human voice in a vertical plane. (Source: From Ref. [61] with permission.)

were conducted by Chu and Warnock in 40 adults, 20 male and 20 female [61]. Directivity effects can become important for audience members seated at the end of the front rows of an auditorium, since they will receive considerably less of the direct sound at high frequencies. This can considerably reduce the intelligibility of speech.

References

1 Palmer, J.W. (1993). *Anatomy for Speech and Hearing*, 4e. Philadelphia, PA: Lippincott Williams & Wilkins.

2 Crocker, M.J. (1974). The ear, hearing, loudness and hearing damage. In: *Reduction of Machinery Noise* (ed. M.J. Crocker), 43. West Lafayette, IN: Purdue University.

3 Crocker, M.J. (1998). *Handbook of Acoustics*, 1083–1138. New York: Wiley.

4 Paul, P.V. and Whitelaw, G.M. (2011). *Hearing and Deafness: An Introduction for Health and Education Professionals*. Sudbury, MA: Jones and Bartlett Publishers.

5 Gulick, W.L., Gescheider, G.A., and Frisina, R.D. (1989). *Hearing: Physiological Acoustics, Neural Coding, and Psychoacoustics*, 2e. New York: Oxford University Press.

6 Tysome, J. and Kanegaonkar, R. (2016). *Hearing: An Introduction & Practical Guide*. Boca Raton, FL: CRC Press.

7 Welch, B.L. and Welch, A.S. (1970). *Physiological Effects of Noise*. New York: Plenum Press.

8 Wada, H. (2007). The ear: its structure and function, related to hearing. In: *Handbook of Noise and Vibration Control* (ed. M.J. Crocker), 277–285. New York: Wiley.

9 Silman, S. (1984). *The Acoustic Reflex: Basic Principles and Clinical Applications*. London: Academic Press.

10 Stevens, S.S. and Warshofsky, F. (1980). *Sound and Hearing*, rev. ed. Morristown, NJ: Time-Life.

11 Fletcher, H. and Munson, W.A. (1933). Loudness, its definition, measurement and calculation. *J. Acoust. Soc. Am.* 5: 82–108.

12 Zwicker, E. (1960). Ein Verfahren zur Berechnung der Lautstarke (A means for calculating loudness). *Acustica* 10: 304–308.

13 Hassall, J.R. and Zaveri, K. (1988). *Acoustic Noise Measurements*. Naerum, Denmark: Bruel & Kjaer.

14 Stevens, S.S. (1961). Procedure for calculating loudness: Mark VI. *J. Acoust. Soc. Am.* 33: 1115.

15 ISO 532 (1975). *Acoustics-Method for Calculating Loudness Level*. Geneva: International Standards Organization.

16 Kryter, K.D. (1985). *The Effects of Noise on Man*, 2e. Orlando, FL: Academic.

17 Fastl, H. (1997). Psychoacoustics of sound-quality evaluation. *Acta Acust. (Stuttgart)* 83 (5): 754–764.

18 Fastl, H. and Zwicker, E. (2007). *Psychoacoustics. Facts and Models*, 3e. Berlin-Heidelberg-New York: Springer.

19 Gelfand, S.A. (2009). *Hearing – An Introduction to Psychological and Physiological Acoustics*, 5e. Boca Raton, FL: CRC Press.

20 Hawkins, J.E. and Stevens, S.S. (1950). The masking of pure tones and of speech by white noise. *J. Acoust. Soc. Am.* 22: 6–13.

21 Blauert, J. and Jekosch, U. (1997). Sound-quality evaluation – a multi-layered problem. *Acta Acust. (Stuttgart)* 83 (5): 747–753.

22 Fastl, H. (2005). Psycho-Acoustics and sound quality. In: *Communication Acoustics* (ed. J. Blauert), 139–162. Berlin: Springer.

23 Moore, B.C.J. (2012). *An Introduction to the Psychology of Hearing*, 6e. Bingley, UK: Emerald Group Publishing.

24 Luce, R.D. (1993). *Sound & Hearing – A Conceptual Introduction*. Hillsdale, NJ: Lawrence Erlbaum Associates.

25 Neuhoff, J.G. (ed.) (2004). *Ecological Psychoacoustics*. London: Elsevier, Academic.

26 Durrant, J.D. and Lovring, J.H. (1995). *Bases of Hearing Science*, 3e. Baltimore, MD: Williams & Wilkins.

27 Moore, B.C.J. (ed.) (1995). *Hearing*. San Diego, CA: Academic.

28 Tobias, J.V. (ed.) (1970). *Foundations of Modern Auditory Theory*, vol. I. New York: Academic.

29 Martin, M. and Summers, I. (1999). *Dictionary of Hearing*. London: Whurr.

30 Schubert, E.D. (ed.) (1979). *Psychological Acoustics, Benchmark Papers in Acoustics*, vol. 13. Stroudsburg, PA: Dowden Hutchinson and Ross.

31 Zwicker, E. (1984). Dependence of post-masking on masker duration and its relation to temporal effect in loudness. *J. Acoust. Soc. Am.* 75: 219–223.

32 Stevens, S.S., Volkmann, J., and Newman, E.B. (1937). A scale for the measurement of the psychological magnitude pitch. *J. Acoust. Soc. Am.* 8: 185.

33 O'Shaughnessy, D. (2000). *Speech Communication: Human and Machine*, 2e. Piscataway, NJ: IEEE Press.

34 Egan, J.P. and Mayer, D.R. (1950). Changes in pitch of tones of low frequency as a function of the pattern of excitation produced by a band of noise. *J. Acoust. Soc. Am.* 22: 827.

35 de Boer, E. (1962). Note on the critical bandwidth. *J. Acoust. Soc. Am.* 34: 985.

36 Beranek, L.L. and Ver, I.L. (1992). *Noise and Vibration Control Engineering*. New York: Wiley.

37 Mills, C.H.G. and D.W. Robinson (1961) The subjective rating of motor vehicle noise. The Engineer 30 June.

38 Shaw, E.A.G. (1996). Noise environments outdoors and the effects of community noise exposure. *Noise Control Eng. J.* 44 (3): 109–119.

39 Priede, T. (1971). Origins of automotive vehicle noise. *J. Sound Vib.* 15: 61–73.

40 Feldtkekker, R. and Zwicker, E. (1956). *Das Ohr als Nachrichtenemofanger*, 82. Stuttgart: S. Hirzel.

41 Zwicker, E. (1961). Subdivision of the audible frequency range into critical bands (Frequenzgruppen). *J. Acoust. Soc. Am.* 33: 248.

42 Buus, S. (1997). Auditory masking. In: *Encyclopedia of Acoustics*, vol. 3 (ed. M.J. Crocker), 1427–1445. New York: Wiley.

43 Buus, S. (1998). Auditory masking. In: *Handbook of Acoustics* (ed. M.J. Crocker), 1147–1165. New York: Wiley.

44 Stevens, S.S. (1955). The measurement of loudness. *J. Acoust. Soc. Am.* 27: 815–829.

45 Yost, W.A. (2007). Hearing thresholds, loudness of sound, and sound adaptation. In: *Handbook of Noise and Vibration Control* (ed. M.J. Crocker), 286–292. New York: Wiley.

46 Fastl, H. (2006) Keynote lecture on sound quality evaluations. Proceedings of the Thirteenth International Congress on Sound and Vibration, ICSV13, Vienna, Austria.

47 Eldridge, D.H. and Miller, J.D. (1969) Acceptable noise exposures – damage risk criteria. In: Noise as a Public Health Hazard, ASHA Report 4. Washington, DC: American Speech and Hearing Association, 110.

48 Le, T.N., Straatman, L.V., Lea, J., and Westerberg, B. (2017). Current insights in noise-induced hearing loss: a literature review of the underlying mechanism, pathophysiology, asymmetry, and management options. *J. Otolaryngol. Head Neck Surg.* 46 (1): 1–15.

49 Rybak, L.P. (1986). Drug ototoxicity. *Annu. Rev. Pharmacol. Toxicol.* 26: 79–99.

50 Erdreich, J. (2007). Hearing conservation programs. In: *Handbook of Noise and Vibration Control* (ed. M.J. Crocker), 383–393. New York: Wiley.

51 Sliwinska-Kowalska, M. (2011). Combined exposures to noise and chemicals at work. In: *Encyclopedia of Environmental Health* (ed. J.O. Nriagu), 755–763. Amsterdam and London: Elsevier Science.

52 Gerges, S.N.Y. and Arenas, J.P. (2010). *Fundamentos y Control del Ruido y Vibraciones*, 2e. Florianópolis, Brazil: NR Editora.

53 Glorig, A. (1958). *Noise and Your Ear*. New York: Grune & Stratton.

54 Hinchcliffe, R. (1969) Report from Wales: Some relations between aging noise exposure and permanent hearing level changes. Committee S3-W-40. American National Standards Institute.

55 Rosen, S. (1970). Noise, hearing and cardiovascular function. In: *Physiological Effects of Noise* (eds. B.L. Welch and A. S. Welch). New York: Plenum Press.

56 Shadle, C.H. (2007). Speech production and speech intelligibility. In: *Handbook of Noise and Vibration Control* (ed. M.J. Crocker), 293–300. New York: Wiley.

57 Crocker, M.J. and Price, A.J. (1975). *Noise and Noise Control*, vol. I, 250. Cleveland, OH: CRC Press.

58 Fletcher, H. (1953). *Speech and Hearing in Communication*, 68. New York: Van Nostran.

59 ANSI S3.5-1997 (R2013) (1997) *Methods for the Calculation of the Speech Intelligibility Index.* New York: American National Standards Institute.

60 Denes, P.B. and Pinson, E.N. (1963). *The Speech Chain*, 139. Baltimore, MD: Williams & Wilkins, Bell Telephone Laboratories.

61 Chu, A.C. and Warnock, A.C.C. (2002) Detailed Directivity of Sound Fields Around Human Talkers. NRC Institute for Research in Construction Research Report RR-104. National Research Council Canada.

5

Effects of Noise, Vibration, and Shock on People

5.1 Introduction

Noise and vibration can have undesirable effects on people. At low sound pressure levels, noise may cause annoyance and sleep disturbance. At increased levels, noise begins to interfere with speech and other forms of communication; at still higher levels that are sustained over a long period of time in industrial and other occupational environments, noise can cause permanent hearing damage. Loud impulsive and impact noise are known to cause immediate hearing damage. Similarly, whole-body vibration experienced at low levels may cause discomfort, while such vibration at higher levels can be responsible for a variety of effects including reduced cognitive performance and interference with visual tasks and manual control. Undesirable vibration may be experienced in vehicles and in buildings and can be caused by road and rail forces, unbalanced machinery forces, turbulent boundary layer pressure fluctuations, and wind forces. High levels of sustained vibration experienced by operators of hand-held machines can cause problems with circulation and neuropathy of the peripheral nerves and can result in chronic diseases of the hand and arm. This chapter discusses some of the main effects of noise and vibration on people.

5.2 Sleep Disturbance

It is well known that noise can interfere with sleep. Not only is the level of the noise important for sleep interference to occur, but so is its spectral content, number and frequency of occurrences, and other factors. Even very quiet sounds such as dripping taps, ticking of clocks, and snoring of a spouse can disturb sleep. One's own whispered name can elicit wakening as reliably as sounds 30 or 40 dB higher in level [1]. Common sources of noise in the community that will interfere with sleep are comprised of all forms of transportation, including road and rail traffic, aircraft, construction, and light or heavy industry.

Noise interferes with sleep in two main ways: (i) it can result in more disturbed, lower-quality sleep, and (ii) it can awaken the sleeper. Sleep has been studied extensively in the laboratory for many years and it has been found that there are several stages of sleep. A person passes from one stage to another during a night's sleep. There are several ways of labeling these sleep stages or states. One common way is (starting from awake): rapid eye movement (REM), 1, 2, 3, and 4. The different stages of sleep can be detected by attaching electrodes to the body and monitoring brain waves and other behaviors. The REM stage has aroused much interest and it is in the REM stage that most dreams occur. Here the person has internalized his/her attention but is not in a deep stage of unconsciousness. However, paradoxically, in the REM stage a person is particularly insensitive to sound. Some studies have reported that it appears that sleep in Stage 2 is most easily disturbed while deep sleep in Stages 3 and 4 is least easily disturbed by noise. Some researchers [2] have

Engineering Acoustics: Noise and Vibration Control, First Edition. Malcolm J. Crocker and Jorge P. Arenas.
© 2021 John Wiley & Sons Ltd. Published 2021 by John Wiley & Sons Ltd.

suggested that while noise undoubtedly produces sleep patterns in people similar to those of a poor sleeper, a person compensates by spending more time in deep sleep, by becoming less responsive to stimuli and by napping. Thus, noise may not deprive a person of efficient sleep to produce an adverse effect on health. We must conclude that without further experimental evidence, we still cannot ascertain for certain that noise-disturbed sleep has an adverse effect on health. However, we do know that sleep disturbance caused by noise will reduce a person's feeling of wellbeing and that if this disturbance is sufficient it could conceivably have an adverse effect on health. Unfortunately, although there has been a considerable amount of research into sleep interference caused by noise, there is no internationally accepted way of evaluating the sleep interference that it causes or indeed of the best techniques to adopt in carrying out research into the effects of noise on sleep.

Also, little is known about the cumulative long-term effects of sleep disturbance or sleep deprivation caused by noise. Fortunately, despite the lack of in-depth knowledge, sufficient noise-induced sleep interference data are available to provide general guidance for land use planning, environmental impact statements for new highways and airports, and sound insulation programs for housing. Some airports restrict nighttime aircraft movements and road traffic is normally reduced at night and is less disturbing but is still potentially a problem. Chapter 6 discusses descriptors used for predicting sleep disturbance caused by noise.

5.3 Annoyance

Noise consists of sounds that people do not enjoy and do not want to hear. However, it is difficult to relate the annoyance caused by noise to purely acoustical phenomena or descriptors. When people are forced to listen to noise against their will, they may find it annoying, and certainly if the sound pressure level of the noise increases as they are listening, their annoyance will likely increase. Louder sounds are usually more annoying, but the annoyance caused by a noise is not determined solely by its loudness.

Very short bursts of noise are usually judged to be not as loud as longer bursts at the same sound pressure level. The loudness increases as the burst duration is increased until it reaches about 1/8 to 1/4 seconds, after which the loudness reaches an asymptotic value, and the duration of the noise at that level does not affect its judged loudness. On the other hand, the annoyance of a noise at a constant sound pressure level may continue growing well beyond the 1/4 seconds burst duration as the noise disturbance continues [1].

Annoyance caused by noise also depends on several other factors apart from the acoustical aspects, which include its spectral content, tonal content, cyclic or repetitive nature, frequency of occurrence, and time of day. Nonacoustical factors include biological and sociological factors and such factors as previous experience and perceived malfeasance. We are all aware of the annoyance caused by noise that we cannot easily control, such as that caused by barking dogs, dripping taps, humming of fluorescent lights, and the like. The fact that the listener does not benefit from the noise, cannot stop the noise, and/or control it is important. For instance, the noise made by one's own automobile may not be judged very annoying while the noise made by other peoples' vehicles, motorcycles, lawnmowers, and aircraft operations at a nearby airport, even if experienced at lower sound pressure levels, may be judged much more disturbing and annoying.

Of course, noise annoyance is difficult to measure and is possibly subject to biasing effects. Whenever a team of people is introduced into a neighborhood to make a social survey, there is a danger that their mere presence may affect people's attitudes. The design of the questionnaire and the way in which it is conducted can also seriously affect the results [2]. The response of a community to noise is even more complicated than the response of individuals. It has been found that individuals who are driven to make a complaint are not always those most annoyed. Also, of those seriously disturbed, only a small fraction will complain. With individuals, noise annoyance is complicated by other factors, e.g. whether a man or woman is annoyed by other factors such as marital quarrels. When concerned with community response, additional factors such as social and political circumstances, noise action groups, and community attitudes toward the noise and even the regulatory authorities may be important. Here the public relations activities of the noise maker are clearly important. If the community is convinced that the noise intrusion is necessary for their economic wellbeing or their safety, then their acceptance of the situation is much more likely to occur (see Chapter 16). Chapter 6 reviews the annoyance caused by noise in more detail and provides references for further reading.

5.4 Cardiovascular Effects

It has been established through laboratory, field and animal experiments that there is a relationship between exposure to noise and cardiovascular diseases, such as ischemic heart disease and hypertension. In particular, a number of studies have shown that exposure to intense noise affects the sympathetic and endocrine system increasing heart rate, blood pressure, peripheral vasoconstriction, and stress hormones [3]. These effects do not only occur in waking hours but also during sleep.

Other effects of noise that have also been reported in subjects exposed to high levels of noise are the release of stress hormones (including adrenaline, noradrenalin, and cortisol).

We may define stress as a state of arousal which can eventually have adverse health effects. The general response to any form of stress is called the *adaptation syndrome*. This consists of three states: alarm, a stage of resistance, and a stage of exhaustion. Miller [2] suggested that it is unlikely that any long-term stress effects are produced unless noise levels are sufficient to produce hearing damage and loss. However, some studies have been published which suggest that in some industrial situations noise-induced stress can produce adverse health effects. Figure 5.1 shows the result of an early study by Jansen [4] on 1005 German industrial workers and Figure 5.2 shows the incidence of hypertension in male and female workers in noisy workshops after Andriukin [5]. As Kryter [6, 7] pointed out, however, these results could be misleading. In the noisy environment other factors may be important such as poor ventilation, heat and light, danger from accidents, anxiety over job security, and the possible selection of people with lower general health and economic or social status for intense-noise jobs. More recent epidemiological studies carried out in the occupational field

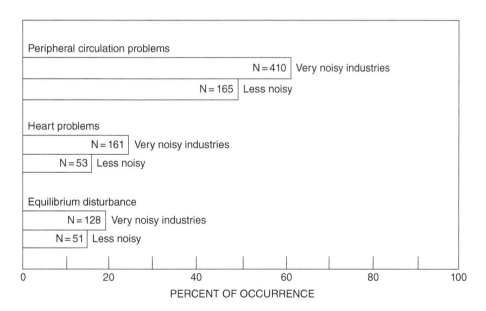

Figure 5.1 Differences in percentages of occurrence of various physiological problems in two different noise levels. Data compiled from 1005 German industrial workers [4].

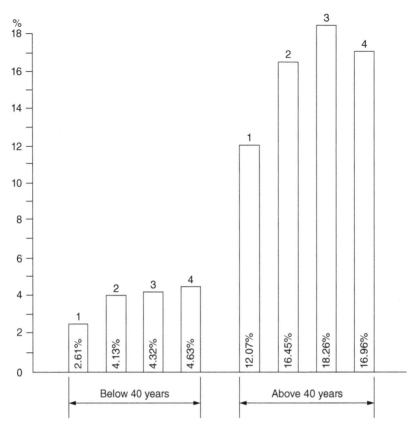

Figure 5.2 Incidence of hypertensive men and women workers (above and below the age of 40) in noisy workshops: (1) tool making; (2) sorting workshop; (3) workshop with automatic lathes; (4) workshop producing ball bearings [5].

have shown that employees working in noisy environments are at a higher risk of high blood pressure and myocardial infarction. It has been observed that workers exposed to A-weighted noise levels greater that 75 dB reported statistically significant increases in blood pressure [8].

Based on a meta-analysis Babisch [9] reported the relationship between road traffic noise and cardiovascular risk. As shown in Figure 5.3, a cubic exposure-response function is given for the increase in relative risk (odds ratio) per increment of the noise level measured as an A-weighted average noise exposure over a 16-hour daytime period, $L_{eq,16h}$ (see Chapter 6). Since this relationship was based on different studies with different A-weighted noise level ranges, it was suggested to use $L_{eq} \leq 60$ dB as a reference category (relative risk = 1). For example, the relative risks for subjects who live in areas where L_{eq} is between 70 and 75 dB would then approximate to 1.15 and 1.29, respectively. A similar study was recently published on the exposure-response relationship of the association between aircraft noise and the risk of hypertension [10].

Mixed epidemiological evidence has been reported for effects of noise on coronary heart disease and coronary risk factors in adults. Clark and Stansfeld [11] have attributed this fact to the difficulty of performing objective measures of blood pressure and to confounding factors associated with coronary heart disease that are difficult to isolate during research. Usually the blood pressure has been estimated indirectly through self-reported measures of hypertension and antihypertensive drug use in many epidemiological studies. However, a unique study recently performed in Europe [12] used reliable measures of blood pressure in the analyses. The research was performed on individuals who had lived near to one of six major airports for five years or more. The study found increased risk of hypertension related to long-term noise exposure, for both night-time aircraft noise and daily average road traffic noise.

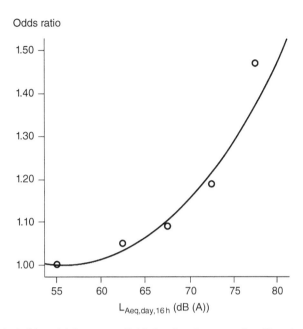

Figure 5.3   Relative risks (odds ratio) for myocardial infarction due to road traffic noise. (*Source:* from Ref. [3]).

5.5   Cognitive Impairment

Several reviews have been written on the possible effects of noise on the performance and efficiency of people at work. Broadbent [13] worked on this topic for many years and has concluded that many investigations both in the laboratory and industry on the effect of noise on mental or motor performance have been inconclusive. Kryter [6, 7] goes even further to suggest that although any mental motor task that requires the perception of any auditory signal for correct performance will undoubtedly be affected by noise, that in his opinion, experimental evidence shows noise has no adverse effect on nonauditory mental or motor tasks. He suggests that adaptation to noise is responsible for this. However, other researchers suggest although many experiments are inconclusive, there is sufficient evidence to conclude that noise does have an adverse effect on work performance. We may summarize some of the generally agreed findings as follows [14]:

a) Steady A-weighted noise levels (without meaning) does not interfere with human performance unless the sound pressure level exceeds 90 dB and even then not consistently.
b) Intermittent or impulsive noises are more disrupting to performance than steady noises. Bursts of such noise can be disruptive even for A-weighted levels below 90 dB.
c) High-frequency noise (above 2000 Hz) appears more disrupting than low-frequency noise (below 2000 Hz).
d) Noise does not seem to change the rate of work but rather increases the variability of the work; there may be work pauses compensated by increases in work rate.
e) Noise increases the number of errors made at work rather than the total amount of work.
f) Complex tasks are more affected by noise than simple tasks.

The importance of these findings has been considered in some countries in their occupational noise legislation, limiting maximum noise levels to enhance concentration and prevent stress during mental tasks [15].

However, many more well-controlled experiments of sufficient duration must be made in industrial work situations before the effects of noise on work performance can be readily established. In particular, the combined effects of noise and other occupational hazards on worker performance, such as work at night, vibrations, heat, illumination, sleep loss, etc. need further research.

Cognitive performance effects due to noise exposure appear to be more significant in children. A number of studies performed on pre-school and primary school children exposed to high levels of environmental noise have found effects on cognitive issues. These studies have suggested that long-term exposure to noise effects on sustained visual attention, cognitive functions involving central processing and language comprehension, speech perception, difficulty in concentrating compared with children from quieter schools, and result in poorer school performance on standardized tests [16]. It has also been suggested that children may adapt to chronic noise exposure by filtering out the unwanted noise stimuli [11]. Although the reported effects are sometimes small in magnitude and the experimental evidence may be debatable, authorities and policy makers responsible for noise abatement have been recommended to become aware of the potential impact of environmental noise on children's development [17].

Table 5.1 summarizes the nonauditory effects of noise on human health and wellbeing [3]. Although not yet conclusive, epidemiological studies report evidence for effects of occupational noise on increased blood pressure and cognitive performance. The evidence for effects of noise on health is more conclusive regarding annoyance and sleep disturbance. The effects of noise are strongest for those effects that, like annoyance, can be classified under wellbeing rather than illness [16].

5.6 Infrasound, Low-Frequency Noise, and Ultrasound

So far we have described the disturbing effects that are produced by noise within the frequency range of human hearing from about 20 to 16 000 Hz. But noise above and below this frequency range can also disturb people. Very low frequency noise or *infrasound* (usually considered to be at frequencies below 20 Hz) may be very intense although "inaudible" in the traditional sense of hearing; but nevertheless it may cause a sensation of pressure or presence in the ears. Low-frequency noise (LFN) between about 20 and 100 Hz, which is within the normally audible range, is usually more disturbing than infrasound.

LFN in the region of 20–50 Hz seems to be the worst problem and more disturbing than infrasound. Infrasound below 20 Hz can excite some of the body organs into resonance as can low-frequency sound in the range 20–75 Hz. Kryter [7] has described some qualitative results obtained by the U.S. Air Force and NASA when people were exposed to intense (above 130 dB) low-frequency (below 100 Hz) pure tones and noise. Some effects observed included chest wall, abdominal and nasal cavity vibration. Additional effects included nausea, giddiness, and coughing. However, these effects only appeared when the levels were above about 130 dB. Such low-frequency levels are only experienced in a very few environments. Such physiological effects can be related to the resonance of organs in the body. This will be discussed further in this chapter. Sources of infrasound and LFN noise include (i) air-conditioning systems in buildings with variable air volume controls or with large-diameter slow-speed fans and blowers and long ductwork in which low-frequency standing waves can be excited, (ii) oil and gas burners, (iii) boilers, and (iv) diesel locomotives and truck and automobile cabins.

Ultrasound covering the range 16000 to 40000 Hz is generated in many industrial and commercial devices and processes. Examples of ultrasonic sources include dental drills, ultrasonic cleaners, humidifiers, ultrasonic welders, air jets, sonic weapons, and jet and rocket engines. Although ultrasound is inaudible, it can produce effects on people. Probably the tinnitus, dizziness, nausea, and headaches sometimes attributed to ultrasound [7] are really caused by parts of the noise which are in the audible range below 20 kHz. However, other studies [7] have suggested that ultrasound becomes audible because subharmonics can be generated in the middle ear by ultrasound. The inner ear will then respond to the subharmonics.

Currently, our best information is that infra- and ultrasound can be tolerated at high sound pressure levels up to 140 dB for very short exposure times, while at levels between 110 and 130 dB, infra- and ultrasound can be tolerated for periods as long as 24 hours without apparent permanent physiological or psychological

Table 5.1 Effects of noise on health and wellbeing with sufficient evidence [3]. Definitions of acoustical indicators can be found in Chapter 6.

Effect	Dimension	Acoustical indicator[a]	Threshold[b]	Time domain
Annoyance disturbance	Psychosocial, quality of life	L_{den}	42	Chronic
Self-reported sleep disturbance	Quality of life, somatic health	L_{night}	42	Chronic
Learning, memory	Performance	L_{eq}	50	Acute, chronic
Stress hormones	Stress Indicator	$L_{max} L_{eq}$	NA	Acute, chronic
Sleep (polysomnographic)	Arousal, motility, sleep quality	$L_{max,indoors}$	32	Acute, chronic
Reported awakening	Sleep	$SEL_{indoors}$	53	Acute
Reported health	Wellbeing clinical health	L_{den}	50	Chronic
Hypertension	Physiology somatic health	L_{den}	50	Chronic
Ischemic heart diseases	Clinical health	L_{den}	60	Chronic

[a] L_{den} and L_{night} are defined as outside exposure levels. L_{max} may be either internal or external as indicated.
[b] Threshold is the level above which effects start to occur or start to rise above background.

effects. There is no current evidence that levels of infrasound, LF, or ultrasound lower than 90–110 dB lead to permanent physiological or psychological damage or other side effects. However, a report recently published in the UK by the Health Protection Agency (HPA) has recommended an exposure limit for the general public to airborne ultrasound sound pressure levels of 70 dB (at 20000 Hz), and 100 dB (at 25000 Hz and above) [18]. For more in-depth discussion on the known effects of infrasound, LFN, and ultrasound, the reader may consult Ref. [19].

5.7 Intense Noise and Hearing Loss

If very intense noise levels of the order of 135 dB or above at any frequency in the hearing range are experienced, immediate hearing damage is likely to result. However, permanent hearing damage is also produced at much lower sound pressure levels if the noise is experienced over much longer periods (weeks, months, or years). This reminds us of the similar phenomenon of metal fatigue where failure can occur at very much lower stress levels than the breaking stress, provided the stress is produced over a sufficiently long time. The problem is that noise-induced hearing loss, or to give it its longer technical name noise-induced permanent threshold shift (NIPTS) is hard to distinguish from presbycusis (the so-called natural hearing loss with age, see Chapter 4).

Since we all suffer from presbycusis in industrialized societies, we should like to be able to subtract the effects of presbycusis from those of NIPTS. The only way we can attempt to do this is to study groups of people who are exposed throughout their lives to intense noise and other control groups who are not. Some attempts have been made to do this for several years. Such studies are very useful but still are troubled by the difficulties in eliminating what Glorig [20] has called sociocusis (losses caused by exposure in the military, recreational, and other nonoccupational activities) and the effects of disease and intermittency in jobs.

Despite the fact that the use of A-weighted sound pressure levels causes problems when noise containing strong pure tones is present, the simplicity afforded is very attractive.

Baughn [21] discusses the construction of risk tables for hearing impairment caused by noise. Here risk is defined as "the difference between the percentage of people with a hearing handicap in a noise-exposed group and the percentage of people with a handicap in a non-noise (but otherwise equivalent) group."

(a)   (b)

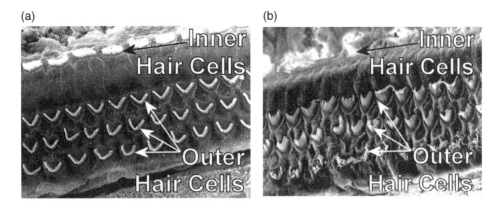

Figure 5.4 Electron microscope image of the hair cells in the cochlea (a) hairs cells before noise exposure and (b) damaged hair cells after intense noise exposure. (*Source:* Courtesy of Pierre Campo, INRS, Department PS, Vandoeuvre, France.)

The level needed to produce hearing loss is much lower than the peak impulsive noise levels required to produce immediate damage. Protective mechanisms exist in the middle ear, which reduce the damaging effect of continuous intense noise. These mechanisms, however, only reduce the noise levels received by the cochlea by about 5 dB and are insufficient to protect the cochlea against most continuous intense noise. Hearing loss can be classified into two main types and is normally measured in terms of the shift of the following hearing threshold shifts: (i) Temporary threshold shift (TTS) is the shift in hearing threshold caused by noise that returns to normal after periods of 24–48 hours. (ii) Permanent threshold shift (PTS) is the shift in hearing threshold that is nonrecoverable even after extended periods of rest.

PTS caused by exposure to intense noise over extended periods of time produces permanent irrecoverable damage to the cochlea (see Chapter 4). Figure 5.4 shows the destruction of the hair cells in the cochlea of a rat caused by intense noise.

Some people have hearing mechanisms that are more sensitive than other people's and are more prone to damage from continuous intense noise and to suffer PTS. Because harmful noise can be quite different in frequency content, often the A-weighted sound pressure level is used as a measure of the intense noise for the prediction of hearing impairment (PTS). Continuous A-weighted sound pressure levels above 75 dB can produce hearing loss in people with the most sensitive hearing if experienced for extended periods of some years. As the A-weighted level increases, an increasing fraction of the workforce experiences PTS. The PTS expected at different levels has been predicted by several organizations including the International Organization for Standardization (ISO), the Environmental Protection Agency (EPA), and the National Institute for Occupational Safety and Health (NIOSH) (see Table 5.2).

5.7.1 Theories for Noise-Induced Hearing Loss

Since World War II, a considerable amount of information has been gathered on hearing loss connected with intense noise. Most of these data have been collected in industry, but there have been other sources (e.g. aircraft pilots and military personnel). Because in most instances the data have been collected for individuals working eight-hour days, little information is available for people working shorter or longer shifts each day and some empiricism is necessary to extrapolate results for use in noise regulations. Two main theories have emerged which are used to aid in these extrapolations: (i) the equal temporary effect hypothesis, and (ii) the equal energy hypothesis.

The equal temporary effect hypothesis assumes that the permanent hearing hazard of noise (NIPTS) can be related to the noise-induced temporary threshold shift (NITTS) caused by the same levels of noise exposure. Some researchers argue that this hypothesis is plausible because it can be related to an observable

Table 5.2 Estimated excess risk of incurring material hearing impairment[a] as a function of average daily noise exposure over a 40-year working lifetime[b]

Reporting Organization	Average Daily A-weighted Noise Level Exposure (dB)	Excess Risk (%)[c]
ISO	90	21
	85	10
	80	0
EPA[d]	90	22
	85	12
	80	5
NIOSH	90	29
	85	15
	80	3

Source: From Ref. [22].

[a] For purposes of comparison in this table, material hearing impairment is defined as an average of the Hearing Threshold Levels for both ears at 500, 1000, and 2000 Hz that exceeds 25 dB.

[b] Adapted from 39 *Fed. Reg.* 43802 [1074b].

[c] Percentage with material hearing impairment in an occupational-noise-exposed population after subtracting the percentage who would normally incur such impairment from other causes in an unexposed population.

[d] EPA = Environmental Protection Agency.

physiological offset in the ear. Some experiments have shown evidence that both the NIPTS and the temporary threshold shift (NITTS) can he related to the same metabolic process occurring in the inner ear (perhaps fatigue through a depletion of nutrients and a build-up of waste products). This hypothesis leads to the conclusion that noise which is intense enough to produce temporary loss (NITTS) will also be intense enough to produce permanent loss (NIPTS).

The equal energy hypothesis, on the other hand, assumes that hearing damage is directly proportional to the acoustic energy an individual has experienced in his or her lifetime. Since acoustic energy is intensity multiplied by the time duration, it is also proportional to the sound pressure level times the duration. This equal energy hypothesis suggests that the sound pressure level may be increased by 3 dB for each halving in exposure time per day. This hypothesis does allow a simple, attractive approach to the regulation problem which also seems thoroughly reasonable. This concept has been used by most of the countries around the world in the development of their noise regulations [15].

5.7.2 Impulsive and Impact Noise

High levels of impulsive and impact noise pose special threats to human hearing. These types of noise can also be very annoying. It is well known now that high levels of such noise damage the cochlea and its hair cells through mechanical processes. Unfortunately, there is currently no commonly accepted definition or recognized standard for what constitutes impulsive noise. However, in general an impulse noise has been described as having a rise time ≤1 second in duration, and if repeated, occurring at intervals >1 second. Impulse noise is also characterized by having a broad spectral content.

Impulsive noise damage cannot be simply related to knowledge of the peak sound pressure level. The duration, number of impulsive events, impulse waveform, and initial rise time are also all important in predicting whether impulsive or impact noise is likely to result in immediate hearing damage.

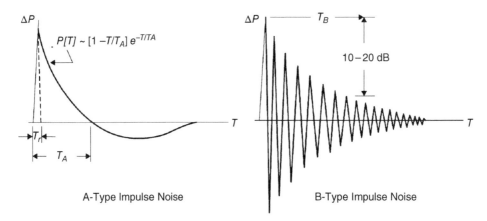

Figure 5.5 Schematic representation of the two basic impulse noise pressure–time profiles, following the simplification of Coles et al. [23]

Impulse noise damage also does not seem to correlate with the noise energy absorbed by the ear. It is known that hearing damage caused by a combination of impulsive noise events during steady background noise cannot be judged by simply adding the noise energy contained in each and may indeed be greater than simple addition would suggest.

There is a huge variation in the characteristics of impulsive and impact noise to which people may be exposed. Peak sound pressure levels can be from 100 dB to as much as 185 dB, and impulse durations may vary from as little as a few microseconds to as much as hundreds of milliseconds. Impulses can be comprised of single events or multiple or repeated events. Coles et al. [23] have classified impulse and impact noise into two main types: (i) single non-reverberant impulses (termed A-waves) and (ii) reverberant impact noise (termed B-waves) (see Figure 5.5). Explosions and the blast waves they create (often commonly called Friedlander waves) can be classified as A-waves in Cole's system. They can have peak sound pressure levels of over 150 dB. They are really shock waves and are not purely sound waves. Riveting and punch press and forge forming and other machining processes in industry can result in reverberant noise caused by ringing of the manufactured parts. Such ringing noise is classified as B-waves in this system.

Although Cole's system includes the effects of peak sound pressure level, duration, and number of impulsive events, it is not completely comprehensive since it neglects the rise times, spectral contents, and temporal patterns of the impulses. Bruel has shown that crest factors (peak value divided by rms value) as high as 50 dB are common in impulsive noises [24]. Unfortunately, there is still no real consensus on how to predict and treat the damaging effects of impulsive noise compared to continuous intense noise. Henderson and Hamernik [25] discuss impulsive and impact noises in more detail.

5.8 Occupational Noise Regulations

According to studies carried out on workers exposed to A-weighted noise levels over 85 dB in the U.S. [26], the types of occupations that present the highest risk for hearing damage in terms of numbers of workers overexposed are: manufacturing and utilities, transportation, military, construction, agriculture, and mining. In 1981, the Occupational Safety and Health Administration (OSHA) estimated that 7.9 million people in the United States working in manufacturing were exposed occupationally to daily A-weighted sound pressure levels at or above 80 dB. In the same year, the U.S. Environment Protection Agency (EPA) estimated that more than nine million U.S. workers were exposed occupationally to daily A-weighted levels above 85 dB. More than half of these workers were involved in manufacturing and utilities. From

Australian data, it has been estimated that around 20.1% of the workforce regularly work in A-weighted noise levels above 85 dB and 9.4% above an exposure of 90 dB [27]. A review on noise-induced hearing loss in countries in Eastern Europe has also been published recently [28].

A World Health Organization study on the burden of disease associated with hearing impairment from occupational noise [29] reported that overall in the Americas, more than 300 000 disability-adjusted life years were lost to noise-induced hearing loss. In this region, Canada and the U.S. accounted for almost half of the years of healthy life lost, as they have large industrial populations.

As previously discussed the hearing loss produced by exposure to hazardous noise depends on many factors, including the sound pressure level, spectral content, exposure duration, and the temporal pattern (continuous, varying, intermittent, or impulsive). The commonly accepted eight-hour average A-weighted sound pressure level (permissible exposure limit, PEL) is based on the exposure during a typical daily work shift. The use of A-weighted sound pressure levels to predict the effect of noise on hearing is based on the response of the human ear at moderate sound pressure levels and is supported by numerous studies (see Chapter 4).

The relationship between noise level and exposure duration is commonly known as the exchange rate (q). This "dose-trading relation" or "trading ratio" is expressed as the number of decibels by which the sound pressure level may be decreased or increased for a doubling or halving of the duration of exposure. For example, the U.S. regulations allow eight hours of exposure at an A-weighted sound pressure level of 90 dB (PEL) and only four hours if the level is 95 dB (exchange rate of 5 dB).

The U.S. was the first western country to introduce occupational noise regulations. In 1969, new standards for industrial noise became effective under a revision of the Walsh-Healey Act [26]. These standards only applied to firms which had federal contracts of $10 000 or more during the course of one year. These standards originally affected only 28 million out of the 40 million U.S. nongovernmental and nonagricultural work forces. It called for a PEL of 90 dB with a q of 5 dB, the reduction of noise levels to the PEL by engineering or administrative controls whenever feasible, the provision and wearing of hearing protection devices above the PEL, and in section (c), the conducting of a "continuing, effective hearing conservation program" for employees exposed above the PEL [30].

Later, in 1971 these standards were extended to apply to the employees of all U.S. companies engaged in interstate commerce when the OSHA of the U.S. Department of Labor increased inspection in subsequent years which encouraged compliance from companies. In 1972 and 1973 OSHA did not classify violations under noise, air, or water, but total OSHA inspections in 1973 averaged approximately 1695 per week, compared to 919 per week a year earlier. In 1973 and 1974, a standards advisory committee reporting to OSHA considered the proposal by the NIOSH that the A-weighted noise exposure level of 90 dB for an eight-hour day should be reduced to 85 dB in the OSHA regulations but the committee did not recommend this reduction. In a 1974 study commissioned by OSHA, the estimated cost to selected U.S. industries most directly affected, of complying with the OSHA regulations was $13.6 billion. If the A-weighted noise exposure level for an eight-hour workday were reduced from 90 to 85 dB, the compliance cost would increase to $31.6 billion [14].

The U.S. noise regulation was amended in 1981 to dictate specific obligations at an A-weighted action level of 85 dB [31]. At this action level, OSHA requires noise measurement, the use and care of hearing protection devices, audiometric testing, employee training and education, and record keeping. The agency promulgated a revision of the hearing conservation amendment in 1983, which is still currently in effect [32]. This amendment replaced section (c) of the noise standard. However, sections (a) and (b) detailing the PEL and the requirement for feasible engineering or administrative controls remained unchanged.

The OSHA regulations permit state agencies to issue their own regulations, as long as these regulations are at least as protective as those promulgated by the federal OSHA. About half of the states have selected to do this, whereas the rest chose to rely on federal enforcement. Some other regulations in the U.S. have been issued by the U.S. Mine Safety and Health Administration, U.S. Departments of Transportation and Energy, U.S. Coast Guard, and NASA. These regulations are discussed in the literature [15].

In 1977, the General Conference of the International Labor Organization adopted Convention 148, regarding the protection of workers against occupational hazards due to air pollution, noise, and vibration in the workplace [33]. This convention established in 24 articles the bases of legislation, considering measures of prevention and protection, the establishment of criteria and exposure limits for occupational noise, the promotion of occupational health research, and official recognition and concern for the health of the exposed workers. The ratification of this convention has generated similar legislation in several different countries. Others have adopted the convention, limiting it to only some of the pollutants such as air and noise, but leaving out vibration [15].

One of the most commonly used standards is ISO 1999 [18]. This international consensus standard can be used to predict the amount of hearing loss expected to occur in various centiles of the exposed population at particular audiometric frequencies as a function of exposure level and duration, age and sex. The standard provides a complete description of NIPTS for various exposure levels and exposure times. It is important to note that ISO 1999 does not explicitly establish limits for the level of occupational exposure, which is defined in the occupational noise legislation or regulation in each particular country. In some countries, technical standards may be typically direct copies of the ISO standards, without the benefit of enabling legislation or regulation [15].

In general, the most relevant acoustical factors in occupational noise regulation in the majority of nations are: The normalized eight-hour PEL, the exchange rate (q), the maximum upper limits for exposure to impulsive sounds, and requirements for engineering controls. Nevertheless, there are notable differences among countries in the defined values for PEL and exchange rate. Of the countries that have regulations, the majority use a PEL of 85 dB with a q value of either 3 or 5 dB. The Canadian federal regulation is one of the few legislations that establishes a PEL of 87 dB with a q of 3 dB. It is important to notice that the U.S. OSHA regulation is one of the few in the world that uses a PEL of 90 dB and the 5-dB exchange rate [34].

Figure 5.6 shows a comparison of duration per day to allowable noise exposure level (100% dose) for different values of PEL and q as used in national standards [15]. This figure shows that the use of an 85 dB PEL with a q of 3 dB, provides the best protection for workers. Legislation using a PEL of 90 dB and q of 3 dB provides better protection than legislation using a PEL of 85 with a q of 5 dB, but only when the A-weighted noise level is above 97.5 dB. Below this level, less protection is provided. Canadian federal law, (PEL 87 dB and q of 3 dB), offers better protection than the regulations using a PEL of 85 dB with a q of 5 dB, at least when the A-weighted noise level is greater than 90 dB.

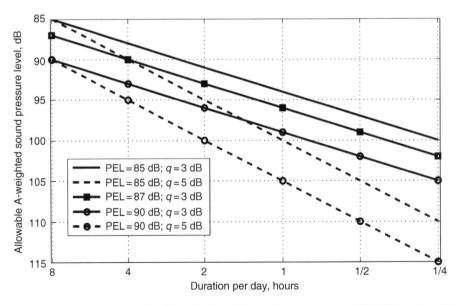

Figure 5.6 Comparison of duration per day for allowable A-weighted noise exposure level (100% dose) for different values of permissible exposure limit and q as used in national standards. (*Source:* From Ref. [15] with permission).

Noise regulations in several nations treat impulse noise separately from continuous noise. A common approach in some countries (usually those using a q of 5 dB) has been to limit the number of impulses at a given peak sound pressure over a workday, although the exact figures vary slightly. Alternatively, other nations have considered impulse noise jointly with any continuous noise present. Most nations limit impulsive noise exposure to a peak unweighted sound pressure level of 140 dB (or C-weighted levels), while a few use slightly lower noise level limits (130 dB or C-weighted noise levels of 120 dB). International legislation and standards on occupational noise have been extensively discussed by Suter [30].

5.8.1 Daily Noise Dose and Time-Weighted Average Calculation

Noise dose (D) is a measure of the exposure to noise to which a person is subjected. It is defined in terms of the A-weighted sound pressure level eight-hour limit (PEL) that represents a 100% dose. For a given duration of the exposure to noise at the constant A-weighted sound pressure level (C) in hours, and the reference duration T_r in hours permitted for exposure to noise at the steady sound level, the dose in percent may be calculated as

$$D = \frac{C}{T_r} \times 100. \tag{5.1}$$

When the daily noise exposure is composed of two or more periods of noise exposure at different levels, their combined effect should be considered, rather than the individual effect of each. Then if D in the equation

$$D = \left(\frac{C_1}{T_1} + \frac{C_2}{T_2} + \frac{C_3}{T_3} + ... + \frac{C_n}{T_n}\right) \times 100 \tag{5.2}$$

exceeds 100%, then, the mixed exposure should be considered to exceed the limit value. C_n indicates the actual total time of exposure at a specified noise level and T_n indicates the total time allowed for exposure at that level. The permissible duration T in hours at any level is calculated as

$$T = \frac{8}{2^{(L_A - PEL)/q}}, \tag{5.3}$$

where L_A is the A-weighted sound pressure level of the noise to which the person is exposed. Note that OSHA establishes a threshold (or cutoff) A-weighted level of 80 dB and exposures that are below this level should not be included in calculating the total dose. NIOSH uses a threshold level of 75 dB.

Example 5.1   Suppose that an unprotected worker is subject to the following four exposure durations and corresponding A-weighted sound pressure levels during an 8-hour workday: 6 hours at 92 dB, 1 hour and 24 minutes at 98 dB, 30 minutes at 107 dB, and 6 minutes at 115 dB. Determine the total daily noise dose according to OSHA regulations.

Solution

Using Eq. (5.3) with PEL = 90 dB and q = 5 dB, the permissible time of exposure at each level is calculated. Therefore,

- 6 hours at 92 dB, then T_1 = 6.06 hours
- 1 hour and 24 minutes (1.4 hours) at 98 dB, then T_2 = 2.64 hours
- 30 minutes (0.5 hours) at 107 dB, then T_3 = 0.76 hours
- 6 minutes (0.1 hours) at 115 dB, then T_4 = 0.25 hours.

Now, the total daily noise dose is calculated using Eq. (5.2) as

$$D = [(6/6.06) + (1.4/2.64) + (0.5/0.76) + (0.1/0.25)] \times 100 = 258\%.$$

Therefore, some corrective action should be taken since the total noise dose considerably exceeds 100%.

Example 5.2 A man works in a large five-star hotel. During a typical working day he spends a total of 2 hours supervising in the kitchen where the average A-weighted noise level is 88 dB. For another 4 hours he carries out administrative tasks in an office where the A-weighted sound pressure level is 77 dB and another 1 hour in the restaurant where the noise level is 66 dB. What is the total noise dose according to both OSHA and NIOSH criteria?

Solution

a) We have to combine the three partial exposures by finding the noise dose for each task. We obtain the permissible time of exposure at each A-weighted level according to OSHA using Eq. (5.3):
- 88 dB to 2 hours, $T_1 = 10.56$ hours
- 77 dB to 4 hours and 66 dB to 1 hour, are not included since the levels are below the OSHA threshold of 80 dB. Therefore using Eq. (5.2)

$$D = [(2/10.56)] \times 100 = 18.9\%.$$

b) In the same way, we obtain the permissible time of exposure at each level according to NIOSH:

- 88 dB to 2 hours, $T_1 = 4$ hours
- 77 dB to 4 hours, $T_2 = 50.8$ hours
- 66 dB to 1 hour, is below the NIOSH threshold of 75 dB and therefore is not included in calculating the total dose. Then

$$D = [(2/4) + (4/50.8)] \times 100 = 57.9\%.$$

From this example we can notice that employees who typically go into high noise level areas for short durations and then return to quiet areas during their work day may exhibit significantly higher daily noise exposures using the NIOSH criteria.

Example 5.3 A driller supervisor in a copper mine works 12 hours shifts. A personal dosimeter registers a noise dose of 70% in 7 hours. What would be the total noise dose if during this time the worker was exposed to a representative sample of the daily noise?

Solution

Considering that the worker was exposed to a representative sample of the daily noise, we can obtain the noise dose per hour directly as 70%/7 hours = 10% per hour. Thus, the total noise dose in a 12 hours shift is $12 \times 10\% = 120\%$, so the worker is overexposed.

OSHA regulations specify a single value called the time-weighted average sound pressure level (*TWA*). *TWA* is that level which, if constant over an eight-hour period of exposure, would result in the same 5-dB-exchange-rate noise dose as is measured. It can be calculated by

$$TWA = 90 + 16.61 \times \log(D/100). \tag{5.4}$$

Note that TWA is a concept similar to a normalized eight-hour average sound level ($L_{eq,8h}$) or the noise exposure level defined in some international occupational noise regulations. In particular, for a PEL = 85 dB and $q = 3$ dB, the noise exposure level recommended in Europe (L_{EX}) is defined as the TWA of the daily noise exposure levels for a nominal week of five 8-hour working days. It is calculated by

$$L_{EX} = 85 + 10\log(D/100). \tag{5.5}$$

Table 5.3 shows the time-weighted average (TWA) noise level limits promulgated by the ACGIH (American Conference of Governmental Industrial Hygienists), NIOSH, and OSHA.

Table 5.3 Time-Weighted Average (*TWA*) noise level limits as a function of exposure duration.

Duration of Exposure [h/day]	A-Weighted Sound Pressure Level (dB)		
	ACGIH	NIOSH	OSHA
16	82	82	85
8	85	85	90
4	88	88	95
2	91	91	100
1	94	94	105
½	97	97	110
¼	100	100	115[a,b]
⅛	103[c]	103	—

[a] No exposure to continuous or intermittent A-weighted sound pressure level in excess of 115 dB.
[b] Exposure to impulsive or impact noise should not exceed a peak sound pressure level of 140 dB.
[c] No exposure to continuous, intermittent, or impact noise in excess of a C-weighted peak sound pressure level of 140 dB

It is possible to make a dosimeter (see Chapter 7) to measure the daily noise dose of a person at work, but since the noise exposures using the OSHA criteria are not added on an energy basis (halving of exposure time for a 5 dB increase, not 3 dB), it is very difficult to make an instrument to do this accurately and also incorporate impulsive noise energy correctly. Dosimeters are easier to make with a 3-dB exchange rate so they incorporate impact noise correctly. In addition, for other exchange rates there is no unique relationship between noise dose and equivalent continuous sound pressure level [35].

Example 5.4   A worker in a pulp mill without using hearing protection has the noise exposure record shown in Table 5.4. Calculate the resulting daily *TWA* and L_{EX}.

Solution

a) We notice that the last two hours of exposure are below the thresholds defined by OSHA and NIOSH. Therefore, we calculate the total A-weighted noise dose according to OSHA as

$$D = [(1/4) + (1/16) + (1/8) + (2/16) + (1/32)] \times 100 = 59.4\%, \text{so}$$
$$TWA = 16.61 \times \log(59.4/100) + 90 = 86.2 \, \text{dB}$$

b) The corresponding total A-weighted noise dose using a PEL = 85 dB and 3-dB exchange rate is

$$D = [(1/0.8) + (1/8) + (1/2.52) + (2/8) + (1/25.4)] \times 100 = 206\%, \text{and}$$

$$L_{EX} = 10 \times \log(206/100) + 85 = 88.1 \, \text{dB}.$$

It is expected a L_{EX} value to be a little higher than the *TWA* value if the noise levels vary during the noise exposure period.

Table 5.4 Worker's A-weighted noise level exposure record.

Level, dB	Exposure, hours
95	1
85	1
90	1
85	2
80	1
70	2

5.9 Hearing Protection

5.9.1 Hearing Protectors

It is best practice to reduce noise through: (i) the use of passive engineering controls such as use of enclosures, sound-absorbing materials, barriers, vibration isolators, etc. and then (ii) using administrative measures such as restricting the exposure of personnel by limiting duration, proximity to noise sources, and the like. In cases where it is not practical or economical to reduce noise exposure to sound pressure levels below that cause hearing hazards or annoyance, then hearing protectors should be used [22]. Hearing protection devices (HPDs) can give noise protection of the order of 30–40 dB, depending on frequency, if used properly. Unfortunately, if they are incorrectly or improperly fitted, then the attenuation they can provide is significantly reduced.

There are four main types of HPDs: earplugs, earmuffs, semi-inserts, and helmets. Figure 5.7 shows examples of the four basic kinds of HPDs [36]. Earplugs are generally low-cost, self-expanding types that are inserted in the ear canal and must be fitted correctly to achieve the benefit. Custom-molded earplugs can be made to fit an individual's ear canals precisely. Some people find earplugs uncomfortable to wear and prefer earmuffs. Earmuffs use a seal around the pinna to protect it and a cup usually containing sound-absorbing material to isolate the ear further from the environment. If fitted properly, earmuffs can be very effective. Unfortunately, earmuffs can be difficult to seal properly. Hair and glasses can break the seals to the head causing leaks and resulting in a severe degradation of the acoustical attenuation they can provide. In addition, they have the disadvantage that they can become uncomfortable to wear in hot weather. The use of earmuffs simultaneously with earplugs can provide some small additional noise attenuation, but not as much as the two individual HPD attenuations added in decibels.

Semi-inserts consist of earplugs are held in place in the ear canals under pressure provided by a metal or plastic band. These are convenient to wear but also have the tendency to provide imperfect sealing of the ear canal. If the plug portion does not extend into the ear canal properly, the semi-insert HPDs provide little hearing protection and can give the user a false sense of security. Helmets usually incorporate semi-inserts and in principle can provide slightly greater noise attenuation than the other HPD types. Attenuation is provided not only for noise traveling to the middle and inner ear through the ear canal, but also for noise reaching the hearing organ through skull bone conduction. Helmets also provide some crash and impact protection to the head in addition to noise attenuation and are often used in conditions that are hazardous not only for noise but potential head injury from other threats. Unfortunately, the hearing protection they provide is also reduced if the semi-inserts are improperly sealed in the ear canals.

In many occupational noise legislations, ear protectors should not be used as a substitute for effective noise control. The use of ear protectors is considered to be an interim measure and must only be used until

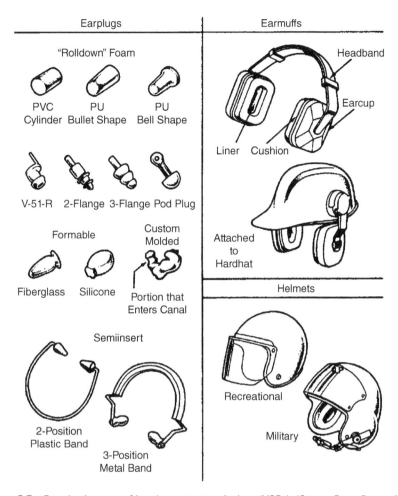

Figure 5.7 Four basic types of hearing protector devices (HPDs). (*Source:* From Berger [36]).

engineering methods successfully bring noise levels below specified limits. The reasons why personal ear protection is not a completely satisfactory solution to the industrial noise problem are not hard to find. Perhaps the overriding reason is discomfort. Most types of ear protectors are uncomfortable to wear. They may cause the ears to get hot and sweaty or the tightness of some types of ear protectors can cause pain or headaches.

Another reason is communication difficulties. Many people claim that ear protectors impair communication which to a certain extent is true. However, in continuous intense noise conditions, hearing protectors actually do not reduce intelligibility of speech for people with normal hearing because noise and speech signals are both reduced equally and the speech signal-to-noise ratio remains the same. Thus, in continuous noise, the Articulation Index should remain constant (see Section 6.5) and speech intelligibility should be unimpaired. Of course, in intermittent or impulsive noise situations, conversation can normally be carried on in the "quiet" intervals and ear protectors will interfere with communication in such instances.

Another communication-related disadvantage of ear protectors is the difficulty in hearing warning signals. In continuous noise, hearing protectors may not offer a disadvantage in this regard except for people with high-frequency hearing loss. Such people, who are often found working in noisy environments, may well have increased difficulty in hearing warning signals and even normal conversation, because ear protectors attenuate high frequencies more and high-frequency signals will then be more easily masked by low-frequency noise (see Section 4.3.3).

Another disadvantage of the use of hearing protectors is that workers may feel that they can no longer judge the performance of their machines or detect the noise caused by wear or some other malfunction. There may be some truth in this, although it is important for people to learn that they can still hear noise when wearing ear protectors, but simply at a reduced level. A final reason is that some ear protectors such as earplugs can exacerbate diseases of the outer ear such as otitis. However, in these cases the use of earmuffs instead or medications will normally overcome this problem.

There are several standards for determining the noise protection produced by ear protectors [37–41]. Noise protection data should be required to be provided by a manufacturer when hearing protectors are purchased. Both the mean noise protection at each frequency band and the standard deviation of the measurements are normally reported.

The NRR (Noise Reduction Rating) is a single-number rating method used to describe HPDs noise attenuation. The NRR is intended to be used for calculating the exposure under the HPD by subtracting it from the C-weighted noise exposure level L_C. An alternative use of the NRR is with the A-weighted noise exposure level, L_A, in which the NRR can be applied if 7 dB is first subtracted from its value. Reference [42] contains more detailed discussion on hearing protectors.

Example 5.5 A hearing protector has an NRR of 17 dB and it used in a factory where one-octave-band sound pressure levels are given in line two of Table 5.5. Calculate the A-weighted noise level entering the ear of a worker in this factory.

Solution

First we need the A-weighted sound pressure level from the one-octave-band levels given in Table 5.5. The A-weighting correction in dB can be taken from Table 4.2 (see Section 4.3.5) and these values are shown in line three of Table 5.5. Line four shows the A-weighted one-octave-band levels. These levels are combined to give

$$L_A = 10 \times \log \left(10^{74/10} + 10^{86/10} + 10^{91/10} + 10^{95/10} + 10^{93.5/10} + 10^{85.5/10} + 10^{76/10} \right)$$

$$= 98.7 \, \text{dB}$$

Therefore, the A-weighted sound pressure level entering at the ear is

$$\text{Level} = L_A - (\text{NRR} - 7) = 98.7 - (17 - 7) = 88.7 \, \text{dB}.$$

Table 5.5 One-octave-band factory noise.

	One-octave-band center frequency, Hz						
	125	250	500	1000	2000	4000	8000
One-octave-band level, dB	92	95	94	95	92	85	78
A-weighting correction, dB	−18	−9	−3	0	+1.5	+0.5	−2
A-weighted one-octave-band levels, dB	74	86	91	95	93.5	85.5	76

5.9.2 Hearing Conservation Programs

Many governments require or mandate hearing conservation programs for workers in industries and in other occupations in which hazardous noise conditions exist. Hearing conservation programs are designed to protect workers from the effects of hazardous noise environments.

Most nations have dictated specific obligations at an action level which corresponds to a dose of 50% of the PEL. Therefore, the OSHA TWA A-weighted action level is 85 dB [31]. The equivalent NIOSH A-weighted action level is 82 dB [22].

At these action levels, regulations require implementing hearing conservation programs.

In general, a hearing conservation program will have at least the following components: noise measurements, the use and care of hearing protection devices, audiometric testing, employee training and education, and record keeping.

Hearing evaluation through audiometric testing is a key component of a hearing conservation program. If employees are working in areas with intense noise levels, it is necessary to monitor their hearing regularly. These hearing tests will help in determining whether or not the administrative and engineering controls and ear protection devices are being adhered to or effectively used. The OSHA 90 dB eight-hour workday limit for the A-weighted sound pressure level only protects a certain percentage of workers (about 93%). Since people's ears are different, some workers (about 77%) will still be suffering appreciable hearing damage in these conditions. Through audiogram results it may be possible to identify individuals with higher than average susceptibility to noise-induced hearing loss (those with "tender" ears). An individual could adhere to the limits of the more conservative occupational noise recommendations and still develop hearing loss due to individual risk factors.

In fact, NIOSH recommendations for a hearing loss prevention program make it mandatory to conduct audiometric tests periodically on all individuals whose A-weighted exposures equal or exceed 85 dB as an eight-hour TWA. It is recommended that the tests should be conducted annually and baseline audiometric tests shall be conducted before employment (conducted upon the initiation of noise exposure) or within 30 days of employment for all workers who must be enrolled in the hearing conservation program [22].

It is extremely helpful to use an educational program to teach management at all levels and all shop personnel an industrial plant the facts concerning occupational hearing loss, noise control, and ear protection devices. With the commitment and example of management, after introducing such a program, the majority of shop personnel will cooperate with a hearing conservation program.

More stringent occupational noise legislations establish that if action levels are exceeded, immediate engineering noise control measures must be implemented within a 6–12 month period as part of a hearing conservation program.

Protection can be provided not only by use of engineering controls designed to reduce the emission of noise sources, control of noise and vibration paths, labeling of noisy areas, and the provision of HPDs, but by the limitation of personnel exposure to noise as well. For instance, arranging for a machine to be monitored with a control panel that is located at some distance from a machine, instead of right next to it, can reduce personnel noise exposure. The rotation of personnel between locations with different noise levels during a workday can ensure that one person does not stay in the same high noise level throughout the workday. Reference [43] describes hearing conservation programs and, in addition, some legal issues including torts, liabilities, and occupational injury compensation.

Example 5.6   As part of a hearing conservation program following the OSHA criteria, we need to determine the maximum time for which a worker may spend in a particular workshop where the A-weighted noise level is 105 dB without using HPD. The rest of the eight-hour working shift the worker is subjected to a constant A-weighted level of 85 dB.

Solution

According to Table 5.3 or Eq. (5.3) the exposure to A-weighted sound pressure levels of 105 and 85 dB is allowed for 1 and 16 hours, respectively. Since the total dose must be less than 100% and if the worker is exposed to 105 dB for t hours and 85 dB for (8-t) hours, then, $1 = t + (8\text{-}t)/16$. Solving for t, we obtain $t = 0.53$ hours or 32 minutes. Therefore, as an administrative measure, the worker should be allowed to work less than 32 minutes in the workshop with high noise level to comply with OSHA.

5.10 Effects of Vibration on People

An extensive discussion of the effects of vibration on people is given by Griffin [44]. People seem most responsive to vertical vibration. The human body can be regarded as a complex nonlinear multi-degree of freedom mechanical system. A simple lumped parameter model of the human body that works well for frequencies below 100 Hz is shown in Figure 5.8. There are relative motions between the body parts that vary with the frequency and the direction of the applied vibration. Several of the human organs have resonances which can produce sensations of pain if exited by vibration or noise at their resonance frequencies. These resonance

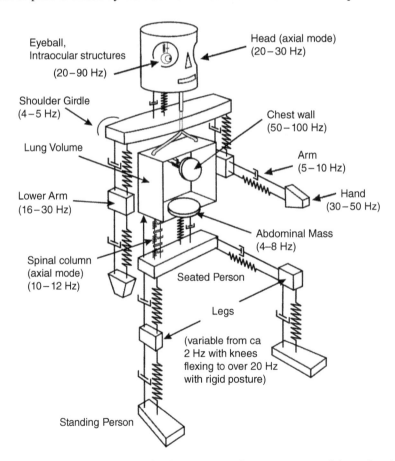

Figure 5.8 Mechanical model of the human body showing resonance frequency ranges of the various body sections. (*Source:* adapted from Bruel and Kjaer [45].)

frequencies are approximately in the range of 5–200 Hz. One important part is the thorax-abdomen system that has a resonance effect in the 3–6 Hz range. A further resonance effect is found in the 20–30 Hz region which is caused by the head–neck-shoulder system [46]. Eyeball resonances are found in the region 60–90 Hz and a resonance effect in the lower jaw-skull system has been determined between 100 and 200 Hz.

In general, it has been shown that vibrations produce both physiological and psychological effects. However, the degree of negative impact of vibration exposure on workers' health is of less importance than in the case of noise, because vibration exposures are less common. It is extremely difficult to conduct epidemiological studies on the long-term effects of vibration exposure on human subjects. Although the scientific evidence on the effects of vibration on human health is limited and not completely conclusive, studies report that the main outcomes of vibration exposure on humans are whole-body effects, hand-arm effects, motion sickness, effects on vision, and annoyance. These effects depend mostly on the magnitude of vibrations, frequency, duration of exposure, and direction of the motion.

Whole-body effects are produced when the body is supported on a vibrating surface, i.e. sitting, standing or lying on a vibrating surface. These effects can affect the performance of activities, comfort and human health. Back pain, displacement of intervertebral disks, degeneration of spinal vertebrae, and osteoarthritis may be associated with vibration exposure, although other causes unrelated to vibration exposure, such as bad sitting posture or heavy lifting may be other sources of disorders of the back. Other disorders that have been associated with occupational whole-body exposure include abdominal pain, digestive disorders, urinary frequency, prostatitis, hemorrhoids, balance, and sleeplessness [44]. Vibration of the whole body is produced by all kinds of transportation and by several types of industrial machinery.

Power tools or workpieces can transmit intense levels of vibration to the hands and arms of their operators. Long-term exposure to these vibrations can cause significant diseases affecting the blood vessels, nerves, bones, joints, muscles, and connective tissues of the hand and forearm. One of the main recognized diseases produced by hand-arm vibration exposure is known as "vibration-induced white finger," Reynaud's disease, and more recently as Hand-Arm Vibration Syndrome (HAVS) [47]. The disease is characterized by intermittent whitening (i.e. blanching) of the fingers because of lack of blood circulation. The effect is amplified in very cold weather. This disease may be common among those workers with long-term exposed to pneumatic drills, pavement breakers, grinding machines, chain saws, etc. Affected people usually report other associated symptoms such as numbness and tingling. Other effects of hand-arm vibration that have been reported include decreased grip strength, reduced tactile discrimination, reduced manipulative dexterity, articular disorders, and some other physiological problems at the wrist and shoulders [44].

Motion sickness, also called *kinetosis*, is caused by real or illusory movements of the body or the environment at low frequency (usually less than 1 Hz). Human reaction to very low frequency vibration is extremely variable and seems to depend on a number of external factors unrelated to the motion. Normal illnesses of people caused by low-frequency vibration include vomiting, nausea, sweating, dizziness, headaches, and drowsiness. Vertical oscillation is the prime cause of sickness in marine craft but usually not in most road vehicles and some environments.

The effect of vibration on the visual abilities is believed to be caused by the resonances of the eyeball. This effect can be very unsafe for helicopter pilots and other occupations where vision-focusing abilities are important. The effect can be appreciated when trying to read a book or newspaper in a moving vehicle subjected to high levels of mechanical vibrations. In addition, the effect has sometimes been associated with headaches.

There is little evidence concerning the possible effects of vibration exposure on simple cognitive tasks (unless the vibration directly affects input and output processes) and fatigue-decreased proficiency.

Vibration has also been reported as a cause of annoyance that includes sleep deprivation, discomfort and interference with activities (e.g. learning, memory, decision making, manual tasks control). Figure 5.9 gives the Reiher-Meister chart showing human response to vibration [48]. Although this was produced in 1931, later measurements with people have produced somewhat similar subjective results. Comparable results

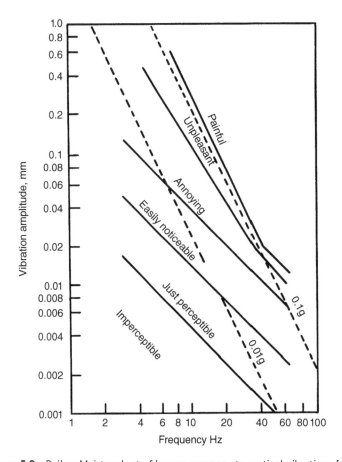

Figure 5.9 Reiher-Meister chart of human response to vertical vibrations [45].

were obtained in a study presented by Dieckmann [49] who extended the frequency range down to 0.1 Hz using higher vibration amplitudes than the Reiher and Meister study. It is rather interesting to note that, for small vibrations, the subjective responses follow constant velocity contours as frequency is changed. For large vibration amplitudes, the responses change to constant acceleration contours with frequency.

In general, rms acceleration values have been found to correlate well with human response for steady, continuous vibration. In the frequency range of 1–100 Hz, the approximate absolute threshold for the perception of vertical vibration is 0.01 m/s^2; a magnitude of 0.1 m/s^2 will be easily noticeable; around 1 m/s^2 will be uncomfortable and acceleration values over 10 m/s^2 are usually dangerous [44]. It has been found that high peak values are often underestimated by rms values. In this case the use of a vibration dose value (VDU) is preferred (see Section 5.10.2).

Vibration is a major source of lost time in occupational environments. It is estimated that about eight million workers in the U.S. are currently exposed to occupational vibration. Of these, an estimated 6.8 million are exposed to whole-body vibration and the remainder to hand-arm vibration [50]. Other authors report that 500 000 construction workers are exposed to whole-body vibration and over 250 000 are exposed to hand-arm vibration above the current international daily exposure limit values [51]. Construction workers are often one of the largest affected groups by population. However, workers in industries like foundries, shipbuilding, forestry, mining, transportation, and defense are also affected. Unlike noise, there is no OSHA regulation for vibration exposure; although standards, directives and best practice guidelines for identifying workers at risk and for taking steps to mitigate the vibration and reduce risk of injury are currently is use in the U.S.A.

5.11 Metrics to Evaluate Effects of Vibration and Shock on People

5.11.1 Acceleration Frequency Weightings

Human beings are more sensitive to some frequencies than to others. As seen in Figure 5.8, each part of the body has a corresponding natural frequency and each part will resonate over a range of exciting frequencies. In addition, this sensitivity depends on the direction in which the vibration is applied. No one person's vibration sensitivity is exactly the same as any other and hence statistical human responses have been determined. These factors need to be considered when assessing harmful effects of vibration on humans. Therefore, different weighting curves have been proposed to account for frequencies to which the body is most and least sensitive. International and national standards define similar acceleration frequency weighting curves for predicting vibration discomfort. Figure 5.10 shows the frequency weighting curves for human response to whole-body vibrations [52]. Appropriate weightings are applied to vibration frequency spectra between 0.1 and 100 Hz to assess whole-body vibration effects. They are designated W_b, W_c, W_d, W_e, W_f, and W_g. The ISO standard defines an additional weighting curve, W_k, which is almost identical to W_b. The choice of the frequency weighting depends on the standard used, the effect of vibration which is being assessed (health, activity, comfort), the vibration input position on the body (seat, seat back, feet), and the direction of vibration (vertical or horizontal). Probably the most widely used of these frequency weightings is W_b which has been designed to reproduce human sensitivity to vertical motions.

In the case of hand-transmitted vibration, all current national and international standards use the same frequency weighting (called W_h) over the frequency range from 8 to 1000 Hz (see Figure 5.11). This acceleration weighting is applied to each of the three axes of vibration at the point of entry of the vibration to the hand.

Human response vibration meters are currently available with these various weightings built in. In practice, instantaneous accelerations are recorded and frequency weightings are applied by means of electronic filters. Orthogonal component accelerations are commonly combined by vector additions.

5.11.2 Whole-Body Vibration Dose Value

Unlike the case of noise exposure, there is no consensus on precise limits needed to avoid the risk of injury expected to occur in the majority of the exposed population. However, some standards have been proposed that provide guidelines as to how to deal with occupational whole-body vibration exposure using dose values [53–56].

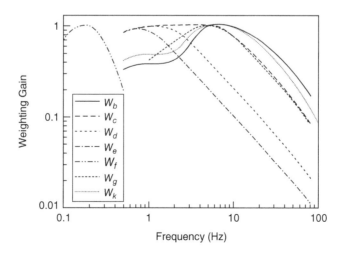

Figure 5.10 Acceleration frequency weightings for whole-body vibration and motion sickness [52].

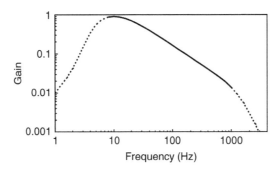

Figure 5.11 Acceleration frequency weighting W_h for the evaluation of hand-transmitted vibration [52].

The *VDV*, is a cumulative exposure-time domain function which is particularly effective when the acceleration crest factor (peak value divided by the rms value) is larger than three. The *VDV* has been defined as the fourth root of the time integral of the fourth power of the frequency-weighted acceleration,

$$VDV = \left[\int_0^T a^4(t)dt \right]^{0.25},$$
(5.6)

where $a(t)$ is the frequency-weighted acceleration and T is the duration of the vibration exposure in seconds. Thus, the units of *VDV* are $ms^{-7/4}$. It is noted that a 16 times increase in exposure duration requires a halving of the vibration magnitude to maintain the same vibration dose value.

For statistically stationary vibration in which both the frequency-weighted rms acceleration a_{rms} and the exposure duration T are known, an approximation to the VDV may be determined from the empirically *estimated vibration dose value, eVDV* as

$$eVDV = 1.4 a_{rms} T^{0.25}.$$
(5.7)

This estimated value is not applicable to cases in which the crest factor is high (transients, shocks, etc.). Values of *VDV* and *eVDV* are usually noted with a subscript as to which weighting has been used, e.g. use of W_d would result in VDV_d and $eVDV_d$.

Example 5.7 Workers in an office close to a subway line are subjected to vibration arising from 20 trains passing during working hours. Vibration measurements are carried out in the office while trains are passing. Each event lasts for 10 seconds and the frequency-weighted rms acceleration, which is constant during each event, is 0.1 m/s². (a) Calculate the estimated VDV. (b) If the number of trains could be reduced, how many would be allowed if an estimated VDV less than $0.4 \, ms^{-7/4}$ is recommended to avoid adverse comments?

Solution

a) We can use Eq. (5.7):

$$eVDV = 1.4 \times 0.1 \times (20 \times 10)^{0.25} = 0.53 \, ms^{-7/4}.$$

b) If N is the number of trains, Eq. (5.7) is written as

$$eVDV = 1.4 \times 0.1 \times (N \times 10)^{0.25} \leq 0.4. \text{ Now we express the condition for } N \text{ as}$$

$$N \leq \left(\frac{0.4}{1.4 \times 0.1} \right)^4 \frac{1}{10} = 6.7.$$

Therefore, if the number of trains is reduced to six,

$$eVDV = 1.4 \times 0.1 \times (6 \times 10)^{0.25} = 0.39 \text{ ms}^{-7/4}.$$

The cumulative effects of individual vibration events can be estimated by combining the dose of each individual event VDV_i according to the fourth-power law

$$VDV = \left(VDV_1^4 + VDV_2^4 + \dots + VDV_N^4\right)^{0.25}. \tag{5.8}$$

Obviously, if there are N identical events each of VDV, the total VDV is $VDV \times N^{0.25}$.

Although no precise limits can be set to prevent disorders caused by whole-body vibration, the British Standard [53] defines an action level in the region of $15 \text{ ms}^{-7/4}$ for vertical vibration. The ISO standard [54] (also adopted by an ANSI standard [55]) considers that health risks are likely for a VDV greater than $17 \text{ ms}^{-7/4}$, and defines a *health guidance caution zone* between 8.5 and $17 \text{ ms}^{-7/4}$. On the other hand, an EC directive [56] defines a VDV exposure action level of $9.1 \text{ ms}^{-7/4}$ and an exposure limit value above which the workers shall not be exposed of $21 \text{ ms}^{-7/4}$.

Example 5.8 The vibration exposure at a particular location consists of four different events having individual VDVs of 0.01, 0.02, 0.05, and $0.03 \text{ ms}^{-7/4}$. What is the total VDV?

Solution

We use Eq. (5.8):

$$VDV = \left[(0.01)^4 + (0.02)^4 + (0.05)^4 + (0.03)^4\right]^{0.25} = 0.05 \text{ ms}^{-7/4}.$$

We observe that the fourth-power law emphasizes the contribution of the largest of the four individual values.

Example 5.9 A sample of the vibration exposure of a forklift operator is measured over a period of 30 minutes. The VDV of the sample is $4.5 \text{ ms}^{-7/4}$. Calculate the exposure over the total working day, of seven hours of duration.

Solution

Assuming that the sample is representative of the entire day, we obtain:

$$VDV = 4.5 \times [7/0.5]^{0.25} = 8.7 \text{ ms}^{-7/4}, \text{ which is inside the health caution zone according to ISO 2631.}$$

5.11.3 Evaluation of Hand-Transmitted Vibration

Occupational exposures to hand-transmitted vibration usually are intermittent and have widely varying daily exposure durations (from a few seconds to many hours). For this, a daily exposure is reported by an energy-equivalent frequency-weighted eight-hour exposure acceleration, $A(8)$, defined as

$$A(8) = \left(\frac{1}{8}\int_0^T [a_{h,w}(t)]^2 dt\right)^{0.5}, \tag{5.9}$$

Table 5.6 ACGIH recommended threshold limit values for hand-transmitted vibration exposure in either *x*, *y*, or *z* directions [59].

Total daily exposure duration	Dominant frequency-weighted component of acceleration which shall not be exceeded (rms)	
	m/s^2	g's
4–8 hours	4	0.4
2–4 hours	6	0.61
1–2 hours	8	0.81
Less than 1 hour	12	1.22

where $a_{h,w}(t)$ is the instantaneous value of the weighted hand-transmitted acceleration and T is the total duration of the working day in hours. The energy-equivalent acceleration over any other period T is related to $A(8)$ by

$$A(8) = A(T)\sqrt{\frac{T}{8}}. \tag{5.10}$$

Therefore, if $A(4) = 10\,\text{m/s}^2$, $A(8) = A(4)/\sqrt{2} = 7.07\,\text{m/s}^2$.

It has been proposed by Griffin [52] that the number of years of exposure (in the range 1–25 years) required for 10% incidence of white-finger disease is $30/A(8)$. The ISO standard [57] states that evidence of HAVS is rare for people exposed to $A(8)$ of less than $2\,\text{m/s}^2$ and unreported for values less than $1\,\text{m/s}^2$. The EU directive on physical agents [56] defines an 8-hour equivalent exposure action value of $2.5\,\text{m/s}^2$ rms and an 8-hour equivalent exposure limit value of $5.0\,\text{m/s}^2$ rms. However, these values do not define safe exposures to hand-transmitted vibration. A 10% probability of white-finger disease is predicted after 12 years at the EU action value and after 5.8 years at the EU exposure limit value. The ANSI standard [58] defines the same action and limit values as the EU directive. The EU directive also states that manufacturers must test and declare vibration levels of their equipment using the action value of $2.5\,\text{m/s}^2$ as a reference. The ACGIH has also recommended threshold limit values (TLV) of frequency-weighted rms acceleration for exposure of the hand to vibration that vary from 4 to $12\,\text{m/s}^2$ depending on the duration of the exposure (see Table 5.6).

Example 5.10 A fettler uses different electric hand tools during a working shift of nine hours making ceramic tiles. The measured frequency-weighted rms accelerations for exposure times of 2 hours, 2½ hours, and 4½ hours are, respectively, 17, 10, and $9\,\text{m/s}^2$. Calculate the equivalent frequency-weighted acceleration over the nine-hour period and the eight-hour equivalent value.

Solution

The total exposure during a working shift is

$$A(9) = \sqrt{\frac{(17^2 \times 2) + (10^2 \times 2.5) + (9^2 \times 4.5)}{9}} = 11.51\,\text{m/s}^2.$$

The eight-hour equivalent acceleration can be determined from Eq. (5.10):

$$A(8) = 11.51\sqrt{\frac{9}{8}} = 12.2\,\text{m/s}^2.$$

Example 5.11   A worker in a machine shop uses three tools during a working day. The frequency-weighted rms accelerations and exposure times are measured resulting in: (a) an angle grinder: 4 m/s^2 for 2½ hours, (b) an angle cutter: 3 m/s^2 for 1 hour, and (c) a chipping hammer: 20 m/s^2, for 15 minutes. Determine the daily vibration exposure and assess the situation using the ANSI S2.70 criteria.

Solution

The partial vibration exposures for the three tasks are:

a) grinder: $A_1(8) = 4 \times \sqrt{2.5/8} = 2.2 \text{ m/s}^2$
b) cutter: $A_2(8) = 3 \times \sqrt{1/8} = 1.1 \text{ m/s}^2$
c) chipper: $A_3(8) = 20 \times \sqrt{15/480} = 3.5 \text{ m/s}^2$.

The daily vibration exposure is then:

$$A(8) = \sqrt{A_1^2 + A_2^2 + A_3^2} = \sqrt{(2.2)^2 + (1.1)^2 + (3.5)^2} = 4.3 \text{ m/s}^2.$$

This value is below the 8-hour equivalent exposure limit value of 5.0 m/s^2 but above the exposure action value of 2.5 m/s^2. Therefore, some actions should be taken to reduce the risks, for example by decreasing the worker's daily exposure time or using vibration protecting gloves.

Like noise, if exposure action values are exceeded, measures intended to reduce whole-body and hand-transmitted vibrations to a safe level and health surveillance should be implemented according to the local legislation.

There are other specific defined metrics, such as the seat effective amplitude transmissibility (SEAT), the motion sickness dose value (MSDV), the head injury criterion (HIC) that are discussed in detail by Griffin [44, 52] and Brammer [60]. Criteria for human comfort and annoyance for vibration in buildings are discussed in Chapters 6 and 12.

References

1 Crocker, M.J. (2007). General introduction to noise and vibration effects on people and hearing conservation. In: *Handbook of Noise and Vibration Control* (ed. M.J. Crocker), 303–307. New York: Wiley.

2 Miller, J.D. (1974). Effects on noise on people. *J. Acoust. Soc. Am.* 56: 729.

3 European Environmental Agency (2010) Good practice guide on noise exposure and potential health effects. Technical Report No. 11/2010. Copenhagen.

4 Jansen, G. (1961). Adverse effects of noise on iron and steel workers. *Stahl Eisen* 81: 217.

5 Andriukin, K.D. (1961). Influence of sound stimulation on the development of hypertension: clinical and experimental results. *Cor. Vasa* 3: 235.

6 Kryter, K.D. (1994). *The Handbook of Hearing and the Effects of Noise: Physiology, Psychology and Public Health*. San Diego, CA: AP Professional.

7 Kryter, K.D. (2012). *Physiological, Psychological, and Social Effects of Noise*. Bodega Bay, CA: NASA Scientific and Technical Information Branch.

8 Chang, T.-Y., Jain, R.-M., Wang, C.-S., and Chan, C.-C. (2003). Effects of occupational noise exposure on blood pressure. *J. Occup. Environ. Med.* 45: 1289–1296.

9 Babisch, W. (2008). Road traffic noise and cardiovascular risk. *Noise Health* 10: 27–33.

10 Babisch, W. and van Kamp, I. (2009). Exposure response relationship of the association between aircraft noise and the risk of hypertension. *Noise Health* 11: 149–156.

11 Clark, C. and Stansfeld, S.A. (2007). The effect of transportation noise on health and cognitive development: a review of recent evidence. *Int. J. Comp. Psychol.* 20: 145–158.

12 Floud, S., Vigna-Taglianti, F., Hansell, A. et al. (2011). *Occup. Environ. Med.* 68: 518–524.

13 Broadbent, D.E. (1958). *Perception and Communication*. Oxford: Pergamon Press.

14 Crocker, M.J. and Price, A.J. (1975). *Noise and Noise Control*, vol. I. Cleveland, OH: CRC Press.

15 Arenas, J.P. and Suter, A.H. (2014). Comparison of occupational noise legislation in the Americas: an overview and analysis. *Noise Health* 16: 306–319.

16 Stansfeld, S.A. and Matheson, M.P. (2003). Noise pollution: non-auditory effects on health. *Br. Med. Bull.* 68: 243–257.

17 Klatte, M., Bergström, K., and Lachmann, T. (2013). Does noise affect learning? A short review on noise effects on cognitive performance in children. *Front. Psychol.* 4: 578.

18 Advisory Group on Non-ionising Radiation (2010) Health Effects of Exposure to Ultrasound and Infrasound. Report RCE-14, 167–170. Health Protection Agency.

19 Broner, N. (2007). Effects of infrasound, low-frequency noise, and ultrasound on people. In: *Handbook of Noise and Vibration Control* (ed. M.J. Crocker), 320–325. New York: Wiley.

20 Glorig, A. (1958). *Noise and Your Ear*. New York: Grune & Stratton.

21 Baughn, W.L. (1972). The risk of hearing impairment as a function of noise exposure. In: *Noise and Vibration Control Engineering* (ed. M.J. Crocker), 133. West Lafayette, IN: Purdue University.

22 Criteria for a recommended standard, Occupational noise exposure, Revised Criteria (1998) U.S. Department of Health and Human Services. www.cdc.gov/niosh/docs/98-126.html (accessed 20 February 2020).

23 Coles, R.R., Garinther, G.R., Hodge, D.C., and Rice, C.G. (1968). Hazardous exposure to impulse noise. *J. Acoust. Soc. Am.* 43: 336–343.

24 Bruel, P.V. (1980). The influence of high crest factor noise on hearing damage. *Scand. Audiol. Suppl.* 12: 25–32.

25 Henderson, D. and Hamernik, R.P. (2007). Auditory hazards of impulse and impact noise. In: *Handbook of Noise and Vibration Control* (ed. M.J. Crocker), 326–336. New York: Wiley.

26 Suter, A.H. (2007). Development of standards and regulations for occupational noise. In: *Handbook of Noise and Vibration Control* (ed. M.J. Crocker), 377–382. New York: Wiley.

27 Williams, W. (2013). The epidemiology of noise exposure in the Australian workforce. *Noise Health* 15: 326–331.

28 Pawlaczyk-Luszczynska, M., Dudarewicz, A., Zaborowski, K. et al. (2013). Noise induced hearing loss: research in Central, Eastern and South-Eastern Europe and Newly Independent States. *Noise Health* 15: 55–66.

29 Concha-Barrientos, M., Campbell-Lendrum, D., and Steenland, K. (2004) Occupational Noise: Assessing the Burden of Disease from Work-related Hearing Impairment at National and Local Levels. WHO Environmental Burden of Disease Series, No. 9. Geneva: World Health Organization.

30 Suter, A.H. (1998). Standards and regulations. In: *Encyclopedia of Occupational Health and Safety*, 4e, vol. 47 (ed. J.M. Stellman), 15–17. Geneva: International Labour Organization.

31 OSHA (1981) *Occupational noise exposure: Hearing conservation amendment*. Occupational Safety and Health Administration. Fed. Reg. 46:4078–179.

32 OSHA (1983) *Occupational noise exposure: Hearing conservation amendment; Final rule*. Occupational Safety and Health Administration, 29 CFR 1910.95. Fed. Reg. 46:9738–85.

33 International Labour Organization (1977) Working environment (air pollution, noise and vibration) Convention 148. Geneva.

34 Suter, A.H. (2009). The hearing conservation amendment: 25 years later. *Noise Health* 11: 2–7.

35 Marsh, A.H. and Richings, W.V. (1998). Measurement of sound exposure and noise dose. In: *Handbook of Acoustical Measurements and Noise Control*, 3e (ed. C.M. Harris). New York: Acoustical Society of America.

36 Berger, E.H. (2003). Hearing protection devices. In: *The Noise Manual*, 5e (eds. E.H. Berger, L.H. Royster, J.D. Royster, et al.), 379–454. Fairfax, VA: American Industrial Hygiene Association.

37 ANSI S12.6-2016 (2016) *Methods for Measuring the Real-Ear Attenuation of Hearing Protectors*. New York: American National Standards Institute.

38 ANSI S3.19-1974 (1974) *Methods for Measuring of Real-Ear Protection of Hearing Protectors and Physical Attenuation of Earmuff.* New York: American National Standards Institute.

39 International Organization for Standardization, ISO 4869-1 (1990) *Acoustics-Hearing protectors- Part 1: Subjective Method for the Measurement of Sound Attenuation.* Geneva: International Standards Organization (reviewed in 2009).

40 EN 24869 (1992/1993) *Acoustics. Hearing protectors. Sound attenuation of hearing protectors.* European Committee for Standardization.

41 EN 352, *Hearing protectors. Safety requirements & testing.* European Committee for Standardization.

42 Gerges, S.N.Y. and Casali, J.G. (2007). Hearing protectors. In: *Handbook of Noise and Vibration Control* (ed. M.J. Crocker), 364–376. New York: Wiley.

43 Erdreich, J. (2007). Hearing conservation programs. In: *Handbook of Noise and Vibration Control* (ed. M.J. Crocker), 383–393. New York: Wiley.

44 Griffin, M.J. (1996). *Handbook of Human Vibration.* London: Academic.

45 Anon (1989) *Human Vibration*, booklet by Bruel & Kjaer, Naerum, Denmark.

46 Broch, J.T. (1984). *Mechanical Vibration Shock Measurements*, 2e. Naerum: Bruel & Kjaer.

47 Janicak, C.A. (2004). Preventing HAVS in the workplace. *Prof. Saf.* 49: 35–40.

48 Reiher, H.J. and Meister, F.J. (1932) Human sensitivity to vibration, *Forsch. Geb. Ingenieurmes*, Stuttgart, Ins. Tech., Nov. 1931. Abstract in Eng. News Rec. 31 March, 470.

49 Dieckmann, D. (1957). A study of the influence of vibration on man. *Ergonomics* 1: 347–355.

50 Bruce, R.D., Bommer, A.S., and Moritz, C.T. (2003). Noise, vibration and ultrasound. In: *The Occupational Environment: Its Evaluation, Control and Management* (ed. S. DiNardi), 435–493. Fairfax, VA: AIHA Press.

51 Brauch, R. (2009) *Vibration Analysis and Standards: A Review of Vibration Exposure Regulations, Standards, Guidelines and Current Exposure Assessment Methods.* AIHA Spring Conference, American Industrial Hygiene Association. Florida.

52 Griffin, M.J. (2007). Effects of vibration on people. In: *Handbook of Noise and Vibration Control* (ed. M.J. Crocker), 343–353. New York: Wiley.

53 British Standard, BS 6841 (1987) *Measurement and evaluation of human exposure to whole-body mechanical vibration and repeated shock.* London: British Standards Institution.

54 International Standard, ISO 2631 (1997) *Mechanical vibration and shock- Evaluation of human exposure to whole-body vibration, Part 1: General requirements.* Geneva: International Standards Organization (reviewed in 2014).

55 ANSI S3.18–2002 (2002) *Guide for the Evaluation of Human Exposure to Whole-body Vibration.* New York: American National Standards Institute.

56 Directive 2002/44/EC (2002) On the minimum health and Safety Requirements Regarding the Exposure of Workers to the Risks Arising from Physical Agents (vibration). *Official J. European Communities*, 6 July; L177/13–19. The European Parliament and the Council of the European Union.

57 International Standard, ISO 5349 (2001) *Mechanical Vibration – Measurement and Evaluation of Human Exposure to Hand-Transmitted Vibration.* Geneva: International Organization for Standardization (reviewed in 2016).

58 ANSI S2.70–2006 (2006) *Guide for the Measurement and Evaluation of Human Exposure to Vibration Transmitted to the Hand.* New York: American National Standards Institute.

59 American Conference of Government Industrial Hygienists (ACGIH) (1984) *Standard for Hand-Arm Vibration.* Cincinnati.

60 Brammer, A.J. (2007). Effects of mechanical shock on people. In: *Handbook of Noise and Vibration Control* (ed. M.J. Crocker), 354–363. New York: Wiley.

6

Description, Criteria, and Procedures Used to Determine Human Response to Noise and Vibration

6.1 Introduction

People are exposed to noise during daytime and nighttime hours. During the day the noise can interfere with various activities and cause annoyance, and at night it can affect sleep. Very intense noise can even lead to hearing damage (see Chapter 4). In the daytime the activities most affected are communications that involve speech between individuals, speech in telephone communications, and speech and music on radio and television. If the noise is more intense, it is normally more annoying, although there are a number of other attitudinal and environmental factors that also affect annoyance.

There are many different ways to measure and evaluate noise, each normally resulting in a different noise measure, descriptor, or scale. The various measures and descriptors mainly result from the different sources (aircraft, traffic, construction, industry, etc.) and the different researchers involved in producing them. From these measures and descriptors, criteria have been developed to decide on the acceptability of the noise levels for different activities. These criteria are useful in determining whether noise control efforts are warranted to improve speech communication, reduce annoyance, and lessen sleep interference. This chapter contains a review and discussion of some of the most important noise measures and descriptors. In the past 20–30 years these measures and descriptors have undergone some evolution and change as researchers have attempted to find descriptors that best relate to different human responses and are more easily measurable with improved instrumentation. For completeness this evolution is traced and some measures and descriptors are described that are no longer in use, since knowledge of them is needed in the study of the results of various noise studies reported in the literature.

6.2 Loudness and Annoyance

As the level of the noise is increased, it is accompanied by an apparent increase in loudness. Loudness may be considered to be the subjective evaluation of the intensity of a noise when this evaluation is divorced from all the attitudinal, environmental, and emotional factors that may affect the listener's assessment of the annoying properties of the noise. Chapter 4 contains a detailed discussion on the loudness of noise. Generally, if a noise is louder, it is judged to be more annoying and vice versa, although there are exceptions. Table 6.1 shows some of the acoustical and nonacoustical factors that can contribute to the annoyance caused by noise. Some of the factors shown in Table 6.1 are also important in considerations of the effects of noise on speech communication (see Sections 6.5 and 6.6) and on sleep (see Section 6.15.1). The annoyance caused by noise is discussed further in Section 6.15.2.

Engineering Acoustics: Noise and Vibration Control, First Edition. Malcolm J. Crocker and Jorge P. Arenas.
© 2021 John Wiley & Sons Ltd. Published 2021 by John Wiley & Sons Ltd.

Table 6.1 Some acoustical and nonacoustical factors that contribute to annoyance caused by noise.

Acoustical Factors	Nonacoustical Factors
Sound pressure level	Time of day
Frequency spectrum	Time of year
Duration	Necessity for noise
Pure-tone content	Community attitudes
Impulsive character	Past experience
Fluctuation in level	Economic dependence on source

6.3 Loudness and Loudness Level

As discussed in Section 4.3.2 which contains an in-depth description of the loudness of sound, the human ear does not have a uniform sensitivity to sound as its frequency is varied. Figure 4.6 in Chapter 4 shows equal loudness level contours. These contours connect together pure-tone sounds that appear equally loud to the average listener. Recently, slightly modified contours have been proposed by Moore and coworkers [1–3] (see Figure 6.1). In 2007, a new American National Standards Institute (ANSI) standard (ANSI S3.4–2007) [4] for the calculation of the loudness of steady sounds was published. This 2007 ANSI standard is based on the loudness model of Moore et al. [1]. The model in ANSI S3.4–2007 gives reasonably accurate predictions of a wide range of data on loudness perception [1]. However, the equal loudness contours predicted by the model shown as the dotted lines in Figure 6.1 differ substantially from those in the International Organization for Standardization (ISO) standard that was applicable in the past [5], which in turn were based on the 1956 data of Robinson and Dadson [6]. Similar contours have been determined experimentally for bands of noise instead of pure tones (see Chapter 4).

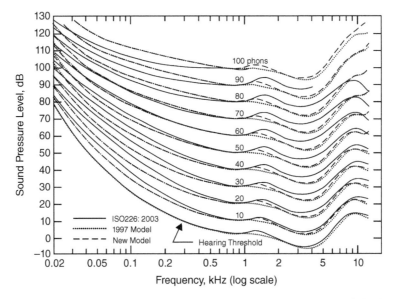

Figure 6.1 The bottom curves (marked Hearing Threshold) show the absolute threshold (free field, frontal incidence, binaural listening) predicted by the original (1997) model [1] (dotted line), and as published in ISO 389-7 [7] (solid line). The other curves show equal loudness contours as predicted by the original model, the modified model, and as published in ISO 226 [8] (dotted, dashed, and solid lines, respectively).

6.4 Noisiness and Perceived Noise Level

6.4.1 Noisiness

Although the level of noise or its loudness is very important in determining the annoyance caused by noise, there are other acoustical and nonacoustical factors that are also important. In laboratory studies, people were asked to rate sounds of equal duration in terms of their noisiness, annoyance, or unacceptability [9, 10]. Using octave bands of noise, Kryter and others have produced *equal noisiness index* contours. These equal noisiness contours are similar to those for equal loudness, except that at high frequency less sound energy is needed to produce equal noisiness and at low frequency more is needed. The unit of noisiness index is the *noy N*. Equal noisiness index curves are shown in Figure 6.2. The procedure to determine the logarithmic measure, the perceived noise level (PNL), is quite complicated and has been standardized [11]. It has also been described in several books [12–14]. Briefly, it may be stated as follows. Tabulate the one-octave

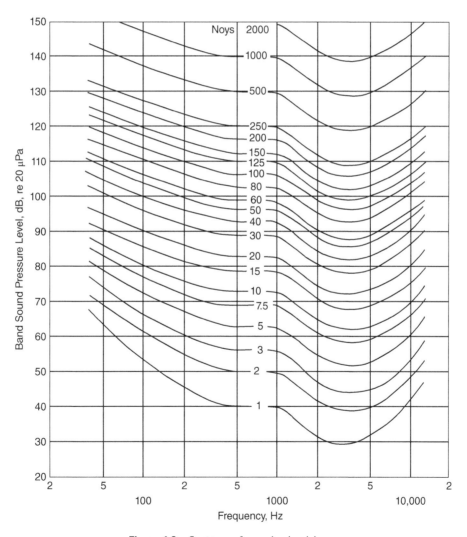

Figure 6.2 Contours of perceived noisiness.

band (or one-third-octave band) sound pressure levels (SPLs) of the noise. Calculate the noisiness index in noys for each band in Figure 6.2. Then calculate the total noisiness index N_t, from

$$N_t = N_{max} + 0.3\left(\sum N - N_{max}\right), \tag{6.1}$$

where N_{max} is the maximum noisiness index and ΣN is the sum of all the noisiness indices. If one-third octave bands are used, the constant 0.3 for octave bands in Eq. (6.1) is replaced by 0.15.

The total perceived noisiness index N_t (summed over all frequency bands) is converted to the PNL or L_{PN} from

$$L_{PN} = 40 + (33.22)\log N_t. \tag{6.2}$$

The procedure is similar to that described in Section 4.3.2 for calculating loudness level (phons) from loudness (sones). Some have questioned the usefulness of this procedure since listeners in laboratory experiments do not seem to be able to distinguish between (i) loudness and (ii) noisiness, and (iii) annoyance. Despite this, the procedure has been widely used in assessing single-event aircraft noise. In the United States the Federal Aviation Administration (FAA) has adopted the effective perceived noise level (EPNL) for the certification of new aircraft. As an example, the noisiness of the spectra given in Figure 16.6 of Chapter 16 can be calculated. The take-off noise shown by the upper dashed line of Figure 16.6 of Chapter 16 has a noisiness of 170 noys and a PNL of 114 PNdB. PNL has received wide acceptance in many countries as a measure of aircraft noisiness with and without tone corrections.

Example 6.1 At one instant in time the sound pressure levels (dB) of a turbojet engine airplane were recorded in one-octave bands (see first and second columns of Table 6.2). The turbojet engine airplane noise spectrum is relatively smooth indicating broadband noise with no appreciable tones. Compute the PNL (PNdB).

Solution

Each sound pressure level is converted to perceived noisiness using Figure 6.2 (see third column of Table 6.2). We observe that the maximum value of noisiness index is at 400 Hz, $N_{max} = 30$. The perceived noisiness values are combined using Eq. (6.1),

$$N_t = 30 + 0.3(3 + 10 + 20 + 30 + 20 + 25 + 19 + 7.5 - 30) = 61.35.$$

The total perceived noisiness index N_t is now converted to the PNL or L_{PN} using Eq. (6.2),

$$L_{PN} = 40 + (33.22)\log(61.35) = 99 \text{ PNdB}.$$

Table 6.2 Sound level values for Example 6.1.

Octave band center frequency, Hz	L_p, dB	Noisiness index, noys
50	75	3.0
100	80	10.0
200	86	20.0
400	89	30.0
800	84	20.0
1600	80	25.0
3150	72	19.0
6300	60	7.5

6.4.2 Effective Perceived Noise Level

Although PNL can be used to monitor the peak noise level of an aircraft pass-by (or flyover or flyby), this measure does not take into account the variation of the noise or its duration. Experiments have shown that annoyance and noisiness increase both with the magnitude and with the duration of a noise event. Noise that is of long duration is normally judged to be more annoying than noise of short duration. Figure 6.3 shows a PNL time history of a flyover of a typical fanjet aircraft. As the airplane approaches, the discrete frequency whine caused by fan and compressor noise radiated from the engine inlets is very evident.

When the airplane is overhead, the noise is dominated by that from the fan exit and is again mostly whine. When the plane has passed, the low-frequency jet rumble is heard. The peaks for each source occur at different times since each source is very directional. The inlet noise is mainly "beamed" forward in the flight direction, while the jet noise is mainly radiated backward about 45° to the jet exhaust direction. In addition to the effects of noise level and duration, the effects of tonal content must be considered. If the noise contains pure tones along with the broadband noise spectrum, it is also judged to be noisier than without such tones. To account for these effects the EPNL or L_{EPN} has been defined as

$$L_{EPN} = L_{PN,\max} + C + D, \tag{6.3}$$

where C is the correction factor for pure tones (between 0 and 6 dB, depending on frequency and the tone magnitude in relation to the broadband noise in adjacent frequency bands) and D is a correction for duration [10–12]. EPNL is simply a time integration of the tone-corrected PNL. Equation (6.3) does not directly reveal the time integration. However, EPNL is defined in Federal Aviation Regulations (FAR) Part 36, and also International Civil Aviation Organization (ICAO) Annex 16 as

$$L_{EPN} = 10 \log \left\{ \sum \left(10^{L_{PN}(k)/10} \Delta t \right) \right\} - 10, \tag{6.4}$$

where $L_{PN}(k)$ is the PNL plus the tone correction of the kth sample in the time history. There are some complexities regarding the maximum value of $L_{PN}(k)$ also known as PNLT(k) if the tone correction at that moment is not as large as the average tone correction for the two preceding and two succeeding samples,

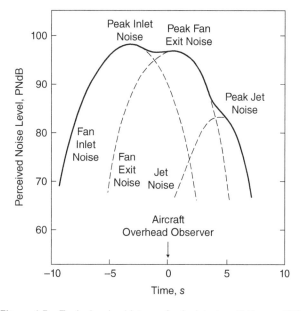

Figure 6.3 Typical noise history of a fanjet aircraft flyover [13].

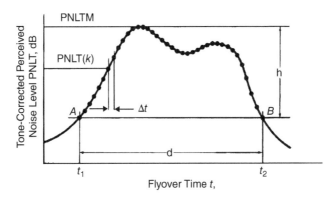

Figure 6.4 How tone-corrected perceived noise level may vary in an aircraft flyover, showing some of the labeling used in the calculation of EPNL [13].

in which case this average replaces the tone correction value for sample k. But essentially Eq. (6.4) states the summation process that is actually taking place, just as in any time-integrated measure of noise level.

The EPNL takes into account tonal content, duration, and the level of the noise by integrating the tone-corrected PNL over the duration of the event. An example of how the tone-corrected PNL varies with an aircraft flyover is shown in Figure 6.4. For aircraft certification purposes h is equal to 10 PNdB, Δt is equal to 0.5 second, and the duration, d, is determined by the 10-dB down points shown as A and B in Figure 6.4.

The procedure for calculating L_{EPN} is quite complicated, and its description is beyond the scope of this book. It is fully described in standards [9, 14, 15] and in some other books [10–12]. There is a useful, approximate relationship between A-weighted SPL, L(A) and PNL. For aircraft noise spectra this is generally taken to be PNL = L(A) + 13, and EPNL = SEL + 3. (SEL is the sound exposure level. See Section 6.9.) The A-weighting filter and A-weighted SPL were discussed in Section 4.3.5.

6.5 Articulation Index and Speech Intelligibility Index

The articulation index (AI) is a measure of the intelligibility of speech in a continuous noise. The AI was first proposed by French and Steinberg [16] and was extended later by Beranek and Ver [17]. Speech has a dynamic range of about 30 dB in each one-third octave band from 200 to 6000 Hz, and the long-term root mean square (rms) overall SPL at the speaker's lips is about 65 dB. In speech, vowels and consonants are joined together to produce not only words but sounds that have a distinctive personal nature as well. The vowels usually have greater energy than consonants and give the speech its distinctive characteristics. This is because vowels have definite frequency spectra with superimposed short-duration peaks. The AI ranges from AI = 0 to 1.0 corresponding to 0 and 100% intelligibility, respectively. If the AI is less than about 0.3, speech communication is unsatisfactory (only about 30% of monosyllabic words are understood); while if the AI is greater than about 0.6 or 0.7, speech communication is generally satisfactory (with more than 80% of monosyllabic words understood). Methods to calculate the AI are somewhat complicated and are given in American National Standard (ANSI S3.5–1969) [18] and explained in several books [10, 13]. Since the calculation of AI is complicated, it will not be explained in detail here. In 1997 the ANSI S3.5–1969 standard was updated and then further revised in 2002, 2007, and 2012 [19]. In this later standard, the AI has been renamed as the speech intelligibility index (SII). As before, the SII is calculated from acoustical measurements of speech and noise. The reader is referred to the new standard for the changes from AI to SII and complete details of the calculation of this index [19]. Because of the complication in the calculation of AI and SII, many favor the use of the speech interference level (SIL), which is easier to calculate. SIL is described in the next section of this chapter.

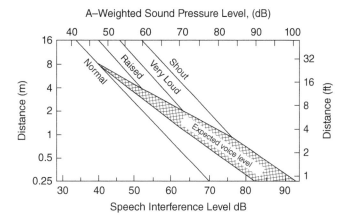

Figure 6.5 Talker-to-listener distances (m) for male speech communication to be just reliable. (*Source*: Reprinted from American Standard ANSI S3.14-1977.)

6.6 Speech Interference Level

The SIL is a measure used to evaluate the effect of background noise on speech communication [17]. The SIL is the arithmetic average of the SPLs of the interfering background noise in the four octave bands with center frequencies of 500, 1000, 2000, and 4000 Hz (see ANSI S3.14–1977(R-1986)). If the SIL of the background noise is calculated, then this may be used in conjunction with Figure 6.5 to predict the sort of speech required for satisfactory face-to face communication with male voices (i.e. for at least 95% sentence intelligibility). As an example, if the SIL is 40 dB and the speakers are males, they should be able to communicate with normal voices at 8 m. If the SIL increases to 50, a raised voice must be used at 8 m. For females the SIL should be decreased by 5 dB (or the *x*-axis moved to the right by 5 dB). The shaded area of Figure 6.5 shows the range of speech levels that normally occur as people raise their voices to overcome the background noise. Because it is simpler to measure than the SIL, the A-weighted SPL is sometimes used as a measure of speech interference, but with somewhat less confidence. Webster has produced a comprehensive diagram (see Figure 6.6) that summarizes speech levels required for communication (at various distances) with 97% intelligibility of sentences for both outdoor and indoor situations [20]. Figure 6.6 is similar to Figure 6.5 but contains some additional information concerning voice levels in different situations. With noise levels above 50 dB, people tend to raise their voice levels as shown by the "expected line" (at the left) for nonvital communication and the "communicating line" (at the right) of the diagonal shaded area for essential communication.

Example 6.2 Background noise levels for an industrial plant were measured to be 45, 54, 56, and 45 dB respectively in the 500-, 1000-, 2000-, and 4000-Hz center frequency bands. What is the SIL and at what distances could people communicate satisfactorily?

Solution

$$SIL = (45 + 54 + 56 + 45)/4 = 50 \, dB.$$

Thus, from Figure 6.5, men could communicate satisfactorily at 2 m (perhaps even 2.5 m) with a normal voice and 4 m (perhaps even 5 or 6 m) with a raised voice. Women could communicate satisfactorily with a normal voice at 1 m (perhaps up to 1.5 m) and with a raised voice at 2 m (perhaps up to 2.5 m).

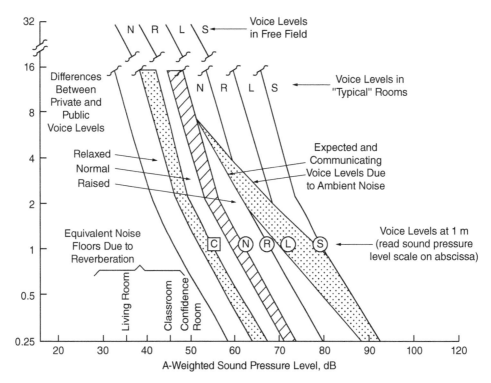

Figure 6.6 Comprehensive diagram summarizing speech levels for communication (at various distances) for 97% intelligibility of sentences on first presentation to listeners for both outdoors (free field) and indoor situations [20].

6.7 Indoor Noise Criteria

The SIL is mainly used to evaluate the effect of noise on speech in situations outdoors or indoors where the environment is not too reverberant. The A-weighted SPL can be used as a guide for the acceptability of noise in indoor situations, but it gives no indication about which part of the frequency spectrum is of concern. A number of families of noise-weighting curves have been devised to evaluate the acceptability of noise in indoor situations. These include *noise criterion* (NC) *curves, noise rating* (NR) *curves, room criterion* (RC) *curves*, and *balanced noise criterion* (NCB) *curves*. The curves have resulted from the need to either specify acceptable noise levels in buildings or determine the acceptability of noise in existing building spaces. A major concern has been to determine the acceptability of air-conditioning noise. Beranek and coworkers [21–24] have been major contributors to the development of the NC and NCB noise criterion curves. Beranek and coworkers [21–24] and Blazier [25] were mainly responsible for the development of the RC room criterion curves. NR curves were devised by Kosten and van Os [26] and are similar to the NC curves. They have been standardized and adopted by the ISO. These noise-weighting curves are now reviewed briefly.

6.7.1 NC Curves

The NC curves (Figure 6.7) were developed from the results of a series of interviews with people in offices, public spaces, and industrial spaces [21, 22]. These results showed that the main concern was the interference of noise with speech communication and listening to music, radio, and television. In order to determine the NC rating of the noise under consideration, the one-octave band SPLs of the noise are measured, and these are then plotted on the family of NC curves (Figure 6.7). The noise spectrum must not exceed the particular NC curve specified in any octave band in order for it to be assigned that particular NC rating [17].

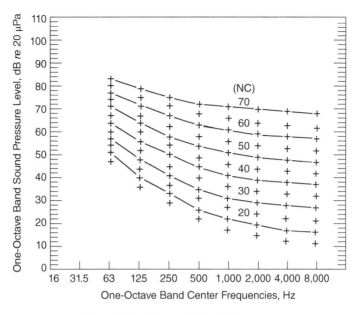

Figure 6.7 Noise criterion (NC) curves.

Example 6.3 In a classroom the one-octave-band noise levels were measured to be 61, 56, 54, 50, 44, 43, 39, and 34 dB for the center frequency bands between 63 and 8000 Hz. Evaluate this room according to the NC curves.

Solution

Using Figure 6.7 we can show that the furthest penetration of the noise data into the NC curves is at 500 Hz where the data are slightly above the NC 45 curve. Therefore, the approximate NC value of the noise would be a rating of NC 46. Usually, for classrooms an NC 30–40 range is recommended, so the background noise of our example would be judged as unacceptable.

6.7.2 NR Curves

The NR curves are similar to the NC curves (see Figure 6.7). They were originally produced to develop a procedure to determine whether noise from factories heard in adjacent apartments and houses is acceptable [26]. The noise spectrum is measured and plotted on the family of NR curves (Figure 6.8) in just the same way as with the NC curves. One difference from the NC curves, however, is the use of corrections for time of day, intermittency, audible pure tones, fraction of time the noise is heard, and type of neighborhood. These corrections are made to the final NR rating and not to the one-octave band levels used to determine the NR. It has been found that in the range of NR or NC of 20 to 50 there is little difference between the results obtained from the two approaches.

6.7.3 RC Curves

NC curves are not defined in the low frequency range (16- and 31.5-Hz one-octave bands) and are also generally regarded as allowing too much noise in the high-frequency region (at and above 2000 Hz). Blazier based his derivation of the RC curves on an extensive study conducted for the American Society of Heating, Refrigeration, and Air-Conditioning Engineers (ASHRAE) by Goodfriend of generally acceptable background spectra in 68 unoccupied offices [25]. The A-weighted SPLs were mostly in the range of 40–50 dB. Blazier [25] found that the curve that he obtained from the measured data had a slope of about −5 dB/octave,

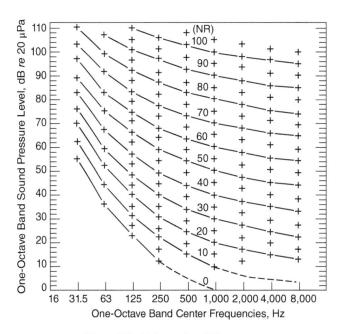

Figure 6.8 Noise rating (NR) curves.

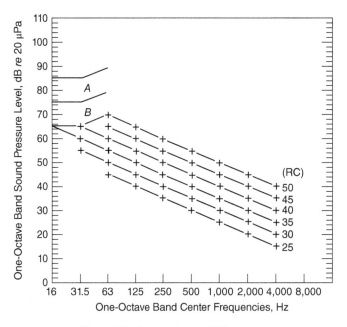

Figure 6.9 Room criterion (RC) curves.

and he thus drew a family of straight lines with this slope (see Figure 6.9). He also found that intense low-frequency noise with a level of 75 dB or more in region *A* is likely to cause mechanical vibrations in lightweight structures (including rattles), while noise in region *B* has a low probability to cause such vibrations. The value of the RC curve is the arithmetic average of the levels at 500, 1000, and 2000 Hz. Since these curves were obtained from measurements made with air-conditioning noise, they are mostly useful in rating the noise of such systems.

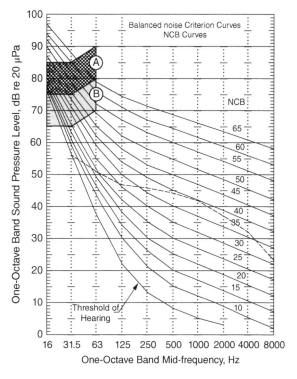

Figure 6.10 Balanced noise criterion (NCB) curves.

6.7.4 Balanced NC Curves

In 1989, Beranek [23, 24] modified the NC curves to include the 16- and 31.5-Hz octave bands and changed the slope of the curves so that it is now −3.33 dB/octave between 500 and 8000 Hz. He also incorporated the A and B regions as specified by Blazier [25] in the RC curves. The rating number of a *balanced noise criterion* NCB curve is the arithmetic average of the octave band levels with mid frequencies of 500, 1000, 2000, and 4000 Hz. The result is a set of rating curves that are useful to rate air-conditioning noise in buildings (see Figure 6.10). Recommended categories of NCB curves for different building interior spaces are presented in Section 6.16.1.

Example 6.4   As an example of the use of the balanced noise criterion, a background noise spectrum from air-conditioning is plotted (as the dashed curve) in Figure 6.10. Evaluate the acceptability of such a noise spectrum according to the NCB curves.

Solution

The NCB is calculated from the formula NCB = ¼ (44 + 42 + 37 + 33) dB. Thus this background noise spectrum can be assigned a rating of NCB 39 dB. Such a noise spectrum might be just acceptable in a general office, and barely acceptable in a bedroom and living room in a house, but not acceptable at all in a church, concert hall, or theater (see Table 6.4).

6.8 Equivalent Continuous SPL

For noise that fluctuates in level with time it is useful to define the *equivalent continuous SPL*, L_{Aeq}, which is the A-weighted SPL averaged over a suitable period, T. This average A-weighted SPL is also sometimes known as the *average sound level* L_{AT} in ANSI documents, so that $L_{Aeq} = L_{AT}$. The equivalent SPL is defined by

$$L_{Aeq} = 10\log\left(\frac{1}{T}\int_0^T p_A^2 dt/p_{ref}^2\right) = 10\log\left(\frac{1}{T}\int_0^T 10^{L(t)/10}dt\right) = 10\log\left(\frac{1}{N}\sum_{i=1}^N 10^{L_i/10}\right), \qquad (6.5)$$

where p_A is the instantaneous sound pressure measured using an A-weighting frequency filter and p_{ref} is the reference sound pressure 20 µPa. The averaging time T can be specified as desired to range from seconds to minutes, hours, weeks, months, and so forth. $L(t)$ is the short-time average. See Figure 6.11a. L_i can be a set of short-time averages for L_p over set periods. If the SPLs, L_i, are values averaged over constant time periods such as one hour, then they can be summed as in Eq. (6.5). See Figure 6.11b.

The average SPL (or the equivalent continuous SPL L_{Aeq}) can be conveniently measured with an integrating sound level meter or some other similar device. Since it accounts both for magnitude and the duration, L_{Aeq} has become one of the most widely used measures for evaluating community (environmental) noise from road traffic, railways, and industry [27–29]. L_{Aeq} has also been found to be well correlated with the psychological effects of noise [30, 31]. For community noise, a long-period T is usually used (often 24 hours). In the literature L_{Aeq} is often abbreviated to L_{eq}.

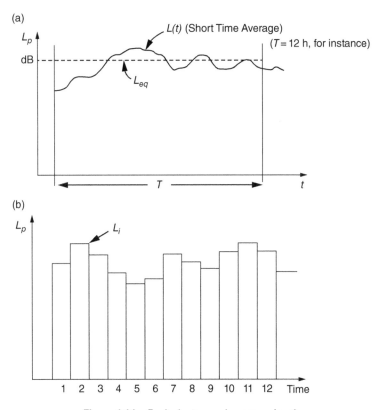

Figure 6.11 Equivalent sound pressure level.

Example 6.5   At one employee's work station, the average A-weighted SPLs, $L_i = 100, 90, 91, 89$, and 90 dB are obtained for five short, equal time intervals. Determine the average SPL L_{Aeq}.

Solution

In this case, Eq. (6.5) gives $L_{Aeq} = 10 \times \log[(10^{10} + 10^9 + 10^{9.5} + 10^{8.9} + 10^9)/5] = 95$ dB.

6.9   Sound Exposure Level

Although L_{eq} measurements provide practical results for fluctuating noise, an L_{eq} measurement does not remove ambiguity in the case of transient noise (e.g. aircraft flyover or a vehicle drive-by). In such a case the SEL is very useful and corresponds to an equivalent SPL L_{eq} normalized to one second, combining both loudness and duration in a single metric. Therefore, SEL is numerically equivalent to the total sound energy while L_{eq} is proportional to the average sound power. For example, an A-weighted noise level of 80 dB lasting one second would have a SEL $= 80$ dB but if the noise event lasted two seconds, the SEL would be 83 dB. SEL can be used to either characterize a single noise event or compare the energy of noise events which have different time durations. Sometimes A-weighted sound exposure level is referred to ASEL and L_{AE}.

Comparing the definition of SEL with Eq. (6.5), we see that

$$\text{SEL} = L_{eq} + 10 \log T, \tag{6.6}$$

where T is the integration time in seconds. Some integrating sound level meters offer an automatic SEL mode.

Example 6.6   What is L_{eq} over eight hours for a machine performing 500 operation cycles each with SEL $= 105$ dB?

Solution

We have that 8 hours $= 8 \times 3600 = 28\,800$ seconds. Therefore, from Eq. (6.6)

$$L_{eq} = \text{SEL} - 10 \log T = 10 \log \left(500 \times 10^{105/10} \right) - 10 \log (28800) = 132 - 44.6 = 87.4 \text{ dB}$$

Example 6.7   For a typical day a lane of traffic in a highway located at a distance of 50 m from a receiver carries a total of 20 000 vehicles of all types. Assume that during the day trucks constitute 10% of the total of vehicles and automobiles 90%. For the example, suppose that the average ASELs at a distance of 50 m are 80 dB for trucks and 65 dB for automobiles. (a) Determine the 24-hour average SPL, and (b) Determine the effect of banning trucks from this highway.

Solution

a) We have that 24 hours $= 24 \times 3600 = 86\,400$ seconds. We obtain a number of 2000 trucks and 18 000 automobiles per day. First, we need to obtain the energy-combined SEL for all the events:

$$L_{AE} = 10 \log \left(2000 \times 10^{80/10} + 18000 \times 10^{65/10} \right) = 10 \log \left(2.57 \times 10^{11} \right) = 114.1 \text{ dB}.$$

Therefore, using Eq. (6.6)

$$L_{eq} = L_{AE} - 10 \log (86400) = 114.1 - 49.4 = 64.7 \, \text{dB}.$$

b) The computation without trucks gives

$$L_{eq} = 10 \log (18000 \times 10^{6.5}) - 49.4 = 58.2 \, \text{dB}.$$

Therefore, banning trucks from this highway will produce an A-weighted noise level reduction in L_{eq} of $64.7 - 58.2 = 6.5 \, \text{dB}$.

6.10 Day–Night Equivalent SPL

In the United States during the 1970s, the Environmental Protection Agency (EPA) developed a measure, from the equivalent SPL, known as the *day–night equivalent level* (DNL) or L_{dn} that accounts for the different response of people to noise during the night [31] given by

$$L_{dn} = 10 \log \frac{15(10^{L_d/10}) + 9(10^{(L_n + 10)/10})}{24}, \tag{6.7}$$

where L_d is the 15-hour daytime A-weighted equivalent SPL (from 07:00 to 22:00 hours) and L_n is the 9-hour nighttime equivalent SPL (from 22:00 to 07:00 hours). The nighttime noise level is subjected to a 10-dB penalty because noise at night is deemed to be much more disturbing than noise during the day. This 10-dB nighttime penalty is analogous to the 10-dB nighttime penalty applied in both the composite noise rating (CNR) and the noise exposure forecast (NEF), as described in Section 6.12. The DNL has become increasingly used in the United States and some other countries to evaluate community noise and in particular airport noise [29, 32]. In 1980 the U.S. Federal Interagency Committee on Urban Noise (FICON) adopted L_{dn} as the appropriate descriptor of environmental noise in residential situations [29–33].

Example 6.8 The energy-equivalent hourly A-weighted SPLs in a residential community are measured using an integrating sound level meter, and the results are presented in Table 6.3. Calculate the daytime and nighttime A-weighted equivalent SPL, and the day-night equivalent level in the community.

Solution

The daytime average A-weighted level is calculated considering measurements 1–15 giving

$$L_d = 10 \times \log \left[(10^6 + 10^{6.5} + \ldots + 10^6)/15 \right] = 63.3 \, \text{dB}.$$

The nighttime average level is calculated considering measurements 16–24 resulting

$$L_n = 10 \times \log \left[(10^6 + 10^6 + \ldots + 10^6)/9 \right] = 58.3 \, \text{dB}.$$

Consequently, the day-night equivalent A-weighted level is

$$L_{dn} = 10 \times \log \left[\left(15 \times 10^{6.33} + 9 \times 10^{(58.3 + 10)/10} \right)/24 \right] = 65.9 \, \text{dB}.$$

Table 6.3 Data used in Example 6.8.

Number	Time	A-weighted sound pressure level, dB
1	07:00	60
2	08:00	65
3	09:00	65
4	10:00	62
5	11:00	63
6	12:00	63
7	13:00	60
8	14:00	60
9	15:00	65
10	16:00	65
11	17:00	65
12	18:00	64
13	19:00	64
14	20:00	62
15	21:00	60
16	22:00	60
17	23:00	60
18	00:00	58
19	01:00	57
20	02:00	57
21	03:00	57
22	04:00	57
23	05:00	57
24	06:00	60

Example 6.9 Assume for the noise profile presented in Table 6.3 that a factory nearby produced an intense venting noise for a period of 10 seconds, at an A-weighted SPL of 95 dB during the 22:00 hours night period. What would be the effect on the day-night equivalent level?

Solution

First we must determine the effective A-weighted hourly level for the 22:00 time period. This level must combine the normal A-weighted background noise of 60 dB that is on 3590 seconds out of the hour (3600 seconds) and 95 dB that is on for the remaining 10 seconds. Therefore, the new A-weighted value for the row 16 in Table 6.3 is

$$L_{16} = 10 \times \log\left[(3590 \times 10^6 + 10 \times 10^{9.5})/3600\right] = 69.9 \text{ dB}.$$

The daytime equivalent SPL L_d is not affected by this venting noise and we have to replace 60 dB by the level including the contribution of the venting noise (69.9 dB) in Table 6.3 and repeat the calculation for L_n. This gives a nighttime A-weighted equivalent level of 62.2 dB. Finally, the day-night A-weighted equivalent level is calculated as in Example 6.8, resulting in 68.8 dB, an increase of almost 3 dB that represents a potential significant source of annoyance to the residents of the community.

6.11 Percentile SPLs

The equivalent SPL discussed above accounts for the fluctuation in noise level of an unsteady noise by forming an average SPL resulting in an equivalent steady A-weighted SPL. There is, however, some evidence that unsteady noise (e.g. from noise sources such as passing road vehicles or aircraft movements) is more disturbing than steady noise. To try to better account for fluctuations in noise level and the intermittent character of some noises, *A-weighted percentile SPLs* are used in some measures, in particular those for community and traffic noise [27, 34, 35]. The level L_n is defined to represent the SPL exceeded $n\%$ of the time, and thus L_{10}, for example, represents the SPL exceeded 10% of the time.

Figure 6.12 gives an example of L_{10}, L_{50}, and L_{90} levels and a cumulative distribution. It is seen in this schematic example figure that the A-weighted level exceeded 10% of the time L_{10} is 85 dB. L_{50} is sometimes termed the *median* noise level, since for half the time the fluctuating noise level is greater than L_{50} and for the other half it is less. L_{50} is used in Japan for road traffic noise. Levels such as L_1 or L_{10} are used to represent the more intense short-duration noise events. L_{10} is used in Australia and the United Kingdom (over an 18-hour 06:00 to 24:00 period) as a target value for new roads and for insulation regulations for new roads. Levels such as L_{90} or L_{99} are often used to represent the minimum noise level, the residual level from a graphic level recorder, or the average minimum readings observed when reading a sound level meter. Figure 6.13 shows the outdoor A-weighted SPLs recorded in 1971 at 18 locations in the United States. Values of L_{99}, L_{90}, L_{50}, L_{10}, and L_1 are shown for the period 07:00 to 19:00 hours [27, 32]. The small range in levels in recordings 1 and 4 (urban situations) and the large range in levels in recordings 6, 11, 13, and 18 (situations involving aircraft overflights) are very evident. Obviously road traffic usually creates more steady noise, while aircraft movements lead to more extreme variations in noise levels.

6.12 Evaluation of Aircraft Noise

The noise levels around airports are of serious concern in many countries. Several attempts have been made to produce measures to predict and assess the annoyance caused by aircraft noise in the community. A study of rating measures in 1994 showed 11 different measures in use in the 16 countries studied [29]. The following measures merit brief discussion.

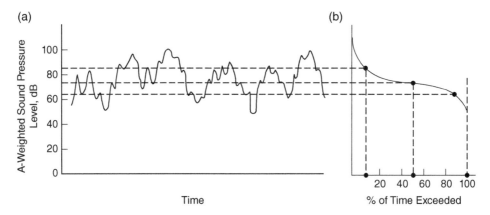

Figure 6.12 (a) Percentile levels and (b) cumulative probability distribution function of percentile levels.

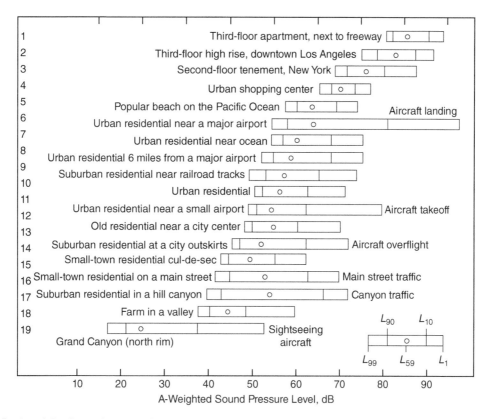

Figure 6.13 A-weighted sound pressure levels measured in 1971 at 18 locations in the United States. Values of the percentile levels L_{99}, L_{90}, L_{50}, L_{10}, and L_1, where L_n is A-weighted sound pressure level exceeded n percent of the time are shown for the period 07:00–19:00 hours [27, 28, 32].

6.12.1 Composite Noise Rating

The CNR has a long history dating back to the early 1950s [36–38]. Originally, the basic measure it used was the level rank – a set of curves placed about 5 dB apart in the midfrequency range, rather similar to the NC and NR curves described earlier. The level rank was obtained by plotting the noise spectrum on the curves and finding the highest zone into which the spectrum protruded. The rank found initially plus the algebraic addition of corrections gave the CNR. The corrections [36] were for spectrum character, peak factor, repetitive character, level of background noise, time of day, adjustment to exposure, and public relations. The value of CNR obtained was associated with a range of community annoyance categories found from case histories – ranging from no annoyance, through mild annoyance, mild complaints, strong complaints, and threats of legal action, to vigorous community response.

In the late 1950s, the CNR was adapted to apply to the noise of military jet aircraft [39] and later of commercial aircraft [40]. The calculation was further modified later when applied to commercial aircraft by using the PNL instead of the level rank or the SPL just referred to.

The final version of CNR does not contain any corrections for background noise, previous experience, public relations, or other factors such as the presence of pure tones. Although CNR is no longer used, it is discussed here for completeness and because it has formed the basis for some of the other noise measures and descriptors, such as NEF, which follow.

6.12.2 Noise Exposure Forecast

The NEF is a similar measure to CNR, but it uses the EPNL instead of PNL [41, 42]. Thus NEF automatically takes account of the annoying effects of pure tones and the duration of the flight events. The use of NEF has been superseded in the United States and most other countries by the day–night level L_{dn} or the day–evening–night level L_{den}.

6.12.3 Noise and Number Index

The *noise and number index* (NNI) is a subjective measure of aircraft noise annoyance first developed and used in the United Kingdom. The NNI was the outcome of surveys in 1961 and 1967 of noise in the residential districts within 10 miles of London (Heathrow) Airport [43, 44]. NNI is based on a summation of $(PNL)_N$ terms weighted by aircraft movements. $(PNL)_N$ is the average peak noise level of all aircraft operating during a day, and N is the number of aircraft movements. Here PNL is the peak PNL produced by an individual aircraft during the day and N is the number of aircraft operations of that type over a 24-hour period.

In 1988 NNI was superseded in the United Kingdom by a measure based on the A-weighted L_{eq}. The L_{eq} is determined over the period 07:00–23:00 hours. Noise at night is evaluated in terms of the size of the 90 SEL footprint of individual aircraft movements, or, less commonly, using L_{eq} determined over the period 23:00–07:00 hours.

6.12.4 Equivalent A-Weighted SPL L_{eq}, Day–Night Level L_{dn}, and Day–Evening–Night Level L_{den}

In recent years some countries have continued to use NEF or NNI or similar noise measures or descriptors related to those that include a weighting based on the number of aircraft movements [29]. However, because they are much simpler to measure and seem to give adequate correlation with subjective response, there has been a move in several countries toward the use of L_{eq} and L_{dn} [29]. In the United Kingdom L_{eq} has been adopted with an 18-hour period only (07:00–23:00) (because nighttime flights are normally restricted). The European Union has specified the use of L_{den} (DENL) to evaluate aircraft noise, which includes three periods: day, evening, and night, from which L_{den} is calculated. In the United States since publication of the EPA's *Levels document* [31] and other similar publications, the use of CNR and NEF has been superseded by the DNL (L_{dn}) for the assessment of the potential impacts of noise and for planning recommendations and land-use management near civilian and military airports.

6.13 Evaluation of Traffic Noise

6.13.1 Traffic Noise Index

In an attempt to develop acceptability criteria for traffic noise from roads in residential areas, Griffiths and Langdon [45] produced a unit for rating traffic noise, the *traffic noise index* (TNI). They measured A-weighted traffic noise levels at 14 sites in the London area and interviewed 1200 people at these sites in the process. Griffiths and Langdon excluded sites with noise sources other than traffic. They then used regression analysis to fit curves to the data. This indicated that L_{10} was better at predicting dissatisfaction than L_{50} or L_{90}, and that TNI was also superior to L_{10}, L_{50}, and L_{90}.

Use of the TNI has not been widespread. The index attempts to make an allowance for the noise variability since fluctuating noise is commonly assumed to be more annoying than steady noise.

Some doubt has been cast on the conclusions of Griffiths and Langdon, and it has been suggested that the very short sample times (100 seconds in each hour) used may have resulted in underestimates of L_{10} and overestimates of L_{90} [46]. TNI is not considered today to be significantly superior to either L_{10} or L_{eq} and has not been widely used.

Instead of TNI, the British government has adopted the A-weighted L_{10}, averaged over 18 hours from 06:00 to 24:00 hours, as the noise index to be used to implement planning and remedial measures to reduce the impact on people of road traffic noise [47–49]. In addition, the British government uses a 16-hour L_{eq} and an 8-hour L_{eq} for the case of land used for residential development.

6.13.2 Noise Pollution Level

In a later survey, Robinson [50] again concluded that, with fluctuating noise, L_{eq}, the equivalent continuous A-weighted SPL on an energy basis, was an insufficient descriptor of the annoyance caused by fluctuating noise. He included another term in his *noise pollution level* (NPL) or L_{NP}.

Robinson examined the available Griffiths and Langdon data. [49] He then examined the aircraft noise experiments of Pearsons [51] and found that L_{NP} predicted very well Pearsons's data points and the tradeoff between duration and level for individual flyover events. A-weighted levels were used in L_{NP} with traffic noise, and PNLs (PNdB) were used with aircraft noise. The superiority of L_{NP} over all other forms of NR has not been proved in practice, and it has not been widely used.

6.13.3 Equivalent SPL

Figure 6.14 shows the annoyance results of Pearsons and coworkers, using six different NRs: L_1, L_{10}, L_{50}, L_{90}, L_{eq}, and L_{NP} [52]. As expected for all of the noise measures, annoyance increases with level. The shapes of the curves, however, do vary considerably when the annoyance is less than very annoying. In particular, the L_{10} and L_{eq} measures exhibit a very steep rise in annoyance from the categories of *slightly* to *moderately annoying* for no increase in noise level, presumably one of their drawbacks. However, except for the case of L_{NP} in the *extremely annoying* category, the standard deviation of L_{eq} for a specified response category is in all cases less than or equal to the standard deviation of the other noise measures. This is an advantage in the use of L_{eq} since there is more confidence in the annoyance scores predicted. There is one clear advantage of L_{eq} over L_{10}, however, in the case of noise containing short-duration, high-level single events. If the events do not occur for

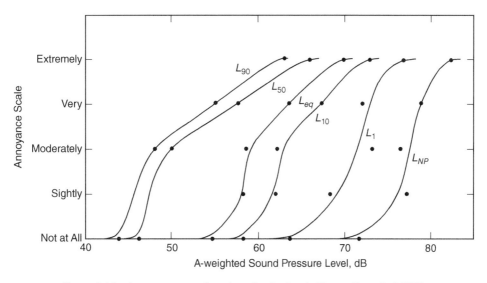

Figure 6.14 Annoyance as a function of noise level. (*Source:* From Ref. [52].)

more than 10% of the time, then L_{10} will be relatively insensitive to these high-level events and will tend to represent the "background" noise. In fact, for a noise measure L_n to be useful, the intruding noise events must be present for *more* than $n\%$ of the time. This suggests that L_{10} would be unsuitable as a measure of aircraft noise annoyance and that it might be a possible source of error in Griffiths and Langdon's results [45] for low traffic flows.

The A-weighted equivalent SPL (often now denoted as L_{eq}) has become the measure most commonly used to assess and regulate road traffic (and railroad noise) [29]. In the United States L_{dn} (a similar measure) is used for road traffic noise assessment.

6.14 Evaluation of Community Noise

In some community noise measures, corrections are applied to community noise levels to account for pure-tone components or impulsive character, seasonal corrections (summer or winter when windows are always closed), type of district (rural, normal suburban, urban residential, noisy urban, very noisy urban), and for previous exposure (such corrections are similar to those for NR).

Figure 6.15 gives three examples of the A-weighted SPLs measured in a community over a 24-hour period [27]. The triangles in the three figures are the maximum levels read from a graphic level recorder. The continuous lines are percentile levels measured for hour-long periods throughout the 24 hours. The highest percentile measure L_1 does not represent the maximum levels well, which are presumably mostly higher level, short-duration sounds (occurring less than 1% of the time). It is observed in Figure 6.15a and 6.15b that although the day–night levels L_{dn} are only 3 dB different (86 and 83 dB), there is a much greater fluctuation in SPL with time in Figure 6.15b. This example is for a location near a major airport, and the fluctuation in level would suggest that the noise environment in location 6 would be much more annoying than in location 1 (near a freeway). Figure 6.15c illustrates another quite different distribution of A-weighted SPLs with time.

In the United States, the A-weighted L_{eq} and L_{dn} are normally used to characterize community noise. However, the Department of Housing and Urban Development (HUD) also recognizes the usefulness of L_{10} in some instances. In other countries the A-weighted L_{eq} is mainly used for community noise studies rather than L_{dn}. International standards on the description and measurement of environmental noise [53] recommend the use of A-weighted L_{eq} and rating levels (which are A-weighted L_{eq} values to which tone and impulse adjustments have been made). These standards also recommend that in some circumstances it may be useful to determine the distribution of A-weighted SPLs by determining percentile levels such as L_{95}, L_{50}, and L_5.

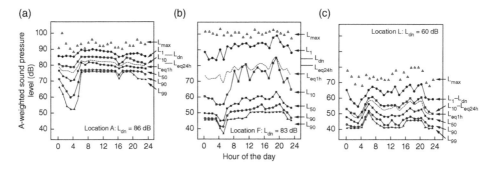

Figure 6.15 Variation of percentile levels throughout 24 hours periods recorded at three residential locations (numbers 1, 6, and 12) in Los Angeles in 1971. (a) third-floor apartment near a freeway, (b) urban residential location near a major airport, and (c) old residential location near a city center. The values of L_n represent the A-weighted SPLs exceeded n percent of the time during 1-hour periods. Hourly maximum levels L_{max}, 1-hour and 24-hour values of L_{eq}, and the day-night levels L_{dn} are also shown. (*Source:* From Refs. [27, 32].)

In California, some state legislation requires the use of the *community noise exposure level* (CNEL) rather than the day–night level L_{dn}. The two descriptors are similar except that the CNEL (L_{CN}) like DENL (L_{den}) has three periods instead of two. Besides the night penalty of 10 dB, an evening penalty of 5 dB is applied with CNEL. CNEL is defined [54] as

$$L_{CN} = 10 \log \frac{12\left(10^{L_d/10}\right) + 3\left(10^{(L_e + 5)/10}\right) + 9\left(10^{(L_n + 10)/10}\right)}{24}, \tag{6.8}$$

where L_d is the average 12-hour day *HNL* (hourly noise level), L_e is the average 3-hour evening *HNL*, and L_n is the average 9-hour night *HNL*.

The hourly noise level is given by

$$HNL = 10 \log \frac{\int 10^{L/10} dt}{3600}, \tag{6.9}$$

where L is the instantaneous A-weighted SPL, and t is the time in seconds. The integral is calculated and summed. *HNL* is usually computed electronically. The CNEL has also been extensively used to evaluate airport noise and assess environmental noise transmission into buildings in California [55].

Example 6.10   Determine the CNEL from the 24 one-hour average sound pressure levels given in Table 6.3.

Solution

In this case the evening penalty of 5 dB is applied to A-weighted SPLs occurring between 19:00 and 22:00 hours. Thus L_d is now calculated considering measurements 1–12 giving

$$L_d = 10 \times \log\left[\left(10^6 + 10^{6.5} + \ldots + 10^{6.4}\right)/12\right] = 63.5 \, \text{dB}.$$

The evening average level is obtained from measurements 13–15 as

$$L_e = 10 \times \log\left[\left(10^{6.4} + 10^{6.2} + 10^6\right)/3\right] = 62.3 \, \text{dB}.$$

The nighttime average level is calculated considering measurements 16–24 resulting

$$L_n = 10 \times \log\left[\left(10^6 + 10^6 + \ldots + 10^6\right)/9\right] = 58.3 \, \text{dB}.$$

Inserting the appropriate values in Eq. (6.8) we obtain

$$L_{CN} = 10 \times \log\left[\left(12 \times 10^{6.35} + 3 \times 10^{6.73} + 9 \times 10^{6.83}\right)/24\right] = 66.4 \, \text{dB}.$$

6.15   Human Response

6.15.1   Sleep Interference

Various investigations have shown that noise disturbs sleep [56–62]. It is well known that there are several stages of sleep and that people progress through these stages as they sleep [56]. Noise can change the progression through the stages and if sufficiently intense can awaken the sleeper. Most studies have been conducted in the laboratory under controlled conditions using brief bursts of noise similar to aircraft flyovers or the passage of heavy vehicles. However, some have been conducted in the participants' bedrooms.

Different measures have been used such as A-weighted *maximum SPL*, L_{Amax}, *ASEL*, *EPNL* (EPN dB), and *day–night SPL*, (DNL (L_{dn})), and most studies have concentrated on the percentage of the subjects that are

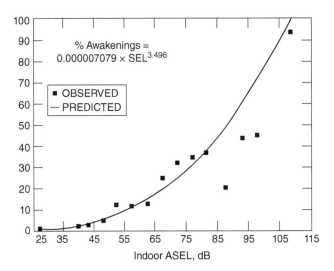

Figure 6.16 Proposed sleep disturbance curve based on data of Pearsons et al. [61]. (*Source:* From Ref. [62] with permission.) Curve represents percentage of subjects awakened against A-weighted sound pressure exposure level ASEL.

awakened. Pearsons et al. [61] have reassessed data from 21 sleep disturbance studies (drawn from the reviews of Lucas [56] and Griefahn [57] and seven additional studies). From these data, Finegold et al. [62] have proposed sleep disturbance criteria based on the indoor ASEL. In the reanalysis, because of the extremely variable and incomplete databases, the data were averaged in 5-dB intervals to reduce variability. A regression fit to these data gave the following expression (which is also shown graphically in Figure 6.16):

$$\% \ Awakenings = 7.1 \times 10^{-6} L_{AE}^{3.5}, \tag{6.10}$$

where L_{AE} is the indoor ASEL. Although the authors recognize that there are concerns with the existing data and there is a recognition that additional sleep disturbance data are needed, Finegold et al. [62] have proposed that Figure 6.16 be used as a practical sleep disturbance curve until further data become available.

6.15.2 Annoyance

In 1978, Schultz published an analysis of 12 major social surveys of community annoyance caused by transportation noise [63]. This Schultz analysis, which relates the percentage of the population that report they are "highly annoyed" by transportation noise to the A-weighted day–night average SPL DNL (L_{dn}), has become widely used all over the world as an important curve for describing the average community response to environmental noise. Since the Schultz curve was published, additional data have become available. Fidell et al. [64] used 453 exposure response data points compared to the 161 data points originally used by Schultz. This resulted in almost tripling the database for predicting noise annoyance from transportation noise. A later study by the U.S. Air Force eliminated 53 of these data points because there was insufficient correlation between the DNL and the percentage of the population that was highly annoyed, $\%HA$ [62]. The results of these two studies and the original Schultz curve are given in Figure 6.17. Finegold et al. [62] recommend a logistic fit,

$$\%HA = \frac{100}{1 + \exp\left(11.13 - 0.14 L_{dn}\right)}, \tag{6.11}$$

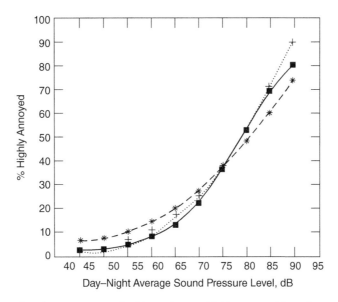

Figure 6.17 Curves representing the percentage of subjects that are highly annoyed by noise against A-weighted day-night average sound pressure level DNL (L_{dn}): (–■–) New logistic USAF curve (400 data points), (···+···) Schultz [63] third-order polynomial (161 data points), and (--*- -) Fidell et al. [64] quadratic curve (453 data points). (*Source:* From Ref. [62] with permission.)

rather than the quadratic fit used by Fidell et al. [64] or the third-order polynomial fit used by Schultz [63]. This results in a very close agreement between the curve obtained with the 400 data points and the original 1978 Schultz curve [63]. See Figure 6.18.

The differences between the curves in Figures 6.17 and 6.18 are not very significant; however, there are several advantages to the use of the logistic fit given in Eq. (6.11), including (i) the same predictive utility

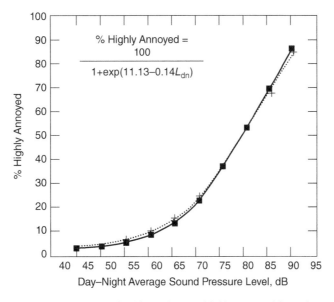

Figure 6.18 Curves representing the percentage of subjects that are highly annoyed by noise against A-weighted day-night average sound pressure level DNL (L_{dn}): (–■–) Logistic fit to 400 community annoyance social survey data points and (···+···) 1978 Schultz [63] curve. (*Source:* From Ref. [62] with permission.)

in both the original Schultz curve and the Fidell et al. curve, (ii) it allows prediction of annoyance to approach but does not reach 0 or 100%, (iii) it approaches a 0% community annoyance prediction for a DNL (L_{dn}) of approximately 40 dB rather than the anomaly of an increase in annoyance for levels of less than 45 dB as predicted by the Fidell et al. curve, (iv) use of a logistic function has had a history of success with U.S. federal environmental impact analyses, and (v) it is based on the most defensible social survey database [62].

Example 6.11   Predict the noise-induced sleep disturbance for a specific receiver due to a single nighttime aircraft flyover of duration 10 seconds and an outdoor measured A-weighted noise level L_{eq} of 95 dB. Assume that the average loss due to the insulation of a typical house is 20.5 dB with windows closed and 15 dB with windows open.

Solution

The outdoor SEL is obtained from Eq. (6.6):

$$SEL = 95 - 10 \log (10) = 85 \, dB.$$

The indoor SEL is estimated from the outdoor SEL by subtracting the insulation values. Therefore,

$$SEL = 85 - 20.5 = 64.5 \, dB \text{ with windows closed, and}$$
$$SEL = 85 - 15 = 70 \, dB \text{ with windows open.}$$

Now, using either Eq. (6.10) or Figure 6.16, we obtain that

$$\% \, Awakenings = 7.1 \times 10^{-6} \times (64.5)^{3.5} = 15.3, \text{ and } \% \, Awakenings = 7.1 \times 10^{-6} \times (70)^{3.5} = 20.4.$$

Therefore, the approximate percentage of people likely to awake due to this single flyover at nighttime with windows closed and open is 15 and 20%, respectively.

Example 6.12   Suppose there are 10 aviation noise events in a period of 24 hours at an airport. Eight events occurred during daytime and two events during nighttime. ASELs of each event at a nearby residential community are 84, 93, 97, 83, 96, 93, 88, and 91 dB during daytime and 99 and 90 dB during nighttime. Determine the percentage of people highly annoyed by the 24 hours airport operation noise.

Solution

We determine the day-night A-weighted equivalent level for 24 hours (86 400 seconds) applying the 10-dB penalty to the two nighttime events:

$$L_{DN} = 10 \log \left[\left(10^{8.4} + 10^{9.3} + 10^{9.7} + 10^{8.3} + 10^{9.6} + 10^{9.3} + 10^{8.8} + 10^{9.1} + 10^{10.9} + 10^{10} \right) / 86400 \right] = 60.8 \, dB.$$

Now, replacing this value in Eq. (6.11), we obtain

$$\%HA = \frac{100}{1 + \exp (11.13 - 0.14 \times 60.8)} = 6.8.$$

Therefore, almost 7% of the residents are expected to be highly annoyed by the airport noise.

Most community noise impact studies since the late 1970s have been based on a combination of aircraft and surface transportation noise sources. However, there has been a continuing controversy over whether all types of transportation should be combined into one general curve for predicting community annoyance to transportation noise [62, 65–71]. Some researchers have suggested that people find aircraft noise more annoying than traffic noise or railroad noise for the same value of DNL [29, 62, 65, 69].

The differences have been discussed in the literature [29, 32, 62] and can perhaps be explained by a variety of causes such as (1) methodological differences, (2) variability in the criterion for reporting high annoyance, (3) inaccuracy in some of the acoustical measurements, (4) community response biases, and (5) aircraft noise entering homes through parts of the building structure with less transmission loss (such as the roof rather than the walls). Figure 6.19 shows logistic fits to 400 final data points from a total of 22 different community annoyance surveys. It can be seen that aircraft noise appears to produce somewhat more annoyance than railroad or traffic noise particularly for the higher DNL (L_{dn}) values. Miedema and Vos have made a reanalysis of data from selected social surveys that also shows that aircraft noise appears to cause more annoyance than other surface transportation noise sources [65]. However, the results from the Miedema and Vos study seem to suggest that for high values of DNL (L_{dn}) (over 60–70 dB), although aircraft noise is by far the most annoying source, railroad noise is also more annoying than traffic noise (in contrast to the results of Finegold et al.) [62].

If the five causes discussed above can be dismissed as responsible for the apparent greater annoyance of aircraft noise, then it may be that aircraft noise is more annoying than surface transportation noise for reasons such as: (i) the higher peak levels, (ii) the greater variation in level with time, and (iii) the different frequency spectra from other types of transportation noise sources. As already discussed in the text accompanying Figure 6.14, aircraft noise does generally exhibit a much greater variation in level from traffic noise and other sources of surface transportation. If such variation is indeed more annoying and is one of the main causes of the difference in annoyance caused by these different forms of transportation noise, this suggests that it may be advisable to reexamine such measures that account for variation in level such as the TNI or the NPL discussed in Sections 6.13.1 and 6.13.2. The annoyance caused by noise is also discussed briefly in Chapter 5.

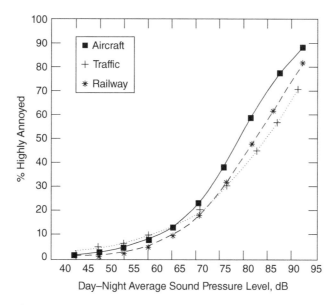

Figure 6.19 Curves representing the percentage of subjects that are highly annoyed by noise against A-weighted day-night average sound pressure level DNL (L_{dn}) for different sources: –■– Aircraft, ···+··· traffic, and --*-- railway. Curves based on data from Fidell et al. [64]. (*Source:* From Ref. [62] with permission.)

6.16 Noise Criteria and Noise Regulations

Using some of the noise measures and descriptors discussed and surveys and human response studies, various criteria have been proposed so that noise environments can be determined that are acceptable for people, for speech communication, for different uses of buildings, for sleep, and for different land uses. In some countries such criteria are used to write noise regulations for new machinery, vehicles, traffic noise, railroad noise, aircraft and airport noise, community noise, and land use and planning. It is beyond the scope of this book to give a comprehensive summary of all these criteria and regulations. Instead just a few are described in this section. The interested reader is referred to the literature [29, 32, 66] for more complete summaries of criteria, regulations, and legislation.

6.16.1 Noise Criteria

An example of noise criteria is given in Table 6.4, which is based on those suggested by Beranek and Ver [17] and gives recommended NCB curve values (and approximate A-weighted levels) for various indoor functional activity areas. The NCB curves are given in Figure 6.10. For example, the air-conditioning unit chosen to supply air to bedrooms (used in residences, apartments, hotels, hospitals, etc.) should have a spectrum corresponding to no more than an NCB curve of 25–40 (or an A-weighted SPL of no more than about 38–48 dB).

Table 6.4 Recommended values of NCB curves for different uses of spaces in buildings[a].

Type of Space (and Acoustical Requirements)	NCB Curve	Approximate L_A
Broadcast and recording studios	10	18
Concert halls, opera houses, and recital halls	10–15	18–23
Large auditoriums, large drama theaters, and large churches	<20	28
Broadcast, television, and recording studios	<25	33
Small auditoriums, small theaters, small churches, music rehearsal rooms, large meeting and conference rooms	<30	38
Bedrooms, sleeping quarters, hospitals, residences, apartments, hotels, motels, etc.	25–40	38–48
Private or semiprivate offices, small conference rooms, classrooms, and libraries	30–40	38–48
Living rooms and drawing rooms in dwellings	30–40	38–48
Large offices, reception areas, retail shops and stores, cafeterias, and restaurants	35–45	43–53
Lobbies, laboratory work spaces, drafting and engineering rooms, general secretarial areas	40–50	48–58
Light maintenance shops, industrial plant control, rooms, office and computer equipment rooms, kitchens, and laundries	45–55	53–63
Shops, garages, etc. (for just acceptable speech and telephone communication)	50–60	58–68
For work spaces where speech or telephone communication is not required, but where there must be *no* risk of hearing damage	55–70	63–78

[a] Also given are the approximate equivalent A-weighted sound pressure levels L_A.
Source: Based in part on Ref. [17].

Table 6.5 Guidelines from EPA [31], WHO [68], FICON [33], and various European agencies [29] for acceptable noise levels.

Authority	Specified A-Weighted Sound Pressure Levels	Criterion
EPA levels Document [31]	$L_{dn} \leq 55$ dB (outdoors) $L_{dn} \leq 45$ dB (indoors)	Protection of public health and welfare with adequate margin of safety
WHO Document (1995) [68]	$L_{eq} \leq 50/55$ dB (outdoors; day) $L_{eq} \leq 45$ dB (outdoors; night) $L_{eq} \leq 30$ dB (bedroom) $L_{max} \leq 45$ dB (bedroom)	Recommended guideline values
U.S. Interagency Committee (FICON) [33]	$L_{dn} \leq 65$ dB $65 \leq L_{dn} \leq 70$ dB	Considered generally compatible with residential development Residential use discouraged
Various European road traffic regulations [29]	$L_{eq} \geq 65$ or 70 dB (day)	Remedial measures required

Source: Based in part on Ref. [32].

Another example of noise criteria are the guidelines recommended by EPA [31], WHO [68], FICON [33], and various European road traffic regulating bodies. See Table 6.5. As already mentioned, L_{eq} is very widely used to evaluate road traffic, railroad, and even aircraft noise [36]. Interestingly, railroad noise has been found to be less annoying than traffic noise in several surveys [29, 72, 73]. This has resulted in noise limits (using L_{eq}) that are 5 dB lower for railroad noise than road traffic noise in Austria, Denmark, Germany, and Switzerland and 3 dB lower in The Netherlands [29]. Gottlob terms this difference the "railway bonus." [29] An example of national noise exposure criteria is the guidance given in the British government guidelines adopted in 1994 for land development. See Table 6.6. This table shows guidelines in A-weighted SPLs, L_{eq}, for four noise exposure categories [74]. The noise exposure categories can be interpreted as follows: [75]

Table 6.6 Guidelines used in the United Kingdom for A-weighted equivalent sound pressure levels for different noise exposures categories.

Noise Source		Noise Exposure Category			
		A	B	C	D
Road traffic	(07:00–23:00)	<55	55–63	63–72	>72
	(23:00–07:00)	<45	45–57	57–66	>66
Rail traffic	(07:00–23:00)	<55	55–66	66–74	>74
	(23:00–07:00)	<45	45–59	59–66	>66
Air traffic	(07:00–23:00)	<57	57–66	66–72	>72
	(23:00–07:00)	<48	48–57	57–66	> 66
Mixed sources	(07:00–23:00)	<55	55–63	63–72	>72
	(23:00–07:00)	<45	45–57	57–66	>66

Source: Based on Ref. [74].

(A) Noise need not be considered as a determining factor in granting planning permission, although the noise level at the high end of the category should not be regarded as a desirable level; (B) noise should be taken into account when determining planning applications, and, where appropriate, conditions should be imposed to ensure an adequate level of protection against noise; (C) planning permission should not normally be granted; where it is considered that permission should be given, for example because there are no alternative quieter sites available, conditions should be imposed to ensure a commensurate level of protection against noise; and (D) planning permission should normally be refused [76, 77].

6.17 Human Vibration Criteria

The effects of vibration on people's health and criteria for protection are discussed in Chapter 5. Criteria for human comfort and annoyance, in particular for vibration in buildings, are discussed in this section.

6.17.1 Human Comfort in Buildings

Vibration discomfort depends on many factors including the characteristics of the vibration. Most standards are designed to assess the relative discomfort of different motions and not necessarily to predict the absolute acceptability of vibration. This is because consistent quantitative data concerning human perception of and reaction to vibration is limited and difficult to measure. However, criteria to assess vibration comfort in buildings have been developed just to give guidance to engineers, local authorities and policy makers.

Construction-site vibrations can have a number of effects including annoyance to residents in neighbors' buildings. Other major sources of vibration causing discomfort in buildings are nearby road traffic (trucks, busses, subways, rail vehicles, etc.), building service machinery (elevators, air-conditioning units, pumps, etc.), internal sources (footsteps, washing machines, door slamming, furniture moving, etc.), and blasting. These sources can produce continuous vibration, intermittent vibration, or repeated shocks.

It has been determined that acceptable magnitudes of vibration in some buildings are close to vibration perception thresholds for frequencies in the range 1–80 Hz. For vertical vibration, the human body is most sensitive in the frequency range 4–8 Hz (see Chapter 5). Studies made by Griffin [78] have shown that the approximate absolute rms acceleration threshold for the perception of vibration is $0.01 \, \text{m/s}^2$ and that the comfortable limit is usually around $0.315 \, \text{m/s}^2$. However, other variables could also cause annoyance depending on the use of the building in addition to the vibration frequency, direction, and duration.

General assessment of human exposure to whole-body vibration has been standardized in the international standard ISO 2631 [79] and the British Standard BS 6841 [80]. The specific British standard BS 6472 [81] has been issued for evaluating disturbance to people in buildings. Ranges of vibration magnitudes associated with varying degrees of discomfort are shown in Figure 6.20 [82]. These values are specified in frequency-weighted rms acceleration (see Chapter 5) measured at the point on the building surface where the vibration enters the human body.

Using the whole-body vibration dose value (discussed in Section 5.11.2) and the guidance in ISO 2631 Part 2 and BS 6472, it is possible to summarize the acceptability of vibration in different types of buildings. See Table 6.7. Human sensitivity to vibration in buildings is such that a doubling of the vibration dose values of the second column of Table 6.7 will result in the possibility of adverse comments from occupants and a further doubling will mean that adverse comments will become probable.

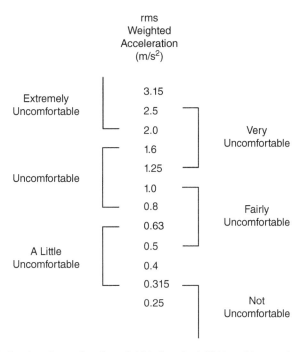

Figure 6.20 Scale of vibration discomfort from British Standard 6841 and International Standard 2631 [82].

Table 6.7 Vibration dose values ($ms^{-7/4}$) at which various degrees of adverse comment may be expected in buildings [82].

Place	Low Probability of Adverse Comment	Adverse Comment Possible	Adverse Comment Probable
Critical working areas	0.1	0.2	0.4
Residential	0.2–0.4	0.4–0.8	0.8–1.6
Office	0.4	0.8	1.6
Workshops	0.8	1.6	3.2

Example 6.13 There is concern that residents in a building may be disturbed by vibration from three types of event on a nearby construction site. The vibration dose values for each type of event were measured having individual *VDV*s of 0.06, 0.04, and 0.05 $ms^{-7/4}$ and they are repeated a number of 4, 20, and 10 events during the day, respectively. Assess the likely effects on the residents of the building.

Solution

First, we need the individual vibration dose values of each event using Eq. (5.8):

$$VDV_1 = 0.06 \times 4^{0.25} = 0.085 \, \text{ms}^{-7/4},$$

$$VDV_2 = 0.04 \times 20^{0.25} = 0.085 \, \text{ms}^{-7/4},$$

$$VDV_3 = 0.05 \times 10^{0.25} = 0.089 \, \text{ms}^{-7/4}.$$

The total vibration dose value is then obtained using Eq. (5.8) as

$$VDV = \left[(0.085)^4 + (0.085)^4 + (0.089)^4\right]^{0.25} = 0.114 \, \text{ms}^{-7/4}.$$

Therefore, since the total *VDV* is below the lower limit in Table 6.7 for a residential building, there is a low probability of adverse comments from the occupants.

6.17.2 Effect of Vibration on Buildings

Many studies have been carried out to define threshold limits for the occurrence of vibration-induced damage to buildings. Interestingly, building response to vibration is somewhere between the constant velocity and constant acceleration contours according to several studies. One of the earliest published by Monk [83] in 1971 covers the frequency range 1–100 Hz and vibration displacement amplitudes in the range 1–1000 μm (see Figure 6.21). Generally, vibration will be insufficient to cause cracks in plaster finishes or in walls until

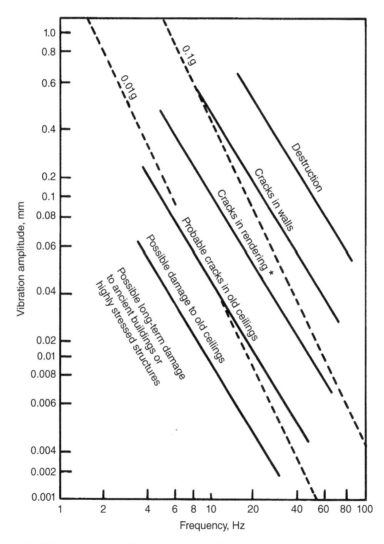

Figure 6.21 Response of building in good condition to vibration (*Rendering is plaster put on stone or brick walls). (*Source:* From Monk [83].)

it becomes unpleasant or painful. Vibration which is only just perceptible to people is unlikely to cause any damage to buildings unless they are very old or highly stressed. In such cases the damage normally appears first as cracks in old ceilings.

Continuous sources of vibration may produce resonances in buildings. Typical values of building natural frequencies are between 0.1 and 10 Hz for very tall and low buildings, respectively. Individual components of the building such as floors, ceilings, windows, etc. also have their own natural frequencies which can be excited at resonance by a number of vibration sources.

In general, vibration-induced damage depends on the characteristics of the vibration, the type of source, the type of building, and the type of soil. It has been identified that man-made sources of ground-borne vibration are in the frequency range 1–150 Hz while natural sources such as wind and earthquakes have vibration energy at frequencies below 0.1 Hz [84].

Although vibration can be expressed in terms of displacement and acceleration, it is common practice to express limits for vibration on buildings in terms of peak vibration velocity. A guide for measurement of vibrations and evaluation of their effects on buildings has been published as an international standard ISO 4866 [85] and an identical British Standard BS 7385 [86–87]. In these standards, vibration-induced damage to buildings has been classified into three categories: cosmetic, minor and major. Cosmetic damage is produced in the range from 15 mm/s at 4 Hz to 50 mm/s at frequencies greater than 40 Hz for measurements taken at the base of the building. Empirical evidence suggests that, below 12.5 mm/s peak vibration velocity, no damage is produced to a building.

Some other guidelines have been presented for assessing the effect of vibration and shock from different sources on structures. The interested reader is referred to these documents [88–90] for more information on the subject.

References

1 Moore, B.C.J., Glasberg, B.R., and Baer, T. (1997). A model for the prediction of thresholds, loudness and partial loudness. *J. Audio Eng. Soc.* 45: 224–240.

2 Moore, B.C.J. (2003). *An Introduction to the Psychology of Hearing*, 5e. San Diego, CA: Academic.

3 Glasberg, B.R. and Moore, B.C.J. (2006). Prediction of absolute thresholds and equal-loudness contours using a modified loudness model (L). *J. Acoust. Soc. Am.* 120 (2): 585–588.

4 ANSI/ASA S3.4–2007 (2007) *Procedure for the Computation of Loudness of Steady Sounds*. New York: American National Standards Institute. (reviewed in 2012).

5 ISO 226 (1987). *Acoustics – Normal Equal-Loudness Contours*. Geneva: International Standards Organization.

6 Robinson, D.W. and Dadson, R.S. (1956). A re-determination of the equal-loudness relations for pure tones. *Br. J. Appl. Phys.* 7: 166–181.

7 ISO 389-7 (2005) *Acoustics – Reference Zero for the Calibration of Audiometric Equipment. Part 7: Reference Threshold of Hearing under Free-Field and Diffuse-Field Listening Conditions*. Geneva: International Standards Organization (reviewed in 2013).

8 ISO 226 (2003) *Acoustics – Normal Equal-Loudness Contours*. Geneva: International Standards Organization (reviewed in 2014).

9 ISO 3891 (1978) *Procedure for Describing Aircraft Noise Heard on the Ground*. Geneva: International Standards Organization (withdrawn in 2010).

10 May, D.N. (1978). Basic subjective responses to noise. In: *Handbook of Noise Assessment* (ed. D.N. May), 3–38. New York: Van Nostrand Reinhold.

11 Harris, C.M. (ed.) (1991). *Handbook of Acoustical Measurements and Noise Control*, 3e. New York: McGraw-Hill.

12 Anderson, J.S. and Bratos-Anderson, M. (1993). *Noise, Its Measurement, Analysis, Rating, and Control*. Aldershot, UK: Avebury Technical.

13 Crocker, M.J. (2007). Rating measures, descriptors, criteria, and procedures for determining human response to noise. In: *Handbook of Noise and Vibration Control* (ed. M.J. Crocker), 394–413. New York: Wiley.

14 Federal Aviation Regulation, FAR Part 36 (2006) *Noise Standards: Aircraft Type and Airworthiness Certification*, 7 February. Federal Aviation Authority.

15 ICA0 *Annex 16 to the Convention on International Civil Aviation, Environmental Protection* (Vol. I and II). International Civil Aviation Organization.

16 French, N.R. and Steinberg, J.C. (1947). Factors governing the intelligibility of speech sounds. *J. Acoust. Soc. Am.* 19: 90–119.

17 Beranek, L.L. and Ver, I.L. (1992). *Noise and Vibration Control Engineering*. New York: Wiley.

18 ANSI S3.5–1969 (R1976) (1976). *Methods for the Calculation of the Articulation Index.* New York: American National Standards Institute.

19 ANSI S3.5–1997 (R2012) (2012). *Methods for the Calculation of the Speech Intelligibility Index.* New York: Acoustical Society of America.

20 Webster, J.C. (1983) Communicating in noise in 1978–1983. *Proceedings of the 4th International Congress on Noise as a Public Health Problem*, Torino, Italy (21–25 June 1983) (ed. E. Rossi), 411–424.

21 Beranek, L.L. (1957). Revised criteria for noise control in buildings. *Noise Control* 3: 19–27.

22 Beranek, L.L., Blazier, W.E., and Figiver, J.J. (1971). *J. Acoust. Soc. Am.* 50: 1223–1228.

23 Beranek, L.L. (1989). Balanced noise criterion (NCB) curves. *J. Acoust. Soc. Am.* 86: 650–664.

24 Beranek, L.L. (1989). Applications of NCB noise criterion curves. *Noise Control Eng. J.* 33: 45–56.

25 Blazier, W.E. (1981). Revised noise criteria for application in the acoustical design and rating of HVAC systems. *Noise Control Eng. J.* 16: 64–73.

26 Kosten, C.W. and van Os, G.J. (1962) Community reaction criteria for external noise. *The Control of Noise, NPL Symposium No. 12*. London: HMSO, 373–382.

27 EPA (1971). *Community Noise*, Report No. NTID 330.3. Washington, DC: U.S. Environmental Protection Agency.

28 Eldred, K.M. (1974). Assessment of community noise. *Noise Control Eng. J.* 3 (2): 88–95.

29 Gottlob, D. (1995). Regulations for community noise. *Noise/News Int.* 3 (4): 223–236.

30 EPA (1973). *Public Health and Welfare Criteria for Noise*, Report No. 550/9-73-002. Washington, DC: U.S. Environmental Protection Agency.

31 EPA (1974). *Information on Levels of Environmental Noise Requisite to Protect Public Health and Welfare with an Adequate Margin of Safety*, Report No. 550/9-74-004. Washington, DC: U.S. Environmental Protection Agency.

32 Shaw, E.A.G. (1996). Noise environments outdoors and the effects of community noise exposure. *Noise Control Eng. J.* 44 (3): 109–119.

33 Federal Interagency Committee on Urban Noise (1980). *Guidelines for Considering Noise in Land Use Planning and Control*, Document 1981-338-006/8071. Washington, DC: U.S. Government Printing Office.

34 Scholes, W.E. and Vulkan, G.H. (1969). A note on the objective measurement of road traffic noise. *Appl. Acoust.* 2: 185–197.

35 Schultz, T.J. (1982). *Community Noise Ratings*, 2e. London: Applied Science.

36 Rosenblith, W.A., Stevens, K.N., and the staff of Bolt, Beranek, and Newman Inc (1953). *Handbook of Acoustic Noise Control, Vol 2, Noise and Man*, WADC TR-52-204, Wright Air Development Center, 181–200. Ohio: Wright Patterson Air Force Base.

37 Stevens, K.N., Rosenblith, W.A., and Bolt, R.H. (1955). A community's reaction to noise: can it be forecast? *Noise Control* 1: 63–71.

38 Galloway, W.J. and Bishop, D.E. (1969) *Noise Exposure Forecasts: Evolution, Evaluation, Extensions, and Land Use Interpretations*. BBN Report No. 1862 for the FAA/DOT Office of Noise Abatement. Washington, DC.

39 Stevens, K.N., Pientrasanta, A.C., and the staff of Bolt, Beranek, and Newman Inc. (1957). *Procedures for Estimating Noise Exposure and Resulting Community Reactions from Air Base Operations*, WADC TN-57-10, Wright Air Development Center. Ohio: Wright Patterson Air Force Base.

40 Bolt, B., and Newman Inc. (1964) *Land Use Planning Relating to Aircraft Noise*. FAA Technical Report, October; also issued as Report No. AFM86–5, TM-5-365, NAVDOCKS P-38. Washington, DC: U.S. Department of Defense.

41 Bishop, D.E. and Horonjeff, R.D. (1967). *Procedures for Developing Noise Exposure Forecast Areas for Aircraft Flight Operations*, FAA Report DS-67-10. Washington, DC: Department of Transportation.

42 Bishop, D.E. (1974) *Community Noise Exposure Resulting from Aircraft Operations: Application Guide for Predictive Procedure*. AMRL-TR-73-105. Canoga Park, CA: Bolt Beranek and Newman Inc.

43 Committee on the Problem of Noise (1963). *Noise – Final Report*. London: HMSO.

44 MIL Research Ltd (1971). *Second Survey of Aircraft Noise Annoyance Around London Heathrow Airport*. London: HMSO.

45 Griffiths, I.D. and Langdon, F.J. (1968). Subjective response to road traffic noise. *J. Sound Vib.* 8 (1): 16–33.

46 Schultz, T.J. (1972). Some sources of error in community noise measurement. *J. Sound Vib.* 6 (2): 18–27.

47 Department of the Environment (1973) Motorway Noise and Dwellings. Digest 153. Garston, Watford, UK: Building Research Establishment.

48 Department of Transport (1988). *Calculation of Road Traffic Noise (CoRTN)*. London: HMSO.

49 Abbott, P.G. and Nelson, P.M. (1989). The revision of calculation of road traffic noise. *Acoust. Bull. (Inst. Acoust.)* 14 (1): 4–9.

50 Robinson, D.W. (1969) The Concept of Noise Pollution Level, NPL Aero Report AC 38. Teddington, Middlesex, UK: National Physical Laboratory, Aerodynamics Division.

51 Pearsons, K.S. (1966) *The Effects of Duration and Background Level on Perceived Noisiness*. Report FAA ADS-78. Washington, DC.

52 Bolt, Beranek, and Newman, Inc. (1974) *Establishment of Standards for Highway Noise Levels (Final Report)*. Vol. 5, Prepared for Transportation Research Board, National Cooperative Highway Research Program, National Academy of Sciences, NCHRP 3–7/3. Washington, DC.

53 ISO 1996-3:1987 (2016) *Description, Measurement and Assessment of Environmental Noise – Part 1: Basic Quantities and Assessment Procedures, ISO 1996-1:2016; Part 2: Determination of Environmental Noise Levels, ISO 1996-2: 2007; Part 3: Application to Noise Limits*. Geneva: International Standards Organization (withdrawn in 2007).

54 Noise Standards for California Airports, California Administrative Code, Title 4, Subchapter 6.

55 Goldstein, J. (1979). Descriptors of auditory magnitude and methods of rating community noise. In: *Community Noise* (eds. R.J. Peppin and C.W. Rodman), 38–72. Philadelphia: ASTM.

56 Lucas, J.S. (1975). Noise and sleep: a literature review and a proposed criterion for assessing effect. *J. Acoust. Soc. Am.* 58 (6): 1232–1242.

57 Griefahn, B. (1980) Research on noise-disturbed sleep since 1973, *Proceedings of the Third International Congress on Noise as a Public Health Problem*. ASHA Report No. 10.

58 Jones, C.J. and Ollerhead, J.B. (1992) Aircraft noise and sleep disturbance: a field study. *Proceedings of EURO-NOISE'92* (ed. R. Lawrence), 119–127. London.

59 Ollerhead, J.B., Horne, J., Pankhurst, F. et al. (1992) Report of a Field Study of Aircraft Noise and Sleep Disturbance. London: Civil Aviation Authority.

60 Pearsons, K.S., Barber, D.S., Tabachnik, B.G., and Fidell, S. (1995). Predicting noise-induced sleep disturbance. *J. Acoust. Soc. Am.* 97: 331–338.

61 Pearsons, K.S., Barber, D.S., and Tabachnik, B.G. (1989) Analyses of the Predictability of Noise-induced Sleep Disturbance. Technical Report HSD-TR-89-029. Brooks Air Force Base, TX: Human Science Division (HSDY/YAH U.S. Air Force Systems Command).

62 Finegold, L.S., Harris, C.S., and von Gierke, H.E. (1994). Community annoyance and sleep disturbance: updated criteria for assessment of the impacts of general transportation noise on people. *Noise Control Eng. J.* 42 (1): 25–30.

63 Schultz, T.J. (1978). Synthesis of social surveys on noise annoyance. *J. Acoust. Soc. Am.* 64: 377–405.

64 Fidell, S., Barber, D.S., and Schultz, T.J. (1991). Updating a dosage effect relationship for the prevalence of annoyance due to general transportation noise. *J. Acoust. Soc. Am.* 89: 221–233.

65 Miedema, H.M.E. and Vos, H. (1998). Exposure response functions for transportation noise. *J. Acoust. Soc. Am.* 104: 3432–3445.

66 Von Gierke, H.E. and Eldred, K. (1993). Effects of noise on people. *Noise/News Int.* 1 (2): 67–89.

67 Passchier-Vermeer, W. (1993). *Noise and Health*. Publication No. A93/02. The Hague: Health Council of Netherlands.

68 Berglund, B., Lindvall, T., and Schwela, D.H. (1999). *Guidelines for Community Noise*. Geneva, Switzerland: World Health Organization.

69 Kryter, K.D. (1982). Community annoyance from aircraft and ground vehicle noise. *J. Acoust. Soc. Am.* 72 (4): 1222–1242.

70 Schultz, T.J. (1982). Comments on K. D. Kryter's paper, community annoyance from aircraft and ground vehicle noise. *J. Acoust. Soc. Am.* 72 (4): 1243–1252.

71 Kryter, K.D. (1982). Rebuttal by Karl D. Kryter to comments by T. J. Schultz. *J. Acoust. Soc. Am.* 72 (4): 1253–1257.

72 Mohler, U. (1988). Community response to railway noise: a review of social surveys. *J. Sound Vib.* 120: 321–332.

73 Lang, J. (1989) Schallimmission von Schienenverkerhstrecken. Forschungsarbeiten aus dem Verkehrswesen 23. Vienna.

74 British Government Planning Policy Guidance PPG (1994). *Planning and Noise*. London: HM.

75 Private communication with Rupert Thornely-Taylor, April 29, 1996.

76 Directive 2002/49/EC (2002) Directive 2002/49/EC of the European Parliament and of the Council of 25 June 2002 Relating to Management of Environmental Noise.

77 IEC 60268-16 (2011) *Sound System Equipment – Part 16, Objective Rating of Speech Intelligibility by Speech Transmission Index*. Geneva: International Electrotechnical Commission.

78 Griffin, M.J. (1996). *Handbook of Human Vibration*. London: Academic.

79 ISO2361 (2016) *Mechanical vibration and shock – Evaluation of human exposure to whole-body vibration, Part 1: General requirements*, ISO2631-1: 1997 (Amended in 2010); *Part 2: Vibration in buildings (1 Hz to 80 Hz)*, ISO2631-2: 2003 (Reviewed in 2013); *Part 4: Guidelines for the evaluation of the effects of vibration and rotational motion on passenger and crew comfort in fixed-guideway transport systems*, ISO2631-4: 2001 (Reviewed in 2016); *Part 5: Method for evaluation of vibration containing multiple shocks*, ISO2631-5: 2004. Geneva: International Standards Organization.

80 British Standard, BS 6841 (1987) *Measurement and Evaluation of Human Exposure to Whole-Body Mechanical Vibration and Repeated Shock*. London: British Standards Institution.

81 British Standard, BS 6472 (2008) *Guide to evaluation of human exposure to vibration in buildings. Part 1: Vibration sources other than blasting*, BS6472-1:2008; *Part 2: Blast-induced vibration*, BS 6472-2:2008. London: British Standards Institution.

82 Griffin, M.J. (2007). Effects of vibration on people. In: *Handbook of Noise and Vibration Control* (ed. M.J. Crocker), 343–353. New York: Wiley.

83 Monk, R.G. (1971). Mechanically induced vibration in buildings. *Environ. Eng.* 51: 912.

84 Smith, B.J., Peters, R.J., and Owen, S. (1996). *Acoustics and Noise Control*, 2e. Harlow: Addison Wesley Longman.

85 International Organization for Standardization (2010) *Mechanical vibration and shock – Vibration of fixed structures – Guidelines for the measurement of vibrations and evaluation of their effects on structures*. ISO 4866:2010. Geneva: International Organization for Standardization (reviewed in 2016).

86 British Standard, BS 7385-1 (1990) *Guide for Measurement of Vibration and Evaluation of their Effects in Buildings, Part 1*. London: British Standards Institution.

87 British Standard, BS 7385-2 (1993) *Guide to Damage Levels from Ground Borne Vibrations*, Part 2. London: British Standards Institution.

88 Jones & Stokes (2004) *Transportation- and construction-induced vibration guidance manual*. Sacramento, CA: California Department of Transportation, Noise, Vibration, and Hazardous Waste Management Office.

89 Hanson, C.E., Towers, D.A., and Meister, L.D. (2006). *Transit Noise and Vibration Impact Assessment*, FTA-VA-1003-9006. Springfield, VA: U.S. Department of Transportation Federal Transit Administration.

90 Siskind, D.E., Stagg, M.S., Koppand, J.W., and Dowding, C.H. (1980). *Structure Response and Damage Produced by Ground Vibration from Surface Mine Blasting*. Report of Investigation 8507. Pittsburgh, PA: U.S. Bureau of Mines.

7

Noise and Vibration Transducers, Signal Processing, Analysis, and Measurements

7.1 Introduction

In the measurement of noise and vibration fields, it is necessary to sense the sound or vibration disturbance with a transducer. The transducer converts some physical property of the sound and vibration field into an electrical signal. This signal is then amplified, attenuated, or transformed in some way so that it can be analyzed and/or processed to provide the data of particular interest. For some cases such as simple measurements of the A-weighted sound pressure level, only limited amounts of processing are needed. In other cases with more sophisticated measurements, special analysis and processing is required. Such examples include modal analysis, sound intensity, wavelet analysis, machinery condition monitoring, acoustical holography, and beamforming, with which quite complicated signal analysis and processing may be needed. In all cases considerable care should be taken to ensure that the transducers together with their measurement systems are calibrated and checked periodically to make sure they are working properly.

7.2 Typical Measurement Systems

It may be necessary to measure noise and/or vibration for various reasons. Before beginning any measurement program, the objectives should be defined. For instance, it may be desired to measure noise to determine if a noise problem exists, whether the noise output of a machine is within its specifications, to determine the main sources of noise on a machine or in a vehicle or building. In the case of vibration, the reasons for measurement can include determining whether vibration of structures may result in unwanted sound generation, or in the case of intense vibration, machine wear and condition, and even the danger of structural fatigue and failure.

Since about 1920, most sound and vibration measurement systems have made extensive use of electrical networks. From that time, the electrical amplification of signals has made possible several measurement techniques that were previously impossible to use. When measurements are made of noise or vibration, it is usually necessary to combine several different types of instruments into one measurement system. The individual components of the measuring system utilized will depend on the particular measurements needed. A generalized system is shown in Figure 7.1. To design a system to make useful measurements, it is desirable for one to have a good understanding of the phenomena being investigated and to have a reasonable understanding of the functioning of the instrumentation and signal processing.

The first item in any noise or vibration-measuring system is the *transducer*. As its name implies, this instrument converts a signal in one physical form into another; that is, a transducer converts a sound pressure signal or a vibration signal into an electrical signal. Normally, the electrical signal obtained from a transducer

Engineering Acoustics: Noise and Vibration Control, First Edition. Malcolm J. Crocker and Jorge P. Arenas.
© 2021 John Wiley & Sons Ltd. Published 2021 by John Wiley & Sons Ltd.

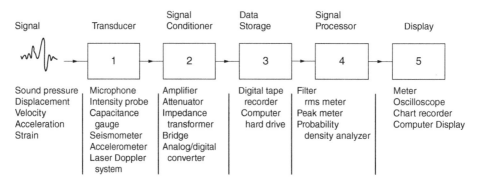

Figure 7.1 Idealized noise or vibration-measuring system.

is not suitable for direct analysis or read-out, and a *signal conditioner* is then used to amplify, attenuate, or transform the signal using analog-to-digital (A/D) conversion. It is optional at this stage to include a *data storage* item in the system before the signal is passed to a signal processor (See Figure 7.1.)

The signal processor may consist of a narrow- or wide-band filter, a root mean square (rms) detector, a probability density analyzer, and the like. The last item in the system is usually the *display unit*, a read-out unit, or a digital computer, which is used to perform some post-processing of the signal. Analog and/or digital meters, oscilloscopes, and other units can all be used to display signals. Also, a data storage item or data distribution system can be included in the main measurement system at this stage instead of earlier, so that the data can be analyzed or shared with others later.

7.3 Transducers

The basis of all noise and vibration measurement systems is the transducer. The microphone is the main transducer used to measure sound, while the accelerometer is the main transducer used to measure vibration. Specialized transducer systems have been developed to measure sound intensity in air and vibration intensity of structural systems. In addition, measurement procedures have been formulated to determine the modes of vibration of structures. Special rooms and systems such as anechoic and reverberant rooms and impedance tubes [1] are now widely used for sound power and noise source identification measurements of machines and the measurement of the acoustical impedance of materials. With any measurement, calibration of the system is essential to obtain reliable results that can be compared with results obtained by others.

Transducers and their associated measuring systems generally suffer from two major shortcomings:

1) A transducer will normally respond to other variables in addition to its response to the variable of interest. For example, a microphone, although being most sensitive to sound pressure, may also be slightly sensitive to variations in temperature, humidity, magnetic fields, and vibration.

2) It is difficult, and in many cases not possible, to introduce a transducer into the measurement medium without disturbing the medium in some way. The transducer will extract some energy from the medium or structural system. In addition, other disturbances will be caused. An accelerometer will add mass to the system and alter the structure's vibration. A pitot tube or hot-wire anemometer will disturb the flow. A microphone will reflect, diffract, and refract the incident sound wave. Some noncontacting vibration transducers and systems will not directly interfere with the vibration of the system to be measured, but they will interfere with any associated sound field generated by the vibrating body.

7.3.1 Transducer Characteristics

An ideal sound or vibration transducer should have the following characteristics [1]:

1) It should cause negligible diffraction of the sound field or structural vibration field (i.e. its dimensions should be small compared with the smallest sound or vibration wavelength of interest).
2) It should have a high acoustic or mechanical (driving point) impedance compared with the fluid medium or structure so that little energy is extracted from the field.
3) It should have low electrical noise.
4) Its output should be independent of temperature, humidity, magnetic fields, static pressure, and wind velocity, and it should be rugged and stable with time.
5) Its sensitivity should be independent of sound pressure or vibration level magnitudes.
6) Its frequency response should be flat.
7) It should introduce a zero phase shift between the sound pressure or structural vibration and the electrical output signal.

No transducer can meet all of the above criteria, and thus different types of transducers and vibration sensors are preferred for different measurements. The microphone is by far the most common form of acoustical transducer, and the piezoelectric accelerometer is the most widely used vibration transducer. Because of their importance the next section of this chapter concerns these devices. But it should be noted that other specialized noise and vibration transducers are used for measurements. For instance, sound intensity probes of different designs can be used for noise source identification and sound power measurements of a source in situ [2, 3], and several other types of vibration-measuring transducers such as strain gauges and laser Doppler interferometer systems are also in use [4].

7.3.2 Sensitivity

An ideal microphone (or accelerometer) together with its measurement system should have an output voltage amplitude E that is proportional to the exciting pressure amplitude p (or acceleration amplitude a) (see Figure 7.2). The ratio of open-circuit output voltage to input pressure (or acceleration) is normally called the sensitivity M_p:

$$M_p = E/p \text{ or } M_p = E/a. \qquad (7.1)$$

The transducer sensitivity [V/(N/m^2) or V/(m/s^2)] depends on the microphone (or accelerometer) design. (Different types are discussed in Refs. [2, 4] and later in Sections 7.4.1, 7.5.1, and 7.5.2 of this chapter).

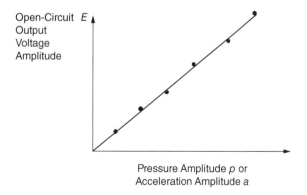

Figure 7.2 Sensitivity of (i) an ideal microphone or accelerometer —— and (ii) an actual transducer ●.

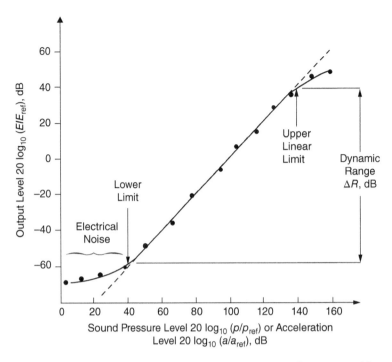

Figure 7.3 Sensitivity of an ideal microphone or accelerometer showing upper and lower limits.

In the case of noise measurements with very low sound pressure amplitudes, electrical noise will exceed the voltage signal generated by the microphone, and this will govern the low signal amplitude limit for measurements. With very high sound pressure amplitudes, the diaphragm displacement may become so large that the voltage generated is no longer proportional to the displacement. Nonlinearity then sets the upper amplitude use limit [2]. For sound pressures somewhat above the nonlinear limit, physical damage to the microphone can occur.

The situation is similar for vibration measurements made with seismic mass accelerometers. For very high vibration amplitudes, the accelerometer displacement may become so large that the voltage generated is no longer proportional to the displacement. Nonlinearity again sets an upper use limit. For displacements above the nonlinear limit, physical damage to the accelerometer can occur. Large accelerometers are normally more sensitive than small ones, and this is advantageous in many applications; but large heavy accelerometers cause more mass loading problems to start to occur at a lower frequency than small lightweight accelerometers [5].

If Figure 7.2 is plotted on a logarithmic scale, Figure 7.3 is obtained. The reference voltage E_{ref} is normally taken as 1 V, while the reference sound pressure p_{ref} is usually taken as 20 μPa, and the reference acceleration is normally taken as 1 μm/s^2.

A microphone or accelerometer has a usable range of operation between its upper and lower amplitude limits. Thus, at any frequency the microphone or accelerometer response magnitude is normally given by subtracting the x- from the y- axis in Figure 7.3, and the range ΔR in which the microphone or accelerometer response R is constant is known as the dynamic range:

$$R = 20 \log \left(\frac{E/p}{E_{ref}/p_{ref}} \right) = 20 \log \left(\frac{E/E_{ref}}{p/p_{ref}} \right) \text{ dB} \qquad (7.2a)$$

or

$$R = 20 \log \left(\frac{E/a}{E_{ref}/a_{ref}} \right) = 20 \log \left(\frac{E/E_{ref}}{a/a_{ref}} \right) \text{ dB.} \qquad (7.2b)$$

Example 7.1   A 1/2 in. microphone has a sensitivity of 12.5 mV/Pa. Calculate the output voltage amplitude produced by a sound pressure level of 74 dB.

Solution

The sound pressure of 74 dB is $p = p_{ref} \times 10^{L_p/20} = 2{\times}10^{-5} \times 10^{74/20} = 0.1 \text{ N/m}^2$.
Since $1 \text{ Pa} = 1 \text{ N/m}^2$, from Eq. (7.1) we obtain the output voltage as

$$E = M_p \times p = 12.5 \times 10^{-3} \times 0.1 = 1.25 \times 10^{-3} = 1.25 \text{ mV.}$$

Example 7.2   A 1-in. microphone has a sensitivity of 50 mV/Pa. What is the sound pressure level that produces an output voltage amplitude of 1 mV?

Solution

From Eq. (7.1): $p = E/M_p = 1/50 = 0.02 \text{ N/m}^2$. Then, the corresponding sound pressure level is

$$L_p = 20 \log \left(p/p_{ref} \right) = 20 \log \left(0.02/(2 \times 10^{-5}) \right) = 60 \text{ dB.}$$

Example 7.3   A sound level meter (SLM) is calibrated to read in decibels (dB) (ref: 20 μPa) when used with a microphone of sensitivity 50 mV/Pa. The microphone is then replaced by an accelerometer, with a sensitivity of 2 mV/m/s^2, attached to a vibrating surface. If the reading is 78 dB, what is the rms acceleration of the surface?

Solution

First we determine the equivalent voltage of 78 dB. If the level were a sound pressure level, $p = 2 \times 10^{-5} \times 10^{78/20} = 0.16 \text{ N/m}^2$, so the output voltage is $E = M_p{\times}p = 0.05 \times 0.16 = 8 \text{ mV}$. Now, considering the sensitivity of the accelerometer, the corresponding rms acceleration is

$$a = E/M_a = 8/2 = 4 \text{ m/s}^2.$$

7.3.3   Dynamic Range

It is seen for the microphone (or accelerometer) shown in Figure 7.3 that the dynamic range is about 100 dB. Most good-quality microphones have a dynamic range of about 100–120 dB (interestingly enough about the same as the human ear) [2]. As the microphone diaphragm diameter (or accelerometer mass) is increased, the transducer sensitivity is normally increased as well so that electrical noise is less of a problem and the lower signal amplitude limit for measurements is decreased. In the case of a microphone, however, a larger diaphragm diameter usually results in a larger deflection for a given sound pressure and a reduced upper sound pressure level limit because of nonlinearity problems. Thus between the upper and lower amplitude limits, the microphone has a usable range of operation.

Small-diameter microphones are not very sensitive but can be used for high-amplitude sound pressures without distortion; their electronic noise floor is quite high, however. (Different types are discussed briefly in this chapter. The fundamentals of the operation of the main types of microphone are described in more detail in Ref. [2].) Large diameter microphones are normally more sensitive and can be used for lower-level noise than small-diameter microphones. Their noise floor is lower, but larger-diameter microphones experience more diffraction problems at low frequencies than small-diameter microphones [5]. Figure 7.4 shows the inherent noise floor plotted against upper limiting frequency for four microphone diameters [6]. Because of the dynamic range problems, the small-diameter microphones cannot be used for very "quiet" sounds, and the large-diameter microphones cannot be used for intense noise.

Figure 7.5 shows the dynamic range for four commercially available microphones. The lower level limit is given in terms of the A-weighted sound pressure level of the internal noise floor of the microphone and associated preamplifier. The upper level limit is set by the sound pressure level at which 3% dynamic distortion occurs. It is observed that the 1-in. and 1/2-in. diameter microphones have the greatest dynamic ranges of about 150 dB. The 1/4-in. microphone has a dynamic range of about 140 dB, while the 1/8-in. diameter microphone has a dynamic range of only about 100 dB [6].

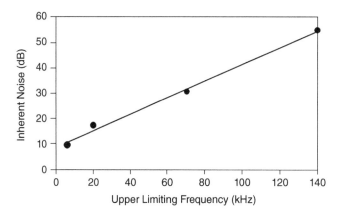

Figure 7.4 Inherent noise floor against upper limiting sound pressure level for four different diameter microphones. The four dots represent the four sizes of microphone (1, 1/2, 1/4, and 1/8 in.) in order from 1 in. (bottom left) to 1/8 in. (top right) [6].

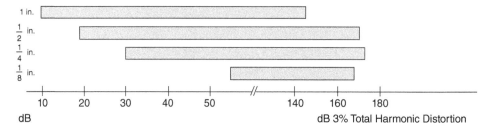

Figure 7.5 Comparison of the dynamic ranges of the same four condenser microphones as shown in Figure 7.4. The x-axis is A-weighted sound pressure level. The upper limit is given in decibels at which 3% total harmonic distortion occurs [6].

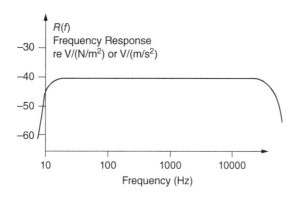

Figure 7.6 Frequency response of an ideal microphone or accelerometer.

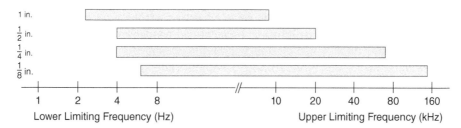

Figure 7.7 Comparison of the frequency response ranges of four different diameter condenser microphones (lower limiting frequency [Hz] and upper limiting frequency [kHz]. Microphone diameter: 1/8, 1/4, 1/2, and 1 in.) [6].

7.3.4 Frequency Response

The *magnitude* of the *frequency response*, $R(f)$, of an ideal microphone is given by Eq. (7.2a) as the frequency of the sound pressure p is changed. The magnitude of the frequency response, $R(f)$, or an ideal accelerometer is given by Eq. (7.2b) as the frequency of the vibration a is changed. Figure 7.6 shows the frequency response of an ideal microphone used for noise measurements or an ideal accelerometer used for vibration measurements. In practice, microphones and accelerometers can only approach the ideal frequency response in Figure 7.6. Resonance peaks of the diaphragm of the microphone or the inertial mass of the accelerometer are usually observed in the high-frequency range. These peaks are normally suppressed by the addition of damping or some other means at the transducer design stage. Note that for some measurements (e.g. explosive blasts with a microphone and shock events with an accelerometer) knowledge of the phase frequency response of the transducer is also important. For such measurements of impulsive phenomena, the additional requirements of zero or linear phase shift with frequency are needed for microphones and accelerometers, respectively.

Figure 7.7 shows the frequency response range of four different diameter condenser microphones [6]. It is observed that the smaller diameter microphones have the largest usable frequency range. The 1/8-in. diameter microphone has a usable range from about 5 Hz to about 150 kHz. The 1-in. microphone on the other hand has a much more restricted usable frequency range from about 2 Hz to only about 10 kHz.

7.4 Noise Measurements

Since the human hearing range extends from about 20 to 20 000 Hz, it is desirable that the frequency response of microphones and noise measurement systems should be "flat" between these limits as shown in Figure 7.6. For certain types of measurements (e.g. for measurement of sonic booms or explosive blasts), it may be necessary to measure sounds that contain frequencies lower than about 20 Hz. For scale-model studies or for the measurement of noise environments or structures subject to fatigue, it may be necessary to measure to frequencies higher than 20 000 Hz. We will first discuss the different types of microphones used in noise measurements and their acoustical properties and then methods by which microphones are calibrated. More details about microphones are given in Ref. [2] and of their calibration in Ref. [7]. Chapter 57 in Ref. [8] also has detailed information about microphone calibration.

7.4.1 Types of Microphones for Noise Measurements

An acoustical transducer is a device that converts some property of a sound field into an electrical signal. The most common device is the microphone designed to measure sound pressure. Some transducers have been designed, however, to measure sound particle velocity, sound pressure gradient, and sound intensity.

Microphones may be divided into three main classes: communication, studio, and measurement microphones [2]. The discussion here mainly concerns noise measurement microphones. There are three main types of microphone used for noise measurements: (i) polarized condenser microphones, (ii) prepolarized condenser microphones (sometimes called electret microphones), and (iii) piezoelectric microphones. The polarized condenser microphone possesses a thin diaphragm under tension and must be provided with an external polarizing voltage that is applied between the diaphragm and the backplate. On the other hand, the prepolarized condenser (electret) microphone avoids the need for a polarizing voltage by the provision of a thin layer of electrically charged material, which is normally deposited on the backplate during manufacture.

Condenser microphones, because of their stability and well-defined mechanical impedance, are the ones mostly preferred for noise measurements. They do have drawbacks of fragility and sensitivity to humidity, however. Piezoelectric microphones are more robust than condenser microphones. They possess a stiff diaphragm that is coupled to a piezoelectric crystal or ceramic element.

Microphones may be divided into directional and nondirectional (or omnidirectional) types. In some cases, directional microphones may be useful, such as in the localization of noise sources. Most noise measurements are made with nominally omnidirectional microphones, although even these types become somewhat directional at high frequency, at which the dimensions of the microphone become comparable with the acoustic wavelength.

Microphones may be further subdivided into three main types: (i) free-field, (ii) pressure-field, and (iii) diffuse-field microphones. The frequency responses of these microphones are adjusted so that they have an essentially flat frequency response when placed in these different sound fields. Today TEDS (transducer electronic data sheet) microphones with built-in sensitivity are also available. These microphones have built-in information about the response required in different sound fields. So such TEDS microphones can be used in all of the three sound fields described above if they are used in conjunction with information loaded into the operating system of the associated analyzer. Reference [2] describes the design and principles of operation of the main types of microphones and furthermore explains how the microphone parameters, physical properties, and design may be chosen to obtain the required microphone sensitivity, frequency response, and dynamic range. Reference [1] also addresses some of the same considerations with respect to noise measurements.

a) Condenser Microphones

Because of its uniform sensitivity, low distortion, and portability, the invention of the condenser microphone in 1917 by E.C. Wente revolutionized electroacoustics, and it soon became an integral part of any high-quality sound system. However, although the polarized condenser microphone has significant advantages,

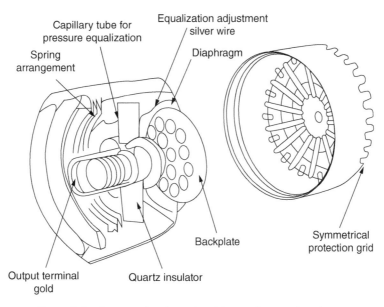

Figure 7.8 Cross-section through a 1-in. condenser microphone.

it also suffers from some disadvantages: specifically its rather low sensitivity, high internal impedance, and the need for a polarizing voltage. The very large, almost distortionless electronic signal amplification that can now be achieved has made the lack of sensitivity of the condenser microphone unimportant, and, because of its smooth sensitivity over a very wide frequency range and its well-defined geometry, the condenser microphone is still preferred for use in many noise measurement applications [9].

Figure 7.8 shows a diagram of a 1-in. condenser microphone. The condenser microphone consists of a thin metal diaphragm stretched under tension and spaced a short distance from an insulated backplate. The diaphragm and backplate constitute the electrodes of the condenser. Holes are drilled in the backplate to provide the required air damping for the diaphragm. A small hole is provided to equalize the static pressure across the diaphragm, provided that it changes slowly. This hole usually determines the lower cutoff frequency of the microphone. Some discussion of the electrical circuit of the system and the theory, design, and construction of condenser microphones is given in Ref. [2] and in Chapters 110 and 112 of the *Handbook of Acoustics* [8]. Other authors have also discussed the electrical theory in some detail [10–12]. It is normally necessary to remove the protection grid during microphone calibration [7].

b) Prepolarized (Electret) Microphones

In the 1970s, a new type of precision microphone became commercially available. This microphone is basically a condenser microphone; however, no direct current (dc) polarization voltage is needed since the electret's foil diaphragm and/or the backplate is permanently polarized during manufacture. The polarization voltage is created by embedding and aligning static electrical charges into a thin layer of material, which is deposited on the microphone diaphragm or backplate.

The first electret microphones used diaphragms made from an insulating material that carried the permanent electrical charge. The diaphragms of electret microphones made in this way have to be quite heavy in order to carry the permanent electrical charge material. Heavy diaphragms have several disadvantages and result in a low resonance frequency peak. Most high-quality electret microphones used for noise studies now have the permanently charged material attached to the stationary backplate instead of the diaphragm. In this

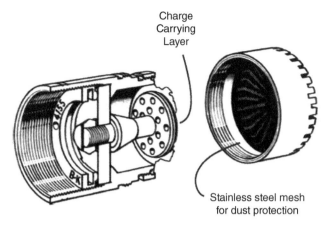

Charge
Carrying
Layer

Stainless steel mesh
for dust protection

Figure 7.9 Electret microphone using a thin electret polymer layer deposited on the perforated backplate.

way, much thinner diaphragms can be used, made of the same metal-coated plastic material used in condenser microphones. Such microphones have a high resonance frequency peak and an overall performance almost rivaling the best condenser microphone. The thin diaphragm and the perforated backplate comprise the two plates of the condenser. Preamplifiers are still needed, and in some recent electret microphones miniature preamplifiers are built into the microphones themselves. See Ref. [2] and Chapters 110 and 112 of the *Handbook of Acoustics* [8]. Figure 7.9 shows a cross-sectional diagram of a typical electret microphone [13].

The electret microphone has the following advantages: (i) no polarization voltage needed, (ii) rugged construction, (iii) large capacitance (about 500 pF), and hence, loading is a lesser problem than with the condenser microphone, and (iv) low cost [11, 14, 15].

c) Piezoelectric Microphones

The design and construction of piezoelectric transducers is discussed in Ref. [2] and in Chapters 110 and 112 in Ref. [8] and in varying detail by several other authors [16, 17].

Piezoelectric crystals may be cut and used in many different orientations. If a slice is cut from a piezoelectric crystal and pressures applied to the opposite faces of the slice causing a deformation, then equal and opposite charges are produced on the opposite faces of the slice with an electric potential developed between the faces. Crystals are often directly exposed to liquids (e.g. as hydrophones) where the high mechanical impedance of the liquid is not a disadvantage. In gases, the large acoustical/mechanical impedance mismatch is a disadvantage, and piezoelectric materials are usually used in conjunction with a diaphragm [2].

Figure 7.10 shows a cross-sectional view of a commercial piezoelectric microphone. Much thicker diaphragms are normally used with piezoelectric than condenser microphones (usually about 50 times greater). This inevitably leads to a lower resonance frequency for the diaphragm (assuming the density of the diaphragm material is the same). To obtain a flat free-field response, it is necessary to damp this resonance overcritically. Hence the upper frequency response is poorer than with a condenser microphone because the mass-controlled region is entered at a lower frequency.

The diaphragm is connected to a ceramic bender element. A bimorph simply supported beam bender element is most often used, although sometimes cantilever benders are used [16, 17]. The force needed to produce a voltage from a crystal or ceramic slice in pure compression is quite large. However, if a thin bar or beam is cut from a crystal in a suitable orientation, a voltage is produced across the beam as it is bent [5].

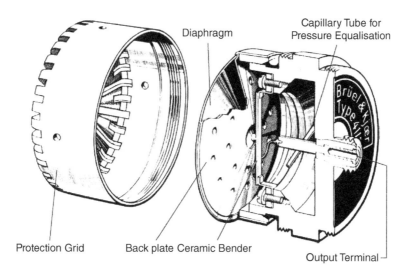

Figure 7.10 Cross-sectional view of piezoelectric microphone.

Metal foil is usually cemented to the outside surfaces of the crystal, and the two foils with the crystal in between form a condenser of the solid dielectric type. It is usual to use two bars in conjunction to produce a *bimorph*. Electrical connections may be applied in two ways to the bimorph to produce either parallel or series connections [5].

Other microphones are used for many communication purposes but not normally for precision noise measurements. The moving coil microphone has been in use for many years but is less valuable in noise work since its frequency response is not very smooth and is poor at high frequencies. Also, its sensitivity is low. However, it does have some advantages that make it ideal for some applications [2].

7.4.2 Directivity

At low frequencies, for example, below about 1000 Hz, the frequency response of a microphone is independent of the angle of incidence of the sound waves. However, at higher frequencies, as the microphone dimensions and the wavelength of the sound become comparable, *diffraction* effects become important and the frequency response of a microphone is strongly dependent on the angle of incidence of the sound (see Figure 7.11). The effect of *directivity* is discussed in detail in Ref. [2] and Chapter 3 in Ref. [5].

Microphones can be designed to have a flat frequency response when exposed to a free progressive wave sound field. They are usually known as free-field microphones. Other microphones can be designed to have a flat frequency response to grazing incidence sound waves and are normally known as pressure-field microphones. Still other microphones are designed to have a flat frequency response to random incidence sound and are known as diffuse-field incidence microphones [1, 2]. Care must be made in ensuring that the right type is used. Figure 7.12 illustrates the use of different types.

7.4.3 Transducer Calibration

It is important to calibrate transducers used for sound, shock, and vibration to ensure the accuracy of measurements made with them. Proper calibration also ensures that the results measured with the transducers are comparable with the results measured by others. The accuracy of the calibration must

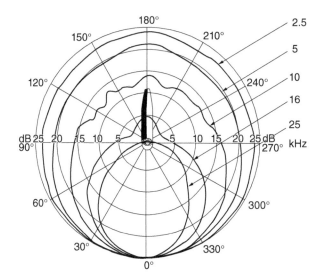

Figure 7.11 Directivity of a microphone with a protection grid at different frequencies.

(a)

(b)

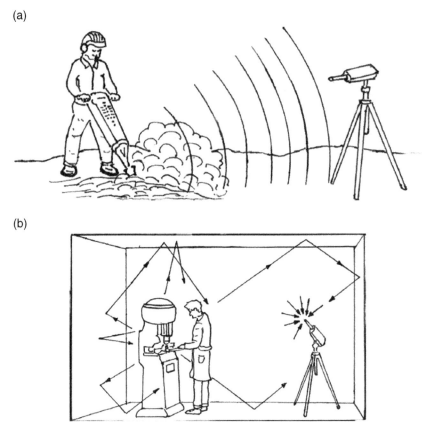

Figure 7.12 Noise measurements using (a) free-field microphone and (b) diffuse-field microphone.

also be known. The transducer system and the calibration method used must perform the calibration with a known accuracy. If these conditions are met, then the calibration is termed as traceable. It should be noted, however, that this traceability is not an indication necessarily of highly accurate measurements, but simply that if the uncertainty is known, the measurements can then be compared with valid measurements made by others, which have also been made using proper calibration procedures. Thus measurements made with transducers and calibration procedures are of no use unless the associated uncertainty in measurement is also known. New and upgraded international standards now provide a variety of calibration methods together with their suitability for ensuring traceability.

Transducers should be calibrated on site before and after each measurement or set of measurements and periodically at service centers or national metrology laboratories. Reference [7] describes calibration procedures for microphones and sound intensity systems. Reference [18] discusses the calibration of shock and vibration transducers. References [7, 18] also describe the traceability of sound and vibration measurements, and Ref. [19] describes the traceability of shock and vibration measurements to national and international standards in detail.

Microphones, calibrators, and SLMs all need to be calibrated. The most common requirement is the calibration of microphones for their sound pressure response, but calibrations for sound intensity, sound power, and acoustic impedance measurements are also commonly needed [7]. Nedelitsky in Chapter 108 of Ref. [8] and others have also discussed the problems of calibrating microphones.

When a microphone is introduced into a sound field, it causes undesirable reflections and hence changes the sound field. This effect is more pronounced at high frequency. Ideally we should like to measure the sound pressure in the unperturbed or "free field," but this is difficult. It is extremely important that manufacturers and users calibrate microphones accurately. Unfortunately, some users do not calibrate microphones sufficiently frequently to ensure the accuracy of sound pressure level readings. As already discussed, microphones may be designed to have three flat frequency responses. TEDS microphones can be used to obtain all of the three flat frequency responses. The two main types of microphones in use are the *free-field* type (primarily used for outdoors measurements) and the *diffuse-field* type (mainly used for indoors measurements). Pressure-field microphones are mostly used for coupler measurements, except for the recently developed flat surface microphone.

As regards calibration of microphones, the pressure response may be considered to be the response of the microphone to a uniform pressure applied over its diaphragm. This is the response of an ideal microphone of *zero size* introduced into a free progressive plane wave field. However, when an actual finite size microphone is introduced into such a field, reflection, and diffraction are caused, which give a different microphone response called *the free-field* response [2]. Because the microphone can be oriented at any arbitrary angle in the plane wave field, perhaps the pressure response is more fundamental. There are several ways of measuring the pressure response of a microphone: [20, 21] (i) pistonphone, (ii) driven-diaphragm-type calibrator, (iii) electrostatic actuator, (iv) reciprocity method, and (v) substitution method. The only absolute method of calibration is the reciprocity method. This is somewhat complicated and time-consuming; and the other four methods, although not giving absolute calibration, are more convenient and normally sufficiently accurate. See Ref. [7] and Chapter 108 in Ref. [8].

a) Pistonphones

The pistonphone is a very accurate, reliable, and simple device for calibrating a microphone that is convenient for use in the field. The principle of operation is quite simple. A small battery-powered electric motor drives a shaft on which is mounted a cam disk. The cam disk drives two pistons symmetrically. The cam gives the pistons a sinusoidal motion at four times the shaft rotational speed. The stroke of the pistons (or peak amplitude from mean position) is thus one quarter of the difference in maximum and minimum diameters of the cam. The pistons vary the cavity volume sinusoidally in time and since for pistonphones a low frequency (e.g. 250 Hz) is normally chosen (for mechanical reasons), a corresponding sinusoidal variation in pressure occurs. (See Ref. [7] and Chapter 3 in Ref. [5] for the theory and further details of its operation.)

Provided the piston stroke is carefully controlled, the sound pressure level produced can be accurately predicted. Good sealing should be maintained when the microphone is fitted into the coupler opening. In principle, this type of pistonphone can be used to calibrate any microphone, provided that good sealing is maintained. If microphones of different diameters are to be calibrated with the pistonphone, and different coupler connections are used to keep the volume unchanged, the sound pressure level is unchanged. If a different volume results, thc changc in sound pressure level can be predicted. (See Ref. [5] for the theory.) A change in atmospheric pressure will alter the calibration; but a correction is simply made by the use of a barometer provided by the manufacturer.

b) Driven-Diaphragm Calibrators

Somewhat simpler, lower cost calibrators are also available that work on the driven-diaphragm principle. Some commercially available types produce 114 dB at 1000 Hz. A stabilized 1000-Hz oscillator feeds a piezoelectric driven element that vibrates the metallic diaphragm creating a pressure in the front coupling. The diaphragm is driven by an oscillator powered by a 9-V battery. Some calibrators generate five selected frequencies: 125, 250, 500, 1000, and 2000 Hz at 94 dB. With such calibrators, corrections should normally be made for changes in ambient temperature and pressure (altitude). Newer types of calibrator have a built-in microphone that measures and controls the sound pressure generated in the coupler cavity via a feedback loop [7].

c) Sound Intensity Probe Calibration

When two phase-matched microphones are used together to form a sound intensity probe, then the two-microphone intensity probe can be calibrated by fitting an intensity coupler attached to the pistonphone [7]. The coupler normally consists of two chambers (upper and lower) connected by a coupling element. When the pistonphone is connected to the coupler, a phase difference is created between the two sound pressures generated in the upper and lower chambers. The sound pressure amplitudes are the same, however, in both chambers. Thus the propagation of a plane progressive sound wave in a free field is simulated. If one microphone of the intensity pair is fitted to the upper chamber and the other is fitted to the lower chamber, then the simulated sound wave produced by the sound in the two chambers can be used for the calibration of the intensity probe for the measurement of both sound intensity and particle velocity. See Chapter 8 and text discussing Figures 8.21 and 8.25.

The intensity coupler and pistonphone can also be used for the sound pressure sensitivity calibration of the two microphones. For this measurement, the two microphones are connected to the upper chamber of the coupler. They are then automatically exposed to the same sound pressure, and the amplitude and phase difference between the two microphones can be measured and checked to see that it is within acceptable limits. If a sound source, which can generate broadband sound, is attached to the coupler, then the pressure residual-intensity index spectrum can be measured as well. This index is used to determine the accuracy of the sound intensity measurements made with the probe [3, 7, 22]. See Chapter 8.

7.5 Vibration Measurements

A vibration sensor is a device that converts some property of the vibration of a structure into an electrical signal. Conversely, a vibration generator works on the opposite principle of converting an electrical signal into a mechanical vibration. Both vibration sensors and generators may be termed transducers, since they convert one physical variable into another [4].

Vibration sensors can be made to work using several different principles and to measure surface displacement, velocity, acceleration, and strain. They may be arbitrarily designated as contacting or noncontacting devices. Contacting sensors are often convenient to use since they can measure vibration at a specific location on a structure. They do have the disadvantage, however, that they can change the vibration of the structure by adding mass, stiffness and damping. This is particularly a problem if the structure is lightweight.

There are three quantities of most interest in vibration studies [5]. These are *displacement*, *velocity*, and *acceleration*. A fourth quantity, *strain*, is also frequently measured. In the early 1900s, most vibration measurements were made using *mechanical or optical devices*. Such devices are still used satisfactorily for low-frequency measurements (a few hertz). With the advent of electronics in the 1920s, transducers that converted mechanical into electrical signals were developed. Before about 1960, displacement and particularly velocity-sensitive transducers were utilized, especially when higher frequency measurements were needed. However, since that time acceleration sensitive transducers (accelerometers) have become preferred. The reason for this is mainly because excellent lightweight accelerometers were developed to measure very high frequency vibrations (5000 Hz or more) in aircraft and spacecraft. Most velocity-sensitive transducers have an upper limiting frequency of about 1000 Hz, while piezoelectric accelerometers can be made to have an upper limiting frequency of 40 000 Hz or more [5].

For many measurements it is unimportant whether displacement, velocity, or acceleration is measured. For simple harmonic motion, the amplitudes of these three quantities are simply related to each other by multiplying or dividing by the angular frequency ω. Even if the vibration is random in nature, if frequency filtration is used, this principle can still be applied. Alternatively, the conversion from acceleration to velocity and displacement may be made by using electronic integration.

7.5.1 Principle of Seismic Mass Transducers

Modern-day piezoelectric accelerometers and some of the earlier displacement transducers [23] work on the same principle: A seismic mass m is supported on a spring of stiffness k and the whole is enclosed in a case (see Figure 7.13). The damping constant R in most applications is small, and it is neglected for simplicity in the following analysis (although it could easily be included).

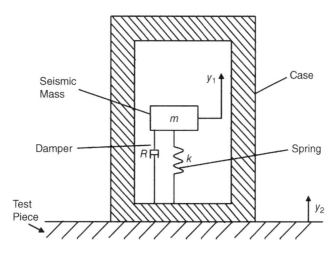

Figure 7.13 Idealized diagram of vibration transducer.

If the vibrating member (test piece) is undergoing a time-dependent displacement, $y_2(t)$, and the mass experiences a time-dependent displacement, $y_1(t)$, then considering the forces acting on the mass m,

$$k(y_2 - y_1) = m\ddot{y}_1. \tag{7.3}$$

The situation is identical to that studied in Chapter 2, except that there is no external force applied to the mass, and instead the test piece receives a displacement input, $y_2(t)$. The easiest variable to measure is the relative displacement $y_2 - y_1$. We will assume the test piece experiences simple harmonic motion:

$$y_2 = \boldsymbol{B}e^{j\omega t}, \tag{7.4}$$

and, consequently in steady-state conditions, the mass experiences simple harmonic motion, y_1, at the same angular frequency ω:

$$y_1 = \boldsymbol{A}e^{j\omega t}. \tag{7.5}$$

Note that $\boldsymbol{A}$ and $\boldsymbol{B}$ are written as complex quantities because y_1 and y_2 will not generally be in phase (see Section 2.4.1). From Eqs. (7.3)–(7.5) we obtain (noting that $j^2 = -1$)

$$k(\boldsymbol{B} - \boldsymbol{A})e^{j\omega t} = -m\omega^2 \boldsymbol{A}e^{j\omega t}, \text{and} \tag{7.6}$$

$$\boldsymbol{B} - \boldsymbol{A} = -\left(m\omega^2/k\right)\boldsymbol{A}. \tag{7.7}$$

Rearranging Eq. (7.7) gives

$$\boldsymbol{A} = -\left(\frac{k}{m\omega^2 - k}\right)\boldsymbol{B}, \tag{7.8}$$

and substituting Eqs. (7.8) into (7.7) gives

$$\boldsymbol{B} - \boldsymbol{A} = \frac{m\omega^2}{m\omega^2 - k}\boldsymbol{B}. \tag{7.9}$$

It is easy to see how this result can be used to design displacement, velocity, or acceleration transducers.

Example 7.4 An accelerometer placed on an object vibrating sinusoidally measures an acceleration amplitude value of $4\,\text{m/s}^2$. If the vibration frequency is known to be 50 Hz, calculate the corresponding velocity and displacement amplitudes.

Solution

We know that $\omega = 2\pi f = 2\pi(50) = 314.16\,\text{rad/s}$.

Since the displacement for harmonic motion is $y(t) = Ae^{j\omega t}$, where A is the displacement amplitude, the velocity is:

$$\dot{y}(t) = j\omega Ae^{j\omega t} = j\omega y(t), \text{and the acceleration is } \ddot{y}(t) = -\omega^2 y(t).$$

Therefore, we can see that the velocity amplitude and the acceleration amplitude are, respectively, $v = \omega A$ and $a = \omega^2 A$. Thus,

$$v = a/\omega = 4/314.16 = 12.7\,\text{mm/s}, \text{and}$$

$$A = a/\omega = 4/(314.16)^2 = 40.5\,\mu\text{m}.$$

a) Seismic Mass Displacement and Velocity Transducers

Seismic mass displacement and velocity transducers were widely used until the 1950s. They required a large seismic mass and a soft spring and were operated well above the system's natural frequency (between about 0.5 and 3 Hz) to obtain a flat frequency response curve. They were heavy and only suitable for use on heavy machinery [24]. The soft spring gave a low resonance frequency between about 0.5 and 3 Hz, but the upper frequency limit was normally about 100 Hz due to limitations in the mechanical linkages used to measure the motion of the mass relative to the system's case. There are now several other displacement and velocity transducers available that do not suffer from this very limited frequency range. For example, the laser Doppler system can be used to measure dynamic displacement and velocity with a very high degree of precision.

b) Seismic Mass Acceleration Transducers

To measure acceleration conveniently with the seismic mass, the system shown in Figure 7.13 must be used as follows. The ratio of relative *displacement* amplitude $|B - A|$ to test piece acceleration amplitude $\omega^2|B|$ is from Eq. (7.9)

$$\frac{|B - A|}{\omega^2|B|} = \frac{1}{|\omega^2 - \omega_n^2|} = \frac{1/\omega_n^2}{\left|1 - \left(\omega^2/\omega_n^2\right)\right|}. \tag{7.10}$$

Except for the constant $(1/\omega_n^2)$ in the numerator, the right-hand side of Eq. (7.10) is identical to that of Eq. (2.20) in Chapter 2 with $\delta = 0$. If we had included damping in our model, the result would have again been identical to Eq. (2.20) in Chapter 2, except for the constant. Thus, we can make use of the results of Figure 2.9 in Chapter 2. This figure shows that we should like to operate the instrument in the stiffness-controlled region, where $\omega/\omega_n \ll 1$.

Again this is obvious from Eq. (7.10), since, if $\omega/\omega_n^2 \ll 1$, the right-hand side becomes a constant (equal to $1/\omega_n^2$). To obtain a flat frequency response over a wide frequency range, ω_n should be made large. But note that in this case the sensitivity decreases, since it is approximately proportional to $1/\omega_n^2$. To obtain a large value of the frequency ω_n, it is necessary to use a very stiff spring and a small mass, which is opposite to what is needed for seismic mass displacement and velocity transducers discussed above [5].

Example 7.5 The spring-mass system of Figure 7.13 with $m = 0.5$ kg and $k = 10\,000$ N/m, with negligible damping, is used as a vibration sensor. When mounted on a machine vibrating with an amplitude of 4 mm, the total displacement of the mass of the sensor is observed to be 12 mm. Find the frequency of the vibrating machine.

Solution

We find that $|B| = 4$ mm and $|A| = 12$ mm. Then, the relative displacement is $|B - A| = |4\text{--}12| = 8$ mm. From Eq. (7.10) we write that:

$$|B - A| = \frac{|B|\left(\omega^2/\omega_n^2\right)}{\left|1 - \left(\omega^2/\omega_n^2\right)\right|}, \text{ i.e. } 8 = \frac{4\left(\omega^2/\omega_n^2\right)}{\left|1 - \left(\omega^2/\omega_n^2\right)\right|}. \text{ Solving the equation gives } \frac{\omega}{\omega_n} = 0.8165.$$

Now, $\omega_n = \sqrt{k/m} = \sqrt{10000/0.5} = 141.42$ rad/s. Therefore,

$$\omega = 0.8165 \times 141.42 = 115.5 \text{ rad/s, i.e. } f = 18.38 \text{ Hz}.$$

Example 7.6 The vibration transducer of Figure 7.13 is used to measure the vibration of an engine whose operating speed range is from 500 to 3000 rpm. The vibration consists of two harmonics. The amplitude distortion must be less than 3%. Find the natural frequency of the transducer if the damping is negligible.

Solution

We select the accelerometer on the basis of the lowest frequency being measured (500 rpm). From Eq. (7.10) and considering 3% of error, we have that

$\dfrac{|B - A|}{|B|} = \dfrac{\omega^2/\omega_n^2}{\left|1 - (\omega^2/\omega_n^2)\right|} = 1.03$. Solving the equation gives $\dfrac{\omega^2}{\omega_n^2} = \dfrac{1.03}{0.03} = 34.33$. Therefore, $\dfrac{\omega}{\omega_n} = \dfrac{2\pi(500)}{60\ \omega_n} = \sqrt{34.33}$. We find that $\omega_n = 8.94$ rad/s.

Thus, the natural frequency of the accelerometer must be $f_n = 1.42$ Hz.

It is possible to measure the relative displacement by attaching a strain gauge to the spring and calibrating the instrument by the change in resistance produced. This is the principle of the *strain gauge* accelerometer, which is normally used with a Wheatstone bridge. Such accelerometers will usually be found to have an upper frequency limit of about 1000 or 2000 Hz. Provided that the accelerometer is dc coupled, it is easy to calibrate the accelerometer by inverting it. If the strain gauge accelerometer is placed horizontally, there is no signal; placed vertically in one direction it gives a constant resistance change equivalent to +1 g and placed in the opposite vertical position the change is equivalent to −1 g. If the accelerometer is alternating current (ac) coupled, such calibration is not possible. However, the most commonly used type of accelerometer is now the *piezoelectric accelerometer* [25, 26].

7.5.2 Piezoelectric Accelerometers

The piezoelectric accelerometer is a seismic mass type of accelerometer that works on the principle described above. The main types include the following: compression, delta shear, planar shear, theta shear, annular shear, and ortho shear. These are all designed to accentuate certain advantages that are useful in different environmental conditions and are described in more detail in Ref. [4]. As already discussed, the heavier accelerometers have the greatest sensitivity but the lowest resonance frequency and vice versa.

Several characteristics of piezoelectric accelerometers need to be considered. The piezoelectric accelerometer produces a charge after deformation. After the signal is passed through a charge converter (incorporated in modern piezoelectric accelerometers), it can be considered as a voltage source. Hence, the *sensitivity* is given in mV/m/s². Besides being sensitive to acceleration in the longitudinal axis, the accelerometer is also slightly sensitive to vibration in the transverse axis due to irregularities in construction and alignment of the piezoelectric disks. (See Figure 7.14.)

Good accelerometers should have a *lateral sensitivity* less than 5% of the longitudinal (or main axis) sensitivity. The transverse sensitivity will vary in the base plane having maximum and minimum values in certain directions. The sensitivity of an accelerometer will be somewhat temperature dependent (one should always consult the manufacturer's instructions), however, provided the maximum (Curie point) temperature is not exceeded, the piezoelectric material will not be damaged and will retain its properties.

Piezoelectric accelerometers produce some output signal when subjected to acoustic signals or base strains. Normally the *acoustic sensitivity* is low (producing a false response output of less than 1 g for a sound pressure

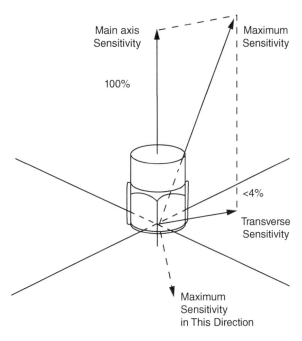

Figure 7.14 Transverse and longitudinal sensitivity of accelerometer.

level input of 160 dB). As the test object vibrates, it will induce strain in the accelerometer base with a consequent output signal. Most accelerometer bases are made thick and rigid to reduce this effect.

As was discussed above (also see Ref. [4]), all accelerometers exhibit a fundamental resonance frequency. The frequency range of an accelerometer is usually assumed to be bounded by an upper frequency limit of one third of the resonance frequency for less than 1-dB error or one fifth the resonance frequency for less than 0.5-dB error. This assumes that the accelerometer design has low damping. The upper frequency limit is extended in some accelerometer designs by using high damping. Today TEDS accelerometers are also available with built-in amplitude response information, so that the upper frequency limit is increased by about 50% to about one half of the resonance frequency.

The lower limiting frequency depends on the type of preamplifier used to follow the accelerometer. Two types may be used and are commercially available. If the preamplifier is designed so that the output voltage is proportional to the input voltage, it is called a voltage amplifier. If the output voltage is proportional to the input charge, it is called a charge amplifier. When a voltage amplifier is used, the accelerometer output voltage is very sensitive to cable capacitance. This is because typically the capacitance of a piezoelectric accelerometer is several hundred picofarads (usually between 100 and 1000 pF). This is somewhere between the very low capacitance of a condenser microphone and the higher capacitance of a piezoelectric microphone (see Section 7.4.1). The effect of cable capacitance on the voltage sensitivity (in mV/m/s^2) of an accelerometer is determined as: [25]

$$M_V = M_{V0} \frac{C_A}{C_A + C_C + C_P},$$ (7.11)

where M_V is the voltage sensitivity, M_{V0} is the open circuit (unloaded) accelerometer voltage sensitivity, and C_A, C_C, and C_P are the accelerometer, cable, and preamplifier capacitances, respectively.

Example 7.7 While taking a vibration measurement with cable A, a requirement has arisen for a longer cable B. Knowing that the capacitance of cable A is 110 pF, the capacitance of the accelerometer (including cable A) is 1117 pF, the capacitance of cable B is 260 pF, and the open circuit voltage sensitivity of the accelerometer is 9.73 mV/m/s^2, calculate the new voltage sensitivity due to the new cable.

Solution

We assume that the preamplifier capacitance can be neglected. The capacitance of the piezoelectric element alone is $C_A = 1117 - 110 = 1007$ pF.

Therefore, using Eq. (7.11) we obtain

$$M_V = 9.73 \times 10^{-3} \frac{1007 \times 10^{-12}}{1007 \times 10^{-12} + 260 \times 10^{-12}} = 7.73 \text{ mV/ms}^{-2}.$$

Therefore, the voltage sensitivity has been reduced simply by changing the cable.

When a charge amplifier is used, the output voltage is not sensitive to changes in cable length. Broch [27] describes the reasons for this in some detail. If a voltage amplifier is used, the lower limiting frequency (3-dB down point) is given by

$$f_\ell = \frac{1}{2\pi RC}, \tag{7.12}$$

where R is the input resistance of the voltage preamplifier and $C = C_A + C_C + C_P$ is the effective circuit capacitance. It should be noted that making the total capacitance as large as possible or designing a preamplifier with a high input resistance may ensure that the low frequency limit is low enough to provide useful operation at frequencies down to 1 Hz or less.

The dynamic range of the piezoelectric accelerometer is determined by the low frequency limit set by electrical noise in the system and the upper frequency limit governed by the preloading of the accelerometer and the mechanical strength of the piezoelectric element. Small accelerometers usually have a higher upper limit [4]. It should also be noted that accelerometers have a main direction of sensitivity. (See Figure 7.14.)

It is also possible to simultaneously measure acceleration in three orthogonal directions. This is the principle of the *triaxial* accelerometer. Each unit incorporates three separate built-in piezoelectric elements that are oriented at right angles with respect to each other. The output signals of the triaxial accelerometer, each representing the vibration for one of the three axes, are connected to a multichannel signal analyzer for processing. This type of accelerometer is particularly suited for application in building vibrations, modal testing in structures, and to assess occupational vibration exposure (see Section 5.10 in Chapter 5 of this book). They can be mounted in a seat pad which can be placed directly on the seat cushion, floor or fixed to the back of a seat to measure whole-body vibration exposure. Through some adaptors, the triaxial accelerometer is also commonly used to assess occupational exposures to hand-transmitted vibration in hand tool operators.

7.5.3 Measurement Difficulties

Piezoelectric accelerometers produce some output signal when subjected to acoustic signals or base strains. Normally, the *acoustic sensitivity* is low (producing an apparent acceleration of smaller than 1 g for a sound pressure level of 160 dB). As the test object vibrates, it will induce strain in the accelerometer base with a consequent output signal. Most accelerometer bases are made thick and rigid to reduce this effect. As was discussed in Section 7.5.2, all accelerometers exhibit a fundamental resonance frequency [4] and the

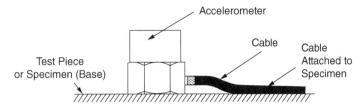

Figure 7.15 Mounting of an accelerometer to reduce cable whip noise.

upper frequency limit is extended in some accelerometer designs by using high damping. However, it should be noted that poor *mounting techniques* can have a marked effect on the frequency response above 2000 or 3000 Hz. Below this frequency, the mounting technique is not important as long as the accelerometer is fixed directly onto the specimen. The best technique is to attach the accelerometer to the specimen with a threaded steel stud. When this mounting method cannot be used, cement is also acceptable, and wax can be used with small to medium-sized accelerometers (up to 20–30 g) [25]. At low frequencies, magnets may be used. *Ground loops* must be avoided, in particular, when low acceleration levels are measured. With higher acceleration levels, ground loops can also cause problems. Measurement errors of 50% or more may be caused by ground loops with acceleration amplitudes of 100 m/s^2 or more. In such cases the accelerometer should be electrically isolated from the specimen and attached with a nonconducting stud or cemented to a mica washer, which is in turn cemented to the specimen. Cable "whip" can produce electrical noise, and cables should be attached to the specimen as shown in Figure 7.15.

In the case of the measurement of high-frequency vibration of thin metal plates, considerable care should be taken when using an accelerometer because of the *mass loading* effect. Well below its resonance frequency, the accelerometer can be assumed to act as a pure mass. The velocity of the test piece $\boldsymbol{u}_0$ can be assumed to be reduced to $\boldsymbol{u}_1$ by the addition of the accelerometer [28]

$$\frac{\boldsymbol{u}_1}{\boldsymbol{u}_0} = \frac{\boldsymbol{Z}}{\boldsymbol{Z} + j\omega m}, \tag{7.13}$$

where $\boldsymbol{Z}$ = mechanical impedance of test piece at the attachment point, ω = angular frequency, and m is the accelerometer mass. The input point impedance for a metal plate can be assumed to be

$$\boldsymbol{Z} = \frac{4}{\sqrt{3}} \rho c_L h^2, \tag{7.14}$$

where ρ = plate density, c_L = longitudinal wave speed in plate, and h = plate thickness. Substituting Eqs. (7.14) in (7.13) and assuming $|\boldsymbol{Z}| = |j\omega m|$, which gives a 3-dB reduction in velocity amplitude (from Eq. (7.13)), gives the frequency f_{3dB} at which the 3-dB reduction occurs as

$$f_{3dB} = \frac{2\rho c_L h^2}{\sqrt{3}\pi m}. \tag{7.15}$$

Equation (7.15) is plotted in Figure 7.16. This figure suggests that even lightweight accelerometers will "load" thin metal plates (e.g. a 10-g accelerometer will mass "load" a 3.2-mm [1/8-in.] aluminum plate at frequencies above about 4000 Hz). A general rule is to ensure that the accelerometer mass is less than one tenth of the mass of the structure [25].

To overcome such mass loading problems, various types of noncontacting gauges have been produced and still are widely used. Particular examples of where noncontacting gauges are invaluable are in the study of the vibration of thin sandwich composite panels and of loudspeaker cones. Examples of noncontacting gauges are given in Ref. [4].

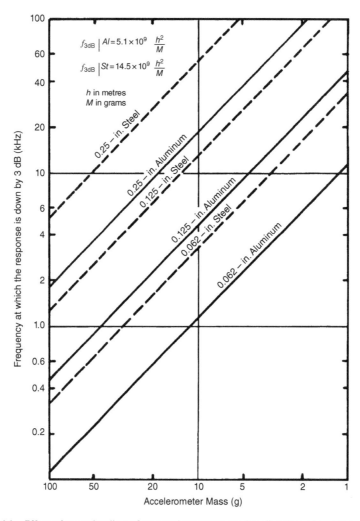

Figure 7.16 Effect of mass loading of an accelerometer on the vibration of a metal plate [28].

Example 7.8 An accelerometer is placed on an aluminum plate of thickness 3 mm. It is required that the frequency at which the response is down by 3-dB be below 9 kHz. Determine the appropriate mass of the accelerometer to comply with this requirement.

Solution

The values of density and longitudinal wave speed in aluminum are $\rho = 2700$ kg/m^3 and $c_L = 5200$ m/s, respectively. Using Eq. (7.15) we find the required mass

$$m = \frac{2\rho c_L h^2}{\sqrt{3}\pi \; f_{3dB}} = \frac{2(2700)(5200)(3 \times 10^{-3})^2}{\sqrt{3}\pi(9000)} = 5.16 \times 10^{-3} \text{ kg.}$$

Therefore, the mass of the accelerometer must be less than 5-g to obtain a good measure of the plate vibration.

7.5.4 Calibration, Metrology, and Traceability of Shock and Vibration Transducers

Vibration measurements of acceleration, velocity, and displacement can be classified in two main ways: (i) relative measurements made between two points with systems such as laser Doppler interferometers and (ii) absolute measurements made with seismic mass transducers (mainly accelerometers). The calibration of transducers used for both types of measurement is described in detail in Ref. [18]. Reference [18] also describes the traceability of vibration measurements and Ref. [19] describes the traceability of shock and vibration measurements to national and international standards in more detail. In many ways the traceability requirements and calibration concerns for shock and vibration transducers are similar to those of microphones.

7.6 Signal Analysis, Data Processing, and Specialized Noise and Vibration Measurements

7.6.1 Signal Analysis and Data Processing

Sound and vibration signals produced by transducers are not normally in a suitable form for the study of noise and vibration problems. Frequency analysis is the most common approach used in the solution of such problems. This is because the human ear acts in many ways like a frequency analyzer [29] and also because frequency analysis can often be used to reveal information that can be related to the operation of machines, in particular rotating machinery and to the properties and the behavior of structures [30].

Until the late 1970s, frequency analysis mostly was carried out with dedicated instruments that incorporated analog filters. Since that time, with the advent of the fast Fourier transform (FFT) algorithm, frequency analysis has been increasingly carried out using digital computers, either as part of a dedicated instrument or after appropriate conversion of the analog signal to digital form (known as A/D conversion). The analog signal is converted into a series of discrete values (also known as a time series). Dual and multichannel analyzers are also in common use for the parallel analysis of the signals from several transducers. Besides carrying out frequency analysis of signals, analyzers have been made to calculate various other functions including cross-spectra, coherence between signals, and the like (see Chapter 1). Reference [31] explains how measured noise and vibration data can be acquired, processed, shared, and stored using the power of modern digital computer systems. The A/D conversion process, digital data storage and retrieval, basic computations used in noise and vibration data analysis, and errors associated with the various calculations are summarized in Ref. [32]. Important considerations in signal processing include sampling rate, sampling interval, resolution, and data format. The analog data must be sampled so that at least two sample values per cycle of the highest frequency of interest in the analog signal are obtained. The format of the data is also important, and it must be ensured that the data produced are in a form that they can be read directly by any computer program written in a common programming language.

7.6.2 Sound Level Meters (SLMs) and Dosimeters

The SLM is an instrument designed to measure sound pressure over a frequency range and a sound pressure level range similar to the human ear. The basic SLM consists of a microphone, amplifier, frequency weighting circuit(s), detector averaging circuit, and an analog or digital read-out device [33–37]. Miniaturization of electrical circuits and components has allowed small lightweight SLMs to be built, which possess considerable sophistication and capabilities. Also, standards for SLMs written by successive national and international standards committees have resulted in different requirements and additional changes in SLM design and manufacture so that there is a wide variety of SLMs in use.

There are four main types of SLM: 0–3. Types 0–2 are the most accurate and type 3 the least. The tolerances for the different types are specified in these standards. The different SLM types are generally used for precision, laboratory, field, and survey measurements. Dosimeters are similar in design to SLMs, except that the time constant, threshold, and exponent circuits are omitted [23, 38]. Dosimeters are worn on the person and are normally designed to measure the so-called noise dose, which is a percentage of a criterion exposure, where exposure is measured in pascal-squared hours (see Chapter 5). The noise dose calculated depends on the exchange rate used. In the United States, the Occupational Safety and Health Administration (OSHA) requires a 5-dB exchange rate while the National Institute for Occupational Safety and Health (NIOSH) and most other countries use the 3-dB (equal energy) exchange rate. Reference [39] reviews the main aspects of SLM design and also briefly describes dosimeter design. Reference [40] reviews dosimeters in more detail.

7.6.3 Sound Power and Sound Intensity

Measurements of the sound power of machines are frequently needed. The European Union requires some machines to be provided with labels giving their sound power output. From knowledge of the sound power level, the sound pressure level can be calculated at a certain distance from a machine, either outside in free space or in a building, provided certain environmental conditions are known. Reference [41] describes several methods that can be used and that have been standardized both nationally and internationally. Also, measurement surfaces including the spherical and box surfaces are described. Reverberation time and comparison methods of measuring sound power are also reviewed in Ref. [41]. Sound intensity measurements described in Ref. [3] provide another method of determining the sound power output of sources, and this approach is particularly useful when the machine cannot be moved into a special facility such as an anechoic room or a reverberation room. In addition, the sound intensity approach can still be used to determine the sound power of a sound source when there is a hostile background noise present. Provided that the background noise is steady, good estimates of the sound power of a source can be made, even if the background noise level is very high. In such situations, the approaches described to obtain the sound power of sources in Ref. [41] normally fail. Sound intensity and sound power measurements are discussed in more detail in Chapter 8 of this book.

7.6.4 Modal Analysis

Modal analysis can be used to provide information concerning the natural frequencies, modal damping factors, and mode shapes of structures and machinery [42]. This information can be obtained either from mathematical analysis of a structure's dynamic response derived from a set of equations and knowledge of its mass and stiffness distributions or from experimental measurements of the structure's response to excitation. Experimental modal analysis (often simply known as modal testing) is discussed in detail in Ref. [43]. As explained in this chapter, however, modal testing cannot be divorced from mathematical models of the structure's dynamical behavior and the objective of modal testing in reality is an attempt to construct or validate a mathematical model of this behavior. Unfortunately, such mathematical models obtained through modal testing can never be quite complete because, in practice, the amount of experimental data obtained is inevitably limited. Usually, the mathematical model to be validated is a finite element model of the structure. Modal analysis is used for several main purposes. These include (i) monitoring the dynamic properties of a structure or machine to obtain early indications of structural deterioration or impending failure (often known as structural or machinery health monitoring), (ii) modification of a structure's mass or stiffness distributions so as to change natural frequencies to avoid high-amplitude resonant vibration, the resulting structural fatigue and the possibility of failure, and (iii) troubleshooting – the display of animated modes of vibration of the structure so that the dynamic behavior of certain critical areas of the structure can be understood, thus making possible practical solutions to vibration problems.

7.6.5 Condition Monitoring

Condition monitoring is discussed in Ref. [44]. Such monitoring of machinery may give sufficient advance warning of wear and possible imminent breakdown and failure so that replacement parts can be obtained in time to avoid costly machine downtime and/or loss of production. Costs savings can be considerable, for instance, in the case of downtime of a city electrical power plant, where unexpected downtime losses can exceed several million U.S. dollars each day. The savings in avoiding unnecessary downtime costs often vastly exceed the costs of replacement machine parts. In cases such as aircraft power plants, predictive maintenance can even prevent the possibility of catastrophic failure of the compressor blades and other components of the engines.

Reference [44] describes two main approaches to condition monitoring: (i) monitoring the relative displacement of a rotating machine shaft or bearing with a proximity probe and (ii) monitoring the vibration of the cover of a machine. Proximity probes are normally built into machines during manufacture; while, on the other hand, accelerometers can be placed on the cover of a machine at any time during service to monitor its vibration. Proximity probes are usually used to monitor changes in absolute shaft displacement and to measure so-called shaft orbits. Chapter 57 in the *Handbook of Acoustics* [8] also contains useful information concerning monitoring changes in shaft motion with the use of proximity probes. Reference [44] is mainly concerned with vibration measurements of machine casings made with accelerometers. If a machine is monitored in its original new condition, measurements made later may reveal changes indicating wear and the possibility of failure. Faults related to the frequency of the shaft speed include: misalignment, imbalance, and cracks in the drive shaft. It is usually difficult to distinguish between these faults. Useful information is given in Ref. [44] about the changes in the vibration signature, which manifest themselves with electrical machines, gears, bearings, and reciprocating machinery such as internal combustion engines, pumps, compressors, and the like.

7.6.6 Advanced Noise and Vibration Analysis and Measurement Techniques

Wavelets are starting to become of practical use in the solution of noise and vibration problems, machinery diagnostics, and health monitoring. Wavelet analysis is concerned with the decomposition of time signals into short waves or *wavelets*. Any waveform may be used for such decomposition, provided that it is localized at a particular time (or position). Reference [45] provides an in-depth discussion of the basics of wavelet analysis of signals, and in this chapter time is taken as the independent variable, although, in practice, any physical variable can be used. The Fourier coefficients of a signal are obtained by averaging over the full length of the signal, and the result is that no information is provided about how the frequency content of the signal may be changing with time.

In principle, the short-time Fourier transform (STFT) does provide this needed frequency time variation information, by dividing the time record of the signal into sections, each of which is analyzed separately. The frequency coefficients of the signal computed by the STFT depend on the length and time (or position) of the short record that the calculation process assumes is one period of an infinitely long periodic signal [32, 42, 46]. The difficulty remains, however, that infinitely long harmonic functions are being used to decompose a transient signal. Wavelets are short functions that avoid this difficulty. A set of wavelet functions is used as the basis for the decomposition of transient time-history signal records.

Near-field *acoustical holography* is a technique that can be used to reconstruct the frequency spectrum of any sound field descriptor (i.e. sound pressure, particle velocity, and sound intensity) at any location in space, normal surface velocity/displacement, directivity patterns, and the total sound power of a source. This is achieved by making a sound pressure measurement on a planar surface located near to the source's surface with either a scanning microphone (or hydrophone) or an array of microphones. If a scanning microphone is used, it should be robot-controlled. The procedure essentially creates a near-field hologram and relies on the measurement of the phase relationship between all of the sensor elements in the array that is used. If a scanning microphone is used instead, then the phase relationship is obtained by comparisons with a reference signal, which is kept stationary throughout the measurement. The approach can be used both with sources that are random in nature (such as fluid flow or boundary layer noise sources) and with deterministic sources (such as those containing pure tones like noise from a propeller). Although the approach is not simple, it eliminates the need for costly facilities and relies on the use of software instead, some of which is becoming commercially available. Reference [47] describes the theory behind this approach and also describes its use in practice to determine the in-flight sound pressure, normal surface velocity, and sound intensity on the interior fuselage surface and the floor of a passenger commuter aircraft.

More recently, a measuring technique called *beamforming* has been used to characterize complex noise sources. The theoretical fundamentals of the technique were developed in the area of antennas but have also been applied to sonar, telecommunications, seismology, and radioastronomy [48]. In acoustical applications, the noise from the source is detected in the far field by an array of standard microphones. The signals from the array are digitized, time-delayed, summed, and digital-signal processed with some beamforming algorithm (under the assumption of a certain source model) in order to produce a very directional sensor [49].

Beamformers have poor spatial resolution at low frequencies but are more accurate than near-field acoustical holography at higher frequencies. The frequency limits of the system are mainly dependent on the number of microphones, the size of the array, and the distance to the source. Several configurations of arrays can be used for this purpose, and examples include linear, random, star, spiral, circular, planar, and spherical arrays [49, 50]. The technique produces acoustical images that can be superimposed to pictures or video where sound pressure levels are represented by colors similar to a thermal camera. Depending on the capabilities of the processing system, the noise source analysis can be done as post-processing or in real-time, which is very useful in the study of moving sources. In this case, however, corrections due to the effect of Doppler shift must be incorporated [51].

Although the signal processing resources to implement beamforming in practice are not simple [52], this technique allows one to get very detailed information about the sound field, performing noise source identification and classification. Applications of beamforming to identification of noise sources include industrial, machinery, automotive, railway, wind turbine blade, and aircraft flying noise [49–51]. At present some of these systems are commercially available.

References

1 Valletta, P.R. and Crocker, M.J. (2007). Noise and vibration measurements. In: *Handbook of Noise and Vibration Control* (ed. M.J. Crocker), 501–525. New York: Wiley.

2 Rasmussen, G. and Rasmussen, P. (2007). Acoustical transducer principles and types of microphones. In: *Handbook of Noise and Vibration Control* (ed. M.J. Crocker), 435–443. New York: Wiley.

3 Jacobsen, F. (2007). Sound intensity measurements. In: *Handbook of Noise and Vibration Control* (ed. M.J. Crocker), 534–548. New York: Wiley.

4 Hansen, C.H. (2007). Vibration transducer principles and types of vibration transducers. In: *Handbook of Noise and Vibration Control* (ed. M.J. Crocker), 444–454. New York: Wiley.

5 Crocker, M.J. and Price, A.J. (1975). *Noise and Noise Control*, vol. I. Boca Raton, FL: CRC Press.

6 Bruel & Kjaer (2003) Catalog. Naerum, Denmark.

7 Frederiksen, E. (2007). Calibration of measurement microphones. In: *Handbook of Noise and Vibration Control* (ed. M.J. Crocker), 612–623. New York: Wiley.

8 Crocker, M.J. (ed.) (1998). *Handbook of Acoustics*. New York: Wiley.

9 Wong, G.S.K. and Embleton, T.F.W. (eds.) (1995). *AIP Handbook of Condenser Microphones. Theory, Calibration, and Measurements*. New York: American Institute of Physics.

10 Kleiner, M. (2013). *Electroacoustics*. Boca Raton, FL: CRC Press.

11 Beranek, L.L. and Mellow, T.J. (2012). *Acoustics: Sound Fields and Transducers*. Oxford: Elsevier.

12 Beranek, L.L. (1954). *Acoustics*. New York: McGraw-Hill; reprinted by the Acoustical Society of America, New York, 1996.

13 Sessler, G.M. and West, J.F. (1963). Electrostatic microphones with electret foil. *J. Acoust. Soc. Am.* 35: 1354–1357.

14 Hansen, K.S. (1969) *Details in the Construction of a Piezoelectric Microphone*, Bruel & Kjaer Tech. Rev. No. 1. Naerum, Denmark.

15 Sessler, G.M. and West, J.F. (1973). Electret transducers: a review. *J. Acoust. Soc. Am.* 53: 1589–1600.

16 Mason, W.P. (1950). *Piezoelectric Crystals and their Application to Ultrasonics*. New York: Van Nostrand.

17 Kinsler, L.E. and Frey, A.R. (1962). *Fundamentals of Acoustics*, 2e, 325. New York: Wiley.

18 Licht, T.R. (2007). Calibration of shock and vibration transducers. In: *Handbook of Noise and Vibration Control* (ed. M.J. Crocker), 624–632. New York: Wiley.

19 von Martens, H.-J. (2007). Metrology and traceability of vibration and shock measurements. In: *Handbook of Noise and Vibration Control* (ed. M.J. Crocker), 633–646. New York: Wiley.

20 Bruel, P.V., and Rasmussen, G. (1959) *Free Field Response of Condenser Microphones*, Bruel & Kjaer Tech. Rev. No. 1, and Rev. No. 2. Naerum, Denmark.

21 ANSI S1.10–1966 (2001) *American National Standard Method for the Calibration of Microphones*. New York: American National Standards Institute (revised in 2001).

22 Sandberg, U. and Ejsmont, J.A. (2007). Tire/road noise-generation, measurement, and abatement. In: *Handbook of Noise and Vibration Control* (ed. M.J. Crocker), 1054–1071. New York: Wiley.

23 ANSI S1.25-1991 (1991) Specification for Personal Noise Dosimeters. New York: Acoustical Society of America.

24 Myklestad, N.O. (1956). *Fundamentals of Vibration Analysis*, 101. New York: McGraw-Hill.

25 Serridge, M. and Licht, T.R. (1986). *Piezo-Electric Accelerometer and Vibration Preamplifier Handbook*. Naerum, Denmark: Bruel and Kjaer.

26 ISO 5348 1998) *Mechanical Vibration and Shock – Mechanical Mounting of Accelerometers*, Geneva: International Standards Organization.

27 Broch, J.T. (1984). *Mechanical Vibration and Shock Measurements*, 2e, 112–114. Naerum, Denmark: Bruel & Kjaer.

28 Starr, E.A. (1971). Sound and vibration transducers. In: *Noise and Vibration Control* (ed. L.L. Beranek). McGraw-Hill. See also Crocker, M.J. and Price, A.J. (1975) *Noise and Noise Control*. Vol. 1. Cleveland, OH: CRC Press.

29 Crocker, M.J. (2007). Psychoacoustics and product sound quality. In: *Handbook of Noise and Vibration Control* (ed. M.J. Crocker), 805–828. New York: Wiley.

30 Crocker, M.J. (2007). Machinery noise and vibration sources. In: *Handbook of Noise and Vibration Control* (ed. M.J. Crocker), 831–846. New York: Wiley.

31 Li, Z. and Crocker, M.J. (2007). Equipment for data acquisition. In: *Handbook of Noise and Vibration Control* (ed. M.J. Crocker), 486–492. New York: Wiley.

32 Piersol, A.G. (2007). Signal processing. In: *Handbook of Noise and Vibration Control* (ed. M.J. Crocker), 493–500. New York: Wiley.

33 IEC 60651-1979 (1979). *Sound Level Meters*. Geneva: International Electrotechnical Commission (IEC).

34 IEC 60804-2000 (2000). *Integrating-Averaging Sound Level Meters*. Geneva, Switzerland: International Electrotechnical Commission (IEC).

35 IEC 61672-1, 2013 (2013). *Electroacoustics – Sound Level Meters – Part 1: Specifications*. Geneva, Switzerland: International Electrotechnical Commission (IEC).

36 ANSI/ASA S1.4–2014 (2014) *American National Standard Electroacoustics – Sound Level Meters – Part 1: Specifications*. New York: American National Standards Institute.

37 ANSI/ASA S1.25–1991 (R2017) (1991) *Specification for Personal Noise Dosimeters*. New York: American National Standards Institute.

38 IEC 61252:1993+AMD1:2000+AMD2:2017 (2017). *Electroacoustics – Specifications for Personal Sound Exposure Meters*. Geneva, Switzerland: International Electrotechnical Commission (IEC).

39 Wong, G.S.K. (2007). Sound level meters. In: *Handbook of Noise and Vibration Control* (ed. M.J. Crocker), 455–464. New York: Wiley.

40 Kardous, C.A. (2007). Noise dosimeters. In: *Handbook of Noise and Vibration Control* (ed. M.J. Crocker), 465–469. New York: Wiley.

41 Jonasson, H.G. (2007). Determination of sound power level and emission sound pressure level. In: *Handbook of Noise and Vibration Control* (ed. M.J. Crocker), 526–533. New York: Wiley.

42 Ewins, D.J. (2000). *Modal Testing, Theory, Practice, and Application*, 2e. Philadelphia: Research Studies Press.

43 Ewins, D.J. (2007). Modal analysis and modal testing. In: *Handbook of Noise and Vibration Control* (ed. M.J. Crocker), 565–574. New York: Wiley.

44 Randall, R.B. (2007). Machinery condition monitoring. In: *Handbook of Noise and Vibration Control* (ed. M.J. Crocker), 575–584. New York: Wiley.

45 Newland, D.E. (2007). Wavelet analysis of vibration signals. In: *Handbook of Noise and Vibration Control* (ed. M.J. Crocker), 585–597. New York: Wiley.

46 Herlufsen, H., Gade, S., and Zaveri, H.K. (2007). Analyzers and signal generators. In: *Handbook of Noise and Vibration Control* (ed. M.J. Crocker), 470–485. New York: Wiley.

47 Williams, E.G. (2007). Use of near-field acoustical holography in noise and vibration measurements. In: *Handbook of Noise and Vibration Control* (ed. M.J. Crocker), 598–611. New York: Wiley.

48 Krim, H. and Viberg, M. (1996). Two decades of array signal processing research. *IEEE Signal Proc. Mag.* 5: 4–24.

49 Bai, M.R., Ih, J.-G., and Benesty, J. (2013). *Acoustic Array Systems: Theory, Implementation, and Application*. Singapore: Wiley.

50 Chistensen, J.J. and Hald, J. (2004) *Beamforming*, Bruel & Kjaer Technical Review No. 1, Naerum, Denmark.

51 Sijtsma, P. (2012) *Acoustic beamforming for the ranking of aircraft noise*. Report NLR-TP-2012-137. National Aerospace Laboratory NRL, Amsterdam.

52 Johnson, D.H. and Dudgeon, D.E. (1993). *Array Signal Processing: Concepts and Techniques*. Englewood Cliffs: Prentice Hall.

8

Sound Intensity, Measurements and Determination of Sound Power, Noise Source Identification, and Transmission Loss

8.1 Introduction

Sound intensity is a measure of the magnitude and direction of the flow of sound energy. Although acousticians have attempted to measure sound intensity as long ago as the early 1870s, the first reliable measurement of sound intensity did not occur until over one hundred years later in the late 1970s. Then the convergence of theoretical and experimental advances, including the derivation of the cross-spectral formulation for sound intensity and developments in digital signal processing, propelled sound intensity measurements from the laboratory into practical use. Most modern measurements of sound intensity are made using the simultaneous measurement of the sound pressure with two closely-spaced microphones and this approach will receive the most attention in this chapter.

Sound intensity $\mathbf{I}$ is a vector quantity and is defined as the time average of the net flow of sound energy through a unit area in a direction perpendicular to the area. The dimensions of the intensity are sound energy per unit time per unit area (watts per square meter). For sound energy to be conserved, the sound power generated by a source must be equal to the normal component of the sound intensity integrated over any surface that completely encloses the source. This holds even in the presence of other sources outside the surface. A central point in noise control engineering is to determine the sound power radiated by sources. The value and relevance of determining the sound power radiated by a source is due to this quantity being largely independent of the surroundings of the source in the audible frequency range. Also, it can be used to predict the sound pressure level at a distance from a source once physical details of the surroundings are known. Sound intensity measurements make it possible to determine the sound power of sources without the use of costly special facilities such as anechoic and reverberation rooms and in the presence of steady background noise.

Sound intensity measurements seem to be most useful for the determination of the sound power of large machinery in situ, for noise source identification, and for measurement of the sound transmission loss of partitions. Care must be taken to sample the sound intensity in the sound field appropriately in order to reduce errors. Calibration should be undertaken as recommended by the manufacturers and standards bodies.

8.2 Historical Developments in the Measurement of Sound Pressure and Sound Intensity

It is hard to realize that engineers and scientists have only been able to make quantitative acoustical field measurements with transducers for the last century. It is true that many scientists had previously been making qualitative acoustical studies. Osborne Reynolds describes the use of bells, the human voice and ear in his studies of acoustical reciprocity in 1875 [1]. Soon after in 1876 Lord Rayleigh used such devices as Tyndall's sensitive flames and smoke jets [2] in his acoustical reciprocity studies [3], (see Figures 8.1 and 8.2). We know

Engineering Acoustics: Noise and Vibration Control, First Edition. Malcolm J. Crocker and Jorge P. Arenas.
© 2021 John Wiley & Sons Ltd. Published 2021 by John Wiley & Sons Ltd.

Figure 8.1 Tyndall's flames [2]. A long flame may be shortened and a short one lengthened, according to circumstances, by these sonorous vibrations. Here, for example, are two flames. The one flame is long, straight, and smoky; the other is short, forked and brilliant. On sounding the whistle, the long flame becomes short, forked, and brilliant; the forked flame becomes long and smoky [2].

Figure 8.2 Tyndall's smoke jets [2]. The amount of shrinkage exhibited by some of these smoke columns, in proportion to their length, is far greater than that of the flames. A tap on the table causes a smoke-jet 18 in. high, to shorten to a bushy bouquet. It dances to the tune of a musical box, (but) as the music continues it consists of a series of rapid leaps from one form to another [2].

that several scientists including Mayer, Koenig, Helmholtz, and Rayleigh were concerned with the development of methods to measure the intensity of sound (throughout the 1870s). (See Figure 8.3, the first page of Rayleigh's copy of a paper by Mayer on which Rayleigh inscribed the title himself.) Mayer in this paper clearly describes experimental studies concerning the cancelation of sound waves of the opposite direction and phase with the same sound intensity. It is not until 1882 that we see the birth of the first quantitative

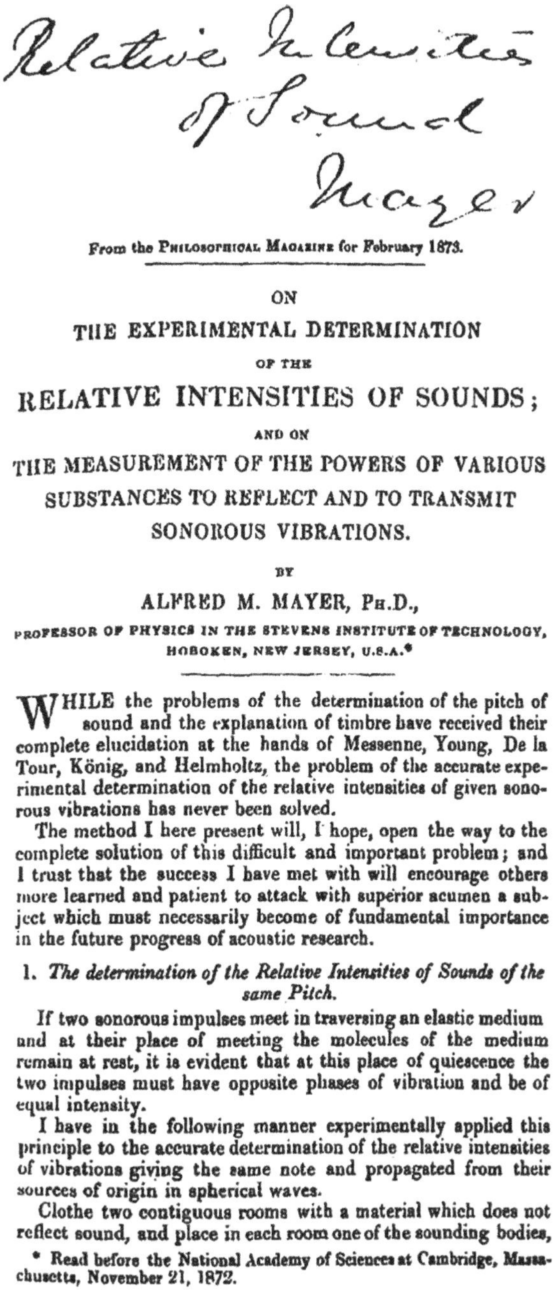

Relative Intensities of Sound
Mayer

From the PHILOSOPHICAL MAGAZINE for February 1873.

ON

THE EXPERIMENTAL DETERMINATION

OF THE

RELATIVE INTENSITIES OF SOUNDS;

AND ON

THE MEASUREMENT OF THE POWERS OF VARIOUS

SUBSTANCES TO REFLECT AND TO TRANSMIT

SONOROUS VIBRATIONS.

BY

ALFRED M. MAYER, Ph.D.,

PROFESSOR OF PHYSICS IN THE STEVENS INSTITUTE OF TECHNOLOGY,
HOBOKEN, NEW JERSEY, U.S.A.*

WHILE the problems of the determination of the pitch of sound and the explanation of timbre have received their complete elucidation at the hands of Messenne, Young, De la Tour, König, and Helmholtz, the problem of the accurate experimental determination of the relative intensities of given sonorous vibrations has never been solved.

The method I here present will, I hope, open the way to the complete solution of this difficult and important problem; and I trust that the success I have met with will encourage others more learned and patient to attack with superior acumen a subject which must necessarily become of fundamental importance in the future progress of acoustic research.

1. *The determination of the Relative Intensities of Sounds of the same Pitch.*

If two sonorous impulses meet in traversing an elastic medium and at their place of meeting the molecules of the medium remain at rest, it is evident that at this place of quiescence the two impulses must have opposite phases of vibration and be of equal intensity.

I have in the following manner experimentally applied this principle to the accurate determination of the relative intensities of vibrations giving the same note and propagated from their sources of origin in spherical waves.

Clothe two contiguous rooms with a material which does not reflect sound, and place in each room one of the sounding bodies,

* Read before the National Academy of Sciences at Cambridge, Massachusetts, November 21, 1872.

Figure 8.3 Rayleigh's copy of Mayer's paper on sound intensity.

(a) (b)

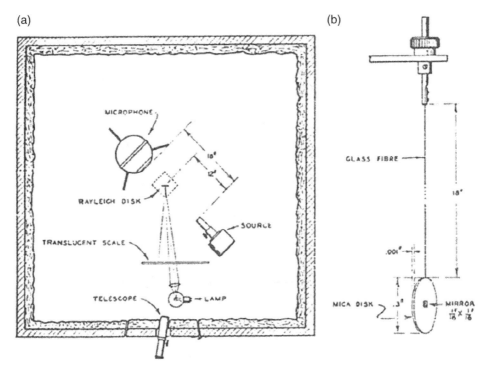

Figure 8.4 (a) Arrangement of apparatus for wave calibration by means of the Rayleigh disk; (b) dimensions of typical Rayleigh disk system [4].

acoustical transducer with the arrival of the now famous Rayleigh disk [3]. (See Figure 8.4.) Surprisingly enough, the design of the disk was conceived accidently during Rayleigh's work on the absolute measurement of the ohm [4, 5].

Although the title of Rayleigh's paper suggests that the Rayleigh disk measures sound intensity, it is actually a device for measuring the mean square acoustic particle velocity, as has been shown by Koenig. Rayleigh's disk was a great step forward, in quantitative acoustical measurement, and was for many years the standard instrument for acoustical calibration – however it is hardly a practical instrument for "field" measurements.

The next important advance was the development of the telephone transmitter into a reliable electroacoustic transducer (microphone) [6]. Several researchers assisted in this development including Wente, Arnold, Crandall, and Fletcher [7]. Wente's papers in 1917 and 1922 describe a sophisticated device which he called a condenser transmitter or electrostatic transmitter from which modern condenser microphones are descended [8, 9]. Although Wente suggests in his titles and text that the condenser microphone measures sound intensity, it is actually a device for measuring sound pressure (a scalar quantity). The sound intensity is of course a vector quantity and may be defined to be the time average rate of flow of sound energy through unit area (with the vector directed perpendicular to the area). Most intensity devices which have been proposed have required the use of two or more similar or different acoustical transducers. Unfortunately until the early 1980s, these intensity devices appear to have suffered from calibration or other problems.

Olson in 1932 was probably the first to describe a device designed to measure the real sound intensity vector [10, 11]. Olson's device consisted of two closely-spaced pressure microphones. In 1941 Clapp and Firestone used a device consisting of two crystal pressure microphones combined with a ribbon velocity microphone [12]. Their device appeared to suffer from temperature instability and internal resonances of the ribbon. Bolt and Petrauskas in 1943 described an acoustic impedance meter using two microphones to measure the sound pressure sum and sound pressure difference [13]. This principle of operation is much the same as that used in most sound intensity devices today. In 1955 Baker described an acoustic intensity meter consisting of a hot-wire anemometer (to measure particle velocity) and a pressure microphone [14]. Because this device requires a steady air flow and is sensitive to undesired air movements it appears unsuitable for practical applications. In 1956 Schultz used an intensity probe made from two electrostatic microphones mounted back-to-back and later discussed problems with intensity measurements in reverberant enclosures [15, 16]. In 1973 Burger et al. and van Zyl and Anderson described sound intensity measurements with a probe made from a pressure sensitive microphone and a velocity-sensitive microphone [17, 18]. Credit for first demonstrating in 1979 the reliability of the two microphone intensity technique for directly measuring sound power of a large machine such as a diesel engine should perhaps go to Chung, Pope, and Feldmaier [19]. Reinhart and Crocker published similar results in 1980 [20]. Also in 1980, Crocker et al. first demonstrated the use of the sound intensity method to measure the sound transmission loss of structures [21, 22].

8.3 Theoretical Background

Sound fields are usually described in terms of sound pressure, which is the quantity we hear. However, sound fields are also energy fields, in which kinetic and potential acoustic energies are generated, transmitted, and dissipated. The acoustic energy in a sound field is not only of interest theoretically, but it is of practical importance as well. Energy considerations are useful in room acoustics and also in noise control engineering problems. Determination of the sound power radiated by a source is a basic quantity needed in the control of noise. At any point in a sound field, in the absence of mean fluid flow, the instantaneous intensity vector $\mathbf{I}(t)$ expresses the magnitude and direction of the instantaneous flow of sound energy as follows:

$$\mathbf{I}(t) = p(t)\mathbf{u}(t), \tag{8.1}$$

where $p(t)$ and $\mathbf{u}(t)$ are the instantaneous sound pressure and vector particle velocity at a point in the sound field.

By combining the fundamental equations that govern a sound field, the equation of mass continuity, the adiabatic relation between sound pressure and density change, and Euler's equation of motion (Newton's second law), one can derive the equation [23, 24]

$$\int_S \mathbf{I}(t) \cdot d\mathbf{S} = -\frac{\partial}{\partial t}\left(\int_V w(t)dV\right) = -\frac{\partial E(t)}{\partial t}, \tag{8.2}$$

in which $w(t)$ is the instantaneous total sound energy density, S is the area of an enclosing surface, V is the volume of fluid contained by the surface S, and $E(t)$ is the instantaneous total sound energy within the surface. The left-hand term is the net outflow of sound energy through the surface, and the middle and right-hand terms are the rate of change of the total sound energy within the surface. This is the equation of conservation of sound energy, which expresses the simple fact that the net flow of sound energy out of a closed surface equals the (negative) rate of change of the sound energy within the surface, because the energy is conserved.

In practice the time-averaged intensity

$$\mathbf{I} = \langle p(t)\mathbf{u}(t)\rangle_t \tag{8.3}$$

is more important than the instantaneous intensity. Examination of Eq. (8.2) leads to the conclusion that the time average of the instantaneous net flow of sound energy out of a given closed surface is zero unless there is generation (or dissipation by sound absorption) of sound power within the surface; in this case the time average of the net flow of sound energy out of a given surface enclosing a sound source is equal to the net sound power of the source. In other words,

$$\int_S \mathbf{I} \cdot d\mathbf{S} = 0, \tag{8.4}$$

unless there is a steady source (or a sink) within the surface, irrespective of the presence of steady sources outside the surface, and

$$\int_S \mathbf{I} \cdot d\mathbf{S} = W_a \tag{8.5}$$

if the surface encloses a steady sound source that radiates the sound power W_a, irrespective of the presence of other steady sources outside the surface. In practice, Eq. (8.5) cannot be used to measure the sound power W_a of a source, if sources situated outside the surface S are unsteady (e.g. time-varying background noise) and/or there is sound-absorbing material inside the surface.

If the sound field is simple harmonic with angular frequency $\omega = 2\pi f$, then from Eq. (8.1), the sound intensity in the r-direction is of the form

$$I_r \approx \langle p \cos(\omega t) u_r \cos(\omega t + \varphi) \rangle_t = \frac{1}{2} p u_r \cos \varphi, \tag{8.6}$$

where φ is the phase angle between the sound pressure $p(t)$ and the particle velocity, $u_r(t)$ in the r-direction. (For simplicity we consider only the component in the r-direction here.) It is common practice for a simple harmonic wave in a stationary sound field to rewrite Eq. (8.3) as

$$I_r = \frac{1}{2} \operatorname{Re} \{p u_r^*\}, \tag{8.7}$$

where both the sound pressure p and the particle velocity u_r here are complex exponential quantities, and u_r^* denotes the complex conjugate of u_r. The time averaging gives the factor of ½. We note that the use of complex notation is mathematically very convenient and that Eq. (8.7) gives the same result as Eq. (8.6). We note that by using the complex notation, we lose the time dependence of the instantaneous intensity, thus considerable care must be taken in making the correct physical interpretation of the mathematical results when using complex notation.

In a plane progressive wave (of any waveform) traveling in the r-direction, the sound pressure p and the particle velocity u_r, are in phase ($\varphi = 0$) and related by the characteristic impedance of the medium, ρc, where ρ is the density and c is the speed of sound:

$$p(t) = \rho c u_r(t). \tag{8.8}$$

Thus, for a plane wave, the time-average sound intensity is

$$I_r = \langle p(t) u_r(t) \rangle = \langle p^2(t)/\rho c \rangle_t = p_{rms}^2/\rho c. \tag{8.9}$$

In this case the sound intensity is simply related to the mean square sound pressure p_{rms}^2, which can be measured with a single microphone. Eq. (8.9) also holds sufficiently far from any source of any arbitrary time history and frequency content since in the far geometric field the wave fronts can be regarded as locally plane surfaces. If the characteristic impedance, ρc, is assumed to be equal to 400 kg/m²s, then the sound intensity level normal to the wavefronts far from any source is

$$L_i = L_p. \tag{8.10}$$

This is a result of the choice of reference values for sound pressure of $p_{ref} = 20 \times 10^{-6}$ Pa (N/m²) and sound intensity $I_{ref} = 10^{-12}$ watt/m². However, in normal ambient conditions in air at sea level,

$$\rho c = 415 \text{ kg/m}^2\text{s}, \tag{8.11}$$

and the error in Eq. (8.10) is about 0.1 dB. In most practical cases, however, the sound intensity is not simply related to the sound pressure. Examination of Eq. (8.1) shows that both the sound pressure and the particle velocity must be estimated simultaneously and that their product must be time-averaged. This requires the use of a more complicated device than a single microphone.

8.4 Characteristics of Sound Fields

Many different types of sound fields are encountered in practice. The sound field near to a simple point sound source has certain well-known characteristics. However, sound fields generated by many simultaneously operating independent sources have much more complicated characteristics. A reverberant sound field, which is created when sources operate in spaces with hard wall surfaces, has even more complicated properties, and so on.

8.4.1 Active and Reactive Intensity

We have seen that the sound pressure and the particle velocity are in phase in a plane propagating wave. This is also the case in a free field, sufficiently far from any source that generates the sound field. Conversely, one of the characteristics of the sound field near a source is that the sound pressure and the particle velocity are partly out of phase (in quadrature). To describe such phenomena, one may introduce the concept of active and reactive sound fields.

In a simple harmonic sound field, the particle velocity may, without loss of generality, be divided into two components: one component in phase with the sound pressure and the other component out of phase with the sound pressure.

The *instantaneous active intensity* is the product of the sound pressure and the in-phase component of the particle velocity. This quantity fluctuates at twice the frequency of the sound wave and has a nonzero time average. The time-averaged quantity is the component that is generally referred to simply as sound intensity. The (active) sound intensity is associated with net flow of sound energy and has a direction normal to the wavefronts [24–27].

The *instantaneous reactive intensity* is the product of the sound pressure and the out-of-phase component of the particle velocity. This quantity fluctuates at twice the frequency of the sound wave and has a time average equal to zero at any point in a sound field [24–27].

8.4.2 Plane Progressive Waves

Consider first the situation in a plane progressive simple harmonic wave traveling in the positive x-direction. The sound pressure $p(x,t)$ may be written:

$$p(x, t) = P \cos (\omega t - kx) \tag{8.12}$$

and since sound pressure and particle velocity are in phase

$$u(x, t) = U \cos (\omega t - kx). \tag{8.13}$$

But from Eq. (8.8) in a sinusoidal progressive plane wave, $p/u = P/U = \rho c$.

The total mechanical energy per unit volume is known as the sound energy density and can be written

$$e_{tot}(x, t) = e_p(x, t) + e_k(x, t). \tag{8.14}$$

The potential energy density takes the form

$$e_p(x, t) = \left(P^2/2\rho c^2\right) \cos^2 (\omega t - kx); \tag{8.15}$$

and the kinetic energy density is given by

$$e_k(x,t) = \left(U^2/2\rho c^2\right) \cos^2(\omega t - kx), \tag{8.16}$$

where P is the sound pressure amplitude, U is the particle velocity amplitude, and k is the wavenumber ω/c. Thus from Eq. (8.8) summing Eqs. (8.15) and (8.16), and using Eq. (8.8), the total instantaneous energy density is

$$e_{tot}(x,t) = \left(P^2/\rho c^2\right) \cos^2(\omega t - kx) \tag{8.17}$$

and the time-average of the total energy density, $\langle e_{tot}(x,t)\rangle$

$$\langle e_{tot}(x,t)\rangle = \tfrac{1}{2}\,P^2/\rho c^2 = p^2_{rms}/\rho c^2 = \langle e(t)\rangle_t, \tag{8.18}$$

which is seen to be independent of x.

The time-average total energy density is given by $\tfrac{1}{2}\,P^2/\rho c^2$ for all values of x and t. The instantaneous intensity is (from Eqs. (8.12) and (8.13)) given by

$$I(x,t) = \left(P^2/\rho c\right) \cos^2(\omega t - kx) = \left(P^2/\rho c\right)\left[1 + \cos 2(\omega t - kx)\right]. \tag{8.19}$$

The time average intensity is given by $\tfrac{1}{2}\,P^2/\rho c$ for all values of x and t. Thus

$$\langle I\rangle_t = c\,\langle e\rangle_t, \tag{8.20}$$

where $\langle e\rangle_t$ is the time-average of the total instantaneous energy density $e(t)$. The result in Eq. (8.20) can also be obtained by considering the sound energy in a progressive wave without reflections in a tube. All of the sound energy in a column of fluid 1 m long in the tube must flow through the cross-section area dS in unit time. Thus $e_{tot}(x,t)\,cdS = I\,dS$, which on rearranging gives Eq. (8.20).

The instantaneous distribution of the sound pressure, particle velocity, sound intensity and energy density $e(x,t)$ are shown at one instant of time in Figure 8.5. It is seen that the energy is concentrated in specific regions where the pressure and particle velocity are maximum and travels in the positive x-direction at the wave speed, c.

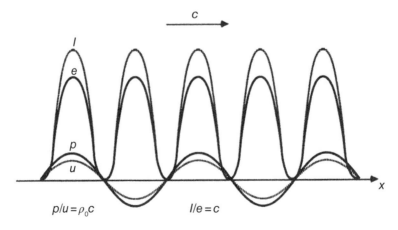

Figure 8.5 Instantaneous spatial distributions of sound pressure, particle velocity, energy density and sound intensity in a plane progressive wave traveling in the positive x-direction at velocity c.

8.4.3 Standing Waves

Consider now the case of a pure tone standing wave with angular frequency $\omega = 2\pi f$. The sound pressure takes the form

$$p(x,t) = 2P\cos(\omega t)\cos(kx). \tag{8.21}$$

The instantaneous distribution of the total energy density (kinetic plus potential) is given by

$$e(x,t) = \left(P^2/\rho c^2\right)\left[1 + \cos(2\omega t)\cos(2kx)\right], \tag{8.22}$$

and the instantaneous sound intensity distribution is given by

$$I(x,t) = P^2/\rho c\left[\sin(2\omega t)\sin(2kx)\right]. \tag{8.23}$$

In this case, the time average sound intensity is seen to be zero for all positions in the sound field in the tube. The instantaneous sound intensity represents an oscillatory flow of sound energy back and forth between regions of high potential energy and kinetic energy. The fluctuations of $2\omega = 4\pi f$ are at twice the angular frequency of the sound wave, ω. See Figure 8.6.

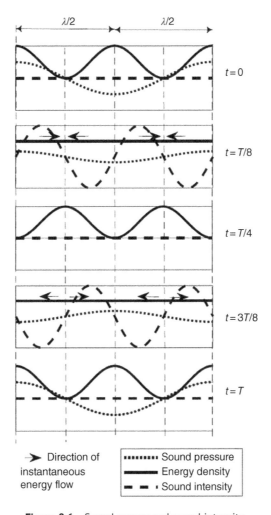

Figure 8.6 Sound energy and sound intensity.

Figure 8.6 illustrates this behavior for one sound energy oscillation, which is equivalent to one half of the period of the sound pressure and particle velocity. The plot shows the standing wave where the wavelength λ is equal to the tube length ℓ and thus $k = \lambda/\ell$.

Very near a sound source, the reactive field is usually stronger than the active field. However, in the absence of reflections, the reactive field diminishes rapidly with increasing distance from sources. Therefore, at a moderate distance from sources, the sound field is dominated by the active field. The extent of the reactive field depends on the frequency and on the dimensions and the radiation characteristics of the sound source; however, in practice, the reactive field may be assumed to be negligible at a distance greater than, say, 0.6 m from the source, provided the reflected sound field is small [27].

8.4.4 Vibrating Piston in a Tube

Consider a sound field created by a vibrating piston [27]. Plane sound waves are traveling along a hard-walled tube, as illustrated in Figure 8.7, when only a single frequency is present.

In Figure 8.7a the tube is terminated at the right end with a perfect absorber; therefore, there is no reflection of sound at the termination of the tube. Under these conditions the pressure and the particle velocity are in phase at every position in the tube, and the spatial distribution of the pressure is in phase with the spatial distribution of the particle velocity, as shown in the figure at two instants of time by the continuous and broken lines. The instantaneous intensity is always positive in the direction toward the termination and is given from Eq. (8.1) by:

$$I_x = P \cos{(\omega t - kx)}\, U\,(\cos{\omega t - kx}) = P^2/\rho c \cos^2{(\omega t - kx)}$$

$$I_x = \tfrac{1}{2}\left(P^2/\rho c\right)\left[1 + \cos{(\omega t - kx)}\right]. \tag{8.24}$$

In Figure 8.7b the tube is terminated with material that is partly absorptive. There will be partial reflection at the termination in this example, so that a weaker wave returns from right to left. The two opposite traveling waves add together, giving the pressure distribution shown at two different instants of time. In this case the spatial distribution of the particle velocity is somewhat out of phase with the spatial distribution of the pressure. The two waves interact to give an active intensity that is less than shown in Figure 8.7a. There is also a reactive component that flows back and forth in the positive then negative direction (to and from locations of high sound pressure and thus potential energy to regions of high particle velocity and thus kinetic energy). The time average of this reactive component is zero at any point in the tube.

In Figure 8.7c the tube is terminated with an infinitely hard material. Therefore, the waves are perfectly reflected at the termination, and the reflected waves traveling back to the left have the same amplitude as the incident waves traveling to the right. The incident and reflected waves combine to give a standing wave pattern. In this sound field the pressure and the particle velocity are in quadrature at all positions; therefore, the time average of the intensity is zero at any point (i.e. the sound field is completely reactive). The spatial distribution of the pressure is 90° out of phase with the spatial distribution of the particle velocity. The magnitude of the reactive intensity varies with the position: at some locations it is maximum, and at other locations it is zero.

In the general case, where the sound field cannot be assumed to be simple harmonic, one cannot make an instantaneous separation of the particle velocity into components in phase and in quadrature with the pressure; under some conditions the particle velocity is not fully correlated with the sound pressure [27–29].

(a)

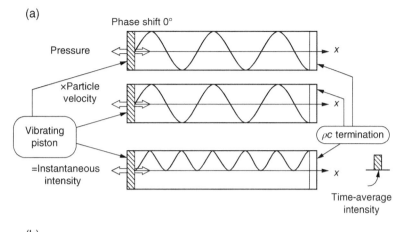

(b)

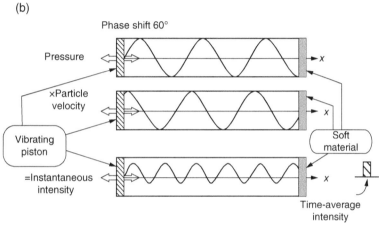

(c)

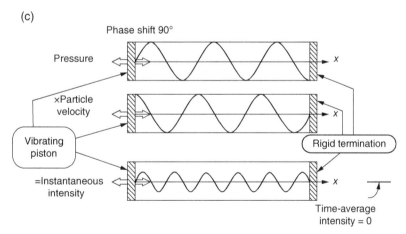

Figure 8.7 Spatial distributions of instantaneous sound pressure, instantaneous particle velocity and instantaneous sound intensity for a pure-tone one-dimensional sound field in a tube. (a) Case with no reflection at the right end of the tube. (b) Case with partial reflection at the right end of the tube. (c) Case with perfect reflection at the right end of the tube [27].

8.5 Active and Reactive Sound Fields

As already discussed, in a one-dimensional pure tone plane progressive wave the sound pressure and particle velocity are everywhere in phase for all values of time. The same situation exists far from idealized point sources of sound (monopoles, dipoles, quadrupoles, etc.) since the wave front curvature decreases and the surface becomes almost planar (flat) far from a source. However, close to the center of a source the situation is quite different and the sound pressure and particle velocity are almost completely 90° out of phase (in quadrature).

The near-field behavior of simple discrete frequency sources is also sometimes studied by introducing the concepts of *active* and *reactive* sound fields. Four sound field quantities, potential energy density, kinetic energy density, active intensity and reactive intensity can be defined. These quantities were already described in the previous discussion of progressive plane wave propagation (Figure 8.5) and standing waves (Figure 8.6).

8.5.1 The Monopole Source

The active and reactive intensity can most easily be understood for a spherically spreading point source (or monopole) using trigonometrical functions. The sound pressure in outward traveling waves may be written (see Eq. (3.34)):

$$p = \frac{\rho c k Q}{4\pi r} \sin{(\omega t - kr)}, \tag{8.25}$$

where Q is the source strength $4\pi a^2 U$ for a simple harmonic spherical source of radius a with normal velocity amplitude U. See Figure 8.8.

Here it is assumed that the source radius is small in wavelengths, $a \ll \lambda$, or $ka \ll 1$. The source strength Q has units of volume flow rate. The particle velocity u is everywhere radial and is given by

$$-\frac{\partial p}{\partial r} = \rho \frac{\partial u}{\partial t}, \tag{8.26}$$

and thus, after integrating the results for $\partial u / \partial t$, we obtain Eq. (8.27):

$$u = \frac{c k^2 Q}{\omega 4\pi r} \sin{(\omega t - kr + \varphi_1)} - \frac{c k Q}{\omega 4\pi r^2} \cos{(\omega t - kr)}. \tag{8.27}$$

We notice that the first term represents the particle velocity, which is in phase with the sound pressure and is dominant far from the origin. The second term represents the particle velocity which is out-of-phase with the pressure and is dominant near to the source origin where r is small.

When $kr \ll 1$, the sound field is known as the *near field*; when $kr \gg 1$, the sound field is known as the *far field*: when $kr = 1$, the in-phase and out-of-phase particle velocity components are equal.

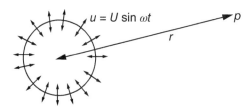

$u = U \sin{\omega t}$

p

r

Figure 8.8 Idealized monopole source of sound.

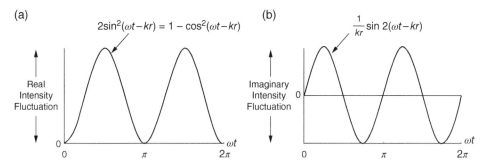

Figure 8.9 (a) Real intensity fluctuations, (b) Imaginary intensity fluctuations. This is the case for $kr = 1$ at the boundary between the near field and the far field.

Upon multiplication of Eq. (8.25) with Eq. (8.27), we obtain the instantaneous sound intensity

$$I\,(r,t) = \left(\rho c k Q^2 / 16\pi^2 r^2\right) \left[k\,\sin^2\,(\omega t - kr) - (1/r)\,\sin\,(\omega t - kr)\,\cos\,(\omega t - kr)\right]. \tag{8.28a}$$

$$= \left(\rho c k Q^2 / 16\pi^2 r^2\right) k\left[\sin^2\,(\omega t - kr) - (1/2kr)\,\sin\,2(\omega t - kr)\right]. \tag{8.28b}$$

We see that the fluctuating intensity at a distance r from the origin is comprised of two terms. The first term is seen to fluctuate at frequency 2ω but is always positive and represents energy flowing away from the source. The second term is also seen to fluctuate at frequency 2ω, but is alternatively positive and negative and represents energy flowing from and back toward the source. (See Figure 8.9 for $kr = 1$.)

When $kr = 1$, the real and imaginary intensity fluctuations are equal in magnitude. See Figure 8.9. When $kr \ll 1$, the imaginary term dominates; when $kr \gg 1$, the real intensity term dominates.

Example 8.1 A sphere of radius 5.5 cm pulsates with a surface velocity amplitude of 1.0 cm/s at the following frequencies: 50 Hz, 500 Hz, and 5000 Hz. Calculate (a) the volume velocity Q, (b) the distance to the near-field/far-field boundary at each frequency, (c) the magnitude of the real and imaginary intensity fluctuations at the boundary, and (d) the time-average of the real and imaginary intensity components at the boundary.

Solution

a) $Q = 4\pi a^2 U = 4\pi(3.025 \times 10^{-3}) \times (1 \times 10^{-2}) = 3.8 \times 10^{-4}\,\mathrm{m^3/s}$.

b) $kr = 1$, so $r = 1/k = c/(2\pi f) = 344/(2\pi f)$. Then, $r = 1.1$ m at 50 Hz, $r = 11$ cm at 500 Hz, and $r = 1.1$ cm at 5000 Hz.

c) From Eq. (8.28), the magnitude of the real intensity fluctuation is $(\rho c k^2 Q^2 / 16\pi^2 r^2)$ and the magnitude of the imaginary fluctuation is $(\rho c k Q^2 / 16\pi^2 r^3)$. Therefore, at $r = 1/k$ both magnitudes are equal to $(\rho c k^4 Q^2 / 16\pi^2)$. Thus, this magnitude is $2.6 \times 10^{-7}\,\mathrm{watt/m^2}$ at 50 Hz, 0.0026 watt/m^2 at 500 Hz, and 26.1 watt/m^2 at 5000 Hz.

d) The imaginary intensity component given by the second term in Eq. (8.28) integrates to be zero over a period $T = 2\pi/\omega$. The time-average of the real intensity becomes:

$$\langle I \rangle_t = \frac{\rho c k^2 Q^2}{16\pi^2 r^2} \frac{1}{T} \int_0^T \sin^2(\omega t - kr)dt = \frac{\rho c k^2 Q^2}{32\pi^2 r^2}. \text{ Now for } kr = 1, \langle I \rangle_t = \frac{\rho c Q^2}{32\pi^2 r^4}. \text{ Thus}$$

$\langle I \rangle_t = 1.3 \times 10^{-7}\,\mathrm{watt/m^2}$ at 50 Hz, $\langle I \rangle_t = 0.0013\,\mathrm{watt/m^2}$ at 500 Hz, and $\langle I \rangle_t = 1.3\,\mathrm{watt/m^2}$ at 5000 Hz.

Example 8.2 Consider the same pulsating sphere of Example 8.1 and calculate its sound power at the frequencies 50 Hz, 500 Hz and 5000 Hz.

Solution

The total sound power emitted by the source is found by integrating the time-average real intensity over a sphere at some radius r (see Eq. (8.5)). Then,

$W = \int_S \langle I \rangle_t dS = \int_0^{2\pi} \int_0^\pi \frac{\rho c k^2 Q^2}{32\pi^2 r^2} r^2 \sin\theta d\theta d\phi = \frac{\rho c k^2 Q^2}{8\pi}$, which is independent of radius as conservation of

energy requires. Thus, the sound power of the pulsating sphere is $W = 1.96 \times 10^{-6}$ watt at 50 Hz, $W = 1.96 \times 10^{-4}$ watt at 500 Hz, and $W = 1.96 \times 10^{-2}$ watt at 5000 Hz.

The fluctuating intensity is much more complicated for a dipole than a monopole, having more different regions and also an angular dependence, θ.

8.5.2 The Dipole Source

As discussed in Chapter 3, unlike the monopole, the sound pressure field produced by a dipole exhibits both near and far field behaviors (see Eq. (3.36)):

$$p = \frac{\rho c k Q l \cos\theta}{4\pi r} \left[\frac{1}{r} \sin(\omega t - kr + \varphi) + k \cos(\omega t - kr + \varphi) \right]. \tag{8.29}$$

The first term represents sound pressure fluctuations in the near field and the second, in the far field. We notice again that the division between the two fields occurs when $kr = 1$.

The radial particle velocity can again be obtained from Eq. (8.26) and has three distinct regions. The radial sound intensity fluctuations are obtained by multiplying the pressure by the velocity (see Eq. (3.40)).

8.5.3 General Case

In general, the active intensity I is defined to be the product of the sound pressure and the in-phase radial particle velocity and the reactive intensity J is defined to be the product of the sound pressure and the out-of-phase component of the radial particle velocity. Using complex notation, the real (active) and imaginary (reactive) components of the sound intensity are

$$I = \frac{1}{2} \operatorname{Re}\left\{ p\mathbf{u}^* \right\}, \tag{8.30a}$$

$$J = \frac{1}{2} \operatorname{Im}\left\{ p\mathbf{u}^* \right\}. \tag{8.30b}$$

The reactive intensity J is thus a vector quantity with a direction (like the active intensity) pointing away from the source in the radial r-direction. However, by definition it is seen that it has a long time average of zero at any point in the sound field. The reactive intensity has been misinterpreted by some making it a somewhat controversial subject. This may be because when it is plotted, it is usually shown as an arrow (of length representing its magnitude) and direction away from the source. Thus, some have assumed it represents a real flow of energy, instead of a fluctuating flow, from and to a point in the sound field. Perhaps a two-headed arrow representation could be better with the arrow heads at each end pointing away from and also back toward the point in the field.

It has been shown mathematically that the real active intensity can be written as [28, 29]

$$\mathbf{I} = -\frac{|p|^2}{2\rho c}\frac{\nabla\varphi}{k}. \tag{8.31}$$

The active intensity has a direction orthogonal (perpendicular) to surfaces of constant phase (wave fronts). It is seen immediately that it is very important to avoid phase errors in the measurement of sound intensity. The reactive intensity can be written as [28, 30]

$$\mathbf{J} = -\frac{\nabla(|p|^2)}{4\rho ck}. \tag{8.32}$$

The reactive intensity is thus seen to be proportional to the gradient of the mean square pressure. It is orthogonal to surfaces of equal sound pressure. Because the active and reactive intensities are given mathematically by the real and imaginary parts of the product of the sound pressure and the complex conjugate of the particle velocity, the term *complex intensity* has been defined and used by some:

$$I_{comp} = I + \mathrm{i}J = (1/2)\,pu^*. \tag{8.33}$$

With the advent of the cross-spectral method of measurement of sound intensity in the late 1970s [31, 32], interest grew in theoretical modeling of sound fields. Pascal published an early analytical study in 1981 concerning active and reactive sound intensity and coined the term *complex intensity* [29], Tichy and his colleagues extended this analytical study in 1984 and 1985 [33–36].

Some have found these theoretical studies not to have been very helpful or even have been confusing for practitioners who use sound intensity in engineering problems for sound power measurement and sound source identification of machinery. There are two main reasons: (i) these studies have mostly concerned idealized single-frequency sound sources (the particle velocity can only be separated into components in phase and out of phase with the sound pressure for discrete frequency sources), and (ii) the plots of reactive sound intensity are represented by arrows of different lengths (normally pointing away from sources): the lengths correspond to the fluctuation magnitude at these locations. Jacobsen has pointed out that the direction of the reactive intensity arrows is related to the spatial derivative of the square of sound pressure only and could equally well be shown in the reverse direction. As already stated, it has been suggested that this confusion would be overcome by replacing the single direction arrows with double headed arrows showing both directions of the pulsating energy flow.

In 1984, Elko and Tichy suggested that the reactive intensity, which is known to be high near sources, could be useful in noise source location [34]. However, Crocker pointed out at that time that in the ideal case of a pure tone standing wave in a tube (see Figure 8.6) the reactive intensity is high, although no source is present [25, 37]. The same situation exists for the ideal case of standing waves in a room with hard walls in which no source is present. Indeed Pascal has emphasized that when sources are present, "(reactive intensity) cannot be used directly to locate sound sources; (this is) because interference situations obliterate the assumption that the decrease of (the) reactive intensity vector indicates source positions" [38]. Pascal gives an example of the interference of a plane wave and a near-field wave [38]. He goes on to say "A general answer to the source identification problem will only be given by a detailed study of active intensity potentials (scalar and vector)."

Although separating the particle velocity in a sound field into components that are in phase and out-of-phase with the sound pressure can strictly be done only for discrete frequency sounds, Jacobsen used this approach with one-third octave band random noise signals [28]. The approach required the use of the Hilbert Transform as formulated by Heyser [40] Jacobsen [28] made measurements in the sound field, near a loudspeaker, a vibrating box and in a reverberation room. The measurements were made in the near field at varying distances from the source: near field, 30 and 50 cm in one-third octave bands centered at 240 Hz, 500 Hz, and 1 kHz. The results are said to represent the running short-time average of the active and reactive components of the complex instantaneous intensity [28]. The time windows used vary from 31.25 to 125 ms. The results obtained are not easy to interpret. By definition, the instantaneous reactive intensity should average to zero if a sufficiently long time window is taken. Since this was not the case, it suggests that sufficiently long averaging times need to be used in practice to obtain a reliable and repeatable value of the mean sound intensity. Fahy also reached this conclusion in evaluating Jacobsen's results [24]. Kutruff and Schmitz commented in 1994 that "the measurement of sound intensity (has) so far not found the practical application it deserves" [41]. They commented that one possible reason is the conceptual difficulty inherent in (interpreting) the intensity (a vector). They also went on to say that "the frequently made distinction between 'active' and 'reactive' intensity may have added to the confusion about intensity" [41].

8.6 Measurement of Sound Intensity

The measurement of sound intensity is much more complicated than the measurement of sound pressure. In general, it requires the simultaneous measurement of sound pressure and particle velocity. This needs the use of at least two transducers. There are currently three main methods in use:

1) The so-called p–p method, in which the particle velocity is determined from the sound pressure difference between two closely-spaced microphones and the sound pressure is taken to be the average of the two microphones signals.
2) The p–u method, in which the particle velocity is determined by a particle velocity transducer and the sound pressure is obtained with a pressure microphone.
3) The p–a method, in which the surface velocity of a vibrating surface is obtained by integrating the signal from a surface-mounted accelerometer and the sound pressure is measured with a pressure microphone located close to the accelerometer.

The first p–p method is well established and has been in use now for almost 40 years and will be discussed in detail first. It is currently the dominant way of measuring sound intensity.

8.6.1 The p–p Method

The most successful measurement principle employs two closely-spaced pressure microphones [26, 27, 30, 37]. Both the International Electrotechnical Commission (IEC) and the American National Standards Institute (ANSI) standards for sound intensity measurements deal exclusively with the two-microphone method [42, 43].

In this approach, the particle velocity is obtained through an elaboration of Newton's Law known as Euler's relation

$$\nabla p(t) + \rho \frac{\partial \mathbf{u}(t)}{\partial t} = \mathbf{0}. \tag{8.34}$$

Use is made of a finite difference approximation of the sound pressure gradient, as

$$\hat{u}_r(t) = -\frac{1}{\rho} \int_{-\infty}^{t} \frac{p_2(\tau) - p_1(\tau)}{\Delta r} d\tau, \tag{8.35}$$

where p_1 and p_2 are the sound pressure signals from the two microphones, Δr is the microphone separation distance, and τ is a dummy time variable. The caret indicates the finite difference estimate obtained from the "two-microphone approach." The sound pressure at the center of the probe is estimated as

$$\hat{p}(t) = \frac{p_1(t) + p_2(t)}{2}, \tag{8.36}$$

and the time-averaged intensity component in the axial direction is, from Eqs. (8.3), (8.35), and (8.36),

$$\hat{I}_r = \langle \hat{p}(t)\hat{u}_r(t) \rangle_t = \frac{1}{2\rho\Delta r} \left\langle (p_1(t) + p_2(t)) \int_{-\infty}^{t} (p_1(t) - p_2(t))d\tau \right\rangle_t. \tag{8.37}$$

The majority of sound intensity measurement systems in commercial production today are based on the two-microphone method and the use of condenser microphones, which are the most reliable and stable types available. Some commercial intensity analyzers use Eq. (8.37) to measure the intensity in one-third-octave frequency bands. See for example Figure 8.10. Another type of analyzer uses Eq. (8.37) to calculate the intensity from the imaginary part of the cross-spectrum G_{12} between the two microphone signals. See for example Figure 8.11.

$$I_r(\omega) = -\frac{1}{\omega\rho\Delta r} \text{Im}\{G_{12}(\omega)\}. \tag{8.38}$$

The time domain formulation and the frequency domain formulation, Eqs. (8.37) and (8.38) respectively, are equivalent. Eq. (8.38), derived originally by Chung [31] for his frequency analysis of engine noise in the mid 1970s, makes it possible to determine sound intensity spectra with a dual-channel Fast Fourier Transform (FFT) analyzer. Fahy [32] published the same intensity expression in his Letter to the Editor in JASA.

Figure 8.12 shows two of the most common microphone arrangements, "side-by-side" and "face-to-face" [44]. The side-by-side arrangement has the advantage that the diaphragms of the microphones can be placed very near a radiating surface, but has the disadvantage that the microphones shield each other. At high

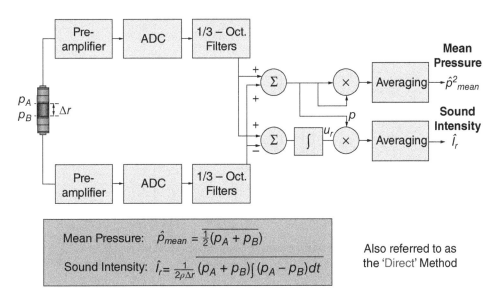

Figure 8.10 Schematic diagram of sound intensity measurements made with the "direct" method. (Figure Courtesy of Brüel & Kjær).

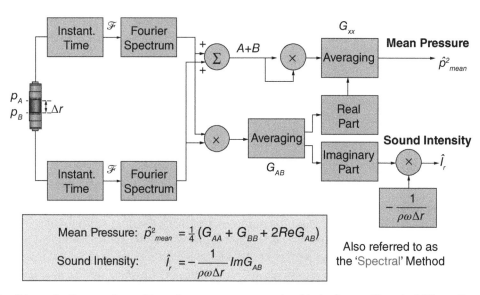

Figure 8.11 Schematic diagram of sound intensity measurements made with the "spectral" method (Figure Courtesy of Brüel & Kjær).

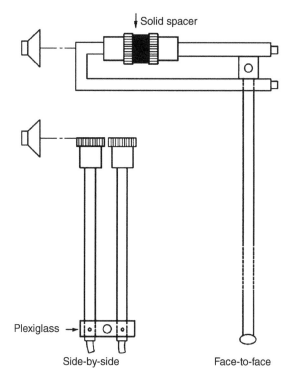

Figure 8.12 Microphones arrangements used to measure sound intensity.

frequencies the face-to-face configuration with a solid spacer between the microphones is normally the superior arrangement [44, 45]. Figure 8.13 shows a sound intensity probe with two microphones in the face-to-face configuration (Bruel & Kjaer, Denmark.) Figure 8.14 shows two examples of commercial three-dimensional sound intensity probes.

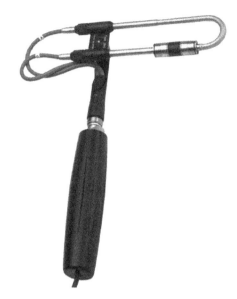

Figure 8.13 Sound intensity probe with the microphones in the face-to-face configuration (*Source:* Bruel & Kjaer, Denmark).

(a) (b)

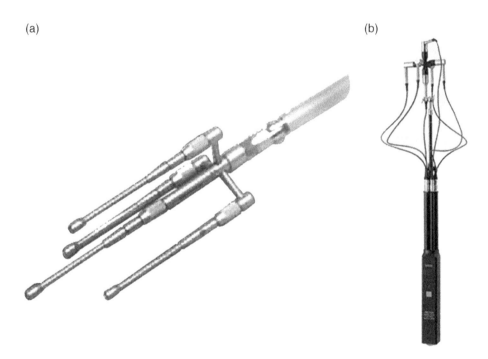

Figure 8.14 Three-dimensional sound intensity probe for vector measurements; (a) Ono Sokki, Japan; (b) G.R.A.S. Sound & Vibration, Denmark.

a) Errors and Limitations in *p–p* Measurements of Sound Intensity

There are many sources of error in the measurement of sound intensity with two microphones, and a considerable part of the sound intensity literature has been concerned with identifying and studying such errors, some of which are fundamental and others, which are associated with technical deficiencies [37, 44–47]. One complication is that the accuracy depends very much on the sound field under study; under certain conditions even minute imperfections in the measuring equipment will have a significant influence. Another complication is that small local errors are sometimes magnified into large global errors when the intensity is integrated over a closed surface [48].

The following is an overview of some of the sources of error in the measurement of sound intensity. A more detailed discussion is given in Ref. [49]. Those who make sound intensity measurements should be aware of the limitations imposed by:

- The finite difference error [50]
- Errors due to scattering and diffraction [45]
- Instrumentation phase mismatch [47].

Other possible sources of error that are usually less serious are:

- Random errors associated with a given finite averaging time [46] and
- Bias errors caused by turbulent airflow [51].

Chung and Pope demonstrated that a two-microphone system (in a side-by-side arrangement) using the cross-spectrum formula (Eq. (8.38)) along with microphone switching could measure the sound intensity [52]. This was proved experimentally by confirming that at 1 m from a loudspeaker the intensity measurements gave the same result as that made with single microphones assuming the sound intensity $I = p^2_{rms}/\rho c$. See Figure 8.15a. Later Blaser and Chung [53] compared the sound power propagating along a tube (measured by two microphones) from a loudspeaker source with the sound power radiated out of the tube through

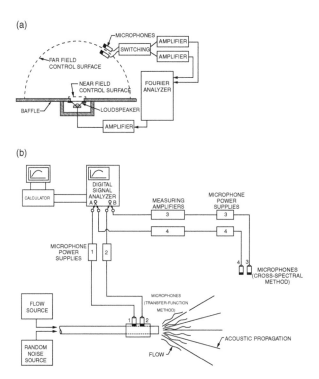

Figure 8.15 Experimental set-up for measuring sound intensity, (a) near field near a loudspeaker and far field [53] and (b) both inside and outside of a duct system [52–54].

a control surface with the two microphone arrangement with the same microphone separation (see Figure 8.15b). By summing the space–time average of the sound intensity measured normal to the control surface (after multiplying by appropriate control area segments) they obtained almost perfect agreement with the in-duct sound power measurements.

After obtaining sound power estimates of sources by making fixed point measurements, Chung went on to demonstrate that continuously scanning the two side-by-side microphone arrangement over a conformal surface around a diesel engine could be used to measure its sound power [19]. Soon after Reinhart and Crocker showed that scanning intensity measurements made on selected parts of a diesel engine gave more accurate results than sound power single microphone measurements on a spherical surface obtained by selective lead wrapping [20].

b) Errors Due to the Finite Difference Approximation

One of the obvious limitations of the measurement principle based on the use of two pressure microphones is the frequency range [25–27, 30]. The finite difference estimate given by Eq. (8.37) is accurate only if the distance between the microphones is much less than the wavelength, as suggested in Figure 8.16; and this clearly implies an upper frequency limit that is inversely proportional to the distance between the microphones. The finite difference error, that is the ratio of the measured intensity $\hat{I}_r$ to the true intensity I_r, can be shown to be

$$\hat{I}_r/I_r = \frac{\sin k\Delta r}{k\Delta r} \tag{8.39}$$

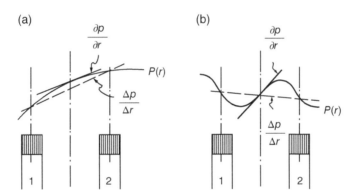

Figure 8.16 Illustration of the error due to the finite difference approximation: (a) good approximation at a low frequency and (b) very poor approximation at a high frequency.

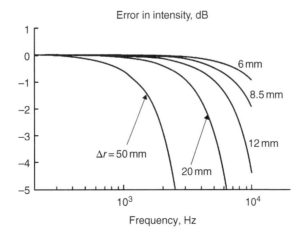

Figure 8.17 Finite difference error of an ideal *p–p* face-to-face sound intensity probe in a plane wave of axial incidence for different values of the separation distance Δr [45].

for an ideal intensity probe in a plane wave of axial incidence. This relation is shown in Figure 8.17 for different values of the microphone separation distance. Although the finite difference error in principle depends on the sound field [37, 44], the upper frequency limit of intensity probes had generally been considered to be the frequency at which the error given by Eq. (8.36) is acceptably small. Note, however, that the interference of the microphones on the sound field had been ignored. A numerical and experimental study of such interference effects has now shown that the upper frequency limit of an intensity probe with the microphones in the usual face-to-face configuration is an octave above the limit determined by the finite difference error if the length of the spacer between the microphones equals the diameter. This is because the resonance of the cavities in front of the microphones gives rise to a pressure increase that to some extent compensates for the finite difference error [45]. See Figure 8.18.

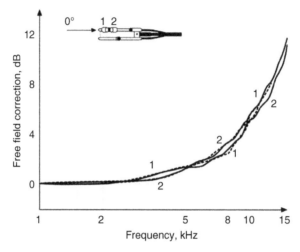

Figure 8.18 Pressure increase on the two microphones of a sound intensity probe with ½ in. microphones separated by a 12 mm spacer for axial sound incidence. Experimental results (solid line); numerical results (dashed line) [56, 57].

Figure 8.19, which corresponds to Figure 8.17, shows the error calculated for a probe with two half-inch microphones. It is apparent that the optimum length of the spacer is about 12 mm and that a probe with this geometry performs very well up to 10 kHz [45].

c) Errors in *p–p* Probe due to Instrumentation Phase Mismatch

Phase mismatch between the two measurement channels is the most serious source of error in the measurement of sound intensity. In the first ground breaking measurements of sound intensity made by Chung and his colleagues in the late 1970s and early 1980s, commercial sound intensity probes with phase-matched microphones were not available. So they used a pair of high-quality microphones, cartridges, and preamplifiers in a side-by-side configuration. This arrangement has the advantage that the phase mismatch between the two measurement systems can be measured and corrected. The side-by-side arrangement of the microphone pair (probe) also allows it to be located closer to noise sources than sound intensity probes now commercially available, which mostly have asymmetric arrangements.

Chung [55] and Chung and Blaser [54] have described three main approaches to make phase-mismatch corrections as follows. Krishnappa has also described another similar method [58].

1) *Transfer Function Approach*

This procedure is presented in detail by Chung and Blaser, and Chung [54, 55]. The calibration transfer function (magnitude and phase difference) between the two systems is normally achieved by mounting the two microphones in a circular plate attached to the one end of a circular tube. The two microphones are subjected to the same broadband frequency sound field produced by a loudspeaker mounted at the other end of the tube.

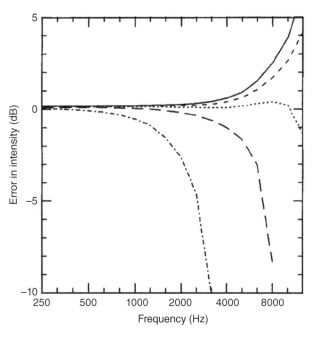

Figure 8.19 Error of a p–p sound intensity probe with half-inch microphones in the face-to-face configuration in a plane wave of axial incidence for different spacer lengths: 5 mm (solid line); 8.5 mm (dashed line); 12 mm (dotted line); 20 mm (long dashes); 50 mm (dash-dotted line) [45].

The transfer function thus measured with an FFT analyzer represents the phase and gain difference between the two measurement systems. If the first microphone (#1) is used as a reference, the phase and gain correction $[G_{12}]_{\text{calibrated}}$ is given by:

$$[G_{12}]_{\text{calibrated}} = G_{12}/\left(H_{12} \cdot |H_1|^2\right), \tag{8.40}$$

where H_{12} is the calibration transfer function and $|H_1|$ is the gain factor of the microphone system #1.

2) *Microphone Switching Approach*

Chung describes the switching procedure as follows, where the intensity in the x-direction, $I_x(\omega)$ is given by:

$$I_x(\omega) = \text{Im}\left\{\left[G_{12} \cdot G_{12}{}^S\right]^{1/2}\right\}/\rho\omega\Delta r \; |H_1||H_2|, \tag{8.41}$$

and where $G_{12}{}^S$ is the cross-spectrum measured with the two microphone sensing locations interchanged and $|H_1|$ and $|H_2|$ are the gain factors of each microphone. It is seen from Eq. (8.41) above that the phase-corrected cross spectrum is obtained from the geometric mean of the original cross-spectrum G_{12} and the switched cross-spectrum $G_{12}{}^S$, or

$$[G_{12}]_{\text{calibrated}} = \left[G_{12} \cdot G_{12}{}^S\right]^{1/2}/|H_1||H_2|, \tag{8.42}$$

3) *Modified Microphone Switching Approach*

The modified switching approach has been described and used by Chung and Blaser [54] and Chung [55]. In this approach, the relative phase between the two microphone systems and the gain factors of each system are obtained by taking the square root of the ratio of the original and switched transfer functions. Any broadband signal can be used in determining the transfer functions and the microphones need not be subjected to the same sound fields. The procedure is given by:

$$[G_{12}]_{\text{calibrated}} = G_{12}/\big(K_{12} \cdot |H_1|^2\big), \tag{8.43}$$

where

$$K_{12} = \big[H_{12}/H_{12}{}^S\big]^{1/2}, \tag{8.44}$$

and where H_{12} and $H_{12}{}^S$ are the original and switched transfer functions described above.

Chung has discussed the advantages of the three methods described above in Ref. [55] to correct for phase mismatch. In the first method, the Transfer Function Approach, a very sophisticated spectrum analyzer is not needed. It is not necessary to repeat measurements with suitable microphones to take the square root of the complex variable in Eq. (8.42). However, it is usually difficult to measure the transfer function, H_{12} over a broad frequency range. Less error is likely with use of the second method, the Microphone Switching Approach, with microphones in the side-by-side arrangement. However, some intensity probes are not symmetrical and this method cannot be used. The third method is a compromise between the first two methods. A good free field condition is not needed and it saves ensemble-averaging time. But it is necessary to take the square root of a complex variable.

Some continue to make phase mismatch corrections in intensity measurements, as explained by Chung, to obviate the need to purchase special probes produced by manufacturers and their associated software. However, many now make use of commercial sound intensity probes. In such commercial probes, microphones, and associated electronics are chosen individually in which the phase mismatch has been carefully minimized. In 1987, a special calibrator was produced by Bruel & Kjaer for the calibration of p–p probes. This calibrator makes it possible to simulate an acoustic field of known intensity by positioning acoustical elements between the microphones thus creating a certain frequency-dependent phase difference between them. The calibrator can then be used to perform particle and sound intensity calibrations of the probe.

Even with the best equipment that is available today, phase mismatch remains a problem. It can be shown that the estimated intensity, subject to a phase error φ_e, is a very good approximation to the true intensity I_r, (which is unaffected by phase mismatch) and it can be written as [59]

$$\hat{I}_r \approx I_r - \frac{\varphi_e}{k\Delta r}\frac{p_{rms}^2}{\rho c} = I_r\left(1 - \frac{\varphi_e}{k\Delta r}\frac{p_{rms}^2/\rho c}{I_r}\right). \tag{8.45}$$

That is, the phase error causes a bias error in the measured intensity that is proportional to the phase error and to the mean square pressure, but inversely proportional to the wave number (and thus frequency) [30, 57].

For a given phase error, φ_e, the bias error in the intensity is proportional to the ratio of the mean square sound pressure to the sound intensity; but it is inversely proportional to the frequency and the microphone separation distance. Ideally the phase error should be zero, of course. In practice one must, even with state-of-the-art equipment, allow for phase errors ranging from about 0.05° at 100 Hz to 2° at 10 kHz. Both the IEC standard and the North American ANSI standards on instruments for the measurement of sound intensity specify performance evaluation tests that ensure that the phase error is within certain limits. Figure 8.20 [60] shows the error in intensity level as a function of frequency for a phase error of 0.3°.

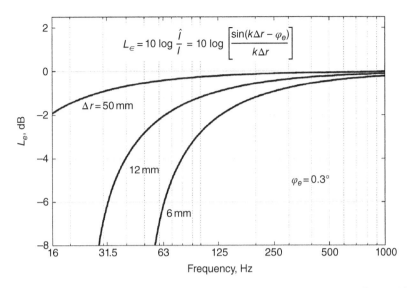

Figure 8.20 Error level L_e due to a phase error φ_e of 0.3° in a plane propagating wave. (*Source:* After Gade [60].)

Eq. (8.45) is often written

$$\hat{I}_r = I_r + \left(\frac{I_0}{p_0^2}\right)p_{rms}^2 = I_r\left(1 + \frac{I_0}{p_0^2/\rho c}\frac{p_{rms}^2/\rho c}{I_r}\right),\tag{8.46}$$

where the residual intensity I_0

$$\frac{I_0}{p_0^2/\rho c} = -\frac{\varphi_e}{k\Delta r}\tag{8.47}$$

has been introduced. The residual intensity, which is normally measured in one-third octave bands, is the "false" sound intensity indicated by the instrument when the two microphones are exposed to the same pressure p_0. Under such conditions the true intensity is zero and the indicated intensity I_0 should be made very small. A commercial instrument for measuring I_0 in which the two microphones are subjected to the same sound pressure p_0 is shown in Figure 8.21.

The right-hand side of Eq. (8.46) shows how the error caused by phase mismatch depends on the ratio of the mean square pressure to the intensity in the sound field, which is governed by the conditions in the sound field.

Example 8.3 A two-microphone sound intensity probe has a total phase mismatch of 1° at 400 Hz and the distance between the microphones is 12 mm. The probe measures a sound intensity level of 76.5 dB in a 400 Hz progressive plane wave. What is the true value of the sound intensity level?

Solution

If a phase mismatch $\varphi_e = 1° = \pi/180$ rad exists between the two measuring channels, we determine the approximation error in a plane sinusoidal wave using Eq. (8.39): $\varepsilon = \hat{I}/I = \sin(k\Delta r \pm \varphi_e)/k\Delta r$. Since $k\Delta r = (2\pi f/c)\,\Delta r = 2\pi(400)(12\times10^{-3})/344 = 0.0279\pi$. Thus, the underestimation is $\varepsilon = \sin(0.0279\pi - \pi/180)/0.0279\pi = 0.8$, so $L_e = 10\log(0.8) = -0.97$ dB. Now, the overestimation is $\varepsilon = \sin(0.0279\pi + \pi/180)/0.0279\pi = 1.197$, and $L_\varepsilon = 10\log(1.197) = +0.78$ dB. Therefore, the true intensity will be $76.5 + 0.78 \approx 75.7$ dB or $76.5 - 0.97 \approx 75.5$ dB.

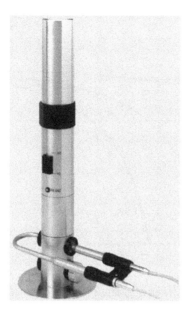

Figure 8.21 Coupler for measurement of the pressure-residual intensity index of p–p sound intensity probes (Bruel & Kjaer, Denmark).

Example 8.4 The two microphones in a sound intensity probe are known to have a response phase difference of $0.3°$ at 63 Hz when the probe is oriented in the direction of propagation of a plane wave. Determine the separation between the microphones such that the maximum approximation error at 63 Hz is -2 dB.

Solution

Since $L_\varepsilon = 10\log\varepsilon$, then $\varepsilon = 10^{-2/10} = 0.63$. Now, $k = (2\pi f/c) = 2\pi(63)/344 = 0.367\pi$, and $\varphi = 0.3° = 1.7\pi \times 10^{-3}$. If $k\Delta r - \varphi$ is small, then $\varepsilon = 1 - (\varphi/k\Delta r)$. Thus, $1 - (1.7\pi \times 10^{-3}/0.367\pi\Delta r) = 0.63$. Solving for Δr gives that the separation between the microphones must be 0.0125 m ≈ 12.5 mm (see Figure 8.20).

Phase mismatch in sound intensity probes is usually described in terms of the so-called pressure-residual intensity index:

$$\delta_{pI_0} = 10\log\left(\frac{p_0^2/\rho c}{I_0}\right), \tag{8.48}$$

which is a common way of describing the phase error. With a microphone separation distance of 12 mm, the typical phase error corresponds to a pressure-residual intensity index of 18 dB over most of the frequency range. Figure 8.21 shows a commercial coupler for the measurement of the pressure-residual intensity index. The error due to phase mismatch is small, provided that the measured pressure-intensity index δ_{pI} is much less than the pressure-residual intensity index δ_{pI_0} (Eq. (8.48)); that is

$$\delta_{pI} \ll \delta_{pI_0}, \tag{8.49}$$

where

$$\delta_{pI} = 10\log\left(\frac{p_{rms}^2/\rho c}{I_r}\right) \approx L_p - L_I \tag{8.50}$$

is the pressure-intensity index of the measurement. The inequality Eq. (8.49) shows that the phase error in the equipment must be much smaller that the phase angle between the two sound pressure signals in the sound field for measurement errors to be minimized. A more specific requirement is given by

$$\delta_{pI} < L_d = \delta_{pI_0} - K, \tag{8.51}$$

where the quantity

$$L_d = \delta_{pI_0} - K \tag{8.52}$$

is called the *dynamic capability index* of the instrument and K is the *bias error index*. The dynamic capability index states the maximum acceptable value of the pressure-intensity index of the measurement for a given grade of accuracy and is used in the ISO standards. The larger the value of K the smaller is the dynamic capability index, the stronger and more restrictive is the requirement, and the smaller is the error. The condition expressed by the inequality Eq. (8.51) and a bias error index of 7 dB guarantee that the error due to phase mismatch is less than 1 dB. This corresponds to the phase error of the equipment being five times less than the actual phase angle in the sound field. Figure 8.22 gives a plot of the maximum error due to phase mismatch as a function of the bias error index K.

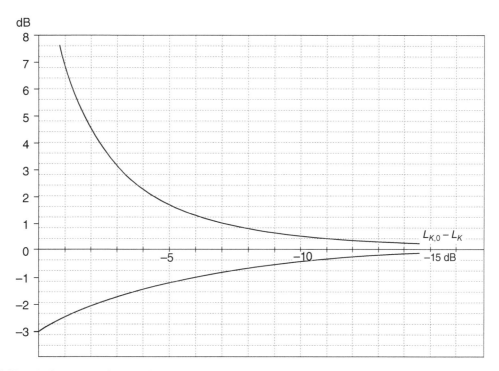

Figure 8.22 Maximum error due to a phase mismatch as a function of the bias error index K for negative residual intensity (*upper curve*) and for positive residual intensity (*lower curve*). (*Source:* After Gade [62].)

Example 8.5 Calculate the pressure-intensity index in a plane traveling wave field with the intensity probe axis at $30°$, $60°$, and $90°$ to the direction of propagation.

Solution

If a plane sound wave is incident at an angle θ to the probe axis, the measured intensity is reduced by a $\cos \theta$ factor, such that $I(\theta) = I_r \cos \theta$. Then, in Eq. (8.50) we obtain that

$$\delta_{pI} = 10 \log \left(\frac{p_{rms}^2 / \rho c}{I_r \cos \theta} \right) \approx L_p - L_I - 10 \log \cos \theta = -10 \log \cos \theta.$$

Therefore, $\delta_{pI} = 0.63$ dB at $30°$, 3 dB at $60°$, and ∞ at $90°$.

Sound power measurements using sound intensity involve integrating the normal component of the intensity over a surface. The global version of Eq. (8.46) has the form [30, 57]

$$\hat{W}_a = W_a \left[1 + \frac{I_0 \rho c}{p_0^2} \frac{\int_S \left(p_{rms}^2 / \rho c \right) dS}{\int_S \mathbf{I} \cdot d\mathbf{S}} \right], \tag{8.53}$$

which shows that the global version of the inequality Eq. (8.51) can be written as

$$\Delta_{pI} < L_d = \delta_{pI_0} - K, \tag{8.54}$$

where

$$\Delta_{pI} = 10 \log \left[\frac{\int_S \left(p_{rms}^2 / \rho c \right) \, dS}{\int_S \mathbf{I} \cdot d\mathbf{S}} \right] \tag{8.55}$$

is the global pressure-intensity index of the measurement. This quantity plays the same role in scanning sound power estimation as the pressure-intensity index does in measurements at discrete points.

It is obvious that the presence of noise sources outside the measurement surface increases the mean square pressure on the surface, and thus the influence of a given phase error; therefore, phase mismatch limits the range of measurement. Most modern sound intensity analyzers can determine the global pressure-intensity index concurrently with the actual measurement. Figure 8.23 shows examples of the index measured under various conditions [59]. In practice one should examine whether the inequality Eq. (8.54) is or is not satisfied if there is significant noise from extraneous sources. If the inequality is not satisfied, it can be recommended to use a measurement surface somewhat closer to the source than is advisable in more favorable circumstances. It may also be necessary to modify the measurement conditions – to shield the measurement surface from strong extraneous sources, for example, or to increase the sound absorption present in the room.

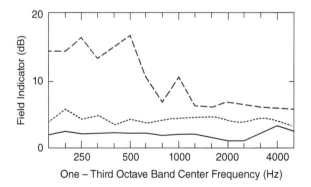

Figure 8.23 The global pressure-intensity index Δ_{pI} determined under three different conditions: sold line, measurement using a "reasonable" surface; dashed line, measurement using an eccentric surface; and dash-dotted line, measurement with strong background noise. (*Source:* After Jacobsen [61].)

d) Calibration of *p–p* Sound Intensity Probes

Several calibration procedures are required for *p–p* sound intensity probes in order to ensure accurate measurements are made. The two microphones should be calibrated with a pistonphone as usual. In addition, the IEC standard and the ANSI counterpart standard specify minimum acceptable values of the pressure-residual intensity index for the probe and the processor system. The test to ensure these minimum values are achieved involves exposing the two microphones to the same sound pressure in a small cavity driven by a broadband noise source. A similar test is carried out on the processor by feeding the same signal to the two channels. As already discussed, the measured pressure-intensity index, which is equal to the pressure-residual intensity index, reveals how well the microphones are matched. According to the results of the test, the probe and its associated system are classified as "class 1" or "class 2."

The pressure and intensity response of the probe should also be checked in a plane propagating wave as a function of frequency. In addition, tests should be performed to ensure that the directional response of the probe follows the ideal cosine law within a specific tolerance. A final test is required for the frequency range below 400 Hz. In this test the intensity probe is exposed to a sound field in a standing wave tube with a prescribed standing wave ratio (24 dB for class 1 probes.) The sound intensity indicated by the system, when the probe is traversed through the resulting interference field, should be within a given tolerance. More details are given by Jacobsen in Refs. [30, 57].

The Bruel and Kjaer Company produces an intensity probe with the microphones arranged in a face-to-face arrangement. The probe when it is supplied has microphones which have been specially selected to have an identical phase performance and effective microphone separation. See Figure 8.24. Bruel and Kjaer also produces three calibrators which can be used in the field and laboratory. See Figure 8.25. The checks to the sound intensity system can be finished using these calibrators and checked using the knowledge that the known sound power source shall be accurately determined in a free field (Figures 8.29 and 8.30).

Calibration of a sound intensity system using a *p–p* intensity probe requires knowledge of:

- Sound pressure transducer properties
  - Sensitivity and gain adjustment
  - Phase matching
  - Effective microphone acoustical separation

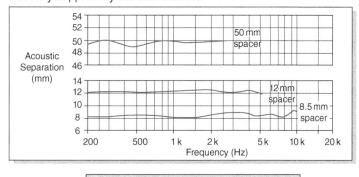

Figure 8.24   Effective separation results performed by the manufacturer of the Bruel and Kjaer probe (Courtesy of Brüel & Kjær).

Calibrators

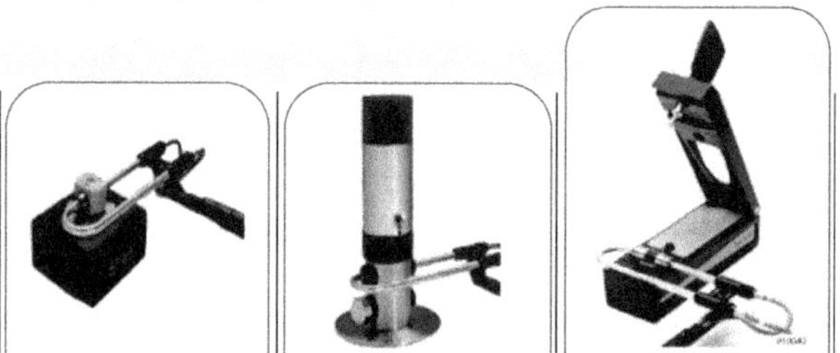

		Sound Level Calibrator 4231	Sound Intensity Calibrator 3541	Sound Intensity Calibrator 4297
Main Application		In the field	In the laboratory	In the field
Dismantling of Probe		Necessary	Necessary	Unnecessary (up to 3 kHz)
Calibration of Sound Pressure Level	L_p	Yes	Yes	Yes
Verification of Sound Intensity Level	L_I	No	Yes	No
Verification of Particle Velocity Level	L_V	No	Yes	No
Pressure-Residual Intensity Index	$L_p - L_I$	No	20 Hz to 5 kHz	20 Hz to 3 kHz* 20 Hz to 6.3 kHz**

* with spacer ** without spacer

Figure 8.25 Calibrators for calibration of sound intensity calibration in the field and laboratory (Courtesy of Brüel & Kjær).

- Density of fluid medium

  – Composition of
  – Temperature
  – Ambient pressure

Using the three calibrators in Figure 8.25, the system can be calibrated as follows:

1) The sensitivity and gain can be checked using the B&K 3541 calibrator, see for example Figure 8.26.
2) The manufacturer's phase matching can be checked using the B&K Type 4297 calibrator (see Figure 8.27) or the B&K Type 3541 calibrator (see Figure 8.28), in which an intensity coupler is used to simulate the microphone separation.

Finally, the sound intensity level is determined in a free field with an omnidirectional sound source and the sound power level in checked and compared with that of a standard sound power level source (Figure 8.29). Also checked are the consistency of the sound pressure, sound intensity and velocity levels (Figure 8.30).

8.6.2 The p–u Method

The p–u method of measuring sound intensity requires the use of two completely different transducers, one to measure the sound pressure and the other the particle velocity. As discussed before, Clapp and Firestone in 1941, Baker in 1955, Burger et al. in 1973, and Van Zyl and Anderson in 1975 all used such devices [12–14, 17, 18]. Unfortunately, all of these devices produced unsatisfactory results.

- Sensitivity is supplied by the manufacturer
- Gain adjustment by
  use of conventional calibration techniques

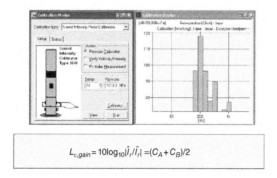

$$L_{c,gain} = 10\log_{10}|\hat{I}_r/\tilde{I}_r| = (C_A + C_B)/2$$

Figure 8.26 Sensitivity and gain adjustment using B&K calibrator 3541 (Courtesy of Brüel & Kjær).

- Use pressure chamber, for example as
  in Type 4297 Sound Intensity calibrator
  - Measure phase by use of a
    dual channel FFT/CPB analyzer
  - Measure $L_p - L_I = PRI = \delta_{PI_0}$ and
    convert to phase using the formula

$$\varphi = \sin^{-1}\left(\frac{I_{ref} \cdot \omega \cdot \rho \cdot \Delta r}{P_{ref}^2 \cdot 10^{\frac{PRI}{10}}}\right)$$

4297
Sound Intensity
Calibrator

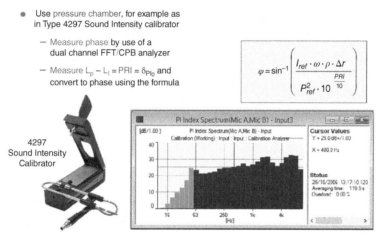

Figure 8.27 Phase difference determination between microphone pair using B&K Type 4297 calibrator and formula to determine phase (Courtesy of Brüel & Kjær).

- Upper chamber for
  - Gain calibration/adjustment,
    » Use Pistonphone Type 4228
  - Phase verification,
    » Use Sound Source Type ZI 0055
- Upper/lower chamber for
  - Intensity verification
  - Particle Velocity verification

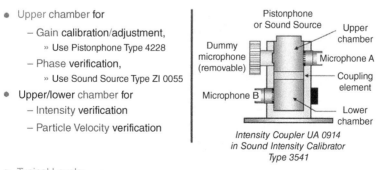

*Intensity Coupler UA 0914
in Sound Intensity Calibrator
Type 3541*

- Typical Levels:
  - Sound Pressure Level: 118.0 dB re 20μPa
  - Particle Velocity Level: 117.8 dB re 50nm/s
  - Sound Intensity Level: 117.9 dB re 1pW/m²

Figure 8.28 Use of B&K coupler UA 0914 and calibrator B&K 3541 to verify phase difference between microphone pairs (Courtesy of Brüel & Kjær).

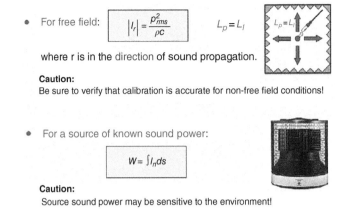

- For free field: $\left|I_r\right| = \dfrac{p_{rms}^2}{\rho c}$ $L_p = L_I$ $L_p = L_I$

where r is in the direction of sound propagation.

Caution:
Be sure to verify that calibration is accurate for non-free field conditions!

- For a source of known sound power: $W = \int I_n ds$

Caution:
Source sound power may be sensitive to the environment!

Figure 8.29 Final verification steps in system calibration using free field check to see that sound pressure level and sound intensity level are the same in a free field using a known value for ρc (Courtesy of Brüel & Kjær).

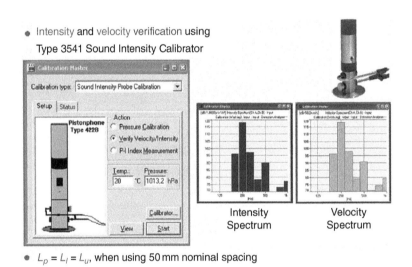

- Intensity and velocity verification using Type 3541 Sound Intensity Calibrator

Intensity Spectrum   Velocity Spectrum

- $L_p = L_I = L_u$, when using 50 mm nominal spacing

Figure 8.30 Final verification check to see that the sound pressure level (Figure 8.26) is equal to the intensity level and spectrum to those for the sound intensity and velocity for 50 mm microphone spacing (Courtesy of Brüel & Kjær).

a) The *p–u* Measurement Principle

In 1984, Norwegian Electronics manufactured a sound intensity probe, which used a microphone in conjunction with a special transducer [63]. The transducer detected the particle velocity based on the displacement of an ultrasound beam. Unfortunately, the device proved to be difficult to calibrate and was rather bulky and was withdrawn from sale after 10 years in the mid-1990s [23].

A new miniaturized sensor device known as a Microflown became available for the measurement of particle velocity in the late 1990s [29, 64, 65]. Subsequently this device, when combined with a small microphone, has become available as a small sound intensity probe. The particle velocity sensor is similar to the hot-wire anemometer employed in the sound intensity probe described by Baker in 1955 [14]. But it is much smaller and does not use wires. Instead two miniature sensors made of silicon nitride on a thin platinum layer are arranged in parallel. They are heated by an electrical current of 10 mA to about 200 °C and sense the fluctuating particle velocity through a cooling effect and related electrical resistance fluctuations. The second sensor is shielded by the first sensor making the intensity device exhibit a cosine directivity in the same way as a p–p probe. Since the sensors are so small (only 1 mm in length) and situated only 40 µm apart, phase shift errors caused by the small spacing can be neglected even at very high frequency. This device is manufactured by Microflown Technologies and probes made for the measurement of intensity in one, two, and three dimensions are available [29, 64–66].

The instantaneous sound intensity is, as before, the product of the sound pressure and particle velocity

$$I_r = p(t)\, u_r(t), \tag{8.56}$$

and for the simple harmonic case, using complex notation (see Eq. (8.7))

$$I_r = \tfrac{1}{2}\, \mathrm{Re}\ (pu_r{}^*). \tag{8.57}$$

Thus, in the frequency domain

$$I_r\,(\omega) = \tfrac{1}{2}\, \mathrm{Re}\ [P\,(\omega)U^*(\omega)], \tag{8.58}$$

$$I_r\,(\omega) = G_{pu}\,(\omega), \tag{8.59}$$

where $P(\omega)$ and $U(\omega)$ are the Fourier transforms of the sound pressure and particle velocity, and G_{pu} is the cross spectrum between sound pressure and particle velocity.

The frequency response of the Microflown probe has been found to be relatively flat up to 1 kHz. Then between 1 kHz and about 10 kHz, the frequency response decreases.

b) Sources of Error in *p–u* Measurement Probes

Some of the sources of error in p–u measurement probes are similar to those in p–p probes and some are different. Phase mismatch is still a problem with p–u systems. With the Microflown probe, phase mismatch in the particle velocity sensor itself is not a problem, but it remains as a problem between the microphone and the sensor. Phase mismatch between the p and u transducers must be corrected otherwise the measurements will be in error. The error is most important in measurements close to a source. This can be illustrated as follows. From

$$I_r = \frac{1}{2} p u_r \cos \varphi_r, \tag{8.60}$$

it can be seen that for simple harmonic progressive plane waves when the sound pressure and particle velocity are in phase, $\varphi_r = 0$ and $\cos \varphi_r = 1$ and small phase mismatch errors will have a negligible effect since $\cos \varphi_r \approx 1$ until the phase angle becomes very large. However, when the sound pressure and particle velocity are nearly out of phase (for example near a source or a hard reflective surface), then $\varphi_r \approx 90°$, and the reactive intensity becomes

$$J_r = \frac{1}{2} p u_r \sin \varphi_r \approx \frac{1}{2} p u_r \varphi_r, \tag{8.61}$$

and even a small mismatch error in the estimation of φ_r will have a large effect on the value of I_r measured. See Figure 8.31.

Using complex notation, it can be seen that by introducing a small mismatch error, in φ_r into Eq. (8.57), that the measured time-average intensity I_r is

$$I_r = \frac{1}{2}\, \mathrm{Re}\left\{p u_r^* e^{-j\varphi_r}\right\} = \mathrm{Re}\left\{[I_r + jJ_r](\cos \varphi_r - j\sin \varphi_r)\right\} \approx I_r + \varphi_r J_r, \tag{8.62}$$

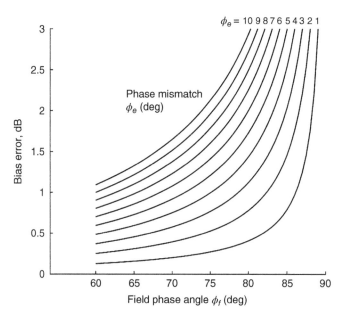

Figure 8.31 Normalized systematic error ϕ_e of a p–u system due to phase mismatch as a function of field phase angle [24].

when φ_r is small and where I_r is the real intensity and J_r is the reactive intensity. Eq. (8.62) again clearly shows that a small phase error can lead to a significant error in the measured intensity I_r when the time-average imaginary intensity is much greater than the real intensity $J_r \gg I_r$ (e.g. near to a source). However, when the real intensity is much greater than the imaginary intensity $I_r \gg J_r$, then a much greater phase error can be tolerated.

Example 8.6 A p–u intensity probe is used to measure a field angle between pressure and particle velocity of 70°. Calculate the maximum instrumentation phase mismatch of the probe to have a systematic error of 1 dB.

Solution

The estimated sound intensity is $\hat{I} = \frac{1}{2}\, pu_r \cos(\phi_f \pm \phi_e)$ and the true intensity is
$I = \frac{1}{2}\, pu_r \cos(\phi_f)$, where ϕ_f is the angle between the sound pressure and particle velocity and ϕ_e is the phase error. Therefore, the error is $10\log(\hat{I}/I) = 10\log[\cos(\phi_f \pm \phi_e)/\cos(\phi_f)] = 10\log[\cos(\phi_e) \mp \tan(\phi_f) \sin(\phi_e)]$. When $\phi_e \ll 1$, the error can be written as $10\log[1 \mp \phi_e \tan(\phi_f)]$. For an error of 1 dB and $\phi_f = 70°$, we obtain that $\mp\phi_e \tan(70°) = 10^{0.1} - 1$. Therefore, $\phi_e = \pm 0.0942\,\text{rad} \approx \pm 5.4°$ (see Figure 8.31).

---

c) Calibration of p–u Measurement Probes

Unfortunately, p–u intensity probes are unlike p–p probes, in which phase mismatch can be corrected in principle by reversing the probe in the sound field. This is because the phase mismatch in this case with a p–u probe does not change sign with probe reversal and thus cannot be canceled out.

Fahy has discussed the importance of correcting for p–u phase mismatch in reference [24]. Unfortunately, currently there are no standardized methods for calibrating p–u probes or for determining and correcting for phase mismatch. A small p–u probe was brought into production about 1990 (see Figure 8.32). The Microflown p–u probe, since it is much smaller, can be put in a standing wave tube. According to the Microflown

Figure 8.32 A *p*–*u* sound intensity probe. (*Source:* by Microflown Technologies, The Netherlands.)

company their probe can be calibrated in a standing wave tube in the frequency range 20 Hz–3.5 kHz. At the end of the tube, the phase shift between sound pressure and particle velocity should be equal to ±90°. Near to the end of a rigidly terminated standing wave tube the phase shift is a simple sine function.

Jacobsen et al. [67, 68] have also investigated the calibration of this *p*–*u* probe using two other approaches. One calibration method is known as the "Piston on a Sphere" and is used in the frequency range 10–400 Hz. The other method requires the use of a special loudspeaker with a known acoustical calibration and a reference microphone placed in an anechoic room. This method is useful in the range 300 Hz–20 kHz. Since the calibration of the Microflown probe is complicated, most users will need to rely on that carried out by the manufacturer [68]. The Microflown probe has now been used to make sound power measurements on large machines [69]. Its uses in scanning and measurement criteria have also been discussed [70, 71].

8.6.3   The Surface Intensity Method

With the surface intensity method, in principle the surface vibration velocity can be measured with a non-contacting device, such as a laser Doppler system, or a contacting device such as an accelerometer. The sound pressure still has to be measured with a microphone.

The disadvantage of this method is that it is laborious. The surface velocity must be measured at numerous locations and the sound pressure must also be sensed close to these locations in which the sound field is normally quite reactive. However, the method does have the advantage that the information about the surface vibration can be also of considerable value independently and allows the radiation efficiency of the surface to be calculated as part of the measurement process. In addition, the microphone measurements must be made in the near field and although this has measurement difficulties, background noise is less of a problem because the source sound is normally dominant. Most attention to date has been given to the *p*–*p* and *p*–*u* approaches described in Sections 8.6.1 and 8.6.2 above.

The surface intensity technique which uses an accelerometer mounted on a vibrating surface and a microphone located close to the accelerometer has been under development since about 1974. Macadam described the use of this technique in the measurement of the sound power radiated from room surfaces in lightweight buildings [72, 73]. Hodgson also discussed this technique and its use on a large centrifugal chiller machine

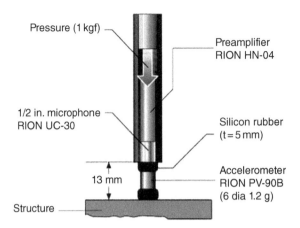

Figure 8.33 Hand-held probe for surface intensity measurement [84].

[74]. Brito investigated theoretically and experimentally the case of a vibrating rectangular flexible panel and obtained good agreement [75, 76]. Kaemmer and Crocker [77–79] measured the sound power of a vibrating cylinder using the surface intensity method and compared it with the reverberation room method and theory. McGary and Crocker continued development of the surface intensity technique [80–83]. They then applied it successfully to the determination of the sound intensity radiated from the different surfaces of a Cummins NTC 350 (260 kW) diesel engine [80–82]. They obtained good results when they carefully accounted for phase shifts between the microphone and accelerometer signals [83]. Although, the surface intensity probe is more complicated and more difficult to calibrate than a sound intensity probe, it is still used by some. See for example the hand-held probe developed by Hirao et al. and shown in Figure 8.33 [84].

a) The Surface Intensity Principle
In the case of surface intensity measurements, the velocity u_n normal to the vibrating surface areas S can be found directly by various noncontacting devices or by integrating the signal from a surface-mounted accelerometer. The pressure p can be measured by a microphone located close to the accelerometer. The finite distance between the noncontacting device or accelerometer and the microphone introduces a time delay Δt between pressure and velocity signals. This can be approximately related to a phase shift ϕ by $\Delta t = \phi/2\pi f$ where f is the frequency (Hz). As with the p–u method, the intensity is usually computed in the frequency domain by feeding the velocity and pressure signals into an FFT analyzer, since the spectral distribution of sound power is usually required. The sound intensity may be shown to be [77, 78]

$$I_n(\omega) = C_{up}(\omega) \cos \phi + Q_{up}(\omega) \sin \phi, \qquad (8.63)$$

where C_{up} is the co-spectrum (real part) and Q_{up} is the quad-spectrum (imaginary part) of the one-sided cross-spectral density between velocity and pressure G_{up}:

$$G_{up}(f) = C_{up}(f) - j\,Q_{up}(f), \qquad (8.64)$$

b) Sources of Error in Surface Intensity Measurements
The phase shift between the surface velocity and the sound pressure must be corrected otherwise large errors will occur. The phase shift can be caused by the finite distance between the microphone and the vibrating surface. This can be approximated by the travel time in a progressive wave (see Figure 8.34). However, a phase error can also be caused by the velocity and pressure instrumentation. The total error E in intensity caused by uncorrected phase shift can be written as [78]

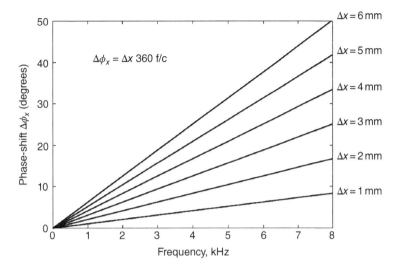

Figure 8.34 Phase shift $\Delta\phi_x$ from the finite distance between the microphone and the vibrating surface [78].

$$E = 10 \log I_t/I = 10 \log \left(\cos \phi + Q_{up} \sin \phi \right). \tag{8.65}$$

The effect on the estimate of the intensity of undetected phase errors is seen to be a function of ϕ and also the ratio of the quad-to co-spectrum (imaginary to real part of the cross-spectrum). The effect of the ratio Q_{up}/C_{up} and uncorrected phase shift ϕ on the intensity is seen to be extremely important. If the quad-spectrum (imaginary part) is much larger than the co-spectrum (real part), then even moderate uncorrected phase shifts can cause large errors in the sound intensity measurements.

c) Calibration of Surface Intensity Measurement Systems.

It is obvious that it is very important to measure and correct for phase shift caused by the propagation of sound waves from the surface velocity probe or accelerometer to the pressure sensing device. Additional phase shifts between the velocity/accelerometer sensor and the pressure sensor can also be caused by the associated electronics. In their work with accelerometer/pressure measurements (see Figure 8.35) Kaemmer and Crocker, and McGary and Crocker studied this problem [78, 79, 82].

McGary and Crocker measured the phase shift caused by the microphone/accelerometer separation and also the instrumentation.

Noting that $G_{ap}(\omega) = -j\omega G_{up}(\omega)$, and that $G_{ap} = G_{pa}^*$, Eq. (8.65) can be written as

$$E = 10 \log \left[\cos \phi + \left(C_{pa}/Q_{pa} \right) \sin \phi \right] \tag{8.66}$$

which is plotted in Figure 8.36. It is observed that the error caused by uncorrected phase shifts depends strongly on the ratio of C_{pa}/Q_{pa}.

8.7 Applications

Some of the most common practical applications of sound intensity measurements are now discussed. In 1978, Chung and Pope demonstrated that using a two-microphone probe that sound intensity could be measured both (i) at fixed points in the near field (2.5 cm) from a loudspeaker source and (ii) at 12 points on a hemisphere of 1 m radius surrounding the source [52]. Then using these two measurements and by integration over the hemisphere they obtained sound power level estimates agreeing within 0.4 dB over the

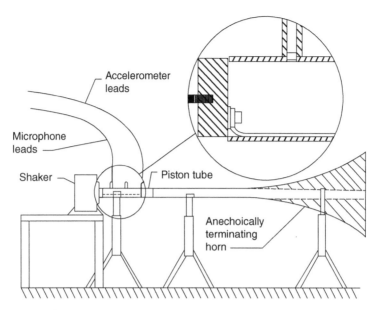

Figure 8.35 Transducer arrangement for phase shift determination [82].

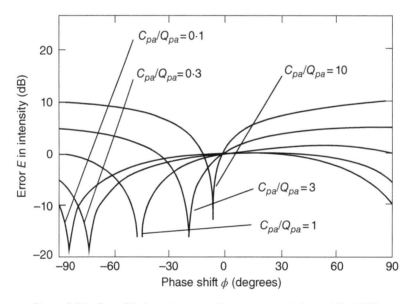

Figure 8.36 Error E in intensity caused by uncorrected phase shift ϕ [83].

frequency range 500–5000 Hz. In 1979, Chung used the scanning procedure to identify parts of a diesel engine [19]. Soon after, Crocker et al. used conformal surface scanning to source-identify diesel engines [20]. They were able to verify that surface scanning on an engine produced results very similar to those made with fixed points using single microphones and making the far field assumption $I = p^2/\rho c$ on a spherical array around the engine. They were also able to demonstrate that sound intensity could be used to determine the transmission loss of partitions by surface scanning as accurately as with single microphone measurements using the two-room method [21, 22].

Sound intensity measurements became widely used in industry in the early 1980s. But a fundamental concern remained whether to measure the intensity at fixed points or to use the faster and more convenient approach of scanning the intensity probe over a control surface encompassing the source. Several researchers conducted theoretical and experimental studies to compare the accuracy of results for the sound power of sources made using sound intensity scans compared with fixed points. In 1984, Bockhoff studied this surface sampling problem using a numerical simulation of continuous scanning (moving the probe over a plane control surface) compared with the use of discrete fixed points measurements on the surface [85]. He concluded that "continuous scanning (even with non-constant (speed) provides a higher accuracy than fixed point measurements" [85].

In 1986 Pope published theoretical and experimental studies of the sound power of a reference source located over a reflecting surface [86]. The spatial sampling was made on a hemispherical control surface. Pope also concluded that for a "given number of measurements, scan averaging can be expected to yield better sound power estimates than point sampling" [86]. In addition, Crocker raised the issue in 1986 with his editorial "To scan or not to scan" [87]. Yang and Crocker also re-examined the questions in 1994 with a computer simulation of sound intensity sampling with fixed points and with scanning of a source in the presence of background noise [88]. They similarly concluded that "the scanning method is more accurate that the fixed point method" [88]. Most engineering measurements of sound power and transmission loss of partitions using sound intensity are made with the scanning method. The rest of Section 8.7 reviews a number of practical results.

8.7.1 Sound Power Determination

One important application of sound intensity measurements is the determination of the sound power of operating machinery in situ. Sound power determination using intensity measurements is based on Eq. (8.5), which shows that the sound power of a source is given by the integral of the normal component of the intensity over a surface that encloses the source. Figure 8.37 shows measurements of the normal intensity measured over hemispherical and rectangular surfaces, where the total sound power of a machine source is desired. Theoretical considerations seem to indicate the existence of an optimum measurement surface that minimizes measurement errors. In practice a surface of a simple shape at some distance, approximately

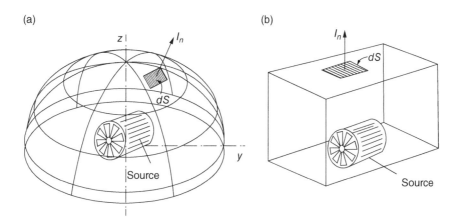

Figure 8.37 Sound intensity measured on a segment of (a) a hemispherical measurement surface and (b) a rectangular measurement surface.

(a)   (b)

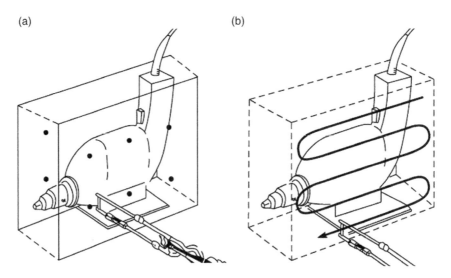

Figure 8.38 Typical box surface used in sound power determination with the intensity method: (a) measurements at discrete points and (b) measurements path used in scanning measurement.

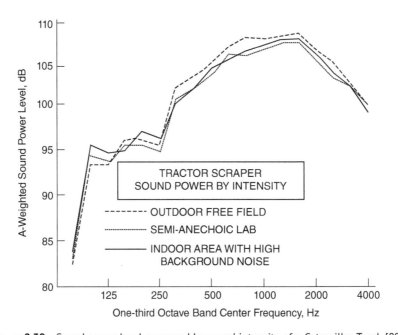

Figure 8.39 Sound power level measured by sound intensity of a Caterpillar Truck [89].

25–50 cm, from the source is recommended by some, as illustrated in Figures 8.38 and 8.39. If there is a strong reverberant field or significant ambient noise from other sources, the measurement surface should be chosen to be somewhat closer to the source under study. Such measurements made closely following the source contour are known as conformal. Sound power measurements using sound intensity were soon put into practice in industry in North America with good results. See, for example, measurements made on earth-moving equipment in Figure 8.39 [89]. The measurements (which have been A-weighted) show surprisingly good

Table 8.1 Sound power measurement by pressure vs. intensity techniques in varying acoustical environments. Identical tests on same crawler tractor [89].

Environment	A-weighted difference from true value sound power dB	
	By Intensity	By Pressure/Hemisphere
Outdoor free field	−1.0	0[a]
Semi-anechoic lab	−1.0	+0.4
Typical service shop	−1.3	+1.8
Reflective cell	—	+4.2
Reverberant room	−1.7	+8.2

[a] Assumed as "true value."

Figure 8.40 Indoor sound power measurement of a reciprocating compressor [90].

agreement between the results obtained in three different environments, one of which had high background noise (see Table 8.1). The outdoor free field result was taken as the reference. Figure 8.40 shows sound power measurements made on a reciprocating compressor using sound intensity [90]. Figures 8.41 and 8.42 show sound power measurements made on vehicles using gated sound intensity [91].

The surface integral can be approximated either by sampling at discrete points or by scanning manually or with a robot over the surface as shown in Figure 8.38. As discussed above, a third approach is to scan the probe over a conformal surface such as closely following the outline of a machine source. This is particularly useful when sound radiation from different parts of a machine must be determined. With the scanning approach, the intensity probe is moved continuously over the measurement surface in such a way that the axis of the probe is always perpendicular to the measurement surface. The scanning procedure, which was introduced in the late 1970s on a purely empirical basis, was regarded by some with much skepticism for more than a decade [57], but is now generally considered to be more accurate and far more convenient than the procedure based on fixed points [57]. A moderate scanning rate, say 0.5 m/s, and a "reasonable" scan line density should be used, say, 5 cm between adjacent lines if the surface is very close to the source, 20 cm if it is farther away.

Figure 8.41 Automated sound-intensity system used to measure the sound power of a passenger car [91].

Figure 8.42 Garden tractor used for tests of gated sound power [91].

Example 8.7 A sound wave traveling in the direction of the probe axis has an intensity level $L_1 = (+)$ 73 dB. Another sound wave travels in the opposite direction with an intensity level $L_2 = (-)$ 70 dB. The signs indicate the direction of the intensity. Determine the total sound intensity level measured by the probe when both waves travel simultaneously.

Solution

We must add intensity levels in different directions. We convert the levels back to intensity by $I = I_{ref} \times 10^{(L/10)}$. Then, $I_1 = I_{ref} \times 10^{(73/10)}$ and $I_2 = I_{ref} \times 10^{(70/10)}$. Now, the total intensity considering the directions is $I_T = I_{ref} \times (10^{7.3} - 10^7) = 9.953 \times 10^{-6}$. Finally, the intensity level is $L_I = 10 \times \log (9.953 \times 10^{-6} / 10^{-12}) = 70$ dB.

Example 8.8 A box-like hypothetical surface of 1 m × 1 m × 0.7 m is constructed around a noisy machine located on a reflective ground to determine its sound power through the scanning procedure. With a suitably long averaging time, a sound intensity probe is swept over each surface. The areas and respective space-average sound intensity levels on the surfaces are: 82 dB on $S_1 = 1$ m^2, 70 dB on $S_2 = 1$ m^2, 80 dB on $S_3 = 0.7$ m^2, 78 dB on $S_4 = 0.7$ m^2, and 81 dB on $S_5 = 0.7$ m^2. Calculate the sound power and sound power level of the machine.

Solution

The partial sound powers can be found from each side of the box and added. Thus, the total power is:

$$W = \Sigma (I_i \times S_i) = I_{ref} \times \left[10^{(82/10)} \times 1 + 10^{(70/10)} \times 1 + 10^{(80/10)} \times 0.7 + 10^{(78/10)} \times 0.7 + 10^{(81/10)} \times 0.7 \right]$$

$$= 10^{-12} \times 3.707 \times 10^8 = 3.707 \times 10^{-4} \text{ watt. Therefore, the total sound power level is}$$

$$L_W = 10 \times \log \left(3.707 \times 10^{-4} / 10^{-12} \right) = 85.7 \text{ dB}.$$

8.7.2   Noise Source Identification

This is perhaps the most important application. Every noise reduction project starts with the identification and ranking of noise sources and transmission paths. In one study of diesel engine noise, the whole engine was wrapped in sheet lead. The various parts of the engine were then exposed, one at a time, and the sound power was estimated with single microphones at discrete positions on the assumption that $I_r \approx p^2_{rms}/\rho c$. Sound intensity measurements make it possible to determine the partial sound power contribution of the various components directly.

Plots of the sound intensity measured on a measurement surface near to a machine can be used in locating noise sources. Figure 8.43a shows measurement surfaces near to a vacuum cleaner. Figures 8.43b and 8.43c show a mesh diagram and a contour plot of the normal intensity, and Figure 8.43d gives a vector flow diagram of the intensity.

Magnitude plots using arrows are useful in visualizing sound fields. Figure 8.44 shows the sound intensity measured in the vicinity of a violoncello [92].

8.7.3   Noise Source Identification on a Diesel Engine Using Sound Intensity

In these experiments the sound intensity was assumed to be given by

$$I_r = \text{Im}\{G_{12}\}/\rho\omega\Delta r. \tag{8.67}$$

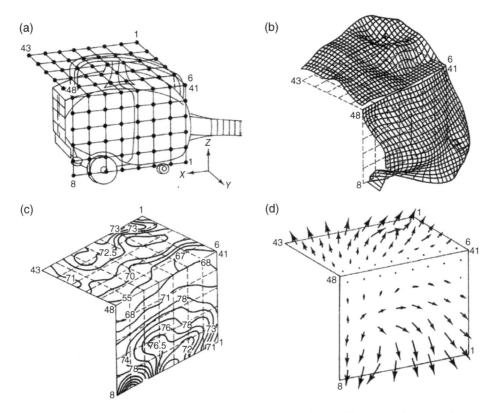

Figure 8.43 Measurements of the sound intensity radiated by a vacuum cleaner: (a) vacuum cleaner and measurement points (two measurement surfaces each with 48 measurements points), (b) mesh diagram showing magnitude of normal intensity on the two surfaces, (c) contour plot showing contours of equal normal intensity on the two measurements surfaces, and (d) vector intensity flow diagram. (*Source:* RION Technical Notes.)

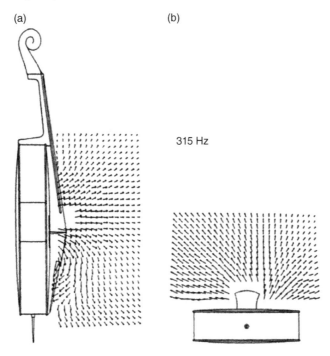

Figure 8.44 Sound intensity vectors measured in two planes near a violoncello producing a note having a fundamental frequency of 315 Hz. (a) in a plane on the axis and (b) in a plane intersecting the bridge. (*Source:* After Tachibana with permission [92].)

Corrections for phase shift between the two microphone channels were made using the approach of Krishnappa [58]. The true cross-spectrum G_{12} between the microphone signals is related to the measured value $\overline{G}_{12}$ by:

$$G_{12} = \overline{G}_{12} T_{12} / |H_2|^2, \tag{8.68}$$

where T_{12} is the transfer function between microphone channel systems 1 and 2 and H_2 is the gain of microphone system 2. In principle, this approach has some advantage over the switching technique suggested by Chung [54, 55]. Since switching is eliminated, measurements can be made about twice as fast after the phase shift is determined and stored on the FFT.

In this phase shift determination, the 1/4 in. microphones were mounted at the same longitudinal position at the end of a small tube which was excited by a white noise source from a small loudspeaker. The two microphones were subjected to the same signal to obtain the transfer function T_{12} below the tube cut off frequency. The gain $|H_2|$ was obtained separately by supplying microphone number 2 with a signal from a pistonphone.

If phase-matched microphones and high-quality microphone amplifiers are used, the standard deviation of the transfer function measured is often greater than the phase error being measured. In such cases it is not necessary to measure phase mismatch errors. Two microphones were used in a side-by-side arrangement for the sound intensity. See Figure 8.12.

The area-averaging intensity technique was used for the engine measurements, with the hand-held microphone pair being moved just over the engine surface by hand. An ensemble average was made over one hundred and fifty samples during each intensity measurement. The dynamic range of the intensity measurement system was 40 dB but improvements in the computer programming have extended the dynamic range to 80 dB.

The microphone spacing of 8 mm finally chosen allowed accurate intensity measurements to be obtained in the frequency range from about 200 to 7000 Hz [20, 39, 93].

a) Comparison between Sound Intensity and Lead-Wrapping Measurements

Using this approach and Eqs. (8.67) and (8.68), Reinhart and Crocker measured the sound power radiated by 103 different sub-areas on the surface on a Cummins NTC 350 (260 kW) diesel engine. A two-microphone probe was moved over the surface of the engine to obtain the sound intensity radiated by a particular surface area of the engine. The results from groups of areas were summed together to duplicate the eight major areas used in the lead-wrapping measurements and five major areas in the surface intensity measurements. The engine was run under the same condition of speed and load so that comparisons could be made between the results. Complete details of the sound intensity engine results and measurements are given in Refs. [20, 39, 93]. Figures 8.45 and 8.46 show the sound power obtained by summing over 25 locations on the oil pan. Figure 8.45 shows narrow-band results. Figure 8.46 shows the results given in Figure 8.45 summed into one-third octave bands.

It is interesting to note some peaks in the narrow-band spectrum in Figure 8.45 which are presumably caused by a forcing frequency being close to a structural resonance frequency. Also of interest is the good agreement between the lead-wrapping and the sound intensity results in Figure 8.46 for frequencies at and above 325 Hz. This result is similar to that for the oil pan when the lead-wrapping and the surface intensity results are compared in the later section of this chapter. The results for the other major engine surfaces are given in Refs. [20, 39, 93].

Table 8.2 shows a comparison of the overall sound power levels of the engine obtained from the sound intensity, surface intensity and lead-wrapping methods [39]. It is seen that the three methods all rank the oil pan as the predominant source. The weaker sources seem to be overestimated by the lead-wrapping approach [39, 93, 94].

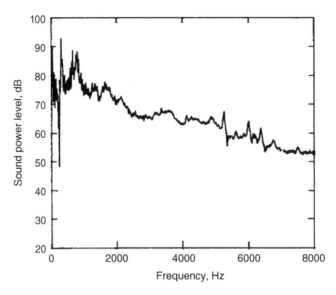

Figure 8.45 Oil pan narrow-band sound power level spectrum determined from two-microphone sound intensity method. (Peaks occur at 300, 650, and 790 Hz.) All intensity measurements were made at an engine speed of 1500 rpm and a load of 540 Nm [39].

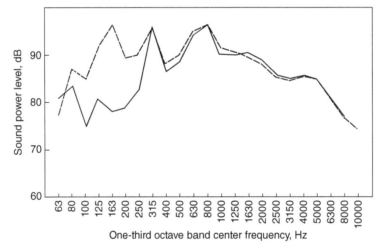

Figure 8.46 Comparison of sound power level determined for the oil pan from the spherical lead-wrapping technique - - -, and the sound power level from the two-microphone sound intensity method — [39].

b) Comparison between Surface Intensity and Lead-Wrapping Measurements

The sound power radiated from five different surfaces of the engine was measured using the surface intensity technique and compared with the sound power radiated from the same surfaces determined from the lead-wrapping approach [80–82, 94]. The five parts chosen for surface intensity measurements were: (i) the oil pan, (ii) the aftercooler, (iii) the left block wall, (iv) the right block wall, (v) the oil filter and cooler. The exhaust manifold and cylinder head were not investigated because of the intense heat radiated by these parts.

Table 8.2 Comparison of overall sound power levels obtained by the sound intensity, surface intensity, and lead-wrapping methods [39].

Engine part	Overall Sound Power Level (dB ref. 1 pW) at 1500 rpm and 540 Nm load		
	Sound intensity	Surface intensity	Lead-wrapping
Oil pan	102.7	103.3	102.6
Exhaust manifold, turbocharger, Cyl. Head, and valve covers	101.4	—	101.6
Aftercooler	100.8	101.9	100.6
Engine front	95.0	—	100.0
Oil filter and cooler	91.1	93.4	98.1
Left block wall	97.4	94.6	97.3
Right block wall	94.8	93.3	97.3
Fuel and oil pumps	91.5	—	96.3
Sum of oil pan, aftercooler, oil filter and cooler, block walls	106.1	106.4	106.7
Sum of all parts	108.1	—	108.8
Bare engine sound power	—	—	109.5

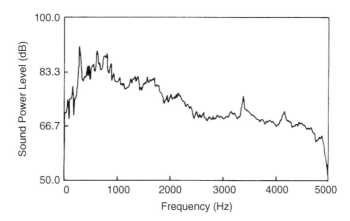

Figure 8.47 Narrow-band sound power level determined from surface intensity method summed over 24 sub-areas of the surface of the oil pan measured at 1500 rpm and 540 Nm load [80–82, 94].

High surface temperatures can make acceleration readings difficult or inaccurate and are somewhat dangerous for investigation when moving the accelerometer from location to location. The front of the engine was not examined either, because pulleys there made it very difficult to mount an accelerometer and locate the microphone. Figures 8.47 and 8.48 show the sound power obtained by summing over 24 locations on the oil pan using Eq. (8.5). Figure 8.49 shows narrow-band results, while Figure 8.48 shows the narrow-band sound power results presented in Figure 8.49 summed into one-third octave bands. The curve with symbols O was obtained from the lead-wrapping method, while the curve with symbols S was obtained from the surface intensity technique. The results for the other four surfaces examined on the engine are given in Refs. [81, 82, 94]. It is seen in Figure 8.48 that the agreement between the two methods is good except at very low frequency (below 315 Hz). At low frequency the lead becomes "transparent" to sound as already discussed and

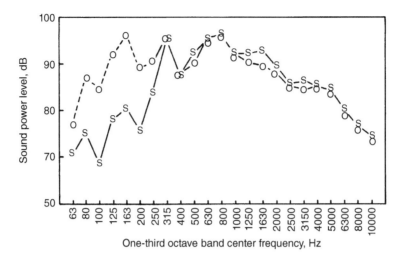

Figure 8.48 Comparison of sound power level determined for the oil pan from the spherical traverse lead-wrapping technique (O) and the sound power level from the surface intensity measurements (S) [80–82, 94].

the surface intensity method is less accurate because of calibration difficulties [82]. The trend for the lead-wrapping results to be higher than the surface intensity results in the lower frequency region was observed for all the five parts examined and is as expected, since lead-wrapping fails at low frequency.

c) Radiation Efficiency of Different Engine Surfaces

Many authors have theoretically investigated the sound radiation from vibrating surfaces. Provided the edge conditions are known, the sound radiation can be predicted theoretically for flat beams and plates in different modes of vibration, see Chapters 3 and 12. However, except at or near a resonance frequency, it is difficult experimentally to excite a single mode of vibration and to measure the sound power radiated. In practice many modes of vibration are usually excited simultaneously. It is also much more complicated theoretically to predict the sound radiated from curved vibrating surfaces or where the geometry is complicated or when the edge conditions are not well known.

The sound power W_{rad} radiated by a vibrating surface of area S is given by:

$$W_{rad} = \rho c S \langle v^2 \rangle \sigma_{rad}, \tag{8.69}$$

where ρc is the air characteristic impedance, $\langle v^2 \rangle$ is the space-average of the mean-square normal surface velocity, and σ_{rad} is the structure radiation efficiency. If a single mode is excited, then the quantities are for that mode. If many modes are excited simultaneously, then Eq. (8.69) still applies, but W_{rad} and $\langle v^2 \rangle$ are the values for the frequency band under consideration and σ_{rad} is a value for the frequency band, averaged over the modes excited. Averaged values of σ_{rad} have been calculated theoretically for idealized flat vibrating structures, but are difficult to calculate for curved structures with ill-defined edge conditions. However, averaged values of σ_{rad} may be determined experimentally for complicated structures by measuring W_{rad} and $\langle v^2 \rangle$.

McGary and Crocker attempted to determine σ_{rad} for the five surfaces of the diesel engine under examination by measuring W_{rad} from the surface intensity method and dividing by $\langle v^2 \rangle$ which is also determined as a by-product of this approach [39, 80–82]. The results for σ_{rad} and for the five surfaces (averaged over one-third octave bands) are given in Ref. [82]. One typical result for the oil pan is presented in Figure 8.49. At high frequency (above about 1000 Hz) all the surfaces had a radiation efficiency approaching one (as theory should predict).

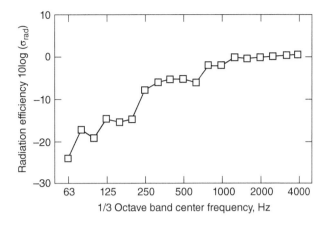

Figure 8.49 Radiation efficiency of the oil pan determined at an engine speed of 1500 rpm and a load of 540 Nm [80–82, 94].

Such radiation efficiency curves, if they can be accurately predicted or measured, can be very useful in engine design in a number of ways. They enable an experimentalist to determine the sound power radiated by each engine surface, simply by measuring the space-averaged mean-square velocity (as a function of frequency) on that surface This can easily be accomplished using an accelerometer rather than making a complicated or sophisticated intensity or sound power measurement. Also, in principle, if theory can be used to predict the vibration of different surfaces of an engine in the design stage, then the sound power radiated by the different surfaces can also be predicted before the engine is constructed.

8.7.4 Measurements of the Transmission Loss of Structures Using Sound Intensity

The transmission loss (sound reduction index) is a measure of the performance of a structure in preventing transmission of sound from one space to another. It is defined as 10 times the logarithm (to base 10) of the ratio of the sound power incident on the dividing structure to the sound power radiated on the receiving side. In principle the receiving space should be anechoic. The evaluation of the transmission loss of panels and partitions is very important in several noise problems, in particular with panels in aircraft, spacecraft, surface vehicles and buildings.

The traditional method of measuring transmission loss requires the use of a very expensive transmission suite consisting of two vibration-isolated reverberation rooms. The incident sound power is deduced from an estimate of the spatial average of the mean square sound pressure in the source room on the assumption that the sound field is diffuse, and the transmitted sound power is determined from a similar measurement in the receiving room where, in addition, the reverberation time must be determined. In 1980, Crocker was the first to describe a new method for the determination of the transmission loss of panels [21, 22].

This method involves the measurement of the incident and transmitted sound intensities. The incident intensity is determined from measurements of the space-averaged mean square sound pressure $\langle p_{rms}^2 \rangle$ in a reverberation room on the source side of the panel. It was checked to see if the space-averaged value of $\langle p_{rms}^2 \rangle_s / 4\rho c$ would be a good estimate of the intensity on the source side of the panel. Figure 8.50 confirms that it is [22]. Here in Figure 8.50, the diffuse field intensity ① is compared with the intensity passing through the open window with the panel absent ②. The transmitted intensity is measured directly, using the two-microphone technique. One advantage of this sound intensity method is that it uses one reverberation room instead of two as used in the conventional transmission suite method. Another advantage is that it makes possible the identification of the energy transmitted through different parts of composite panels (for example, panels made from a heavy solid wall with lightweight windows) [22]. Measurements of transmission loss

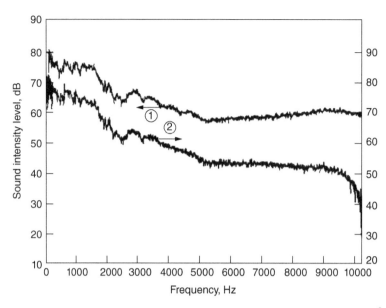

Figure 8.50 Comparison of diffuse field intensities averaged over narrow frequency bands. ① from $p_{rms}^2/4\rho c$; ② Two-microphone technique [22].

made with this method compare favorably with those made using the conventional method and the simple mass law theory [22]. This method has now been standardized. See Section 8.9 and ISO standards 15 186-1 and 15 186-2 and ASTM standard E 2249-02 (2009).

It would be attractive to eliminate the use of a reverberation room on the source side of the panel. Unfortunately, this is not normally possible because direct measurements of intensity on the source side give the vector sum of the incident and reflected intensities. What is required on the source side is solely the incident intensity. Ideally the receiving space should be very quiet and non-reverberant as in the case of an anechoic room. In practice, almost any relatively non-reverberant environment on the receiving side of the panel is sufficient, provided the background noise levels are steady and low. The transmitted sound intensity is measured directly using a sound intensity probe.

When a reverberation room is available on the source side of a panel, it is normally best to estimate the incident sound intensity from measurements of the space-average mean square pressure $\langle p_{rms}^2/4\rho c\rangle_s$. However, there are other possible ways to overcome the difficulty in measuring the incident sound intensity. If the source and receiving rooms are completely anechoic, the insertion loss can be measured by placing loudspeakers in the source space and first measuring the space-average normal intensity in the area when the panel is removed, and then the transmitted intensity with the panel in place. If the source and receiving rooms are completely anechoic, then the transmission loss (TL) is equal to the insertion loss (IL). See Chapter 12 and also Chapter 10 where this is proved for the case of the insertion of a muffler in the exhaust system. If two anechoic rooms are available on the source and receiving side of a panel under test, then the panel can be placed in a special "window" opening between the two rooms. Then this approach can also be used in practice to determine the IL and thus the TL of doors or windows in the open and shut case when they are placed in the opening. Crocker and Valc have used this approach with patches of sound-absorbing material placed in the source room of a two-reverberation room suite and obtained encouraging results [96].

In real field measurements in buildings under construction or completed, a similar approach can be used to test doors, provided the source and receiving spaces are made as absorbing as possible by including fiberglass, carpets, drapes, curtains, etc. in the two spaces. It is also possible to determine the TL of leaks around doors by first measuring the incident intensity and then the transmitted intensity through the air gap around doors in a like manner. Another way to obtain an estimate of the TL of a wall or door between two rooms in a built condition is to place a highly absorbing piece of sound-absorbing material on or very near to the panel under test and measure the intensity normal to the surface. The TL determined in this way will only be useful when the absorption coefficient is greater than 0.2. Material placed on the test side of the object (wall or door) must be removed before the transmitted intensity is measured. But absorbing material placed nearby can be left in place. By measuring the incident and transmitted intensity in this way, the TL of the object under test can be determined in the field for engineering purposes. Again, the source and receiving spaces should be made as anechoic as possible by placing sufficient absorbing material in both spaces. A number of loudspeakers should be used in the source room to obtain a reasonably uniform field there.

a) Experimental Measurements

Experiments were conducted to measure the diffuse field intensity in the reverberation room, and the sound intensity transmitted by the panel by the traditional and new techniques discussed previously. The panel was clamped within a frame in an opening of one wall in the reverberation room. The panel edge conditions were intended to be fully-fixed. The edges between the panel and the frame, as well as the edges between the frame and the wall of the opening, were fully sealed with modeling clay to minimize any leaks. A schematic diagram of the experimental set-up is shown in Figure 8.51. The diffuse sound field inside the reverberation room was produced by the use of an air jet noise source supplemented with a loudspeaker driven by a random noise generator. An overall level of at least 110 dB was produced throughout the experimental investigation.

Figure 8.52 shows the transmission loss of an aluminum panel determined with the intensity approach and with the conventional two-room method, and Figure 8.53 shows measured and calculated transmission loss values of a composite panel, an aluminum aircraft panel with a plexiglass window [21, 22]. Figure 8.54 shows measurement of the transmitted intensity with an intensity probe with microphones in a side-by-side arrangement. This arrangement allows the TL of small areas of the partition to be estimated by scans made by scanning the probe over the small area under investigation.

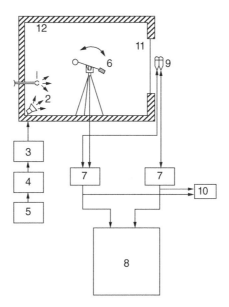

Figure 8.51 Experimental set-up for measurement of transmission loss of a panel by sound intensity method. 1. Jet Noise Source, 2. Loudspeaker, 3. Power Amplifier, 4. Band-pass Filter, 5. Random Noise Generator, 6. Rotating Boom and, Microphone, 7. Measuring Amplifier, 8. FFT Analyzer, 9. Microphones, 10. Oscilloscope, 11. Panel, 12. Reverberation Room [21, 22].

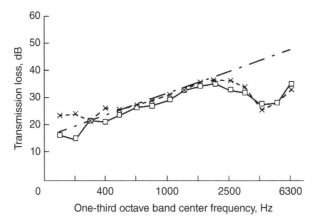

Figure 8.52 Transmission loss of a 3.2 mm thick aluminum panel: --□--, Sound intensity method; ×-×-×, conventional method: –·–·–, mass law. (*Source:* After Crocker et al. [22, 26, 95].)

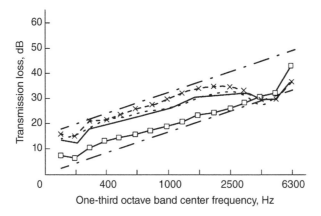

Figure 8.53 Measured and calculated transmission loss of a composite aluminum-plexiglass panel. Measured values: × ×-×, aluminum 3.2 mm thick; –□–, plexiglass 1.6 mm thick; - - - -, total transmission loss. Calculated values; — - —, mass law, aluminum; — - —, mass law, plexiglass; —— total transmission loss. (*Source:* After Crocker et al. [22, 26, 95].)

Figure 8.54 The transmitted sound intensity measured with a probe with microphones in a side-by-side arrangement.

b) Round Robin Comparison of Transmission Loss

Figures 8.55 and 8.56 show the transmission loss results of a round robin test on two single-leaf and double-leaf constructions using the conventional two-room method (Figure 12.30) and the sound intensity method (Figure 8.51).

c) Transmission Loss of Aircraft Structures

Sound intensity can also be used to study the transmission loss and radiation efficiency of engineering structures that are not flat and cannot be located in the window opening between two reverberation of transmission suite. Studies have been made on real aircraft structures such as a single engine Piper Cherokee [93, 99] (Figure 8.57) and a simulated half-scale model of a light aircraft fuselage.

The sound transmission loss of different aircraft panels was measured in a semi-anechoic chamber as well as in a reverberation chamber. These measurements represent normal incidence sound transmission loss and random incidence sound transmission loss, respectively.

In one set of measurements, the fuselage of the Piper Cherokee aircraft in Figure 8.57 was the subject of the tests. Four areas of the starboard fuselage sidewall were chosen for measurement studies; two single layer plexiglass windows and two aluminum panels with standard trim. One of the two plexiglass windows was part of the door unit, while the other was the passenger window behind the door. Similarly, one of the aluminum panels was located on the door under the window, while the other was located beneath the back passenger window.

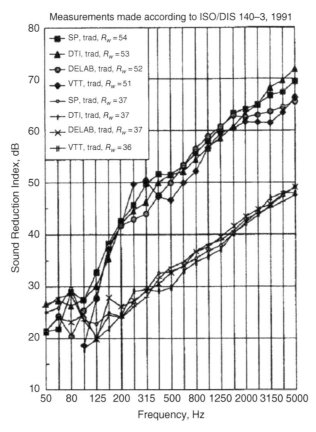

Figure 8.55 Interlaboratory comparisons according to ISO 140-3 for a single metal leaf window (lower curves) and a double metal leaf window (upper curves) using conventional reverberation room method [97].

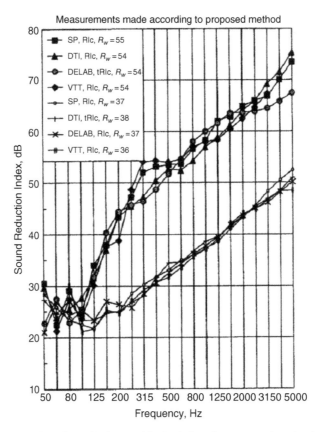

Figure 8.56 Interlaboratory comparison for a single metal leaf window (lower curves) and a double metal leaf window (upper curves) using sound intensity method [97].

Figure 8.57 Photograph of the aircraft fuselage in the semi-anechoic chamber [99].

Semi-Anechoic Chamber

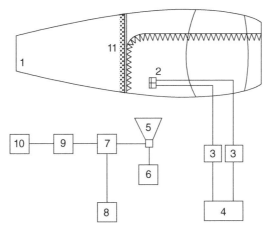

Figure 8.58 Instrumentation used for the measurement of sound transmission loss in the semi-anechoic chamber: 1. Fuselage; 2. Phase-matched microphones; 3. Microphone amplifiers; 4. Fast Fourier Transform Analyzer (FFT); 5 and 6. Horn and driver; 7. Amplifier; 8. Voltmeter; 9. Filter; 10. White noise generator; 11. Plywood-Fiberglass divider [99].

For the measurement of normal incidence sound transmission loss, the fuselage was suspended from three points in the semi-anechoic chamber with the bottom of the fuselage 1.37 m above the floor. The chamber itself is 12.5 m × 8.2 m × 5.5 m and has a concrete floor. To reduce reflections between the floor and the fuselage and to make the environment essentially anechoic, fiberglass sheets were placed on the floor beneath the fuselage. The acoustical excitation used was emitted from a pneumatic driver with a rectangular horn attached.

The instrumentation used to measure the sound transmission loss in the semi-anechoic chamber is shown in Figure 8.58. To measure the sound intensity, half-inch pairs of field-incidence microphones (phase-matched by Bruel and Kjaer on the basis of measurements in a pressure chamber), were arranged side-by-side with a spacing of 13.2 mm [99].

The space-averaged transmitted sound intensity for each panel and respective horn position was measured by sweeping the two-microphone array over the interior of each panel of interest. Sweeping measurements were made as close as possible to each panel. A large amount of fiberglass was placed in the interior of the fuselage to make it reasonably anechoic for the sound intensity measurements. During the measurements inside the fuselage, the difference between the interior sound pressure level and the sound intensity level was monitored, and provided this difference did not exceed 12 dB, the interior transmitted sound intensity measurements were judged to be accurate. To prevent flanking of the panel under study, all panels having similar or lower sound transmission loss than the panel under study were covered with sheets of lead-vinyl. The incident sound intensities for each panel and respective horn position were measured by removing the fuselage and sweeping the two-microphone array over the same area as was previously occupied by the fuselage panel under investigation.

The random incidence sound transmission loss of the same four panels was measured with the fuselage suspended from three points in a reverberation chamber having dimensions 7.5 m × 6.2 m × 5.5 m. With the addition of the fuselage, the chamber was no longer truly reverberant. However, for the purpose of these tests, the sound field was considered to be sufficiently diffuse in the frequency range of interest, namely 100–1250 Hz [98].

The transmitted sound intensity was measured using a Bruel and Kjaer sound intensity probe (Type 3519) with a face-to-face microphone arrangement. To prevent sound entering the cabin via other parts of the fuselage, all parts except the starboard side containing the panels were covered with lead-vinyl. The interior of the

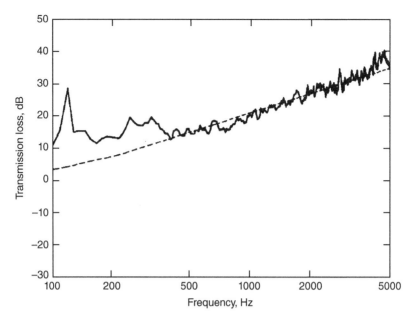

Figure 8.59 Transmission loss versus frequency for the back passenger window at normal incidence —— experimental data with horn and driver source and - - - theoretical mass law [99].

fuselage was made reasonably anechoic, and care was taken to prevent flanking of neighboring panels [98]. The sound intensity incident on the fuselage was assumed to be given by the diffuse field sound intensity:

$$I_i = \frac{p^2_{rms}}{4\rho c},$$
(8.70)

where p^2_{rms} is the space-average mean square sound pressure measured in the reverberation chamber using a half-inch microphone rotating on a boom.

Figure 8.59 shows the comparison between the semi-anechoic experimental results and mass law predictions in narrow 10 Hz bandwidths for the back window. The mass law predictions of the transmission loss seem to agree very well with those found experimentally. In the low frequency range, the transmission loss of a panel of this size is no longer mass-controlled, as assumed in the mass law predictions, which may partly explain the observed discrepancy at low frequencies.

Figure 8.60 shows the one-third octave band results for the sound transmission loss of the various panels under study. The two different aluminum panels with trim and the plexiglass windows seem to have nearly the same transmission loss up to about 400 Hz. It should be remembered, however, that at very low frequencies, (below 100 or 200 Hz) phase errors may lessen the accuracy of the results. Above the frequency of 400 Hz, the aluminum panels are seen to transmit considerably less sound power than the plexiglass windows.

d) Transmission Loss of Cylindrical Structures

Studies were made on aircraft models. In this study the fuselage structure was idealized as a cylindrical shell [95, 100]. An aluminium cylindrical shell of 0.76-m diameter and 1.67-m length was built. The boundary conditions of the cylindrical shell were intended to be fully clamped at the two ends of the shell and both ends were closed with massive end plates.

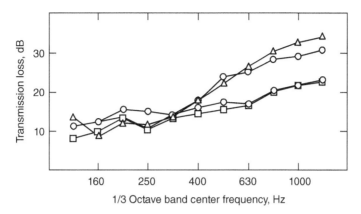

Figure 8.60 Sound transmission loss versus frequency for different aircraft panels in a diffuse sound field. --□-- rear window (panel 1), --○-- door window (panel 3), --△-- rear aluminum panel (panel 2), and --○-- door aluminum panel (panel 4) [99].

The dimensions of the cylinder were chosen to simulate a half-scale model of a light aircraft fuselage. The incident intensity is obtained through the measurement of the diffuse field intensity in the transmission room or air space. Two cases have been studied. (i) The sound source was located inside the cylinder which was placed in an anechoic room. The interior sound field is assumed to be diffuse. (ii) The source was outside the cylinder which was placed in a reverberation room. In this second case, the interior of the cylinder was filled with fiberglass making it anechoic. In both the cases, the sound intensity transmitted from the exterior into the interior receiving space was measured using the two-microphone sound intensity method. Thus, in these two cases studied, the transmitting space was assumed to be reverberant and the receiving space anechoic.

To rate the sound transmission property of a cylindrical shell, the difference between the incident intensity level and the transmitted intensity level is desired. Figure 8.61 shows the set-up with the sound source inside. In the second case the sound source was located outside in a reverberant room and the intensity was measured inside.

In the first case, the incident intensity was evaluated from Eq. (8.70) assuming the interior sound field to be diffuse. The average sound pressure inside the cylinder p^2_{rms} was measured by moving a microphone probe longitudinally and circumferentially inside the cylinder. The intensity transmitted through the cylinder from the interior to the exterior was measured outside using the sound intensity technique. The experimental results agree fairly well with the theoretical prediction in the higher frequency region. In the transmission suite method, it is assumed that the sound fields on the incident and transmitted sides of the structure are diffuse in nature. The assumption that the field inside the cylinder is diffuse in the present case cannot be satisfied in the low-frequency range. This may explain the disagreement at low frequencies.

In the second case, the transmission loss was investigated by suspending the cylinder in a reverberation room. A very intense sound field was generated in the reverberation room. The central interior of the cylinder was filled with a large amount of $48 \, kg/m^3$ ($3.0 \, lb./ft^3$) wedge-shaped pieces of fiberglass to make the interior sound field as anechoic as possible. The two microphones were held so that the line joining their centers was in the radial direction and the microphone closer to the shell was 10 mm away from the surface. During the measurement the microphones were rotated circumferentially inside the cylindrical shell to obtain a spatially-averaged sound intensity [95, 100].

The incident intensity in this case was measured by a single microphone mounted on a rotating boom in the reverberation room. Figure 8.62 shows narrow-band plots of the transmission loss of the cylindrical shell as obtained in the two cases studied; (i) with the noise source in the cylinder and the transmitted intensity measured on the exterior surface of the cylinder when it was situated in an anechoic room and (ii) with the noise source outside the cylinder when it was situated in a reverberation room and the transmitted intensity

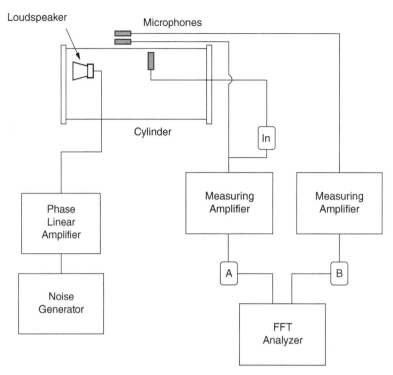

Figure 8.61 Schematic diagram of the experimental set-up for the measurement of transmission loss of a cylinder with the source inside [95, 100].

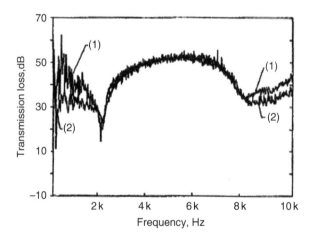

Figure 8.62 Transmission loss of the cylindrical shell measured by two-microphone sound intensity method in (1) outward and (2) inward directions [95, 100].

measured on the interior surface of the cylinder which was filled with fiberglass and thus anechoic inside. Good agreement is found between the two plots in the high-frequency region above 1500 Hz. Two dips in the transmission loss curve, one at the ring frequency (2100 Hz) and the other at the critical coincidence frequency (7500 Hz) are clearly observed.

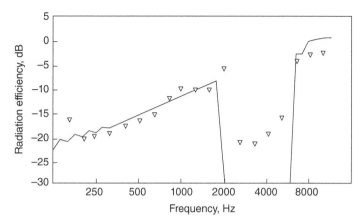

Figure 8.63 Experimental results of radiation efficiency at one-third octave center frequencies (∇) compared to theoretically predicted values of radiation efficiency (−) [100].

The agreement between the transmission loss measurements when the noise source is either in the cylinder or exterior to it in the high-frequency region also indicates that measurements can be done in either direction on a real aircraft fuselage providing that the incident field is diffuse and the receiving space is anechoic for both situations. For example, if only an anechoic chamber or a space out of doors without reflections is available, one can use an interior sound source and measure the transmitted intensity with the two- microphone method outside the fuselage, even though the real aircraft has noise sources outside the cabin.

The radiation efficiency indicates how efficiently sound is radiated from a structure to the acoustic field. It is defined as (see Chapter 3)

$$\sigma_{rad} = \frac{W}{\rho c S \langle v^2 \rangle},$$ (8.71)

where W is the sound power radiated from the structure, ρc is the characteristic impedance of the air, S is the surface area, and $\langle v^2 \rangle$ is the space-averaged mean-square velocity on the radiating surface.

The radiation efficiency of the cylinder can be determined by measuring the surface velocity of the vibrating structure and the sound power it radiates. The cylinder was excited by an external airborne noise source. The spatially averaged velocity was taken at 60 randomly distributed locations over the cylindrical shell. Figure 8.63 presents the one-third octave band plot of the radiation efficiency measured over a frequency range of 50–10 000 Hz. Also shown in Figure 8.63 are the theoretical results of the radiation efficiency obtained from the wavenumber diagrams.

The agreement between the theoretical and experimental results seems to be fairly good – especially since the two peaks, one at the critical coincidence frequency and the other at the ring frequency, can be seen in the experimental radiation efficiency measurements.

8.8 Comparison Between Sound Power Measurements Using Sound Intensity and Sound Pressure Methods

It has been found that, in field determinations of sound power, the sound intensity method is superior to the sound pressure method because it is affected less by background noise and the measurement environment. This section describes real sound power measurements on an automated packaging machine using both sound intensity and sound pressure methods.

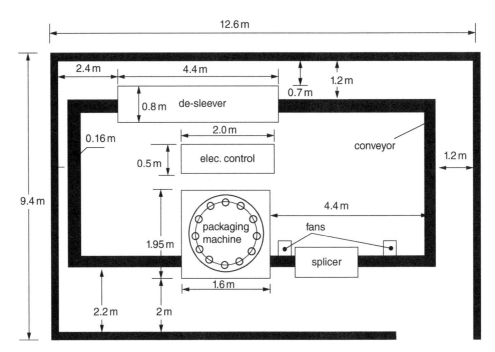

Figure 8.64 Layout of the packaging machine [101].

Sound power level measurements using the sound intensity method (ANSI standard Sl2.l2-1992) and the sound pressure method (ISO 3744) were made on a packaging machine [101]. The packaging machine was located inside a big building in which other machines were operating nearby. Figure 8.64 shows the layout of the packaging machine.

The sound power measurements were made with a real-time analyzer and a sound intensity probe with two phase-matched 1/4 in. microphones. A measurement surface was defined and both sound intensity measurements and sound pressure measurements were performed on the measurement surface. Figures 8.65 and 8.66 show the sound power measurement set-ups for both the sound intensity and the sound pressure measurements.

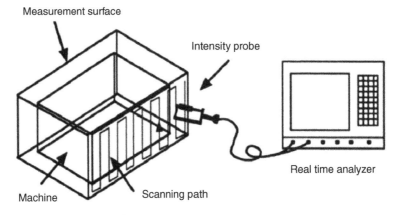

Figure 8.65 Sound intensity measurement set-up.

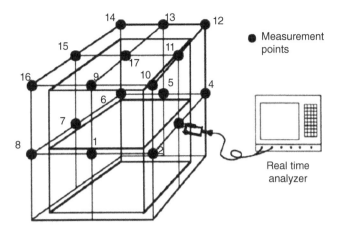

Figure 8.66   Sound pressure measurement set-up [101].

8.8.1   Sound Intensity Method

The sound power of the packaging machine was determined twice using the sound intensity method [101]. A measurement surface was first defined which had five plane rectangular surfaces and a total of five sound intensity scans was performed on these five surfaces; then each rectangular surface was subdivided into two halves, and a total often sound intensity scans was performed on the 10 surfaces. The sound power levels determined from the 5 scan and the 10 scan measurements are shown in Figure 8.67. Good agreement between the two sound power level results is seen. The final sound power level result for the packaging machine is the one obtained by the 10 scan method, according to the ANSI S12.12-1992 standard.

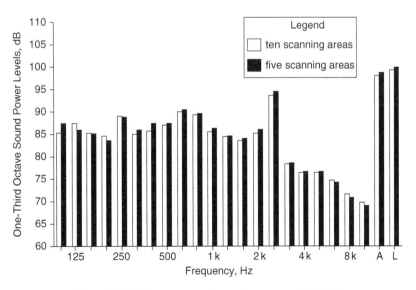

Figure 8.67   Convergence of sound power results [101].

8.8.2 Sound Pressure Method

The sound power of the packaging machine was also determined using a 17 point sound pressure method (ISO 3744.) Two environmental correction factor methods were used: (i) reference sound source method, and (ii) reverberation time (T_{60}) method. Both methods gave acceptable sound power level estimates, except in the very low frequency bands where the sound pressure method failed because of the poor sound signal/background noise ratio. Only after the background noise was reduced by stopping other machines nearby could the sound pressure method be used. Figure 8.68 shows the signal/noise ratio condition in the in-situ measurement. Figure 8.69 shows the sound power level results obtained from the sound intensity method and the

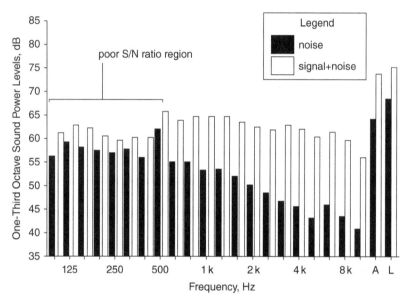

Figure 8.68 Signal/noise ratio [101].

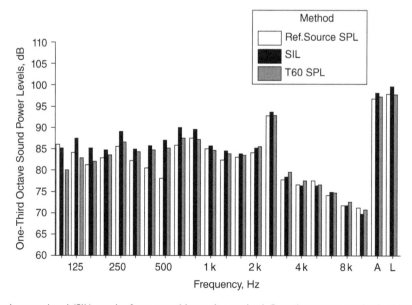

Figure 8.69 Sound power level (SIL) results from sound intensity method. Sound pressure method with (a) reference sound source correction method (Ref. Source SPL) and (b) T_{60} correction method (T_{60} SPL) [101].

sound pressure method using: (i) the reference sound source correction factor method and (ii) the reverberation time (T_{60}) correction factor method.

For the reference sound source correction factor method, the environmental correction factor K was determined by placing a reference sound power source, on the center of the top surface of the machine and measuring the difference between the one-third octave band sound power levels obtained from the 17 point sound pressure method and its known sound power level results obtained from sound intensity measurements in an ideal quiet anechoic environment.

For the reverberation time (T_{60}) correction factor method, the environmental correction factor K was determined by measuring the reverberation time of the room in which the packaging machine was housed. Because of the poor signal/noise ratio and the impossibility of shutting off all the sound sources simultaneously, the reverberation time T_{60} was determined by making an impulsive sound and recording the decay in the sound pressure level with a tape recorder. Thus, the environmental correction factor K was calculated by:

$$K = 10 \log \left(1 + \frac{4S}{A}\right), \tag{8.72}$$

where K is the sound power correction factor in decibels, S is the total surface area of the measurement surface, and A is the total sound absorption area of the room and is given by (see Chapter 3):

$$A = 0.161(V/T_{60}), \tag{8.73}$$

where V is the total volume of the test room, and T_{60} is the measured reverberation time.

Reasonably good agreement was achieved between the two methods except for the very low frequency bands where the T_{60}, method failed because of the lack of enough sound energy decay due to a poor impulsive sound signal/background noise ratio.

A) The sound power determined from sound intensity measurements is less affected by environmental conditions because of the cancelation of the positive flow of background noise sound intensity into one measurement surface and the corresponding equal negative flow out of the other surface. If a sound source can be treated as a constant velocity source, as is usually the case for massive machinery sound sources, the background noise in theory has no effect on the sound power estimate of the source. Reflected sound changes the sound power radiated by the sound source at low frequency because of the "acoustical image effect." In both cases Gauss's law guarantees that the sound intensity method gives a true sound power estimate of a sound source, provided there is no energy sink inside the measurement surface.

B) In general, the sound pressure method overestimates the sound power of a source if other significant background noise sources exist at the measurement site. If the sound pressure level, due to the background noise only, is at least 6 dB lower than the total sound pressure level due to the machine and background noise, the sound pressure method can be used as an alternative way to determine the sound power level of a noise source or to verify the sound power measurements obtained from the sound intensity method. In real situations, however, it is not always possible to meet this 6 dB criterion because sometimes it is impossible to shut down other operating machinery nearby. In the sound power measurements described in Section 8.82 of this chapter, some of the nearby machines were shut doom to meet this 6 dB criterion. Figure 8.68 shows the signal/noise ratio obtained after some operating machinery nearby was shutdown. Notice that the 6 dB criterion was not met in some low frequency bands.

C) There are basically two types of environmental corrections, the reference sound source method, and the reverberation time (T_{60}) method. Environmental corrections are possible only when the requirement concerning the signal/noise ratio stated in paragraph B) above is satisfied. The reference sound source method is more acceptable because it takes care of both the reflected sound and the background noise, and therefore gives a better sound power correction. The reverberation time T_{60} method tends to fail because it is not suitable for strong background noise situations.

The sound intensity method to determine the sound power of machinery noise sources is more accurate and useful in most real situations since the sound power estimates are much less affected by background noise and reverberation. It is not always possible to use the sound pressure method in practice. The 6 dB criterion must be met before the sound pressure method can be used. Once this criterion is met, the sound pressure method can give reasonably good sound power estimates. If the environmental correction factor is obtained for use in the sound pressure method, the reference sound source correction factor method is better than the reverberation time correction factor method because it takes into account both background noise and reverberation effects.

8.9 Standards for Sound Intensity Measurements

During the 1980s and early 1990s, extensive efforts were made in North America by an ANSI committee and in Europe by an ISO committee to develop a standard for the determination of the sound power of sources using sound intensity measurements. The ANSI Committee was concerned with developing an engineering grade standard to assist users in industrial situations. This ANSI standard, published in 1992, allows either scans or fixed points to be used on a surface around the source. The standard requires either the number of scans or the number of fixed points to be doubled until the sound power estimates in each one-third octave band does not differ from the previous estimate by more than a given tolerance.

On the other hand, the ISO committee initially concentrated on developing a standard requiring measurements of sound intensity made at fixed points. During the ISO committee's work, studies were made on the number of fixed points needed to give certain grades of accuracy in complicated sound fields with extraneous noise present. Crocker and Fahy reviewed the progress with these standards in that period [102, 103].

Unfortunately, during some of the ISO studies it was assumed that the error in the sound power estimate is inversely proportional to the square root of the number of measurement points, N; that is the error in the true power W is

$$E\{W\} = \frac{s/\mu}{\sqrt{N}}, \tag{8.74}$$

where s^2 is the sample variance of the measured intensity components and μ is the sample mean. Jacobsen states categorically that this reasoning is "specious" and "fallacious" and moreover can lead to the requirement of a prohibitively large number of discrete measurement points [104]. Jacobsen states that "point measurements cannot possibly be modeled as independent repeated trials (in a probabilistic sense) and therefore (such) results cannot be regarded as independent random variables" [104]. In fact, Jacobsen states that the sample values become progressively more correlated as N increases, and the suggestion by Hubner [105] on reducing the number of measurement positions does not sufficiently address the fundamental problem with Eq. (8.74). Jacobsen also states that the modification suggested by Hubner [105] of bisecting the enveloping surface to reduce the number of measurement points, or the similar modification suggested by Bockhoff and Taillifet [106] do not overcome the fundamental problem with Eq. (8.74). Jacobsen concludes that the "scanning technique is at least as accurate as sampling at discrete points, or indeed more accurate" [104].

The ANSI S12.12 Committee initiated a round robin of its draft standard ANSI S12.12-198X in 1988. Measurements were made of the sound power of a reference sound power source (Acculab RSS-101) by 16 different laboratories [107]. The standard deviation results are plotted in Figure 8.70. In general, it was found that the estimates of sound power obtained from scanning had a smaller standard deviation that these obtained

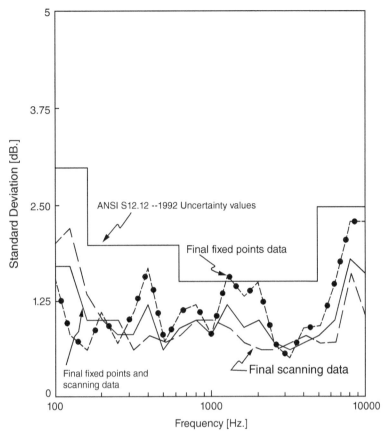

Figure 8.70 Comparison of final fixed points and scanning values, final fixed points values, final scanning values and uncertainty values from table 1 of the ANSI document [107].

with fixed points in most one-third octave bands, except those below 200 Hz. It is also observed that the standard deviations in the final scanning sound power results are less than the uncertainty values stated in the ANSI standard. Also, the standard deviation in the A-weighted sound power level determinations was 0.4 dB using both fixed points and scanning measurements [105].

The ANSI S12.12 standard allowing both scanning and fixed point measurements was published in 1992. The ISO 9614-1 standard for fixed point measurements was published in 1993 and the ISO 9614-2 standard for scanning was published in 1996.

Several international and national standards for the measurement of sound intensity have been completed:

- ISO (International Organization for Standardization) 9614-1 Acoustics – Determination of Sound Power Levels of Noise Sources Using Sound Intensity – Part 1: Measurement at Discrete Points, 1993.
- ISO (International Organization for Standardization) 9614-2 Acoustics – Determination of Sound Power Levels of Noise Sources Using Sound Intensity – Part 2: Measurement by Scanning, 1996.
- ISO (International Organization for Standardization) 9614-3 Acoustics – Determination of Sound Power Levels of Noise Sources Using Sound Intensity – Part 3: Precision Method foe Measurements by Scanning, 2002.

- IEC (International Electrotechnical Commission) 1043 Electroacoustics – Instruments for the Measurement of Sound Intensity, 1993.
- ISO (International Organization for Standardization) 15 186 – 1 Acoustics-Measurements of Sound Insulation in Buildings and Building Elements Using Sound Intensity – Part 1: Laboratory Measurements, 2000
- ISO (International Organization for Standardization) 15 186 – 2 Acoustics-Measurements of Sound Insulation in Buildings and Building Elements Using Sound Intensity – Part 2: Field Measurements, 2003.
- ANSI (American National Standards Institute) S12.12-1992 Engineering Method for the Determination of Sound Power Levels of Noise Sources Using Sound Intensity.
- ANSI (American National Standards Institute). S1-12-1994 Instruments for the Measurement of Sound Intensity.
- ASTM International Standard Test Method for Laboratory measurement of Airborne Transmission Loss of Building partitions and Elements using Sound Intensity – ASTM E2249-02 (2008), 2009.

References

1 Osborne Reynolds to Lord Rayleigh (J. W. Strutt), unpublished letter on the reciprocity of sound, Fallowfield, Manchester dated 2 October 1875.

2 Tyndall, J. (1867). Sound. In: *A Course of Eight Lectures*, Lecture VI, 217–254. London: Longmans, Green and Co.

3 Lord Rayleigh (J. W. Strutt) (1876). On the application of the principle of reciprocity to acoustics. *Proc. R. Soc. Lond.* 25: 118–122. (See also Paper 36 in *Acoustics: Historical and Philosophical Developments*, R. B. Lindsay, ed., Dowden, Hutchinson & Ross, Stroudsburg, PA, 402-406).

4 Lord Rayleigh (J. W. Strutt) (1882). On an instrument capable of measuring the intensity of aerial vibrations. *Philos. Mag.* 14: 186–187.

5 Lord Rayleigh (J. W. Strutt) and Schuster, A. (1881). On the determination of the ohm (B. A. Unit) in absolute measure. *Proc. R. Soc. Lond.* 32: 104–141.

6 Miller, H.B. (ed.) (1982). *Acoustical Measurements-Methods and Instrumentation*, 237. Stroudsburg, PA: Hutchinson Ross Publishing Co.

7 Miller, H.B. (ed.) (1982). *Acoustical Measurements-Methods and Instrumentation*, 114. Stroudsburg, PA: Hutchinson Ross Publishing Co.

8 Wente, E.C. (1917). A condenser transmitter as a uniformly sensitive instrument for the absolute measurement of sound intensity. *Phys. Rev.* 10: 39–63.

9 Wente, E.C. (1922). The sensitivity and precision of the electrostatic transmitter for measuring sound intensities. *Phys. Rev.* 19: 498–503.

10 Olson, H.F. (1932) System response to the energy flow of sound waves. US Patent No. 1,892,644.

11 Olson, H.F. (1975). Acoustic wattmeter. *J. Audio Eng. Soc.* 22 (5): 321–328.

12 Clapp, C.W. and Firestone, F.A. (1941). The acoustic wattmeter, an instrument for measuring sound energy flow. *J. Acoust. Soc. Am.* 13: 124–136.

13 Bolt, R.H. and Petrauskas, A.A. (1943). An acoustic impedance meter for rapid field measurements. *J. Acoust. Soc. Am.* 15: 79(A).

14 Baker, S. (1955). Acoustic intensity meter. *J. Acoust. Soc. Am.* 27: 269–273.

15 Schultz, T.J. (1956). Acoustic wattmeter. *J. Acoust. Soc. Am.* 28: 693–699.

16 Schultz, T.J., Smith, P.W., and Malme, C.I. (1975). Measurement of acoustic intensity in reactive sound field. *J. Acoust. Soc. Am.* 57 (6), Pt. 1): 1263–1268.

17 Burger, J.F., van der Merwe, G.J.J., van Zyl, B.G., and Joffe, L. (2005). Measurement of sound intensity applied to the determination of radiated sound power. *J. Acoust. Soc. Am* 53 (4): 1167–1168.

18 van Zyl, B.G. and Anderson, F. (1975). Evaluation of the intensify method of sound power determination. *J. Acoust. Soc. Am.* 57 (3): 682–686.

19 Chung, J.Y., Pope, J., and Feldmaier, D.A. (1979) Application of acoustic intensity measurement to engine noise evaluation. Proceedings of S.A.E. Diesel Noise Conference, SAE Paper No. 790502.

20 Reinhart, T. and Crocker, M.J. (1980) A comparison of source identification techniques on a diesel engine. Proceedings of Inter-Noise 80 (ed. G.C. Maling, Jr), 1129–1132. New York: Noise Control Foundation.

21 Crocker, M.J., Forssen, B., Raju, P.K., and Mielnicka, A. (1980) Measurement of transmission loss of panels by an acoustic intensity technique. Proceedings of Inter-Noise 80, 741–746.

22 Crocker, M.J., Raju, P.K., and Forssen, B. (1981). Measurement of transmission loss of panels by the direct determination of transmitted acoustic intensity. *Noise Control Eng. J.* 17: 6–11.

23 Pierce, A.D. (1989). *Acoustics: An Introduction to Its Physical Principles and Applications*, 2e. New York: Acoustical Society of America.

24 Fahy, F.J. (1995). *Sound Intensity*, 2e. London: E&FN Spon.

25 Crocker, M.J. (1984) Direct measurement of sound intensity and practical application in noise control engineering. Proceedings of Inter-noise 84, 19–36.

26 Crocker, M.J. (1998). Measurement of sound intensity. In: *Handbook of Acoustical Measurements and Noise Control* (ed. C.M. Harris), 14.1–14.17. New York: Acoustical Society of America.

27 Crocker, M.J. and Jacobsen, F. (1998). Sound intensity. In: *Handbook of Acoustics* (ed. M.J. Crocker), 1327–1340. New York: Wiley.

28 Jacobsen, F. (1991). A note on instantaneous and time-averaged active and reactive sound intensity. *J. Sound Vib.* 147: 489–496.

29 Pascal, J.C. (1981) Mesure de l'intensité active et réactive dans différents champs acoustiques [active and reactive intensity measurements in various acoustics fields]. Proceedings of the International Congress on Recent Developments in Acoustic Intensity Measurements Centre Technique des Industries Mecaniques (CETIM), Senlis, France, 11–19.

30 Jacobsen, F. (2007). Sound intensity measurements. In: *Handbook of Noise and Vibration Control* (ed. M.J. Crocker), 534–548. New York: Wiley.

31 Chung, J.Y. (1978). Cross-spectral method of measuring acoustic intensity without error caused by instrument phase mismatch. *J. Acoust. Soc. Am.* 64: 1613–1616.

32 Fahy, F.J. (1977). Measurement of acoustic intensity using the cross-spectral density of two microphone signals. *J. Acoust. Soc. Am.* 62 (5): 1057–1059.

33 Elko, G.W. (1984) Frequency domain estimation of the complex acoustic intensity and acoustic energy density. Ph.D. thesis in Acoustics. Penn State University.

34 Elko, G.W. and Tichy, J. (1984) Measurement of the complex acoustic intensity and the acoustic energy density. Proceedings of Inter-Noise 84, Honolulu, USA, 1061–1067.

35 Tichy, J. and Adin Mann III, J. (1985) Use of the complex intensity for sound radiation and sound field studies. Proceedings of the Second International Congress on Acoustic Intensity Centre Technique des Industries Mecaniques (CETIM), Senlis, France, 113–120.

36 Elko, G.W. (1985) Simultaneous measurement of the complex acoustics intensity and the acoustic energy density. Proceedings of the Second International Congress on Acoustic Intensity Centre Technique des Industries Mecaniques (CETIM), Senlis, France, 69–78.

37 Waser, M.P. and Crocker, M.J. (1984). Introduction to the two-microphone gross-spectral method of determining sound intensity. *Noise Control Eng. J.* 22 (3): 76–85.

38 Pascal, J.C. (1985) Structure and patterns of acoustic intensity fields. Proceedings of the Second International Congress on Acoustic Intensity Centre Technique des Industries Mecaniques (CETIM), Senlis, France, 97–104.

39 Reinhart, T.E. and Crocker, M.J. (1982). Source identification on a diesel engine using acoustic intensity measurements. *Noise Control Eng.* 18 (3): 84–92.

40 Heyser, R.C. (1986) *Instantaneous Intensity*. Preprint 2399, 81st Conf Audio Eng. Soc.

41 Kutruff, H. and Schmitz, A. (1994). Measurement of sound intensity by means of multi-microphone probes. *Acta Acust. Acust.* 80: 386–396.

42 IEC 1043 (1993) *Electroacoustics – Instruments for the Measurement of Sound Intensity – Measurements with Pairs of Pressure Using Microphones*. Geneva: International Electrotechnical Commission.

43 ANSI S1.9 – 1996 (1996) *Instruments for the Measurement of Sound Intensity*. Washington, DC: American National Standards Institute.

44 Rasmussen, G. and Brock, M. (1981) Acoustic intensity measurement probe. Proceedings of Recent Development in Acoustic Intensity. Senlis, France, 81–88.

45 Jacobsen, F., Cutanda, V., and Juhl, P.M. (1998). Numerical and experimental investigation of the performance of sound intensity probes at high frequencies. *J. Acoust. Soc. Am.* 103 (3): 953–961.

46 Jacobsen, F. (1989). Random errors in sound intensity estimation. *J. Sound Vib.* 128: 247–257.

47 Ren, M. and Jacobsen, F. (1991). Phase mismatch errors and related indicators in sound intensity measurement. *J. Sound Vib.* 149: 341–347.

48 Pope, J. (1989) Qualifying intensity measurements for sound power determination. Proceedings of Inter-Noise 89, 1041–1046.

49 Jacobsen, F. (1997). An overview of the sources of error in sound power determination using the intensity technique. *Appl. Acoust.* 50: 155–156.

50 Shirahatti, U. and Crocker, M.J. (1992). Two-microphone finite difference approximation errors in the interference fields of point dipole sources. *J. Acoust. Soc. Am.* 92: 258–267.

51 Jacobsen, F. (1994) Intensity measurements in the presence of moderate airflow. Proceedings of Inter-Noise 94, 1737–1742.

52 Chung, J.Y. and Pope, J. (1978) Acoustical measurement of acoustic intensity – the two microphone cross-spectral method. Proceedings of Inter-Noise 78, 893–900.

53 Blaser, D.A. and Chung, J.Y. (1978) A transfer function technique for determining the acoustic characteristics of duct systems with flow. Proceedings of Inter-Noise 78, 901–908.

54 Chung, J. and Blaser, D.A. (1980). Transfer function method of measuring acoustic intensity in a duct-system with flow. *J. Acoust. Soc. Am.* 68 (6): 1570–1577.

55 Chung, J.Y. (1981) Fundamental aspects of the cross-spectral method of measuring acoustic intensity. Proceedings of Rec. Develop. Acoustic Intensity, Senlis, France: CETIM, 1–10.

56 Juhl, P. and Jacobsen, F. (2004). A note on measurement of sound pressure with intensity probes. *J. Acoust. Soc. Am.* 116 (3): 1614–1620.

57 Jacobsen, F. (2007). Sound intensity. In: *Springer Handbook of Acoustics* (ed. T.D. Rossing), 1053–1075. New York: Springer.

58 Krishnappa, G. (1981). Cross spectral method of measuring acoustic intensity by correcting phase and gain mismatch errors by microphone calibration. *J. Acoust. Soc. Am.* 69: 307–310.

59 Jacobsen, F. (1991). A simple and effective correction for phase mismatch in intensity probes. *Appl. Acoust.* 33: 165–180.

60 Gade, S. (1982). Sound intensity (part I. theory). *Bruel Kjaer Tech. Rev.* 3: 3–39.

61 Jacobsen, F. (1990). Sound field indicators: useful tools. *Noise Control Eng. J.* 35: 37–46.

62 Gade, S. (1985). Validity of intensity measurements in partially diffuse sound field. *Bruel Kjaer Tech. Rev.* 4: 3–31.

63 Nordby, S.A., Bjor, O.H. (1984) Measurement of sound intensity by use of a dual channel real-time analyzer and a special sound intensity microphone. Proceedings of Inter-Noise 84, Honolulu, 1107–1110.

64 Druyvesteyn, W.F. and de Bree, H.E., A new sound intensity probe: comparison to the Bruel & Kjaer p-p probe. Audio Eng. Soc. Conv. 104, 1998.

65 de Bree, H.-E. (2003). The microflown: an acoustic particle velocity sensor. *Acoust. Aust.* 31: 91–94.

66 Raangs, R., Druyvesteyn, W.F., and de Bree, H.E. (2003). A low-cost intensity probe. *J. Audio Eng. Soc.* 51: 344–357.

67 Jacobsen, F. and de Bree, H.-E. (2005). A comparison of two-deferent sound intensity measurement principles. *J. Acoust. Soc. Am.* 118: 1510–1517.

68 Jacobsen, F. and Jaud, V. (2006). A note on the calibration of pressure-velocity sound intensity probes. *J. Acoust. Soc. Am.* 120: 830–837.

69 Tijs, E. and de Bree, H.-E. (2009) PU sound power measurements on large turbo machinery equipment. Proceedings of 16th International Congress on Sound and Vibration, ICSV16.

70 Comesana, D.F., Wind, J., Grosso, A., and Holland, K. (2011) Performance of p-p and p-u intensity probes using scan and paint. Proceedings of 18th International Congress on Sound and Vibration, ICSV18, Rio de Janeiro.

71 Comesana, D.F., Peksel, B.O., and de Bree, H.-E. (2014) Expanding the sound power measurement criteria for sound intensity probes. Proceedings of 18th International Congress on Sound and Vibration, ICSV21, Beijing.

72 Macadam, J.A. (1974) *The Measurement of Sound Power Radiated by Individual Room Surfaces in Lightweight Buildings*, Building Research Establishment Current Paper 22174.

73 Macadam, J.A. (1976). The measurement of sound radiation from room surfaces in lightweight buildings. *Appl. Acoust.* 9 (2): 103–118.

74 Hodgson, T. (1977). Investigation of the surface acoustical intensity method for determining the noise sound power of a large machine in-situ. *J. Acoust. Soc. Am.* 61 (2): 487–491.

75 Brito, J.D. (1976) Sound intensity patterns for vibrating surfaces. Ph.D. thesis. Massachusetts Institute of Technology.

76 Brito, J.D. (1984) Machinery noise source analysis using surface intensity measurements. *Proceedings of 1979 National Conference on Noise Control Engineering* (ed. J.W. Sullivan and M.J. Crocker), 137–142. Poughkeepsie, NY: Institute of Noise Control Engineering.

77 Kaemmer, N. (1978) Determination of sound power from intensity measurements on a cylinder. MSME thesis. Purdue University.

78 Kaemmer, N. and Crocker, M.J., Sound power determination from surface intensity measurements on a vibrating cylinder. *Proceedings of 1979 National Conference on Noise Control Engineering* (ed. J.H. Sullivan and M.J. Crocker), 153–160. Poughkeepsie, NY: Institute of Noise Control Engineering.

79 Kaemmer and Crocker, M.J. (1998). Sound power determination from surface intensity measurements on a vibrating cylinder. *J. Acoust. Soc. Am.* 73 (3): 856–866.

80 McGary, and Crocker, M.J. (1980) Noise Source Identification of Diesel Engines Using Surface Intensity Measurements. EPA Contract 68-01-4907, Report No. 5, HL80-2.

81 McGary, M., and Crocker, M.J. (1980) Surface Acoustical Measurements on a Diesel Engine NASA Technical Memorandum 81807.

82 McGary, M. and Crocker, M.J. (1981). Surface intensity measurements on a diesel engine. *Noise Control Eng.* 16 (1): 26–36.

83 McGary, M. and Crocker, M.J. (1982). Phase shift errors in the theory and practice of surface acoustical intensity measurements. *J. Sound Vib.* 82: 275–288.

84 Hirao, Y., Yamamoto, S., Nakamura, K., and Boha, S. (2004). Development of a hand-held sensor probe for detection of sound components radiated from a specific device using surface intensity measurements. *Appl. Acoust.* 65 (7): 719–735.

85 Bockhoff, M. (1984) Some remarks on the continuous sweeping method in sound power determination. Proceedings of Inter-Noise 84, 1173–1176.

86 Pope, J. (1986) Intensity measurements for sound power determination over a reflecting plane. Proceedings of Inter-Noise 86, 1115–1120.

87 Crocker, M.J. (1986). Sound power determination from sound intensity – to scan or not to scan. *Noise Control Eng. J.* 27: 66–69.

88 Yang, S. and Crocker, M.J. (1994) Errors in sound power measurements due to finite number of samples. Third International Congress on Air-and Structure Borne Sound and Vibration, Montreal, Canada, 1079–1084.

89 Pope, J., Hickling, R., Feldmaier, D. and Blaser, D. (1981) The use of acoustic intensity scans for sound power measurement and for noise source identification in surface transportation vehicles. SAE Paper 81401.

90 Roozen, B.N., van den Oetelaar, J., Geerlings, A., and Vliegenthart, T. (2009). Source identification and noise reduction of a reciprocating compressor; a case history. *Int. J. Acoust. Vib.* 14 (2): 90–98.

91 Hickling, R., Lee, P., and Wei, W. (1997). Investigation of integration accuracy of sound power measurement using an automated sound-intensity system. *Appl. Acoust.* 50 (2): 125–140.

92 Tachibana, H. (1987) Visualization of sound fields by the sound intensity technique (in Japanese). Proceedings of Second Symp. Acoustic Intensity, Tokyo, 117–127.

93 Reinhart, T.E., and Crocker, M.J. (1980) Noise source identification of diesel engines using acoustic intensity measurements. EPA Contract 68-01-4907, Report No. 7, HL80-39.

94 Crocker, M.J. (1982). Comparison between surface intensity, acoustic intensity and selective wrapping noise measurements on a diesel engine. In: *Engine Noise, Excitation Vibration and Radiation* (eds. R. Hickling and M.M. Komal), 279–311. New York: Plenum Press.

95 Crocker, M.J., Forssen, B., Raju, P.K. and Wang, Y.S. (1981) Application of acoustic intensity measurement for the evaluation of transmission loss of structures. Proceedings of the International Congress on Recent Developments in Acoustic Intensity Measurements Centre Technique des Industries Mecaniques (CETIM), Senlis, France, 161–169.

96 Crocker, M.J. and Valc, Z. (1990) New techniques to measure the transmission loss of partitions using sound intensity and single-microphone sound pressure measurements. Proceedings of Inter-Noise 90, 103–106.

97 Jonasson, H.G. (1993). Sound intensity and sound reduction index. *Appl. Acoust.* 40: 281–293.

98 Wang, Y.S. and Crocker, M.J. (1982). Direct measurement of aircraft structures using the acoustic intensity approach. *Noise Control Eng. J.* 19 (3): 80–85.

99 Lyle, K.H., Atwal, M.S., and Crocker, M.J. (1988). Light aircraft sound transmission studies: the use of the two-microphone sound intensity technique. *Noise Control Eng. J.* 31 (3): 145–153.

100 Wang, Y.S., Crocker, M.J. and Raju, P.K. (1983) Theoretical and experimental evaluation of transmission loss of cylinders. Presented as Paper 81–1971 at the AIAA 7th Aeroacoustics Conference, Palo Alto, October 5–7, 1981 and published in AIAA Journal 21(2): 186–192.

101 Crocker, M.J. and Yang, S. (1993) Sound power measurements using sound intensity and sound pressure methods. Proceedings of Inter-Noise 93, Leuven, Belgium, 377–382.

102 Crocker, M.J. (1984) Standards for the determination of the sound power of sources from the measurement of sound intensity. Proceedings of Inter-Noise 84, 1305–1310.

103 Fahy, F.J. (1997). International standards for the determination of sound power levels of sources using sound intensity measurement an exposition. *Appl. Acoust.* 50 (2): 97–109.

104 Jacobsen, F., Some observations on the approximation of the surface integrate in sound power determination based on intensity. Proceedings of 13th International Congress on Acoustics, Belgrade, Yugoslavia, 1989.

105 Hubner, G. (1988) Sound power determination of machines using sound intensity measurements – Reduction of number of measurement positions in cases of "hot areas". Proceedings of Inter-Noise 88, 1113–1116.

106 Bockhoff, M. and Taillifet, D. (1988) Sound power determination of machines by intensity technique, Proceedings of Inter-Noise 88, 1125–1128.

107 Shirahatti, U.S., Crocker, M.J. and Peppin, R.I. (1989) Sound power determination from sound intensity: results of the ANSI round robin test. Proceedings of Inter-Noise 89, 1029–1034.

9

Principles of Noise and Vibration Control

9.1 Introduction

The noise and vibration of some sources can be just annoying. Some sources, however, can be intense enough to permanently hurt people. Research on noise and vibration control is therefore very important and it has produced the development of multiple control approaches. It is widely accepted that noise and vibration control is most effective when it is considered at the design stage. Proper design and construction of machinery, buildings, transportation vehicles, airports, and highways can eliminate potential noise and vibration problems. Before selecting noise and vibration control measures, sources must be identified and evaluated. Although each noise and vibration problem is somewhat different, a systematic approach and use of several well-known methods often produce sufficient reduction and acceptable conditions. Implementation of noise and vibration control measures and predictions of their performance are therefore important issues. This chapter begins with a discussion of the source–path–receiver model, continues with a description of the most useful passive noise and vibration control approaches, and concludes with a discussion of the main applications of the relatively new topic of active noise reduction. Several individual worked examples are provided of the use of passive noise and vibration control approaches.

9.2 Systematic Approach to Noise Problems

Noise and vibration control should always be incorporated at the design stage wherever possible because there are more low-cost options and possibilities then to make completed machines or installations quieter [1–3]. After machines are built or installations completed, noise control approaches can still be achieved through various modifications and add-on treatments, but these are frequently more difficult and expensive to implement. Several books deal with the fundamentals of noise control and with practical applications of noise control techniques [4–12]. Other books deal with acoustics and noise theory [13, 14].

Noise and vibration problems can be described using the simple *source–path–receiver* model [4] shown in Figure 9.1. The *sources* are of two main types: (i) airborne sound sources caused by gas fluctuations (as in the fluctuating release of gas from an engine exhaust) or (ii) structure-borne machinery vibration sources that in turn create sound fields (e.g. engine surface vibrations). Moreover, these sound pressure and vibration sources are of two types: (i) steady-state and (ii) impulsive. Both steady-state and impulsive vibrations (caused by impacting parts) are commonly encountered in machines. The *paths* may also be airborne or structure-borne in nature.

Engineering Acoustics: Noise and Vibration Control, First Edition. Malcolm J. Crocker and Jorge P. Arenas.
© 2021 John Wiley & Sons Ltd. Published 2021 by John Wiley & Sons Ltd.

Figure 9.1 Source-path-receiver model for noise and vibration problems.

Source modifications are the best practice but are sometimes difficult to accomplish. Often changes in the path or at the *receiver* may be the only real options available. The model shown in Figure 9.1 is very simple. In reality there will be many sources and paths. The dominant source should be treated first, then the secondary one, and so on. The same procedure can also be applied to the paths. Finally, when all other possibilities are exhausted, the receiver can be treated. If, as in most noise problems, the receiver is the human ear, earplugs or earmuffs or even complete personnel enclosures can be used.

Measurements, calculations, and experience all play a part in determining the dominant noise and vibration sources and paths. The dominant sources (and paths) can sometimes be determined from careful experiments. In some cases, parts of a machine can be turned off or disconnected to help identify sources. In other cases, parts of a machine can be enclosed, and then sequential exposure of machine parts can be used to identify major sources. Frequency analysis of machines can also be used as a guide to the causes of noise, as with the case of the firing frequency in engines, the pumping frequency of pumps and compressors, and the blade passing frequency of fans. More sophisticated methods are also available involving the use of coherence, cepstrum, and sound intensity methods. Methods for determining the sources of noise and vibration in machinery are discussed in Section 9.2.1.

9.2.1 Noise and Vibration Source Identification

In most machinery noise and vibration control problems, knowledge of the dominant noise and vibration sources in order of importance is very desirable, so that suitable modifications can be made. In a complicated machine, such information is often difficult to obtain, and many noise and vibration reduction attempts are made based on inadequate data so that frequently expensive or ineffective noise and vibration reduction methods are employed. Machine noise and vibration can also be used to diagnose increased wear. The methods used to identify noise and vibration sources will depend on the particular problem and the time and resources (personnel, instrumentation, and financial resources) and expertise available and on the accuracy required. In most noise and vibration source identification problems, it is usually the best practice to use more than one method in parallel to ensure greater confidence in the results of the identification procedure.

Noise and vibration source and path identification methods of differing sophistication have been in use for many years [15]. In recent years, a considerable amount of effort has been devoted in the automobile industry to devise better methods of noise and vibration source and path identification [16–32]. Most effort has been expended on interior cabin noise and separating airborne [16–20] and structure-borne [21–23] paths. Engine and power train noise and vibration sources and paths [24–26] and tire noise [27, 28] have also received attention. This effort has been expended not only to make the vehicles quieter but to give them a distinctive manufacturer sound quality [33]. More recently these methods have been extended and adapted in a variety of ways to produce improved methods for source and path identification [34–49]. One example is the so-called transfer path analysis (TPA) approach [43] and variations of it, which to some extent are based on an elaborated version of the earlier coherence approach for source and path modeling [15]. Commercial softwares are now available using the TPA and other related approaches for noise and vibration source and path identification. Pass-by exterior noise sources of automobiles and railroad vehicles have also been studied [35].

Noise and vibration energy must flow from a *source* through one or more *paths* to a *receiver* (usually the human ear). Figure 9.1 shows the simplest model of a *source–path–receiver* system. Noise sources may be *mechanical* in nature (caused by impacts, out-of-balance forces in machines, vibration of structural parts) or *aerodynamic* in nature (caused by pulsating flows, flow–structure interactions, jet noise, turbulence). Noise and vibration energy can flow through a variety of airborne and structure-borne paths to the receiver.

Figure 9.2 shows an example of a propeller-driven airplane. This airplane situation can be idealized as the much more complicated source–path–receiver system shown in Figure 9.3. In some cases the distinction between the source(s) and path(s) is not completely clear, and it is not easy to neatly separate the sources from the paths. In such cases, the sources and paths must be considered in conjunction. However, despite some obvious complications, the source–path–receiver model is a useful concept that is widely used. In this chapter we will mainly concern ourselves with machinery noise sources. We note that *cutting* or *blocking* one

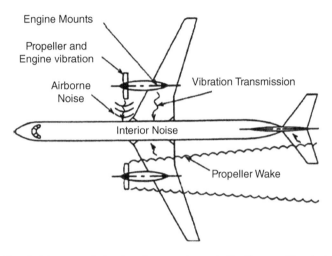

Figure 9.2 Sources and paths of airborne and structure-borne noise and vibration resulting in interior noise in an airplane cabin [15].

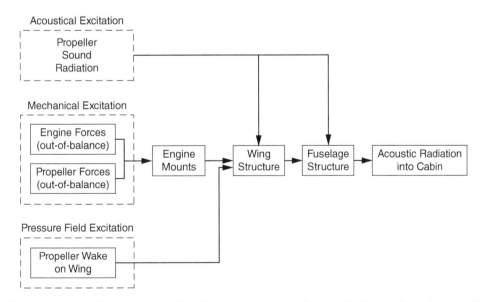

Figure 9.3 Source–path–receiver system showing airborne and structure-borne paths for a twin-engine propeller-driven airplane [15].

or more noise and/or vibration *path* often gives invaluable information about the noise sources. Reference [15] provides a comprehensive discussion on the methods for noise and vibration source identification.

9.2.2 Noise Reduction Techniques

A study of the literature reveals many successful well-documented methods used to reduce the noise of machines. These can be classified using the source–path–receiver model. Some of the most useful approaches can generally be used only at the source or in the path. Others, such as *enclosure*, can be adapted for use at any location. For instance, a small enclosure can be built inside a machine around a gear or bearing, or a larger enclosure or room can be built around a complete machine. Finally, an enclosure or personnel booth can be built for the use of a machine operator. Table 9.1 summarizes a large number of approaches that have been found useful in practice.

The main passive noise control approaches are discussed in the following sections. These include the use of (i) vibration isolators, (ii) vibration damping materials (iii) sound absorption, (iv) enclosures, and (v) barriers. The use of mufflers and silencers is discussed in Chapter 10.

9.3 Use of Vibration Isolators

Vibration isolation has been discussed frequently in the literature [50]. Vibration isolators are used in two main situations: (i) where a machine source is producing vibration and it is desired to prevent vibration energy flowing to supporting structures and (ii) where a delicate piece of equipment (such as an electronics

Table 9.1 Passive noise control approaches that may be considered for source, path, or receiver.

Source	Choose quietest machine source available
	Reduce force amplitudes
	Apply forces more slowly
	Use softer materials for impacting surfaces
	Balance moving parts
	Use better lubrication
	Improve bearing alignment
	Use dynamic absorbers
	Change natural frequencies of machine elements
	Increase damping of machine elements
	Isolate machine panels from forces
	Reduce radiating surface areas (by adding holes)
	Stagger time of machine operations in a plant
Path	Install vibration isolators
	Use barriers
	Install enclosures
	Use absorbing materials
	Install reactive or dissipative mufflers
	Use vibration breaks in ductwork
	Mismatch impedances of materials
	Use lined ducts and plenum chambers
	Use flexible ductwork
	Use damping materials
Receiver	Provide earplugs or earmuffs for personnel
	Construct personnel enclosures
	Rotate personnel to reduce exposure time
	Locate personnel remotely from sources

package or precision grinder) must be protected from vibration in the structure. It is the first case that will receive attention here. Primary emphasis is placed on reducing the force transmitted from the machine source to the supporting structure, but a secondary consideration is to reduce the vibration of the machine source itself.

It is often found that machines are attached to metal decks, grills, and sometimes lightweight wood or concrete floors. The machine on its own is usually incapable of radiating much noise (particularly at low frequency). The supporting decks, grills, and floors, however, tend to act like sounding boards, just as in musical instruments, and amplify the machine noise. Properly designed vibration isolators can overcome this noise problem.

Vibration isolators are of three main types: (i) spring, (ii) elastomeric, and (iii) pneumatic. Spring isolators are durable but have little damping. Elastomeric isolators are less durable and are subject to degradation due to corrosive environments. They have higher damping and are less expensive. Pneumatic isolators are used where very low frequency excitation is present.

Often the exciting forces are caused by rotational out-of-balance forces in machines or machine elements or by magnetic or friction effects. Usually, these forces are simple harmonic in character. Such forces occur in electric motors, internal combustion engines, bearings, gears, and fans. Sometimes, however, the exciting forces may be impulsive in nature (e.g. in the case of punch presses, stamping operations, guillotines, tumblers, and any machines where impacts occur). The design of vibration isolators for a machine under the excitation of a simple harmonic force is considered below.

9.3.1 Theory of Vibration Isolation

A machine may be considered, for simplicity, to be represented by a rigid mass m. If the machine is attached directly to a large rigid massive floor, as shown in Figure 9.4, then all the periodic force $F_m(t)$ applied to the mass is directly transmitted to the floor. We will assume that the force on the mass is vertical and $F_m(t) = F_m \sin 2\pi f t$, where F_m is the amplitude of this force, and f is its frequency (Hz). If a vibration isolator is now placed between the machine and the floor, we can model this system with the well-known single-degree-of-freedom system shown in Figure 9.5. The following discussion assumes that the isolators have a constant stiffness and that the damping is viscous in nature. Suppose, for the moment, that the periodic force is stopped and that the mass m is brought to rest. If the mass m is displaced from its equilibrium position and released, then it will vibrate with a *natural frequency of vibration f_n* given by

$$f_n = \left(\frac{1}{2\pi}\right)\sqrt{\frac{K}{m}} \text{ Hz,} \tag{9.1}$$

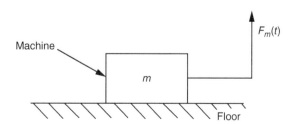

Figure 9.4 Rigid machine of mass m attached to a rigid massive floor.

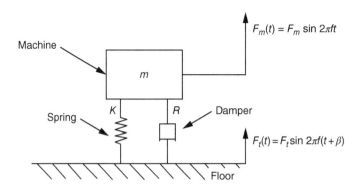

Figure 9.5 Rigid machine of mass m separated from rigid massive floor by vibration isolator of stiffness *K* and damping *R*.

where, in International System (SI) units, K is the stiffness (N/m) and m is the mass (kg). (see Chapter 2 of this book.)

If the exciting force $F_m(t)$ is now resumed, then a force $F_t(t)$ will be transmitted to the rigid floor. This force $F_t(t)$ will be out of phase with $F_m(t)$, but it is simple to show that the ratio of the amplitudes of the forces in the steady state is given by

$$T_F = \frac{F_t}{F_m} = \sqrt{\frac{1 + 4(f/f_n)^2(R/R_c)^2}{\left[1 - (f/f_n)^2\right]^2 + 4(f/f_n)^2(R/R_c)^2}}, \tag{9.2}$$

where f is the frequency of the exciting force (Hz), f_n is the natural frequency of vibration of the mass m supported by the spring (Hz), R is the coefficient of damping (Ns/m), and R_c is the coefficient of critical damping ($R_c = 2\sqrt{mK}$). See Chapter 2 of this book for further discussion on vibration. The ratio F_t/F_m is known as the force transmissibility T_F.

The vibration amplitude A of the machine mass m is given by

$$\frac{A}{F_m/K} = \frac{1}{\sqrt{\left[1 - (f/f_n)^2\right]^2 + 4(f/f_n)^2(R/R_c)^2}}. \tag{9.3}$$

The ratio $A/(F_m/K)$ is known as the *dynamic magnification factor* (DMF). This is because F_m/K represents the static displacement of the mass m if a static force of value F_m is applied, while A represents the dynamic displacement amplitude that occurs due to the periodic force of amplitude F_m. Note that the ratio R/R_c is known as the *damping ratio δ*. If $δ = 1.0$, the damping is called *critical damping*. With most practical vibration isolators, $δ$ may be in the range from about 0.01 to 0.2. Equations (9.2) and (9.3) are plotted in Figures 2.9 and 2.10, respectively.

If the machine is run at the natural frequency f_n, we see from Figure 2.9 that f/f_n is 1.0 and the force amplitude transmitted to the floor is very large, particularly if the damping in the isolator support is small. If the machine is operated much above the natural frequency, however, then the force amplitude transmitted to the floor will be very small.

We define the efficiency $η$ of the isolator as

$$η = (1 - T_F) \times 100\%. \tag{9.4}$$

> **Example 9.1**   Suppose we wish to isolate the 120-Hz vibration of an electric motor so that the system has a natural frequency of 12 Hz. Calculate a) the force transmissibility and b) the isolator efficiency of the resulting system.
>
> **Solution**
>
> If we choose isolators so that the system has a natural frequency of 12 Hz, then the ratio $f/f_n = 10$. If the damping in the isolator system is $R/R_c = 0.1$, the force transmissibility will be only about 0.025, or 2.5% (see Figure 2.10). Thus the isolator efficiency is 97.5%. To reduce the force transmissibility still further, we could use softer isolators and choose a still lower resonance frequency. There is some danger in doing this, however, because the static deflection of the machine will naturally increase if we use softer isolators. Since a large static deflection may be undesirable (it may interfere with the operation of the machine), this restricts the softness of the isolator and thus how low we can make the natural frequency f_n. The static deflection d produced in the isolator by the gravity force on the mass m is given by $d = mg/K$.

We have already seen that the natural frequency is related to K and m by $f_n = (1/2\pi)\sqrt{K/m}$. Hence we can relate the static deflection d to the natural frequency f_n:

$$d = \frac{g}{4\pi^2 f_n^2}, \tag{9.5}$$

where the static deflection d is given in centimetres (or inches) and g is the acceleration of gravity, 981 cm/s². The relationship given in Eq. (9.5) has been plotted in Figure 9.6. The greatest static deflection d that can be allowed from operational considerations should be chosen. This will then allow a determination of the lowest allowable f_n. It should be noted that excessive static deflection may interfere with the operation of the machine by causing alignment problems. Also, using isolators with a small vertical stiffness usually means that they may have a small horizontal stiffness, and there can be stability problems. Both these considerations limit the lowest allowable f_n.

With springs, the damping is small (usually $\delta < 0.1$), and we may use Eqs. (9.2) and (9.3) and Figures 2.9 and 2.10, assuming that $\delta = 0$. With $\delta = 0$ (or, equivalently, $R = 0$), Eqs. (9.2) and (9.5) give

$$T_F = \frac{F_t}{F_m} = \frac{1}{1 - 4\pi^2 f^2 d/g}. \tag{9.6}$$

This result is plotted in Figure 9.7. The suggested procedure is as follows.

1) Establish the total weight of the machine and the lowest forcing frequency experienced.
2) From Figure 9.7 select the force transmissibility allowable (this determines the static deflection, given the lowest forcing frequency.
3) From the spring constants given by the manufacturer, the machine weight, and the static deflection, choose the appropriate vibration isolator.

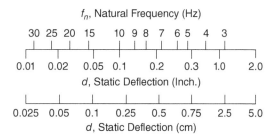

Figure 9.6   Relationship between natural frequency f_n of machine-isolator-floor system and static deflection d of machine.

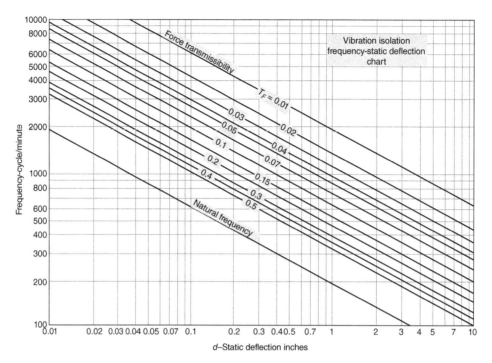

Figure 9.7 Relationship (for a linear isolator) between forcing frequency f, force transmissibility T_F, and static deflection d of machine-isolator-floor system, without damping ($R = 0$). (Note deflection given in inches. 1 in. = 2.54 cm).

Example 9.2 An electric motor of mass 100 kg and a reciprocating compressor of mass 500 kg are mounted on a common support. The motor runs at 2400 revolutions/minute (rpm) and by a belt drives the compressor at 3000 rpm. The vertical force fed to the support is thought to be excessive. Choose six equal spring mounts to provide a force transmissibility not exceeding 5%.

Solution

1) The total machine weight is 600 kg, and the lowest forcing frequency is 2400 rpm (or 40 Hz).
2) From Figure 9.7, the static deflection required is 0.33 cm (0.13 in.).
3) Thus, since there are six spring isolators, each must support a mass of 100 kg. This requires a spring constant for each isolator of 981/0.33 = 3000 N/cm.

   As a check on the calculation, we see from Figure 9.6 that a static deflection of 0.33 cm. requires a natural frequency of about 8.6 Hz. Thus, the ratio of forcing frequency to natural frequency $f/f_n = 40/8.6 = 4.65$. From Figure 2.10, with zero damping, $T_F = 0.05$, which agrees with the design requirement.

9.3.2 Machine Vibration

There are several other factors that should be considered in isolator design. First, we notice from Figures 2.9 and 2.10 that when a machine is started or stopped, it will run through the resonance condition and then, momentarily, $f/f_n = 1$. When passing through the resonance condition, large vibration amplitudes can exist on the machine, and the force transmitted will be large, particularly if the damping is small. Assuming viscous damping, to provide small force transmissibility at the operating speed, small damping is required; however, to prevent excessive machine vibration and force transmission during stopping and starting, large damping is needed. These two requirements are conflicting. Fortunately, some forces, such as out-of-balance forces, are much reduced during starting and stopping. However, it is normal to provide a reasonable amount of damping ($\delta = 0.1$–0.2) in spring systems to reduce these starting and stopping problems. The severity of the machine vibration problem can be gauged from Figure 9.8.

9.3.3 Use of Inertia Blocks

Inertia blocks are normally made from concrete poured onto a steel frame. If the mass supported by the vibration isolators is increased by mounting a machine on an inertia block, the static deflection will be increased. If the isolator stiffness is correspondingly increased to keep the static deflection the same, then there is no change in the system's natural frequency or the force transmissibility T_F. The use of an inertia block does, however, result in reduced vibration amplitude of the machine mass. It also has additional advantages including (i) *improving stability* by providing vibration isolator support points that are farther apart, (ii) *lowering the center of gravity* of the system, thus improving stability and reducing the effect of coupled modes and rocking natural frequencies, (iii) *producing more even weight distribution* for machines often enabling the use of symmetrical vibration isolation mounts, (iv) *functioning as a local acoustical barrier* to shield the floor of an equipment room from the noise radiated from the bottom of the machine and reducing its transmission to rooms below, and (v) *reducing the effect on the machine of external forces* such as transient loads or torques caused by operation of motors or fans or rapid changes in machine load or speed.

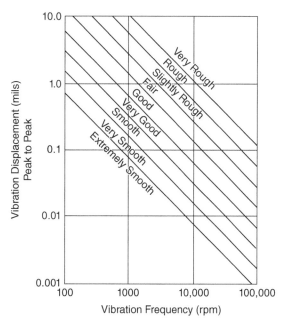

Figure 9.8 Machine vibration severity chart showing peak-to-peak- vibration (2A) mils (1 mil = 0.001 in. = 0.0254 mm).

9.3.4 Other Considerations

The performance of vibration isolators in the high-frequency range ($f/f_n \geq 1$) is often disappointing. There are several possible reasons; but often they all result in an increase in force transmissibility T_F for $f/f_n \geq 1$. Usually, the reasons are deviations from the simple single-degree-of-freedom model as mentioned below [5].

a) Support Flexibility

If the assumption that the support is rigid does not hold, then the support will also deflect (e.g. because of floor flexibility). If the isolator stiffness is similar in magnitude to that of the supporting floor, then additional resonances in the floor–machine system will occur for $f/f_n > 1$. It is normal practice to choose isolators (assuming a rigid support or floor) to have a natural frequency well below the fundamental natural frequency of the floor itself. (If possible there should not be any machine-exciting frequencies in the frequency range 0.8–1.3 times the floor fundamental natural frequency.) The fundamental natural frequency of a wood floor is usually in the 20–30 Hz frequency range, while that of a concrete floor is in the 30–100 Hz range.

b) Machine Resonances

Internal resonances in the machine structure will also increase the force transmissibility T_F in a similar manner to support flexibility. Increasing the stiffness and/or the damping of machine members can help to reduce this effect.

c) Standing-Wave Effects

Standing-wave effects in the vibration isolators can significantly decrease their performance and increase force transmissibility for $f/f_n \geq 1$, but this typically occurs only at high frequency. This effect can be reduced by increasing the damping in the isolators. Springs particularly suffer from this problem because of their low inherent damping. Soft materials (such as felt or rubber) placed between the spring and the support can alleviate the problem. In elastometric isolators the effects of standing wave resonances are small because of their higher damping.

d) Shock Isolation

If the forces in the machine are impulsive in character (such as caused by repeated impacts), then the discussion for isolation of the single degree-of-freedom model presented so far can still be used. It is normal practice to choose an isolator that provides a natural period T ($T = 1/f_n$), which is much greater than the shock pulse duration but less than the period of repetition of the force.

> **Example 9.3** A nail-making machine cuts nails five times each second ($T = 0.2$ second). The shock pulse duration is approximately 0.015 second. What is a suitable choice of vibration isolator for the machine?
>
> **Solution**
>
> Elastometric isolators that provide a natural period of vibration of 0.1 second (natural frequency of 10 Hz) would be a good choice. They have high damping. An elastomer is a type of viscoelastic damping material. In the following section the use of damping materials will be discussed.

9.4 Use of Damping Materials

Load-bearing and non-load-bearing structures of a machine (panels) are excited into motion by mechanical machine forces resulting in radiated noise. Also, the sound field inside an enclosure excites its walls into vibration. When resonant motion dominates the vibration, the use of damping materials can result in significant noise reduction. In the case of machinery enclosures, the motion of the enclosure walls is normally mass-controlled (except in the coincidence frequency region), and the use of damping materials is often disappointing. Damping materials are often effectively employed in machinery in which there are impacting parts since these impacts excite resonant motion. A structure that vibrates in flexure can be damped by the appropriate addition of a layer of damping material.

Damping involves the conversion of mechanical energy into heat. Damping mechanisms include friction (rubbing) of parts, air pumping at joints, sound radiation, viscous effects, eddy currents, and magnetic and mechanical hysteresis. Rubbery, plastic, and tarry materials usually possess high damping. During compression, expansion, or shear, these materials store elastic energy, but some of it is converted into heat as well. The damping properties of such materials are temperature dependent. Damping materials can be applied to structures in a variety of ways, for example by the application of coatings made of some viscoelastic material usually marketed in the form of tapes, sheets or sprays which may be applied like paint.

An essential parameter for measuring the amount of damping energy lost in a harmonically vibrating structure is the system damping loss factor, η, which is defined as the ratio between the energy dissipated within the damping layer and the energy stored in the whole structure, per cycle of vibration. Figure 9.9 shows typical ranges of the loss factors reported for materials at small strains, near room temperature, and at audio and lower frequencies. The range indicated for plastics and rubbers is large because it includes many different materials and because the properties of individual materials of this type may vary considerably with both frequency and temperature. On the other hand, since measurement of Poisson's ratio of viscoelastic materials is very difficult to obtain experimentally, data are not available for most damping materials. Often, viscoelastic materials are assumed to be incompressible in regions of rubbery behavior and about 0.3 in regions of glassy behavior [52]. In general, the damping loss factors of high-strength materials (e.g. metals) tend to be much smaller than those of plastics and rubbers, which are of lower strength [51].

The viscoelastic materials of greatest practical interest for damping applications are plastics and elastomers. An elastomer is a soft substance that exhibits thermo-viscoelastic behavior. Viscoelastic materials possess both elastic and viscous properties. For a purely elastic material, all the energy stored in a sample during

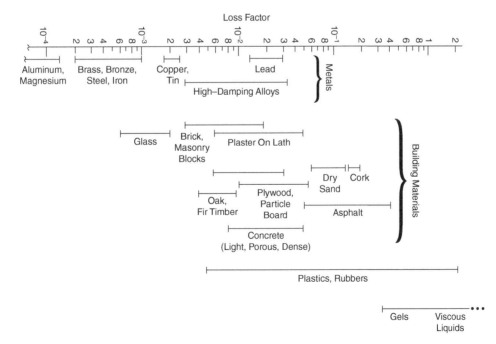

Figure 9.9 Typical ranges of material damping loss factors at small strains and audio frequencies, near room temperature [51].

loading is returned when the load is removed. Furthermore, the displacement of the sample responds immediately, and in-phase, to the cyclic load. Conversely, for a purely viscous material, no energy is returned after the load is removed. The input stress is lost to pure damping as the vibration energy is transferred to internal heat energy. All the materials that do not fall into one of the above extreme classifications are called viscoelastic materials. Some of the energy stored in a viscoelastic system is recovered upon removal of the load, and the remaining energy is dissipated by the material in the form of heat [53].

Undamped metal structures normally have a very low loss factor, typically in the range 0.001–0.01. Using a viscoelastic layer can increase this loss factor. This means that the amplitude of the resonant vibration when the structure is subjected to structure-borne sound or vibration will be much lower than for an undamped structure. Reduced amplitude of vibration means less radiation of sound, and also a reduced risk of fatigue failure. In addition, use of a viscoelastic coating method for damping the vibration of plates has proven to significantly reduce dynamic stresses in the structure as a whole.

A characteristic of viscoelastic materials is that their Young's modulus is a complex quantity, having both a real and imaginary component. Furthermore, this complex modulus varies as a function of many parameters, the most important of which are the frequency and temperature of a given application. Consequently, this results in a corresponding eigenvalue problem in which the stiffness matrix depends on both the frequency and the temperature. The moduli typically possess relatively high values at low temperatures and/or high frequencies (rubbery behavior regions) but take on comparatively small values at high temperatures and/or low frequencies (glassy behavior regions) [54].

In general, the vibration analysis of a system that is frequency independent can be accurately achieved by classical techniques. It is much more difficult to obtain accurate predictions when the equations of motion are frequency-dependent. This is because the solution of the corresponding eigenvalue problem is difficult to compute. Methods based on the modal strain energy have been used to approximate the solution of the problem [55]. However, they are not accurate when the frequency and temperature ranges are increased, and when they include the *transition region*, where the variations of the dissipation and the stiffness of the viscoelastic material are quite pronounced. The greatest loss factors occur in the transition region at intermediate frequencies and temperatures [53].

Although there are several ways of applying damping materials to structures, the use of unconstrained and constrained damping layers are the most common.

9.4.1 Unconstrained Damping Layer

If a free viscoelastic layer is attached to a panel (glued on one or two faces of the panel), which otherwise has a very small damping, then bending produces both flexure and extension of the two layers. As one of the faces of the coating material is free, the added rigidity is due to bending deformation of the material. In this case, shear has little effect on energy storage of the composite panel since the viscoelastic layer is unconstrained. This treatment is usually called extensional damping (see Figure 9.10a).

Unconstrained metal-elastomer composite structures are an important tool for the reduction of mechanical vibrations. Since the first successful modeling of a metal-elastomer composite presented by Ross et al. [56], (known as RUK theory), considerable attention has been paid to the prediction of the dynamic behavior of such structures. The damping properties of a plate are influenced by the stiffness of the unconstrained viscoelastic coating, its dissipation loss factor, and by the thickness of the dissipating layer. These properties of a two-layer plate may be represented by an equivalent plate accounting for mass, stiffness and viscoelastic damping added on the plate by means of the RUK theory [57, 58]. Unconstrained free layer coating treatment is sometimes preferred since it is economic, stable and easy to apply for in situ corrective measures.

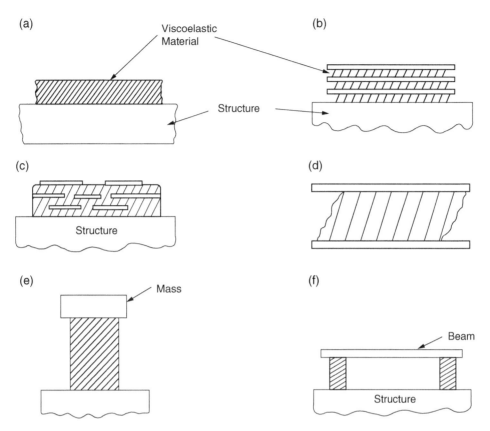

Figure 9.10 Different ways of using vibration damping materials: (a) free (unconstrained) layer, (b) multiple constrained layer, (c) multilayer tile spaced treatment, (d) sandwich panel, (e) tuned damper, and (f) resonant beam damper. Shaded elements represent viscoelastic material.

9.4.2 Constrained Damping Layer

More efficient and effective damping can be achieved by using a laminated composite (sandwich plate) made of one or more sheet metal layers each separated by a viscoelastic layer and the whole being bonded together. Many practical applications operate on the principle of constrained layer damping. The bending of the composite produces not only bending and extensional strains in all layers but also shears. The shear-strain energy storage tends to dominate the damping action of the constrained viscoelastic layers. Therefore, it is possible to use very thin layers of viscoelastic material achieving very high values of total damping [58] (see Figure 9.10b). This treatment is called shear damping and, for example, it has been used in the design of vehicle windscreens and to improve sound insulation of windows in buildings. Constrained metal-vicoelastic-metal structures have been widely used to provide vibration reduction in structures. In general, finite element methods have been used to model their behavior [53, 55].

Figure 9.10 shows some common ways of applying damping materials and systems to structures. Reference [51] describes damping mechanisms in more detail and how damping materials can be used to reduce vibration in some practical situations.

9.5 Use of Sound Absorption

Sound-absorbing materials have been found to be very useful in the control of noise. They are used in a variety of locations: close to sources of noise (e.g. close to sources in electric motors), in various paths, (e.g. above barriers in buildings or inside machine enclosures), and sometimes close to a receiver (e.g. inside earmuffs). When a machine is operated inside a building, the machine operator usually receives most sound through a direct path, while people situated at greater distances receive sound mostly through reflections (see Figure 9.11).

The relative contributions of the sound reaching people at a distance through direct and reflected paths are determined by how well the sound is reflected and absorbed by the walls in the building.

9.5.1 Sound Absorption Coefficient

When sound waves strike a boundary separating two media, some of the incident energy is reflected from the surface and the remaining energy is transmitted into the second medium. Some of this energy is eventually converted by various processes into heat energy and is said to have been absorbed by that medium. The fraction of the incident energy absorbed is termed the *absorption coefficient* $\alpha(f)$, which is a function of frequency and already defined in Chapter 3 as

$$\alpha(f) = \frac{\text{sound intensity absorbed}}{\text{sound intensity incident}}. \tag{9.7}$$

The absorption coefficient theoretically ranges from zero to unity. In practice, values of $\alpha > 1.0$ are sometimes measured. This anomaly is due to the measurement procedures adopted to measure large scale building materials. One sabin is defined as the sound absorption of one square metre of a perfectly absorbing surface, such as an open window. The sound absorption of a wall or some other surface is the area of the surface, in square metres, multiplied by the absorption coefficient.

9.5.2 Noise Reduction Coefficient

Another parameter, which is often used to assess the performance of an acoustical absorber, is the single number known as the noise reduction coefficient (NRC). The NRC of a sound-absorbing material is given by the average of the measured absorption coefficients for the 250-, 500-, 1000-, and 2000-Hz one-octave bands rounded off to the nearest multiple of 0.05. This NRC value is often useful in the determination of the applicability of a material to a particular situation. However, where low or very high frequencies are involved, it is usually better to consider sound absorption coefficients instead of NRC data.

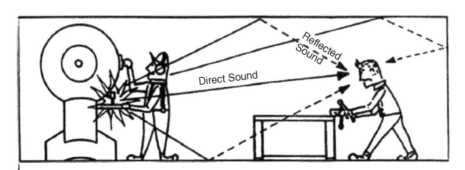

Figure 9.11 Paths of direct and reflected sound emitted by a machine in a building.

Example 9.4 If $\alpha_{250} = 0.25$, $\alpha_{500} = 0.45$, $\alpha_{1000} = 0.65$, and $\alpha_{2000} = 0.81$, what is the NRC?

Solution

$$\text{NRC} = (0.25 + 0.45 + 0.65 + 0.81)/4 = 0.54.$$

A wide range of sound-absorbing materials exist that provide absorption properties dependent upon frequency, composition, thickness, surface finish, and method of mounting. They can be divided into several major classifications. Materials that have a high value of α are usually porous and fibrous. Fibrous materials include those made from natural or artificial fibers including glass fibers. Porous materials made from open-celled polyurethane are also widely used.

9.5.3 Absorption by Porous Fibrous Materials

Porous materials are characterized by the fact that the nature of their surfaces is such that sound energy is able to enter the materials by a multitude of small holes or openings. They consist of a series of tunnel-like pores and openings formed by interstices in the material fibers or by foamed products. (Usually, within limitations, the more open and connecting these passages are, the larger are the values of the sound-absorbing efficiency of the material.) If, on the other hand, the pores and penetrations are small and not joined together, then the material becomes substantially less efficient as a sound absorber. Included in this broad category of porous absorbers are fibrous blankets, hair felt, wood-wool, ceramics, foams, acoustical plaster, a variety of spray-on products, and certain types of acoustical tiles [59].

When a porous material is exposed to incident sound waves, the air molecules at the surface of the material and within the pores of the material are forced to vibrate and in the process lose energy. This is caused by the conversion of sound energy into heat due to thermal and viscous losses of air molecules at the walls of the interior pores and tunnels in the sound-absorbing material. At low frequencies these changes are isothermal, while at high frequencies they are adiabatic. In fibrous materials, much of the energy can also be absorbed by scattering from the fibers and by the vibration caused in the individual fibers. The fibers of the material rub together under the influence of the sound waves and lose energy due to work done by the frictional forces. Figure 9.12 shows the two main mechanisms by which the sound is absorbed in materials. For this reason, high values of absorption coefficient in excess of 0.95 can be observed. Depending upon how α is determined

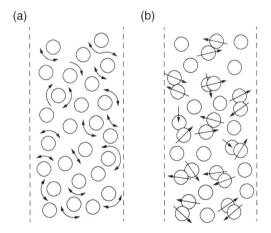

(a)　　　　　　　(b)

Figure 9.12 The two main mechanisms believed to exist in sound-absorbing materials: (a) viscous losses in air channels and (b) mechanical friction caused by fibers rubbing together.

experimentally, values in excess of unity can also be measured [60]. The values of α observed are usually strongly dependent upon (i) frequency, (ii) thickness, and (iii) method of measurement. These should always be considered in the choice of a particular material.

Figure 9.13 shows typical sound absorption characteristics for a blanket-type fibrous porous material placed against a hard wall, and with a 10-cm airspace between the material and the wall. In both cases the absorption properties are substantially better at high frequencies than low. When the same material is backed by an airspace, the low-frequency absorption is improved without significantly changing the high-frequency characteristics. Hanging drapes and curtains a few centimeters away from walls and windows can improve their sound absorption properties, particularly at low frequencies. Figure 9.14 shows the effect of increasing the thickness of sound-absorbing materials mounted on solid backings. Increased low-frequency absorption is also observed for increased material thickness.

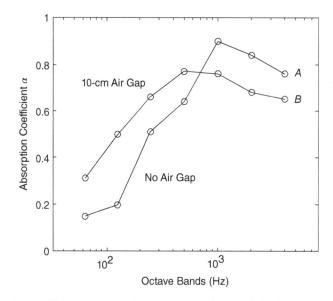

Figure 9.13 Typical absorption coefficient vs. octave band frequency characteristics for a 25-mm-thick fibrous absorbing material. Curve *A* is for the material laid directly on a rigid backing, while curve *B* shows the effect of introducing a 10-cm air gap.

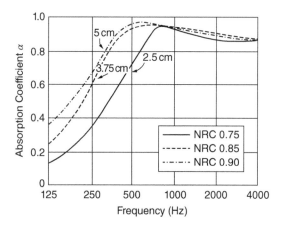

Figure 9.14 Sound absorption coefficient α and noise reduction coefficient (NRC) for typical fiberglass foamboard.

An additional effect to consider is that of surface treatment, since most materials will become discolored or dirty after prolonged exposure for several years and may require cleaning or refinishing. Because the surface pores must be open to incident sound for the porous material to function, it is essential that they should not be blocked with paint or any other surface-coating treatment. The effects of brush painting porous materials are usually more severe than spray painting; however, the usual effect is to lower the absorption coefficient to about 50% of its unpainted value particularly at high frequencies. In addition, the absorption peak is shifted downward in frequency. As more coats of paint are applied, the paint membrane becomes denser and more pores are sealed, the result of which is to shift the absorption peak even lower in frequency and magnitude.

It is useful now to describe briefly the physical parameters used to account for the sound-absorbing and attenuating properties of porous materials. These include flow resistivity, porosity, volume coefficients of elasticity of both air and the skeleton, structural form factor (tortuosity), and specific acoustic impedance. These will each be described separately.

a) Flow Resistivity R

This accounts for the resistance offered to airflow through the medium. It is defined in the meter–kilogram–second (mks) system as

$$R = \frac{\Delta p}{U} \frac{1}{t},$$

(9.8)

where R is the flow resistivity (mks rayls/m), Δp is the differential sound pressure created across a sample of thickness t, measured in the direction of airflow (N/m^2), U is the mean steady flow velocity (m/s), and t is the thickness of the porous material sample (m). Typical values for porous fibrous materials vary from 4×10^3 to 4×10^4 mks rayls/m for a density range of 16 kg/m^3–160 kg/m^3. Generally speaking, if the flow resistivity R becomes very large, then most of the incident sound falling on the material will be reflected; while if R is too small, then the material will offer only very slight viscous losses to sound passing through it, and so it will provide only little sound absorption or attenuation. Although the absorption is proportional to thickness, it is generally found that for a given flow resistivity value R, the optimum thickness of material is approximately given by $t = 100/\sqrt{R}$.

b) Porosity ε

The porosity of a porous material is defined as the ratio of the void space within the material to its total displacement volume as

$$\varepsilon = \frac{V_a}{V_m},$$

(9.9)

where V_a is the volume of air in the void space in the sample and V_m is the total volume of the sample. It has to be noticed that the void space is only that accessible to sound waves. For a material composed of solid fibers the porosity can be estimated from

$$\varepsilon = 1 - \frac{M_s}{V_s \rho_F},$$

(9.10)

where M_s is the total mass of sample (kg), V_s is the total volume of sample (m^3), and ρ_F is the density of fibers (kg/m^3). Typical values of porosity for acceptable acoustical materials are greater than 0.85.

c) Volume Coefficient of Elasticity of Air K

The bulk modulus of air is derived from

$$\Delta p = -K \frac{\Delta V}{V},$$

(9.11)

where Δp is the change in pressure required to alter the volume V by an increment ΔV (N/m^2), ΔV is the incremental change in volume (m^3), and K is the volume coefficient of elasticity for air (N/m^2).

d) Volume Coefficient of Elasticity of the Skeleton Q

This is defined in a similar way to the bulk modulus. It is derived from the change in thickness of a sample sandwiched between two plates as the force applied on them is increased, that is,

$$\delta F = -Q \frac{\delta t}{t} S, \tag{9.12}$$

where δF is the incremental force applied to the sample (N), δt is the incremental change in thickness of the sample (m), t is the original thickness (m), Q is the volume coefficient of elasticity of the skeleton (N/m^2), and S is the sample area (m^2).

e) Structural Form Factor k_s

It is found that in addition to the flow resistance, the composition of the inner structure of the pores also affects the acoustical behavior of porous materials. This is because the orientation of the pores relative to the incident sound field has an effect on the sound propagation. This has been dealt with by Zwikker and Kosten [61] and treated as an effective increase in the density of the air in the void space of the material. Beranek [62] reports that flexible blankets have structure factors between 1.0 and 1.2, while rigid tiles have values between 1.0 and 3.0. He also shows that, for homogeneous materials made of fibers with interconnecting pores, the relationship between the porosity ε and structure factor k_s is $k_s \approx 5.5 - 4.5\varepsilon$. In the technical literature it is possible to find the self-explanatory term *tortuosity* used instead of structural form factor. For most fibrous materials the structural form factor is approximately unity.

f) Specific Acoustic Impedance z_0

This is defined as the ratio of the sound pressure p to particle velocity u at the surface of the material, for a sample of infinite depth, when plane sound waves strike the surface at normal incidence. This is a complex quantity and it is defined mathematically as

$$z_0 = \frac{p}{u} = \rho c (r_n + j x_n), \tag{9.13}$$

where z_0 is the normal specific acoustic impedance, ρc is the characteristic impedance of air (415 mks rayls), and $j = \sqrt{-1}$; r_n and x_n are the specific acoustic resistance and specific acoustic reactance, respectively.

It is useful to briefly examine some of the interesting relationships that exist between the specific impedance and the absorption coefficient α under certain circumstances. For example, if a porous material is composed so that $r_n \gg 1$ and $r_n \gg x_n$, the absorption coefficient α_θ for a sound wave striking a surface at angle θ to the normal is given by

$$\alpha_\theta = \frac{4 r_n \cos \theta}{(1 + r_n \cos \theta)^2}. \tag{9.14}$$

Furthermore, for a diffuse sound field, the random incidence absorption coefficient α is given by [14]

$$\alpha = \frac{8}{r_n} \left[1 + \frac{1}{1 + r_n} - \frac{2}{r_n} \ln (1 + r_n) \right]. \tag{9.15}$$

If $r_n > 100$ (materials with small absorption coefficients), this equation can be substantially simplified to $\alpha = 8/r_n$.

The use of the complex acoustic impedance, rather than absorption coefficient, allows for a much more rigorous treatment of low-frequency reverberation time analysis for rooms. Although often considerably more complex than the classical theory, this approach does predict more accurate values of reverberation times in rooms containing uneven distribution of absorption material, even if either pair of opposite walls are highly absorbing or if opposite walls are composed of one very soft and one very hard wall [14]. Beranek shows [62, 63] that the specific acoustic impedance of a rigid tile can be written in terms of the previously defined fundamental parameters as

$$z_0 = \left(\rho \frac{k_s K}{\varepsilon} \left[1 - j \frac{R}{\rho k_s \omega} \right] \right)^{1/2}, \tag{9.16}$$

where ρ is the density of air (kg/m^3) and ω is the angular frequency (rad/s). It can therefore be seen from Eqs. (9.13), (9.15), and (9.16) that the absorption coefficient is proportional to the porosity and flow resistance of the material (provided $r_n > 100$ and $r_n \gg x_n$), and is inversely proportional to the density and structure factor. For soft blankets, the expressions for the acoustic impedance become very much more complex and the reader is directed to Refs. [61–63] for a much more complete analysis.

The basic theory used to model the sound propagation within porous absorbents assumes that the absorber frame is rigid and the waves only propagate in the air pores. This is the typical case when the porous absorber is attached to a wall or resting on a floor that constrains the motion of the absorber frame. Neglecting the effect of the structural form factor, it can be shown that plane waves in such a material are only possible if the wavenumber is given by [64]

$$k_p = k \sqrt{1 + j \frac{R \varepsilon}{\omega \rho}}, \tag{9.17}$$

where k is the free-field wavenumber in air (m^{-1}) and ρ is the density of air (kg/m^3). In addition, the wave impedance of plane waves is

$$z_0 = \frac{\rho c}{\varepsilon} \frac{k_p}{k}. \tag{9.18}$$

In another method of analysis, a fibrous medium is considered to be composed of an array of parallel elastic fibers in which a scattering of incident sound waves takes place resulting in their conversion to viscous and thermal waves by scattering at the boundaries. This approach was first used by Rayleigh [65] and has since been refined by several researchers. Attenborough and Walker included effects of multiple scattering in the theory [66] that is able to give good predictions of the impedance of porous fibrous materials. However, even more refined phenomenological models have been presented in recent years [67]. On the other hand, very useful empirical expressions for the prediction of both the propagation wavenumber and characteristic impedance of a porous absorbent have been developed by Delany and Bazley [68]. The expressions as functions of frequency f and flow resistivity R are

$$k_p = k \left(1 + 0.0978 \, X^{-0.7} - j0.189 \, X^{-0.595} \right) \tag{9.19}$$

and

$$z_0 = \rho c \left(1 + 0.0571 \, X^{-0.754} - j0.087 \, X^{-0.732} \right), \tag{9.20}$$

where $X = \rho f / R$. Equations (9.19) and (9.20) are valid for $0.01 < X < 1.0$, $10^3 \le R \le 5 \times 10^4$, and $\varepsilon \approx 1$. Additional improvements to the Delany and Bazley empirical model have been presented by other authors [69, 70]. If the frame of the sound absorber is not constrained (elastic-framed material), a more complete "poroelastic" model of sound propagation can be developed using the Biot theory [71]. In addition, both guidelines and charts for designing absorptive devices using several layers of absorbing materials have been presented [72, 73].

9.5.4 Panel or Membrane Absorbers

When a sound source is turned on in a room, a complex pattern of room modes is set up, each having its own characteristic frequency. These room modes are able, in turn, to couple acoustically with structures in the room, or even the boundaries of the room, in such a way that sound power can be fed from the room modes to other structural modes in, for example, a panel hung in the room. A simply-supported thin plate or panel can only vibrate at certain allowed natural frequencies $f_{m,n}$ and these are given by (see Eq. (2.70b))

$$f_{m,n} = 0.453 c_L h \left[\left(\frac{m}{a} \right)^2 + \left(\frac{n}{b} \right)^2 \right], \tag{9.21}$$

where $f_{m,n}$ is the characteristic modal frequency, m and n are integers (1, 2, 3, ...), c_L is the longitudinal wave speed in the plate material (m/s), h is the thickness of the plate (m), and a and b are the dimensions of the plate (m).

If $m = n = 1$, then this gives the first allowable mode of vibration along with its fundamental natural frequency $f_{1,1}$. The above equation is valid only for plates with simply supported edges. For a plate with clamped edges, the fundamental mode occurs at approximately twice the frequency calculated from Eq. (9.21). Thus a room mode, at or close to the fundamental frequency of a plate hung in the room, will excite the plate fundamental mode. In this way the plate will be in a resonant condition and therefore have a relatively large vibration amplitude. In turn, this will cause the plate to dissipate some of its energy through damping and radiation. Therefore, the plate can act as an absorber having maximum absorption characteristics at its fundamental frequency (and higher order modes), which will depend upon the geometry of the plate and its damping characteristics. In all practical cases this effect takes place at low frequencies, usually in the range 40–300 Hz. Particular care has to be taken, therefore, that any panels that may be hung in a room to improve reflection or diffusion are not designed in such a manner that they act as good low-frequency absorbers and have a detrimental effect on the acoustics of the space.

If a panel is hung in front of a hard wall at a small distance from it, then the airspace acts as a compliant element (spring) giving rise to a resonant system comprised of the panel's lumped mass and the air compliance. The resonance frequency f_r (Hz) of the system is given by

$$f_r \approx \frac{59.5}{\sqrt{Md}}, \tag{9.22}$$

where M is the mass surface density (kg/m^2), and d is the airspace depth (m). Hence, a thin panel of 4 kg/m^2 placed a distance of 25 mm from a rigid wall will have a resonance frequency of 188 Hz. As in the case of a simply supported panel, a spaced panel absorbs energy through its internal viscous damping [74, 75]. Since its vibration amplitude is largest at resonance, its sound absorption is maximum at this frequency. Usually this absorption can be both further increased in magnitude and extended in its effective frequency range (i.e. giving a broader resonance peak) by including a porous sound-absorbing material, such as a fiberglass blanket, in the airspace contained by the panel. This effectively introduces damping into the resonant system.

Figure 9.15 shows the effect of introducing a 25-mm-thick fibrous blanket into the 45-mm airspace contained by a 3-mm plywood sheet. The change is quite significant at low frequencies in the region of the resonance peak where both the magnitude and width of the peak are increased. On the other hand, there is little effect at high frequencies.

Membrane absorbers are one of the most common bass absorbers used in small rooms. In addition, their nonperforated surfaces are durable and can be painted with no significant effect on their acoustical properties. It is important that this type of sound absorber be recognized as such in the design of an auditorium. Failure to do so, or to underestimate its effect, will lead to excessive low-frequency absorption, and the room will have a relatively short reverberation time. The room will then be considered to be acoustically

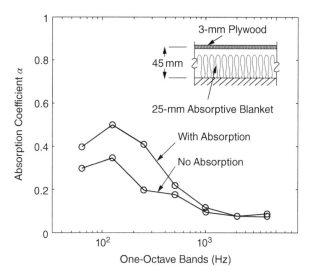

Figure 9.15 Effect on the sound absorption coefficient α of placing a 25-mm-thick sound-absorptive blanket in the airspace (45-mm deep) behind a flexible 3-mm plywood panel.

unbalanced and will lack warmth. Typical panel absorbers found in auditoriums include gypsum board partitions, wood paneling, windows, wood floors, suspended ceilings, ceiling reflectors, and wood platforms. Porous materials also possess better low-frequency absorption properties when spaced away from their solid backing (see Figure 9.13) and behave in a manner similar to the above solid panel absorbers. When the airspace equals one-quarter wavelength, maximum absorption will occur, while when this distance is a one-half wavelength, minimum absorption will be realized. This is due to the fact that maximum particle velocity in the air occurs at one-quarter wavelength from a hard wall and hence provides the maximum airflow through the porous material. This, in turn, provides increased absorption at that frequency. Such an effect can be useful in considering the performance of curtains or drapes hanging in an auditorium. A similar effect is observed for hanging or suspended acoustical ceilings [4].

To achieve well-balanced low-frequency absorption, a selection of different size and thickness spaced panels can be used and indeed have been used successfully in many auditoriums. Combinations of resonant panels have also been suggested [76].

9.5.5 Helmholtz Resonator Absorbers

A Helmholtz resonator, in its simplest form, consists of an acoustical cavity contained by rigid walls and connected to the exterior by a small opening called the neck, as shown in Figure 9.16. Incident sound causes air molecules to vibrate back and forth in the neck section of the resonator like a vibrating mass while the air in the cavity behaves like a spring. As shown in the previous section, such an acoustical mass–spring system has a particular frequency at which it becomes resonant. At this frequency, energy losses in the system due to frictional and viscous forces acting on the air molecules in and close to the neck become maximum, and so the absorption characteristics also peak at that frequency. Usually there will be only a very small amount of damping in the system, and hence the resonance peak is normally very sharp and narrow, falling off very quickly on each side of the resonance frequency. This effect can be observed easily if we blow across the top of the neck of a bottle, we hear a pure tone developed rather than a broad resonance.

If the neck is circular in cross-section and if we neglect boundary layer effects, the undamped resonance frequency f_r is given by

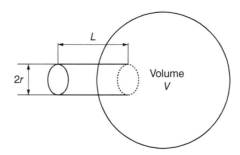

Figure 9.16 A Helmholtz resonator consists of a neck of radius *r*, length *L* and cross-sectional area *S*, and backed by a closed volume *V*.

$$f_r = \frac{c}{2\pi} \sqrt{\frac{S}{(L + 1.7r)V}}, \tag{9.23}$$

where S is the area of the neck (m^2), L is the length of neck (m), r is the radius of neck (m), V is the cavity volume (m^3), and c is the speed of sound (m/s). The factor $(L + 1.7r)$ in Eq. (9.23) gives the effective length of the baffled neck. The factor $1.7r$ is sometimes called the *end correction*. If the neck is unflanged then the end correction is $1.5r$.

Although Helmholtz-type resonators can be built to be effective at any frequency, their size is such that they are used mainly for low frequencies in the region 20–400 Hz. Because of their sharp resonance peaks, undamped resonator absorbers have particularly selective absorption characteristics. Therefore, they are used primarily in situations where a particularly long reverberation time is observed at one frequency. This frequency may correspond to a well-excited low-frequency room mode, and an undamped resonator absorber may be used to reduce this effect without changing the reverberation at other, even nearby, frequencies. They are also used in noise control applications where good low-frequency sound absorption is required at a particular frequency. In this respect, special Helmholtz resonators constructed from hollow concrete blocks with an aperture or slit in their faces, are used in transformer rooms and electrical power stations to absorb the strong 120-Hz noise produced therein. Helmholtz resonators are also used in mufflers. See Chapter 10, Sections 10.6.2 c and 10.6.2 d.

This concept is well known and such resonators were used hundreds of years ago in some churches built in Europe. Some damping may be introduced into such resonators by adding porous material either in the neck region or to a lesser extent in the cavity. The effect of increased damping is to decrease the absorption value at resonance and also considerably to broaden the absorption curve over a wider frequency range. Figure 9.17 shows measured absorption coefficients for slotted concrete blocks filled with porous material. It can be seen that they offer especially good low-frequency absorption characteristics. In addition, they can be faced with a thin blanket of porous material and covered with perforated metal sheeting, as shown in Figure 9.18, to improve their high-frequency absorption with only little effect on their low-frequency performance.

It is useful to note that there is a limit to the absorption any given undamped resonator of this type can provide. According to Zwikker and Kosten [61] the maximum absorption possible, $A_{\max}$ sabins, is given by

$$A_{\max} = 1.717 \left(\frac{c}{f_r}\right)^2, \tag{9.24}$$

where c is the speed of sound (m/s) and f_r is the resonance frequency (Hz). Therefore, if $f_r = 150$ Hz, for example, then we cannot expect to realize more than $A_{\max} = 8.5$ sabins per unit. Equation (9.24) also shows that as the resonance frequency becomes lower, more absorption can be obtained from the resonator.

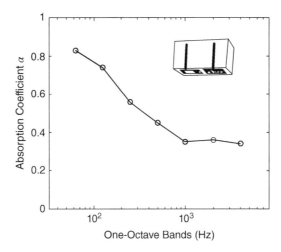

Figure 9.17 Sound absorption coefficient vs. frequency for a slotted 20-cm concrete block filled with an incombustible fibrous material.

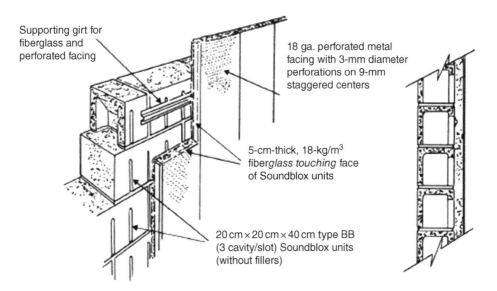

Supporting girt for fiberglass and perforated facing

18 ga. perforated metal facing with 3-mm diameter perforations on 9-mm staggered centers

5-cm-thick, 18-kg/m³ fiber*glass touching* face of Soundblox units

20 cm × 20 cm × 40 cm type BB (3 cavity/slot) Soundblox units (without fillers)

Figure 9.18 Slotted concrete blocks faced with fiberglass and covered with a perforated metal provide good low- and high-frequency absorption characteristics.

Example 9.5   A Helmholtz resonator is made of a sphere of 10 cm internal diameter. If it is to resonate at 400 Hz in air, what is the hole diameter that should be drilled into the sphere?

Solution

In this case we have $L = 0$ and the surface of the sphere acts as a flange. Since $S = \pi r^2$, from Eq. (9.23) we obtain that $r = 1.7 V \pi (2 f_r / c)^2$. The cavity volume can be calculated as $V = 4 \pi r^3 / 3 = 4 \pi (0.05)^3 / 3 = 5.24 \times 10^{-4}$ m³. Therefore, $r = 1.7 (5.24 \times 10^{-4}) \pi (2 \times 400 / 344)^2 = 1.5$ cm, or 3 cm diameter.

In certain circumstances, the performance of a Helmholtz-type resonator can be drastically influenced by the effect of its surrounding space. This is found particularly in the case for such resonators mounted in a highly absorbing plane, when interference can take place between the sound radiated from the resonator and reflected from the absorbing plane. In such circumstances special care has to be taken.

9.5.6 Perforated Panel Absorbers

As described above, single Helmholtz resonators have very selective absorption characteristics and are often expensive to construct and install. In addition, their main application lies at low frequencies. Perforated panels offer an extension to the single resonator absorber and provide a number of functional and economic advantages. When spaced away from a solid backing, a perforated panel is effectively made up of a large number of individual Helmholtz resonators, each consisting of a "neck," constituted by the perforation of the panel, and a shared air volume formed by the total volume of air enclosed by the panel and its backing. The perforations are usually holes or slots and, as with the single resonator, porous material may be included in the airspace to introduce damping into the system. Perforated panels are mechanically durable and can be designed to provide good broadband sound absorption. The addition of a porous blanket into the airspace tends to lower the magnitude of the absorption maximum but, depending on the resistance of the material, generally broadens the effective range of the absorber [77]. At low frequencies, the perforations become somewhat acoustically transparent because of diffraction, and so the absorbing properties of the porous blanket remain almost unchanged. This is not so at high frequencies at which a reduction in the porous material absorption characteristics is observed. The resonance frequency f_r for a panel perforated with holes and spaced from a rigid wall may be calculated from

$$f_r = \frac{c}{2\pi} \sqrt{\frac{P}{Dh'}} \,, \tag{9.25}$$

where c is the speed of sound (m/s), D is the distance from wall (m), h' is the thickness of the panel with the end correction (m), r is the radius of hole or perforation (m), and P is the open area ratio (or perforation ratio). For a panel made up of holes of radius r metres and spaced s metres apart, the open area ratio P is given by $\pi(r/s)^2$ (see Figure 9.19).

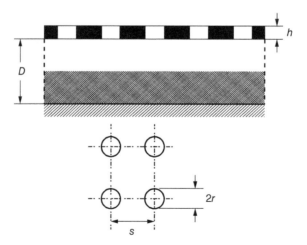

Figure 9.19 Geometry for a typical perforated panel absorber.

Table 9.2 Different formulas for the end correction factor. P is the open area ratio.

δ	Notes
0.85	Single hole in a baffle
$0.8(1 - 1.4\sqrt{P})$	For $P < 0.16$
$0.8\left(1 - 1.47\sqrt{P} + 0.47\sqrt{P^3}\right)$	Includes $P = 1$
$0.85(1 - 1.25\sqrt{P})$	Square apertures; for $P < 0.16$
$-\ln[\sin(\pi P/2)]/\pi$	Slotted plate; in Eq. (9.26) $v = 0$ and $r = $ width of slots

A full expression for h' in Eq. (9.25) that takes into account the boundary layer effect is given by [78]

$$h' = h + 2\delta r + \left[\frac{8\nu}{\omega}\left(1 + \frac{h}{2r}\right)\right]^{1/2}, \tag{9.26}$$

where h is the thickness of panel (m), ω is the angular frequency (rad/s), ν is the kinematic viscosity of air (15×10^{-6} m^2/s), and δ is the end correction factor. For panel hole sizes not too small we can write $h' \approx h + 2\delta r$. To a first approximation the end correction factor can be assumed to be 0.85, as was done in the previous section for a single hole. However, more accurate results that include the effects of the mutual interaction between the perforations have been predicted. Table 9.2 presents some of these results.

We can see that the resonance frequency increases with the open area ratio (i.e. the number of holes per unit area) and is inversely proportional to the thickness of the panel and its distance from the solid backing.

Example 9.6 A perforated panel with a 10% open area ($P = 0.10$) and thickness 6 mm is installed 15 cm in front of a solid wall. Determine the resonance frequency of the system.

Solution

Assuming that the holes are not too small and if we neglect the end effect and put $h' = h + 2\delta r = 6$ mm and $D = 15$ cm, then the resonance frequency $f_r = 560$ Hz. If the percentage open area is only 1% ($P = 0.01$), then in the same situation, $f_r = 177$ Hz.

In practical situations, air spaces up to 30 cm can be used with open areas ranging from 1 to 30% and thicknesses from 3 mm to 25 mm. These particular restrictions would allow a resonance frequency range of 60–4600 Hz. Many perforated panels and boards are commercially available and can readily be used to make perforated absorbers. These include: hardboards, plastic sheets, wood and plywood panels, and a variety of plane and corrugated metal facings. Some perforated sheets are available that consist of a number of different size perforations on one sheet. This can be useful to give broader absorption characteristics.

More recently, microperforated panels have been developed, which means that the diameter of the perforations is very small (less than a millimeter). In this case, the diameter is comparable to the thickness of the boundary layer, resulting in high viscous losses as air passes through the perforations and, consequently, achieving absorption without using a porous material [79, 80]. Some commercial microperforated panels take advantage of this fact, allowing the construction of a transparent absorbent device similar to a double-glazing unit. However, obtaining broadband sound absorption using microperforated panels is difficult and requires the use of multiple layers, increasing the depth and cost of the device.

Another technique commonly used to achieve broader sound absorption characteristics is to use a variable, often wedge-shaped, airspace behind the perforated panel [81], as shown in Figure 9.20. One of the

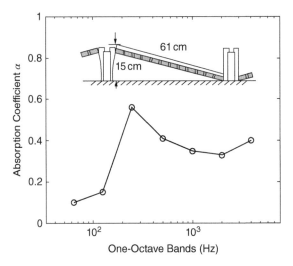

Figure 9.20 Variable airspace perforated panels give broader absorption characteristics than those with a constant air depth. Here the panel is 16 mm thick and has holes of 9.5 mm in diameter spaced 3.5 cm between centers.

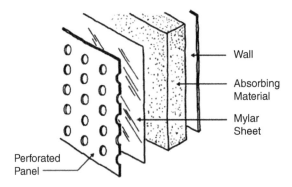

Figure 9.21 Porous absorbing material protected by a thin Mylar (polyester) sheet behind the perforated panel.

commercial types of perforated absorber has been referenced in the literature as the "Kulihat." [82] This is a conical absorber consisting of two or three perforated aluminum sectors held together by steel clips. The interior of the Kulihat is lined with mineral wool.

It is common practice to install a thin (1 mil to 2 mil; 1 mil = 0.0254 mm) Mylar (polyester) film or sheet behind the perforated panel to cover and protect the porous absorbing material, as shown in Figure 9.21. As long as this sheet is thin, its effect is usually only observed at high frequencies where the effective absorption can be reduced by approximately 10% of its value with no covering sheet present.

9.5.7 Slit Absorbers

The slit or slat absorber is another type of perforated resonator. These are made up of wooden battens fixed fairly closely together and spaced at some distance from a solid backing. Porous material is usually introduced into the air cavity (see Figure 9.22). This type of resonator is fairly popular with architects since it can be constructed in many different ways, offering a wide range of design alternatives. Equation (9.25) does not hold for slat absorbers since the perforations are very long. For a detailed discussion of the properties of such absorbers, the reader is advised to consult Refs. [83] and [84]. The principle of the slit resonator is,

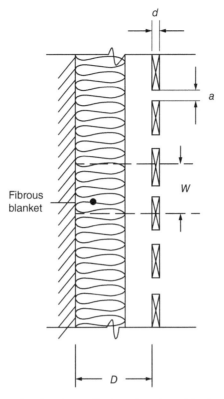

Figure 9.22 Slat type of resonator absorber (normally the mineral wool is placed immediately behind the slots).

however, the same as for a general Helmholtz resonator. The resonance frequency f_r (Hz) is given by the solution to the following equation:

$$f_r = \frac{c}{2\pi} \sqrt{\frac{a}{S}\left(d + \frac{2a}{\pi}\left[1.12 + \ln\left\{\frac{c}{\pi a f_r}\right\}\right]\right)^{-1/2}}, \tag{9.27}$$

where c is the speed of sound (m/s), a is the slit width (m), d is the slit depth (m), and S is the cross-sectional area (m^2) of space behind the slats formed by each slat (i.e. slat width $W \times$ air space D). A typical acoustical design procedure would be to choose a resonance frequency f_r and a slat of width W and thickness d and then determine the required value of air space from $S = W \times D$ for a chosen value of slit width a.

Example 9.7 Design a slit resonator having $f_r = 200$ Hz and with the resonator constructed with slats measuring 25×84 mm.

Solution

We have $W = 84$ mm and $d = 25$ mm. Since the slit width should be approximately 10–20% the width of the slat, let us try a slat spacing (i.e. slit width) of 8 mm. Then, all the known values in Eq. (9.27) are $f_r = 200$ Hz, $W = 0.084$ m, $d = 0.025$ m, $a = 0.008$ m, and $c = 343$ m/s. Equation (9.27) can be rewritten by multiplying and dividing both sides of the equation by $\sqrt{S}$ and f_r, respectively. Then, S can be obtained by squaring both sides of the resulting equation. Replacing the known values gives $S = 0.01$ m^2 and since $W = 0.084$ m, we require $D = 0.13$ m.

We see, therefore, that for the above example of a slit resonator, an air space of 13 cm is required to provide a resonance frequency of 200 Hz. If the calculation is repeated for a 100-Hz resonance frequency, it is found that an air space of some 48 cm is required.

9.5.8 Suspended Absorbers

This class of sound absorbers is known by several names including *suspended absorbers*, *functional absorbers*, and *space absorbers*. They generally refer to sound-absorbing objects and surfaces that can be easily suspended, either as single units or as a group of single units within a room. They are particularly useful in rooms in which it is difficult to find enough surface area to attach conventional acoustical absorbing materials either through simple lack of available space or interference from other objects or mechanical services such as ducts and pipes, in the ceiling space (see Figure 9.23). It is relatively easy and inexpensive to install them, without interfering with existing equipment. For this reason they are often used in noisy industrial installations such as assembly rooms or machine shops. An advantage of flat suspended panels is that they absorb sound on both sides.

Functional absorbers are usually made from highly absorbing materials in the form of a variety of three-dimensional shapes, such as spheres, cones, double cones, cubes, and panels. These are usually filled with a porous absorbing material. Since sound waves fall on all their surfaces and because of diffraction, they are able to yield high values of effective absorption coefficients. It is, however, more usual to describe their absorption characteristics in terms of their total absorption in sabins per unit, as a function of octave band frequencies, rather than to assign an absorption coefficient to their surfaces. One also finds that when a group of functional absorbers are installed, the total absorption realized from the group depends upon the spacing between the individual units to a certain extent. Once a certain optimum spacing has been reached the effective absorption per unit does not increase [85, 86].

9.5.9 Acoustical Spray-on Materials

Acoustical spray-on materials consist of a range of materials formed from mineral or synthetic organic fibers mixed with a binding agent. This holds the fibrous content together in a porous manner and also acts as an adhesive. During the spraying application, the fibrous material is mixed with a binding agent and water to produce a soft lightweight material of coarse surface texture with high sound-absorbing characteristics. Due to the nature of the binding agent used, the material may be easily applied directly to a wide number of surface types including wood, concrete, metal lath, steel, and galvanized metal. When sprayed onto a solid backing, this type of spray-on material usually exhibits good mid- and high-frequency sound absorption and when

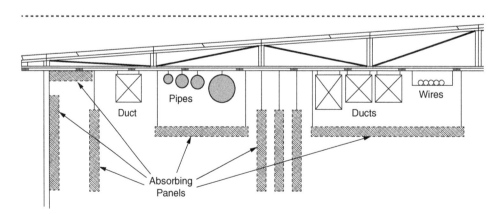

Figure 9.23 Sound-absorbing material placed on the walls and under the roof and suspended as panels in a factory building.

applied to, for example, metal lath with an airspace behind it, the material then exhibits good low-frequency absorption as well. As would be expected from previous discussions of porous absorbers, the absorption values increase with greater depth of application, especially at low frequencies. Spray-on depths of up to 5 cm are fairly common [59].

The sound absorption characteristics of such materials are very much dependent upon the amount and type of binding agent used and the way it is mixed during the spray-on process. If too much binder is used, then the material becomes too hard and therefore a poor absorber, while on the other hand, if too little binder is employed, the material will be prone to disintegration; and, since it will have a very low flow resistance, it will also not be a good absorber [4]. Some products use two binders. One is impregnated into the fibrous material during manufacture, while the other is in liquid form and included during application. The fibrous material (containing its own adhesive) and the liquid adhesive are applied simultaneously to the surface using a special nozzle. The material is particularly resistant to disintegration or shrinkage and, furthermore, is fire-resistant and possesses excellent thermal insulation properties.

If visually acceptable, spray-on materials of this type can be successfully used as good broadband absorbing materials in a variety of architectural spaces including schools, gymnasiums, auditoriums, shopping centers, pools, sports stadiums, and in a variety of industrial applications such as machine shops and power plants. Disadvantages generally include the difficulty to clean and redecorate the material, although some manufacturers claim that their product can be spray-painted without loss of acoustical performance. The data for such claims should be checked carefully before one proceeds to the painting of a surface.

9.5.10 Acoustical Plaster

The term acoustical plaster has been applied to a number of combinations of vermiculite and binder agents such as gypsum or lime. They are usually applied either to a plaster base or to concrete and must have a solid backing. Because of this, acoustical plasters have poor low-frequency sound absorption characteristics (also due to the thickness of application, which rarely exceeds 13 mm). This can be slightly improved by application to metal lath. The material may be applied by a hand trowel or by machine. However, the latter tends to compact the material and give lower sound absorption characteristics. The surface of acoustical plasters can be sprayed with water-thinned emulsion paint without any significant loss of absorption, although brushed oil-based paints should never be used. Figure 9.24 shows typical sound absorption characteristics for a 13-mm-thick hand trowel applied acoustical plaster taken from a variety of published manufacturer's data [4].

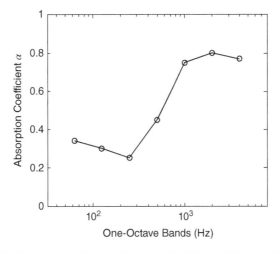

Figure 9.24 Sound absorption coefficient α of a 13-mm thick acoustical plaster.

9.5.11 Measurement of Sound Absorption Coefficients

Although some approximate values for sound absorption coefficients and resonance frequency characteristics can be estimated from the geometry, flow resistance, porosity, and other physical factors, it is clearly necessary to have actual measured values of absorption coefficients for a variety of materials and different constructions. There are two standardized methods to measure the sound absorption coefficient: using either a reverberation room or using an impedance tube [60, 87–90].

Tables of absorption coefficients are published for a variety of acoustical materials in which numerous mounting systems are employed. Sound absorption coefficients and corresponding values of NRC of some common sound-absorbing materials and construction materials are shown in Table 9.3. More extensive tables of sound absorption coefficients of materials can be found in the literature. The book by Trevor Cox provides a good review on sound-absorbing materials [91]. The tables are very often made up after receiving absorption results from a number of independent testing laboratories for the same sample of material. It is found that even when all of the requirements of either the International Organization for Standardization (ISO) or ASTM standards are met, there can still be quite large differences observed between the values obtained for the absorption coefficients measured in different laboratories for the same sample. These differences can be due to variations in the diffuseness of the testing rooms and to human error and bias in measuring the reverberation times.

The main applications of sound-absorbing materials in noise control are [11, 92]: (i) incorporation in noise control enclosures, covers, and wrappings to reduce reverberant build-up of sound and hence increase insertion loss, (ii) incorporation in flow ducts to attenuate sound from fans and flow control devices, (iii) application to the surfaces of rooms to control reflected sound (e.g. to reduce steady-state sound pressure levels in reverberant fields), (iv) in vehicles (walls, engine compartments, and engine exhausts), (v) in lightweight walls and ceilings of buildings, and (vi) on traffic noise barriers to suppress reflections between the side of the vehicle and the barrier or to increase barrier performance by the presence of absorbent on and around the top of the barrier.

However, sound-absorbing materials are most commonly used to optimize the reverberation time in rooms and to reduce the sound pressure level in reverberant fields. These applications will now be described separately.

9.5.12 Optimization of the Reverberation Time

When sound is introduced into a room, the reverberant field level will increase until the rate of sound energy input is just equal to the rate of sound energy absorption. If the sound source is abruptly turned off, the reverberant field will decay at a rate determined by the rate of sound energy absorption.

As mentioned before, the reverberation time of a room depends upon the sound absorption present within it. In addition, the reverberation time has become recognized as the most important single parameter used to describe the acoustical performance of auditoriums. Changes in the total absorption can be made by modifying the values of areas and sound absorption coefficients of the room surfaces. Therefore, it is possible to change the values of reverberation time to provide performers and listeners with a high degree of intelligibility throughout the room and optimum sound enrichment.

In a room, for speech to be understood fully, each component of the speech must be heard by the listener. If the room has a long reverberation time, then the speech components overlap and a loss of intelligibility results. Similarly, for music enjoyment and appreciation, a certain amount of reverberation is required to obtain the quality and blend of the music. It is, however, much more difficult to describe the acoustical qualities of a room used for music since many are subjective and therefore cannot be described in measurable physical quantities. Optimum reverberation times are not only required for good subjective reception but also for efficient performances.

Table 9.3 Sound absorption coefficient $\alpha(f)$ and corresponding NRC of common materials [60].

Material	Frequency (Hz)						NRC
	125	250	500	1000	2000	4000	
Fibrous glass (typically 64 kg/m³) hard backing							
25 mm thick	0.07	0.23	0.48	0.83	0.88	0.80	0.61
50 mm thick	0.20	0.55	0.89	0.97	0.83	0.79	0.81
10 cm thick	0.39	0.91	0.99	0.97	0.94	0.89	0.95
Polyurethane foam (open cell)							
6 mm thick	0.05	0.07	0.10	0.20	0.45	0.81	0.21
12 mm thick	0.05	0.12	0.25	0.57	0.89	0.98	0.46
25 mm thick	0.14	0.30	0.63	0.91	0.98	0.91	0.71
50 mm thick	0.35	0.51	0.82	0.98	0.97	0.95	0.82
Hair felt							
12 mm thick	0.05	0.07	0.29	0.63	0.83	0.87	0.46
25 mm thick	0.06	0.31	0.80	0.88	0.87	0.87	0.72
Brick							
Unglazed	0.03	0.03	0.03	0.04	0.04	0.05	0.04
Painted	0.01	0.01	0.02	0.02	0.02	0.02	0.02
Concrete block, painted	0.01	0.05	0.06	0.07	0.09	0.03	0.07
Concrete	0.01	0.01	0.02	0.02	0.02	0.02	0.02
Wood	0.15	0.11	0.10	0.07	0.06	0.07	0.09
Glass	0.35	0.25	0.18	0.12	0.08	0.04	0.16
Gypsum board	0.29	0.10	0.05	0.04	0.07	0.09	0.07
Plywood, 10 mm	0.28	0.22	0.17	0.09	0.10	0.11	0.15
Soundblox concrete block							
Type A (slotted), 15 cm (6 in.)	0.62	0.84	0.36	0.43	0.27	0.50	0.48
Type B, 15 cm (6 in.)	0.31	0.97	0.56	0.47	0.51	0.53	0.63
Spray-acoustical (on gypsum wall board)	0.15	0.47	0.88	0.92	0.87	0.88	0.79
Acoustical plaster (25 mm thick)	0.25	0.45	0.78	0.92	0.89	0.87	0.76
Carpet							
On foam rubber	0.08	0.24	0.57	0.69	0.71	0.73	0.55
On concrete	0.02	0.06	0.14	0.37	0.60	0.66	0.29

Optimum values of the reverberation time for various uses of a room may be calculated approximately by [93]

$$T_R = K\left(0.0118V^{1/3} + 0.1070\right), \tag{9.28}$$

where T_R is the reverberation time (s), V is the room volume (m³), and K is a constant that takes the following values according to the proposed use: For speech $K = 4$, for orchestra $K = 5$, and for choirs and rock bands $K = 6$. It has been suggested that, at frequencies in the 250-Hz octave band and lower frequencies, an increase is needed over the value calculated by Eq. (9.28), ranging from 40% at 250 Hz to 100% at 63 Hz. Other authors have suggested optimum T_R values for rooms for various purposes [14, 62].

However, achieving the optimum reverberation time for a room may not necessarily lead to good speech intelligibility or music appreciation. It is essential to adhere strictly to the other design rules for shape, volume, and time of arrival of early reflections [94, 95].

Example 9.8   Consider an auditorium of dimensions 15 m × 20 m × 4 m, with an average Sabine absorption coefficient $\bar{\alpha} = 0.24$ for the whole room surface at 500 Hz. Is this room appropriate to be used as an auditorium?

Solution

The total surface area of the room is calculated as $S = 2(20 \times 4 + 15 \times 4 + 15 \times 20) = 880 \text{ m}^2$ and its volume is $V = 1200 \text{ m}^3$. Assuming that there is a uniform distribution of absorption throughout the room, the reverberation time can be obtained as given by Sabine $T_R = 0.161 \, V/(S\bar{\alpha}) = (0.161 \times 1200)/(880 \times 0.24) = 0.9 \text{ s}$. Now substituting the value of room volume in Eq. (9.28), and considering $K = 4$ for using the room for speech, we obtain $T_R = 0.9 \text{ s}$. Therefore, the total absorption of such a room is optimum for its use as an auditorium, at least at 500 Hz. This calculation may be repeated at each frequency for which absorption coefficient data are available.

9.5.13 Reduction of the Sound Pressure Level in Reverberant Fields

When a machine is operated inside a building, the machine operator usually receives most sound through a direct path, while people situated at greater distances receive sound mostly through reflections (see Figure 9.11).

In the case of machinery used in reverberant spaces such as factory buildings, the reduction in sound pressure level L_p in the reverberant field caused by the addition of sound-absorbing material, placed on the walls or under the roof (see Figure 9.23), can be estimated for a source of sound power level L_W from the so-called room equation:

$$L_p = L_W + 10 \log \left(\frac{D}{4\pi r^2} + \frac{4}{R} \right), \tag{9.29}$$

where the *room constant* $R = S\bar{\alpha}/(1-\bar{\alpha})$, $\bar{\alpha}$ is the surface average absorption coefficient of the walls, D is the source directivity, and r is the distance in metres from the source. The surface average absorption coefficient $\bar{\alpha}$ may be estimated from

$$\bar{\alpha} = \frac{S_1\alpha_1 + S_2\alpha_2 + S_3\alpha_3 + \cdots}{S_1 + S_2 + S_3 + \cdots}, \tag{9.30}$$

where $S_1, S_2, S_3, \ldots$ are the surface areas of material with absorption coefficients $\alpha_1, \alpha_2, \alpha_3, \ldots$, respectively. For the suspended absorbing panels shown in Figure 9.23, both sides of the panel are normally included in the surface area calculation. If the sound absorption is increased, then from Eq. (9.29) the change in sound pressure level ΔL in the reverberant space (beyond the critical distance r_c) is

$$\Delta L = L_{p1} - L_{p2} = 10 \log \frac{R_2}{R_1}. \tag{9.31}$$

If $\bar{\alpha} \ll 1$, then the reduction in sound pressure level (sometimes called *the noise reduction*) is given by

$$\Delta L \approx 10 \log \frac{S_2\bar{\alpha}_2}{S_1\bar{\alpha}_1}, \tag{9.32}$$

where S_2 is the total surface area of the room walls, floor, and ceiling and any suspended sound-absorbing material, $\bar{\alpha}_2$ is the average sound absorption coefficient of these surfaces after the addition of sound-absorbing material, and S_1 and $\bar{\alpha}_1$ are the area and the average sound absorption coefficient before the addition of the material.

Example 9.9 A machine source operates in a building of dimensions 30 m × 30 m with a height of 10 m. Suppose the average absorption coefficient is $\bar{\alpha} = 0.02$ at 1000 Hz. What would be the noise reduction in the reverberant field if 100 sound-absorbing panels with dimensions 1 m × 2 m each with an absorption coefficient of $\bar{\alpha} = 0.8$ at 1000 Hz were suspended from the ceiling (assume both sides absorb sound)?

Solution

The room surface area = $2(900) + 4(300) = 3000$ m^2, therefore $R_1 = (3000 \times 0.02)/0.98 = 60/0.98 = 61.2$ sabins (m^2). The new average absorption coefficient $\bar{\alpha}_2 = (3000 \times 0.02 + 200 \times 2 \times 0.8)/3400 = 0.11$. The new room constant is $(3400 \times 0.11)/0.89 = 420$ sabins (m^2). Thus from Eq. (9.31) the predicted noise reduction $\Delta L = 10 \log(420/61.2) = 8.4$ dB. This calculation may be repeated at each frequency for which absorption coefficient data are available. It is normal to assume that about 10 dB is the practical limit for the noise reduction that can be achieved by adding sound-absorbing material in industrial situations.

9.6 Acoustical Enclosures

Acoustical enclosures are used wherever containment or encapsulation of the source or receiver is a good, cost-effective, feasible solution. They can be classified in four main types: (i) large loose-fitting or room-size enclosures in which complete machines or personnel are contained, (ii) small enclosures used to enclose small machines or parts of large machines, (iii) close-fitting enclosures that follow the contours of a machine or a part, and (iv) wrapping or lagging materials often used to wrap pipes, ducts, or other systems.

The performance of such enclosures can be defined in three main ways [5]: (i) *noise reduction* (*NR*), the difference in sound pressure levels between the inside and outside of the enclosure, (ii) *transmission loss* (*TL*, or equivalently the *sound reduction index*), the difference between the incident and transmitted sound intensity levels for the enclosure wall, and (iii) *insertion loss* (*IL*), the difference in sound pressure levels at the receiver point *without* and *with* the enclosure wall in place. Enclosures can either be complete or partial (in which some walls are removed for convenience or accessibility). Penetrations are also often necessary to provide access or cooling.

9.6.1 Reverberant Sound Field Model for Enclosures

In the energy model for an enclosure it is assumed that the reverberant sound field produced within the enclosure is added to the direct sound field produced by the sound source being enclosed. The sum of the two sound fields gives the total sound field within the enclosure, which is responsible for the sound radiated by the enclosure walls.

If the smallest distance ℓ between the machine surface and the enclosure walls is greater than a wavelength λ ($\ell > \lambda$), for the lowest frequency of the noise spectrum of the machine (noise source), then the enclosure can be considered large enough to assume that the sound field within the enclosure is diffuse (the sound energy is uniformly distributed within the enclosure). Another criterion [96] used to assume the diffuse sound field condition requires that the largest dimension of the interior volume of the enclosure be less than $\lambda/10$.

Therefore, according to classical theory, the reverberant sound pressure level within the enclosure, L_{prev}, is given by [3, 97]

$$L_{prev} = L_W + 10\log T - 10\log V + 14, \tag{9.33}$$

where T is the reverberation time within the enclosure in seconds, and V is the internal volume of the enclosure in cubic metres. Then, the reverberant sound intensity incident on the internal enclosure walls can be estimated from (See Chapter 3)

$$I_i = \frac{\langle p^2 \rangle_{rev}}{4\rho c}, \tag{9.34}$$

where $\langle p^2 \rangle_{rev}$ is the reverberant mean-square sound pressure (space–time average) within the enclosure, ρ is the density of the medium (air) within the enclosure, and c is the speed of sound.

9.6.2 Machine Enclosure in Free Field

The sound field immediately inside of an enclosure consists of two components: (i) the internal reverberant sound field and (ii) the direct sound field of the machine noise source [3]. The fraction of sound energy that is incident on the interior of the enclosure wall that is transmitted outside depends on its transmission coefficient. The transmission coefficient τ of a wall may be defined as

$$\tau = \frac{\text{sound intensity transmitted by wall}}{\text{sound intensity incident on wall}}, \tag{9.35}$$

and this coefficient τ is related to the transmission loss TL (or sound reduction index) by

$$TL = 10\log\frac{1}{\tau}. \tag{9.36}$$

If the enclosure is located in a free-field, as shown in Figure 9.25, the sound pressure level immediately outside the enclosure is [3, 97]

$$L_{pext} = L_{prev} - TL - 6. \tag{9.37}$$

Therefore, the enclosure with its noise source inside can be considered to be an equivalent sound source placed in free-field conditions. Now, if the floor is hard and highly reflective, the sound pressure level at a distance r from an enclosure wall can be estimated by

$$L_p(r) = L_{pext} + 10\log S - 10\log\left(2\pi r^2\right), \tag{9.38}$$

where S is the total outer surface area of the enclosure walls.

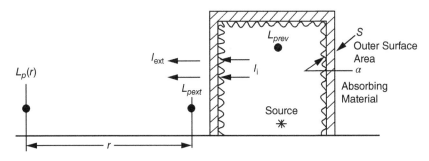

Figure 9.25 Acoustical enclosure placed in free field.

9.6.3 Simple Enclosure Design Assuming Diffuse Reverberant Sound Fields

If the sound fields are assumed to be reverberant both inside and outside a completely sealed enclosure (typical of a machine enclosure in a machine shop), then the noise reduction *NR* is given by

$$NR = L_{p1} - L_{p2} = TL + 10 \log \frac{A_2}{S_e}, \tag{9.39}$$

where L_{p1} and L_{p2} are the reverberant sound pressure levels inside and outside the enclosure, $A_2 = S_2 \bar{\alpha}_2$ is the *absorption area* in square metres of the outside receiving space where $\bar{\alpha}_2$ is the surface average absorption coefficient of the absorption material in the receiving space averaged over the area S_2, and S_e is the enclosure surface area in square metres.

Example 9.10 A noisy diesel generator is operated in the middle of the floor of a shop measuring 20×20 m by 4 m high. The shop has an average surface absorption coefficient $\bar{\alpha} = 0.1$ at 1000 Hz. The machine is placed inside a completely sealed enclosure of 10 m^2 surface area. The enclosure is constructed of steel panels having $TL = 24.5$ dB at 1000 Hz. Determine the noise reduction provided by the enclosure.

Solution

The room surface area $= 2(20 \times 20 + 20 \times 4 + 20 \times 4) = 1120$ m^2, therefore the absorption area of the outside receiving space is $A_2 = 1120 \times 0.1 = 112$ sabins (m^2). The noise reduction is obtained from Eq. (9.39): $NR = 24.5 + 10\log(112/10) = 35$ dB. This calculation may be repeated at each frequency for which absorption coefficient and transmission loss data are available.

---

a) Personnel Booth or Enclosure in a Reverberant Sound Field

Equation (9.39) can be used to design a personnel booth or enclosure in which the noise source is external and the reason for the enclosure is to reduce the sound pressure level inside (see Figure 9.26). If the enclosure is located in a factory building in which the reverberant level is L_{p1}, then the enclosure wall TL and interior absorption area A_2 can be chosen to achieve a required value for the internal sound pressure level L_{p2}. In the case that the surface area S_2 of the internal absorbing material $S_2 = S_e$, the enclosure surface area, then Eq. (9.39) simplifies to

$$NR = TL + 10 \log \bar{\alpha}_2. \tag{9.40}$$

The *NR* achieved is seen to be less than the *TL* in general. When $\bar{\alpha}_2$, the average absorption coefficient of the absorbing material in the receiving space approaches 1, then *NR* approaches *TL* (as expected), although when $\bar{\alpha}_2$ approaches 0, then the theory fails.

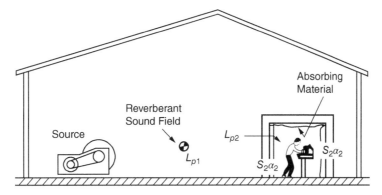

Figure 9.26 Personnel enclosure placed in a reverberant sound field.

Example 9.11 It is desired to build a personnel enclosure in the assembly area of a manufacturing shop. The reverberant level in the manufacturing shop is 85 dB in the 1600-Hz one-third octave band. It is required to provide values of *TL* and $\bar{\alpha}$ to achieve an interior level inside the personnel enclosure of less than 60 dB.

Solution

Assuming that $S_2 = S_e$, then if *TL* is chosen as 30 dB and $\bar{\alpha} = 0.2$, $NR = 30 + 10 \log 0.2 = 30 - 7 = 23$ dB, and $L_{p2} = 62$ dB. Now, if $\bar{\alpha}$ is increased to 0.4, then $NR = 30 + 10 \log 0.4 = 30 - 4 = 26$ and $L_{p2} = 59$ dB, meeting the requirement. Since *TL* varies with frequency (see Eq. (9.43)), this calculation would have to be repeated for each one-third octave band center frequency of interest. In addition, at low frequency, some improvement can be achieved by using a thicker layer of sound-absorbing lining material.

b) Machine Enclosure in a Reverberant Space

When an enclosure is designed to contain a noise source (see Figure 9.27), it operates by reflecting the sound back toward the source, causing an increase in the sound pressure level inside the enclosure. From energy considerations the insertion loss is zero if there is no sound absorption inside. The effect is an increase in the sound pressure at the inner walls of the enclosure compared with the sound pressure resulting from the direct field of the source alone. The build-up of sound energy inside the enclosure can be reduced by placing sound-absorbing material on the walls inside the enclosure. It is also useful to place sound-absorbing materials inside personnel noise protection booths for similar reasons.

The internal surfaces of an enclosure are usually lined with glass or mineral fiber or an open-cell polyurethane foam blanket. However, the selection of the proper sound-absorbing material and its containment will depend on the characteristics of each noise source. Sound absorption material also requires special protection from contamination by oil or water, which weakens its sound absorption properties. If the noise source enclosed is a machine that uses a combustible liquid, or gas, then the material should also be fire-resistant. Of course, since the sound absorption coefficient of linings is generally highest at high frequencies, the high-frequency components of any noise source will suffer the highest attenuation.

The effect of inadequate sound absorption in enclosures is very noticeable. Table 9.4 shows the reduction in performance of an ideal enclosure with varying degrees of internal sound absorption [3]. The first column of Table 9.4 shows the percentage of internal surface area that is treated. The sound power of the source is assumed to be constant and unaffected by the enclosure.

For an enclosure that is installed around a source with a considerable amount of absorbing material used inside to prevent any interior reverberant sound energy build-up, then from energy considerations

$$IL \cong TL,$$

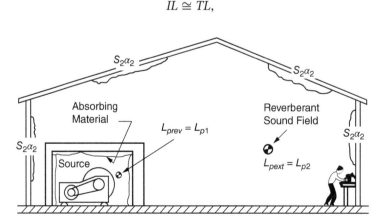

Figure 9.27 Machine enclosure placed in a reverberant environment.

Table 9.4 Change in noise reduction performance of enclosure as function of percentage of internal surface covered with sound absorptive material.

Percent Sound-Absorbent (%)	Change in NR (dB)
10	−10
20	−7
30	−5
50	−3
70	−1.5

if the receiving space is quite absorbent and if IL and TL are averaged over wide frequency bands (e.g. at least one octave). If insufficient sound-absorbing material is put inside the enclosure, the sound pressure level will continue to increase inside the enclosure because of multiple reflections, and the enclosure effectiveness will be reduced. From energy considerations it is obvious that if there is no sound absorption inside ($\bar{\alpha} = 0$), the enclosure will be ineffective, and its insertion loss will be zero.

An estimate of the insertion loss for the general case $0 < \bar{\alpha} < 1$ can be obtained by assuming that the sound field inside the enclosure is diffuse and reverberant and that the interior surface of the enclosure is lined with absorbing material of surface-averaged absorption coefficient $\bar{\alpha}$. Let us assume that (i) the average absorption coefficient in the room containing the noise source is not greater than 0.3, (ii) the noise source does not provide direct mechanical excitation to the enclosure walls, and (iii) the noise source volume is less than about 0.3–0.4 of the enclosure volume, then we may show that the insertion loss of a loose-fitting enclosure made to contain a noise source situated in a reverberant space is [98]

$$IL = L_{P1} - L_{P2} = TL + 10 \log \frac{A_e}{S_e}, \tag{9.41}$$

where L_{p1} is the reverberant level in the room containing the noise source (with no enclosure), L_{p2} is the reverberant level at the same location (with the enclosure), A_e is the absorption area inside the enclosure $S_i \bar{\alpha}_i$, and $\bar{\alpha}_i$ is the surface average absorption coefficient of this material. If the surface area of the enclosure $S_e = S_i$, the area of the interior absorbing material, then Eq. (9.41) simplifies to

$$IL = TL + 10 \log \bar{\alpha}_i. \tag{9.42}$$

We see that normally the insertion loss of an enclosure containing a noise source is less than the TL. If $\bar{\alpha}_i$, the average absorption coefficient of the internal absorbing material approaches 1, the IL approaches TL (as expected), although when $\bar{\alpha}_i$ approaches 0, then this theory fails.

Example 9.12 Consider that the sound pressure level L_{p1} in a reverberant woodworking shop caused by a wood-planing machine in the 1600-Hz one-third octave band is 90 dB. Then what values of TL and $\bar{\alpha}$ should be chosen for a loose-fitting enclosure made to contain the machine to guarantee that the reverberant level will be less than 60 dB?

Solution

Assuming that $S_i = S_e$, then if TL is chosen to be 40 dB and $\bar{\alpha} = 0.1$, $IL = 40 + 10 \log 0.1 = 40 - 10 = 30$ dB and $L_{p2} = 60$ dB. Now, if $\bar{\alpha}$ is increased to 0.2, then $IL = 40 + 10 \log 0.2 = 40 - 7 = 33$ dB and $L_{p2} = 57$ dB, thus meeting the requirement. As in Example 9.10, we can observe that since TL varies with frequency f (see Eq. (9.43)), this calculation should be repeated at each frequency of interest. At low frequencies, since large values of TL and $\bar{\alpha}$ are difficult to achieve, it may not be easy to obtain large values of IL.

Thus the *TL* can be used as an approximate guide to the *IL* of a sealed enclosure only when allowance is made for the sound absorption inside the enclosure. The transmission loss of an enclosure is usually mostly governed by the mass/unit area ρ_s of the enclosure walls (except in the coincidence-frequency region). The reason for this is that when the stiffness and damping of the enclosure walls are unimportant, the response is dominated by inertia of the walls $\rho_s(2\pi f)$ where *f* is the frequency in hertz. The transmission loss of an enclosure wall for sound arriving from all angles is approximately

$$TL = 20 \log (\rho_s f) - C, \tag{9.43}$$

where ρ_s is the surface density (mass/unit area) of the enclosure walls and $C = 47$ if the units of ρ_s are kg/m^2 and $C = 34$ if the units are lb./ft^2. Equation (9.43) is known as the *field-incidence mass law* (see Chapter 12 of this book). The transmission loss of a wall (given by Eq. (9.43)) theoretically increases by 6 dB for each doubling of frequency or for each doubling of the mass/unit area of the wall.

Where the enclosure surface is made of several different materials (e.g. concrete or metal walls and glass), then the average transmission loss TL_{ave} of the composite wall is given by

$$TL_{ave} = 10 \log \frac{1}{\bar{\tau}}, \tag{9.44}$$

where the average transmission coefficient $\bar{\tau}$ is

$$\bar{\tau} = \frac{S_1\tau_1 + S_2\tau_2 + \cdots + S_n\tau_n}{S_1 + S_2 + \cdots + S_n}, \tag{9.45}$$

where τ_i is the transmission coefficient of the *i*th wall element and S_i is the surface area of the *i*th wall element (m^2).

In general, it is difficult to predict the insertion loss of an enclosure with a high degree of accuracy. This is because both the sound field inside and outside the enclosure cannot always be modeled using simple approaches. The models discussed so far are valid for large enclosures where resonances do not arise. However, in the design of enclosures at least two types of enclosure resonances have to be taken into account: (i) structural resonances in the panels that make up the enclosure and (ii) standing-wave resonances in the air gap between the machine and the enclosure. At each of these resonance frequencies, the insertion loss due to the enclosure is significantly reduced and in some instances can become negative, meaning that the machine with the enclosure may radiate more noise than without the enclosure. Therefore, the enclosure should be designed so that the resonance frequencies of its constituent panels are not in the frequency range where high insertion loss is required.

If the sound source being enclosed radiates predominantly low-frequency noise, then the enclosure panels should have high natural frequencies, that is, the enclosure should be stiff and not massive. These requirements are very different from those needed for good performance of a single-leaf partition at frequencies below the critical frequency. To achieve a high insertion loss in the stiffness-controlled region, benefit can be gained by employing small panel areas, large panel aspect ratios, clamped edge conditions, and materials having a high bending stiffness [99].

On the other hand, a high-frequency sound source requires the use of an enclosure with panels having low natural frequencies, implying the need for a massive enclosure. Additionally, the panels of the enclosure should be well damped. This would increase *IL*, in particular for frequencies near and above the critical frequency and at the first panel resonance. Mechanical connections between the sound source and the walls of the enclosure, air gap leaks, structure-borne sound due to flanking transmission, vibrations, and the individual radiation efficiency of the walls of the enclosure will reduce the *IL* in different frequency ranges.

For a sealed enclosure without mechanical connections between the source and the enclosure walls, the problem can be divided in three frequency ranges: (i) the low-frequency range, where the insertion loss is frequency independent and corresponds to frequencies below air gap or panel resonances, (ii) the intermediate-frequency range, where the insertion loss is controlled by panel and/or air gap resonances that do not overlap so that statistical methods are not applicable, and (iii) the high-frequency range, where high modal densities exist, consequently, statistical energy methods can be used for modeling. The insertion loss in the intermediate frequency region fluctuates widely with frequency and position and thus is very difficult to analyze by theoretical methods [96, 100]. Reference [101] discusses other models for estimating insertion loss of enclosures.

9.6.4 Close-Fitting Enclosures

In many cases when machines are enclosed it is necessary to locate the enclosure walls close to the machine surfaces, so that the resulting air gap is small. Such enclosures are termed close-fitting enclosures. In such cases the sound field inside the enclosure is neither reverberant nor diffuse, and the theory discussed at the beginning of this section can be used to calculate a first approximation of the insertion loss of an enclosure.

There are several effects that occur with close-fitting enclosures. First, if the noise source has a low internal impedance, then in principle the close-fitting enclosure can "load" the source so that it produces less sound power. However, in most machinery noise problems, the internal impedance of the source is high enough to make this effect negligible. Second, and more importantly, reductions in the IL occur at certain frequencies (when the enclosure becomes "transparent"). These frequencies f_0 and f_{sw} are shown in Figure 9.28. When an enclosure is close-fitting, then to a first approximation the sound waves approach the enclosure walls at normal incidence instead of random incidence. When the air gap is small, then a resonant condition at frequency f_0 occurs where the enclosure wall mass is opposed by the wall and air gap stiffness. This resonance frequency can be increased by increasing the stiffness, as seen in Figure 9.28. In addition, standing-wave resonances can occur in the air gap at frequencies f_{sw}. These resonances can be suppressed by the placement of sound-absorbing material in the air gap [102, 103].

Jackson has produced simple theoretical models for close-fitting enclosures that assume a uniform air gap [102, 103]. He modeled the source enclosure problem in terms of two parallel infinite panels separated by an

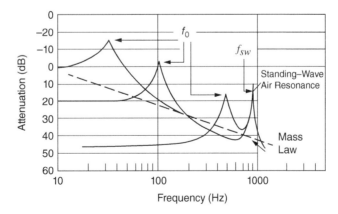

Figure 9.28 Close-fitting enclosure attenuation in sound pressure level for different values of panel stiffness. The resonance frequency f_0 is increased by increasing panel stiffness [101].

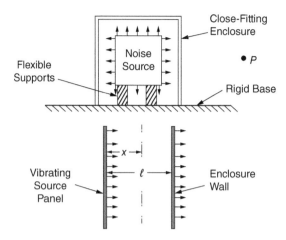

Figure 9.29 Simplified one-dimensional model for a close-fitting enclosure [101].

air gap, as shown in Figure 9.29. One panel is assumed to be vibrating and to be the noise source, and the second panel is assumed to be an enclosure panel. Then, the enclosure performance is specified in terms of the relative vibration levels of the two panels. Later, Junger considered both the source panel and enclosure panel to be of finite area [104]. He assumed that the source panel vibrates as a uniform piston and the enclosure panel vibrates as a simply supported plate excited by a uniform sound pressure field. Comparisons of the Jackson, Junger, and Ver [96] models have been presented by Tweed and Tree [105].

Fahy has presented details of an enclosure prediction model based upon a one-dimensional model similar to that of Jackson [106]. It is assumed that the enclosure panel is a uniform, nonflexible partition of mass per unit area ρ_s, and it is mounted upon viscously damped, elastic suspensions, having stiffness and damping coefficients per unit area s and r, respectively. The insertion loss of the enclosure in this one-dimensional case is [106]

$$IL = 10\log\left\{\left[\cos k\ell - \frac{(\omega\rho_s - s/\omega)}{\rho c}\sin k\ell\right]^2 + \left[1 + \frac{r}{\rho c}\right]^2 \sin^2 k\ell\right\}, \tag{9.46}$$

where $k = \omega/c$ is the wavenumber, and ℓ is the separation distance between the source and enclosure panel. From an examination of Eq. (9.46) it is clear that the insertion loss will be zero at frequencies when the cavity width ℓ is equal to an integer number of half-wavelengths and the panel enclosure velocity equals the source surface velocity. The insertion loss will also have a minimum value at the frequency ω_0

$$\omega_0^2 \approx \frac{\rho c^2}{\rho_s \ell} + \frac{s}{\rho_s}, \tag{9.47}$$

where $\omega_0 = 2\pi f_0$. Figure 9.30 shows a generalized theoretical insertion loss performance of a close-fitting enclosure, according to Eq. (9.46).

Other theoretical models to predict the acoustical performance of close-fitting enclosures have been reported in the literature [107, 108]. However, in practice, the real source panel exhibits forced vibrations in a number of modes and the air gap varies with real enclosures. Thus these simple theoretical models and some later ones can only be used to give some guidance of the insertion loss to be expected in practice. Finite element and boundary element approaches can be used to make insertion loss predictions for close-fitting enclosures with complicated geometries and for the intermediate frequency region [109].

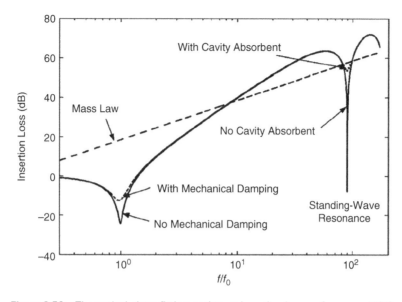

Figure 9.30   Theoretical close-fitting enclosure insertion loss performance [101].

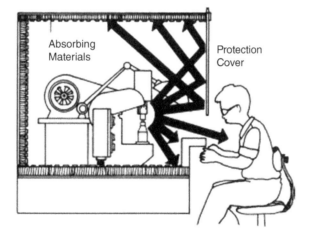

Figure 9.31   Partial enclosure.

9.6.5   Partial Enclosures

When easy and continuous access to parts of a machine is necessary or the working process of the machinery or the safety or maintenance requirements do not allow a full enclosure, a partial enclosure is usually used to reduce the radiated noise [110]. Figure 9.31 shows an example of a partial enclosure used in machinery noise control. The noise reduction produced by a partial enclosure will depend upon the particular geometry. Most of the time, the available attenuation will be limited by diffractive scattering and mechanical connections between the partial enclosure and the vibrating machine. It is recommended that partial enclosures be fully lined with sound absorption material. As a general rule, the enclosure walls of a partial enclosure should have a transmission loss of at least 20 dB. The maximum sound power reduction that can be achieved for such an enclosure is about 10 dB. However, in some cases, the noise levels radiated may be reduced more. Table 9.5

Table 9.5 Effectiveness of a partial enclosure.

Sound Energy Enclosed and Absorbed (%)	Maximum Achievable Noise Reduction (dB)
50	3
75	6
90	10
95	13
98	17
99	20

shows the effectiveness of partial enclosures, where the values are based on the assumption that the noise source radiates uniformly in all directions and that the partial enclosure that surrounds the source is fully lined with sound-absorptive material [111].

9.6.6 Other Considerations

Most equations presented in this chapter will give good estimates for the actual performance of an enclosure. However, some guidelines should be followed in practice to avoid degradation of the effectiveness of an enclosure. In addition, when designing an enclosure, care should be taken so that production costs and time, operational cost-effectiveness, and the efficiency of operation of the machine or equipment being enclosed are not adversely affected.

Most enclosures will require some form of ventilation through duct openings. Such necessary permanent openings must be treated with some form of silencing to avoid substantially degrading the performance of the enclosure. For a good design, it is required that the acoustical performance of silencing of the permanent duct openings will match the performance of the enclosure walls. The usual techniques employed to control the sound propagation in ducts can be used for the design of silencers [112].

When ventilation for heat removal is required but the heat load is not large, then natural ventilation with silenced air inlets low down close to the floor and silenced outlets at a greater height, well above the floor, will be adequate. If forced ventilation is needed to avoid excessive heat build-up in the enclosure, then the approximate amount of airflow needed can be determined by [3]

$$\rho C_p V = \frac{H}{\Delta T}, \tag{9.48}$$

where V is the volume flow rate of the cooling air required (m³/s), H is the heat input to the enclosure (W), ΔT is the temperature differential between the external ambient and the maximum permissible internal temperature (°C), ρ is the air density (kg/m³), and C_p is the specific heat of the air (m²C⁻¹/s²). When high-volume flow rates of air are required, the noise output of the fan that provides the forced ventilation should be considered very carefully, since this noise source can degrade the performance of the enclosure. In general, large slowly rotating fans are always preferred to small high-speed fans since fan noise increases with the fifth power of the blade tip speed.

The effectiveness of an enclosure can be very much reduced by the presence of leaks (air gaps). These usually occur around removable panels or where ducts or pipes enter an enclosure to provide electrical and cooling air services and the like. If holes or leaks occur in the enclosure walls (e.g. cracks around doors or around the base of a cover) and if the *TL* of the holes is assumed to be 0 dB (as is customary), then the reduction in insertion loss as a function of the *TL* of the enclosure walls, with the leak ratio factor β as the parameter, is given by Figure 9.32 [96]. The leak ratio factor β is defined as the ratio of the total face area of the leaks and gaps to the surface area (one side) of the enclosure walls.

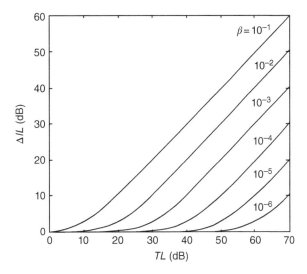

Figure 9.32 Decrease of enclosure insertion loss, ΔIL, as a function of the wall sound transmission loss TL with the leak ratio factor β as parameter.

If the penetrations in the enclosure walls are lined with absorbing materials as shown in Figure 9.33, then the degradation in the enclosure IL is much less significant. Some studies indicate that leaks not only degrade the noise reduction but also can introduce resonances when the leak ratio factor is not very large [113].

It is necessary to provide sufficient vibration isolation to reduce the radiation of noise from the surface on which the machinery is mounted, particularly if low-frequency noise is the main problem. Therefore, it is advisable to mount the machine and/or the enclosure itself on vibration isolators that reduce the transmission of energy to the floor slab. In doing so, control of both the airborne and the structure-borne sound transmission paths between the source and receiver will be provided. Great care is necessary to ensure that the machine will be stable and its operation will not be affected adversely. Insertion of flexible (resilient) connectors between the machine and conduit, cables, piping, or ductwork connected to it must be provided to act as vibration breaks.

In addition, proper breaking of any paths that permit noise to "leak" through openings in the enclosure must be provided. Then all joints, seams, and penetrations of enclosures should be sealed using a procedure such as packing the leaks with mineral wool, which are closed by cover plates and mastic sealant [3].

Any access doors to the enclosure must be fitted tightly and gasketed. Locking handles should be provided that draw all such doors tightly to the gasketed surfaces so as to provide airtight seals. Inspection windows should be double-glazed and the glass thicknesses and pane separations should be chosen carefully to avoid degradation by structural/air gap resonances. Placing porous absorbing material in the reveals between the two frames supporting the glass panes can improve the transmission loss of a double-glazed inspection window. Figure 9.34 shows an enclosure in which some basic noise control techniques have been applied. Figure 9.35 shows an enclosure for a compressor driven by a diesel engine. In this design, sound-absorbing material is located inside the enclosure, and air paths are provided for the passage of cooling air. Reference [114] gives extensive details of a theoretical and experimental acoustical design study of the inlet cooling duct for the compressor enclosure.

Figure 9.36 shows an enclosure built for a bandsaw. A sliding window is provided that can be opened for access to controls. A sliding door is also provided that can be closed as much as possible around items fed to the bandsaw. The interior of the enclosure is lagged with 50 mm of sound-absorbing material. Enclosures are available from manufacturers in a wide variety of ready-made modular panels. Figure 9.37 shows how these

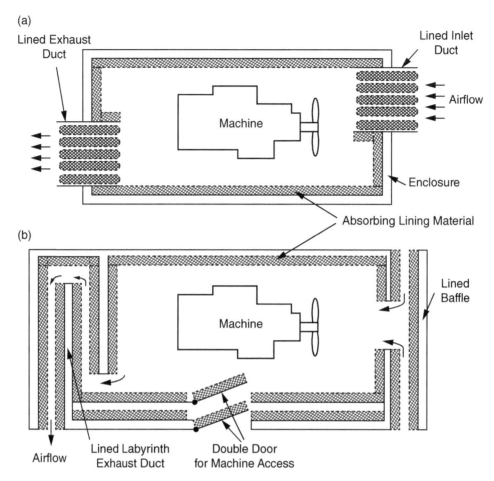

Figure 9.33 Enclosures with penetrations (for cooling) lined with absorbing materials: (a) lined ducts and (b) lined baffles with double-door access provided to interior of machine.

panels can be built into a variety of complete machine enclosures and partial and complete personnel enclosures. Information about cost, construction details, and performance of several enclosures can be found in the manufacturers' literature and in some books [115].

9.7 Use of Barriers

An obstacle placed between a noise source and a receiver is termed a *barrier* or *screen*. When a sound wave approaches the barrier, some of the sound wave is reflected and some is transmitted past (see Figure 9.38). At high frequency, barriers are quite effective, and a strong acoustic "shadow" is cast. At low frequency (when the wavelength can equal or exceed the barrier height), the barrier is less effective, and some sound is diffracted into the shadow zone. Indoors, barriers are usually partial walls. Outdoors, the use of walls, earth berms, and even buildings can protect residential areas from traffic and industrial noise sources [116, 117].

The use of barriers to control noise problems is an example of a practical application of a complicated physical theory: the theory of diffraction, a physical phenomenon that corresponds to the nonspecular reflection

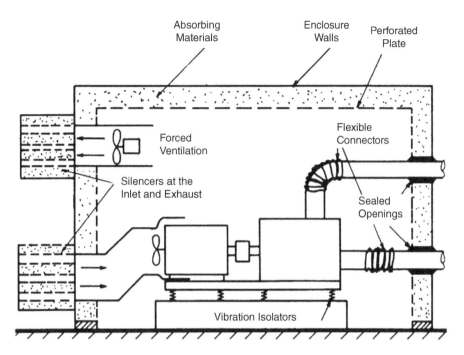

Figure 9.34 Basic elements of an acoustical enclosure used for machinery noise control.

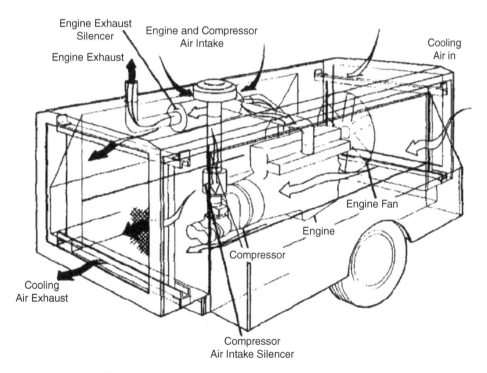

Figure 9.35 Major components and cooling airflow of an air compressor [114].

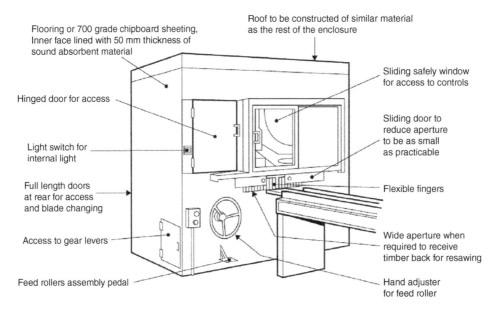

Flooring or 700 grade chipboard sheeting, Inner face lined with 50 mm thickness of sound absorbent material

Roof to be constructed of similar material as the rest of the enclosure

Sliding safely window for access to controls

Hinged door for access

Sliding door to reduce aperture to be as small as practicable

Light switch for internal light

Flexible fingers

Full length doors at rear for access and blade changing

Access to gear levers

Wide aperture when required to receive timber back for resawing

Feed rollers assembly pedal

Hand adjuster for feed roller

Figure 9.36   Enclosure for a bandsaw.

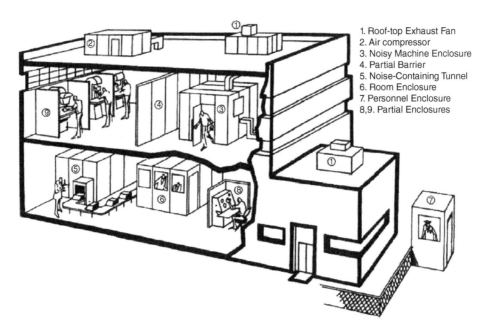

1. Roof-top Exhaust Fan
2. Air compressor
3. Noisy Machine Enclosure
4. Partial Barrier
5. Noise-Containing Tunnel
6. Room Enclosure
7. Personnel Enclosure
8,9. Partial Enclosures

Figure 9.37   Ready-made modular materials used to make enclosures and barriers in a factory building. (Source: Courtesy of Lord Corporation, Erie, PA.)

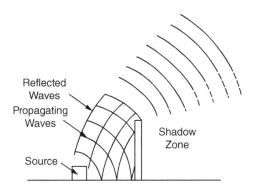

Figure 9.38 Sound waves reflected and diffracted by barrier and acoustical shadow zone.

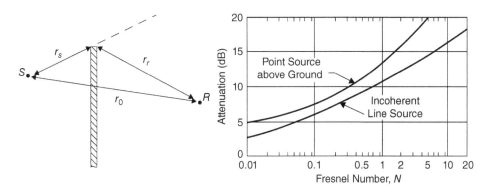

Figure 9.39 Attenuation of a barrier as a function of Fresnel number *N* for point and incoherent line sources.

or scattering of sound waves by an object or boundary. Most of the theories of diffraction were originally formulated for optics, but they find many applications in acoustics. As in the diffraction of light waves, when the sound reaches a listener by an indirect path over a barrier, there is a shadow zone and a bright zone, as shown in Figure 9.38. However, the diffracted wave coming from the top edge of the barrier affects a small transition region close to the shadow zone by interfering with the direct wave [118].

Certainly, from a practical point of view, most of the applications of the physical and geometrical theory have been difficult to use due to the complexity of the analysis, which does not permit fast calculation for design purposes. Because of this, several algorithms, charts, and plots have been developed from time to time to determine the insertion loss or reduction in sound pressure level expected after installation of a barrier both indoors and outdoors.

Figure 9.39 shows the *insertion loss* or reduction in sound pressure level expected after installation of a semi-infinite barrier in free space between a source and receiver. In Figure 9.39, *N* is the dimensionless Fresnel number related to the shortest distance over the barrier $r_s + r_r$ and the straight line distance r_0 between the source *S* and receiver *R*:

$$N = 2(r_s + r_r - r_0)/\lambda, \tag{9.49}$$

where λ is the wavelength and r_s and r_r are shown in Figure 9.39. Usually $\delta = r_s + r_r - r_0$ is called the path length difference. It can be observed in Figure 9.39 that when a noise source approximates to an incoherent line source (e.g. stream of traffic), then the attenuation is lower than the one calculated for a point source.

9.7.1 Transmission Loss of Barriers

Barriers are a form of partial enclosure (they do not completely enclose the source or receiver) to reduce the direct sound field radiated in one direction only. The barrier edges diffract the sound waves, but some waves can pass through the barrier according to the sound transmission laws. All the theories of diffraction have been developed assuming that the transmission loss of the barrier material is sufficiently large that transmission through the barrier can be ignored. Obviously, the heavier the barrier material, or the higher the frequency, the greater is the transmission loss for sound traveling through the barrier. A generally applicable acoustical requirement for a barrier material is to limit the component of sound passing through it to 10 dB less than the predicted noise level due to sound diffracted over the barrier.

Evidently, this is not a governing criterion for concrete or masonry, but can be important for light aluminum, timber, and for glazing panels. In addition, this may be an important consideration when designing "windows" in very tall barriers.

In a study on barriers used indoors, Warnock compared the transmitted sound through a barrier with the diffracted sound over the barrier [119]. He found that the transmitted sound is negligible if the surface density of a single screen satisfies the criterion $\rho_s = 3\sqrt{\delta}\,\text{kg/m}^2$. The minimum acceptable value of ρ_s corresponds to the transmission loss at 1000 Hz being 6 dB greater than the theoretical diffraction loss. A formula for calculating the minimum required surface density for a barrier is [120]

$$\rho_s = 3 \times 10^{(A-10/14)}\,\text{kg/m}^2,\tag{9.50}$$

where A is the A-weighted potential attenuation of the barrier in decibels when it is located outdoors.

As a general rule, when the barrier surface density ρ_s exceeds 20 kg/m^2, the transmitted sound through the barrier can be ignored, and then the diffraction sets the limit on the noise reduction that may be achieved.

According to the discussion above, when butting or overlapping components are used to assemble a noise barrier, it is important that the joints be well sealed to prevent leakage. As an indication, it is common for timber barriers to be manufactured from 19-mm-thick material. As indicated by the mass law, this provides a sound reduction index of 20 dB if joints are tight, which is quite sufficient for barriers designed to provide an attenuation of 10 dB. In some countries, the legislation requires a sample of barrier to be tested in accordance with the local standard for sound insulation of partitions in buildings.

9.7.2 Use of Barriers Indoors

Single-screen barriers are widely used in open-plan offices (or landscaped offices) to separate individual workplaces to improve acoustical and visual privacy. The basic elements of these barriers are freestanding screens (partial-height partitions or panels). However, when a barrier is placed in a room, the reverberant sound field and reflections from other surfaces cannot be ignored.

The diffraction of the sound waves around the barrier boundaries alters the effective directivity of the source. For a barrier placed in a rectangular room, if the receiver is in the shadow zone of the barrier and the sound power radiated by the source is not affected by insertion of the barrier, the approximate insertion loss (the reduction of the sound pressure level before and after the erection of the barrier) can be calculated by [121]

$$IL = 10\log\left(\frac{\left(\dfrac{Q_\theta}{4\pi r^2} + \dfrac{4}{S_0\alpha_0}\right)}{\dfrac{Q_B}{4\pi r^2} + \dfrac{4\Gamma_1\Gamma_2}{S(1-\Gamma_1\Gamma_2)}}\right),\tag{9.51}$$

where Q_θ is the source directivity factor, r is the distance between the source and receiver without the barrier, $S_0\alpha_0$ is the room absorption for the original room before placement of the barrier, S_0 is the total room surface area, α_0 is the mean room Sabine absorption coefficient, S is the open area between the barrier perimeter and the room walls and ceiling,

$$Q_B = Q_\theta \sum_{i=1}^{n} \left(\frac{1}{3 + 10N_i} \right), \tag{9.52}$$

is the effective directivity, n is the number of edges of the barrier (e.g. $n = 3$ for a freestanding barrier, see Figure 9.40). The parameters Γ_1 and Γ_2 are dimensionless numbers related to the room absorption on the source side ($S_1\alpha_1$) and the receiver side ($S_2\alpha_2$) of the barrier, respectively, as well as the open area, and are given by

$$\Gamma_1 = \frac{S}{S + S_1\alpha_1} \text{ and } \Gamma_2 = \frac{S}{S + S_2\alpha_2}. \tag{9.53}$$

Note that $S_1 + S_2 = S +$ (area of two sides of the barrier) and α_1 and α_2 are the mean Sabine absorption coefficients associated with areas S_1 and S_2, respectively.

It is seen that when the barrier is located in a highly reverberant field the IL tends to zero. This means that the barriers are ineffective in highly reverberant environments. Consequently, in this case the barrier should be treated with sound-absorbing material, increasing the overall sound absorption of the room.

The approximation for the effective directivity given in Eq. (9.52) is based on Tatge's result [122]. In deriving Eq. (9.51) the interference between the sound waves has been neglected, so Eq. (9.51) predicts the insertion loss accurately when octave-band analysis is used. However, the effects of the reflections by the floor and the ceiling are not taken into account. This effect will be discussed later. In general, the ceiling in an open-plan office must be highly sound absorptive to ensure maximum performance of a barrier. This is particularly important at those frequencies significant for determining speech intelligibility (500–4000 Hz).

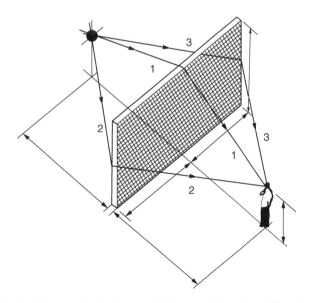

Figure 9.40 Freestanding barrier used indoors and the three diffraction paths.

A more general model for calculating the insertion loss of a single-screen barrier in the presence of a floor and a ceiling has been presented by Wang and Bradley [123]. Their model was developed using the image source technique. More recently, Lau and Tang [124] have presented a study on the insertion loss provided by rigid noise barriers in an enclosed space using different approaches.

Example 9.13   A machine source operates on the floor of a building of dimensions 30 m × 30 m with a height of 10 m. Suppose the average absorption coefficient is $\bar{\alpha} = 0.02$ at 500 Hz. A receiver is located 15 m away from the machine. A wall-to-wall 5-m high barrier is built right in the middle between the source and the receiver. The barrier is made of metal plates with an absorption coefficient of $\alpha = 0.02$ at 500 Hz. Consider the case that the receiver and the acoustical center of the machine are 1 m above the floor. Estimate the insertion loss of the barrier for the 500-Hz one-octave band.

Solution

The room surface area $S_0 = 2(900) + 4(300) = 3000 \text{ m}^2$. The surface of the barrier is 150 m^2 and the open area between the barrier perimeter and the room walls and ceiling is also 150 m^2. Due to the symmetry of the problem, the surfaces forming the room on the source and receiving side of the barrier are $S_1 = S_2 = 150 + S_0/2 = 1650 \text{ m}^2$. In addition, $\alpha_1 = \alpha_2 = 0.02$ and $S_1\alpha_1 = S_2\alpha_2 = 33 \text{ m}^2$. We assume that the machine is an omnidirectional source situated on a hard floor, i.e. $Q_\theta \approx 2$. Now we determine the geometrical distances:

$$r_s = r_r = \sqrt{4^2 + (7.5)^2} = 8.5; r_0 = 15; \delta = 8.5 + 8.5 - 15 = 2 \text{ m}.$$

The wavelength $\lambda = 344/500 = 0.688$ m, and the Fresnel number $N = 2\delta/\lambda = 5.81$. Now, as the barrier extends all the way across the room, the sound is diffracted over the top of the barrier only, so

$$Q_B = \frac{Q_\theta}{3 + 10N} = \frac{2}{3 + 10(5.81)} = 0.033, \text{ and } \Gamma_1 = \Gamma_2 = \frac{150}{150 + 33} = 0.82.$$

Therefore, inserting these values in Eq. (9.51), we obtain

$$IL = 10 \log \left(\frac{\dfrac{2}{4\pi(15)^2} + \dfrac{4}{3000(0.02)}}{\dfrac{0.033}{4\pi(15)^2} + \dfrac{4(0.82)^2}{150\left[1 - (0.82)^2\right]}} \right) = 10 \log (1.237) = 0.9 \text{ dB}.$$

We observe that, although the receiver is deep in the shadow zone of the barrier ($N = 5.81$), the IL is insignificant as compared to the open space situation. If barriers are used inside buildings, their performance is often disappointing because sound can propagate into the shadow zone by multiple reflections. To produce acceptable attenuation, it is important to suppress these reflections by the use of sound-absorbing material, particularly on ceilings just above the barriers.

An ISO standard has been published with guidelines for noise control in offices and workrooms by use of acoustical screens [125]. The standard specifies the acoustical and operational requirements to be agreed upon between the supplier or manufacturer and the user of acoustical screens. In addition, the standard is applicable to (i) freestanding acoustical screens for offices, service areas, exhibition areas, and similar rooms, (ii) acoustical screens integrated in the furniture of such rooms, (iii) portable and removable acoustical screens for workshops, and (iv) fixed room partitions with more than 10% of the connecting area open and acoustically untreated.

9.7.3 Reflections from the Ground

When considering the reflections of sound from the ground, extra propagation paths are created that can result in increased sound pressure at the receiver. The geometry showing reflections from acoustically hard ground with an infinite barrier is shown in Figure 9.41. Application of the image source method indicates that a total of four diffraction paths must be considered: *SOR*, *SAOBR*, *SAOR*, and *SOBR*. Therefore, the attenuation and expected sound pressure level at the receiver has to be calculated for each of the four paths. Then, the four expected sound pressure levels are combined logarithmically to obtain the sound pressure level at the receiver. The process is repeated for the case without the barrier (which has just two paths) to calculate the combined level at the receiver before placement of the barrier. Then, the insertion loss is determined as usual. If the barrier is finite, eight separate paths should be considered since the diffraction around the ends involves only one ground reflection.

Usually, the ground is somewhat absorptive. Therefore, the amplitude of each reflected path has to be reduced by multiplying its sound pressure amplitude by the reflection coefficient of the ground. Other effects can reduce the performance of a barrier. Such is the case if the barrier is close to the noise source, as illustrated in Figure 9.42. The surface of a noise barrier can be treated with an acoustically absorbing material to reduce the effect of multiple reflections. Reduction in the performance of a barrier may also be observed

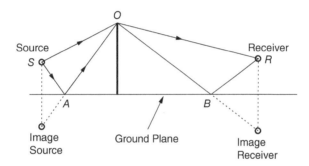

Figure 9.41 Image method for reflections on the ground.

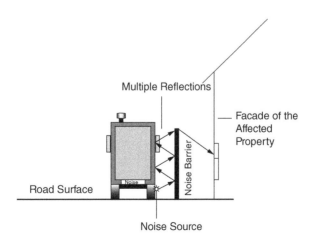

Figure 9.42 Effect of multiple reflections on the acoustical performance of a road noise barrier [117].

when dealing with parallel barriers. This is the case when barriers are constructed on both sides of a road or when the road is depressed with vertical retaining walls. To overcome this problem of multiple reflections and insertion loss degradation, it is possible to:

1) Increase the height of the barriers.
2) Use barriers with sound-absorbing surfaces facing the traffic (an NRC greater than 0.65 is recommended).
3) Simply tilt the barriers outward (a tilt of 5° to 15° is usually recommended).

9.7.4 Use of Barriers Outdoors

Although barriers outdoors are used to reduce the noise from different types of sources, their use to control the noise from highways is surely the most well-known application. While noise barriers do not eliminate all highway traffic noise, they do reduce it substantially and improve the quality of life for people who live adjacent to busy highways. Noise barriers include walls, fences, earth berms, dense plantings, buildings, or combinations of them that interrupt the line of sight between source and observer. It appears that construction of barriers is the main method used for the reduction of noise, although quiet road surfaces, insulation of residential dwellings, and tunnels have also been used for this purpose.

The ISO describes a general method for the calculation of attenuation of sound during its propagation outdoors [126]. This standard has been adopted widely for making predictions of barrier insertion loss using the empirical formula:

$$IL = 10 \log \left[3 + \left(C_2 \frac{\delta}{\lambda} \right) C_3 K_{met} \right] \tag{9.54}$$

where $C_2 = 20$ and includes the effect of ground reflections; $C_2 = 40$ if ground reflections are taken into account elsewhere. C_3 is a factor to take into account of a double diffraction or finite barrier effects, $C_3 = 1$ for a single diffraction, $\delta = (r_s + r_r) - r_0$, and

$$C_3 = \frac{1 + \left(\dfrac{5\lambda}{w} \right)^2}{\dfrac{1}{3} + \left(\dfrac{5\lambda}{w} \right)^2} \tag{9.55}$$

for double diffractions, $\delta = (r_s + r_r + w) - r_0$, where w is the width of the barrier.

When barriers are used outdoors in practice, they are inevitably affected by meteorological conditions which can substantially reduce barrier performance. The term K_{met} in Eq. (9.54) is a correction factor for average downwind meteorological effects, and is given by

$$K_{met} = \begin{cases} \exp \left(-\dfrac{1}{2000} \sqrt{\dfrac{r_s r_r r_0}{2\delta}} \right) & \text{for } \delta > 0 \\ 1 & \text{for } \delta \le 0. \end{cases} \tag{9.56}$$

It can be seen that Eq. (9.54) reduces to the simple formula $IL = 10 \log(3 + 20 N)$ when the barrier is thin, there is no ground, and when meteorological effects are ignored. This formula is identical to that suggested by Tatge [122] and that derived from the empirical results of Maekawa [127] and depends only on the Fresnel number, N.

Example 9.14

A long acoustical barrier of 1 m width and 4 m height is installed to reduce the noise of a source outdoors. The source and the receiver are both located 3 m above the ground and 6 m from the barrier, respectively. Estimate the insertion loss of the barrier in the 500-Hz one-octave band when the effect of ground reflections is included.

Solution

The geometrical distances are:

$r_s = \sqrt{3^2 + 4^2} = 5$ m, $r_r = \sqrt{6^2 + 4^2} = 7.2$ m, $r_0 = 10$ m, and $w = 1$ m. Then
$\delta = (r_s + r_r + w) - r_0 = 5 + 7.2 + 1 - 10 = 3.2$ m. The wavelength $\lambda = c/f = 344/500 = 0.688$ m.

Now, $(5\lambda/w)^2 = (3.44)^2 = 11.83$. We use $C_2 = 20$ and $C_3 = \dfrac{1 + 11.83}{1/3 + 11.83} = 1.054$.

The meteorological effect is determined as

$K_{met} = \exp - \dfrac{1}{2000} \sqrt{\dfrac{5 \times 7.2 \times 10}{2 \times 3.2}} = 0.996$. Therefore,

$IL = 10 \log \left[3 + \left(20\dfrac{3.2}{0.688} \right) \times 1.054 \times 0.996 \right] = 20$ dB in the 500-Hz octave band.

Example 9.15 A noisy machine is located 30 m away from a receiver. A long and thin 2 m high barrier is placed 5 m from the machine to reduce the noise. The center of the machine is 0.5 m above the ground and the receiver is located 1.5 m above the ground. The machine produced a sound pressure level of 69.6 dB at the receiver position in the 250-Hz one-octave band without the barrier. Calculate the sound pressure level at the receiver with the barrier, ignoring meteorological effects. Consider that the ground between the machine and the receiver is soft, level, flat and unobstructed.

Solution

We first calculate the geometrical distances involved in the problem:

$$r_s = \sqrt{5^2 + (2-0.5)^2} = 5.220 \text{ m}; r_r = \sqrt{25^2 + (0.5)^2} = 25.005 \text{ m}; r_0 = \sqrt{30^2 + 1^2} = 30.017 \text{ m}.$$

Then, $\delta = 5.220 + 25.005 - 30.017 = 0.208$ m. $\lambda = 344/250 = 1.376$ m, and

$$N = 2\delta/\lambda = 2(0.208)/1.376 = 0.3.$$

Since the ground is soft and the barrier is thin, we can use $IL = 10\log(3 + 20 N)$. Therefore,

$$IL = 10 \log [3 + 20(0.3)] = 9.6 \text{ dB}.$$

Finally, the sound pressure at the receiver with the barrier is
$L_p = L_p$ (without the barrier) $- IL = 69.6 - 9.6 = 60$ dB in the 250-Hz octave band.

Other empirical charts, algorithms, and theories have been presented for predicting the noise attenuation provided by barriers [127–130]. References [116, 117] discuss some of these approaches.

9.8 Active Noise and Vibration Control

The passive noise and vibration control techniques discussed above work well at mid and high frequencies or in a narrow frequency range but often have the disadvantage of added weight and poor low-frequency performance. Active noise and vibration control has proved useful in the solution of many of these problems.

Although the principle of active noise control has been known since the patent filed by Lueg in 1936 [131], advances in electronic and fast digital signal processing have only made these systems possible since the 1980s.

Active control systems reduce sound and vibration by applying the principle of destructive interference between the fields of the original (*primary*) source and a number of controllable (*secondary*) sources. The term *active* refers to the use of an external source of energy in the reduction process as compared to traditional passive methods such as those discussed in previous sections of this chapter.

When explaining the physical principles of active noise control it is easier to discuss the one-dimensional problem of plane waves propagating in a duct. If the wavelength of the sound is large compared to the diameter of a duct we can assume that only plane waves will travel along the duct. If the primary source produces pure tones only, then a single secondary sound source can be used to cancel the sound by producing a waveform that is equal in amplitude but opposite in phase to the sound detected by a microphone placed downstream from the noise source (see Figure 9.43a). Although the principle is simple, the performance of the system will eventually be limited by small changes in temperature, airflow, etc. At higher frequencies, higher duct modes can be excited so multiple sound waves can propagate and then a large number of sources would need to be used to cancel the sound waves from the noise source.

If the sound from the primary source has a more complicated frequency content and varies with time, the system must be able to adapt itself to changes in operating conditions using the signals from additional error microphones that detect some undesirable state in the canceling process. These error signals are fed back to an electronic controller that implements a control algorithm to minimize a cost function and drive the secondary acoustic sources in order to either reduce the noise or create zones of quiet (see Figure 9.43b). In this case, the controller makes use of adaptive digital filtering techniques in which the weightings of a digital filter are updated with time. Electrical signal processing controllers required to drive the secondary sources can either be feedback or adaptive feedforward controllers. Feedforward control can be applied when the

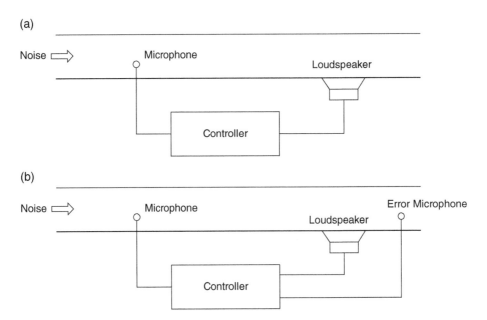

Figure 9.43 Elements of an active noise control system in a duct: (a) simple active noise canceling system, (b) adaptive active noise canceling system.

designer has direct access to information about the primary field. On the other hand, feedback control has primarily been applied when the disturbance cannot be directly observed [132].

The number of secondary sources needed depends on the complexity of the sound field. To control a source in a free field, this number increases as the square of the distance of the secondary source away from the primary source. Acoustical modal overlap determines the number of secondary sources in active noise control applications in enclosures [132]. Zones of quiet of a useful size can be achieved in practice up to several hundred hertz. The theory and practical implementation of active noise and vibration control systems are rather complicated and they are beyond the scope of this book, but the reader is referred to Refs. [132, 133] and some textbooks [134–137].

Examples of current applications of active noise control include the control of sound in ventilation and exhaust ducts, active headsets, and the interiors of propeller aircraft and cars.

Active vibration control can also be used for a number of vibration problems using exactly the same principles as with active noise problems. In active vibration control, secondary vibration actuators provide the necessary inputs applied to a structure to modify its response. Error sensors, such as accelerometers or other vibration transducers, are needed to measure the system response and an electric controller implements the chosen control algorithm to reduce the unwanted vibration. Reference [138] discusses some of the principles of actuators used in active vibration control. Examples of practical active vibration control systems include active engine supports in vehicles and active vibration isolation systems for propeller aircraft power-plants [133].

Vibrating structures often radiate or transmit unwanted sound. An effective technique to control the structural sound radiation is called active structural acoustical control (ASAC). In this technique control inputs are applied directly to a structure in order to minimize the radiated sound or components of structural motion associated with the sound radiation [133]. ASAC systems have been successfully applied to propeller aircraft. In a commercial application, electrodynamic actuators are attached to the aircraft fuselage and a feedforward controller is used. Attenuations from about 10 dB at the propeller fundamental tone to 3 dB at the third harmonic have been reported [139]. In general, it is found to become less efficient and ineffective to use active noise and vibration control methods as the noise or vibration source frequency increases.

A very successful application of active noise control is in hearing protectors and communication headsets. In the case of hearing protectors, the cancelation is established at or very near the outer ear. The entrance of the ear canal is kept very close to the position of a secondary loudspeaker. The fundamental acoustical limitations of active noise control can generally be avoided up to a frequency of about 1 kHz. Active noise control has been incorporated into two types of at-the-ear systems: (i) those designed solely for hearing protection and (ii) those designed for one- or two-way communications. Both types are further dichotomized into open-back (or supraaural) and closed-back (or circumaural earmuff) variations. In the former, a lightweight headband connects active noise control microphone/earphone assemblies that are surrounded by foam pads that rest on the pinnae. In that there are no earmuff cups to afford passive protection, the open-back devices provide only active noise reduction [140].

Closed-back devices, which represent most active noise control-based hearing protection devices, are typically based on a passive noise-attenuating earmuff having good sound attenuation at high frequencies that houses the transducers and, in some cases, the electric signal processing controller. If backup attenuation is needed in the event of an electronic failure of the active noise reduction circuit, the closed-back hearing protector is advantageous due to the passive attenuation established by its earcup. The main features of an active headset are illustrated in Figure 9.44, in which an analog feedback controller is used to control the low-frequency sound pressure at a microphone located close to the ear. Some commercial noise reduction headsets include an input for external audio signals, such as the audio entertainment system in airplanes.

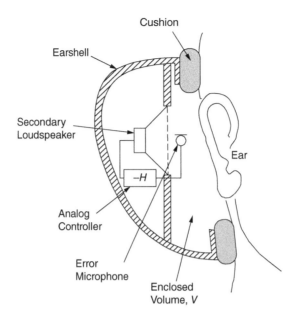

Figure 9.44 Active headset in which the sound inside the headset is detected by the microphone and typically fed back via an analog controller to the loudspeaker to reduce the low-frequency sound pressure level [132].

Fan noise control does not usually involve modifying the fan itself because such noise control modifications almost always result in a degradation of the aerodynamic performance of the fan [141]. Thus, active noise control has been used in reducing tonal noise from fans. Two approaches are possible: (i) create sound-canceling zones from a loudspeaker, or an array of speakers, placed near the fan, or (ii) create an out-of-phase force with a mechanical shaker that is assumed to cancel the unsteady aerodynamic forces created by the fan [142–145]. Both approaches utilize a feedforward controller that requires a fan blade rotational sensor, like a tachometer. Both approaches have been effective in reducing the lower-order blade passing frequency (BPF) tonal noise. Sound pressure level reductions of the BPF noise as much as 22 dB has been shown, while global sound power reductions of up to 14 dB have been reported. Reduction of broadband fan noise by active means is still in the developmental stage [141].

The application of active noise control to reduce turbofan noise has proven to be very challenging because of the complexity of noise sources, which typically consist of the propagation of many circumferential and radial modes for even a single tone. However, some research progress has been made in recent years in applications ranging from single-frequency/mode cancelation to multiple frequency/mode cancelation involving both inlet and exhaust duct radiated fan noise. A typical turbofan active noise control system consists of ring(s) of actuators used to provide the source cancelation, a set of error microphones to monitor the sound pressure level, and a control algorithm to provide real-time optimization of the noise cancelation. The secondary actuators have been placed in various locations the inlet and exhaust fan ducts and sometimes imbedded in the acoustical treatment [146].

Bolt, Beranek & Newman and NASA performed a test [147] using actuators embedded in the stators to provide more control over the radial spinning modes (see Figure 9.45). A difficulty is that the number of error microphones and source actuators required to reduce fan noise sources (in particular at higher frequencies) is too large for practical applications. In addition, the actuators need to be robust, produce high-amplitude sound, and be effective over a broad range of frequencies [146]. It is expected that further developments will reduce the system requirements and cost to make active noise control possible for aircraft engines. An overview of active noise control research for fans has been presented by Envia [148].

Figure 9.45 Active noise control for fan noise reduction [146].

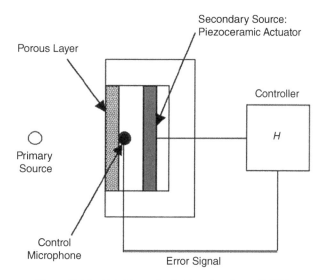

Figure 9.46 Hybrid passive/active absorber cell [60].

The use of active noise control has also been combined with passive control to develop hybrid absorbers. Active control technologies appear to be the only practical way to attenuate the low-frequency turbofan noise components. A hybrid passive/active absorber can absorb the incident sound over a wide frequency range. Figure 9.46 shows the principle of such a system, which combines passive absorbent properties of a porous layer and active control at its rear face, and where the controller can be implemented using digital techniques [149, 150]. The use of piezoelectric actuators as a secondary sources and wire mesh as porous material has allowed the design of thin active liners, composed of several juxtaposed cells of absorbers, to be used in the reduction of noise in flow ducts [150]. The combination of active and passive control using microperforated panels has produced promising results for the application in absorbing systems [151]. This principle has been used in turbofan liners to make them more effective in reducing noise. An example of this hybrid active/passive approach has been developed by Northrop Grumman [152].

Active control systems for reducing tonal airplane cabin noise are generally based on either active noise control or ASAC. Broadband cabin noise may be attenuated with the aid of active headsets or silent seats. An application of active control of propeller aircraft cabin noise and vibration is presented in Chapter 15 of this book. Reference [153] presents a comprehensive discussion of active noise control in aircraft.

Other applications of active noise control have included active mufflers located in the gas exhaust systems for exterior noise reduction. Such systems provide effectiveness of 5 dB to 15 dB at some low frequencies (lower than 200 Hz) [154]. Practical application of active control systems for reducing the response of buildings to wind and earthquake loading still remains very limited and exists mainly only in full-scale applications [155].

References

1 Bolt, R.H. and Ingard, K.U. (1957). System considerations in noise control problems. In: *Handbook of Noise Control* (ed. C.M. Harris), 21-1–21-20. New York: McGraw-Hill.

2 Lyon, R.H. (1987). *Machinery Noise and Diagnostics*. Boston, MA: Butterworths.

3 Bies, D.A. and Hansen, C.H. (2009). *Engineering Noise Control: Theory and Practice*, 4e. Abingdon: Spon Press.

4 Crocker, M.J. and Price, A.J. (1975). *Noise and Noise Control*, vol. I. Cleveland, OH: CRC Press.

5 Crocker, M.J. and Kessler, F.M. (1982). *Noise and Noise Control*, vol. II. Boca Raton, FL: CRC Press.

6 Sharland, I. (1972). *Woods Practical Guide to Noise Control*. Waterlow, London: Woods of Colchester.

7 Irwin, J.D. and Graf, E.R. (1979). *Industrial Noise and Vibration Control*. Englewood Cliffs, NJ: Prentice-Hall.

8 Bell, L.W. and Bell, D.H. (1993). *Industrial Noise Control – Fundamentals and Applications*, 2e. New York: Marcel Dekker.

9 Barron, R.F. (2003). *Industrial Noise Control and Acoustics*. New York: Marcel Dekker.

10 Wilson, C.E. (2006). *Noise Control*, rev. ed. Malabar, FL: Kreiger.

11 Fahy, F.J. and Walker, J.G. (eds.) (1998). *Fundamentals of Noise and Vibration*. London: E&FN Spon, Routledge Imprint.

12 Ver, I.L. and Beranek, L.L. (2005). *Noise and Vibration Control Engineering – Principles and Applications*, 2e. Hoboken, NJ: Wiley.

13 Fahy, F. (2001). *Foundations of Engineering Acoustics*. London: Academic.

14 Kinsler, L.E., Frey, A.R., Coppens, A.B., and Sanders, J.V. (1999). *Fundamentals of Acoustics*, 4e. New York: Wiley.

15 Crocker, M.J. (2007). Noise and vibration source identification. In: *Handbook of Noise and Vibration Control* (ed. M.J. Crocker), 668–684. New York: Wiley.

16 Huertas, J.I., Duque, J.C., Zuluaga, J.P., and Parra Mariño, D.F. (2003) Identification of annoying noises in vehicles. SAE Noise and Vibration Conference and Exposition. Warrendale, PA: Society of Automotive Engineers.

17 Uchida, H. and Ueda, K. (1998) Detection of transient noise of car interior using non-stationary signal analysis. SAE International Congress and Exposition. Warrendale, PA: Society of Automotive Engineers.

18 Alt, N.W., Wiehagen, N., and Schlitzer, M.W. (2001) Interior noise simulation for improved vehicle sound. SAE Noise and Vibration Conference and Exposition. Warrendale, PA: Society of Automotive Engineers.

19 Eisele, G., Wolff, K., Alt, N., and Huser, M. (2005) Application of vehicle interior noise simulation (VINS) for NVH analysis of a passenger car. Noise and Vibration Conference and Exhibition. Warrendale, PA: Society of Automotive Engineers.

20 Unruh, J.F., Till, P.D., and Farwell, T.J. (2000) Interior noise source/path identification technology. SAE General Aviation Technology Conference and Exposition. Warrendale, PA: Society of Automotive Engineers.

21 Chae, C.-K., Bae, B.-K., Kim, K.-J. et al. (2000). Feasibility study on indirect identification of transmission forces through rubber bushing in vehicle suspension system by using vibration signals measured on links. *Veh. Syst. Dyn.* 33 (5): 327–349.

22 Browne, M. and Pawlowski, R. (2005) Statistical identification and analysis of vehicle noise transfer paths. SAE Noise and Vibration Conference and Exhibition. Warrendale, PA: Society of Automotive Engineers.

23 Rust, A. and Edlinger, I. (2001) Active path tracking. A rapid method for the identification of structure borne noise paths in vehicle chassis. SAE Noise and Vibration Conference and Exposition. Warrendale, PA: Society of Automotive Engineers.

24 Steel, J.A. (1998). Study of engine noise transmission using statistical energy analysis. *Proc. Inst. Mech. Eng. Part D: J. Automobile Eng.* 212 (3): 205–213.

25 Alt, N.W., Nehl, J., Heuer, S., and Schlitzer, M.W. (2003) Prediction of combustion process induced vehicle interior noise. SAE Noise and Vibration Conference and Exposition. Warrendale, PA: Society of Automotive Engineers.

26 Diemer, P., Hueser, M.G., Govindswamy, K., and D'Anna, T. (2003) Aspects of powerplant integration with emphasis on mount and bracket optimization. SAE Noise and Vibration Conference and Exposition. Warrendale, PA: Society of Automotive Engineers.

27 Kim, G.J., Holland, K.R., and Lalor, N. (1997). Identification of the airborne component of tyre-induced vehicle interior noise. *Appl. Acoust.* 51 (2): 141–156.

28 Constant, M., Leyssens, J., Penne, F., and Freymann, R. (2001) Tire and car contribution and interaction to low frequency interior noise. SAE Noise and Vibration Conference and Exposition. Warrendale, PA: Society of Automotive Engineers.

29 Christensen, J.J., Hald, J., Mørkholt, J. et al. (2003) A review of array techniques for noise source location. SAE Noise and Vibration Conference and Exposition. Warrendale, PA: Society of Automotive Engineers.

30 Martins, R., Pinto, M., Mendonca, C., and Ribeiro Morais, A. (2003) Pass by noise analysis in a commercial vehicle homologation. 12th SAE Brazil Congress and Exposition. Warrendale, PA: Society of Automotive Engineers.

31 Genuit, K., Guidati, S., and Sottek, R. (2005) Progresses in pass-by simulation techniques. noise and vibration conference and exhibition. Warrendale, PA: Society of Automotive Engineers.

32 Wu, S.F., Rayess, N.E., and Shiau, N.-M. (2001). Visualizing sound radiation from a vehicle front end using the Hels method. *J. Sound Vib.* 248 (5): 963–974.

33 Crocker, M.J. (2007). Psychoacoustics and product sound quality. In: *Handbook of Noise and Vibration Control* (ed. M.J. Crocker), 805–828. New York: Wiley.

34 Frid, A. (2000). Quick and practical experimental method for separating wheel and track contributions to rolling noise. *J. Sound Vib.* 231 (3): 619–629.

35 de Beer, F.G. and Verheij, J.W. (2000). Study of engine noise transmission using statistical energy analysis Experimental determination of pass-by noise contributions from the bogies and superstructure of a freight wagon. *J. Sound Vib.* 231 (3): 639–652.

36 Mas, P., Sas, P., and Wyckaert, K. (1994) Indirect force identification based upon impedance matrix inversion: a study on statistical and deterministical accuracy. Nineteenth ISMA Conference, Leuven, 12–14 September.

37 Otte, D. (1995). The use of SVD for the study of multivariate noise & vibration problems. In: *SVD and Signal Processing III: Algorithms, Architectures and Applications* (eds. M. Moonen and B. De Moor), 357–366. Amsterdam, Netherlands: Elsevier Science.

38 Otte, D., Leuridan, J., Grangier, H., and Aquilina, R. (1991) Prediction of the dynamics of structural assemblies using measured FRF data: some improved data enhancement techniques. Proceedings Ninth International Modal Analysis Conference, Florence, 909–918.

39 Starkey, J.M. and Merrill, G.L. (1989). On the ill conditioned nature of indirect force measurement techniques. *J. Modal Anal.* 4 (3): 103–108.

40 Verheij, J.W. (1992) Experimental procedures for quantifying sound paths to the interior of road vehicles. Proceedings of the Second International Conference on Vehicle Comfort, Bologna, 14–16 October.

41 Wyckaert, K. and Van der Auweraer, H. (1995) Operational analysis, transfer path analysis, modal analysis: tools to understand road noise problems in cars. SAE Noise & Vibration Conference, Traverse City, 139–143.

42 van der Linden, P.J.G. and Varet, P. (1996) Experimental determination of low frequency noise contributions of interior vehicle body panels in normal operation. SAE Paper 960194, Detroit, February 26–29, 61–66.

43 LMS, Transfer Path Analysis – The Qualification and Quantification of Vibro-acoustic Transfer Paths, see http://www.lmsintl.com/downloads/cases.

44 Tournour, M., Cremers, L., and Guisset, P. (2000) Inverse numerical acoustics based on acoustic transfer vectors. 7th International Congress on Sound and Vibration, Garmisch, Partenkirchen, Germany, 2069–1076.

45 Fahy, F.J. (1995). The Vibro-acoustic reciprocity principle and applications to noise control. *Acustica* 81: 544–558.

46 Veronesi, W.A. and Maynard, J.D. (1989). Digital holographic reconstruction of sources with arbitrarily shaped surfaces. *J. Acoust. Soc. Am.* 85: 588–598.

47 Bai, M.R. (1992). Application of BEM-based acoustic holography to radiation analysis of sound sources with arbitrarily shaped geometries. *J. Acoust. Soc. Am.* 92: 533–549.

48 Kim, B.K. and Ih, J.G. (1996). On the reconstruction of the Vibro-acoustic field over the surface enclosing an interior space using the boundary element method. *J. Acoust. Soc. Am.* 100: 3003–3015.

49 Seybert, A.F. and Martinus, F., Forward and inverse numerical acoustics for NVH applications. 9th International Congress on Sound and Vibration, P714–1, Orlando, FL, July 8–11, 2002.

50 Ungar, E.E. (2007). Use of vibration isolation. In: *Handbook of Noise and Vibration Control* (ed. M.J. Crocker), 725–733. New York: Wiley.

51 Ungar, E.E. (2007). Damping of structures and use of damping materials. In: *Handbook of Noise and Vibration Control* (ed. M.J. Crocker), 734–744. New York: Wiley.

52 Lumsdaine, A. and Scott, R.A. (1998). Shape optimization of unconstrained viscoelastic layers using continuum finite elements. *J. Sound Vib.* 216: 29–52.

53 Floody, S.E., Arenas, J.P., and de Espíndola, J.J. (2007). Modelling metal-elastomer composite structures using a finite-element-method approach. *J. Mech. Eng.* 53 (2): 66–77.

54 Arenas, J.P. and Hornig, K.H. (2008) Sound power radiated from rectangular plates with unconstrained damping layers. *Proceedings of the Ninth International Conference on Computational Structures Technology* (ed. B.H.V. Topping and M. Papadrakakis). Stirlingshire, Scotland: Civil-Comp Press.

55 Soni, M.L. (1980) Finite element analysis of viscoelastically damped sandwich structures. Proceedings of 51st Shock and Vibration Symposium, San Diego, 97–109.

56 Ross, D., Ungar, E.E., and Kerwin, E.M. Jr. (1959). Damping of plate flexural vibrations by means of viscoelastic laminate structural damping. In: *Structural Damping* (ed. J.E. Ruzicka), 49–88. New York: ASME.

57 Bespalova, E.I. and Kitaigorodskii, A.B. (1999). Damping of the vibrations of plates by the coating method. *Int. Appl. Mech.* 35: 1167–1172.

58 Kerwin, E.M. (1959). Damping of flexural waves by a constrained viscoelastic layer. *J. Acoust. Soc. Am.* 31: 952–962.

59 Arenas, J.P. and Crocker, M.J. (2000). Recent trends in porous sound-absorbing materials. *Sound Vib.* 44 (7): 12–17.

60 Crocker, M.J. and Arenas, J.P. (2007). Use of sound-absorbing materials. In: *Handbook of Noise and Vibration Control* (ed. M.J. Crocker), 696–713. New York: Wiley.

61 Zwikker, C. and Kosten, C.W. (1949). *Sound Absorbing Materials*. Amsterdam: Elsevier.

62 Beranek, L.L. (1971). *Noise and Vibration Control*. New York: McGraw-Hill.

63 Beranek, L.L. (1947). Acoustical properties of homogeneous isotropic rigid tiles and flexible blankets. *J. Acoust. Soc. Am.* 19: 556–568.

64 Crighton, D.G., Dowling, A.P., Ffowcs Williams, J.E. et al. (1992). *Modern Methods in Analytical Acoustics*. London: Springer.

65 Strutt, J.W. (Lord Rayleigh) (1945). *The Theory of Sound*. New York: Dover.

66 Attenborough, K. and Walker, L.A. (1971). Scattering theory for sound absorption in fibrous media. *J. Acoust. Soc. Am.* 49: 1331–1338.

67 Allard, J.F. and Atalla, N. (2009). *Propagation of Sound in Porous Media: Modeling Sound Absorbing Materials*, 2e. Chichester: Wiley.

68 Delany, M.E. and Bazley, F.N. (1970). Acoustical properties of fibrous materials. *Appl. Acoust.* 3: 105–116.

69 Allard, J.F. and Champoux, Y. (1992). New empirical equation for sound propagation in rigid frame fibrous materials. *J. Acoust. Soc. Am.* 91: 3346–3353.

70 Mechel, F.P. (2002). *Formulas of Acoustics.* Berlin: Springer.

71 Biot, M.A. (1956). Theory of propagation of elastic waves in a fluid-saturated porous solid. *J. Acoust. Soc. Am.* 28: 168–191.

72 Simon, F. and Pfretzschner, J. (2004). Guidelines for the acoustic design of absorptive devices. *Noise Vib. Worldwide* 35: 12–21.

73 Mechel, F.P. (1988). Design charts for sound absorber layers. *J. Acoust. Soc. Am.* 83: 1002–1013.

74 Ford, R.D. and McCormick, M.A. (1969). Panel sound absorbers. *J. Sound Vib.* 10: 411–423.

75 Mechel, F.P. (2001). Panel absorber. *J. Sound Vib.* 248: 43–70.

76 Becker, E.C.H. (1954). The multiple panel sound absorber. *J. Acoust. Soc. Am.* 26: 798–803.

77 Ingard, U. (1954). Perforated facing and sound absorption. *J. Acoust. Soc. Am.* 26: 151–154.

78 Guess, A.W. (1975). Result of impedance tube measurements on the acoustic resistance and reactance. *J. Sound Vib.* 40: 119–137.

79 Maa, D.Y. (1998). Potential of microperforated panel absorber. *J. Acoust. Soc. Am.* 104: 2861–2866.

80 Dupont, T., Pavic, G., and Laulagnet, B. (2003). Acoustic properties of lightweight micro-perforated plate systems. *Acta Acust.* 89: 201–212.

81 Jordan, V.L. (1947). The application of Helmholtz resonators to sound absorbing structures. *J. Acoust. Soc. Am.* 19: 972–981.

82 Ginn, K.B. (1978). *Architectural Acoustics.* Naerum: Bruel & Kjaer.

83 Smith, J.M.A. and Kosten, C.W. (1951). Sound absorption by slit resonators. *Acustica* 1: 114–122.

84 Kristiansen, U.R. and Vigran, T.E. (1994). On the design of resonant absorbers using a slotted plate. *Appl. Acoust.* 43: 39–48.

85 Cook, R.K. and Chrazanowski, P. (1949). Absorption by sound absorbent spheres. *J. Acoust. Soc. Am.* 21: 167–170.

86 Moulder, R. (1998). Sound-absorptive materials. In: *Handbook of Acoustical Measurements and Noise Control*, 3e1991, reprinted (ed. C.H. Harris), 30.1–30.27. New York: Acoustical Society of America.

87 ASTM C423-17 (2017) *Standard Test Method for Sound Absorption and Sound Absorption Coefficients by the Reverberation Room Method.* American Society for Testing and Materials.

88 ISO 354 (2003) *Acoustics – Measurement of Sound Absorption in a Reverberation Room.* Geneva: International Standards Organization

89 ASTM C384-04 (2016): *Standard Test Method for Impedance and Absorption of Acoustical Materials by Impedance Tube Method.* American Society for Testing and Materials.

90 ISO 10534 (1998) *Acoustics – Determination of Sound Absorption Coefficient and Impedance in Impedance Tubes, Part 1: Method Using Standing Wave Ratio, 1996; Part 2: Transfer-function method.* Geneva: International Standards Organization.

91 Cox, T.J. and D'Antonio, P. (2009). *Acoustic Absorbers and Diffusers: Theory, Design, and Application*, 2e. London and New York: Taylor & Francis.

92 Fuchs, H. (2013). *Applied Acoustics: Concepts, Absorbers, and Silencers for Acoustical Comfort and Noise Control.* Berlin: Springer.

93 Stephens, R.W.B. and Bate, A.E. (1950). *Wave Motion and Sound.* London: Edward Arnold.

94 Beranek, L.L. (1996). *Concert and Opera Halls – How they Sound.* New York: Acoustical Society of America.

95 Barron, M. (1993). *Auditorium Acoustics and Architectural Design.* London: E&FN Spon.

96 Ver, I.L. (1992). Enclosures and wrappings. In: *Noise and Vibration Control Engineering: Principles and Applications* (eds. L.L. Beranek and I.L. Ver). New York: Wiley.

97 Schultz, T.J. (1971). Wrapping, enclosures, and duct linings. In: *Noise and Vibration Control* (ed. L.L. Beranek). New York: McGraw-Hill.

98 Smith, B.J., Peters, R.J., and Owen, S. (1996). *Acoustics and Noise Control*, 2e. Harlow: Addison Wesley Longman Limited.

99 Hillarby, S.N. and Oldham, D.J. (1983). The use of small enclosures to combat low frequency noise sources. *Acoust. Lett.* 6 (9): 124–127.

100 Cole, V., Crocker, M.J., and Raju, P.K. (1983). Theoretical and experimental studies of the noise reduction of an idealized cabin enclosure. *Noise Control Eng. J.* 20 (3): 122–133.

101 Arenas, J.P. and Crocker, M.J. (2007). Use of enclosures. In: *Handbook of Noise and Vibration Control* (ed. M.J. Crocker), 685–695. New York: Wiley.

102 Jackson, R.S. (1962). The performance of acoustic hoods at low frequencies. *Acustica* 12: 139–152.

103 Jackson, R.S. (1966). Some aspects of the performance of acoustic hoods. *J. Sound Vib.* 3 (1): 82–94.

104 Junger, M.C. (1970) Sound transmission through an elastic enclosure acoustically closely coupled to a noise source. ASME Paper No. 70-WA/DE-12. New York: American Society of Mechanical Engineers.

105 Tweed, L.W. and Tree, D.R. (1978). Three methods for predicting the insertion loss of close fitting acoustical enclosures. *Noise Control Eng.* 10 (2): 74–79.

106 Fahy, F.J. (1985). *Sound and Structural Vibration*. London: Academic.

107 Byrne, K.P., Fischer, H.M., and Fuchs, H.V. (1988). Sealed, close-fitting, machine-mounted acoustic enclosures with predictable performance. *Noise Control Eng. J.* 31 (1): 7–15.

108 Oldham, D.J. and Hillarby, S.N. (1991). The acoustical performance of small close fitting enclosures, part 1: theoretical models and part 2: experimental investigation. *J. Sound Vib.* 50 (2): 261–300.

109 Agahi, P., Singh, U.P., and Hetherington, J.O. (1999). Numerical prediction of the insertion loss for small rectangular enclosures. *Noise Control Eng. J.* 47 (6): 201–208.

110 Alfredson, R.J. and Seow, B.C. (1976). Performance of three sided enclosures. *Appl. Acoust.* 9 (1): 45–55.

111 Gordon, C.G. and Jones, R.S. (1998). Control of machinery noise. In: *Handbook of Acoustical Measurements and Noise Control*, 3e 1991, reprinted (ed. C.M. Harris), 40.1–40.16. New York: Acoustical Society of America.

112 Keith, R.H. (2007). Noise control for mechanical and ventilation systems. In: *Handbook of Noise and Vibration Control* (ed. M.J. Crocker), 1328–1347. New York: Wiley.

113 Moreland, J.B. (1984). Low frequency noise reduction of acoustic enclosures. *Noise Control Eng. J.* 23 (3): 140–149.

114 Wang, Y. and Sullivan, J.W. (1984). Baffle-type cooling system: a case study. *Noise Control Eng. J.* 22 (2): 61–67.

115 Miller, R.K. and Montone, W.V. (1977). *Handbook of Acoustical Enclosures and Barriers*. Atlanta, GA: Fairmont Press.

116 Arenas, J.P. (2007). Use of barriers. In: *Handbook of Noise and Vibration Control* (ed. M.J. Crocker), 714–724. New York: Wiley.

117 Horoshenkov, K., Lam, Y.W., and Attenborough, K. (2007). Noise attenuation provided by road and rail barriers, earth berms, buildings, and vegetation. In: *Handbook of Noise and Vibration Control* (ed. M.J. Crocker), 1446–1457. New York: Wiley.

118 Kurze, U., J. and Beranek, L.L. (1971). Sound propagation outdoors. In: *Noise and Vibration Control* (ed. L.L. Beranek), 164–193. New York: McGraw-Hill.

119 Warnock, A.C.C. (1974) Acoustical Effects of Screens in Landscaped Offices, Canadian Building Digest Vol. 164. National Research Council of Canada.

120 Department of Transport, Noise Barriers – Standards and Materials (1976) Technical Memorandum H14/76. London: Department of Transport.

121 Moreland, J. and Minto, R. (1976). An example of in-plant noise reduction with an acoustical barrier. *Appl. Acoust.* 9: 205–214.

122 Tatge, R.B. (1973). Barrier-Wall attenuation with a finite-sized source. *J. Acoust. Soc. Am.* 53: 1317–1319.

123 Wang, C. and Bradley, J.S. (2002). A mathematical model for a single screen barrier in open-plan office. *Appl. Acoust.* 63: 849–866.

124 Lau, S.K. and Tang, S.K. (2009). Performance of a noise barrier within an enclosed space. *Appl. Acoust.* 70 (1): 50–57.

125 ISO 17624 (2004) *Acoustics – Guidelines for Noise Control in Offices and Workrooms by Means of Acoustical Screens*. Geneva: International Standards Organization.

126 ISO 9613-2 (1996) *Acoustics – Attenuation of Sound during Propagation Outdoors – Part 2: General Method of Calculation*. Geneva: International Standards Organization.

127 Maekawa, Z. (1968). Noise reduction by screens. *Appl. Acoust.* 1: 157–173.

128 Kurze, U.J. and Anderson, G.S. (1971). Sound attenuation by barriers. *Appl. Acoust.* 4: 35–53.

129 Barry, T.M. and Reagan, J. (1978) FHWA Highway Traffic Noise Prediction Model. Report No. FHWA-RD-77-108. Washington, DC: Federal Highway Administration.

130 Department of Transport and Welsh Office (1988). *Calculation of Road Traffic Noise*. London: HMSO.

131 Lueg, P. (1936) Process of Silencing Sound Oscillations, U.S. Patent, No. 2,043,416.

132 Elliott, S.J. (2007). Active noise control. In: *Handbook of Noise and Vibration Control* (ed. M.J. Crocker), 761–769. New York: Wiley.

133 Fuller, C. (2007). Active vibration control. In: *Handbook of Noise and Vibration Control* (ed. M.J. Crocker), 770–784. New York: Wiley.

134 Nelson, P.A. and Elliott, S.J. (1992). *Active Control of Sound*. London: Academic.

135 Kuo, S.M. and Morgan, D.R. (1996). *Active Noise Control Systems, Algorithms and DSP Implementations*. New York: Wiley.

136 Fuller, C.R., Elliott, S.J., and Nelson, P.A. (1996). *Active Control of Vibration*. London: Academic.

137 Hansen, C.H. and Snyder, S.D. (1997). *Active Control of Sound and Vibration*. London: E&FN Spon. Elliott, S.J. (2001). *Signal Processing for Active Control*. London: Academic.

138 Hansen, C.H. (2007). Vibration transducer principles and types of vibration transducers. In: *Handbook of Noise and Vibration Control* (ed. M.J. Crocker), 444–454. New York: Wiley.

139 Ross, C.F. and Purver, M.R.J. (1997) Active Cabin Noise Control, Proceedings of Active 97, Budapest, Hungary.

140 Gerges, S.N.Y. and Casali, J.G. (2007). Hearing protectors. In: *Handbook of Noise and Vibration Control* (ed. M.J. Crocker), 364–376. New York: Wiley.

141 Lauchle, G.C. (2007). Centrifugal and axial fan noise prediction and control. In: *Handbook of Noise and Vibration Control* (ed. M.J. Crocker), 868–884. New York: Wiley.

142 Koopmann, G.H., Neise, W., and Chen, W. (1988). Active noise control to reduce the blade tone noise of centrifugal fans. *J. Vib. Acoust.* 110: 377–383.

143 Quinlan, D.A. (1992). Application of active control to axial flow fans. *Noise Control Eng. J.* 39: 95–101.

144 Gee, K.L. and Sommerfeldt, S.D. (2003). A compact active control implementation for axial cooling fan noise. *Noise Control Eng. J.* 51: 325–334.

145 Lauchle, G.C., MacGillivray, J.R., and Swanson, D.C. (1997). Active control of axial-flow cooling fan noise. *J. Acoust. Soc. Am.* 101: 341–349.

146 Huff, D.L. and Envia, E. (2007). Jet engine noise generation, prediction, and control. In: *Handbook of Noise and Vibration Control* (ed. M.J. Crocker), 1096–1108. New York: Wiley.

147 Curtis, A.R.D. (1999) Active Control of Fan Noise by Vane Actuators. NASA/CR–1999–209156.

148 Envia, E. (2001). Fan noise reduction: an overview. *Int. J. Aeroacoust.* 1 (1): 43–64.

149 Furstoss, M., Thenail, D., and Galland, M.A. (1997). Surface impedance control for sound absorption: direct and hybrid passive/active strategies. *J. Sound Vib.* 203: 219–236.

150 Galland, M.A., Mazeaud, B., and Sellen, N. (2005). Hybrid passive/active absorbers for flow ducts. *Appl. Acoust.* 66: 691–708.

151 Cobo, P., Pfretzschner, J., Cuesta, M., and Anthony, D.K. (2004). Hybrid passive-active absorption using microperforated panels. *J. Acoust. Soc. Am.* 116: 2118–2125.

152 Parente, C.A., Arcas, N., Walker, B.E. et al. (1999) Hybrid Active/Passive Jet Engine Noise Suppression System. NASA/CR–1999–208875.

153 Johansson, S., Hakansson, L., and Claesson, I. (2007). Aircraft cabin noise and vibration prediction and active control. In: *Handbook of Noise and Vibration Control* (ed. M.J. Crocker), 1207–1215. New York: Wiley.

154 Drozdova, L., Ivanov, N., and Kurtsev, G.H. (2007). Off-road vehicle and construction equipment exterior noise prediction and control. In: *Handbook of Noise and Vibration Control* (ed. M.J. Crocker), 1490–1500. New York: Wiley.

155 Crocker, M.J. (2007). Vibration response of structures to fluid flow and wind. In: *Handbook of Noise and Vibration Control* (ed. M.J. Crocker), 1375–1392. New York: Wiley.

10

Mufflers and Silencers – Absorbent and Reactive Types

10.1   Introduction

A muffler (also known as a silencer) is any section of pipe or duct which has been treated or profiled to reduce the propagation of sound from a source, while allowing the free flow of gas. The performance of a muffler is normally strongly dependent on frequency.

A well-designed muffler should have (i) an acoustical performance providing a minimum acceptable noise reduction (NR) as a function of frequency, (ii) a maximum allowable pressure drop, (iii) an acceptable shape and volume, (iv) suitable durability, and (v) a reasonable cost.

Ducted sources are commonly found in mechanical systems. Typical ducted-source systems include engines and mufflers, and fans and air-moving devices (flow ducts and fluid machines and associated piping). In these systems, the source is the active component and the load is the path, which consists of passive elements such as mufflers, ducts, and end terminations. Active mufflers, which can suppress the noise of sources particularly with strong low-frequency pure tones are beyond the scope of this chapter.

The designs of both reactive mufflers and passive mufflers and combined types are discussed. Finite element and boundary element models are useful in the study of muffler elements. The acoustical performance of a system with a muffler as a path element is described in terms of the transmission loss (TL), insertion loss (IL), and radiated sound pressure. The acoustical performance of ducted muffler systems incorporating reactive mufflers depends strongly on the source-load interactions.

This chapter presents a system for reactive muffler modeling based on electrical analogies that has been found useful in predicting the acoustical performance of such systems. The various methods for determining the impedance of a ducted source are described. Theoretical models for the acoustical performance of dissipative mufflers are complicated, particularly above the cutoff frequency, and often practitioners must rely on design charts or mufflers designed and tested by manufacturers. Different theoretical models are reviewed, and some empirical methods are described for simple predictions of the IL of lined ducts.

10.2   Muffler Classification

Mufflers can be classified into two main types: reactive and dissipative. The design of reactive and dissipative mufflers is reviewed in a number of practical and theoretical books and book chapters [1–15]. Reactive mufflers are composed of one or more chambers and resonators of different volumes and shapes linked together by sections of pipe. They work by reflecting sound waves back from the muffler elements toward the source and back and forth in some of the chambers. This results in an impedance mismatch between the source and the muffler

Engineering Acoustics: Noise and Vibration Control, First Edition. Malcolm J. Crocker and Jorge P. Arenas.
© 2021 John Wiley & Sons Ltd. Published 2021 by John Wiley & Sons Ltd.

at certain frequencies and also some absorption of energy by nonlinear behavior of the gas. Dissipative mufflers on the other hand are lined with acoustical material which absorbs the sound energy and converts it into heat. It should be noted that some question the common classification of mufflers into reactive and dissipative (reflecting and absorbing types). Moser states that the dominating effect of so-called dissipative mufflers is not necessarily absorption, since reflection is often of considerable importance [16]. Mufflers can be designed to be partly reactive and partly dissipative and in fact some internal combustion engine mufflers (particularly those for motorcycles) do sometimes incorporate absorbing materials. However, this material can deteriorate because of the severe temperature conditions and become clogged, melt, or fatigue. Thus, most automobile mufflers manufactured today are of the reactive type and do not incorporate absorbing materials. Nevertheless, some dissipation can still occur in a reactive muffler due to viscous dissipation.

10.3 Definitions of Muffler Performance

The definitions of muffler performance in most common use will be given here [1, 5–8]. It should be noted, however, that some authors use different nomenclature and confusion can sometimes arise.

A) **Insertion Loss** (IL). This is the difference in the sound pressure level (SPL) measured at one point in space with and without the muffler inserted between that point and the source [1, 5–8]. IL is a convenient quantity to measure and its use is favored by manufacturers.
B) **Transmission Loss** (TL). This is defined as $10 \log_{10}$ of the ratio of the sound power incident on the muffler to the sound power transmitted. This is the quantity which is most easily predicted theoretically, and its use is favored by those engaged in research.
C) **Noise Reduction** (NR). This is the difference in SPLs measured upstream and downstream of the muffler.
D) **Attenuation**. This is the decrease in propagating sound power between two points in an acoustical system. This quantity is often used in describing absorption in lined ducts where the decrease in SPL per unit length is measured [7, 8].

The first three definitions are used frequently in work on mufflers for automobile engines and they are illustrated in Figure 10.1. It is of interest to note that these definitions are also used with similar meanings to describe sound transmission through air-conditioning ducts, walls of buildings, aircraft cabins, or machine enclosures.

In general, IL, TL, and NR are not simply related, since, except for the TL, they depend either on the termination impedance or on the internal impedance of the source (radiation impedance of the tail pipe). However, if the source and termination impedances are equal to $\rho c/S$ (i.e. the source and the termination are nonreflecting), then,

$$IL = TL < NR,$$

and usually,

$$NR - TL \approx 3 \text{ dB}.$$

10.4 Reactive Mufflers

Reactive mufflers can be subdivided into straight-through and reverse-flow types [4, 5]. Figure 10.2 shows some typical straight-through types. These mufflers are usually comprised mainly of expansion chambers (chambers in which the area is suddenly increased then decreased) and concentric tube resonators (side-

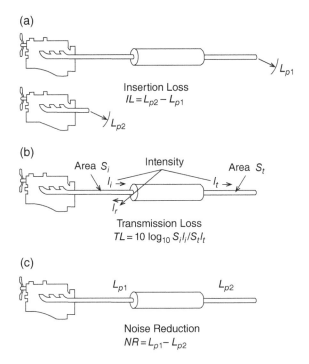

Figure 10.1 Definitions of muffler performance.

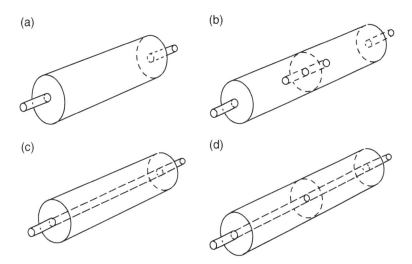

Figure 10.2 Typical straight-through reactive mufflers: (a) single expansion chamber; (b) double expansion chamber with internal connecting tubes; (c) single-chamber resonator; (d) double-chamber resonator.

branch Helmholtz resonators). Reverse-flow types can be built in many different configurations. A typical reverse-flow muffler is shown in Figure 10.3. Figure 10.4 shows two commercial perforated mufflers. As shown, such mufflers consist of several chambers connected by straight pipes. There are usually two end chambers in which the flow is reversed and one or more large low-frequency Helmholtz resonators. Sometimes louver patches are used to produce side-branch Helmholtz resonators (which reflect high-frequency

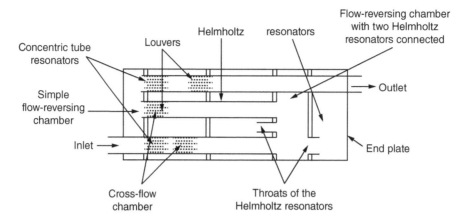

Figure 10.3 Cross-section of typical U.S. automobile muffler with flow-reversing chambers indicated.

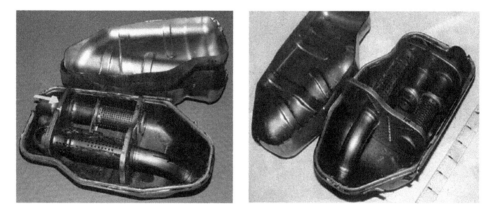

Figure 10.4 Examples of common commercial automobile mufflers [13].

noise). In addition, cross flow is often allowed to occur and attenuation is then created by interference of sound traveling over different path lengths. Most automobile mufflers are of the reverse-flow type, although trucks can use either reverse-flow or straight-through mufflers.

10.5 Historical Development of Reactive Muffler Theories

Although Quincke in the nineteenth century studied the interference of sound propagation through different length pipes, theory of real use in muffler design was not developed until the 1920s. This was probably partly because prior to this time it was difficult (if not impossible) to measure sound pressure quantitatively due to the lack of suitable microphones and partly, because there was less need to reduce the noise produced by less powerful engines.

In 1922 Stewart, in the USA, began developing acoustical filter theory using a lumped parameter approach [17]. In 1927, Mason developed this theory further [18]. In Britain and Germany in the 1930s, work was conducted on designing mufflers for aircraft [19] and single-cylinder engines [20].

However, it was not until the 1950s when another significant improvement in muffler theory occurred. Davis and his coworkers [21, 22] then developed theory for plane wave propagation in multiple expansion chambers and side-branch resonators. They made many experiments and found that in general their predictions for TL were good, provided the cutoff frequency in the pipes and chambers was not exceeded in practice. Above this frequency, cross modes, in addition to plane waves, can exist and one of their theoretical assumptions was violated.

When Davis et al. tried to use their theories to design a helicopter muffler, their predictions were very disappointing, since they only measured IL values of about 10 dB, compared with the 20 dB they had expected from their TL theory. Davis et al. tried to explain this by saying that finite amplitude wave effects must be important. However, although these effects are important with concentric tube resonators, a more likely reason is their neglect of mean flow, which can be of particular importance in IL predictions. For a more complete discussion of the assumptions made by Davis et al. in their theory, see Ref. [22].

In the late 1950s Igarashi et al. began to calculate the transmission properties of mufflers using equivalent electric circuits [23–25]. This approach is very convenient. The total acoustic pressure and total acoustic volume velocity are related before and after the muffler by using the product of four-terminal transmission matrices for each muffler element [5]. The equivalent electrical analog for a muffler is quite convenient since electrical theory and insight may be brought to bear. This approach does, however, assume linearity.

The four-terminal transmission matrices are useful since it is only necessary to know the four parameters A, B, C, D, which characterize the system to define its acoustical performance. The parameter values are not affected by connections to elements upstream or downstream as long as the system elements can be assumed to be linear and passive.

Several transmission matrices have been evaluated for various muffler elements by Igarashi et al. [23–25] and Fukuda et al. [26–28]. Parrott [29] also gives results for transmission matrices, some of which include the effects of a mean flow. However, note that in Ref. [29], Parrott presents a matrix (Eq. (10.28)) for a straight pipe carrying a mean flow of Mach number M, which is in error. Sullivan has given the corrected result in Ref. [30]. Ingard, and Bender and Bammer, recognized the importance of including both the source and tailpipe radiation impedance in muffler system modeling. Crocker [5] and Sullivan [30] continued with such modeling studies.

In the middle and late 1960s and early 1970s several workers including first Davies [31, 32] and then Blair, Goulbourn, Benson, Baites, and Coates [33–38] developed an alternative method of predicting muffler performance based on shock wave theory. Perhaps this work was inspired by Davis's belief [21] that the failure of his helicopter muffler design was caused by the fact that exhaust pressures are much greater than those normally assumed in acoustical theory so that finite amplitude affects become important. This alternative approach involves the use of the method of characteristics and can successfully predict the pressure–time history in the exhaust system. Also, one-third octave spectra of the acoustic noise have been predicted [38]. However, the method is time-consuming and expensive and has difficulties in dealing with complex geometries and some boundary conditions. Although such an approach is probably necessary and useful with the design of mufflers for single-cylinder engines, so far this method has found little favor with manufacturers of mufflers for multicylinder engines. It appears furthermore that Davis' belief [21] may have been incorrect. There are several other possible reasons why Davis failed to obtain better agreement between theory and experiment, each of which can be important. These include [39]: neglect of source impedance effects, neglect of mean gas flow (and its effect on net energy transport), incorrect boundary conditions for exhaust ports and tail pipe, neglect of interaction between mean gas flow and sound in regions of disturbed flow, and, neglect of mean temperature gradients in the exhaust system.

In 1970 Alfredson and Davies published work which shed new light on the acoustical performance of mufflers [39–43]. Alfredson mainly considered the design of long expansion chamber type mufflers commonly used on diesel engines. Alfredson's work has been important since he has shown that (at least with the mufflers and engine he studied) that acoustical theory could be used to predict the radiated exhaust sound and the

TL of a muffler and that finite amplitude effects can be neglected, provided that mean gas flow effects are included in the theory. Alfredson concluded that as the mean flow Mach number approached $M = 0.1$ or 0.2 in the tail pipe, the zero flow theory overpredicted the muffler effectiveness by 5–10 dB or more. The most serious discrepancy occurred for values of reflection coefficient $R \rightarrow 1$. This would occur for low frequencies (large wavelengths). Alfredson computed this error to be

$$Error = 10 \log \left\{ \frac{(1 + M)^2 - (1 - M)^2 R^2}{1 - R^2} \right\}$$

(10.1)

and the result is plotted in Figure 10.5.

As a check on his acoustical theory and on Eq. (10.1), Alfredson later measured the attenuation of an expansion chamber and compared it with theory [41]. The result is shown in Figure 10.6. The good agreement between theory (with flow included) and experiment and the poor agreement with theory when flow was neglected seem to confirm that acoustical theory is probably adequate in many instances in muffler design, provided the effects of mean flow are included in the model where necessary. These conclusions are very important.

Another development occurred in 1970 when Young and Crocker began the use of finite elements to analyze the TL of automobile muffler elements [44]. The reason for the use of finite elements is that some chambers in reverse-flow mufflers (e.g. flow-reversing end chambers and end-chamber/Helmholtz-resonators combinations) are not axisymmetric. Thus, it is difficult, if not impossible, to analyze these chambers using classical assumptions of continuity of pressure and volume velocity at discontinuities, even in the plane wave region. The use of a numerical technique such as finite element analysis makes the acoustical performance of complicated-shaped chambers possible to predict even in the higher frequency cross-mode region. The work of Young and Crocker [44–48] is described in some detail later in this chapter.

Other investigators have since used finite elements in muffler design. Kagawa and Omote [49] have used two-dimensional triangular ring elements. Craggs [50] has used isoparametric three-dimensional elements, while Ling [51], using a Galerkin approach, included mean flow in his acoustical finite element model. However, Ling's work was mainly concentrated on propagation in ducts rather than muffler design.

Side-branch resonators (known by manufacturers as bean cans or spit chambers), see Figures 10.2 and 10.4, have also been studied by Sullivan and Crocker [52, 53]. In practical situations, axial standing waves can exist in the outer concentric cavity of the resonator. Previous theories had been unable to account for this phenomenon (assuming the cavity acts like a lumped parameter stiffness.) Sullivan's work is also described

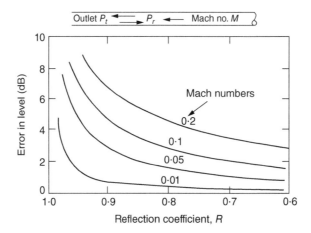

Figure 10.5 Radiated sound pressure level error due to neglect of mean flow, $P_R = R \times Pe^{j\Theta}$. Radiated sound pressure level calculated neglecting mean flow is low by the amount given here as error [39].

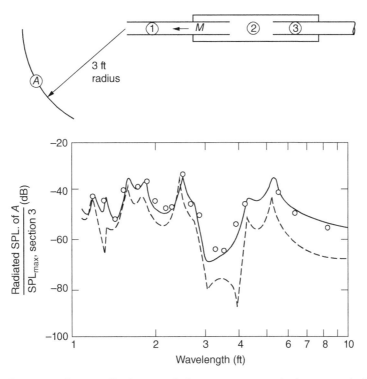

Figure 10.6 Influence of mean gas flow on effectiveness of silencer. ○, measured values: — calculated, *M* = 0.15; - - - -, calculated, *M* = 0.0 [41].

in more detail later in this chapter [53–55]. Other developments in muffler design have included the Bond Graph approach by Karnopp [56, 57]. It is claimed that this approach can extend the frequency range of lumped parameter filter elements.

Another important topic little touched on until the 1980s is the effect of flow in mufflers. Various phenomena can occur. Noise can be generated by the flow process. Interactions can occur between the flow and sound waves. Fricke and Crocker found that the TL of short expansion chambers could be considerably reduced [58]. The effect appears to be amplitude dependent and a feedback mechanism was postulated. Kirata and Itow [59] have studied the influence of air flow on side-branch resonators and concluded that the peak attenuation is considerably reduced by flow. Anderson [60] has concluded that a mean air flow causes an increase in the fundamental resonance frequency of a simple single side-branch Helmholtz resonator connected to a duct.

Another important development, which occurred in the 1970s, is the two-microphone method for determining acoustical properties described by Seybert and Ross [61]. White noise is used as a source. Two flush-mounted wall microphones are used and measurements of the auto- and cross-spectra enable incident and reflected wave spectra and the phase angle between the incident and reflected waves to be determined. The method can be used to measure impedance and TL. Agreement between this two-microphone random noise method and the traditional standing wave tube method is very good and the method is much faster. Figure 10.7 shows a comparison between theory and experiment for the power reflection coefficient *R* for an open end tube and the phase angle [61]. Figure 10.8 shows the TL, of a prototype automobile muffler with a comparison between this method and the classical standing wave ratio (probe tube) method (SWR). For TL measurements, a third microphone was used downstream of the muffler.

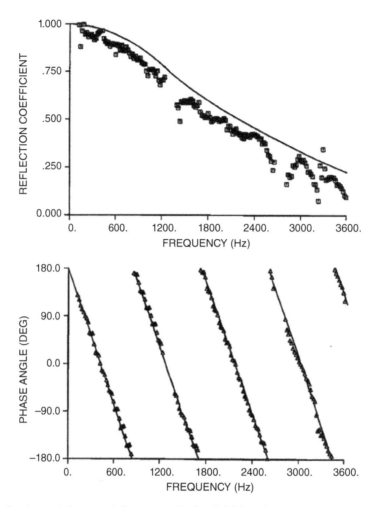

Figure 10.7 Power reflection and phase angle for open end tube. Solid line: theory; open square and open triangle: experiment [61].

Ingard was one of the first researchers to recognize the importance of including both the source and tailpipe radiation impedance in muffler system modeling [62]. Later Bender and Brammer [63], Crocker [5], and Sullivan [30] also included source and radiation impedance in their models.

10.6 Classical Reactive Muffler Theory

10.6.1 Transmission Line Theory

We will first make some simplifying assumptions:

a) Sound pressures are small compared with the mean pressure
b) There are no mean temperature gradients or mean flow, and
c) Viscosity can be neglected.

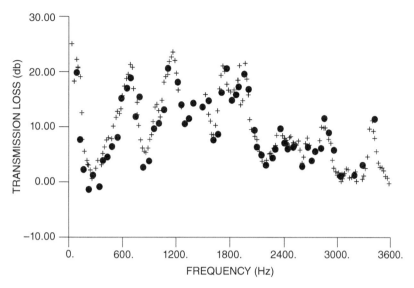

Figure 10.8 Measured values of transmission loss for prototype automotive muffler. ● : SWR method; +: two microphone random excitation method [61].

If plane waves are assumed to exist in a muffler element (see Figure 10.9) then the sound pressure p anywhere in the muffler element can be represented as the sum of right and left traveling waves p^+ and p^- respectively

$$p = p^+ + p^-, \tag{10.2a}$$

$$p = P^+ e^{-jkx} + P^- e^{jkx}, \tag{10.2b}$$

$$V = V^+ + V^-, \tag{10.3a}$$

$$V = (S/\rho c)\left(P^+ e^{-jkx} - P^- e^{jkx}\right), \tag{10.3b}$$

$$V = (S/\rho c)\left(p^+ - p^-\right). \tag{10.3c}$$

Note that p and V represent the magnitude (and phase) of the total sound pressure and volume velocity. The time dependence (constant multiplying factor $e^{j\omega t}$) has been omitted for brevity. The right and left traveling sound waves are represented by the $+$ and $-$ superscripts, respectively, while P represents the pressure amplitude; S the cross-sectional area, $\rho c/S$ the characteristic acoustic impedance (traveling wave pressure divided by traveling wave volume velocity), $k = \omega/c$, the acoustic wavenumber, ω the angular frequency, c the speed of sound, and ρ the fluid density.

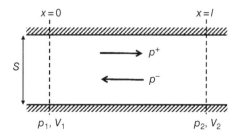

Figure 10.9 Muffler element.

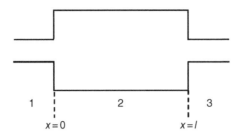

Figure 10.10 Expansion chamber with inlet pipe 1 and outlet pipe 3, both of cross-sectional area S_1. The expansion chamber pipe 2 has cross-section area S_2 at $x = 0$ and $x = l$.

10.6.2 TL of Resonators

Davis et al. used theory such as this to predict the TL of various resonator chamber type mufflers [21, 22] by assuming: (i) continuity of pressure and (ii) continuity of volume velocity at discontinuities.

a) TL of Expansion Chambers

For example, if there is a sudden increase in area at station 1 and a sudden decrease in area (see Figure 10.9) at station 2, then the resulting chamber formed is known as an expansion chamber (see Figure 10.10) and its TL can be found as follows.

The chamber TL is easily derived from Eqs. (10.2a) to (10.3c) above by assuming that the sudden area changes occur at $x = 0$ and $x = l$ (resulting in the area change ratio, m) and by assuming the continuity of pressure and volume velocity at the area discontinuities [21, 22]. In the inlet pipe, P_1^+ and P_1^- are the pressure amplitudes of the right and left traveling waves incident and reflected at the expansion chamber entrance.

Using Eqs. (10.2a) through (10.3c), the sound pressure and volume velocity in the inlet pipe 1 are:

$$p = P_1^+ e^{-jkx} + P_1^- e^{jkx}, \tag{10.4a}$$

and

$$V = (S_1/\rho c)\left(P_1^+ e^{-jkx} - P_1^- e^{jkx}\right). \tag{10.4b}$$

The sound pressure and volume velocity in the expansion chamber 2 are:

$$p = P_2^+ e^{-jkx} + P_2^- e^{jkx}, \tag{10.5a}$$

and

$$V = (S_2/\rho c)\left(P_2^+ e^{-jkx} - P_2^- e^{jkx}\right). \tag{10.5b}$$

The sound pressure and volume velocity in the outlet pipe 3 (assuming anechoic conditions with no reflections) are:

$$p = P_3^+ e^{-jkx}, \tag{10.6a}$$

and

$$V = (S_1/\rho c)\left(P_3^+ e^{-jkx}\right). \tag{10.6b}$$

The boundary conditions of continuity of sound pressure p and volume velocity V at the pipe junctions $x = 0$ and $x = l$ give:

At $x = 0$:

$$P_1^+ + P_1^- = P_2^+ + P_2^-, \tag{10.7a}$$

and

$$V = (S_1/\rho c)\left(P_1^+ - P_1^-\right) = (S_2/\rho c)\left(P_2^+ - P_2^-\right). \tag{10.7b}$$

At $x = l$:

$$p = P_2^+ e^{-jkl} + P_2^- e^{jkl} = P_3^+ e^{-jkl}, \tag{10.8a}$$

and

$$V = (S_2/\rho c)\left(P_2^+ e^{-jkl} - P_2^- e^{-jkl}\right) = (S_1/\rho c) P_3^+ e^{-jkl} \ . \tag{10.8b}$$

For a given expansion chamber length l and wavenumber k, there are four Eqs. (10.7a), (10.7b), (10.8a), and (10.8b) and five unknown complex sound pressure amplitudes P_1^+, P_1^-, P_2^+, P_2^-, and P_3^+. These equations can be solved simultaneously (putting $S_2/S_1 = m$), to give:

$$P_1^+/P_3^+ = \tfrac{1}{4}\left[(1 + m)(1 + 1/m)\,e^{jkl} + (m - 1)(1/m - 1)\,e^{-jkl}\right] = \cos kl + (j/2)(m + 1/m)\sin kl. \tag{10.9}$$

For plane waves, the ratio of the incident sound intensity at the expansion chamber entrance to the transmitted intensity at the outlet (in the tailpipe) assuming anechoic conditions in the tailpipe given by:

$$\left(p^+_{1rms}\right)^2/\left(p^+_{3rms}\right)^2 = \left(P_1^+\right)^2\left(P_3^+\right)^2 = \cos^2 kl + \tfrac{1}{4}(m - 1/m)^2\sin^2 kl, \tag{10.10}$$

and thus, the TL is

$$\mathrm{TL} = 10\log\left(P_1^{+2}/P_3^{+2}\right), \tag{10.11}$$

which from Eq. (10.10) becomes

$$\mathrm{TL} = 10\log\left\{1 + \tfrac{1}{4}(m - 1/m)^2\sin^2 kl\right\}. \tag{10.12}$$

Figure 10.11 gives a plot of Eq. (10.12)

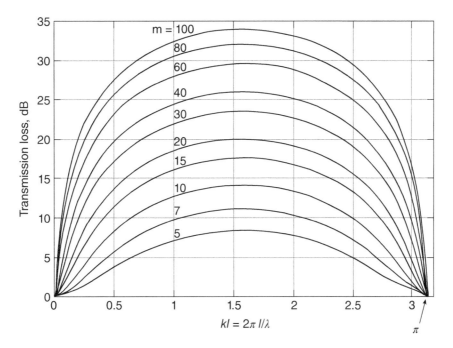

Figure 10.11 Transmission loss TL of an expansion chamber of length l and $S_2/S_1 = m$. The cross-section of the muffler need not be round, but its greatest transverse dimension should be less than $0.8\ \lambda$ (approximately) for the graph to be valid. For values of kl between π and 2π, subtract π and use the scale given along the abscissa. Similarly, for values between 2π and 3π, subtract 2π; etc. Note that when $kl = \pi$, then $l = \lambda/2$; when $kl = 2\pi$, then $l = \lambda$; etc.

Example 10.1   A compressor is pumping air with a gas temperature of 84 °C. The fundamental pumping frequency is 6000 rpm. Noise at the fundamental frequency is a problem. Design an expansion chamber with the shortest length suitable to suppress the exhaust noise. The inlet and outlet tube diameters are 5.0 cm (≈2.0 in.)

Solution

The speed of sound c at 84 °C is (see Eq. (3.3)): $c = 331.6 + 0.6 \times 84 = 382$ m/s.

Assume the pump noise is predominantly at the pumping frequency of 6000/60 = 100 Hz.

The wavelength $\lambda = 382/100 = 3.82$ m. For the shortest length chamber, $kl = \pi/2$ and so $l = \pi/2\,k = \pi/2\,(2\pi/\lambda) = \lambda/4 = 3.82/4 = 0.955$ m. Interpolating between the curves in Figure 10.11, select $m = 12$.

To calculate the TL for $m = 12$, Eq. (10.12) is used, with $\sin kl = \sin \pi/2 = 1$. Thus, TL for $(m = 12) =$ 10 log[1 + ¼ (12 – 1/12)²] = 10 log[1 + ¼ (12 – 0.083)²] = 10 log[1 + ¼ (11.92)²] = 10 log[1 + 142/4] = 10 log[36.5] = 15.6 dB. For a circular cross-section chamber with $m = 12$, the chamber diameter $d = 5.0 \times \sqrt{12} = 5 \times 3.46 = 17.3$ cm.

In order to obtain exact values for the expansion ratio m, and chamber diameter d, a quadratic equation must be solved. Writing Eq. (10.12) with $kl = \pi/2$

$$\text{TL} = 10 \log \left[1 + \tfrac{1}{4}\,(m - 1/m)^2\right],$$

and $m - 1/m = n$, which gives $(1/m)\,[(m^2 - nm) - 1] = 0$. Writing $n = 2\,(10^{\text{TL}/10} - 1)^{1/2}$, we have the solution for n in the quadratic

$$m = \tfrac{1}{2}\left[n + \left(n^2 + 4\right)^{1/2}\right]$$

and for TL = 15 dB,

$$n = 2\left(10^{15/10} - 1\right)^{1/2} = 11.07,$$

$$m = \tfrac{1}{2}\left[11.07 + \left(11.07^2 + 4\right)^2\right] = 11.16.$$

The negative root to the quadratic has no practical meaning, so the chamber diameter becomes:

$$d = 5.0 \times \sqrt{11.16} = 5 \times 3.34 = 16.7 \text{ cm}.$$

An expansion chamber diameter of 17.3 cm instead of 16.7 cm will give a small increase in TL of 0.6 dB.

b) TL Caused by a Side-Branch

Figure 10.12 shows a pipe with a side-branch located at $x = 0$.

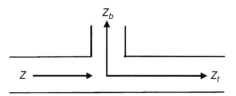

Figure 10.12   Acoustical conditions at the side-branch of input acoustical impedance Z_b.

In the inlet pipe (to the left of the side-branch), the pressure and volume velocity are

$$p = p_1^+ + p_1^-, \tag{10.13}$$

$$p = P_1^+ \, e^{-jkx} + P_1^- \, e^{jkx}, \tag{10.14}$$

and

$$V = V_1^+ + V_1^-, \tag{10.15}$$

$$V = (S_1/\rho c) \left(P_1^+ \, e^{-jkx} + P_1^- \, e^{jkx} \right). \tag{10.16}$$

At the branch junction, where $(x = 0)$, the sound pressure and volume velocity are assumed to be continuous, so that

$$p_1^+ + p_1^- = p_b = p_t, \tag{10.17}$$

where p_b is the pressure at the mouth branch and p_t is the transmitted pressure in pipe 3. Continuity of volume velocity gives:

$$V = V_1^+ + V_1^- = V_b + V_t, \tag{10.18}$$

where V_t is the transmitted volume velocity in pipe 3, assuming anechoic conditions in the tailpipe, pipe 3. Writing $S_1 = S_3 = S$,

$$V_1^+ = p_1^+ /(\rho c/S), V_1^+ = -p_1^- /(\rho c/S), V_t = p_t/(\rho c/S), \tag{10.19}$$

and $V_b = p_b/Z_b$ and dividing Eqs. (10.18) by (10.17), and using the relations in Eq. (10.19), we obtain:

$$1/Z = 1/Z_t + 1/Z_b, \tag{10.20}$$

which gives

$$Z = Z_b Z_t/(Z_b + Z_t), \tag{10.21}$$

where the Z, Z_t, and Z_b represent the system input, output, and branch impedances respectively (ratios of sound pressure to volume velocity). Figure 10.13 shows an electrical analogy of Eq. (10.21).

$$Z = (\rho c/S) \left(P_1^+ + P_1^- \right)/\left(P_1^+ - P_1^- \right), \tag{10.22}$$

and

$$Z_t = \rho c/S. \tag{10.23}$$

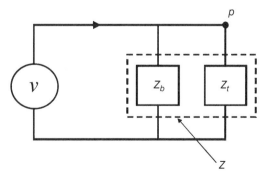

Figure 10.13 Electrical analogy of the side-branch system shown in Figure 10.12.

After rearrangement, we write the transmission coefficient α_t as

$$\alpha_t = (P_t/P_1^+)^2 = \left(R_b^2 + X_b^2\right) / \left[\,\left[\,(\rho c/2S) + R_b\,\right]^2 + X_b^2\,\right], \tag{10.24}$$

where $Z_b = R_b + jX_b$ and R_b and X_b are the real and imaginary parts of the impedance Z_b. The TL is then:

$$\text{TL} = 10 \log\,(1/\alpha_t) = 10 \log\,\left[\,\left[\,(\rho c/2S + R_b)^2 + X_b^2\,\right] / \left(R_b^2 + X_b^2\right)\,\right]. \tag{10.25}$$

c) Helmholtz Resonator

Consider a container of volume V_{hr} connected to a neck of length l. This is called a Helmholtz resonator (see Figure 10.14).

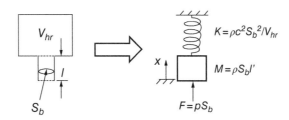

Figure 10.14 Helmholtz resonator with equivalent simple mechanical system.

The acoustical behavior of Helmholtz resonators is well known. Their use in noise control applications where good low-frequency sound absorption is required at a particular frequency is discussed in Chapter 8 of this book. The gas in the neck acts as a mass at low frequency and vibrates against the stiffness of the gas in the volume. Since the gas at each end of the neck of length l will also move, it is normal to add an end connection Δl at each end making the effective neck length $l' = l + 2\Delta l$.

The acoustic mass can thus be written as $M = \rho S_b l'$. The acoustic stiffness K can be written as $K = \rho c^2 S_b^2 / V_{hr}$, where S_b is the neck cross-sectional area, and V_{hr} is the resonator volume. The equation of motion of the gas in the neck is

$$M_b \ddot{x} + R_b \dot{x} + K_b x = p S_b, \tag{10.26}$$

and rearranging and assuming simple harmonic motion (but omitting the time variation $e^{j\omega t}$),

$$R_b + j\,(\omega M - K_b/\omega)u_b = p S_b. \tag{10.27}$$

Thus, the impedance is

$$Z_b = R_b + jX_b = p/S_b\,u_b = R_b + j\,\omega M - jK_b/\omega. \tag{10.28}$$

Setting $R_b = 0$, the impedance becomes:

$$Z_b = jX_b = p/Su_b = \left(\,j/S_b^2\,\right)(\omega M - K/\omega). \tag{10.29}$$

The natural frequency of the Helmholtz resonator is obtained when the reactive impedance is zero, thus

$$\omega = \sqrt{K/M} = c\sqrt{S_b/l'V_{hr}}, \tag{10.30}$$

where V_{hr} is the volume of the Helmholtz resonator.

Example 10.2 A 400 ml bottle has a 2.5 cm ($\approx$1 in.) diameter neck, which has an effective length l' of 6 cm. What is its resonance frequency when it is excited by blowing?

Solution

$$f_r = \frac{c}{2\pi}\sqrt{\frac{S_b}{l'V_{hr}}} = \frac{343}{2\pi}\sqrt{\frac{\pi(0.025)^2/4}{(0.06)(400\times10^{-6})}} = 247 \text{ Hz.}$$

It is seen that the resonance frequency is just 9 Hz lower than lower C on the piano or violin (256 Hz). This frequency is one octave lower than middle C (512 Hz). We note that the resonance frequency can be increased by reducing the volume in the bottle by adding water. To reach a note close to middle C (512 Hz), the volume would need to be reduced to 100 ml by adding 300 ml of water.

d) Helmholtz Resonator as a Side-Branch

Substituting $R_b = 0$ and the reactance X_b given by Eq. (10.29) into Eqs. (10.24) and (10.25) gives the TL:

$$TL = 10\log\left[1 + \left(\frac{c/2S}{\omega l'/S_b - c^2/\omega V_{hr}}\right)^2\right]. \tag{10.31}$$

We note that the TL $\to \infty$ at the resonance frequency $f_r = (c/2\pi)\sqrt{(S_b/l'V_{hr})}$. This type of side-branch resonator produces a useful broadband attenuation around this frequency. Davis et al. [22, 23, 64] give a series of curves for the side-branch Helmholtz resonator after Eq. (10.31) is reorganized with f/f_r as a parameter.

Example 10.3 Calculate the TL of the Helmholtz resonator as a side-branch for a volume of $V_{hr} = 0.00112$ m^3, $l = 6$ mm ($\approx$1/4 in.), $S_b = 7.5\times10^{-4}$ m^2, $S = 2.8\times10^{-3}$ m^2 over the range from 0 to 500 Hz. Assume $l' = l + 0.85a$, where a is the branch tube radius.

Solution

If $S_b = 7.5\times10^{-4}$ m^2 = 7.5 cm^2, then $a^2 = 7.5/\pi$, $a = 1.55$ cm. Now, $l' = 0.006 + 2\times(0.85)\times0.0155 = 3.23$ cm (assuming two end corrections). Substituting the given values for S, S_b, and V_{hr} into Eq. (10.31) and assuming $c = 343$ m/s and $l' = 3.23$ cm, we obtain Figure 10.15. From Eq. (10.30) we obtain $f_r = 249$ Hz. We observe that the TL is greater than 10 dB over a 75 Hz frequency range.

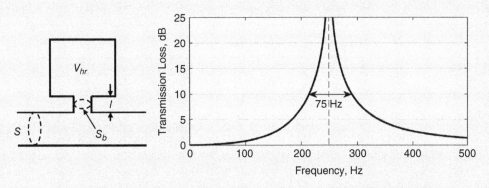

Figure 10.15 Calculated transmission loss of Helmholtz resonator in Example 10.3.

e) Quarter-Wave Resonator as a Side-Branch

We obtain the input impedance Z_b of a quarter-wave resonator tube by considering simple harmonic waves traveling in and out of the tube (length l, see Figure 10.16). Waves will be reflected at the rigid hard end at which the impedance $p/S_b u_b$ will be infinity. The input impedance (provided losses are ignored and $R_b = 0$) is

$$Z_b = jX_b = -(j\rho c/S_b)\cot(\omega l/c). \tag{10.32}$$

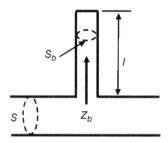

Figure 10.16 Quarter-wave resonator as a side-branch.

The input impedance is zero when $\omega l/c = n\pi/2$, $n = 1, 3, 5, \ldots$ thus, when $\omega_n = n\pi c/2l$, or $f_n = nc/4l$ and $l = nc/4f = n\lambda/4$. Thus the input impedance of a quarter-wave resonator as a side-branch has multiple values of high TL (unlike the Helmholtz resonator). These are given by $f_n = nc/4l$, $n = 1, 3, 5, \ldots$

The TL of a quarter-wave resonator as a side-branch is obtained by substituting Z_b in Eq. (10.32) into Eq. (10.25) giving

$$TL = 10\log\left[\frac{\tan^2(kl) + 4(S/S_b)^2}{4(S/S_b)^2}\right]. \tag{10.33}$$

The TL is seen to depend upon frequency $\omega = kc$ or $f = kc/2\pi$ and the ratio S/S_b.

Example 10.4   Calculate the length l of the quarter-wave resonator to make the first standing quarter-wave frequency 700 Hz. Also calculate the frequency corresponding to three and five standing quarter-waves.

Solution

If $f_n = nc/4l$, and putting $n = 1$, $l = (1/f_n)\times(1\times343)/4 = 343/(700\times4) = 0.1225$ m. f_n for $n = 3$ is $3\times700 = 2100$ Hz and f_n for $n = 5$ is $5\times700 = 3500$ Hz.

Example 10.5   Calculate the TL for the side-branch quarter-wave resonator for $l = 0.2$ m. Assume $S_b = S = 5\times10^{-5}$ m^2.

Solution

If $S_b = 2\times10^{-3}$ m$^2 = 20$ cm^2, then $a^2 = 20/\pi$, $a \approx 2.5$ cm. Assuming an effective length $l + 0.85a$, where a is the side-branch tube radius, then $l' = 20 + 2.5(0.85) = 22.1$ cm. Substituting the value for $l = l'$ and $S/S_b = 1$ into Eq. (10.33) gives the result in Figure 10.17. We note that when $kl' = n\pi/2$, $n = 1, 3, 5$, then the TL $\to \infty$.

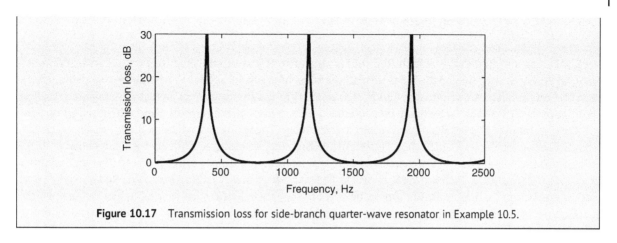

Figure 10.17 Transmission loss for side-branch quarter-wave resonator in Example 10.5.

f) Orifice as a Side-Branch

Consider a short length of pipe of length l as a branch. The pipe radius a and length l are assumed to be small compared to the wavelength λ. The real and imaginary parts of the branch impedance of the orifice pipe are

$$Z_b = \rho c k^2/2\pi + j\rho l'\omega/\pi a^2, \tag{10.34}$$

where the effective length of the short pipe is $l' = l + 2 \times (0.85)a = = l + 1.7a$.

The TL is

$$TL = 10 \log \left[\frac{\left(\rho c/2S + \rho c k^2/2\pi\right)^2 + \left(\rho l'\omega/\pi a^2\right)^2}{\left(\rho c k^2/2\pi\right)^2 + \left(\rho l'\omega/\pi a^2\right)^2} \right]. \tag{10.35}$$

The ratio of acoustic resistance R_b to acoustic reactance X_b is

$$R_b/X_b = \left(\rho c k^2/2\pi\right)/\left(\rho l'\omega/\pi a^2\right) = k a^2/2\, l'. \tag{10.36}$$

Since the tube length l and radius a are assumed to be much less than the wavelength λ, then $ka \to 0$. By neglecting losses in the orifice and setting $R_b << X_b$, the term $\rho c k^2/2\pi$ can be neglected. The TL becomes:

$$TL = 10 \log \left[1 + \left(\pi a^2/2Sl'k\right)^2 \right]. \tag{10.37}$$

Example 10.6 A small tube-axial fan has eight blades and is running at a constant speed of 750 rpm. The fan is operating in a tube of inside diameter 100 mm and thickness 2 mm. A hole of diameter 30 mm is drilled in the tube's wall far from the fan. If the speed of sound inside the tube is $c = 344$ m/s, determine the TL provided by this orifice as a side-branch at the blade passing frequency (BPF) of the fan.

Solution

The BPF of the fan is calculated using Eq. (11.7) as $f = 8 \times 750/60 = 100$ Hz. Then, $k = 2\pi(100)/344 = 1.8265\ \text{m}^{-1}$. Now, $S = \pi(0.1/2)^2 = 0.0079\ \text{m}^2$ and $a = 0.03/2 = 0.015\ \text{m}$. The effective length of the short pipe is $l' = 2 + 1.7\,(15) = 27.5\ \text{mm} = 0.0275\ \text{m}$. Since $ka = 0.0274 << 1$, we can use Eq. (10.37): $TL = 10 \log [1 + (\pi \times 0.015^2/2 \times 0.0079 \times 0.0275 \times 1.8265))^2] = 10 \log [1 + 0.7933] = 10 \log (1.7933) = 2.5\ \text{dB}$.

10.6.3 NACA 1192 Study on Reactive Muffler TL

The extensive study on reactive mufflers published by Davis et al. as NACA 1192 is a very important seminal work [21]. See also Ref. [64] in which Ref. [21] is reprinted. It was republished in Ref. [22] with additional material and discussion included. It has formed the basis of much subsequent research on reactive mufflers used on automobiles and in industry. Davis et al. studied 77 different mufflers. Measurements of TL (called attenuation by Davis et al.) were made on each one with a loudspeaker as a source and with a reflection-free tailpipe termination. The results were compared with theory. The mufflers studied included expansion chambers, side-branch resonators including Helmholtz resonators and combinations of these elements.

Figure 10.18a shows the effect of increasing the expansion ratio m from 4 to 16. It is seen that increasing m (see Eq. (10.12)) causes an increase in the peak TL. The plane wave theory fails above about 690 Hz, where cross modes begin to appear. Figure 10.18b shows the effect of keeping the ratio m constant, but increasing the muffler length. It is noted that the TL drops to zero when the muffler length ℓ is equal to one half a wavelength $\lambda/2$ or for frequencies $f = nc/2\ell$. As the length ℓ increases, the first peak in the TL keeps constant at about 20 dB, but continually moves to a lower frequency. Further length increase results in additional peaks occurring.

Figure 10.19 shows the effect of connecting together two expansion chamber mufflers of the same expansion ratio $m = 16$. Figure 10.19a shows the effect of different length external connection tubes, while Figure 10.19b shows the effect of using different length internal connection tubes. Muffler 12, which appears in both Figure 10.19a and b, can be considered to have connection tubes of zero length. If a muffler 48 in. (1.25 m) long could be acceptable, muffler 19 could be used to provide about 30–40 dB attenuation over a broad frequency range up to 500 Hz.

Figure 10.20a shows the effect of using connecting tubes of length equal to the individual chamber lengths. The connecting tubes in each case are terminated at the center of each chamber.

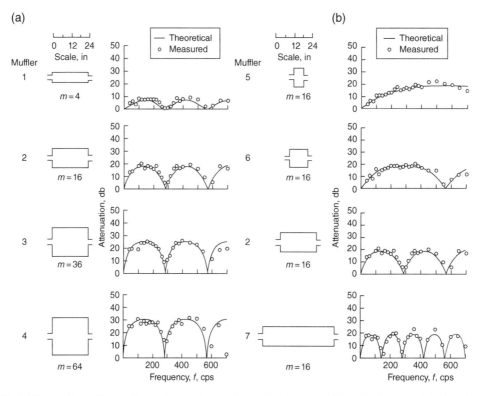

Figure 10.18 (a) Comparison of theoretical and experiment attenuation characteristics – single expansion chamber mufflers – showing effect of expansion ratio m; (b) effect of length, keeping m expansion ratio constant $m = 16$ [21, 64].

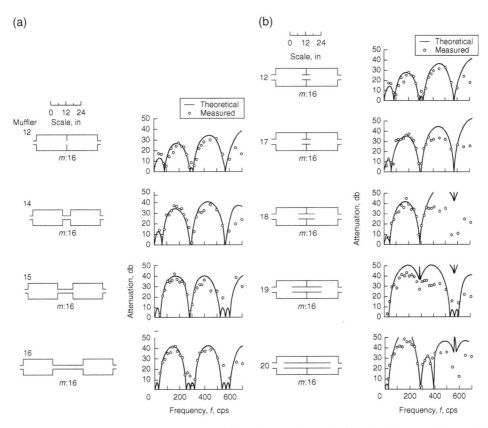

Figure 10.19 Multiple-expansion-chamber mufflers. (a) Effect of connecting-tube length with external connecting tubes. (b) Effect of connecting-tube length with internal connecting tubes [21, 64].

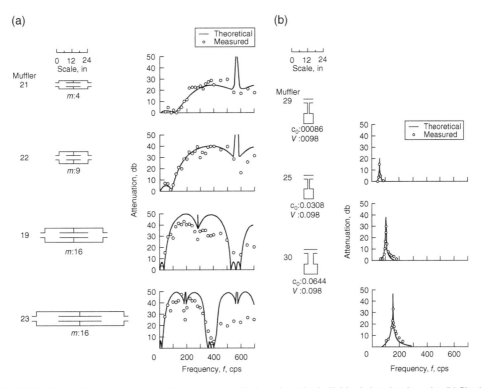

Figure 10.20 (a) Mufflers with internal connecting tubes equal in length to the individual chamber lengths. (b) Single-chamber resonators with resonator chambers separate from tailpipe [21, 64].

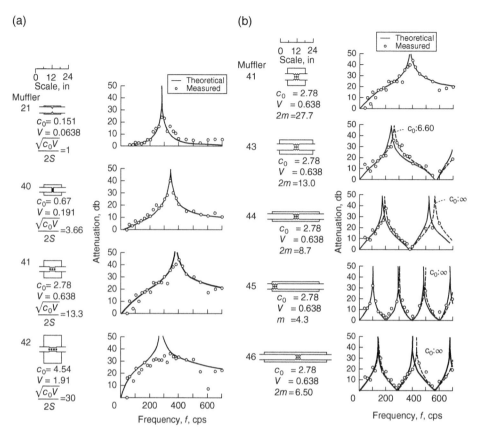

Figure 10.21 Effect of varying the conductivity c_0 and the tube length of concentric tube resonators. (a) variation in attenuation with changes in conductivity, (b) variation in attenuation with changes in length for a constant conductivity [21, 64].

Muffler 19 is shown in both Figures 10.19b and 10.20a. The mufflers in Figure 10.20a can all produce high broadband attenuation of 30 dB or more, but the shorter mufflers have a poorer performance than muffler 19 at frequencies below 200 Hz. Figure 10.20b shows the attenuation provided by side-branch Helmholtz resonators of the same volume but with different diameter necks.

In Figure 10.21, Davis et al. investigated the performance of single-chamber concentric resonators. In Figure 10.21a, the effect of varying the attenuation parameter $\sqrt{c_0 V}/2S$ while keeping the chamber length constant is shown. Here c_0 is the conductivity, V the volume of the concentric resonator, and S the cross-sectional area of the connecting tubes. Figure 10.21b shows the effect of varying the chamber length while keeping both the conductivity c_0 and the resonator volume constant.

Figure 10.22 shows the effect of varying the conductivity c_0 of the connecting elements (tubes or orifices) of double resonator mufflers while the volume V is kept constant. Figure 10.22a shows the effect of varying the conductivity c_0 when small tubes are used as the connecting elements between the exhaust tube and the concentric resonators. Figure 10.22b shows the effect when orifices are used as the connecting elements instead of small tubes. The considerable increase in the attenuation (TL) that occurs when orifices are used instead of small tubes is observed when the results of Figure 10.22b are compared with those in Figure 10.22a.

Figure 10.23 shows the attenuation (TL) of combination mufflers comprised of expansion chambers and concentric tube Helmholtz resonators. Mufflers 73 and 73R (shown in Figure 10.23) show that the order of the double chamber resonators is relatively unimportant. The two peaks are caused by the two different types of chamber. The attenuation caused by these mufflers is over a broad frequency range.

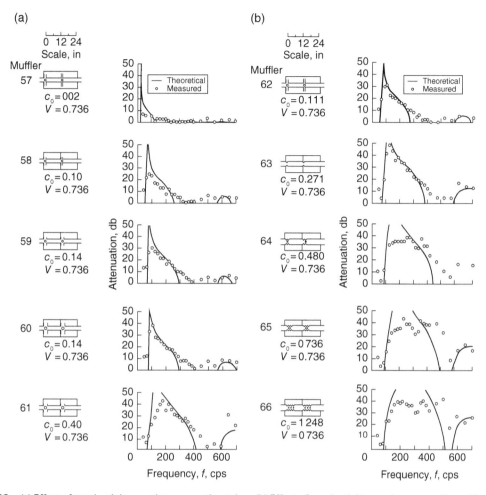

Figure 10.22 (a) Effect of conductivity c_0 using connecting tubes. (b) Effect of conductivity c_0 using connecting orifices [21, 64].

10.6.4 Transfer Matrix Theory

An alternative approach is to assume that the sound pressure p and volume velocity V at stations 1 and 2 in Figure 10.9 can be related by:

$$p_1 = Ap_2 + BV_2, \tag{10.38}$$

and

$$V_1 = Cp_2 + DV_2. \tag{10.39}$$

An electrical circuit analogy can be used where the pressure p is analogous to voltage and volume velocity V to current. This is known as the impedance analogy. Note that an alternative mobility analogy is sometimes used [5]. The circuit element can be represented by the four-pole element shown in Figure 10.24. If the muffler section is simply a rigid straight pipe (See Figure 10.9.) of constant cross-section, then from Eqs. (10.2b) and (10.3b), the sound pressure and volume velocity at stations 1 and 2 ($x = 0$ and $x = L$) are:

$$p_1 = p^+ + p^-, \tag{10.40}$$

$$p_2 = P^+ e^{-jkL} + P^- e^{jkL}, \tag{10.41}$$

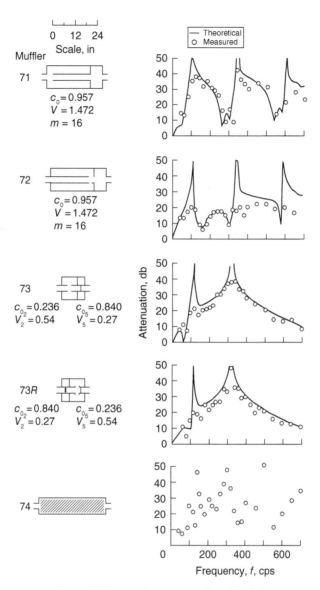

Figure 10.23 Combination mufflers [21, 64].

Figure 10.24 Four-pole representation of muffler element.

$$V_1 = (S/\rho c)\,(p^+ - p^-),\qquad\qquad (10.42)$$

$$V_2 = (S/\rho c)\,\left(P^+\, e^{-jkL} - P^-\, e^{jkL}\right).\qquad\qquad (10.43)$$

The parameters A, B, C, and D may be evaluated using a "black box" system identification technique. To evaluate A and C, assume that the matrix output terminals are open circuit, or $V_2 = 0$.

Then Eq. (10.43) gives $P^+/P^- = e^{j2kL}$ and Eqs. (10.38) and (10.39) give

$$A = p_1/p_2 \text{ and } C = V_1/p_2.$$

Using this result for P^+/P^-, and Eqs. (10.40)–(10.42) after some manipulation, it is found that

$$A = \cos kL \text{ and } C = j(S/\rho c) \sin kL.$$

Similarly, to evaluate B and D assume that the matrix output terminals are short-circuited and $p_2 = 0$. Then Eq. (10.43) gives $P^+/P^- = e^{j2kL}$ and Eqs. (10.38) and (10.39) give

$$B = p_1/V_2 \text{ and } D = V_1/V_2.$$

Using this result for P^+/P^- and Eqs. (10.40), (10.42) and (10.43), it is found that

$$B = j(\rho c/S) \sin kL \text{ and } D = \cos kL.$$

Substituting these results for A, B, C, and D into Eqs. (10.38) and (10.39) and writing them in matrix form gives

$$\begin{bmatrix} p_1 \\ V_1 \end{bmatrix} = \begin{bmatrix} A & B \\ C & D \end{bmatrix} \begin{bmatrix} p_2 \\ V_2 \end{bmatrix}, \tag{10.44}$$

where the four-pole constants (for a straight pipe of length L) are:

$$\begin{bmatrix} A & B \\ C & D \end{bmatrix} = \begin{bmatrix} \cos kL & j(\rho c/S) \sin kL \\ j(S/\rho c) \sin kL & \cos kL \end{bmatrix}. \tag{10.45}$$

Note that $AD - BC = 1$. This is a useful check on the values derived of the four-pole parameters and is a consequence of the fact that the system obeys the reciprocity principle [5]. The matrix in Eq. (10.45) relates the total sound pressure and volume velocity at two stations in a straight pipe.

If several component systems are connected together in series, as in Figure 10.25 then the transmission matrix of the complete system is given by the product of the individual system matrices:

$$\begin{aligned} \begin{bmatrix} p_1 \\ V_1 \end{bmatrix} &= \begin{bmatrix} A_1 & B_1 \\ C_1 & D_1 \end{bmatrix} \begin{bmatrix} p_2 \\ V_2 \end{bmatrix} \\ &= \begin{bmatrix} A_1 & B_1 \\ C_1 & D_1 \end{bmatrix} \begin{bmatrix} A_2 & B_2 \\ C_2 & D_2 \end{bmatrix} \begin{bmatrix} p_3 \\ V_3 \end{bmatrix} \\ &= \begin{bmatrix} A_1 & B_1 \\ C_1 & D_1 \end{bmatrix} \begin{bmatrix} A_2 & B_2 \\ C_2 & D_2 \end{bmatrix} \begin{bmatrix} A_3 & B_3 \\ C_3 & D_3 \end{bmatrix} \begin{bmatrix} p_4 \\ V_4 \end{bmatrix}. \end{aligned} \tag{10.46}$$

This matrix formulation is very convenient particularly where a digital computer is used. The four-pole constants A, B, C, and D can be found easily for simple muffler elements such as expansion chambers and straight pipes as has just been shown (see Eq. (10.45)). They can also be found in a similar manner for more complex muffler shapes (reversing end chambers and reversing end-chamber/Helmholtz resonator combinations) by the finite element method using the same black box identification technique mentioned above (by alternately setting $p_2 = 0$ and $V_2 = 0$).

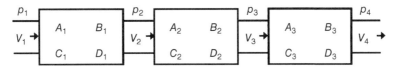

Figure 10.25 Series connection of transmission matrices.

10.7 Exhaust System Modeling

It will now be shown that for any linear passive muffler element that the TL is a property only of the muffler geometry (i.e. four-terminal constants *A*, *B*, *C*, and *D*) and unaffected by connection of subsequent muffler elements or source or load impedances. On the other hand, it will be shown that the IL is affected by the source and load impedances. Finally, if it is desired to predict the SPL outside of the tail pipe it is necessary to have a knowledge not only of the source impedance and load impedance but also of the source strength – either sound pressure or volume velocity.

The TL of a muffler is the quantity most easily predicted theoretically and is certainly of guidance in muffler design. However, either IL or a prediction of the sound pressure radiated from the tail pipe is much more useful to the muffler designer and these are now discussed.

10.7.1 Transmission Loss

The source-muffler-termination system may be modeled as an equivalent electrical circuit [5, 30, 62, 63, 65]. The velocity source model in Figure 10.26b will be used in the derivations of TL (although the pressure source model gives the same result). For simplicity, the mean flow is ignored and the Mach number set to zero, $M = 0$. The cross-sectional areas of the muffler inlet and outlet pipes S_0 are assumed to the equal and there is no mean temperature gradient in the muffler system. To determine the TL, the incident and transmitted pressure amplitudes $|p_1^+|$ and $|p_2^+|$ are needed. The transmitted pressure $|p_2^+|$ is most easily determined by making the tail pipe non-reflecting ($Z_r = \rho c/S_0$). Thus $|p_2^-| = 0$.

From Figure 10.26b (see Eqs. (10.2a), (10.3a), and (10.3b)):

$$p_1 = p_1^+ + p_1^-, \tag{10.47}$$

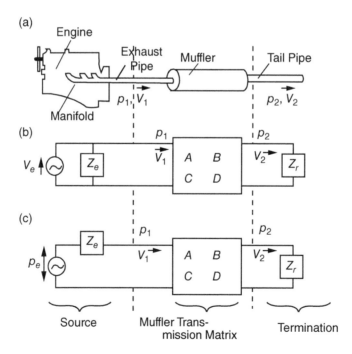

Figure 10.26 (a) Real engine-muffler exhaust system; (b) Volume velocity analog; (c) Pressure source analog [5].

$$V_1 = (S_0/\rho c)\left(p_1^+ - p_1^-\right), \tag{10.48}$$

$$V_2 = (S_0/\rho c)p_2^+, \tag{10.49}$$

and from Eqs. (10.38), (10.39) and (10.44):

$$p_1^+ + p_1^- = Ap_2^+ + Bp_2^+(S_0/\rho c), \tag{10.50}$$

$$(S_0/\rho c)\left(p_1^+ + p_1^-\right) = Cp_2^+ + Dp_2^+(S_0/\rho c). \tag{10.51}$$

From the definition in Figure 10.4b:

$$TL = 10\log\frac{\left|p_1^+\right|^2/\rho c}{\left|p_2^+\right|^2/\rho c} = 20\log\frac{\left|p_1^+\right|}{\left|p_2^+\right|}. \tag{10.52}$$

Then eliminating p_1^- in Eqs. (10.50) and (10.51) and substituting into Eq. (10.52) gives:

$$TL = 20\log\left\{\frac{1}{2}\left|A + B(S_0/\rho c) + C/(S_0/\rho c) + D\right|\right\}. \tag{10.53}$$

Equation (10.53) is a similar result to that obtained by Young and Crocker [46]. Except note that in Ref. [46] particle velocity was used instead of volume velocity and so A, B, C, and D have slightly different definitions. Sullivan [30] has also derived a result similar to Eq. (10.53) in which the mean temperature, cross-sectional area and mean flow in pipes 1 and 2 are different.

The TL is convenient to predict but somewhat inconvenient to determine experimentally. With some care it is possible to construct an anechoic termination from an absorbently-lined horn or use of absorbent packing [47] enabling $\left|p_2^+\right|$ to be measured directly. The quantity p_1^+ can also be determined when the source (in Figure 10.26) is a loudspeaker, by measuring the standing wave in the exhaust pipe, using a microphone probe tube (although it is a laborious process). The TL can be measured for an absorbent silencer in this way. However, if the TL is determined in the "real-life" situation with an automobile engine as a source, the microphone probe tube is placed under severe environmental conditions of high temperature and moisture condensation. Alternatively, as suggested by Seybert and Ross, the TL can be measured using two microphones instead of a probe tube [61]. However, if an automobile engine tail pipe anechoic termination is used, it must be of special design to withstand the high temperature. The IL is of much more practical interest and much easier to measure when an engine is a source. It is also easy to measure with absorbent silencers in ducted systems. The measurement of the IL is discussed next.

10.7.2 Insertion Loss

Using Figure 10.26b again gives:

$$V_1 = V_e - p_1/Z_e, \tag{10.54}$$

$$V_2 = p_2/Z_r, \tag{10.55}$$

where Z_e and Z_r are the source internal impedance and tail pipe or exhaust duct radiation impedance, respectively. Then from Eq. (10.44):

$$p_1 = Ap_2 + Bp_2/Z_r, \tag{10.56}$$

$$V_1 = Cp_2 + Dp_2/Z_r. \tag{10.57}$$

Substituting for V_1 from Eqs. (10.54) into (10.57) and combining Eqs. (10.56) and (10.57) to eliminate p_1 gives:

$$p_2 = \frac{Z_e Z_r V_e}{A Z_r + B + C Z_e Z_r + D Z_e}. \tag{10.58}$$

If a different muffler with four-terminal parameters A', B', C', and D' is now connected to the engine, a new pressure p_2' results:

$$p_2' = \frac{Z_e Z_r V_e}{A' Z_r + B' + C' Z_e Z_r + D' Z_e}. \tag{10.59}$$

Thus

$$\frac{p_2'}{p_2} = \frac{A Z_r + B + C Z_e Z_r + D Z_e}{A' Z_r + B' + C' Z_e Z_r + D' Z_e}. \tag{10.60}$$

This result is similar to that obtained by Sullivan [30]. If p_2' is measured with no muffler in place and only a short (in wavelengths) exhaust pipe, then $A' = D' = 1$, and $B' = C' = 0$,

$$\frac{p_2'}{p_2} = \frac{A Z_r + B + C Z_e Z_r + D Z_e}{Z_e + Z_r}. \tag{10.61}$$

This result is similar to that obtained in Eq. (10.53). Since

$$IL = 20 \log |p_2'/p_2|,$$

it is seen from either Eqs. (10.60) or (10.61) that unlike the *TL*, *IL* depends on both the internal impedance of the source and the tail pipe radiation impedance, besides the transmission characteristics of the muffler itself. Several researchers in the 1970s predicted the *IL* of mufflers installed on engines, e.g. Young [45] and Davies [66]. However, they have normally had to rely on assumed values of engine impedance (e.g. $Z_e = 0$, $\rho c/S_0$, etc.), since measured values were not available. Young's results for *IL* [45], will be discussed later in Section 10.10.

In prediction of IL, Z_r must also be known. Discussion on the problems of estimating Z_e and Z_r follows in a later section.

If the source and radiation impedances are assumed to be $Z_e = Z_r = \rho c/S_0$, then Eq. (10.61) becomes:

$$\frac{p_2'}{p_2} = \frac{A(\rho c/S_0) + B + C(\rho c/S_0)^2 + D(\rho c/S_0)}{2\rho c/S_0}, \tag{10.62}$$

$$IL = 20 \log |p_2'/p_2|,$$

$$IL = 20 \log \left\{ \frac{1}{2} |A + B(S_0/\rho c) + C/(S_0/\rho c) + D| \right\}, \tag{10.63}$$

a result identical to Eq. (10.53). This demonstrates the fact that, in the general case, the muffler TL is not equal to the IL except when the IL is measured with source and termination impedances equal to the characteristic acoustic impedance $\rho c/S_0$. The same conclusion can be reached intuitively or theoretically. However, it is more difficult to reach this conclusion by studying traveling wave solutions (transmission line theory) in mufflers and the exhaust and tail pipes than by using the transmission matrix theory.

10.7.3 Sound Pressure Radiated from Tailpipe

A prediction of this quantity is of probably more importance to muffler designers than a knowledge of either TL or IL. After all, the radiated SPL is the quantity which finally determines the acceptability of a muffler. Examining Eq. (10.58), shows that if the source volume velocity source strength V_e, source impedance Z_e, radiation impedance Z_r and muffler four-terminal (four-pole) parameters A, B, C, and D are known, then the total sound pressure amplitude (and phase) at the end of the tail pipe p_2 can be calculated. It is a fairly simple matter to calculate the radiated pressure amplitude p_r at distance r from the tail pipe or flow duct outlet [39, 40, 42]. For engine exhaust systems, the method used is to assume monopole radiation from the tail pipe so that the net sound intensity transmitted out of the tail pipe is equal to the intensity in the diverging spherical wave at radius r. This gives:

$$2\pi a^2 \left(\frac{|p_2^+|^2}{2\rho_2 c_2} \right) \{(1 + M)^2 - (1 - M)^2 R^2(M)\} = 4\pi r^2 |p_r|^2 / 2\rho_0 c_0, \qquad (10.64)$$

where a is the tail pipe radius and $R(M)$ the tail pipe reflection coefficient (dependent on Mach number M) of the mean flow. Subscript 2 refers to conditions just inside the tail pipe. From Eqs. (10.40) and (10.42), at any station in the muffler:

$$2p^+ = p + (\rho c / S_0) V, \qquad (10.65)$$

and at the tail pipe exit:

$$p_2 = V_2 Z_r. \qquad (10.66)$$

Thus, at the tail pipe or duct exhaust exit, from Eqs. (10.65) and (10.66):

$$p_2 = \frac{2p_2^+}{1 + (\rho c / S_0) / Z_r} \qquad (10.67)$$

and substituting Eqs. (10.67) into (10.58) gives:

$$p_2^+ = \frac{V_e Z_e (Z_r + \rho c / S_0)}{2(A Z_r + B + C Z_e Z_r + D Z_e)}. \qquad (10.68)$$

Taking the modulus of Eq. (10.68) and substituting it into Eq. (10.64) eliminates p_2^+ and gives the pressure $|p_r|$ in terms of the source volume velocity, V_e, the engine and tail pipe radiation impedances, Z_e and Z_r, the muffler four-pole parameters, the tail pipe reflection coefficient $R(M)$ and the mean-flow Mach number in the tail pipe, M.

10.8 Tail Pipe Radiation Impedance, Source Impedance and Source Strength

10.8.1 Tail Pipe Radiation

Early work on engine exhaust mufflers was hampered by a lack of knowledge of the reflection of waves at the end of the tail pipe. As Alfredson and Davies discuss [39], various assumptions have been made in the past about the magnitude and phase of the reflection (some researchers assumed the reflection coefficient R was zero and some, one). In 1948, Levine and Schwinger [67] published a rigorous, lengthy theoretical derivation of the reflected wave from an unflanged circular pipe. The solution assumes plane wave propagation in the pipe and no mean flow. In 1970, Alfredson measured the reflection coefficient R and phase angle θ of waves in an engine tail pipe using the engine exhaust as the source signal. The motivation was to determine if a mean flow and an elevated temperature had a significant effect on the zero flow reflection coefficient and phase calculated by Levine and Schwinger. Both the theoretical results of Levine and Schwinger and Alfredson's experimental results are given in Figure 10.27. Alfredson's experimental results show only a 3–5% increase in the reflection coefficient and virtually no change in the phase angle, as the flow and temperature increase to those conditions found in a typical engine tail pipe. Either Alfredson's or Levine and Schwinger's results for R

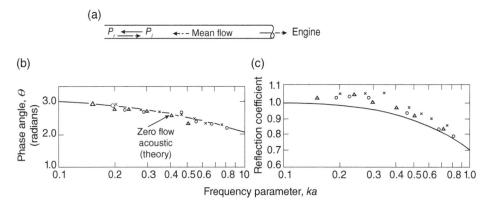

Figure 10.27 Reflection of sound. (a) exhaust tail pipe, $P_r = R \times P_i \, e^{j\theta}$. (b) phase angles vs. frequency parameter. (c) reflection coefficient vs. reflection parameter. —, theoretical zero flow acoustic; ○, $M = 0.078$; △, $M = 0.11$; ×, $M = 0.17$ [39].

and θ can be used to determine the tail pipe radiation impedance Z_r used in IL or sound pressure predictions (Eqs. (10.60) and (10.61) or (10.58) and (10.68)).

The sound pressure and volume velocity can be written as follows:

$$p_2 = p_2^+ + p_2^- = p_2^+ \left(1 + R\, e^{j\theta}\right), \tag{10.69}$$

and

$$V_2 = (S_0/\rho_2 c_2)\left(p_2^+ - p_2^-\right) = p_2^+ \, (S_0/\rho_2 c_2)\left(1 - R\, e^{j\theta}\right). \tag{10.70}$$

Then the ratio of the pressure and volume velocity at the tail pipe exit is given by dividing Eqs. (10.69) by (10.70). The radiation impedance Z_r is calculated from Eq. (10.71).

$$Z_r = p_2/V_2 = (\rho_2 c_2/S_0)\frac{\left(1 + R\, e^{j\theta}\right)}{\left(1 - R\, e^{j\theta}\right)}. \tag{10.71}$$

10.8.2 Internal Combustion Engine Impedance and Source Strength

Until the late 1970s, values of engine impedance used in predictions had been completely speculative. Values of Z_e, of ∞, $\rho c/S$, and 0 were assumed by various workers in making IL calculations. Other experimenters tried to simulate these different values in their idealized experimental arrangements. Values of $Z_e = \infty$, and 0, correspond to constant-volume velocity (current) and constant pressure (voltage) sources, respectively. Suppose the muffler and termination impedances shown in Figure 10.26 are lumped together as a load impedance, then Figure 10.26b and 10.26c reduce to Figure 10.28a and 10.28b, respectively.

In the case of a volume velocity source, $V_1 = V_e Z_e/(Z_e + Z_l)$ and if the internal impedance $Z_e \to \infty$, $V_1 \to V_e$. Thus, a constant-volume velocity is supplied to the load, independent of its value (provided it remains finite). When $Z_e \to \infty$, this source is known as a *constant-volume velocity source*.

In the case of a pressure source, $p_1 = p_e Z_l/(Z_e + Z_l)$ and if the internal impedance $Z_e \to 0$, $p_1 \to p_e$. Thus, a constant sound pressure is supplied to the load terminals, independent of the load impedance value (provided it remains finite also). When $Z_e \to 0$, this source is known as a *constant pressure source*. Note that if $Z_e = \rho c/S$ in either model, that constant sources are not obtained in either model. These constant-volume velocity and constant pressure sources are equivalent to constant current and constant voltage sources which are well known in electrical circuits (see, e.g. Reference [68]).

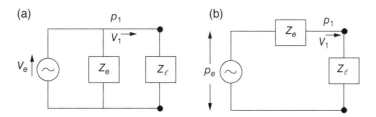

Figure 10.28 (a) Volume velocity source; (b) Pressure source.

In the case of automobile engines, it is of course unlikely, in practice, that the engine impedance is approximately 0, $\rho c/S$, or ∞. However, it could approach one of these values in certain frequency ranges. Some have even questioned the meaning of engine impedance since it must vary with time as exhaust ports open and close. There are at least three approaches to model the automobile engine source characteristics. Without directly using the concept of engine impedance as such, Mutyala and Soedel [69, 70], have used a mathematical model of a single-cylinder two-stroke engine connected to a simple expansion chamber muffler. The passages and volumes are treated as lumped parameters and kinematic, thermodynamic and mass balance equations are used. Good agreement between theory and experiment was obtained for the radiated exhaust noise.

Galaitsis and Bender [71] used an empirical approach to measure engine impedance directly. Using an electromagnetic pure tone source and by measuring standing waves in an impedance tube connected to a running engine they were able to determine the engine internal impedance. At low rpm the impedance fluctuated with frequency. However, at high rpm the impedance approached $\rho c/S$ at high frequency. Ross [72] has also used a similar technique.

A third approach to the determination of engine impedance (and source strength) is the two-load method. This method is well known in electrical circuit theory.

Using the pressure source representation (see Figure 10.28b) and two different known loads Z_l and Z'_l, two simultaneous equations are obtained:

$$p_1 = \frac{p_e Z_l}{Z_e + Z_l},$$ (10.72)

$$p'_1 = \frac{p_e Z'_l}{Z_e + Z'_l}.$$ (10.73)

Eliminating p_e in Eqs. (10.72) and (10.73) gives:

$$Z_e = \frac{p_1 - p'_1}{p'_1/Z'_l - p_1/Z_l}$$ (10.74)

Substitution of Z_e in Eqs. (10.72) or (10.73) now gives the source strength p_e. Kathuriya and Munjal suggest using two different length pipes so that there is little change in back pressure and so that (presumably) the load impedances, and Z_l (comprised of straight pipe and radiation impedance) are well known [65]. In order to remove the necessity to measure p_l inside the tail pipe (where the exhaust gas is hot) it should be possible to measure the sound pressure radiated from the tail pipe p_r since this can be related to the pressure p_l in the straight pipe by equations, such as (10.64) and (10.67). Egolf [73] has used this two-load method in the design of a hearing aid. Sullivan [30] has discussed the limitations of the method. Methods of measuring the internal impedance of sources are discussed further in Section 10.11.

10.9 Numerical Modeling of Muffler Acoustical Performance

10.9.1 Finite Element Analysis

Young and Crocker [44–48] were the first to use finite element analysis in muffler design. So far in this chapter it has been assumed that the mufflers are axi-symmetric in shape and acoustic filter theory [21, 22] provides a sufficient theoretical explanation for the behavior of muffler elements. This filter theory is normally based on the plane wave assumption. However, when a certain frequency limit is reached (known as the cutoff frequency), the filter ceases to behave according to plane wave theory. (This cutoff frequency is usually proportional to the pipe or chamber diameter.) In addition, if the muffler element shape is complicated, the simple plane wave assumptions and the boundary conditions are difficult to apply.

In Young and Crocker's work, first provided to a sponsor in 1971 [44], a numerical method was produced to predict the TL of complicated-shaped muffler elements. In this approach, variational methods were used to formulate the problem instead of using the wave equation. The theoretical approach is described in detail in [44–48] and will not be given here. It is assumed that the muffler element is composed of a volume V of perfect gas with a surface area S. The surface S is composed of two parts: one area over which the normal acoustic displacement is prescribed and the other area over which the pressure is prescribed. The pressure field in the muffler element is solved by making the Langrangian function stationary [44–46]. Thus, this approach is essentially an approximate energy approach. The muffler element is divided into a number of subregions (finite elements). At the corners of the elements the acoustic pressure and volume velocity are determined. The four-pole parameters A, B, C, and D, relating the pressure and volume velocity before and after the muffler element, are obtained in a similar manner to that described above assuming that the matrix output terminals are alternately open-circuited or short-circuited [44–46].

At the corners of the elements the sound pressure and volume velocity are determined. The four-pole parameters A, B, C, D, relating the pressure and volume velocity before and after the muffler element are obtained in a similar manner to that described above assuming that the matrix output terminals are alternately open-circuited or short-circuited [44–46].

In order to check the finite element approach and computer program, it was first applied to the classical expansion chamber case. See Figure 10.29. The dimensions of the simple expansion chamber used are given

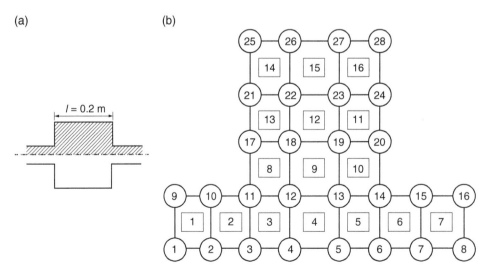

Figure 10.29 (a) simple expansion chamber (b) simple expansion chamber, showing division into 16 finite elements and 28 node points.

in Figure 10.29. The chamber was 8 in. (0.20 m) long and 10 in. (0.25 m) in diameter. Since the chamber was symmetrical, only half the chamber was represented with finite elements. Three finite element models were studied. The first had 8 elements with 16 nodal points, the second had 16 elements with 28 nodal points (see Figure 10.29). The third had 24 elements with 38 nodal points.

Figure 10.30 shows the TL predicted by the three finite element models and by the classical theory for an expansion chamber (see Eq. (10.12)). It shows the rapid convergence of the finite element approximation. Eight elements are insufficient to predict the TL, although the TL predicted by 16 or 24 elements is about the same. Note, however, that above about 1100 Hz, the classical theory and the finite element *TL* predictions diverge. Above this frequency the chamber-diameter-to-wavelength-ratio becomes less than 0.8 and higher modes are present in addition to plane waves, can exist in the expansion chamber. However, the classical theory (Eq. (10.12)) only predicts the plane wave performance.

Having shown that the finite element program could be used to predict TL successfully on known chambers, it was now used to examine chambers such as reversing flow end chambers (see Figure 10.31). Then it

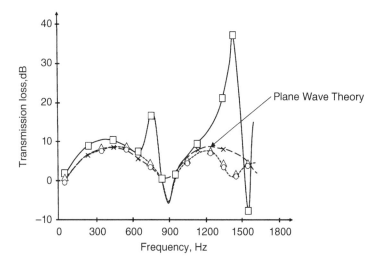

Figure 10.30 Transmission loss of simple expansion chamber [44–46]. ×, Plane wave acoustical filter theory. Finite element theory: □, 8 elements; △, 16 elements; ○, 24 elements.

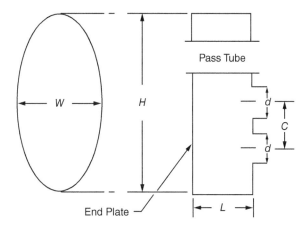

Figure 10.31 Flow-reversing chamber with pass tube and end plate. c: distance between centers of inlet and outlet tubes; H: height of chamber; L: length of chamber; W: width of chamber; and d: pipe diameter [44–46].

was used with end-chamber Helmholtz-resonator combinations and finally mufflers comprised of combinations of straight pipes, end chambers and up to two Helmholtz resonators.

A typical end chamber examined is shown in Figure 10.31. The measurement of TL was based on the standing wave method, see Figure 10.32. An acoustical driver (*H*) was used to supply a pure tone signal and the standing wave in the test section (*J*) was measured with the microphone probe tube (*I*). Using standing wave theory, the amplitude of the incident wave was determined by measuring the maxima and minima of the standing wave at different frequencies. The transmitted wave was determined by a single microphone (*M*), since the reflections were minimized by the anechoic termination (*L*). A steady mean air flow could be supplied to the plenum chamber (*G*) and was used to investigate flow effects on TL in some experiments.

Figures 10.33 and 10.34 show the predicted and measured TL of two different shape reversing end chambers, with and without a mean air flow of 110 ft/s (33.5 m/s). Neither end chamber examined had a pass tube. The first chamber has side-in side-out (SI-SO) tubes and the second side-in center out (SI-CO) tubes. It is

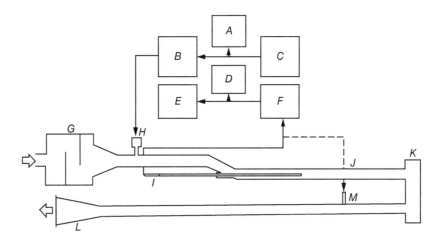

Figure 10.32 Experimental system for measuring the transmission loss. *A*: frequency counter; *B*: amplifier; *C*: frequency oscillator; *D*: oscilloscope; *E*: level recorder; *F*: spectrometer; *G*: plenum chamber; *H*: acoustical driver; *I*: microphone probe; *J*: standing wave tube; *K*: flow-reversing chamber; *L*: anechoic termination and *M*: microphone port [44–46].

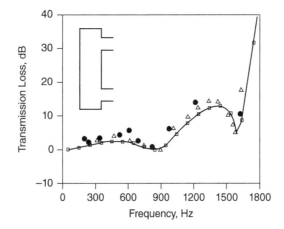

Figure 10.33 Transmission loss for SI-SO flow-reversing chamber (*L* = 2.0 in., *H* = 9.0 in. *W* = 4.75 in.). □, predicted by theory for no flow condition; Δ, measured without flow; ●, measured with flow at 110 ft/s (33.5 m/s) [47].

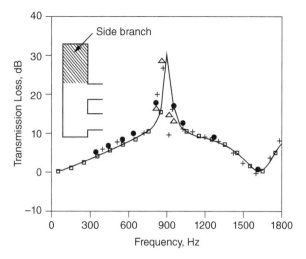

Figure 10.34 Transmission loss for SI-CO flow-reversing chamber (*L* = 2.0 in., *H* = 9.0 in. *W* = 4.75 in.) ─☐─, predicted by theory; +, measured without flow condition; Δ, measured without flow (end plate vibration eliminated); ●, measured with flow at 110 ft/s (33.5 m/s) [47].

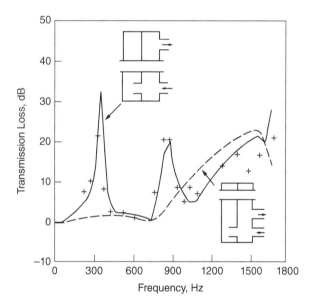

Figure 10.35 Transmission loss characteristics for SI-SO muffler chambers: ─ ─ ─ ─ flow-reversing chamber only – predicted; ──── flow-reversing chamber and resonator – predicted; and + are the measured data for the flow-reversing chamber and resonator [48].

observed that experimental agreement with theory is good and that flow effects appear small at the mean flow velocity (Mach number) used. Part of the volume appeared to act as a side-branch with the SI-CO chamber (Figure 10.34). The theory developed was then used to conduct a theoretical parametric study on reversing end chambers as dimensions, and locations of inlet, outlet and pass tubes were changed. The results are given in Ref. [47].

Figures 10.35 and 10.36 show the predicted and measured TL of similar SI-SO and SI-CO end chambers both of which have pass tubes. Both the cases when the end chambers have Helmholtz resonators attached

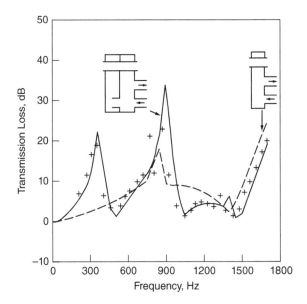

Figure 10.36 Transmission loss characteristics for SI-SO muffler chambers: – – – – flow-reversing chamber only – predicted; —— flow-reversing chamber and resonator – predicted; and + are the measured data for the flow-reversing chamber and resonator [48].

(solid line) and when there are no resonators (broken line) are shown. The no-resonator cases are similar to Figures 10.33 and 10.34, except that here pass tubes are present. It should be noted that the experimental points were measured without flow but with resonators attached. The predictions were made by dividing both the end chamber and the resonator into finite elements [47]. Although only two-dimensional finite elements were used, the third dimension and the elliptical cross-sectional shape were allowed for by varying the mass of the elements corresponding to their thickness [44–48]. It is noted in Figures 10.35 and 10.36 that the addition of the Helmholtz resonators produces sharp attenuation peaks in the TL curves. The first resonance frequency peak at 350 Hz agrees well with the value of 356 Hz calculated for the resonance frequency of a Helmholtz resonator using lumped parameter (mass-spring) theory. The higher frequency peak must be produced by a higher mode resonance caused by interactions between the Helmholtz resonators and the end chambers.

Figure 10.37 shows that the positioning of the resonator neck is theoretically an important factor in determining the TL curve [48].

Figures 10.38–10.40 show the predicted and measured TL for three different muffler combinations. The predictions were made by combining the predicted four-pole parameters of the end-chamber systems with those of the straight pipes using the matrix multiplication method discussed earlier (see Eq. (10.46)). The muffler combinations shown in Figures 10.38–10.40 are typical of automobile reverse-flow mufflers used in the USA except that cross-flow elements and side-branch concentric resonators are absent. It is seen that, at least at the low Mach number used (flow velocity of 32 m/s), that there is very little difference in the TL measured with or without flow. Flow effects may be more important at higher flow rates (corresponding to higher engine loads). Also, flow is expected to have a greater effect on the radiated sound (see Eq. (10.1) and Figure 10.5).

The finite elements (FEM) approach continues to be used to analyze the acoustical performance of mufflers. An important development is the availability of in-house and commercial FEM (and boundary element

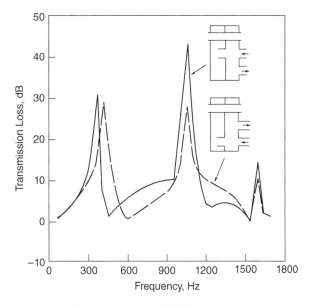

Figure 10.37 Predicted transmission losses for combination of SI-CO flow-reversing chamber and Helmholtz resonator, with different throat locations: – – – – Side-located throat and —— Centrally located throat [48].

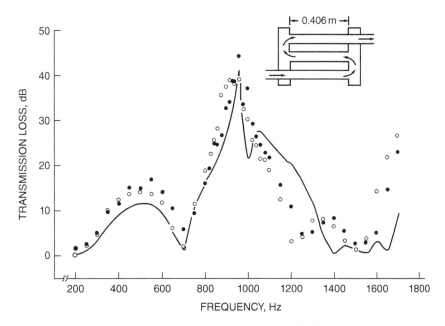

Figure 10.38 Transmission loss characteristics for combination of SI-CO and CI-SO flow-reversing chambers; —— is predicted; ○ is measured without flow and ● is measured with a flow speed of 32 m/s [48].

method [BEM]) software programs and computer codes. The BEM is discussed in the next Section 10.9.2. The first FEM code developed for muffler design was produced in-house on punched computer cards for a large digital computer at Purdue University in 1970 by Young and Crocker [44, 45]. The development of this in-house code continued until the end of the 1970s under several contracts for Arvin Industries [74].

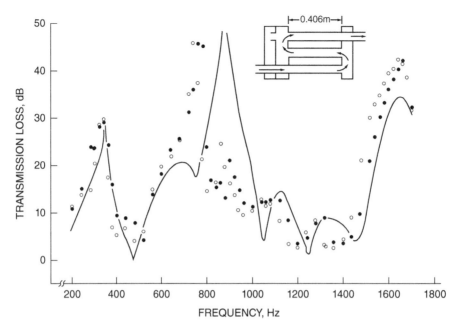

Figure 10.39 Transmission loss characteristics for combination of SI-CO and CI-SO flow-reversing chambers; — is predicted; ○ is measured without flow and ● is measured with a flow speed of 32 m/s [48].

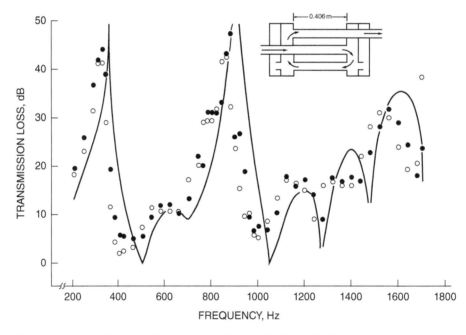

Figure 10.40 Transmission loss characteristics for combination of CI-CO and SI-SO flow-reversing chambers with two resonators; — is predicted; ○ is measured without flow and ● is measured with a flow speed of 32 m/s [48].

Interestingly, the first commercial software contained both FEM and BEM softwares and was produced by SYSNOISE (now called SIMCENTER) in 1988. A number of other such codes are now available such as, ABAQUS, ACTRAN, ANSOL, ANSYS, COMET, COMSOL, LS-DYNA, MUMPS, VA-One, and VNOISE

[75–81]. The well-known FEM code, NASTRAN, was originally created for structural analysis, but now marketed by MSC, it can also be used for acoustical problems.

Although, the FEM and BEM numerical methods are now well developed, muffler manufacturers still rely to a large extent on modifying existing muffler designs and testing them to obtain the acoustical performance desired. Bilawchuk and Fryfe discuss the use of FEM and BEM to predict the TL and IL of mufflers at the design stage [75]. They also describe the so-called "traditional" laboratory method the four-pole transfer matrix method and the three-point method used in the calculation of TL. They conclude that the FEM is better suited than the BEM in the TL determination of mufflers. The BEM is, however, superior in determining the IL since it does not have the problem of the infinite medium domain at the exhaust tube outlet. In addition, they state that the three-point method is faster and easier to use than the four-pole method [75]. Figure 10.41 shows the results obtained with the traditional method for both FEM and BEM [75].

Zhou and Copiello have used FEM modeling coupled with the transfer matrix approach (TMM) to examine complicated muffler systems [76]. In the motorcycle systems they have analyzed, both pipe transmitted noise and muffler shell radiated noise are of concern. In such cases, the acoustical FEM can be easily coupled to the structural FEM. FEM modeling appears to have an advantage over BEM, since mean flow and temperature gradient effects are easier to incorporate. In their studies they have used both ACTRAN and MUMPS (http://mumps.enseeiht.fr).

Using the acoustics module of the COMSOL Multiphysics software, Vasile and Gillich have predicted the TL of an expansion chamber with and without a 15 mm layer of glass wool sound-absorbing material [81]. Figure 10.42 shows the TL without glass wool and with 15 mm glass wool. It is seen that the addition of a 15 mm glass wool layer smooths the TL behavior and increases the TL particularly in the high frequency range. However, sound-absorbing materials cannot always be used since they can absorb moisture, oil, and particles, which can severely degrade their performance.

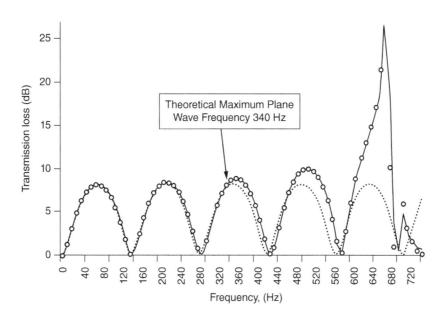

Figure 10.41 Transmission loss for single expansion chamber with the *Traditional* method [75]. ···, theoretical; —— FEM; –O–O– BEM.

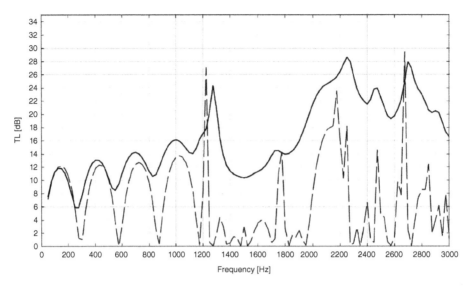

Figure 10.42 Transmission loss of muffler as a function of frequency: – – – without glass wool (solution 1) and —— with 15 mm glass wool (solution 2) [81].

10.9.2 Boundary Element Analysis

The BEM can be used for interior, exterior, and multidomain problems. It was developed subsequently to the FEM approach during the late 1980s by Seybert and his colleagues [82]. Its application to interior domain muffler analysis will be reviewed here. BEM and FEM share a number of assumptions and features, in particular the use of element division of the domain and pre- and post-processing of the information generated [82]. An advantage of the BEM is that the boundary surface needs to be modeled with mesh elements. Another advantage is that for infinite domain problems (such as the sound radiation from a muffler tailpipe), the so-called Sommerfield radiation condition. The advent of the AML capabilities has helped to overcome this infinite domain problem. Thus, there is no need to create a mesh to approximate the radiation condition.

Figure 10.43 shows the boundary element mesh created for a simple expansion chamber muffler. A series of points (called nodes) are selected on the interior muffler surface which are connected together to form the elements. The elements can be either of quadrilateral or triangular shape. The elements must be chosen to be

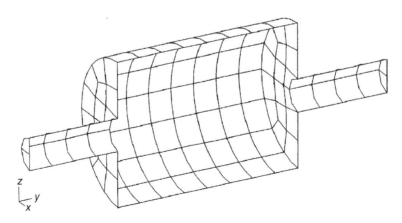

Figure 10.43 Boundary element mesh for a simple expansion chamber muffler [83].

small enough to produce an accurate solution, but not so small than an excessive amount of computer time is required. Figure 10.44 shows the TL made with experiment.

Figure 10.45 shows the SPL contour plot at 2900 Hz for this muffler. This frequency is seen in Figure 10.44 to be where the TL is almost zero and where a cross mode is excited making the muffler performance very poor. The SPL variation from 65 dB to 95 dB is shown and the variation of SPL in the radial direction is clearly seen.

Figure 10.46 shows a muffler in which a multidomain model is used. The BEM allows for temperature variation. The first volume is filled with bulk-reacting absorbing material housed under a perforated inlet tube. The BEM modeling also allows for inclusion of the perforations.

Figure 10.47 shows the SPL contour plot for this muffler at 700 Hz. The SPL varies from 85 to 35 dB. The reduction in the SPL from the inlet to the outlet of the muffler is clearly seen [83].

Seybert and Cheng published a study in 1987 concerning the use of the BEM for the solution of interior acoustics problems [82]. As they state, the BEM approach is ideally suited for use in the TL predictions. They found that predictions they made using the BEM are in excellent agreement with those using the FEM.

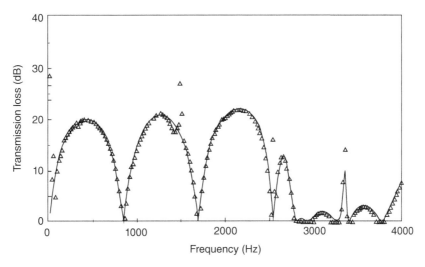

Figure 10.44   TL for simple expansion chamber muffler in Figure 10.43: ——, BEM; Δ, experiment [83].

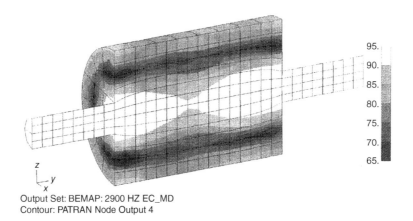

Output Set: BEMAP: 2900 HZ EC_MD
Contour: PATRAN Node Output 4

Figure 10.45   Sound pressure level (SPL) contour plot for the expansion chamber muffler in Figure 10.43 at 2900 Hz [83].

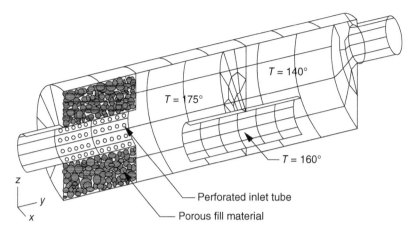

Figure 10.46 Muffler model using the multidomain BEM [83].

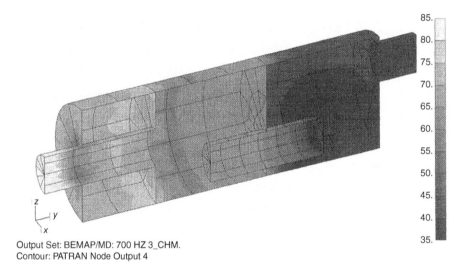

Output Set: BEMAP/MD: 700 HZ 3_CHM.
Contour: PATRAN Node Output 4

Figure 10.47 The SPL contour plot for multidomain muffler at 700 Hz [83].

Figure 10.48 presents their results for the TL of a simple expansion chamber of length 6 in (15.2 cm) expansion ratio $m = 36$. Figure 10.49 shows the result for two expansion chambers each 6 in (15.2 cm) long coupled together by a 4 in (10.16 cm) long tube.

It is seen that the plane wave solution is adequate at frequencies of kl_e less than about 0.6. But above that value of kl_e, the plane wave theory fails because higher modes can exist in the expansion chamber(s) as well as plane waves. The authors also show how the four-pole parameters, A, B, C, and D can be evaluated using two different boundary conditions in the BEM solution at the pipe system exit. In the first, the volume velocity is set to zero. In the second, the pressure is set to zero.

Wang, Tse, and Chen presented a detailed study using BEM to determine the TL of concentric perforated tube resonators [84]. They calculated the TL of a long and a short resonator with the same dimensions as these reported by Sullivan and Crocker [53]. Using the same porosity values and formula for the acoustic impedance used in Ref. [53] they were able to make predictions to compare with the experimental results

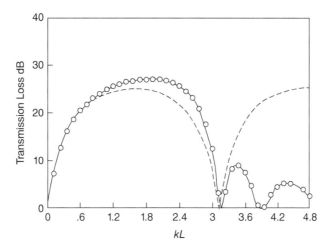

Figure 10.48 The transmission loss of the simple expansion chamber with length L = 6 in. (15.2 cm) and m = 36 (– – – –, plane-wave solution; ——, BEM solution; ○, 2-D FEM solution) [82].

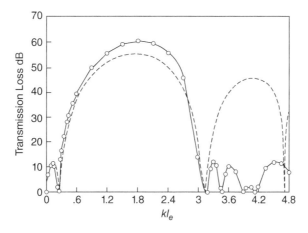

Figure 10.49 The transmission loss of the double expansion chamber with l_e = 6 in. (15.2 cm), l_c = 2 in. (5.08 cm) and m = 36 (– – – –, plane-wave solution; –○–, BEM solution) [82].

of Sullivan and Crocker. The agreement is generally good, except for the inaccurate prediction of the first resonance peak of the short resonator [53]. They used their own set-up to measure the TL of a variety of resonators without and with flow up to a Mach number of $M = 0.15$. As the authors point out, the BEM analysis is not restricted to acoustically long resonators and can include complex-shaped boundaries.

In 1996, Wu and Wan used a direct mixed-body BEM together with a three-point method to evaluate the TL of three concentric tube resonators [85]. They repeated the TL evaluations of the long and short resonators made by Wang, Tse, and Chen, but also examined a long concentric tube resonator with a flow plug. These resonators are exactly the same as the models examined by Sullivan and Crocker [53] and Sullivan [54, 55]. The results are shown in Figures 10.50–10.52. Their results are generally quite good, except for the first peak in the TL and the inaccurate prediction of the TL in the region between the two peaks in Figure 10.50.

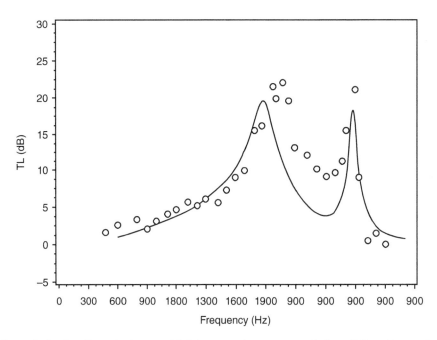

Figure 10.50 Transmission loss for a short concentric tube resonator; —, numerical prediction; ○, experimental data by Sullivan and Crocker (1978) [85].

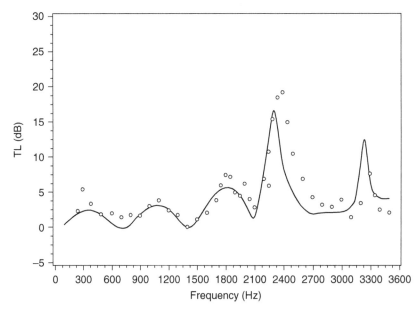

Figure 10.51 Transmission loss for a long concentric tube resonator; —, numerical prediction; ○, experimental data by Sullivan and Crocker (1978) [85].

In 1998, Wu, Zhang, and Cheng used the BEM to improve the derivation of the four-pole parameters [86]. Although their method still requires two BEM runs at each frequency, it is only necessary to solve the BEM matrix once at each frequency. Their improved method uses the velocity boundary condition at the system inlets and outlet. The authors present TL results for a concentric tube plug-flow muffler and also one with two

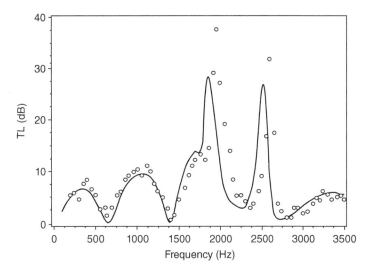

Figure 10.52 Transmission loss for a long concentric tube resonator with a flow plug; —, numerical prediction; ○, experimental data by Sullivan (1979) [85].

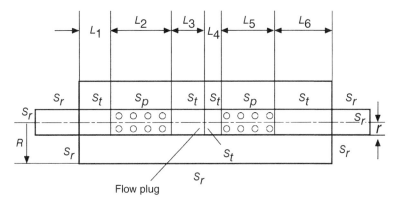

Figure 10.53 Concentric tube muffler with a flow plug (L_1 = 0.0317 m, L_2 = 0.0952 m, L_3 = 0.0255 m, L_4 = 0.0063 m, L_5 = 0.0952 m, L_6 = 0.0953 m, R = 0.0538 m, r = 0.254 m). For perforations in L_2: σ = 0.2168, d_h = 0.00635 m, t = 0.0011938 m; for perforations in L_5: σ = 0.1354, d_h = 0.00635 m, t = 0.0011938 m. (d_h is the hole diameter) [86].

parallel perforated tubes. Wu, Zhang, and Cheng first present predictions for the TL of a concentric tube muffler with a flow plug inserted in the middle of the perforated tube (See Figure 10.53.) [86].

The geometry of the muffler is axisymmetric. The porosity in the first region is 21.68% and in the second region it is 13.54%. Figure 10.54 shows the TL predicted by the BEM (when Sullivan and Crocker's impedance formula is used) compared with the experimental results of Wu et al.

Figures 10.54 and 10.55 show the TL results predicted when either the impedance formula of Sullivan and Bento Coelho are used.

In 2000, Wu edited a book on boundary elements, which included a CD-ROM providing FORTRAN codes [88]. He authored Chapter 5 and included Figures 10.53 and 10.54 in this chapter. This example of a perforated muffler design along with its experimental results were used later by Gunda [79].

The TL predictions for the muffler with two parallel tubes (see Figure 10.56) were made first using Sullivan and Crocker's formula for the perforate transfer impedance (see Figure 10.57a) and then also with an impedance formula proposed by Bento Coelho [87]. See Figure 10.57b.

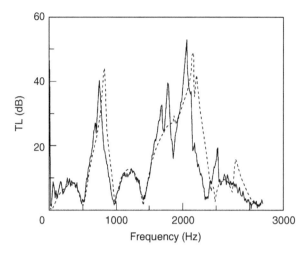

Figure 10.54 Comparison between the experimental data (——) and the BEM prediction using Sullivan and Crocker's formula [53] (- - - -) for the concentric tube muffler with a flow plug [86].

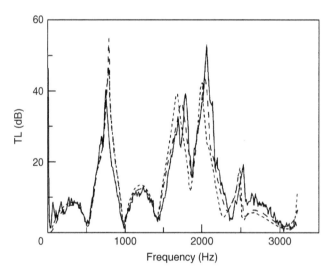

Figure 10.55 Comparison between the experimental data (——) and the BEM predictions using Sullivan's formula [15] (- - -) and Bento Coelho's formula [87] (- - - -) for the concentric tube muffler with a flow plug [88].

It is seen in Figure 10.57b that Bento Coelho's impedance formula gives some improvement in the TL prediction in the region of the large peak at 1800 Hz.

Gunda proposed an analysis method that uses BEM and a Fast Multiple Method (FMM). The FMM speeds up solution times considerably. However, most users still prefer FEM [79]. The analysis is performed using a commercial code formulation called Coustyx developed by ANSOL. This BEM/FMM approach reduces the number of nodes required by BEM allowing more complicated models to be solved at higher frequencies. Figure 10.58 shows the TL prediction for the perforated plug-flow muffler made using the Coustyx code compared with experimental results from Wu [88].

Although in the low frequency region, Gunda's BEM predictions seem to be as good as those of Wu. Also, in the mid- and high-frequency ranges they are no better.

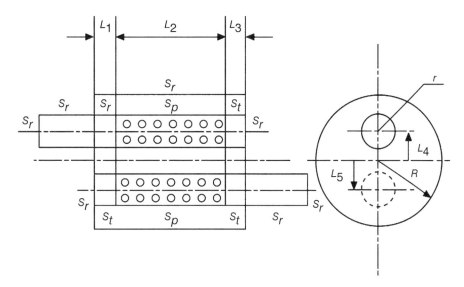

Figure 10.56 Muffler with two parallel perforated tubes (L_1 = 0.0245 m, L_2 = 0.4572 m, L_3 = 0.0254 m, L_4 = 0.0381 m, L_5 = 0.0381 m, R = 0.1016 m, r = 0.0254 m). For both tubes: σ = 0.144, d_h = 0.003175 m, t = 0.0011938 m [86].

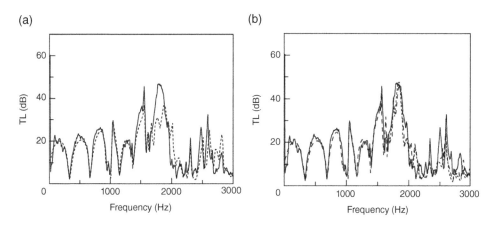

Figure 10.57 (a) Comparison between the experimental data (—) and the BEM prediction using Sullivan's and Crocker's formula [53] (----) for the muffler with two parallel perforated tubes shown in Figure 10.56. (b) Comparison between the experimental data (----) and the BEM prediction using Sullivan's formula [15] (- - -) and Bento Coelho's formula [87] (– – – –) for the muffler with two parallel perforated tubes [86].

Recent developments have included extending the BEM approach to include mean flow [89]. Yang, Ji, and Wu have discussed the use of the BEM in conjunction with a point collocation approach to reduce the number of degrees of freedom [90]. Wu and Wang use a mixed-body approach to evaluate TL using BEM and a three-point method [85]. Hua et al. have explained that different methods can be used such as multiple acoustical loads and different sources to simplify TL determinations [91]. Wu, Zhang, and Cheng have described an improved method for deriving the four-pole parameters using BEM [86]. Banerjee and Jacobi use a method based on Green's Functions to predict TL [92]. Although, most modeling approaches assume that acoustical models are adequate in automobile muffler design, some researchers continue to use fluid mechanics and computational fluid dynamic (CFD) approaches [93–95]. Others have considered nonlinear effects as is also discussed in Section 10.12 [96].

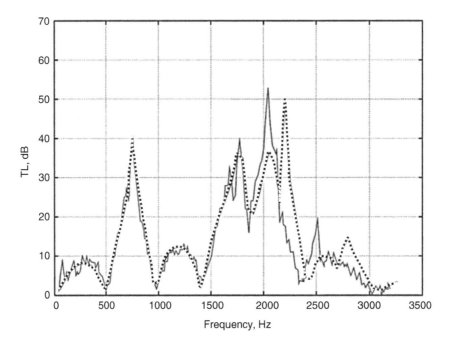

Figure 10.58 Transmission loss of a perforate muffler with flow plug [79]. -----, COUSTYX – FMM; ——, Wu et al. experiment [88].

10.9.3 TL of Concentric Tube Resonators

Although Davis et al. examined concentric tube resonators, their studies were not extended to the case where the center tubes have multiple perforations (or louvers) along their entire lengths. These are known in the automobile industry as "spit chambers" or "bean cans."

Sullivan and Crocker have examined the TL of such concentric tube resonators. (See Figure 10.3.) These resonators, which are often used to provide broad added attenuation in the high-frequency range, are constructed by placing a rigid cylindrical shell around a length of the perforated tube, thus forming an unpartitioned cavity. Sullivan and Crocker used a one-dimensional control volume approach to derive a theoretical model which accounted for the longitudinal wave motion in the cavity and the coupling between the cavity and the tube via the impedance of the perforate [52, 53].

The impedance of the perforate needed for the computer program to predict the TL was obtained by measurement [53]. The instrumentation used is shown in Figure 10.59. The perforate sample was mounted flush with the face of a small anular cavity and the sound pressures at both the face and at the back of the cavity were measured simultaneously. Figure 10.60 presents an example of the specific resistance and reactance for a *single* orifice at room temperature without flow as a function of the peak orifice velocity. A linear behavior is observed in the resistance for SPLs in the 100–120 dB range and is largely independent of frequency. At levels between 120 and 150 dB, the behavior of both the resistance and reactance becomes nonlinear.

Figures 10.61 and 10.62 show the TL for both *short* and *long* resonators [52, 53]. In *short* resonators the primary resonance frequency f_r is less than the first axial modal frequency f_1 of the cavity, $(f_1 = c/2l)$ where c is the speed of sound and l the length. If $f_r > f_1$, then the cavity is said to be *acoustically long*. The TL of *acoustically short* resonators (Figure 10.61) is characterized by two peaks. The first resonance peak results from the coupling of the center tube with the concentric cavity and its frequency f_r can be calculated approximately from the branch Helmholtz equation [52, 53]. However in Figure 10.61, the Helmholtz frequency f_0 is less than the fundamental frequency f_r by 27%. The frequency of the second peak in Figure 10.59 is related to but not equal to the first axial cavity modal frequency $f_1 = c/2l$.

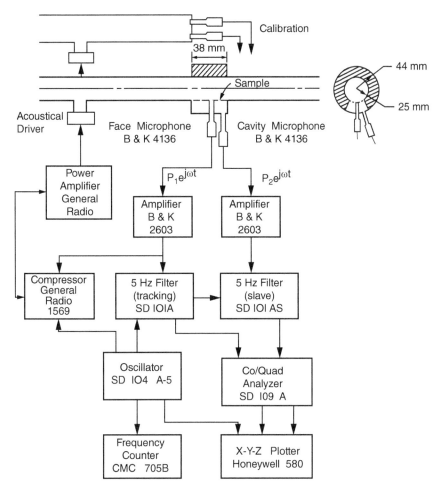

Figure 10.59 Instrumentation for impedance measurements [53].

The performance of concentric tube resonators is dependent on the parameter $k_0 l$, where $k_0 = 2\pi f_0/c$ $= \sqrt{C/V}$. Here k_0 is the wave number of the Helmholtz resonance frequency f_0, c is the speed of sound, and C, V, and l are the conductivity, volume and axial length of the resonator respectively.

In Figure 10.62, the TL of an *acoustically* long resonator is shown. Here the primary resonance frequency f_r occurs above the first and several other longitudinal standing wave mode frequencies in the cavity. Figure 10.63 shows the theoretical effect of changing the porosity of a resonator of constant length 66.7 mm so that as the porosity is increased from 0.5 to 5.0%, the primary resonance frequency f_r and the first axial mode frequency f_1 are gradually merged to provide a wide band of high TL [52, 53].

Even before the studies were published by Davis et al. in 1954, it had become known that the porosity and length of concentric tube resonators was important. Ingard and Pridmore-Brown showed the resonator length was important in 1951 [97]. In 1958, Igarashi and Toyama demonstrated the importance of porosity [23]. However neither of the 1951 and 1958 experimental studies included the effects of mean flow and finite amplitude exhaust noise. Although, Sullivan and Crocker's 1978 paper demonstrated the importance of the length, these other effects were not included in that paper. Sullivan demonstrated these effects clearly in important papers published in 1979. In the 1978 study, mean flow can be included, but it is assumed that the whole louver pattern is uniform and that the impedance of the perforations is constant along its length.

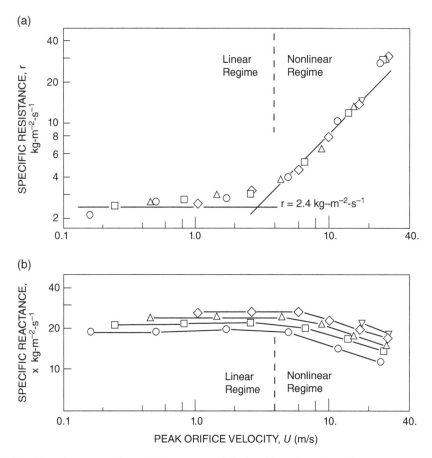

Figure 10.60 (a) Specific resistance and (b) specific reactance of single orifice of sample as function of theoretical peak orifice velocity in the absence of mean flow with frequency as variable: ○ 1000, □ 1100, △ 1200, ◇ 1300, and ▽ 1400 Hz [53].

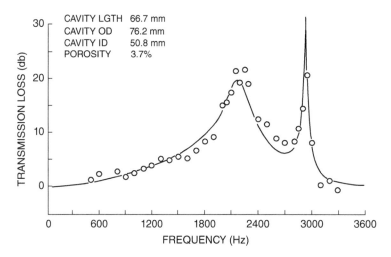

Figure 10.61 Transmission loss for a short resonator: ——, predicted (solid line), ○, measured [53].

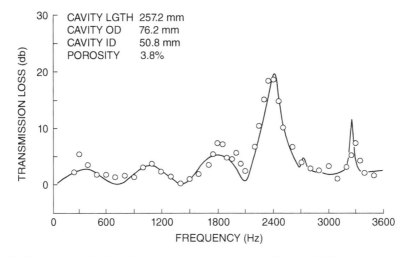

Figure 10.62 Transmission loss for a long resonator: ——, predicted (solid line), ○, measured [53].

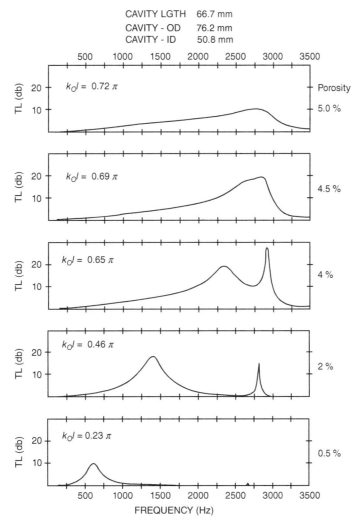

Figure 10.63 Effect of porosity on transmission loss for a short resonator, predicted from mathematical model [53].

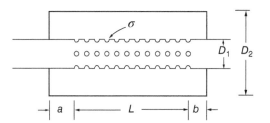

Figure 10.64 Resonator configuration [54].

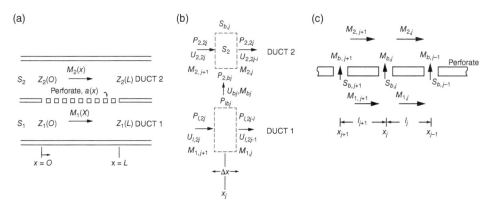

Figure 10.65 (a) The basic two-duct element; (b) Branch point model of perforate; (c) Control volume representation of jth branch, two-duct element [55].

Sullivan continued his studies to consider mathematical modeling and experimental modeling of perforated two-duct through flow, cross-flow and reverse-flow elements (Table 10.1). Subsequently he studied duct cross-flow and reverse-flow three-duct systems (Table 10.2). The focus of his work was on such configurations that are acoustically long in the flow direction and short in the cross dimension. Davies et al. and Igavashi and Toyama [23] had studied the effect of porosity of perforated concentric tube elements experimentally. However, Sullivan was the first to provide theoretical models with experimental confirmation for such elements including mean flow and finite amplitude effects [52, 53].

Figure 10.64 shows the resonator Sullivan studied. (See the top row of Table 10.1.) Using a segmentation approach, the resonator is divided into a number of elements; Figure 10.65a shows the basic two-duct element; Figure 10.65b the branch point model of the perforate, and Figure 10.65c the control volume representation of the jth branch at which the acoustic properties are matched.

Figure 10.66 shows the TL of the resonator operating in the linear and nonlinear regimes. This clearly shows a 10 dB reduction in TL, which occurs in the nonlinear regime.

Figure 10.67 shows the cross-flow chamber studied. Figure 10.68a shows the basic three-duct element and Figure 10.68b the control volume representation of the jth branch. Figure 10.69 shows the effect of a mean flow of Mach number 0.05 on the TL. In this case, the peak TL at 2100 Hz is reduced by more than 10 dB and shifted to a higher frequency. However, the major effect is that the TL is increased by 10 dB or more over most of the frequency range.

Table 10.1 Basic elements and corresponding muffler components of two-duct model [55].

Basic Element	Muffler Component
Through Flow Element $Z_{2,2N}$ Z_{21}	**Resonator** a b DUCT 2 / DUCT 1 $Z_{2,2N} = jZ_2 \cot ka$ $Z_{21} = -jZ_2 \cot kb$
Cross Flow Element $Z_{2,2N}$ Z_{11}	**Flow Expansion Chamber** a b DUCT 2 / DUCT 1 **Flow Contraction Chamber** a b DUCT 1 / DUCT 2 $Z_{2,2N} = jZ_2 \cot ka$ $Z_{11} = -jZ_2 \cot kb$
Reverse Flow Element Z_{21} Z_{11}	**Flow Reversing Chamber** b DUCT 2 / DUCT 1 $Z_{21} = -jZ_2 \cot kb$ $Z_{11} = -jZ_1 \cot kb$

Table 10.2 Basic elements and corresponding muffler components for three-duct model [55].

Basic Element	Muffler Component
Cross Flow Element	**Cross Flow Expansion Chamber**
Reverse Flow Element	**Reverse Flow Expansion Chamber**

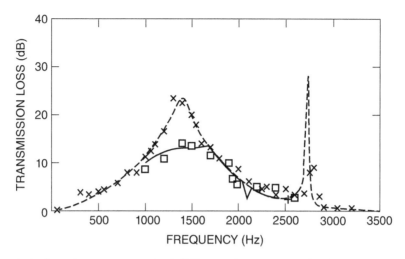

Figure 10.66 Transmission loss of resonator operating in (i) linear regime: ×, measured and, ‒ ‒ ‒ ‒, predicted; and (ii) nonlinear regime: □, measured and, ——, predicted [54].

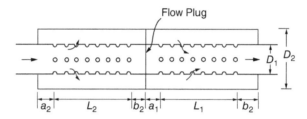

Figure 10.67 Cross-flow chamber configuration [54].

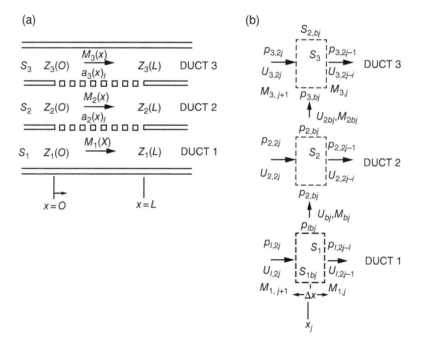

Figure 10.68 (a) The basic three-duct element; (b) control volume of jth branch, three-duct model [55].

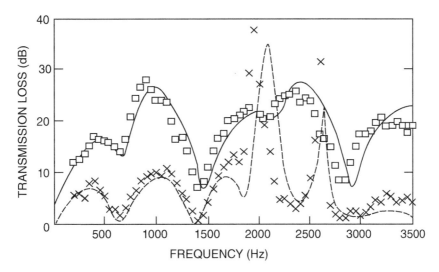

Figure 10.69 Transmission loss of cross-flow chamber operating with (i) $M_\infty = 0$: X, measured and, – – – –, predicted; and (ii) $M_\infty = 0.05$: □, measured and, ——, predicted [54].

10.10   Reactive Muffler IL

The effect of source impedance on IL of an automobile muffler was investigated theoretically in 1970 by Young [45]. Some results are shown in Figures 10.70 and 10.71. In Figure 10.70 it is seen that there is a large difference between the IL curves for the muffler for the three different source impedances investigated: $Z_e = 0$, $\rho c/S$, and ∞, when the prediction is made for discrete frequencies. However, Figure 10.71 shows that if the IL is averaged on an energy basis (with a theoretical 25 Hz filter) that the differences in IL predictions are much less. Note that the vertical scales in Figures 10.70 and 10.71 are different and that a different engine firing frequency is chosen. Also, of considerable interest is the fact that in both figures the TL curve passes through the middle of the IL curves. In Figure 10.70, the hills and valleys in the IL curves are thought to be caused by standing waves in the lengths of straight (exhaust and tail) pipes in the muffler systems.

10.11   Measurements of Source Impedance

Figure 10.72 shows that it is very important to have knowledge of the source impedance if narrow-band predictions of the IL of an automobile muffler are desired. In many cases, such as automobile engines, compressors and blowers, the mechanical sound source is not constant, but varies over restricted frequency range. Figure 10.73 shows that the IL is largely independent of the source impedance if frequency-averaged results over a bandwidth of such as 25 Hz are sufficient. If frequency averaging over a 200 Hz bandwidth is acceptable, then the IL results can be approximated by the TL results.

In some cases, it is important to have detailed information of muffler IL (for instance when the source operates at a fixed speed.) Various approaches have been used to determine the impedance of mechanical sources. The termination of a duct system represents one of the impedance boundary conditions. It is more difficult to characterize the impedance of a mechanical source compared to the impedance of the termination because of the dynamic nature of the source [98, 99].

Direct methods for the measurement of source impedance are based on either the standing-wave technique or the transfer function technique [99]. These techniques have been successfully used to measure the

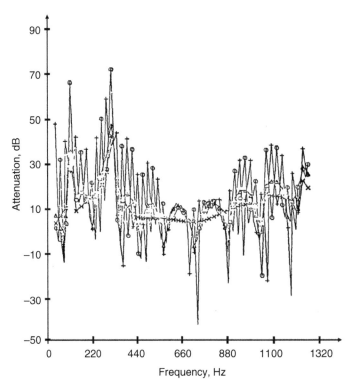

Figure 10.70 Theoretical insertion losses and transmission loss for an automobile engine exhaust muffler system with actual exhaust temperature profile at firing frequency 100 Hz [45]. ○, I.L. with zero source impedance; Δ, I.L. with anechoic source termination; +, I.L. with infinite source impedance; X, Transmission loss (without filter).

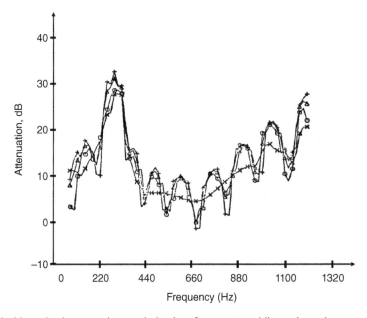

Figure 10.71 Theoretical insertion losses and transmission loss for an automobile engine exhaust muffler system at elevated temperature at firing frequency 70 Hz [45]. ○, I.L. with zero source impedance; Δ, I.L. with anechoic source termination; +, I.L. with infinite source impedance; X, Transmission loss (with 25 Hz filter).

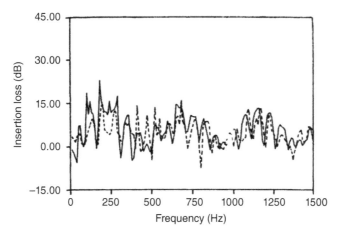

Figure 10.72 Insertion loss of an expansion chamber of the engine operating at 2000 rpm, 10 in. Hg: —, predicted (using measured engine impedance); – – – –, measured [98].

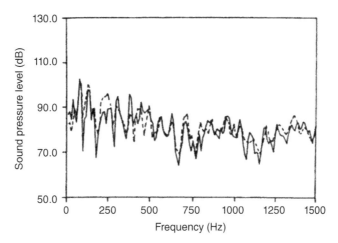

Figure 10.73 Radiated sound pressure level with the expansion chamber of the engine operating at 2000 rpm, 10 in. Hg: —, predicted (using measured engine impedance); – – – –, measured [98].

impedance of primary active mechanical sources such as engines, compressors, fans, and blowers [100, 101]. In this context, the signal refers to the SPL of the secondary measurement source and noise refers to the SPL of the primary source in operation. A minimum signal-to-noise ratio of 10 dB of the secondary source above the primary one is required. Pneumatic and/or electroacoustic drivers are usually needed to achieve a sufficient signal-to-noise ratio. In direct methods, the microphones are placed inside the duct. A modified transfer function method for a low signal-to-noise ratio has been developed that only requires an additional measurement of a calibration transfer function [102].

The advantages of indirect methods are that no secondary source is required as the response is measured from the test source and the microphone can be placed outside the duct. However, the methods are sensitive even to slight errors in the measured response, which strongly influences the numerical aspects. The direct and indirect methods are essentially experimental methods based on frequency domain analysis [98].

Indirect methods are based on the use of different loads and their corresponding responses [103–105]. The pressure response is measured for a known load of impedance Z_L. Assuming that the source pressure is invariant, the system of equations for two, three, four, or more load impedances is solved to calculate the source impedance. With two and three loads, the complex pressure response needs to be used; whereas with four loads only the SPL response is needed [98].

There has been extensive work in recent years continuing determination of the source characteristics of internal combustion engines [106–111]. Desmons and Auregan used several calibrated loads [106]. Liu and Herrin used a two-load wave decomposition approach to measure source impedance [109]. Other researchers have concentrated on evaluating the complete exhaust systems [112–116]. Munjal provided an extensive review in 2004 [115]. Others have made in-depth studies of the exhaust and intake systems of particular engines [117–119]. Hynninen produced a useful review in 2015 as part of a doctoral thesis [120]. Whether linear acoustics theory is adequate for engine exhaust system design continues to be under discussion [121–123]. Hynninen has suggested use of capsules to identify source characteristics of large engines in situ [124].

10.12 Dissipative Mufflers and Lined Ducts

Porous sound-absorbing materials are used to control the noise of heavy equipment and machinery in industry and to line air-conditioning ducts in buildings. Both intake and exhaust noise of the equipment normally needs to be muffled. Improved sound-absorbing materials are also now being used to increase the NR of reactive mufflers.

10.13 Historical Development of Dissipative Mufflers and Lined Duct Theories

Rayleigh made the earliest study of the acoustical performance of sound-absorbing materials in 1883 [125]. That was followed by a theoretical study of the propagation of sound in rectangular lined ducts by Sivian in 1937 who used the impedance of the lining material as the acoustical boundary condition for his model [126]. Morse [127] extended Sivian's model in 1939. Sabine produced a simple empirical model for the attenuation of lined ducts in 1940 [128]. Many other researchers have continued these studies since then [130–148]. Most of these studies involve propagation in rectangular ducts. Scott continued these early studies in 1946 and made measurements of sound absorption in circular ducts [129]. Zwicker and Kosten published an important book on sound-absorbing materials in 1949 [130]. Cremer went on to develop Morse's model in 1953 [131].

The theoretical models are based on the assumptions that the sound absorption in ducts depends on lining thickness, material flow resistivity (flow resistance per unit length), lining thickness, wavelength of the sound, porosity (fractional volume of fluid in the porous material) and duct length. Lined ducts are normally used where mean flow is present and sound attenuation is found to depend not only on flow velocity magnitude but also on direction. (Such lined ducts are usually placed on the intakes and exhausts of fans and blowers, compressors, power plant turbines, etc.) Meyer et al. [132] produced one of the first studies on the influence of flow on sound attenuation in absorbing ducts in 1958.

Most early theoretical studies assumed that sound cannot propagate in the material and that the lining material can be completely described by its wall impedance [127, 131]. The material is then known as "locally reacting." In Scott's analytical model, sound can propagate in the lining in the direction parallel to the duct axis and the material is known as "non-locally reacting" or "homogeneous." Kurze and Ver [133] have developed a theory for the attenuation of ducts lined with non-isotropic sound-absorbing materials, which in the extreme cases gives the results of models for the locally reacting lining models of Morse [127] and Cremer [131] and the homogenous (non-locally reacting) lining model of Scott [129]. A number of theoretical and experimental studies on lined ducts and ducts with multiple parallel absorbing elements (usually termed baffles) have been published in recent years [134–148].

In 1978, Ver published a review of the attenuation of sound in lined and unlined ductwork of rectangular cross-section [134]. The review was mostly concerned with rectangular ducts lined on four sides used in air-conditioning systems. In 1982, Ver published another paper, but this time it was concerned with parallel-baffle mufflers such as those used to quiet turbines in power plants or large fans or blowers [136]. Several other authors since then have conducted theoretical studies to predict the sound attenuation of lined ducts [137–140]. The paper by Astley and Cummings [140] in 1987 used a finite element scheme. They provided a series of design charts for air-conditioning ducts, giving the attenuation per unit length of the ducts lined on four sides for a range of geometries and flow resistivities. They also included the effect of mean flow [140].

10.14 Parallel-Baffle Mufflers

Power generation plants around the world often employ gas turbines and other industrial equipment, which need to be muffled. The inlets and exhausts of large fans and blowers in buildings must also be muffled. Very often large industrial mufflers have to be designed individually for specific applications to achieve optimum performance and acceptable physical size. Normally high sound attenuation over a wide frequency band is required. In addition, the pressure drop in the flow along the muffler should be small and the muffler's physical dimensions made as small as possible. To meet each of these technical requirements individually is not technically difficult, but to meet all three requirements at once is very difficult. For example, a small muffler designed to have a large attenuation will result in one with a large pressure drop, or one with attenuation in only a narrow frequency band. The amount of attenuation required depends not only on the size and type of the industrial equipment to be muffled, but on the proximity of fans and blowers to occupants of buildings or those living in residential communities nearby. Thus, it is important that the properties and size of such mufflers be determined properly in the design stage. As discussed by Vanderburgh [141], designs must be concerned with the increased diversity of equipment and air-conditioning systems to be muffled. There is now more awareness of the problems of low-frequency noise and sound spectrum imbalance in buildings.

Figure 10.74a gives a sketch of a duct lined on two sides along with the symbols and terminology normally used. Parallel-baffle mufflers consist of multiple side-by-side arrangements of ducts lined on two sides. See Figure 10.74b. Figure 10.74c shows a circular muffler. There is not one universally-accepted method to predict the attenuation of parallel-baffle and circular-section mufflers. If a duct is lined on all four sides, the total attenuation may be assumed to be the sum of the attenuation of each pair of sides evaluated independently [14, 15]. The attenuation of a circular duct can be assumed to be equal to a square duct lined on all four sides and having a cross-sectional area equal to that of the circular duct. There are several empirical schemes with associated charts and theoretical studies, which can be used to determine the attenuation of parallel-baffle mufflers.

The studies assume that the attenuation in a duct can be related to the percentage of open area, (% open area/duct cross-sectional area). The early study of Sabine [128] found that the attenuation could be related to the Sabine absorption coefficient α_{sab} (raised to the power 1.4) and the ratio of the perimeter P of the duct cross-section to the open area A. This empirical result may be explained by assuming the acoustic energy in the duct length l is exposed to the surface area Al of the absorbing material which has an absorption

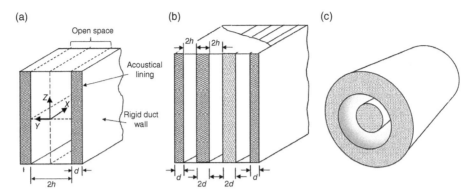

Figure 10.74 (a) A sketch of a lined duct, showing the nomenclature used in describing the direction of a sound wave in the duct. (b) A sketch of a parallel-baffle muffler. (c) A sketch of a circular muffler with annular passage.

coefficient α_{sab}. Although this result becomes increasingly inaccurate at the frequency increases, it is still useful at very low frequency (e.g. less than 200 Hz.)

Embleton was one of the first to publish a chart in 1971 to predict the attenuation of parallel-baffle mufflers [8]. Embleton's chart was based on the then unavailable results of Ingard and his unpublished computer program, which remained unpublished until 1994 [149]. Ver also produced design charts in 1982 for the attenuation of parallel-baffle mufflers [136]. His results were republished in the book edited by Beranek and Ver in 1992 and also in the later version of this book edited by Ver and Beranek in 2006. Bies, Hansen, and Bridges also produced a family of parallel-baffle attenuation prediction charts, which were also republished in the Bies and Hansen book [14, 150]. Bogdanovic [145] has written a useful review of all of these parallel-baffle muffler attenuation prediction methods with the exception of the ones by Bies and Hansen [14] and Ramakrishnan and Watson [151].

Various authors have used different symbols for flow resistance, baffle thickness, baffle spacing, etc. The symbols used are summarized in Table 10.3. In Sections 10.14.1–10.14.7 the original symbols used by these authors in the figures and text in this chapter have not been altered. To avoid any confusion, the reader should consult Table 10.3 and/or the original references.

10.14.1 Embleton's Method [8]

Embleton's chart is given in Figure 10.75. The parameters required to use the chart are the % open area, the spacing between the baffles l_y and the wavelength of sound λ. The airflow temperature is assumed to be 25 °C. The curves give values of the normalized flow resistance parameter $Rt/\rho c$, where R is the flow resistance per unit thickness, t is the material thickness and ρc the air characteristic impedance. However, Embleton states that the open area has the dominant effect and that the curves are not very sensitive to changes in the flow resistance per unit thickness, R. Thus, Embleton suggests that the curves in Figure 10.75 may be used over an appreciable range of values of R (from one half to twice the nominal values of $Rt/\rho c$ given in Figure 10.75) [8]. The attenuation parameter Al_y given in Figure 10.75 is the attenuation per length of duct equal to the duct width l_y. Thus, the attenuation of a duct of length l_e is $Al_y (l_e/l_y)$ dB.

Embleton's chart remains useful because of its simplicity. However, several other authors have now produced attenuation versus frequency prediction curves for parallel-baffle mufflers [136] in which baffle thickness and spacing, mean flow and material resistivity are considered separately. The prediction schemes normally present results for the attenuation for a muffler having a length equal to the spacing between baffles (termed duct "width" or "height") as ordinate. The abscissa is usually the non-dimensionalized frequency, (half duct width h divided by wavelength λ). The authors' own symbols and terminology are used in the following sections of this review.

Table 10.3 Symbols used by various authors in parallel-baffle mufflers.

Symbol	Flow Resistance	Flow Resistivity	Normalized Flow Resistance	Liner Thickness, m	Gap Between Liners, m	Frequency Parameter
Author						
Embleton	R_l	R	$Rt/\rho c$	t	l_y	l_y/λ
Ver	R	R_1	$R_1 d/\rho c$	d	$2h$	$2hf/c$
Ingard	—	—	R	d	D	D/λ
Bies and Hansen	R	R_1	$R_1 d/\rho c$	t	$2h$	$2h/\lambda$
Mechel	$r_0 d$	r_0	$r_0 d/Z_0$	$2d$	$2h$	fh
Ramakrishnan and Watson	$r_c d$	r_0	$r_0 d/\rho c$	d	h	$\mu = 2hf/c$

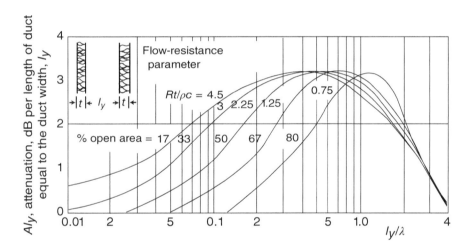

Figure 10.75 Graph used to predict attenuation values. (Source: After Embleton [8].)

10.14.2 Ver's Method [11, 12, 136]

In 1982, Ver published a series of five charts for open areas ranging from 16.6 to 83% to predict the attenuation of parallel-baffle mufflers [136]. Three of these curves with % open areas ranging from 33 to 66% were subsequently republished in two books [11, 12]. See Figure 10.76. Figure 10.76 shows that the attenuation does not depend strongly on the normalized flow resistance $R = R_1 d/\rho c$ in the range from $R = 1$ to 5. Although designers may be tempted to reduce the % open area to broaden the high attenuation region, this can cause problems. When the % open area is reduced, the mean flow velocity will need to increase correspondingly unless the total muffler size is increased. Higher mean flow velocity causes increased flow noise and greater pressure drop with an increased loss in system efficiency.

Ver [11, 12, 136] also published a chart showing the attenuation TL_h (equivalent to the Al_y of Embleton) against the frequency parameter $2hf/c$ (equivalent to $2h/\lambda$) illustrating the effect of variations in the % open area. Figure 10.77 shows the very strong dependence of the normalized attenuation on the % open area for a

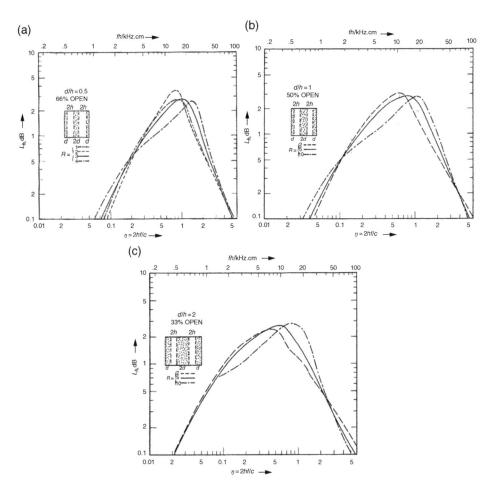

Figure 10.76 Normalized attenuation-versus-frequency curves for parallel-baffle silencers with normalized baffle flow resistance $R = R_1 d/\rho c$ as parameter: (a) 66% open area; (b) 50% open area; (c) 33% open area (after Ver) [11, 12, 136].

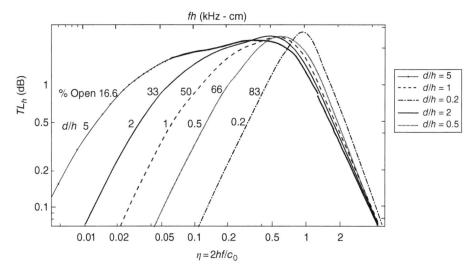

Figure 10.77 Normalized attenuation-versus-frequency curves for parallel-baffle mufflers, illustrating the effect of percentage open area on attenuation bandwidth for $R = R_1 d/\rho c = 5$ (after Ver) [11, 12, 136].

value of the normalized flow resistance parameter $R = 5$. As the value of the % open area is reduced from 83% to 16.6% the region of high attenuation is broadened substantially.

10.14.3 Ingard's Method [149]

As discussed before, Embleton states that his prediction scheme for parallel-baffle mufflers is based on unpublished results of Ingard. Ingard later published his results for such mufflers in 1994 and it is of interest to compare them with Figure 10.75 published by Embleton. Ingard's results given here are for a rectangular duct with a non-locally reacting porous layer on one side only [149]. See Figures 10.78 and 10.79. They are

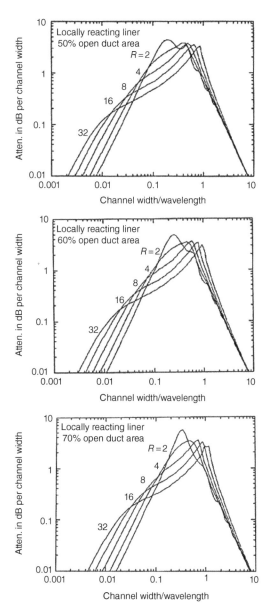

Figure 10.78 Attenuation of the fundamental mode in a rectangular duct with one side lined with a *locally reacting* porous layer with a total normalized flow resistance R. Channel width $= D$ and % open duct area $= 100\, D/(d + D) = 50$, 60, and 70% (after Ingard) [149].

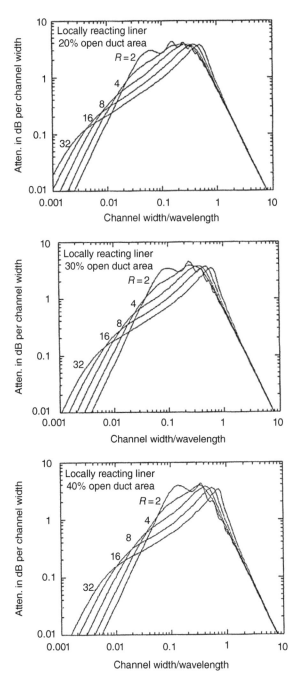

Figure 10.79 Attenuation of the fundamental mode in a rectangular duct with one side lined with a *locally reacting* porous layer with a total normalized flow resistance R. Channel width = D and fraction open duct area = 100 $D/(d + D)$ = 20, 30, and 40% (after Ingard) [149].

with the percentage open area varying from 20 to 70%. Ingard has also similar results for locally reacting ducts lined on one side only but with the open area varying only from 20 to 40%. The theoretical model for the non-locally reacting duct is more complicated than for a locally reacting duct. Ingard provides a computer program along with his book, but unfortunately the programs are designed to run using the DOS language and are difficult to implement with modern computers. Although Ingard's results are for ducts lined on one side only, those for parallel-baffle mufflers may be inferred by assuming his result applies to a duct lined on both sides of double the width (i.e. 2D.) Locally reacting linings are difficult to implement in practice since multiple partitions need to be embedded in the absorbing layer periodically. However, Ingard's results for non-locally and locally reacting liners for the same % open area and normalized flow resistance differ from each other only marginally in most cases.

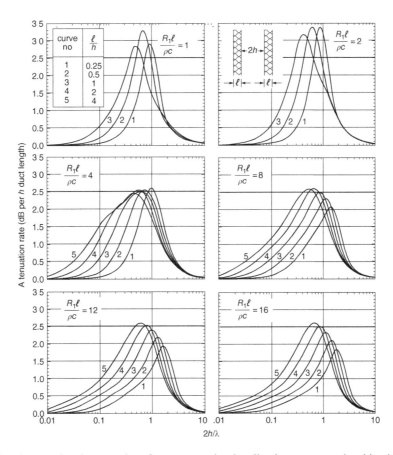

Figure 10.80 Predicted octave band attenuations for a rectangular duct lined on two opposite sides. Lined circular ducts or square ducts lined on all four sides give twice the attenuation shown here. The quantity ρ is the density of fluid flowing in the duct, c is the speed of sound in the duct and ℓ is the liner thickness, h is the half width of the airway, σ is the surface density of a limp membrane covering the liner, R_1 is the liner flow resistivity. Bulk-reacting liner with no limp membrane covering (density ratio $\sigma/\rho h = 0$). Zero mean flow ($M = 0$) (after Bies and Hansen) [14, 15]. (Note the authors use ℓ for the lining thickness instead of d used by Ingard.)

10.14.4 Bies and Hansen Method [14]

Bies and Hansen give a series of curves for the attenuation rate of ducts lined on two sides. They assume two cases: (i) locally reacting absorbent liners, and (ii) non-locally reacting liners (which they term bulk-reacting.) They also give results for flow ($M = 0.1$) in the direction of sound propagation and flow in the opposite direction to the sound propagation ($M = -0.1$). Their results suggest that the assumption of locally reacting liners leads to a peak attenuation somewhat higher than for non-locally (or bulk) liners. The positive or negative flow direction seems to affect the magnitude of the peak attenuation much more than shifting the frequency of the peak attenuation (See Figures 10.80 and 10.81.). One explanation is that when the flow is in the same direction as the sound propagation, the convection reduces the time the sound has to be absorbed and so the attenuation is also reduced accordingly. An alternative explanation preferred by Hansen is that due to friction along the duct walls, the speed of the flow is maximum in the duct center and reduces as the duct wall is approached. This results in sound rays being refracted away from the duct walls when sound is traveling in the same direction as the flow and refracted into the walls when sound is traveling in the opposite direction to the flow, resulting in more sound energy being absorbed.

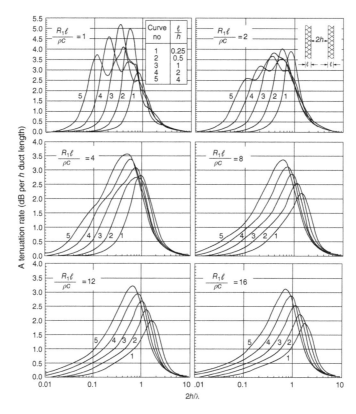

Figure 10.81 Predicted octave band attenuations for a rectangular duct lined on two opposite sides. Lined circular ducts or square ducts lined on all four sides give twice the attenuation shown here. The quantity ρ is the density of fluid flowing in the duct, c is the speed of sound in the duct, ℓ is the liner thickness, h is the half width of the airway, R_1 is the liner flow resistivity. Locally reacting liner with no limp membrane covering (density ratio $\sigma/\rho h = 0$). Zero mean flow ($M = 0$) (after Bies and Hansen) [14, 15].

10.14.5 Mechel's Design Curves [152]

Mechel has published a series of design curves for parallel-baffle mufflers. See Figures 10.82 and 10.83. His predictions are assumed to be for locally reacting lining material since partitions are shown in the linings. Figure 10.82a shows the effect of varying the normalized flow resistance $r_0 d/Z_0$ while the open area ratio OA is held constant at 50% ($d/h = 1$). Here the characteristic impedance ρc is replaced with the symbol Z_0. The first thickness resonance of the baffle is observed clearly for a small normalized flow resistance $r_0 d/Z_0 = 1.5$. It is smoothed out as the normalized resistance is increased to 3.0 and disappears entirely when the resistance is increased to 6.0. As the flow resistance increases, so does the attenuation increase at low frequency, which is important in real applications. Figure 10.82b shows the effect of varying the normalized flow resistance with the open area ratio OA kept at 33% ($d/h = 2$). The first thickness resonance is moved to a lower frequency and

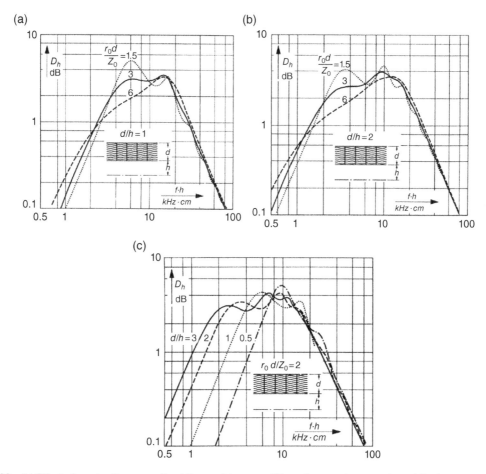

Figure 10.82 (a) Effect of varying the normalized flow resistance $r_0 d/Z_0$ on the sound attenuation while the open area ratio OA is kept constant, $OA = 50\%$ ($d/h = 1$); (b) Effect of varying normalized flow resistance, while open area ratio is kept constant, $OA = 33\%$ ($d/h = 2$); (c) Effect on varying open area ratio from 66 to 25% ($d/h = 0.5$ to 3) for normalized flow resistance $r_0 d/Z_0 = 2$ (after Mechel) [152].

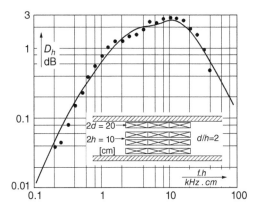

Figure 10.83 Comparison of theoretical values with experiment for given baffle configuration (OA = 33%, d/h = 2), $2d$ = 20 cm, $2h$ = 10 cm (after Mechel) [152].

the low frequency attenuation is increased compared with the $d/h = 1$ case. Figure 10.82c shows the effect of keeping the normalized flow resistance constant $r_0d/Z_0 = 2$ while the open area ratio is varied from $OA = 66$ to 25%. As predicted by other researchers, the increase in open area tends to broaden the attenuation curve while causing only a small reduction in peak attenuation. Increasing the thickness ratio d/h is the most effective way of obtaining high values of attenuation at low frequency [152]. However, care must be taken, since increasing d/h will also lead to higher values of pressure drop or increase in muffler size. Figure 10.83 provides a very encouraging comparison between Mechel's prediction and measurements. Unfortunately, the flow resistance is not given for this case.

10.14.6 Ramakrishnan and Watson Curves [151]

Ramakrishnan and Watson present a series of curves for the attenuation rates for lined parallel-baffle rectangular mufflers [151]. See Figures 10.84 and 10.85. The authors define the normalized frequency $\mu = 2fh/c$ and normalized flow resistance $R = r_0d/\rho c$. The finite element approach which they use also considers multimodal acoustic propagation. The sound-absorbing liners are assumed to be bulk-reacting (non-locally reacting) and thus the model allows for sound propagation in the lining material itself. The authors also compare their predictions with experiment and obtain good agreement except in some cases where $N_1 = d/h$ is small. (See their published tabulated results.)

In the example following, the attenuations predicted by the different design curves in Sections 10.14.1–10.14.6 are compared. Some of the design methods can include the effect of temperature and mean flow (usually in terms of Mach number M). Since other methods do not make allowance for mean flow, it will be assumed to be zero in this example. The main effect of the temperature of the mean flow is to change the speed of sound c and thus the wavelength $\lambda = c/f$ used in the non-dimensional frequency calculation.

The number of baffles and parallel duct sections needed depends on providing sufficient ducting to accommodate the volume flow rates of the fans, blowers, turbines etc. The number of parameters required includes the mean flow resistance of the absorbing material, the thickness of the baffles, d, and the spacing between them, $2h$.

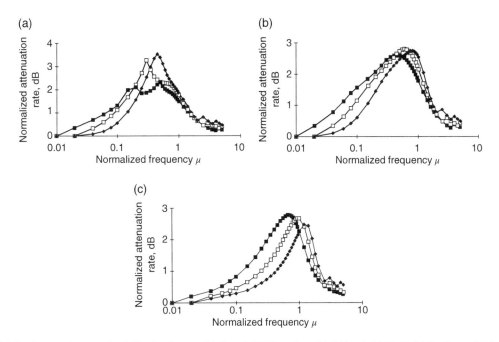

Figure 10.84 Attenuation rate for full unit silencer, (■) $N_1 = 5$, (□) $N_1 = 2$, and (◆) $N_1 = 1$. (a) $R = 2$, (b) $R = 5$, and (c) $R = 20$ (after Ramakrishnan and Watson) [151].

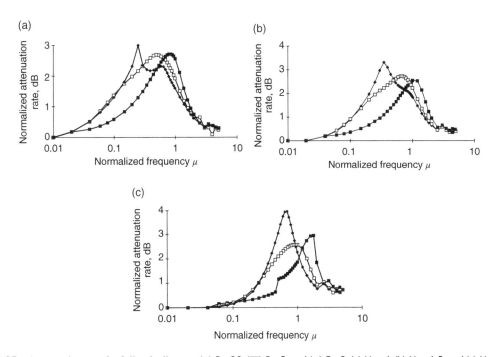

Figure 10.85 Attenuation rate for full unit silencer, (■) $R = 20$, (□) $R = 5$, and (◆) $R = 2$. (a) $N_1 = 4$, (b) $N_1 = 1.5$, and (c) $N_1 = 0.5$ (after Ramakrishnan and Watson) [151].

Example 10.7 A parallel-baffle muffler consists of baffles each 0.1 m thick with a spacing of 0.1 m between each set of baffles. Calculate the attenuation of a muffler of length 1 m. Assume the normalized flow resistance of the absorbing material is 5.0.

Solution

The results are shown in Table 10.4.

Table 10.4 Calculated values of muffler attenuation from methods given in Sections 10.14.1–10.14.6.

| Method | Frequency, Hz | 125 | 250 | 500 | 1000 | 2000 | 4000 | 8000 |
	$2h/\lambda$	0.036	0.073	0.146	0.292	0.584	1.167	2.334
Embleton	Attenuation, dB	2	9	19	27	32	29	13
Ver	Attenuation, dB	1	3	8	18	26	20	4
Ingard	Attenuation, dB	3	7	12	20	22	3	1
Bies/Hansen	Attenuation, dB	1	3	7	15	27	20	5
Mechel	Attenuation, dB	1	3	10	20	28	24	6
Ramakrishnan/Watson	Attenuation, dB	1	2	10	20	25	24	6

The results of the worked Example 10.7 show that the six different methods illustrated in Table 10.4 produce similar results with the greatest predicted attenuations in the 1000–4000 Hz range. Choosing accurate values from the curves for attenuation at low frequencies in the range 125–250 Hz and at high frequencies in the range 4000–8000 Hz is difficult because the curves are steeply sloping and interpolation must be used with the non-dimensional frequency scale. Mechel's curves [152] can be used with some confidence since they have experimental confirmation with measurements made in a very high-quality facility in Stuttgart, Germany. The Ramakrishnan/Watson curves agree very closely with Mechel's curves except for the peak values at 2000 Hz. Although Bogdanovic presents one experimental result [145], which was in good agreement with Embleton's curves, Embleton's method seems to overpredict the attenuation compared with the five other methods. Ingard's curves are very steeply sloping at high frequency, leading to the prediction of very small attenuation values for 4000 and 8000 Hz.

10.14.7 Finite Element Approach for Attenuation of Parallel-Baffle Mufflers

Astley and Cummings [140] presented a family of attenuation prediction curves for lined ducts and parallel-baffle mufflers using finite elements in 1987. However, these curves are not particularly easy to use. More recently, in 2012, Borelli and Schenone [143] also used FEM to predict the attenuation of parallel-baffle mufflers. They used a commercial software for the FEM using a brick mesh and the PARDISO solver which allowed analysis up to 8000 Hz much faster than with the use of traditional solvers. They claim very good agreement between company catalog attenuations given for commercial mufflers and their predictions. Figure 10.86 shows one comparison between their predictions and their own measurements. The muffler that they tested using EN ISO 11691 and EN ISO 7235 standards is given in Figure 10.87. However, another

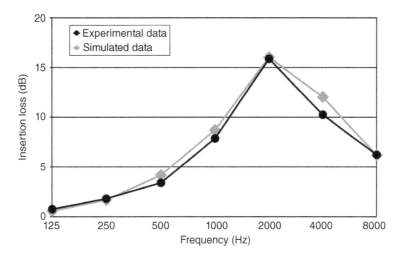

Figure 10.86 Comparison between experimental and simulated data for silencer shown in Figure 10.87 [143].

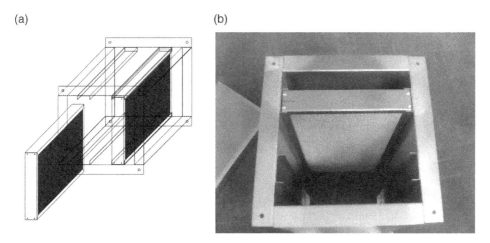

Figure 10.87 Tested silencer: (a) sketch; (b) picture [143].

muffler model FCR SQ-A-110-600 they tested produced somewhat less satisfactory prediction results with an error in their predictions of the order of about 3 dB at 2000 Hz.

Three additional effects should be accounted for:

1) *Discontinuities at the beginning and end of the lined duct section of the muffler.* Although normally the muffler entrance and exit is made continuous with the main ductwork, the soft lining presents a sudden expansion in cross-section, a similar effect happens at the exit, where the sound field experiences a sudden contraction. These effects can be accounted for by considering the muffler as a lined expansion chamber. A correction for parallel-baffle muffler attenuation can be made using the curves given by Embleton [8]. Embleton's correction curves are repeated in the books by Bies and Hansen [14] and Munjal [15].
2) *Pressure drop.* Embleton, Bies and Hansen, and Mechel discuss the problems of pressure drop (sometimes called back pressure) in muffler designs [8, 14, 15]. Embleton describes three main sources, including the inlet and exhaust junction reflections, turbulent-flow creation, and additional end expansion effects [8]. Mechel describes the same three sources of pressure drop and characterizes them in terms of drag

coefficients relative to the pressure. Mechel also stresses that it is important to streamline the leading edges of the baffles [152]. Ver and Beranek discuss how the pressure drop can be estimated [15].

3) *Flow effects including flow temperature.* As already discussed, flow effects the effective sound wavelength in the muffler like a Doppler shift affecting the absorption of the sound [8, 12, 15, 152]. The temperature of the mean flow effects the speed of sound and thus also the effective frequency. The reader is directed to other references for more detailed discussions of these effects [8, 12, 15, 152].

References

1 Crocker, J. (1972) Mufflers. In: Tutorial Papers on Noise Control (ed. Crocker M.J.). Proceedings of Inter-Noise 72, 40–44. Also published in (1975) Reduction of Machinery Noise, rev. ed. (ed. M.J. Crocker), 112–116. Purdue University.

2 Dyer, I. (1956). Noise attenuation of dissipative mufflers. *Noise Control* 2 (3): 50–57.

3 Sanders, G.J. (1968). Silencers: their design and application. *Sound Vib.*: 6–13.

4 Nelson, C.E. (1956). Truck muffler design. *Noise Control* 2 (3): 24–27. and 77.

5 Crocker, M.J. (1977) Internal Combustion Engine Exhaust – Muffling. NOISE-CON 77 Proceedings, October 17–19, 331–358.

6 Franken, P.A. (1960). Reactive mufflers. In: *Noise Reduction* (ed. L.L. Beranek), 414–453. New York: McGraw-Hill Book Co.

7 Doelling, N. (1960). Dissipative mufflers. In: *Noise Reduction* (ed. L.L. Beranek), 454–465. New York: McGraw-Hill Book Co.

8 Embleton, T.F.W. (1971). Mufflers. In: *Noise and Vibration Control* (ed. L.L. Beranek), 362–405. New York: McGraw-Hill Book Co.

9 Bell, L.H. and Bell, D.H. (1993). *Industrial Noise Control.* New York: Marcel Dekker.

10 Barron, R.F. (2003). *Industrial Noise Control and Acoustics.* New York: Marcel Dekker.

11 Galaitsis, A.G. and Ver, I.L. (1992). Passive silencers and lined ducts. In: *Noise and Vibration Control Engineering* (eds. L.L. Beranek and I.L. Ver), 367–427. New York: Wiley.

12 Munjal, M.L., Galaitsis, A.G., and Ver, I.L. (2006). Passive silencers. In: *Noise and Vibration Control Engineering* (eds. I.L. Ver and L.L. Beranek), 279–343. New York: Wiley.

13 Boden, H. and Glav, R. (1998). Exhaust and intake noise and acoustical design of mufflers and silencers. In: *Handbook of Acoustics* (ed. M.J. Crocker), 1034–1053. New York: Wiley.

14 Bies, D.A., Hansen, C.H., and Howard, C. (2018). *Engineering Noise*, 5e. CRC Press.

15 Munjal, M.L. (2014). *Acoustics of Ducts and Mufflers*, 2e. Chichester: Wiley.

16 Moser, M. (2004). *Engineering Acoustics – An Introduction to Noise Control*, (Silencers). Berlin: Springer.

17 Stewart, G.W. (1922). Acoustic wave filters. *Phys. Rev.* 20: 528. See also Stewart, G.W. (1924). Phys. Rev. 23: 520, and Stewart, G.W. (1925), Phys. Rev. 25: 90.

18 Mason, W.P. (1927). A study of the regular combination of acoustic elements, with applications to recurrent acoustic filters, tapered acoustic filters, and horns. *Bell Syst. Tech. J. Q.* 2((3): 258–294.

19 Morley, A.W. (1937) Progress in experiments in aero-engine exhaust silencing. R&M No. 1760. British A.R.C.

20 Martin, H., Schmidt, V. and Willins, W. (1940) The present stage of development of exhaust silencers. RTP TIB Translation No. 2596. British Ministry of Aircraft Production (from MTZ, No. 12).

21 Davis, D.D., Jr., Stokes, G.M., Moore, D. and Stevens, G.L., Jr (1954) *Theoretical and Experimental Investigation of Mufflers with Comments on Engine Exhaust Muffler Design.* NACA 1192.

22 Davis, D.D. Jr. (1957). Acoustical filters and mufflers. In: *Handbook of Noise Control* (ed. C.M. Harris), 21.1–21.44. New York: McGraw-Hill Book Co.

23 Igarashi, J. and Toyama, M. (1958) Fundamentals of Acoustical Silencers (I). Report No. 339. Aeronautical Research Institute, University of Tokyo, December, 223–241.

24 Miwa, T. and Igarashi, J. (1959) Fundamentals of Acoustical Silencers (II). Report No. 344. Aeronautical Research Institute, University of Tokyo, May, 67–85.

25 Igarashi, J. and Arai, M. (1960) Fundamentals of Acoustical Silencers, (III). Report No. 351. Aeronautical Research Institute, University of Tokyo, February, 17–31.

26 Fukuda, M. (1963). A study on the exhaust muffler of internal combustion engines. *Bull. JSME* 6 (22): 255–269.

27 Fukuda, M. (1969). A study on characteristics of cavity-type mufflers, (1st report). *Bull. JSME* 13 (50): 333–349.

28 Fukuda, M. and Okuda, J. (1970). A study on characteristics of cavity-type mufflers (2nd report). *Bull. JSME* 13 (55): 96–104.

29 Parrott, T.L. An improved method for design of expansion – chamber mufflers with application to an operational helicopter. NASA TN D-7309.

30 Sullivan, J.W. (1977) Modelling of engine exhaust system noise. Paper presented at ASME Winter Conference, Atlanta, Georgia, November.

31 Davies, P.O.A.L. (1964). The design of silencers for internal combustion engine. *J. Sound Vib.* 1 (2): 185–201.

32 Davies, P.O.A.L. and Dwyer, M.J. (1964). A simple theory for pressure pulses in pipes. *Proc. Inst. Mech. Eng.* 179 (1): 365–394.

33 Blair, G.P. and Goulbourn, J.R. (1967) Pressure-time history in the exhaust system of a high speed reciprocating internal combustion engine. *SAE Transactions*, 76, paper 670477.

34 Benson, R.S. and Foxcroft, J.S. (1970) Nonsteady flow in internal combustion engine inlet and exhaust systems. *Inst. Mech. Eng.* (London), paper no. 3, 2–26.

35 Blair, G.P. and Spechko, J.A. (1972) Sound pressure levels generated by internal combustion engine exhaust systems. SAE Paper 720155.

36 Blair, G.P. and Coates, S.W. (1973) Noise produced by unsteady exhaust efflux from an internal combustion engine. *SAE Transactions*, Vol. 82, paper 730160.

37 Coates, S.W. (1974) The prediction of exhaust noise characteristics of internal combustion engines, Ph.D. thesis. Queen's University of Belfast, Northern Ireland.

38 Coates, S.W. and Blair, G.P. (1974) Further studies of noise characteristics of internal combustion engine exhaust systems. SAE Paper 740713.

39 Alfredson, R.J. and Davies, P.O.A.L. (1970). The radiation of sound from an engine exhaust. *J. Sound Vib.* 13 (4): 389–408.

40 Alfredson, R.J. (1970) The design and optimization of exhaust silencers. Ph.D. thesis. University of Southampton.

41 Alfredson, R.J. and Davies, P.O.A.L. (1971). Performance of exhaust silencer components. *J. Sound Vib.* 15 (2): 175–196.

42 Alfredson, R.J. (1971). The design of exhaust mufflers using linearized theoretical models. *SAE Paper* 719139: 1048–1052.

43 Davies, P.O.A.L. and Alfredson, R.J. (1972) Design of silencers for internal combustion engine exhaust systems. Proc. J. Mech. Eng., Conference Vibration and Noise in Motor Vehicles, 17–23.

44 Young, C.-I.J. and Crocker, M.J. (1971) Muffler analysis by finite element method. Ray W. Herrick Laboratories. Purdue University Report No. HL 71-33.

45 Young, C.-I.J. (1973) Acoustic analysis of mufflers for engine exhaust systems. Ph.D. thesis. Purdue University.

46 Young, C.-I.J. and Crocker, M.J. (1975). Prediction of transmission loss in mufflers by the finite-element method. *J. Acoust. Soc. Am.* 37 (1): 144–148.

47 Young, C.-I.J. and Crocker, M.J. (1976). Acoustical analysis, testing, and design of flow – reversing muffler chambers. *J. Acoust. Soc. Am.* 60 (5): 1111–1118.

48 Young, C.-I.J. and Crocker, M.J. (1977). Finite-element acoustical analysis of complex muffler systems with and without wall vibrations. *Noise Control Eng.* 9 (2): 86–93.

49 Kagawa, Y. and Omote, T. (1976). Finite-element simulation of acoustic filters of arbitrary profile with circular cross section. *J. Acoust. Soc. Am.* 60 (5): 1003–1013.

50 Craggs, A. (1976). A finite element method for damped acoustic systems: an application to evaluate the performance of reactive mufflers. *J. Sound Vib.* 48 (3): 377–392.

51 Ling, S.F. (1976) A finite element method for duct acoustics problems. Ph.D. thesis. Purdue University.

52 Sullivan, J.W. (1974) Theory and methods for modelling acoustically- long, unpartitioned cavity resonators for engine exhaust systems. Ph.D. thesis. Purdue University.

53 Sullivan, J.W. and Crocker, M.J. (1978). A mathematical model for concentric tube resonators. *J. Acoust. Soc. Am.* 64 (207): 207–215.

54 Sullivan, J.W. (1979). A method for modeling perforated tube muffler components. II. Applications. *J. Acoust. Soc. Am.* 66 (772): 772–788.

55 Sullivan, J.W. (1979). A method for modeling perforated tube muffler components. I. Theory. *J. Acoust. Soc. Am.* 66 (779): 779–778.

56 Karnopp, D. (1975). Lumped parameter models of acoustic filters using normal modes and bond graphs. *J. Sound Vib.* 42 (4): 437–446.

57 Karnopp, D., Reed, D., Margolis, D., and Dwyer, H. (1975). Computer – aided design of acoustic filters using bond graphs. *Noise Control Eng.* 4 (3): 114–118.

58 Fricke, F.R. and Crocker, M.J. (1975) Sound amplification in expansion chambers. Proceedings Inter-Noise 75.

59 Kirata, Y. and Itow, T. (1973). Influence of air flow on the attenuation characteristics of resonator type mufflers. *Acustica* 28 (2): 115–120.

60 Anderson, J.S. (1977). The effect of an air flow on a single side branch Helmholtz resonator in a circular duct. *J. Sound Vib.* 52 (3): 423–431.

61 Seybert, A.F. and Ross, D.F. (1977). Experimental determination of acoustic properties using a two-microphone random-xcitation technique. *J. Acoust. Soc. Am.* 61 (5): 1362–1370.

62 Ingard, K.U. (1957) unpublished information. In: *Handbook of Noise Control* (ed. C.M. Harris), Fig. 21.47. New York: McGraw-Hill Book Co.

63 Bender, E.K. and Brammer, A.J. (1975). Internal combustion engine intake and exhaust system noise. *J. Acoust. Soc. Am.* 58 (1): 22–30.

64 Crocker, M.J., ed. (1984) Noise control. Benchmark Papers in Acoustics Series, paper #2. Van Nostrand Reinhold Company, 25–71.

65 Kathuriya, M.L. and Munjal, M.L. (1976). A method for the experimental evaluation of the acoustic characteristics of an engine exhaust system in the presence of mean flow. *J. Acoust. Soc. Am.* 60 (3): 745–751.

66 Davies, P.O.A.L. (1972) Exhaust System Silencing. The Institution of Marine Engineers, 46–51.

67 Levine, M. and Schwinger, J. (1948). On the radiation of sound from an unflanged circular pipe. *Phys. Rev.* 73 (4): 383–406.

68 Skilling, H.H. (1957). *Electrical Engineering Circuits*, 23–24. New York: Wiley.

69 Mutyala, B.R.C. (1975) A mathematical model of Helmholtz resonator type gas oscillation discharges of two-cycle engines, Ph.D. thesis. Purdue University.

70 Mutyala, B.R.C. and Soedel, W. (1976). A mathematical model of Helmholtz resonator type gas oscillation discharges of two-stroke cycle engines. *J. Sound Vib.* 11 (4): 479–491.

71 Galaitsis, A.G. and Bender, E.K. (1975). Measurement of the acoustic impedance of an internal combustion engine. *J. Acoust. Soc. Am.* 58 Supplement No. (1).

72 Ross, D. (1976) Experimental determination of the normal specific acoustic impedance of an internal combustion engine Ph.D. thesis. Purdue University.

73 Egolf, D.P. (1976) A mathematical scheme for predicting the electro-acoustic frequency response of hearing aid receivers – earmold – ear systems. Ph.D. thesis. Purdue University.

74 Ross, D.F. and Crocker, M.J. (1979) Measurement of acoustic parameters for automotive exhaust systems. Proceedings Noise-Con 79: 235–244.

75 Bilawchuk, S. and Fyfe, K.R. (2003). Comparison and implementation of the various numerical methods used for calculating transmission loss in silencer systems. *Appl. Acoust.* 64: 903–916.

76 Zhou, Z. and Copiello, D. (2013) Simulation of exhaust line noise using FEM and TMM. *Sound and Vibration*, September, 10–13.

77 Copiello, D., Zhou, Z., and Lielens, G. (2015). Acoustic simulation of vehicle exhaust system using high order transfer matrix method coupled with finite element method. *SAE Int. J. Engines* 8 (1): 258–265.

78 Jena, D.P. and Panigrahi, S.N. (2015). Estimating acoustic transmission loss of perforated filters using finite element method. *Measurement* 73: 1–14.

79 Gunda, R. (2008) Boundary element acoustics and the fast multipole method (FMM). *Sound and Vibration*, 12–16.

80 Siano, D. (2011). Three-dimensional/one-dimensional numerical correlation study of a three-pass perforated tube. *Simul. Model. Pract. Theory* 19: 1143–1153.

81 Vasile, O. and Gillich, G.R. (2012) Finite element analysis of acoustic pressure levels and transmission loss of a muffler, advances in remote sensing. *Finite Differences and Information Security*, 43–48.

82 Seybert, A.F. and Cheng, C.Y.R. (1987). Application of the boundary element method to acoustic cavity response and muffler analysis. *J. Vib. Acoust. Stress Reliab. Des. ASME* 109/15: 15–21.

83 Seybert, A.F. and Wu (1997). Acoustic modeling: boundary element methods. In: *Encyclopedia of Acoustics* (ed. M.J. Crocker), 173–183. Wiley.

84 Wang, C.N., Tse, C.C., and Chen, Y.-N. (1993). A boundary element analysis of a concentric-tube resonator. *Eng. Anal. Bound. Elem.* 12: 21–27.

85 Wu, T.W. and Wan, G.C. (1996). Muffler performance studies using a direct mixed-body boundary element method and a three-point method for evaluating transmission loss. *J. Vib. Acoust., ASME* 118: 479–484.

86 Wu, T.W., Zhang, P., and Cheng, C.Y.R. (1998). Boundary element analysis of mufflers with an improved method for deriving the four-pole parameters. *J. Sound Vib.* 217 (4): 767–779.

87 Bento Coelho, J.L. (1984) Modeling of cavity backed perforate liners in flow ducts. Proceedings of 1984 Nelson Acoustics Conference.

88 Wu, T.W. (2000). *Boundary Element Acoustics: Fundamentals and Computer Codes*. WIT Press/Computational Mechanics.

89 Ji, Z., Ma, Q., and Zhang, Z. (1994). Application of the boundary element method to predicting acoustic performance of expansion chamber mufflers with mean flow. *J. Sound Vib.* 173 (1): 57–71.

90 Yang, L., Ji, Z.L., and Wu, T.W. (2015). Transmission loss prediction of silencers by using combined boundary Element method and point collocation approach. *Eng. Anal. Bound. Elem.* 61: 265–271.

91 Hua, X., Jiang, C., Herrin, D.W., and Wu, T.W. (2014). Determination of transmission and insertion loss for multi-inlet mufflers using impedance matrix and superposition approaches with comparisons. *J. Sound Vib.* 333: 5680–5692.

92 Banerjee, S. and Jacobi, A.M. (2013). Transmission loss analysis of single-inlet/double-outlet (SIDO) and double-inlet/single-outlet (DISO) circular chamber Mufflers by using Green's function method. *Appl. Acoust.* 74 (12): 1499–1510.

93 Broatch, A., Serrano, J.R., Arnau, F.J., and Moya, D. (2007). Time-domain computation of muffler frequency response: comparison of different numerical schemes. *J. Sound Vib.* 305: 333–347.

94 Yasuda, T., Wu, C., Nakagawa, N., and Nagamura, K. (2010). Predictions and experimental studies of the tail pipe noise on an automotive muffler using a one dimensional CFD model. *Appl. Acoust.* 71: 701–707.

95 Mishra, P.C., Kar, S.K., Mishra, H., and Gupta, A. (2016). Modeling for combined effect of muffler geometry modification and blended fuel use on exhaust performance of a four stroke engine: a computation fluid dynamics approach. *Appl. Therm. Eng.* 108: 1105–1118.

96 Montenegro, G., Onorati, A., and Della Torre, A. (2013). The prediction of silencer acoustical performances by 1D, 1D-3D and quasi-3D non-linear approaches. *Comput. Fluids* 71: 208–223.

97 Ingard, U. and Pridmore-Brown, D. (1951). The effect of partitions in the absorption lining of sound absorbing ducts. *J. Acoust. Soc. Am.* 23: 589–590.

98 Prasad, M.G. and Crocker, M.J. (1997). Acoustic modeling ducted-source systems. In: *Encyclopedia of Acoustics* (ed. M.J. Crocker), 185–190. New York: Wiley.

99 Prasad, M.G. (1991) Characterization of acoustical sources in duct systems – progress and future trends. Proceedings of Noise-Control, Tarrytown NY, 213–220.

100 Prasad, M.G. and Crocker, M.J. (1983). Acoustical source characterization studies on a multi-cylinder engine exhaust system, and studies of acoustical performance of a multi-cylinder engine exhaust muffler system. *J. Sound Vib.* 90 (4): 479–508.

101 Doige, A.G. and Alves, H.S. (1989). Experimental characterization of noise sources for duct acoustics. *Trans. ASME: J. Vib. Acoust.* 111: 108–114.

102 Kim, W. and Prasad, M.G. (1993) A modified transfer function method for acoustic source characterization duct systems. Presented at the ASME Winter Annual Meeting, Paper 93-WA/NCA-5.

103 Prasad, M.G. (1987). A four load method for evaluation of acoustical source impedance in a duct. *J. Sound Vib.* 180 (5): 347–355.

104 Sridhara, B.S. and Crocker, M.J. (1992). Error analysis for the four-load method used to measure the source impedance in ducts. *J. Acoust. Soc. Am.* 92: 2924–2931.

105 Boden, H. (1995). On multi-load methods for determination of the source data of acoustic one-port sources. *J. Sound Vib.* 180 (5): 725–743.

106 Desmons, L. and Auregan, Y. (1995). Determination of the acoustical source characteristics of an internal combustion engine by using several calibrated loads. *J. Sound Vib.* 179 (5): 869–878.

107 Jang, S.H. and Ih, J.G. (2002). On the selection of loads in the multiload method for measuring the acoustic source parameters of duct systems. *J. Acoust. Soc. Am.* 111: 1171–1176.

108 Rämmal, H. and Bodén, H. (2007). Modified multi-load method for nonlinear source characterization. *J. Sound Vib.* 299 (4–5): 1094–1113.

109 Zhang, Y., Liu, J., Kadlaskar, G. et al. (2018). Using the Moebius transformation to predict the effect of source impedance on insertion loss. *Noise Control Eng. J.* 66 (2): 105–116.

110 Zheng, S., Liu, H., Dan, J., and Lian, X. (2015). Analysis of the load selection on the error of source characteristics identification for an engine exhaust system. *J. Sound Vib.* 344: 126–137.

111 Tanttari, J., Isomoisio, H., and Hynninen, A. (2016) On power plant IC-engine source characterization in-situ. Proceedings of Baltic Nordic Acoustic Meeting – BNAM.

112 Davies, P.O.A.L. (1991). Transmission matrix representation of exhaust system acoustic characteristics. *J. Sound Vib.* 151 (2): 333–338.

113 Desmons, L. and Hardy, J. (1994). A least squares method for evaluation of characteristics of acoustical sources. *J. Sound Vib.* 175 (3): 365–376.

114 Boden, H. and Abom, M. (1995). Modeling of fluid machines as sources of sound in duct and pipe systems. *Acta Acust.* 3: 549–560.

115 Munjal, M.L. (2004) Acoustic characterization of an engine exhaust source – a review, Proceedings of Acoustics 2004.

116 Dokumaci, E. (2005). Prediction of source characteristics of engine exhaust manifolds. *J. Sound Vib.* 280 (3–5): 925–943.

117 Ih, J.-G., Kim, H.-J., Lee, S.-H., and Shinoda, K. (2009). Prediction of intake noise of an automotive engine in run-up condition. *Appl. Acoust.* 70 (2): 347–355.

118 Hynninen, A., Turunen, R., Åbom, M., and Bodén, H. (2012). Acoustic source data for medium speed IC-engines. *J. Vib. Acoust.* 134: 051008.

119 Macián, V., Torregrosa, A.J., Broatch, A. et al. (2013). A view on the internal consistency of linear source identification for IC engine exhaust noise prediction. *Math. Comput. Model.* 57 (7): 1867–1875.

120 Hynninen, A. (2015) Acoustic in-duct characterization of fluid machines with applications to medium speed IC-engines, PhD thesis. KTH Royal Institute of Technology, Stockholm.

121 Jones, A.D., Moorhem, W.K.V., and Voland, R.T. (1986). Is a full nonlinear method necessary for the prediction of radiated engine exhaust noise? *Noise Control Eng. J.* 26 (2): 74–80.

122 Bodén, H. and Albertson, F. (1998) In-duct acoustic one-port sources, linear or non-linear. Proceedings of Inter-Noise 98. 1: 227–230.

123 Payri, F., Torregrosa, A.J., and Payri, R. (2000). Evaluation through pressure and mass velocity distributions of the linear acoustical description of IC engine exhaust systems. *Appl. Acoust.* 60: 489–504.

124 Hynninen, A., Isomoisio, H., and Tanttari, J. (2017). IC-engine acoustic source characterization in-situ with capsule tube method. *Appl. Acoust.* 126: 1–18.

125 Lord Rayleigh, J.W.S. (1883). On porous bodies in relation to sound. *Phil. Mag.* XVI: 181–186. *Sci. Pap., I* 103: 203–225. See also (1945). *Theory of Sound*. Vol. II., 2e. New York: The Macmillan Company.

126 Sivian, L.J. (1937). Sound propagation in ducts lined with absorbing materials. *J. Acoust. Soc. Am.* 9: 135–140.

127 Morse, P.M. (1939). The transmission of sound inside pipes. *J. Acoust. Soc. Am.* 11: 205–210.

128 Sabine, H.J. (1940). The absorption of noise in ventilating ducts. *J. Acoust. Soc. Am.* 12: 53–57.

129 Scott, R.A. (1946). The propagation of sound between walls of porous material. *Proc. R. Soc.* 58: 358–368.

130 Zwikker, E. and Kosten, C.W. (1949). *Sound Absorbing Materials*. New York: Elsevier Publishing Company Inc.

131 Cremer, L. (1953). Theory of acoustics damping in rectangular ducts with an absorbing wall and the resulting highest attenuation. *Acustica* 3: 249–263.

132 Meyer, E., Mechel, F., and Kurtze, G. (1958). Experiments on the influence of flow on sound attenuation in absorbing ducts. *J. Acoust. Soc. Am.* 30: 165–174.

133 Kurze, U.J. and Ver, I.L. (1972). Sound attenuation in ducts lined with non-isotropic material. *J. Sound Vib.* 24: 177–187.

134 Ver, I.L. (1978). A review of the attenuation of sound in straight lined and unlined ductwork of rectangular cross section. *ASHRAE Trans.* 84: 122–149.

135 Kuntz, H.L. and Hoover, R.M. (1987) The interrelationships between the physical properties of fibrous duct lining materials and lined ducts sound attenuation. ASHRAE No. 3082 (RP-478), 449–470.

136 Ver, I.L. (1982) Acoustical design of parallel baffle muffler, Proceedings Inter-noise 82: 381–384, also published in Nelson Industries Conference Proceedings, 1980.

137 Howe, M.S. (1983). The attenuation of sound in a randomly lined duct. *J. Sound Vib.* 87: 83–103.

138 Reinstra, S.W. (1985). Contributions to the theory of sound propagation in ducts with bulk-reacting lining. *J. Acoust. Soc. Am.* 77: 1681–1685.

139 Galaitsis, A.G. (1986). Predicted sound interaction with a cascade of porous resistive layers in a duct. *Noise Control Eng. J.* 26: 62–67.

140 Astley, R.J. and Cummings, A. (1987). A finite element scheme for attenuation in ducts lined with porous material: comparison with experiment. *J. Sound Vib.* 116 (2): 239–263.

141 Vanderburgh, C.R. (1993) How in-duct silencing can be changed to better match the acoustic and aerodynamic needs of HVAC systems, NOISE-93 Proceedings, St. Petersburg, Russia, 217–222.

142 Grey, A. (2004) Induct dissipative bar-silencer design, Master's thesis. University of Canterbury, New Zealand.

143 Borelli, D. and Schenone, C. (2012). A finite element model to predict sound attenuation in lined and parallel-baffle rectangular ducts. *HVAC Res.* 18 (3): 390–405.

144 Ray, E.F. (2013) *Industrial Noise Series*, Part VIII: Absorptive silencer design, www.universalaet.com.

145 Bogdanovic, D. (2014) Calculation methods for predicting attenuation of parallel baffle type silencers, Master's thesis. Chalmers University of Technology, Sweden.

146 Binois, R., Perrey-Debain, E., Dauchez, N. et al. (2015). On the efficiency of parallel baffle-type silencers in rectangular ducts: prediction and measurement. *Acta Acust. Acust.* 101 (3): 520–530.

147 Denia, F.D., Selamet, A., Fuenmayor, F.J., and Kirby, R. (2007). Acoustic attenuation performance of perforated dissipative muffler with empty inlet/outlet extensions. *J. Sound Vib.* 302 (4–5): 1000–1017.

148 Denia, F.D., Antebas, A., Martinez-Cabas, J., and Fuenmayor, F.J., Transmission loss calculations for dissipative mufflers with temperature gradients, Proceedings of 20th International Congress on Acoustics, ICA 2010, August 2010, Sydney, Australia.

149 Ingard, K.U. (1994). *Notes on Sound Absorption Technology*. Poughkeepsie, NY: Noise Control Foundation.

150 Bies, D.A., Hansen, C.H., and Bridges, G.E. (1991). Sound attenuation in rectangular and circular cross-section ducts with flow and bulk reacting liner. *J. Sound Vib.* 146: 47–80.

151 Ramakrishnan, R. and Watson, W.R. (1992). Design curves for rectangular splitter silencers. *Appl. Acoust.* 35: 1–24.

152 Mechel, F.P. (1975) Design criteria for industrial mufflers. Proceedings Inter-Noise 75, 751–760.

11

Noise and Vibration Control of Machines

11.1 Introduction

Many different machine components are used in appliances, vehicles, aircraft, and industry. Some machine components including bearings, gears, fans, burners, cutters, and valves are used in machines built from several of these components such as pumps, compressors, electric motors, and internal combustion (IC) engines. It is impossible to discuss every type of machine and machine component here; so, the discussion is mainly concentrated on common machine components and complete machines. These machine components and machines do work, transfer energy, or convert one form of energy into another. In these processes some undesirable vibration and acoustical energy or noise is produced as an unwanted by-product. It is the purpose of this chapter to review briefly the functioning of these machine components and machines, noise and vibration generation mechanisms, and methods of control. Several other books, book chapters, and articles also discuss machinery noise mechanisms and noise and vibration sources in more detail than is possible here [1–15].

11.2 Machine Element Noise and Vibration Sources and Control

11.2.1 Gears

Gears are used in a number of applications where mechanical power is transmitted. Gears can emit annoying and harmful noise levels when a fraction of the transmitted power is converted to noise [16]. Most modern gear teeth have an involute profile, although some have circular-arc profiles [1, 2]. Figure 11.1a shows some of the terms used with involute gears and Figure 11.1b shows two parallel-axis spur-type gears meshing together.

There are several different types of gears in common use, as shown in Figure 11.2. They are of two main classes, either having (i) parallel axes (spur, helical) or (ii) nonparallel axes (straight bevel, spiral bevel, hypoid). Spur gears and straight bevel gears are usually the noisiest, while helical spiral bevel gears are usually the quietest in their respective classes. This is because the load between gear wheel teeth is transferred more gradually with helical and spiral-bevel gears and rather more abruptly with spur and straight-bevel gears.

Gear noise can arise from a variety of causes. As the gear teeth mesh, a pulsating force occurs at the gear tooth meshing frequency f_m and its harmonics. Harmonics are present because the pulsating force on the gear teeth is not purely sinusoidal. The strength of the harmonics depends on the force pulse shape and also on other impulsive forces caused by tooth deformation, production machinery errors, bearing misalignment,

Engineering Acoustics: Noise and Vibration Control, First Edition. Malcolm J. Crocker and Jorge P. Arenas.
© 2021 John Wiley & Sons Ltd. Published 2021 by John Wiley & Sons Ltd.

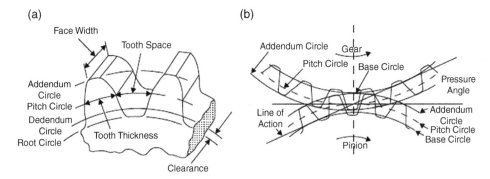

Figure 11.1 Terms used with gears: (a) involute gear, (b) meshing of two parallel-axis spur gears. (*Source:* Based in part on figures in Refs. [1–3].)

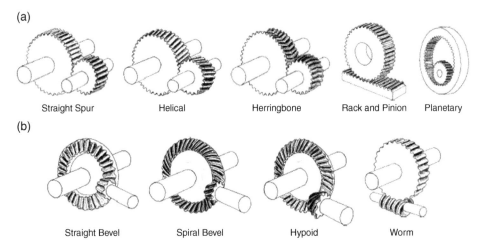

Figure 11.2 The main types of gear in use: (a) parallel axis (straight spur, helical, herringbone, rack and pinion, and planetary); (b) nonparallel axis (straight bevel, spiral bevel, hypoid, and worm) [18].

pinion wheel deformation, and so on. At high gear speeds, air or lubricating fluids can be expelled from between the meshing teeth at supersonic speeds and can even become the dominant source of noise. At low speed and load, the sound pressure level produced by a gear increases by about 3 dB for a doubling of load; while at higher speeds and loads the sound pressure level increases by about 6 dB for a doubling of speed or load [17]. Houser [16] discussed the energy flow in gears in schematic form (see Figure 11.3). This figure is useful to identify the steps in developing models to predict noise and vibration and to identify transducer locations along the gear noise and vibration paths.

The noise of a gear set is quite dependent on the quality of manufacture and the tolerances achieved. Theoretical and experimental studies have been made that attempt to relate gear surface deformation and profiles to noise [19–21]. Precision gear sets can now be manufactured [22], which make very low noise levels, although the cost is higher. In some cases, where low noise levels are required, it is more cost-effective to choose a gear set of moderate cost and to vibration-isolate the pinion and gear bearings, apply damping to the gear housing, and, if necessary, completely enclose it. With gears used to transmit only small loads (e.g. those in electric clocks), very low noise levels can be achieved by the use of soft plastic gears (which reduce the gear tooth force impulses) and by other means [23]. Manufacturing deficiencies can result in variations in pitch and profile from tooth to tooth and eccentricity of the gear wheel, which causes increased

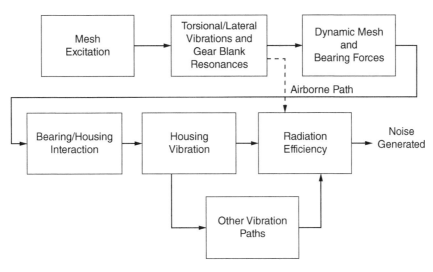

Figure 11.3 Gear noise energy flow diagram [16].

noise and vibration [1–4, 17]. There would, however, be some noise generated even if the gears were without any imperfections. The frequency of this noise (and vibration) would only occur at the gear meshing frequency f_m and its harmonics:

$$f_m = \frac{N_p n_p}{60} \text{ Hz,} \tag{11.1}$$

where N_p is the number of pinion teeth and n_p is the pinion speed in revolutions/minute (rpm). The mesh frequency computation for planetary gears is more complex and depends upon the gear configuration. Noise and vibration measurements have also been used to produce a gear noise and vibration rating index in an attempt to avoid the necessity for using a jury test for this purpose [24].

Example 11.1 Calculate the gear meshing frequency for a spur gear having 36 teeth rotating at 2000 rpm.

Solution

Using Eq. (11.1), the gear mesh frequency is $f_m = 36 \times 2000/60 = 1200$ Hz.

Example 11.2 Determine the mesh frequency and two higher harmonic frequencies of a 25-tooth gear rotating at 1200 rpm.

Solution

The mesh frequency is determined using Eq. (11.1) as $f_m = 25 \times 1200/60 = 500$ Hz. Gear whine noise also occurs at harmonics of mesh frequency ($2 f_m$, $3 f_m$, etc.). Therefore, two higher harmonic frequencies are 1000 and 1500 Hz. In addition, modulations due to tooth spacing errors, eccentricities, and torsional vibrations often create sidebands that are spaced at shaft rotational frequency ($f_s = n_p/60$) intervals on either side of the mesh frequency and its harmonics. This is clearly observed in Figure 11.4.

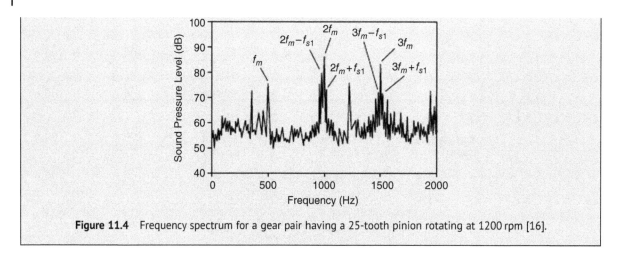

Figure 11.4 Frequency spectrum for a gear pair having a 25-tooth pinion rotating at 1200 rpm [16].

Table 11.1 Effects of different gear design and manufacturing parameters on gear noise [16].

	Direction to reduce noise	Noise reduction (dB)	Comments
Number of teeth	Decrease	0–6	Lowers mesh frequency
Contact ratio	Increase	0–20	Requires accurate lead and profile modifications
Helix angle	Increase	0–20	Machining errors have less effect with helical gears. Little improvement above about 35°
Surface finish	Reduce	0–7	Depends on initial finish – reduces friction excitation
Profile modification		4–8	Good for all types of gears
Lapping		0–10	Very effective for hypoid gears
Pressure angle	Reduce	0–3	Reduces tooth stiffness, reduces eccentricity effect, and increases contact ratio
Face width	Increase		Increases contact ratio for helical gears; reduces deflections

Some of the gains that might be achieved by altering transmission gear designs have been summarized by Houser in Table 11.1 [16]. However, the information in Table 11.1 must be used with care since (i) the many reductions that are possible in the table are not additive in nature and (ii) changing one design quantity inevitably changes other quantities.

Reference [16] provides a detailed discussion of the sources of noise and vibration in the main types of gears in common use. Some information on the noise of gearboxes and transmissions used in vehicles can be found in Ref. [25].

11.2.2 Bearings

There are two main types of bearings: (i) rolling contact and (ii) sliding contact [6, 7]. Rolling contact bearings are more commonly used, but sliding contact bearings are usually quieter than rolling contact bearings, if properly manufactured, installed, and maintained. Proper lubrication is essential for both rolling and sliding contact bearings. Reference [26] presents a detailed review of the noise of bearings.

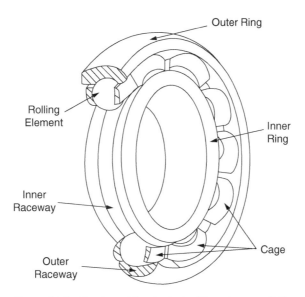

Figure 11.5 Bearing with spherical rolling elements [18].

Rolling contact bearings consist of the rolling elements contained between the *inner* and *outer raceways*. The rolling elements are normally kept from touching each other by a *cage*. The rolling elements may be spherical, cylindrical, tapered, or barrel shaped [6]. Figure 11.5 shows a bearing with spherical rolling elements. The noise made by a rolling contact bearing is normally caused by vibration from two main sources: (i) rotation of bearing elements and (ii) resonances in the elements, raceways, or cage. Reference [3] has likewise identified discrete frequencies (and their harmonics) that are related to bearing geometry and rotational speed. The fundamental frequency is the shaft rotational frequency f_s:

$$f_s = N/60 \text{ Hz}, \tag{11.2}$$

where N is shaft rotational speed in rpm.

The other frequencies are related to the shaft frequency f_s by factors that depend on the roller diameter (d_{rol}), the pitch diameter of the bearing (d_{cg}), the contact angle between the roller element and the raceway (ϕ), and the number of rolling elements (Z_{rol}). Manufacturing imperfections and misalignment cause bearing noise. This noise can be increased further by wear. Approximate formulas for the calculation of the main harmonic frequencies signifying rolling bearing defects have been proposed [26]. These main harmonic frequencies are the cage rotational frequency (*FTF*, Fundamental Train Frequency), the frequency of tumbling of rolling elements over the outer race (*BPFO*, Ball Pass Frequency of Outer ring), the frequency of tumbling of rolling elements over the inner race (*BPFI*, Ball Pass Frequency of Inner ring), and the rotational frequency of rolling elements (*BSF*, Ball Spin Frequency). These formulas, in Hz, are given by:

$$FTF = \frac{f_s}{2}\left(1 - \frac{d_{rol}}{d_{cg}}\cos\phi\right), \tag{11.3}$$

$$BPFO = \frac{f_s}{2}Z_{rol}\left(1 - \frac{d_{rol}}{d_{cg}}\cos\phi\right), \tag{11.4}$$

$$BPFI = \frac{f_s}{2}Z_{rol}\left(1 + \frac{d_{rol}}{d_{cg}}\cos\phi\right), \tag{11.5}$$

$$BSF = \frac{d_{cg}}{d_{rol}} \frac{f_s}{2} Z_{rol} \left(1 - \left[\frac{d_{rol}}{d_{cg}} \cos\phi \right]^2 \right). \tag{11.6}$$

Bearing manufacturers normally provide tabulated values of *FTF*, *BPFO*, *BPFI*, and *BSF* normalized by the shaft rotational frequency. Thus, the main harmonic frequencies are found by multiplying these tabulated values by the rotational speed of the shaft.

Example 11.3 A ball bearing is to operate at 3000 rpm. The bearing has 20 rolling elements, a roller diameter of 0.531 in., the pitch diameter of the bearing is 4.036 in., and the contact angle between the roller element and the raceway is 30°. What are the main harmonic frequencies signifying rolling bearing defects?

Solution

First, we determine the shaft rotational frequency f_s using Eq. (11.2) as $f_s = 3000/60 = 50$ Hz. Then, the *FTF* is calculated from Eq. (11.3) as

$FTF = (50/2)[1 - (0.531/4.036) \times \cos(30)] = 22.15$ Hz. Now, the *BPFO* ring can be determined by multiplying *FTF* by the number of rolling elements, thus $BPFO = 20 \times 22.15 = 443$ Hz. Finally, using Eqs. (11.5) and (11.6) we get that $BPFI = 557$ Hz and $BSF = 187.5$ Hz.

If the inner ring of the rolling bearing is fixed and the outer ring is rotational (e.g. wheel pairs), then the cage rotational frequency (*FTF*) in Eq. (11.3) should be changed by the frequency $(f_s - FTF)$. If the inner ring of the rolling bearing rotates and the outer ring is fixed (e.g. midshaft bearing in aviation engines), then the shaft rotational frequency (f_s) in Eqs. (11.3)–(11.6) should be changed by the differential frequency $(f_{s1} - f_{s2})$ (frequencies of rotation of rotors f_{s1} and f_{s2}), while the cage rotational frequency should be changed by the sum frequency $(f_{s2} + FTF)$.

Even a perfect bearing will make noise when loaded [2]. If a rolling bearing is manufactured to a higher grade of precision (smaller tolerance), then it normally becomes quieter (and more expensive). Classes of tolerance are specified in ISO Standard 492–1986. Methods to test bearings for noise and vibration are given in ANSI/AFBMA Standard 13–1970. The (American) Military Specification MIL-B-17913D defines permissible vibration limits for bearings.

Sliding contact bearings can be divided into three main types: (i) journal, (ii) thrust, and (iii) guide. Journal bearings are cylindrical in shape and allow rotation (see Figure 11.6). Thrust bearings are used to prevent motion along a shaft axis, while guide bearings are normally used for motion of a part in one direction without rotation (e.g. a piston sliding in a cylinder of an IC engine).

When shaft rotation occurs with a journal bearing, the shaft rides on a film of lubricant. Under some conditions, however, this film can break down, causing metal-to-metal contact and consequently wear, noise, and vibration. A well-known instability condition called *oil whirl* can occur causing noise at a frequency of about half of the shaft rotational speed. This is because the average speed of the lubricant film is

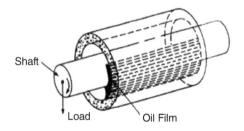

Figure 11.6 Sliding contact bearing [18].

about half of the shaft speed. Oil-whirl noise can be magnified if a shaft resonance frequency occurs near to half the shaft rotational speed.

To minimize sliding bearing noise, proper attention should be paid to lubricant viscosity, pressure, alignment, and structural stiffness. Proper installation of bearings is important to achieve low vibration and noise [27]. Increased bearing noise and vibration are an indication of wear and/or misalignment, and, if correctly analyzed, these can be indicative of potential bearing failure [28, 29]. Active control of special magnetic bearings for a fan has been undertaken in an attempt to set the fan blades into vibration so that they become a secondary out-of-phase sound source to cancel the primary fan noise. Noise reductions of up to 4 dB have been reported [30]. Reference [26] contains more detailed discussion on bearing noise. Further discussion can also be found in Refs. [2, 3, 6, 7].

11.2.3 Fans and Blowers

Fans and blowers are used in appliances, in buildings, in air distribution systems for heating and cooling, and in industry for a variety of purposes. Fan noise, its generation, and control are discussed in detail in Ref. [31] and Chapter 13 in the present book. There are two main types of fan designs: axial and centrifugal (see Figure 11.7). The first three centrifugal fan types – airfoil, backward-curved, and radial – are mostly used in industrial applications. The airfoil fan is the most efficient but it is only suitable for clean-air industrial applications because dust and other particles can adhere to the fan blades and cause malfunctioning.

The forward-curved fan is usually made of lightweight low-cost materials and is generally the least efficient. The radial fan is the noisiest and least efficient but is useful for dirty or corrosive gas flows. Axial fans have the disadvantage that the discharged air rotates, unless downstream or upstream guide vanes (stators) are installed; these fans are mainly used in low- or medium-pressure air-conditioning systems. The vane-axial fan is the most efficient axial type and can generate high pressures. The propeller type is the least expensive of the axial type. It has the lowest efficiency and is normally limited to low-pressure, high-volumetric flow applications. The primary purpose of a fan is to move a required volumetric flow rate of air against a given back pressure with maximum efficiency and low cost and noise. There may be additional requirements such as high resistance to abrasion, ability to transport dusty air, ease of manufacture, maintenance and repair, and noise restrictions.

Fan noise has pure-tone and broadband frequency components. Although the mathematical theory of fan noise is well developed, it is beyond the scope of this chapter, and physical explanations are presented instead. Noise from fans is caused by several mechanisms that have been summarized by Lauchle in Figure 11.8 [31].

Each time a blade passes a point in space or a solid-body obstruction, an impulsive force fluctuation is experienced by the fluid or solid body at the point. If a fan has n equally-spaced blades and the rotational speed is N rpm, then the number of impulses experienced per second f_B is

$$f_B = nN/60 \text{ Hz.} \tag{11.7}$$

The frequency f_B is known as the *blade passing frequency (BPF)* or often the *blade frequency* for short. Since the time history of the impulsive force on the fluid or solid-body obstruction will not be completely sinusoidal, harmonics will appear. The strength of the harmonics is affected by upstream or downstream solid-body flow obstructions.

The noise generated by a fan depends primarily on its design features, geometrical dimensions, and operating speed and load. Both broadband and pure-tone fan noise normally increase with increasing fan speed N. The frequency of the pure-tone noise generated by a fan increases with fan speed N as shown by Eq. (11.7).

The overall sound power in watts of a fan of rotor diameter D can be approximately predicted by [31]

Fan Type	Description	Design Details

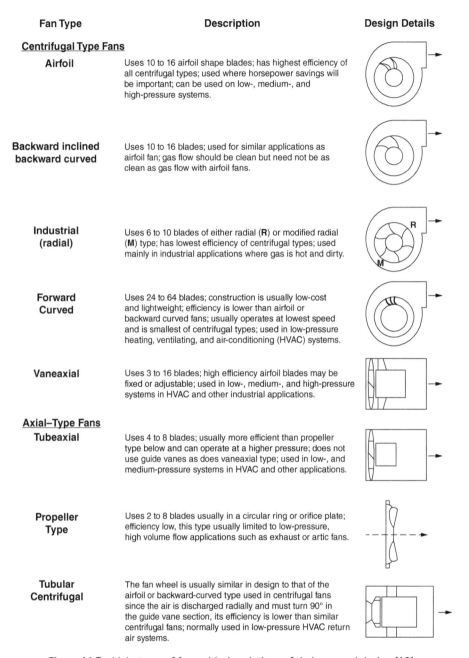

Centrifugal Type Fans

Airfoil — Uses 10 to 16 airfoil shape blades; has highest efficiency of all centrifugal types; used where horsepower savings will be important; can be used on low-, medium-, and high-pressure systems.

Backward inclined backward curved — Uses 10 to 16 blades; used for similar applications as airfoil fan; gas flow should be clean but need not be as clean as gas flow with airfoil fans.

Industrial (radial) — Uses 6 to 10 blades of either radial (**R**) or modified radial (**M**) type; has lowest efficiency of centrifugal types; used mainly in industrial applications where gas is hot and dirty.

Forward Curved — Uses 24 to 64 blades; construction is usually low-cost and lightweight; efficiency is lower than airfoil or backward curved fans; usually operates at lowest speed and is smallest of centrifugal types; used in low-pressure heating, ventilating, and air-conditioning (HVAC) systems.

Vaneaxial — Uses 3 to 16 blades; high efficiency airfoil blades may be fixed or adjustable; used in low-, medium-, and high-pressure systems in HVAC and other industrial applications.

Axial–Type Fans

Tubeaxial — Uses 4 to 8 blades; usually more efficient than propeller type below and can operate at a higher pressure; does not use guide vanes as does vaneaxial type; used in low-, and medium-pressure systems in HVAC and other applications.

Propeller Type — Uses 2 to 8 blades usually in a circular ring or orifice plate; efficiency low, this type usually limited to low-pressure, high volume flow applications such as exhaust or artic fans.

Tubular Centrifugal — The fan wheel is usually similar in design to that of the airfoil or backward-curved type used in centrifugal fans since the air is discharged radially and must turn 90° in the guide vane section, its efficiency is lower than similar centrifugal fans; normally used in low-pressure HVAC return air systems.

Figure 11.7 Main types of fans with descriptions of their use and design [18].

$$W = D^7 \times (N/60)^5. \tag{11.8}$$

Equation (11.8) indicates that the overall sound power at a constant point of operation is proportional to D^7 and increases as the fifth power of the fan speed. Other important dimensionless parameters in fan design are the flow coefficient ϕ and the static pressure coefficient ψ, given by

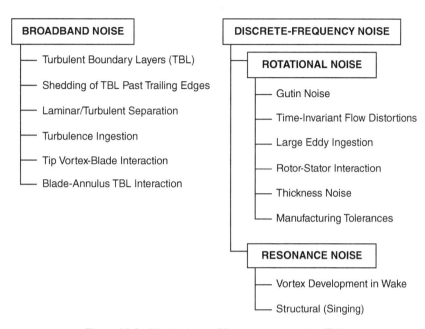

Figure 11.8 Mechanisms of fan noise generation [31].

$$\phi = \frac{4Q}{\pi^2 D^3 (N/60)},$$ (11.9)

and

$$\psi = \frac{2\Delta P}{\rho_0 \pi^2 D^2 (N/60)^2},$$ (11.10)

where Q is the volumetric flow rate (m^3/s), ΔP is the static pressure rise (Pa), and ρ_0 is the air density (kg/m^3).

Example 11.4   Consider a fan of diameter 0.4 m operating at 3000 rpm. Find either a new diameter or new operating speed to achieve a 10-dB noise reduction.

Solution

The noise reduction is $\Delta L_w = 10 \log (W_1/W) = -10$ dB, where W_1 is the overall sound power of the modified fan (see Eq. (11.8)).

a) At constant operating speed, we have: $10 \log \left(\frac{D_1^7 (N/60)^5}{D^7 (N/60)^5}\right) = 70 \log (D_1/D) = -10$. Therefore, $D_1 = D \times 10^{-1/7} = 0.72 \times D = 0.72 \times 0.4 = 0.29$ m.

b) At constant diameter: $10 \log \left(\frac{D^7 (N_1/60)^5}{D^7 (N/60)^5}\right) = 50 \log (N_1/N) = -10$, so $N_1 = N \times 10^{-1/5} = 0.63 \times N = 0.63 \times 3000 = 1893$ rpm.

  Thus, we can change the diameter to 29 cm or change the fan speed to 1893 rpm to obtain a 10-dB overall noise reduction while maintaining the same value of φ and ψ. However, the performance of the fan will suffer because of the noise reduction enforced. Thus, one needs to decide from the application whether these decreases in performance are justified. Equations (11.9) and (11.10) show that at constant speed, the flow rate is proportional to D^3 and the pressure rise to D^2. The new flow rate would be $Q_1 = (0.72)^3 Q = 0.37Q$ and the new static pressure rise would be $\Delta P_1 = (0.72)^2 \Delta P = 0.52\Delta P$. On the other hand, at constant diameter, these new values are $Q_1 = (0.63)^3 Q = 0.25Q$ and $\Delta P_1 = (0.63)^2 \Delta P = 0.40\Delta P$. Therefore, if flow rate is more important than static pressure rise, one might choose to lower the fan speed to obtain the required noise reduction [31].

Structural resonances can also be excited in a fan. These resonance frequencies are largely independent of fan speed. If the fan is operated at off-design conditions, its noise can also be higher than normal. If the fan is operated at reduced volumetric flow rates, the A-weighted sound pressure level can be as much as 15 dB higher than normal because of fan surge and rotating fan stall.

Fan sound power level data are normally provided by manufacturers. Table 11.2 gives specific sound power levels (ref: 10^{-12} watts) in eight octave bands from 63 to 8000 Hz for different types of fans and blowers [31]. This table, based on the research of Graham and Hoover [32] and others, provides an empirical method for predicting the noise of axial and centrifugal fans. In order to account for flow rate and static pressure rise, simply add $10\log Q + 20 \log \Delta P$ (with Q in m^3/s and ΔP in kPa) to the tabulated values. The last column in Table 11.2 gives the value to be added to the level of the particular octave band in which the BPF falls.

Table 11.2 Specific sound power levels, dB, in eight lowest one-octave bands for a variety of axial and centrifugal fans [31].

Fan type	Rotor diameter (m)	63	125	250	500	1000	2000	4000	8000	Add for BPF
Backward-curved	>0.75	85	85	84	79	75	68	64	62	3
centrifugal	<0.75	90	90	88	84	79	73	69	64	3
Forward-curved	All	98	98	88	81	81	76	71	66	2
centrifugal										
Low-pressure radial	>1.0	101	92	88	84	82	77	74	71	7
$996 \le \Delta P \le 2490$	<1.0	112	104	98	88	87	84	79	76	7
Midpressure radial	>1.0	103	99	90	87	83	78	74	71	8
$2490 \le \Delta P \le 4982$	<1.0	113	108	96	93	91	86	82	79	8
High-pressure radial	>1.0	106	103	98	93	91	89	86	83	8
$4982 \le \Delta P \le 14\,945$	<1.0	116	112	104	99	99	97	94	91	8
Vane-axial										
$0.3 \le D_h/D \le 0.4$	All	94	88	88	93	92	90	83	79	6
$0.4 \le D_h/D \le 0.6$	All	94	88	91	88	86	81	75	73	6
$0.6 \le D_h/D \le 0.8$	All	98	97	96	96	94	92	88	85	6
Tube-axial	>1.0	96	91	92	94	92	91	84	82	7
	<1.0	93	92	94	98	97	96	88	85	7
Propeller	All	93	96	103	101	100	97	91	87	5

Example 11.5 Estimate the total sound power level of a 24-bladed forward-curved centrifugal fan running at 2400 rpm. The fan delivers a volume flow rate $Q = 20\,m^3/s$ against a static pressure of 5 kPa.

Solution

We must calculate the BPF using Eq. (11.7), BPF = $24 \times 2400/60 = 960$ Hz. Now, we calculate $10 \log Q + 20 \log \Delta P = 10 \log (20) + 20 \log (5) = 13 + 14 = 27$ dB.

Thus, referring to the third row of Table 11.2 we must add 27 dB to the values of sound power levels for all one-octave bands, and we must also add 2 dB at the 1000 Hz one-octave band where the BPF falls. Therefore, the sound power levels in eight one-octave bands from 63 to 8000 Hz are: 125, 125, 115, 108, 110 (108 + 2), 103, 98, and 93 dB, respectively. The overall sound power level is obtained by logarithmically adding the previous one-octave band levels. Then,

$$L_w = 10 \log \left(2 \times 10^{12.5} + 10^{11.5} + 10^{10.8} + 10^{11} + 10^{10.3} + 10^{9.8} + 10^{9.3}\right) = 128 \, dB.$$

As discussed in Example 11.4 it is difficult to reduce fan noise by changing fan design parameters since such changes may adversely affect fan performance as well. Most fan noise reduction is achieved by proper use of well-known passive control methods. If noise is of major concern and the fan has been properly installed, it will probably be necessary to install intake and discharge sound attenuators (with flexible vibration breaks at attachment points to duct systems, if present). It is also possible to provide further noise attenuation by the use of ducting, elbows, and plenum chambers lined with sound-absorbing material. Care must be taken that significant noise is not generated in this ductwork. Chapter 13 in the present book and Ref. [33] describe the prediction and control of the noise and vibration in ducted heating, ventilation, and air-conditioning systems, and Ref. [34] discusses the aerodynamic sound generated in low-speed flow ducts.

Use of tuned resonators such as cavities for controlling tones is another passive noise control technique. For axial fans, the cavity can be integrated into the hub. In centrifugal fans a cavity can be installed in the cutoff region. The cavity must have rigid walls and designed to be a one-quarter wavelength long at the BPF. The application of this method to a 44-blade centrifugal blower is shown in Figure 11.9.

Considerable efforts continue to be made to reduce fan noise since fans are used in most computers, electric motors, household appliances, vehicles, trucks, and many other items of machinery. Numerous studies have been conducted on the noise control of fans [30, 35–49]. Both passive [30, 35–43] and active [44–48] noise control methods continue to be studied. Experimental and theoretical studies have been made to aid in machinery noise reduction and noise predictions. Sound quality studies have also been conducted on the acceptability of fan noise [49].

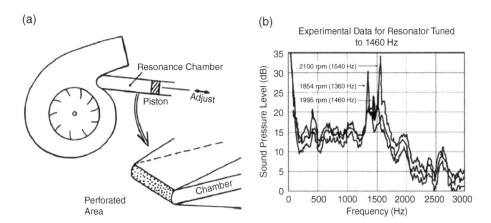

Figure 11.9 Application of a quarter-wavelength resonator for blower BPF noise control: (a) Sketch of the quarter-wavelength resonator installed in the cutoff; (b) Sound pressure level spectra measured for a 44-blade forward-curved centrifugal blower at different speeds. A cavity resonator is mounted in the cutoff of the volute. When the length of the cavity corresponds to the one-quarter wavelength at the frequency of the BPF, tonal noise reduction is observed. The frequencies in parentheses are the BPFs for the noted rotor speed [31].

11.2.4 Metal Cutting

Many industrial processes involve cutting metals. Metal-cutting processes can either be *continuous* or *impulsive* in character. Examples of continuous processes include sawing, drilling, milling, and grinding. Additional continuous cutting processes include use of water jets for cutting steel plates up to 300-mm thickness and plasma and laser cutting techniques. Examples of impulsive processes include punching, piercing, and shearing [50].

The following noise sources can be observed in continuous metal-cutting processes: (i) aerodynamic noise usually caused by vortex shedding from a spinning tool such as from the teeth of a high-speed rotating saw, (ii) noise caused by vibrations of the cutting tool, (iii) noise due to structural vibration and radiation from the cutting tool and the workpiece such as regenerative chatter, and (iv) noise due to material fracture in which energy is built up until it is released at fracture as acoustic emission.

The noise level radiated by continuous cutting processes depends on the feed rate of the workpiece, the depth of cut, the resonance frequencies of the cutting tool and the workpiece, the geometry of the cutting tool and the workpiece, and the radiation efficiencies of the resonant modes of vibration. In impulsive cutting processes such as stamping, forging, punching, piercing, and shearing, impulsive forces are involved. The noise generated by these operations includes acceleration noise, ringing noise, noise due to fracture of the feedstock material (cutting noise), and other machinery noise sources. Acceleration noise is generated by the impact between the cutting blade and the feedstock, causing the air around to be compressed due to the rapid surface deformations. This noise is usually low frequency in nature and is normally much less than the ringing noise caused by flexural vibrations.

Ringing noise is generated from vibrations of the feedstock and machine structure including the machine foundations; it usually makes a significant contribution to the overall noise generated during the metal-cutting process. The magnitude of the ringing noise depends on the radiation efficiency, and spatial averaged mean-square normal vibration velocity of the surface and its area, the air density, and speed of sound in air. Figure 11.10 shows the noise signature at various stages of the operation of a roll former production line. Different sources of noise radiated can be identified from this sound pressure time-history, such as the noise

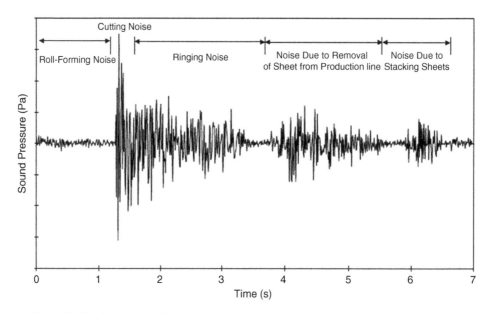

Figure 11.10 Typical sound pressure trace during the operation of a roll former shear [50].

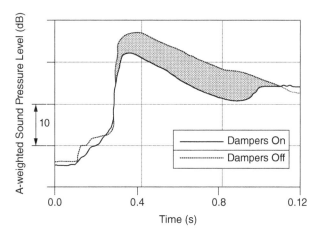

Figure 11.11 Comparison of the sound pressure level during cutting with and without sheet dampers [50].

due to roll forming a flat metal sheet into a profiled sheet, cutting the sheet to the required length, removing the sheet from the production line and dropping the sheet onto a stack. It is noted that high impulsive noise levels are produced by the cutting action (fracturing the metal) and the resulting impact-induced vibration of the product and the surrounding structure (ringing noise). The noise due to removal of the product from the production line and stacking the product may be reduced by changing the operator's work practice or by installing an automatic stacking machine [50].

In some cases, the ringing noise can be effectively reduced by the judicious use of damping materials. Figure 11.11 shows the sound pressure level as a function of time of a damping system used to reduce the vibrations transmitted to a sheet from the impact of the shear blades during the cutting action. The damping system was designed to clamp the sheet prior to and just after the cutting of the sheet [50]. Figure 11.11 shows that a noise reduction of over 5 dB has been achieved by the installation of sheet dampers at the operator position.

Richards et al. have published extensive discussions on acceleration noise and ringing noise [51, 52]. Reference [50] discusses noise sources due to *continuous* metal-cutting processes and to *impulsive* impact/shearing processes. The reference also covers basic theory for the noise emission caused in the cutting of metals. In addition, various noise control approaches (such as use of enclosures, damping materials, sound absorption materials, barriers, and vibration isolation) to reduce metal-cutting noise are presented. Reference [50] reviews numerical methods to predict metal-cutting noise and vibration as well. Research continues on reducing metal-cutting noise [53–60].

11.2.5 Woodworking

Woodworking machinery includes a wide variety of equipment, ranging from off-road forest equipment to simple circular saws, band saws, and jig saws used in industrial and residential workshops. Extremely high noise levels produced by these woodworking machines during both operating and machine idle conditions pose a severe threat to the hearing of the operators of such machinery [61]. Many industrial woodworking machinery noise sources result from the use of saw blades and cutters.

The main types of sawing operations in woodworking machinery are circular sawing and band sawing. In both cases, noise is produced by aerodynamic sources involving the tooth and gullet area of the blade during idle, blade structural vibration noise, and structural vibration noise that is produced by workpiece-related

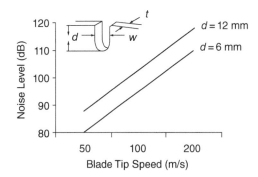

Figure 11.12 Effect of tip speed and gullet depth (*d*) on aerodynamic noise generation for circular saw blades [61].

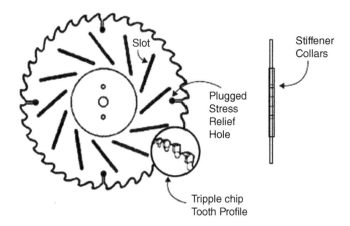

Figure 11.13 Slotted saw blade [63].

sources. Large numbers of circular saws are used all over the world to cut metal, stone, and wood. They are an effective but very noisy form of cutting tool. Circular saws of the carbide-tipped design operating at relatively high peripheral speed (greater than 50 m/s) are usually responsible for the great majority of employee over-exposures found in the woodworking industry. Circular saw blades are commonly found on cutoff/trim saws, single- and multiple-blade rip saws, and panel saws [61].

Rotating rigid disks with various types of openings (gullets) cut into the rim have been used to study airflow disturbances and to predict aerodynamic noise of circular saw blades. The aerodynamic source mechanism involves fluctuating forces set up near the blade periphery. However, details of how this fluctuating force is created or the effect of gullet geometry on the noise produced is still unclear. Figure 11.12 shows the experimental observations made by Stewart of the effect of tip speed and gullet depth on aerodynamic noise generation for circular saw blades [62].

The results of Figure 11.12 indicate that doubling of tip speed results in a 15–18-dB increase in aerodynamic noise level. Noise levels also depend on details of gullet geometry and saw blade plate thickness. In particular, the gullet depth has a direct effect on noise level, as shown in Figure 11.12.

Many different passive methods have been used to reduce the noise of circular saws. These range from source, path, to receiver approaches. The most effective approach is to control the saw noise at the source. One approach reported by Bobeczko involves minimizing the kerf (tooth width), slotting the blade body and adding collars to stiffen and damp the blade vibration (see Figures 11.13 and 11.14) [63]. Figure 11.14 shows that an A-weighted noise reduction of 19 dB was achieved in this particular case. This particular type of blade

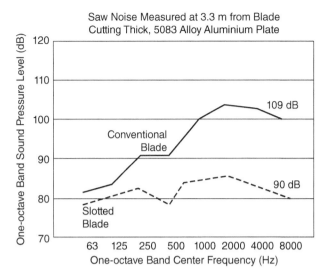

Figure 11.14 One-octave band noise levels of saw blades. Overall A-weighted levels also shown 109 dB – conventional blade, 90 dB – slotted blade [63].

was used extensively over seven years without mechanical problems [63]. To avoid problems, however, it should be made by a skilled machinist [63].

It is believed that the blade shown in Figure 11.13 is substantially quieter than similar blades, because (i) the natural modes of vibration have been changed, (ii) the slots allow some cancelation of sound radiated from the front and back of the blade, and (iii) the added collars and plugged stress release holes add some stiffness and damping, thus reducing the blade vibration and consequent sound radiation.

Ringing noise can also be reduced by mechanical clamping of the workpiece as shown, for example, in Figure 11.15. The blade-radiated noise is also reduced because the saw blade is "buried" beneath the work table for much of the time, and the operator is shielded from its noise. Acoustical shielding such as this is particularly effective for the high-frequency saw blade noise.

Cutters are widely used in the woodworking industry on machines such as planers, molders, cutters, routers, and the like to smooth and shape wood. Most cutters have several rows of knives that protrude above the cutter body, which is normally cylindrical. In many cases the machining process involves a peripheral milling in which each knife removes a "chip," leaving a relatively smooth surface. The smoothness of the cut depends on the machine feeds and speeds used. Most cutters have straight knives (aligned parallel to the cutter rotational axis). This design of cutter is used because it is easy to manufacture and sharpen as needed. Although this design of cutter is effective in working the wood, it is inherently noisy.

The compression of the air, which is caused when cutters rotate near to a stationary surface, results in a siren-type noise mechanism. This mechanism is very efficient in generating sound. This noise is pure tone in nature and occurs at the knife blade passing frequency (number of knives × rpm/60 Hz). The rotating cutter A-weighted noise level increases with cutter rpm and also increases with decreasing distance (clearance) between the knives and the stationary surfaces [61]. The A-weighted noise level produced by rotating cutters during idle also increases with increasing length of the cutter.

The noise during idle for machines equipped with cutters also increases with increased height of the knives and increased open area of the gullet. During cutting, the periodic impact of the cutter knives excites vibration in the wood workpiece (board) which usually becomes the dominant source of operator noise exposure. This noise is also pure tone in nature and takes place at knife BPF and harmonics of this frequency. Another factor that determines the noise radiated during cutting is the workpiece geometry.

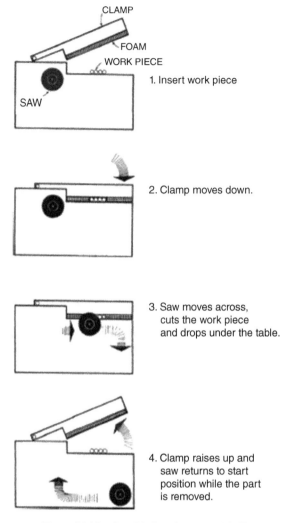

Figure 11.15   Saw blade noise control [63].

Many different techniques have been used to reduce the noise of cutters, including making changes in cutter design. It has been shown that continuous helical cutter design is the most effective way for reducing both idling and cutter noise. This design effectively smoothes out the force input which dramatically reduces the vibration of the workpiece. Figure 11.16 shows the cutter force–time history for straight knife and helical cutters.

In some cases, noise reduction of cutters is accomplished by reducing the width of the workpiece, using slotted tables and air guides, performing machine design changes, modification of feed rates and rotating speeds, and using acoustical enclosures to control the noise in the path [61].

Complete durable panel enclosures such as those illustrated in Figure 11.17a can have an A-weighted insertion loss of as much as 20 dB, provided considerable care is taken to seal the enclosure wherever it is possible and feasible. The enclosure shown in Figure 11.17a was built to house a wood molder and is provided with a viewing window and an entry door to provide access to the tool inside. An additional barrier/guard at the wood output location is provided to accommodate varying sizes of wood parts being molded [65]. The

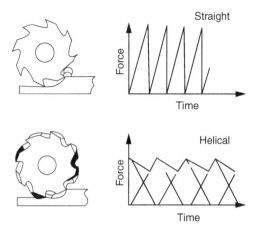

Figure 11.16 Smoothing of force–time history for helical vs. straight knife cutter [64].

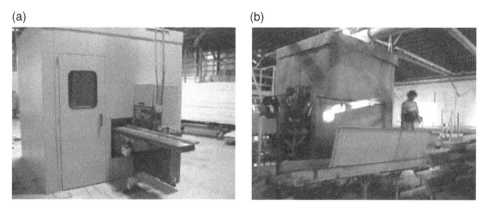

Figure 11.17 (a) Use of complete panel enclosure for molder noise and (b) use of mass-loaded vinyl "curtains" to control rip saw noise [65].

insertion loss is limited in practice by the need for the penetration provided for the workpiece. Figure 9.36 in Chapter 9 of this book shows another similar example of a woodworking machine enclosure for a band saw. The A-weighted insertion loss of mass-loaded vinyl barrier quilted blankets (commonly called "curtains"), as shown in Figure 11.17b, is typically only about 5–7 dB. However, the assembly of the curtain system is relatively simple and the cost of the mass-loaded curtain system shown in Figure 11.17b is less than that of the complete enclosure shown in Figure 11.17a. Curtains, however, are not as durable as panel enclosures and normally must be replaced periodically.

11.3 Built-up Machines

11.3.1 Internal Combustion Engines

The IC engine is a major source of noise in transportation and industrial use. The intake and exhaust noise can be effectively silenced. However, the noise radiated by engine surfaces is more difficult to control. In gasoline engines, a fuel–air mixture is compressed to about one-eighth to one-tenth of its original volume and ignited by a spark plug. In diesel engines air is compressed to about one-sixteenth to one-twentieth of its original volume and liquid fuel is injected in the form of a spray; then spontaneous ignition and combustion occurs. Because the rate of pressure rise is initially more abrupt with a diesel engine than with a gasoline engine, diesel engines tend to be noisier than gasoline engines. The noise of diesel engines has consequently received more attention from both manufacturers and researchers. The noise of IC engines is discussed in detail in Ref. [66].

The noise of engines can be divided into two main parts: combustion noise and mechanical noise. The combustion noise is caused mostly by the rapid pressure rise caused by ignition, and the mechanical noise is caused by a number of mechanisms, with perhaps piston slap being one of the most important, particularly in diesel engines. The noise radiated from the engine structure has been found to be almost independent of load, although dependent on cylinder volume and even more dependent on engine speed [67]. Priede has given a good review of IC engine noise with an emphasis on diesel engine noise [68]. Reference [14] also includes a detailed discussion on various aspects of engine noise. Further discussion and worked examples on automotive engine noise can be found in Chapter 14 of this book.

11.3.2 Electric Motors and Electrical Equipment

Examples of electrical equipment that cause noise and vibration include motors, generators, and alternators [5, 13], transformers, relays, solenoids, and circuit breakers. Electric motors are used widely in appliances, vehicles, and industry in a variety of types and sizes. They may be commutated, synchronous, or induction types. Electrical energy is converted into mechanical energy, and in the process some heat is produced. Fans are often provided to remove the heat and are the main sources of noise in electric motors. Because of the requirement for most motors that they should operate in either direction of rotation, they are usually provided with axial or tubular centrifugal fans, which can be quite noisy (see Section 11.2.3).

The sources of noise and vibration in electrical equipment are mostly aerodynamic, mechanical, and electromagnetic in nature. They are summarized in Table 11.3. Electric motors and the noise they generate are discussed in detail in Ref. [69].

Some empirical studies have been developed to give a rough estimate of the sound pressure level of different types of electric motors [70, 71]. For totally enclosed, fan cooled (TEFC) small motors, the following equation can be used to give a conservative estimate of the overall sound pressure level at 1 m:

$$L_p = A + B \log(W_R) + 15 \log(N) \text{ dB}, \tag{11.11}$$

where N is the motor speed in rpm, W_R is the motor power rating (in kW), and A and B are constants that depend on the electrical power of the motor. For motors under 40 kW, $A = B = 17$. For motors over 40 kW, $A = 28$ and $B = 10$. It has been observed that the sound pressure level produced by Drip-Proof (DRPR) motors is approximately 5 dB lower than the sound pressure level produced by TEFC motors. A TEFC motor with a quiet fan is likely to be 10 dB quieter than the overall value estimated by Eq. (11.11). The one-octave band sound pressure levels can be determined by subtracting the adjustments listed in Table 11.4 from the overall values given by Eq. (11.11). The estimation of the noise levels produced by large electric motors is discussed in Refs. [70, 71].

Table 11.3 Main sources of noise in electric motors [18].

Mechanical	Excessive bearing clearance
	Nonround bearings
	Rotor unbalance
	Rotor eccentricity
	Crooked shaft
	Brush and brush holder vibration
	Misalignment
	Loose laminations
Electromagnetic	Magnetostriction
	Torque pulsations
	Air gap eccentricity
	Air gap permeance variation
	Dissymmetry
	Sparking or arcing
Aerodynamic	Fan blade-passing frequency
	Turbulence
	Noise due to airflow path restrictions

Table 11.4 One-octave band level adjustments (dB) for small electric motors [71].

	One-octave band center frequency (Hz)								
Type of motor	**31.5**	**63**	**125**	**250**	**500**	**1000**	**2000**	**4000**	**8000**
TEFC	14	14	11	9	6	6	7	12	20
DRPR	9	9	7	7	6	9	12	18	27

Example 11.6 Estimate the overall sound pressure level produced by a small 30 kW TEFC motor operating at 2000 rpm at a receiver located at a distance of 10 m from the motor. Both the motor and the receiver are situated on hard ground in open space. For simplicity consider that the motor radiates sound as an omnidirectional source.

Solution

We use Eq. (11.11) with $A = B = 17$ to estimate the overall sound pressure level produced by the motor at 1 m. Thus, $L_p = 17 + 17 \times \log(30) + 15 \times \log(2000) = 92$ dB. If the motor radiates only to a half space and no sound power is absorbed by the hard ground, then we can use Eq. (3.52) to determine the sound power level L_w of the motor. Rearranging Eq. (3.52) we get $L_w = L_p + 20 \times \log(r) + 8$ dB $= 92 + 20 \times \log(1) + 8 = 100$ dB. Therefore, the sound pressure level at 10 m is: $L_p = L_w - 20 \times \log(r) - 8$ dB $= 100 - 20 \times \log(10) - 8 = 72$ dB. Alternatively, since the sound pressure level at 1 m is known, the level at 10 m can be calculated directly from the inverse square law without using the sound power level. The difference in levels at 1 and 10 m is $20 \times \log(10/1) = 20$ dB. Therefore, the sound pressure level produced by the motor at 10 m is: $92 - 20 = 72$ dB.

Example 11.7 Estimate the sound pressure level produced at 1 m for each one-octave band from 31.5 to 8000 Hz, by a 50 kW DRPR motor running at 3000 rpm.

Solution

First we use Eq. (11.11) to estimate the overall sound pressure level at 1 m with $A = 28$ and $B = 10$. Therefore, $L_p = 28 + 10 \times \log(50) + 15 \times \log(3000) = 97$ dB. Now we subtract the values given in the last row of Table 11.4 from 97 dB for each one-octave band. Then, the sound pressure levels at 1 m, in the nine one-octave bands from 31.5 to 8000 Hz are: 88, 88, 90, 90, 91, 88, 85, 79, and 70 dB, respectively.

Electric motor noise is normally controlled by passive means (use of enclosures, sound-absorbing materials, vibration isolation, etc.) The pure-tone vibration and noise produced at twice line frequency and multiples can also be reduced by active control methods, although active control of electrical equipment so far has received limited attention. One exception is the active control of the vibration and noise of large electrical transformers, which has been successfully reduced by active vibration control approaches [70].

11.3.3 Compressors

Compressors can be considered to be pumps for gases. Although there are some differences in construction details between compressors and pumps, their principles of operation, however, are essentially the same (see Section 11.3.4). Since gases normally have much lower densities than liquids, it is possible to operate compressors at much higher speeds than pumps. However, gases have lower viscosities than most liquids and so leakage with compressors is more of a problem than with pumps. Thus, this requires tighter manufacturing tolerances in the moving parts of compressors. Due to the low viscosity of gases and their compression, the temperature of the working medium and of the machine itself increases during compression. As a consequence, when the pressure ratio from after to before compression is more than about five, additional cooling is needed for the compressor and lubrication system. This makes compressors more complicated, as regards maintenance and servicing, than pumps used for liquids. In general compressors are more expensive to operate than pumps.

Compressors are used for many different applications, and there are a large number of quite different designs. Rather like pumps, there are two basic types of compressor: (i) positive-displacement compressors including reciprocating piston and rotary types and (ii) dynamic compressors including axial and centrifugal types. Like pumps, there are a few special-effect compressor types as well.

Positive-displacement compressors operate by increasing the pressure of the gas by reducing its volume in a compression chamber through work applied to the compressor mechanism. Very large numbers of small positive-displacement compressors have been mass produced around the world for use in household refrigerators and domestic and automobile air-conditioning systems. Considerable efforts, which are driven in part by market forces, have been expended to produce quieter small positive-displacement compressors. Positive-displacement compressors can be characterized by the location of the motor: (i) External-drive (open-type) compressors have a shaft or other moving part extending through the casing and are driven by an external power source, thus requiring a shaft seal or equivalent rubbing contact between fixed and moving parts; (ii) hermetic compressors have the compressor and motor enclosed in the same housing shell without an external shaft or shaft seal and the motor operates in the compressed gas; and (iii) semi-hermetic refrigerant compressors have the compressor directly coupled to an electric motor and contained within a gas-tight bolted casing. Positive-displacement compressor mechanisms can

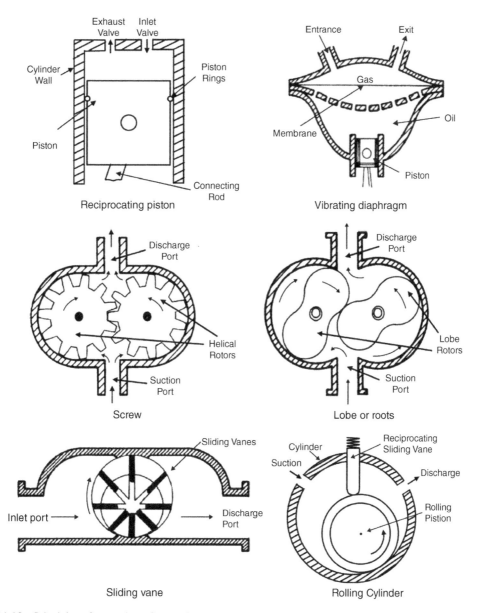

Figure 11.18 Principles of operation of several positive-displacement compressors. *Source:* Adapted from Ref. [72].

be further subdivided into (i) reciprocating types: piston, diaphragm, or membrane, and (ii) rotary types: rolling piston, rotary vane, single-screw, twin-screw, scroll and, throchoidal (lobe). Figure 11.18 shows the principles of operation of several positive-displacement compressors.

Dynamic compressors, on the other hand, work on the principle of using bladed impellors on continuously flowing gas to increase its kinetic energy, which is eventually converted into potential energy and gas of higher pressure. They can be made to be of low weight and generally have higher efficiencies than positive-displacement types. Their operational principles are very similar to fans (see Section 11.2.3).

Centrifugal compressors are widely used in large buildings, offices, factories, and industrial plants that require large central air-conditioning and cooling systems [72, 73]. Such centrifugal compressors eliminate the need for valves. The number of parts with sliding contact and close clearances are reduced, compared with positive-displacement types. Thus, maintenance costs are reduced, although operating costs may be increased due to their somewhat lower efficiency than comparable positive-displacement compressors. They are smaller and lighter in weight, and generally have lower original equipment and installation costs than equivalent reciprocating types. The noise and vibration characteristics are quite different, however, due to the higher speed and the lack of out-of-balance machine parts. The main components of a centrifugal compressor include (i) an inlet guide vane, (ii) an impellor, (iii) a diffuser, and (iv) a volute.

Axial and centrifugal compressors are competitive for volume flow rates from about 25 to 90 m^3/s, but when volume flow rates higher than about 60 m^3/s are needed, axial-flow compressors are normally used instead of centrifugal-flow machines [72]. This is because they are more efficient, smaller in size and weight, and require lower installation costs. They have several disadvantages, however, including generally more complex control systems, a narrower range of available flow rates, and surge and ingestion protection requirements. They also produce higher noise levels than centrifugal types, thus requiring more extensive acoustical treatment. Currently, axial compressors have their greatest use in aircraft and air transportation systems.

The ejector compressor is the simplest form of dynamic compressor [73]. It has no moving parts and is thus low cost. It is inexpensive, but has a low efficiency, however, and thus sees use mostly for vacuum applications. It requires a high-pressure source and transfers the momentum of the high-pressure jet stream to the low-pressure process gas.

The noise generated by the piston type of compressors depends upon several factors, the most important being the reciprocating frequency and integer multiples, number of pistons, valve dynamics, and acoustical and structural resonances. The noise produced by the rotary types depends upon rotational frequency and multiples, numbers of rotating elements, flow capacity, and other flow factors. The noise generated by centrifugal and axial compressors also depends upon rotational frequency and the number of rotating compressor blade elements. Flow speed and volume flow rate, however, are also important factors. Such dynamic compressors are used in aircraft jet engines and large commercial electricity generating plants.

A number of empirical equations have been presented to estimate the overall sound power levels of different types of compressors. For large centrifugal compressors (greater than 75 kW), the overall sound power level generated within the exit piping can be estimated from [70]

$$L_w = 20 \log (W_R) + 50 \log U - 46 \text{ dB}, \tag{11.12}$$

where W_R is the power of the driver motor in kW, and U is the impeller tip speed in m/s. Equation (11.12) is valid for $30 < U < 230$. The impeller tip speed $U = \pi D \times N/60$, where D is the diameter in metres of the impeller and N is the impeller rotary speed in rpm. The frequency in hertz at which the maximum noise level is produced is

$$f_p = 4.1U. \tag{11.13}$$

The one-octave band sound power levels can be determined by subtracting 4.5 dB from the level calculated by Eq. (11.12) for the one-octave band containing f_p. Then, the spectrum rolls off at the rate of 3 dB per octave above and below the one-octave band of maximum noise level.

A similar procedure can be used to estimate the overall sound power level within the exit piping of a reciprocating compressor using

$$L_w = 106.5 + 10 \log (W_R) \text{ dB}, \tag{11.14}$$

but now $f_p = N_c \times N/60$, where N_c is the number of cylinders of the compressor.

The overall sound power radiated by the compressor casing and exit pipe walls can be roughly determined by subtracting the transmission loss provided by the casing and exit piping from the values determined from Eqs. (11.12)–(11.14).

Example 11.8 Estimate the overall sound power level of a 120 kW centrifugal compressor with impeller diameter of 1 m and operating at 2400 rpm.

Solution

The impeller tip speed is $U = \pi(2400)/60 = 125.7$ m/s. Then, use of Eq. (11.13) yields $f_p = 4.1 \times 125.7 = 515$ Hz. Therefore, we must subtract 4.5 dB at the 500 Hz one-octave band in which the frequency f_p falls. Now using Eq. (11.12), $L_w = 20 \times \log(120) + 50 \times \log(125.7) - 46 \approx 100.5$ dB. Thus, the sound power level at the 500 Hz one-octave band would be $100.5 - 4.5 = 96$ dB. The levels in higher and lower one-octave bands decrease by 3 dB per octave. Then, the sound power levels in nine one-octave bands from 31.5 to 8000 Hz are: 84, 87, 90, 93, 96, 93, 90, 87, and 84 dB, respectively. The overall sound power level is obtained by logarithmically adding the previous one-octave band levels. Then,

$$L_w = 10 \log \left(2 \times 10^{8.4} + 2 \times 10^{8.7} + 2 \times 10^9 + 2 \times 10^{9.3} + 10^{9.6}\right) = 100.6 \text{ dB}.$$

An alternative equation to calculate the overall external sound power levels directly for rotary, reciprocating, and centrifugal compressors is [70]

$$L_w = A + 10 \log (W_R) \text{ dB}, \tag{11.15}$$

where A is a constant that depends upon the type of compressor (see Table 11.5). The one-octave band sound power levels can be determined by subtracting the adjustments listed in Table 11.5 from the overall values of Eq. (11.15).

Example 11.9 Estimate the exterior casing sound power level spectrum for a 100 kW centrifugal compressor.

Solution

Use of Eq. (11.15) with $A = 79$ yields $L_w = 79 + 10 \times \log(100) = 99$ dB. Now we subtract the values given in the second row of Table 11.5 from 99 dB for each one-octave band. Then, the sound power levels in the nine one-octave bands from 31.5 to 8000 Hz are: 89, 89, 88, 86, 86, 88, 92, 91, and 87 dB, respectively.

Table 11.5 One-octave band level adjustments (dB) for exterior noise levels radiated by compressors [70].

Type of compressor	One-octave band frequency (Hz)								
	31.5	63	125	250	500	1000	2000	4000	8000
Rotary and reciprocating (including partially muffled inlets); A = 90	11	15	10	11	13	10	5	8	15
Centrifugal (casing noise excluding air inlet noise); A = 79	10	10	11	13	13	11	7	8	12
Centrifugal (unmuffled air inlet noise excluding casing noise); A = 80	18	16	14	10	8	6	5	10	16

For noise control it is normal to classify noise problems in terms of the source–path–receiver framework (See Figure 9.1). In the case of compressors, it is not always so easy to make a distinct division between sources and paths. With positive-displacement compressors, the main noise source is the time-varying pressure pulsation created between the suction (inlet) and discharge manifolds of the compressor. This fluctuating pressure forces the compressor casing and any connecting structures into vibration, which consequently results in sound radiation. Many compressors have housing shell structures, and it is normal to vibration-isolate the compressor casing from the compressor shell housing, which itself may or may not be completely hermetically sealed. Suction and discharge piping must be provided, and care must be taken to reduce vibration transmission paths from the casing to the shell housing through this piping. It is not possible to completely eliminate the vibration transmission through the compressor vibration isolation system, and the piping and the gas between the casing and housing also provides another path for sound energy transmission to the shell. Reference [72] describes some design, operational noise, and vibration features of several categories of dynamic compressors in use and also gives some examples of how noise and vibration problems have been overcome in practice.

11.3.4 Pumps

Pumps are used to transport liquids and suspensions of solid particles in hydraulic systems. The noise generated in such systems is produced not only by the pump but also by the driving motor (usually an electric motor with its cooling fan). The noise and vibration created is carried throughout the hydraulic system as fluid-borne noise and mechanically by the pipe system itself as structure-borne noise [74, 75] (personal communication with Cesare Angeloni, 9 February 2006).

There are three main types of pumps: (i) positive displacement, (ii) kinetic (i.e. dynamic), and (iii) special effect type. Some of the main pump designs are illustrated in Figure 11.19 [76] and are similar in many ways to compressors used to pump or transport gases as described in Section 11.3.3. Positive-displacement pumps work by periodically adding energy to the fluid by one or more elements moving within a cylinder or pump case. In a similar way to positive displacement compressors, positive displacement pumps can be further subdivided into reciprocating and rotary types. The operation of a pump causes mechanical forces resulting in vibration, and even more important, pressure pulsations in the fluid, both of which cause noise. Rotary pumps can be divided into two main types: those possessing single rotors and those with multiple rotor elements. There are many different types of single- and multiple-type pumps including vane, gear, lobe, and screw types as shown in Figure 11.19 [76].

The main sources of noise in reciprocating pumps include: (i) flow and pressure pulsations sometimes called "ripple," turbulent and separated flows, and vortex formation in the fluid and (ii) unbalanced mechanical forces, inlet and discharge valve impacts, and piston slap. The main sources of noise in rotary pumps depend upon the pump design and include (i) fluid pressure pulsations, turbulence, and vortices and (ii) mechanical forces resulting from impacts between the teeth of gear pumps, the sliding of the vanes of vane pumps, and friction forces created in screw types of pumps. Gear pumps are noisy since they trap and compress the working fluid between the gear teeth and expel it in a direction perpendicular to the axis of revolution. They can operate at pressures of 150 bars or more. Screw pumps are generally much quieter and move the fluid in a direction parallel to the screw axis but are limited to lower pressures, usually less than about 40 bars.

Empirical expressions for the overall A-weighted sound power level of different types of pumps are listed in Table 11.6 [76]. Here Q_{BEP} corresponds to the flow at the design or best efficient point of the pump. Equations in Table 11.6 are valid for a limited range of the power consumption within which the measurements were done, and by taking into account a certain degree of uncertainty.

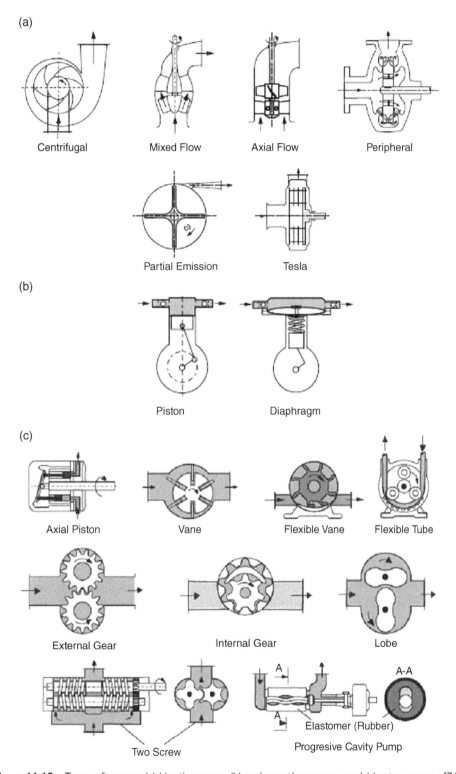

Figure 11.19 Types of pumps: (a) kinetic pumps, (b) reciprocating pumps, and (c) rotary pumps [76].

Table 11.6 Prediction of the A-weighted sound power level generated by different pumps [76].

Type of pump	A-weighted sound power level, (P in kW, P_{ref} = 1 kW) in dB	Valid for power consumption P
Centrifugal pumps (single stage)	$L_{WA} = 71 + 13.5 \log P \pm 7.5$	4 kW $\leq P \leq$ 2,000 kW
Centrifugal pumps (multistage)	$L_{WA} = 83.5 + 8.5 \log P \pm 7.5$	4 kW $\leq P \leq$ 20,000 kW
Axial-flow pumps	$L_{WA} = 78.5 + 10 \log P \pm 10$ at Q_{BEP}	10 kW $\leq P \leq$ 1,300 kW
	$L_{WA} = 21.5 + 10 \log P + 57 Q/Q_{BEP} \pm 8$	$0.77 \leq Q/Q_{BEP} \leq 1.25$
Multipiston pumps (inline)	$L_{WA} = 78 + 10 \log P \pm 6$	1 kW $\leq P \leq$ 1,000 kW
Diaphragm pumps	$L_{WA} = 78 + 9 \log P \pm 6$	1 kW $\leq P \leq$ 100 kW
Screw pumps	$L_{WA} = 78 + 11 \log P \pm 6$	1 kW $\leq P \leq$ 100 kW
Gear pumps	$L_{WA} = 78 + 11 \log P \pm 3$	1 kW $\leq P \leq$ 100 kW
Lobe pumps	$L_{WA} = 84 + 11 \log P \pm 5$	1 kW $\leq P \leq$ 10 kW

Example 11.10 A small screw pump of power consumption 2 kW is installed outdoors on the floor (assumed to be reflecting). If the pump ordinarily radiates equally well in all directions (omnidirectional), what is the A-weighted sound pressure level to be expected at 2 m from the pump?

Solution

First, we determine the A-weighted sound power level from the respective equation in Table 11.6 as $L_{WA} = 78 + 11 \times \log(2) \pm 6 = 81 \pm 6$ dB. Therefore, the expected sound power level would be between 75 dB and 87 dB. Since the pump is an omnidirectional source on a reflecting floor, use of Eq. (3.52) yields the A-weighted sound pressure level at 2 m: $L_p = L_{WA} - 20 \times \log(2) - 8$ dB $= 81 \pm 6 - 20 \times \log(2) - 8 = 81 \pm 6 - 14 = 67 \pm 6$ dB. Thus, the expected A-weighted sound pressure level at 2 m from the pump is predicted to be between 61 and 73 dB.

Various designs of pump have been produced to reduce ripple [75] (personal communication with Cesare Angeloni, 9 February 2006). One novel design has gear/screw elements that makes it a cross between a gear and screw pump (see Figure 11.20). The two contra-rotating elements tend to balance internal forces and torques thus reducing vibration and noise. It is claimed that this type of pump, termed a Continuum® pump, virtually eliminates the trapping of fluid between gear teeth by using special helical gears, which results in much less pressure ripple and pump noise. The pump can operate at speeds up to 5000 rpm and pressures as high as 240 bars. Figure 11.21 shows the pressure pulsations (ripples) for two different designs of gear pump

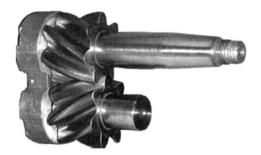

Figure 11.20 Continuum continuous-contact pumps feature helical gears that do not trap fluid as they rotate, as is the case with conventional gear pumps. This minimizes pressure ripple and gives high efficiency and quiet operation at speeds to 5000 rpm [75] (personal communication with Cesare Angeloni, 9 February 2006).

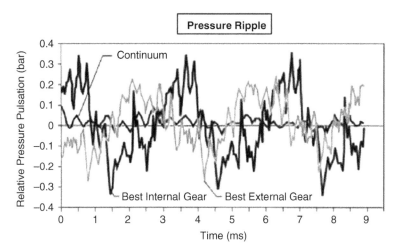

Figure 11.21 Laboratory tests that show reduced pulsation pressure (ripple) for the continuous-contact Continuum pump compared with similar gear pumps. The pulsation tests were conducted at 100 bars, 1500 rpm, and with pressure sampling at 100 kHz [75] (personal communication with Cesare Angeloni, 9 February 2006).

and the Continuum pump when they were all operated at 1500 rpm and a pressure of 100 bars. It is seen that these continuous-contact pumps exhibit a much smaller magnitude of ripple. Figure 11.22 presents the A-weighted sound pressure levels measured using the ISO 4412 method for two gear pumps and the continuous-contact pump when they were tested under similar operating conditions. The overall A-weighted sound pressure level of the continuous-contact pump does appear to be of the order of 5–8 dB lower than the noise levels of the two other pumps and supports the idea that the pressure ripple is a major source of noise in most pump types.

The flow and pressure ripple effects can also be reduced by the use of dampers or accumulators and/or by very careful and precise manufacturing and grinding of the parts of the pump. Dampers can be similar to muffler expansion chambers in design or have special design features including flexible membrane parts to absorb the pressure pulsations. Efforts continue to be made to reduce pump ripple [77] and cavitation [78, 79], which can cause serious damage to pumps and hydraulic systems. Different methods are used to improve pump design [80–82] and also for diagnostics and health maintenance of pumps and hydraulic

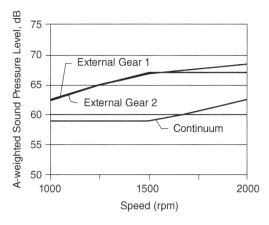

Figure 11.22 Laboratory measurements of the noise of two external gear pumps and the Continuum pump. The measurements were made when all of the pumps were running at 1500 rpm. The pressure produced by all the pumps in the noise test was 150 bars [75] (personal communication with Cesare Angeloni, 9 February 2006).

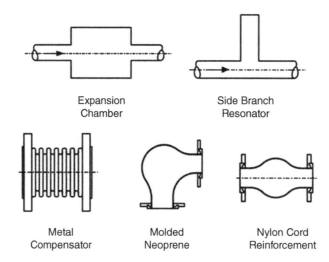

Figure 11.23 Methods for reduction of pressure pulsating in pipes [76].

systems [83, 84]. Precision grinding of parts does add cost and this must be considered in choosing a pump system.

It should be noted that pump pressure pulsations produce vibrations of the pump case, pipe wall or any other structure to which the pump system is attached. These pulsations may also be transported through the pipe structure and the fluid, resulting in noise radiation to the surrounding air. Reduction of these pulsations can be achieved by installing flexible connectors, mufflers, dampers, expansion chambers, side-branch tuned resonators or using flexible couplings, among others. Figure 11.23 shows some devices for reducing structure-borne noise produced by pressure pulsation in piping. Reference [76] presents a detailed review of the noise of pumps and pumping systems.

11.4 Noise Due to Fluid Flow

11.4.1 Valve-Induced Noise

Control valves are used in industrial plants to control the rate of fluid flow by creating a pressure drop across the valve. The flow is first accelerated by this process and then the kinetic energy is converted into thermal energy or heat through turbulence and/or shock waves. In the process a small fraction of the energy is converted into acoustical energy and thus noise. The aerodynamic noise generated by control valves, regulators, and orifices is a major noise source in piping systems. In some cases, exterior noise levels as high as 130 dB can be produced. In addition, in liquid flow systems, hydrodynamic noise (which is mostly due to cavitation) is of some importance. However, in most industrial situations, with high-speed gas flows, the aerodynamic noise is dominant [85]. The main noise-generating mechanisms include turbulent mixing, turbulence interaction with boundaries, cavitation in liquids, shock waves, interaction of turbulence with shocks, cavity resonances, flow separation, vortex shedding, "whistling," and resonant mechanical vibration of valve components [86, 87]. Figure 11.24 presents a simplified view of a control valve and some of the noise sources [87]. See Ref. [88] for an extensive discussion on this subject. Comprehensive reviews of control valve noise have also been given by Reethof [86, 87]. Seebold has reviewed control valve noise sources and approaches for noise control [89].

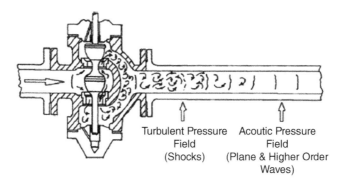

Figure 11.24 Schematic representation of control valve noise generation and propagation [87].

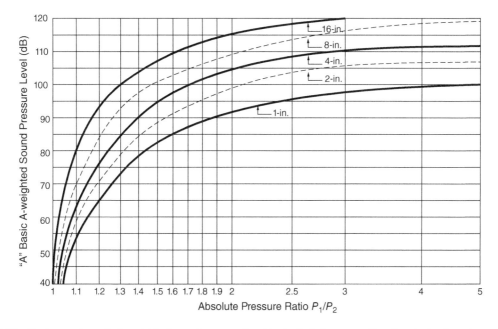

Figure 11.25 Basic aerodynamic A-weighted sound pressure level in decibels for conventional control valves at approximately 70% of rated flow capacity, at 667-kPa inlet pressure and Schedule 40 downstream pipe, measured 1 m from the downstream pipe wall [88].

The major noise-generating processes can be divided into two regimes: *subsonic*, consisting mostly of turbulence–boundary interaction noise, and *supersonic*, consisting mainly of broadband shock noise. Although the flow is confined by the piping, the turbulent mixing noise is similar to the noise of a free jet and is essentially quadrupole in nature [90, 91]. The shocks are caused by abruptly expanded flow after the valve, when it is operating above the critical pressure ratio. The shock noise has two main parts: screech and broadband noise. The screech is discrete in nature and is caused by a feedback mechanism and is not often encountered with valves and regulators. The broadband shock noise is common, however, and has been shown to be mostly independent of flow velocity and to be a function of the pressure ratio across the valve.

Valve noise prediction is quite complex and computational tools are often the only way to compute the noise radiated. However, although not as accurate as a computerized method some simplified noise prediction graphical techniques are available. Figure 11.25 shows a graphical technique from The International

Society of Measurement and Control (ISA). Use of the method is described by Baumann and Abom [88] as: First, find the P_1/P_2 ratio, that is, the absolute inlet pressure divided by the absolute outlet pressure. Next read up to the given valve size and obtain the corresponding "basic sound pressure level" A from the scale on the left. The next step is to correct for the actual inlet pressure. This is given by $B = 12\log(P_1/667)$ where the pressure is in kilopascals. Finally, add C, a correction for the pipe wall if it is other than Schedule 40. Here $C = +1.4$ for Schedule 20, 0 for Schedule 40, -3.5 for Schedule 80, and -7 for Schedule 160. The total A-weighted sound pressure level now is the sum of $A + B + C$. Subtract another 3 dB in case of steam. The following example shows how to use this method [88].

Example 11.11 An 80-mm (3-in.) globe valve with a parabolic plug is reducing steam pressure from 3600 to 2118 kPa. Considering that the pipe Schedule is 80, determine the total A-weighted sound pressure level at 1 m from the pipe.

Solution

The pressure ratio is $3600/2118 = 1.7$. Going to Figure 11.25, we do not find a 3-in. valve. However, we can extrapolate between the lines and find the A factor to be 98 dB for the pressure ratio of 1.7. Factor B is calculated to be $12\log(3600/667) = 8.8$ dB. Finally, we add -3.5 dB for Schedule 80 and we subtract 3 dB for steam. This results in a total A-weighted sound pressure level of 100.3 dB at 1 m from the pipe wall.

Example 11.12 Using a computational method, the A-weighted sound pressure level produced by a valve at 1 m from pipe wall is estimated to be 105 dB. What would the sound pressure level be for a worker 30 m away from the downstream pipe (centerline)?

Solution

Noise produced by valves radiates to the environment largely through the pipe for great distances downstream of a valve. Therefore, this type of noise source is usually treated as a line source (see Section 3.10 of this book). Line sources radiate noise in a cylindrical pattern and the sound pressure is reduced by 3 dB per doubling of the distance. Then, the reduced A-weighted sound pressure level at 30 m from the line source is $L_p(30\text{ m}) = L_p(1\text{ m}) - 10 \times \log(30\text{ m}/1\text{ m}) = 105 - 15 = 90$ dB.

Valve noise can be reduced by several approaches including design of valves with multiple streams, arranging for the pressure drop to occur through several stages, and using absorptive silencers, thicker pipe walls, and pipe lagging. Multiple stream valve designs and absorptive silencers are effective at reducing noise. Absorptive silencers can produce noise reductions of as much as 15–30 dB. However, such silencers can become plugged by solid particles and moisture can be a problem in the silencer material so that effectiveness is lost. Reference [88] deals in detail with noise-generating mechanisms and the prediction and control of control valve noise. Several other authors have also given helpful reviews of valve and piping system noise, and the reader is referred to these as well for more detailed discussions [15, 85–87, 89].

11.4.2 Hydraulic System Noise

Noise in a hydraulic circuit is usually produced by the pumps and motors, although valves are also important noise sources. Many high-pressure hydraulic fluid power systems use positive-displacement pumps. The diameter of pipes used in such systems is usually quite small (of the order of 10–50 mm). The flow is normally single phase with negligible gas bubble and solid particle content. The usual operating pressures are of the order of 100–300 bars. There are systems, however, which operate at pressures of up to 500 or even 650 bars. Hydraulic systems often produce very high noise levels that can limit the range of applications of fluid power systems. Potential high-level noise problems can result in the selection of an alternative means of power transmission in cases where low-noise systems are required.

Hydraulic system noise sources may be categorized as follows: (i) *Airborne noise* originates from the vibration of components, piping, and housings, and is transmitted directly through the air. (ii) *Structure-borne noise* is caused by the mechanical operation of pumps and motors and is transmitted from pumps directly through mounts, drive shafts, and pipes. Structure-borne noise can also arise from the pressure "ripple" in the hydraulic system. (iii) *Fluid-borne noise* is caused primarily by unsteady flow from the pumps and motors but can also be caused by valve instability, cavitation, and/or turbulence.

Fluid-borne noise can be transmitted over considerable distances through pipework with little attenuation. Figure 11.26 gives an illustration of (a) structure-borne noise and (b) fluid-borne noise (commonly called ripples) created by a multipiston axial-flow pump [92].

The main sources of noise in a hydraulic system are usually the pumps and motors. Valves are also important noise generators. See Ref. [88] for further discussion on valve noise. Positive-displacement pumps are mainly used in fluid power applications. The most common types are piston pumps, gear pumps, and vane pumps. See Ref. [76] for more detailed discussion on pump noise. Positive-displacement pumps produce a steady fluid flow rate on which is superimposed a flow fluctuation or ripple, which is caused by the cyclic nature of the pump operation. The flow ripple is also manifested as a pressure fluctuation or ripple. The magnitude of the flow ripple depends on the pump type and its operating conditions [92, 93]. Flow ripple normally occurs in both the suction and the discharge lines. Usually the discharge flow ripple is the most important noise source. Fluid-borne noise in the suction line may cause noise problems especially when it causes vibration and noise radiation from a large-surface reservoir.

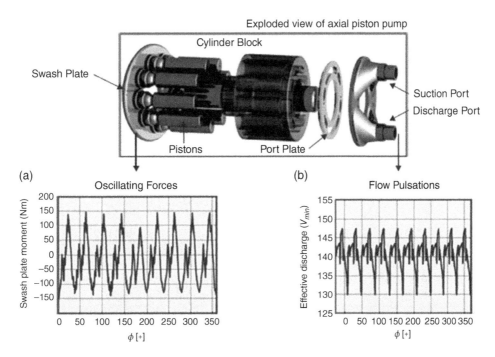

Figure 11.26 Pump forces result as airborne noise (not shown) and (a) structure-borne noise through the piping system and (b) fluid-borne noise (flow and pressure fluctuations called ripples) [18].

There are several ways of reducing the noise of hydraulic systems [90, 92] These include (i) reduction of the pump or motor flow ripples [75] (personal communication with Cesare Angeloni, 9 February 2006), (ii) tuning of the circuit in order to avoid resonant conditions, (iii) use of silencers or pulsation dampers, (iv) use of flexible hoses [74], (v) vibration isolation [74], and (vi) use of enclosures or pipe cladding.

11.4.3 Furnace and Burner Noise

The phenomenon of thermoacoustically-induced oscillations and combustion noise, or "roar," in combustion systems is complicated and usually only qualitative explanations can be given [94]. Various empirical methods have been developed to predict combustion noise. In practice, engineering solutions are mainly used to control combustion oscillations and noise.

Combustion systems consist of two main fuel and air delivery system components: (i) burner and (ii) combustor. The fuel injection component is usually called the burner. The element within which the heat release takes place is normally referred to as the combustor or furnace. The purpose of combustion systems is to add heat to an airstream. The main mechanisms that produce combustion noise are common to all combustion systems. The purpose of the fuel burner is to mix and direct the flow of fuel and air to ensure rapid ignition and complete combustion within the furnace or combustor [94]. The combustion of gaseous fuels takes place mainly in two ways: (i) when the gas and air are mixed before ignition, known as premix flames, and (ii) when the gas and air are mixed after the fuel has been heated, known as diffusion flames.

The noise generated by the combustion can be considered to be either (i) combustion roar or (ii) combustion-driven oscillations or thermoacoustic instabilities. The latter type of noise is observed when pressure oscillations are induced by heat release oscillations. Although thermoacoustic instabilities are closely related to flame instabilities, their instability criteria are different. The main focus of Ref. [94] is on thermoacoustic instabilities since they can produce higher sound pressure levels than combustion roar and since they also have a greater potential for structural damage. It should be noted that some researchers further subdivide combustion noise into four categories [95, 96]. Figure 11.27 shows two types of flame. Type 1 flames tend to be relatively quiet and type 2 flames tend to be relatively noisy. It has been found by several researchers that if sufficient swirl is introduced into the flow, flames change from type 1 to type 2 [95–98].

There are two main passive noise control measures available: (i) reducing the combustion-induced sources and (ii) use of traditional noise control approaches including use of absorptive mufflers, acoustical

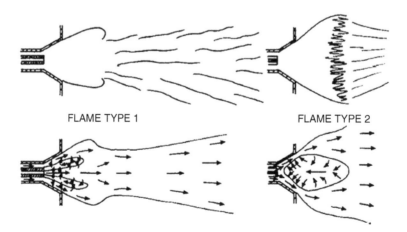

FLAME TYPE 1   FLAME TYPE 2

Figure 11.27 Flame types in burners incorporating swirl [96].

resonators, enclosures for the burner, furnace, or boiler or other units to which they are attached. Active approaches to control thermoacoustic oscillations have also been attempted [99].

11.5 Noise Control of Industrial Production Machinery

Production machinery and equipment that generate intense noise include machines that operate with impacts such as forging hammers [100, 101], cold headers, stamping presses, riveters, jolting tables, some machine tools, and impact-generating assembly stations. With the exception of forging plants, in which forging hammers are the dominant source of noise, noise sources in manufacturing plants can be typically classified in order of intensity/annoyance as: (i) *compressed air* (leakages, air exhaust, air blowing nozzles), (ii) *in-plant material handling systems*, and (iii) *production and auxiliary machinery and equipment*. Reference [102] discusses noise in production areas of manufacturing plants.

The two basic noise reduction techniques are (i) use of acoustical enclosures, which are expensive to build and maintain, may reduce efficiency of the enclosed equipment, and are not always feasible, and (ii) noise reduction at the source, which is effective but often requires a research and development effort. Reference [103] contains a large number of case histories of successful noise reduction projects carried out in industry.

11.5.1 Machine Tool Noise, Vibration, and Chatter

A common problem in the manufacturing industry today is the vibrations or chatter induced in machine tools during machining. Such problems occur, for example, in turning, milling, boring, and grinding. Chatter is very undesirable. Not only does it create noise but the vibration that it produces also results in an uneven cut and undesirable cut quality. The vibration of machine tools may be divided into three different classes: (i) free or transient vibrations of machine tools excited by other machines or engagement of the cutting tool and the like, (ii) forced vibrations usually associated with periodic forces within the machine tool, for example, unbalanced rotating masses, and (iii) self-excited chatter that may be explained by a number of mechanisms. These mechanisms include, among others, the regenerative effect, the mode coupling effect, the random excitation of the natural frequencies of the machine tool caused by the plastic deformation of the workpiece material, and/or friction between the tool and the cut material. Vibration in machine tools affects the quality of machining, particularly the surface finish. Furthermore, machine tool life can be correlated with the vibration and noise levels produced. Machine tool chatter may be reduced by selective passive or active modification of the dynamic stiffness of the tooling structure and/or by the control of cutting data and its use to maintain stable cutting. Forced unbalance vibration in rotating tooling structures may be reduced by passive balancing or active online balancing.

Free or transient vibrations of machine tools may be excited by other machines in the environment via the machine tool base or/and by rapid movements of machine tables, engagement of the cutting tool, and the like. The forced vibrations are usually associated with periodic forces within the machine tool, for example, unbalanced rotating masses, or the intermittent tooth pass frequency excitation in milling. This type of vibration may also be excited by other machines in the environment of the machine tool via its base.

Machine tool vibrations during machining operations are usually denoted *self-excited chatter* or *tool vibration*. Depending on the driving force of the tool vibration, the vibration is generally divided into one of two categories: regenerative chatter (secondary chatter) and nonregenerative chatter (primary chatter). See Refs. [104–106] for examples. Extensive research has been carried out on the mechanisms that control the induction of vibrations in the cutting process. The majority of this research has involved the dynamic modeling of cutting dynamics focusing on analytical or numerical models. Usually, the purpose of this work is to produce dynamic models for the prediction of cutting data that ensure stable cutting and maximize the material removal rate. Active control has also been investigated for turning and boring operations [107, 108]. Reference [109] discusses the problem in the manufacturing industry of the chatter-induced vibration in machine tools during machining.

11.5.2 Sound Power Level for Industrial Machinery

Although the present chapter has discussed a number of procedures for calculating the sound power and sound pressure levels of industrial machinery, more information can be found in Ref. [110]. The measured or calculated sound power levels can be used for predicting the sound pressure levels in a space or developing purchase specifications for new equipment. With any project, acoustical data measured and calculated in accordance with recognized standards should be obtained. Many manufacturers provide sound power levels or measured sound pressure levels at 1 m from their equipment, and some offer special low-noise options. In the European Community (EEC) it is required to determine the sound power level of some items of machinery and to provide a label on the machine giving this information. If manufacturers' data are unavailable, efforts should be made to measure a similar unit in operation. If this is not practical, then theoretical estimations can be used.

It has to be noted that most of the equations presented in this chapter are based on measured data and tend to be conservative, usually predicting somewhat higher sound pressure levels than are measured in the field. Due to recent efforts at reducing equipment noise, sound pressure levels for some equipment may be significantly (10 dB) quieter than the levels calculated in this chapter. Some equipment consists of several different sound-producing components such as motors, pumps, blowers, and the like. The sound power levels for each component should be determined and then combined (using correct decibel addition) to get the total sound power levels.

References

1 Bell, L.H. and Bell, D.H. (1994). *Industrial Noise Control*, 2e. New York: Marcel Dekker.

2 Harris, C.M. (ed.) (1991). *Handbook of Acoustical Measurement and Noise Control*, 3e. New York: McGraw-Hill.

3 Shahan, J.E. and Kamperman, G. (1976). Machine element noise. In: *Handbook of Industrial Noise Control* (ed. L.L. Faulkner), 329–385\. New York: Industrial Press.

4 Smith, J.D. (2003). *Gear Noise and Vibration*, 2e, rev. and expanded. New York: Marcell Dekker.

5 Timar, P.L. (ed.) (1989). *Noise and Vibration of Electrics Machines*. Amsterdam: Elsevier.

6 Eschmann, P., Hasbargen, L., and Weigand, K. (1985). *Ball and Roller Bearings: Theory, Design, and Application*. Chichester: Wiley.

7 Rowe, W.B. (1983). *Hydrostatic and Hybrid Bearings Designs*. London: Butterworths.

8 Fuller, D.D. (1984). *Theory and Practice of Lubrication for Engineers*, 2e. Chichester: Wiley.

9 Skaistis, S. (1988). *Noise Control of Hydraulic Machinery*. New York: Marcel Dekker.

10 Wilson, G. (1989). *Noise Control*. Prentice Hall new edition, reprinted by Kreiger, New York, 2000.

11 Norton, M.P. and Karczub, D.G. (2003). *Fundamentals of Noise and Vibration Analysis for Engineers*, 2e. Cambridge: Cambridge University Press.

12 Ver, I.L. and Beranek, L.L. (eds.) (2005). *Noise and Vibration Control Engineering*, 2e. New York: Wiley.

13 Gieras, J.F., Wang, C., and Lai, J.C. (2006). *Noise of Polyphase Electric Motors*. New York: Taylor and Francis.

14 Hickling, R. and Kamal, M.M. (eds.) (1982). *Engine Noise, Excitation, Vibration and Radiation*. New York: Plenum.

15 Blake, W.F. (1986). *Mechanics of Fluid-Induced Sound and Vibration*. New York: Academic Press.

16 Houser, D.R. (2007). Gear noise and vibration prediction and control methods. In: *Handbook of Noise and Vibration Control* (ed. M.J. Crocker), 847–856. New York: Wiley.

17 Mitchell, L.D. (1971) Gear noise: the purchaser's and the manufacturer's views. Noise and vibration control engineering. *Proceedings of the Purdue Noise Control Conference* (ed. M.J. Crocker), 95–106. Purdue University.

18 Crocker, M.J. (2007). Machinery noise and vibration sources. In: *Handbook of Noise and Vibration Control* (ed. M.J. Crocker), 831–846. New York: Wiley.

19 Amini, N. (1999) Gear surface machining for noise suppression. Ph.D thesis. Chalmers University, Sweden, No. 1498, 1–70.

20 Lian, Q. and Dafoe, R. (2005). Using advanced theory, geometry modification, manufacturing realization to localize bearing contact and reduce cylindrical gear noise. *VDI Ber.* 19041: 73–90.

21 Chen, Y. and Ishibashi, A. (2006). Investigation of the noise and vibration of planetary gear drives. *Gear Tech.* 23 (1): 48–55.

22 Palmer, D. (2005) Unique gear grinding process offers low noise and high precision. *Eureka* January: 13–14.

23 Viebrock, W.M. and Crocker, M.J. (1973). Noise reduction of a consumer electric clock. *Sound Vib.* 7 (3): 22–26.

24 Blankenship, G.W. and Singh, R. (1992). New rating indices for gear noise based on vibro-acoustic measurements. *Noise Control Eng. J.* 38 (2): 81–92.

25 Tuma, J. (2007). Transmission and gearbox noise and vibration prediction and control. In: *Handbook of Noise and Vibration Control* (ed. M.J. Crocker), 1086–1095. New York: Wiley.

26 Zusman, G. (2007). Types of bearings and means of noise and vibration prediction and control. In: *Handbook of Noise and Vibration Control* (ed. M.J. Crocker), 857–867. New York: Wiley.

27 Zandbergen, T. and van Nijen, G. (2000). Less noise and vibration by proper rolling bearing accuracy design and installation. *Shock Vib. Dig.* 32 (1): 40.

28 Hui, Z., Shu-Juan, W., and Guo-fia, Z. (2005) Extraction of failure characteristics of roller element bearing based on wavelet transform under strong noise, J. Harbin Inst. Tech. (New Series) 12(2): 169–172.

29 Lyon, R.H. (1987). *Machinery Noise and Diagnostics.* Boston: Butterworths.

30 Piper, G.E., Watkins, J.M., and Thorp, O.G. (2005). Active control of axial-flow fan noise using magnetic bearings. *J. Vib. Control.* 11 (9): 1221–1232.

31 Lauchle, G.C. (2007). Centrifugal and axial fan noise prediction and control. In: *Handbook of Noise and Vibration Control* (ed. M.J. Crocker), 868–884. New York: Wiley.

32 Graham, J.B. and Hoover, R.M. (1991). Fan noise. In: *Handbook of Acoustical Measurements and Noise Control*, 3e (ed. C.M. Harris), 42.1–42.18. New York: McGraw-Hill.

33 Kingsbury, H.F. (2007). Noise sources and propagation in ducted air distribution systems. In: *Handbook of Noise and Vibration Control* (ed. M.J. Crocker), 1316–1322. New York: Wiley.

34 Oldham, D.J. and Waddington, D.D. (2007). Aerodynamic sound generation in low speed flow ducts. In: *Handbook of Noise and Vibration Control* (ed. M.J. Crocker), 1323–1327. New York: Wiley.

35 Huang, L. and Wang, J. (2005). Acoustic analysis of a computer cooling fan. *J. Acoust. Soc. Am.* 118: 2190.

36 Huang, L. (2003). Characterizing computer cooling fan noise. *J. Acoust. Soc. Am.* 114: 3189.

37 Huang, L., Zou, Z.P., and Xu, L. (2003) Prediction of computer cooling fan noise using a 3D unsteady flow solver. Proceedings of Noise-Con, 84.

38 Nijhof, M.J.J., Wijnant, Y.H., de Boer, A., and Beltman, W.M. (2004) optimizing circular side-resonators to reduce computer fan noise. Proceedings of Noise-Con, 815.

39 Laage, J.W., Mahendra, P., Melzer, P.J., et al. (2004) Measurement tools for studying fan noise. Proceedings of Noise-Con, 593.

40 Shin, H., Lee, S., and Kim, K.B. (2004) The axial fan noise simulation by the freewake method and acoustic analogy. Proceedings of Noise-Con, 2515.

41 Jeon W.-H., Lee, D.-J., and Rhee, H. (2002) An application of the acoustic simularity law to the numerical analysis of centrifugal fan noise. Proceedings of Noise-Con, 154.

42 Bobeczko, M.S. (2000) Condenser fan noise reduction solutions. Proceedings of Noise-Con, 133.

43 Langford, M., Burdisso, R., and Ng, W. (2005) An overview of flow control for fan noise reduction. Proceedings of Noise-Con, 289.

44 Jiricek, O. and Konicek, P. (1998). Active attenuation of fan noise in an air-conditioning duct. *J. Acoust. Soc. Am.* 103: 2836.

45 Gee, K.L. and Sommerfeldt, S.D. (2002) Multi-channel active control of axial cooling fan noise. Proceedings of Noise-Con, 880.

46 Homma, K., Fuller, C., and Man, K.X. (2003) Broadband active-passive control of small axial fan noise emission. Proceedings of Noise-Con, 410.

47 Gee, K.L. and Sommerfeldt, S.D. (2004). Application of theoretical modeling to multichannel active control of cooling fan noise. *J. Acoust. Soc. Am.* 115: 228.

48 Wang, J., Huang, L., and Cheng, L. (2005). A study of active tonal noise control for a small axial flow fan. *J. Acoust. Soc. Am.* 117: 734.

49 Minorikawa, G., Irikado, H., and Ito, T. (2005) Study on sound quality evaluation of fan noise. Proceedings of Noise-Con, 36.

50 Lai, J.C.S. (2007). Metal-cutting machinery noise and vibration prediction and control. In: *Handbook of Noise and Vibration Control* (ed. M.J. Crocker), 966–974. New York: Wiley.

51 Richards, J.E., Westcott, M.E., and Jeyapalan, R.K. (1979). On the prediction of impact noise, I: acceleration noise. *J. Sound Vib.* 62 (4): 547–575.

52 Richards, J.E., Westcott, M.E., and Jeyapalan, R.K. (1979). On the prediction of impact noise, II: ringing noise. *J. Sound Vib.* 65 (3): 419–453.

53 Trabelsi, H. and Kannatey-Asibu, E. (1990) Tool wear and sound radiation in metal cutting. Production Engineering Division (Publication) PED, Vol. 45, Modeling of Machine Tools: Accuracy, Dynamics, and Control, Winter Annual Meeting of the American Society of Mechanical Engineers, Nov. 25–30 1990, Dallas, TX, 121–131.

54 Wu, Y., Ke, S., Yang, S. et al. (1991) Experimental study of cutting noise dynamics. American Society of Mechanical Engineers, Design Engineering Division (Publication) DE, Vol. 36, Machinery Dynamics and Element Vibrations, 13th Biennial Conference on Mechanical Vibration and Noise presented at the 1991 ASME Design Technical Conferences, Sept. 22–25 1991, Miami, FL, 313–318.

55 Ling, C.C., Cheng, C., Chen, J.S., and Morse, I.E. (1995) Experimental investigation of damping effects on a single point cutting process. American Society of Mechanical Engineers, Manufacturing Engineering Division, MED, Vol. 2–1, Manufacturing Science and Engineering, Proceedings of the 1995 ASME International Mechanical Engineering Congress and Exposition. Part 1 (of 2), Nov. 12–17 1995, San Francisco, CA, 149–164.

56 Bahrami, A. and Williamson, H.M. (1997) Effect of blade profile on sheet metal shear noise. Proceedings of the 1997 National Conference on Noise Control Engineering, NOISE-CON. Part 2, University Park, PA, June 15–17, 471–476.

57 Burgess, M.A., Williamson, H.M., and Kanapathipillai, S. (1998). Noise reduction for friction saws. *Acoust. Aust.* 26 (1): 9–12.

58 Bahrami, A., Williamson, H.M., and Lai, J.C.S. (1998). Control of shear cutting noise: effect of blade profile. *Appl. Acoust.* 54 (1): 45–58.

59 Lai, J.C.S., Speakman, C., and Williamson, H.M. (2002). Control of shear cutting noise: effectiveness of passive control measures. *Noise Vib. Worldw.* 33 (7): 6–12.

60 Bahrami, A., Lai, J.C.S., and Williamson, H. (2003) Noise from shear cutting of sheet metal. Proceedings of the Tenth International Congress on Sound and Vibration, Stockholm, Sweden, July 7–10, 4069–4076.

61 Nielsen, K.S. and Stewart, J.S. (2007). Woodworking machinery noise. In: *Handbook of Noise and Vibration Control* (ed. M.J. Crocker), 975–986. New York: Wiley.

62 Stewart, J.S. (1978). An investigation of the aerodynamic noise generation mechanism of circular saw blades. *Noise Control Eng. J.* 11 (1): 5–11.

63 Bobezko, M. (1979) Quieter metal cutting methods and machinery. Proceedings of Noise Con, 163–169.

64 Stewart, J.S. and Hart, F.D. (1976). Control of industrial wood planer noise through improved cutterhead design. *Noise Control Eng. J.* 7 (1): 4–9.

65 Cmar, D.S. (2005) Noise control in the woodworking industry. www.phaseto.com/Noise%20Control%20in%20Woodworking%20Industry.htm (accessed 20 February 2020).

66 Reinhart, T.E. (2007). Internal combustion engine noise prediction and control – diesel and gasoline engines. In: *Handbook of Noise and Vibration Control* (ed. M.J. Crocker), 1024–1033. New York: Wiley.

67 White, R.G. and Walker, J.G., ed. (1982) Noise and Vibration. Ellis Horwood Publishers, New York: Wiley.

68 Priede, T. (1992). *Noise and Vibration Control Engineering* (eds. L.L. Beranek and I.L. Ver). New York: Wiley.

69 Zusman, G. (2007). Types of electric motors and noise and vibration prediction and control methods. In: *Handbook of Noise and Vibration Control* (ed. M.J. Crocker), 885–896. New York: Wiley.

70 Bies, D.A. and Hansen, C.H. (2009). *Engineering Noise Control*, 4e. London: Spon Press.

71 Joint Department of the Army, Air Force and Navy, USA. (1995) Noise and vibration control for mechanical equipment. Technical manual TM 5-805-4/AFJMAN 32-1090. Washington, DC.

72 Crocker, M.J. (2007). Noise control of compressors. In: *Handbook of Noise and Vibration Control* (ed. M.J. Crocker), 910–934. New York: Wiley.

73 Bloch, H.P., Cameron, J.A., Danowski, F.M. et al. (1982). *Compressors and Expanders: Selection and Application for the Process Industry*. New York: Dekker.

74 Becker, R.J. (1979) Noise control in hydraulic equipment. Proceedings of Noise-Con., 307–310.

75 Continuum® pumps. www.settimafm.com (accessed 19 January 2019).

76 Cudina, M. (2007). Pumps and pumping system noise and vibration prediction and control. In: *Handbook of Noise and Vibration Control* (ed. M.J. Crocker), 897–909. New York: Wiley.

77 Harrison, A.M. and Edge, K.A. (2000). Reduction of axial piston pump pressure ripple. *Proc. Inst. Mech. Eng. I: J. Syst. Control Eng.* 214 (1): 53–63.

78 Cudina, M. (2003). Detection of cavitation phenomenon in a centrifugal pump using audible sound. *Mech. Syst. Signal Process.* 17 (6): 1335–1347.

79 Al-Hashmi, S., Gu, F., Li, Y. et al. (2004) Cavitation detection of a centrifugal pump using instantaneous angular speed. Proceedings of the 7th Biennial Conference on Engineering Systems Design and Analysis – 2004, Manchester, UK, Vol. 3, 185–190.

80 Johansson, A. and Palmberg, J.O. (2003) Design aspects for noise reduction in fluid power systems. Proceedings of the Tenth International Congress on Sound and Vibration, July 7–10 2003, Stockholm, Sweden, 3975–3982.

81 Mohamed, O.A., Charley, J., and Caignaert, G. (2002) Vibroacoustical analysis of flow in pipe system. *Am. Soc. Mech. Eng., Appl. Mech. Div.*, 253(2). Proceedings of the 5th International Symposium on Fluid-Structure Interaction, Aeroelasticity, Flow-Induced Vibration and Noise, Vol. 2: Part B, 881–888.

82 Carletti, E. and Vecci, I. (1990). Acoustical control of external pump gear noise by intensity measuring techniques. *Noise Control Eng. J.* 35 (2): 53–59.

83 Parrondo, J.L., Velarde, S., Pistono, J., and Ballesteros, R. (1998) Diagnosis based on condition monitoring of fluid-dynamic abnormal performance of centrifugal pumps. Proceedings of the 23rd International Conference on Noise and Vibration Engineering, ISMA, 309–316.

84 Gao, Y., Zhang, Q., and Kong, X. (2003). Wavelet-based pressure analysis for hydraulic pump health diagnosis. *Trans. Am. Soc. Agric. Eng.* 46 (4): 969–976.

85 Ng, K.M. (1994). Control valve noise. *ISA Trans.* 33: 275–286.

86 Reethof, G. (1978). Turbulence-generated noise in pipe flow. *Annu. Rev. Fluid Mech.*: 333–367.

87 Reethof, G. (1977). Control valve and regulator noise generation, propagation, and reduction. *Noise Control Eng. J.* 9 (2): 74–85.

88 Baumann, H.D. and Abom, M. (2007). Valve-induced noise: its cause and abatement. In: *Handbook of Noise and Vibration Control* (ed. M.J. Crocker), 935–945. New York: Wiley.

89 Seebold, J.G. (1985). Control valve noise. *Noise Control Eng. J.* 24 (1): 6–12.

90 Morris, P.J. and Lilley, G.M. (2007). Aerodynamic noise: theory and applications. In: *Handbook of Noise and Vibration Control* (ed. M.J. Crocker), 128–158. New York: Wiley.

91 Huff, D.L. and Envia, E. (2007). Jet engine noise generation, prediction, and control. In: *Handbook of Noise and Vibration Control* (ed. M.J. Crocker), 1096–1108. New York: Wiley.

92 Johnston, N. (2007). Hydraulic system noise prediction and control. In: *Handbook of Noise and Vibration Control* (ed. M.J. Crocker), 946–955. New York: Wiley.

93 Zanetti-Rocha, L., Gerges, S.N.Y., Nigel-Johnston, D., and Arenas, J.P. (2013). Rotating group design for vane pump flow ripple reduction. *Int. J. Acoust. Vib.* 18 (4): 192–200.

94 Putnam, R.A., Krebs, W., and Sattinger, S.S. (2007). Furnace ad burner noise control. In: *Handbook of Noise and Vibration Control* (ed. M.J. Crocker), 956–965. New York: Wiley.

95 Putnam, A.A. (1976). Combustion noise in industrial burners. *Noise Control Eng. J.* 7 (1): 24–34.

96 Cabelli, A., Pearson, I.G., Shepherd, I.C., and Collins, D.H. (1987). Control of noise from an industrial gas-fired burner. *Noise Control Eng. J.* 29 (2): 38–44.

97 Putnam, A.A. (1971). *Combustion Driven Oscillation in Industry*. American Elsevier.

98 Leuckel, W. and Fricker, N. (1976). The characteristics of swirl stabilized natural gas flames. Part 1: different flame types and their relation to flow and mixing patterns. *J. Inst. Fuel* 49: 103–112.

99 McManus, K.R., Poinsot, T., and Candel, S.M. (1993). A review of active control of combustion instabilities. *Prog. Energy Combust. Sci.* 19: 1–29.

100 Stewart, N.D., Daggerhart, J.A., and Bailey, J.R. (1974). Experimental study of punch press noise. *J. Acoust. Soc. Am.* 56 (6): S8.

101 Shinaishin, O.A. (1972). Punch press noise, a program for analysis and reduction. *J. Acoust. Soc. Am.* 52: 156.

102 Rivin, E. (2007). Noise abatement of industrial production equipment. In: *Handbook of Noise and Vibration Control* (ed. M.J. Crocker), 987–994. New York: Wiley.

103 Jensen, P., Jockel, C.R., and Miller, L. (1978) *Industrial Noise Control Manual*, rev. ed. DHEW NIOSH Publication No. 79–117. www.cdc.gov/niosh/pdfs/79-117-a.pdf (accessed 20 February 2020).

104 Lai, H.-Y. (1987) Nonlinear Adaptive Modeling for Machining Chatter Identification and Monitoring, Vol. 7, Presented at the ASME Design Technology Conference – 11th Biennial Conference on Mechanical Vibration and Noise., Boston, MA, 181–187.

105 Shiraishi, M., Yamanaka, K., and Fujita, H. (1991). Optimal control of chatter in turning. *Int. J. Mach. Tools Manuf.* 31 (1): 31–43.

106 Pan, J. and Su, C.-Y. (2001) Modeling and chatter suppression with ultra-precision in dynamic turning metal cutting process. Proceedings of the ASME Design Engineering Technical Conference, Vol. 6 B, 18th Biennial Conference on Mechanical Vibration and Noise, Pittsburgh, PA, Sept. 9–12, 1125–1132.

107 Andren, L., Hakansson, L., and Claesson, I. (2003). Active control of machine tool vibrations in external turning operations. *Proc. Inst. Mech. Eng. B J. Eng. Manuf.* 217 (6): 869–872.

108 Andren, L., Hakansson, L., and Claesson, I. (2003) Performance evaluation of active vibration control of boring operations using different active boring bars. Proceedings of the Tenth International Congress on Sound and Vibration, Stockholm, Sweden, July 7–10, 3749–3756.

109 Hakansson, L., Johansson, S., and Claesson, I. (2007). Machine tool noise, vibration, and chatter prediction and control. In: *Handbook of Noise and Vibration Control* (ed. M.J. Crocker), 995–1000. New York: Wiley.

110 Bruce, R.D., Moritz, C.T., and Bommer, A.S. (2007). Sound power level predictions for industrial machinery. In: *Handbook of Noise and Vibration Control* (ed. M.J. Crocker), 1001–1009. New York: Wiley.

12

Noise and Vibration Control in Buildings

12.1   Introduction

Buildings are very complex mechanical structures and are subjected to many external and internal sources of noise and vibration. The requirements of lightweight construction in new buildings have made these buildings more susceptible to noise and vibration-related problems. Thus, occupants frequently complain about the levels of noise and vibration they experience. Internal sources of noise and vibration in modern buildings are mechanical equipments, such as roof-mounted heating, ventilation, and air-conditioning (HVAC) units, boilers and elevators. The building occupants themselves produce noise in many ways (e.g. speaking with loud voices and their footsteps). In this chapter, we will start discussing the transmission of airborne sound through structures such as walls and floors. Sound transmission is of concern in many different noise problems, although the transmission of sound through walls of buildings is the topic which seems to have received the most attention. This is not surprising since in buildings (houses, multifamily apartments, and industrial buildings) we are mainly concerned with reducing the sound transmitted from one room to another. However, with the continued mechanization of transportation and other equipment, we are becoming more and more concerned with keeping exterior noise out of buildings (sometimes known as noise "immission") – e.g. from surface transportation vehicles (mainly road traffic), aircraft noise, and industrial noise. Sometimes the prevention of industrial noise from reaching a community (sometimes known as noise "emission") by the industrial building erected around the process must be considered. This problem, however, seems to be less frequently encountered. It is important to notice that consideration to noise and vibration insulation during the design and building stages is by far less expensive than taking corrective actions once the building is constructed. For more extensive discussions of noise and vibration in building structures, the reader is referred to more detailed treatments available in several books [1–10]. Sections 12.2–12.4 present a review of sound transmission theories for single and double walls and composite partitions. Section 12.4 reviews the use of Statistical Energy Analysis (SEA) to predict the transmission of sound and vibration through structures. The effects of leaks and flanking of a partition are discussed in Section 12.25. Sections 12.6 and 12.7 discuss single-number ratings and test methods for airborne sound transmission. The transmission of impact sound (e.g. footsteps) and floor/ceiling assembly performance are reviewed in Section 12.8. Sections 12.9 and 12.10 review measured sound transmission loss data and sound insulation requirements for buildings. The chapter ends with a discussion on the vibration response of buildings to fluid flow and wind.

Engineering Acoustics: Noise and Vibration Control, First Edition. Malcolm J. Crocker and Jorge P. Arenas.
© 2021 John Wiley & Sons Ltd. Published 2021 by John Wiley & Sons Ltd.

12.2 Sound Transmission Theory for Single Panels

12.2.1 Mass-Law Transmission Loss

We will consider the case of a plane sound wave incident on a partition. For simplicity, we shall assume that the partition is thin (in wavelengths) and infinite in extent. For convenience, we assume that the x- and y-axes are in the plane of the paper (see Figure 12.1) and that the direction of propagation of the wave is in the x-y plane, so that there is no variation in the z-direction (perpendicular to the paper).

Thus, the three-dimensional wave equation (see Eq. (3.29)) reduces to the two-dimensional wave equation:

$$\frac{\partial^2 p}{\partial x^2} + \frac{\partial^2 p}{\partial y^2} - \frac{1}{c^2}\frac{\partial^2 p}{\partial t^2} = 0. \tag{12.1}$$

We may assume plane waves p_i, p_r, p_t, incident, reflected, and transmitted by the panel (Figure 12.1b). If the panel stiffness and damping are negligible, it will act like a limp mass, and the panel will be forced into motion. If we assume that the incident wave has a frequency f, then the panel will be forced into vibration at this frequency, and the reflected and transmitted waves will also have the same frequency.

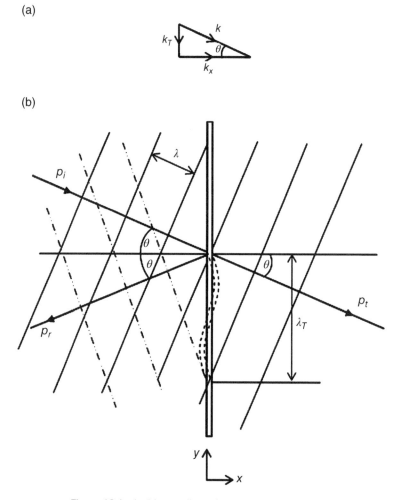

Figure 12.1 Incident, reflected, and transmitted waves.

The wavelengths and wave numbers of the incident, reflected, and transmitted waves must be the same, provided the fluid medium is the same on each side of the panel. By Snell's law, the angles of incidence, reflection, and transmission will also be the same (equal to θ). Provided the fluid remains in contact with the surface of the panel everywhere, the trace wavelength of the vibration on the panel $\lambda_T = \lambda/\sin\theta$, where λ is the wavelength in the fluid.

Equations for the plane waves in the fluid which satisfy Eq. (12.1) are given by

$$p_i = A_1 e^{j(\omega t - k_x x + k_y y)}, \tag{12.2a}$$

$$p_r = B_1 e^{j(\omega t + k_x x + k_y y)}, \tag{12.2b}$$

$$p_t = A_2 e^{j(\omega t - k_x x + k_y y)}. \tag{12.2c}$$

The amplitudes of the waves A_1, B_1, and A_2 are shown as complex quantities since the waves are not necessarily in phase with each other (see Chapter 3). The wave number components k_x and k_y, in the x- and y-directions are related to the wave number k by

$$k_x = k \cos \theta, \tag{12.3a}$$

and

$$k_y = k \sin \theta. \tag{12.3b}$$

See Figure 11.1a, which shows the situation for the incident wave. Squaring Eqs. (12.3a) and (12.3b) and adding gives

$$k^2 = k_x^2 + k_y^2. \tag{12.4}$$

This result can also be seen from Figure 11.1a or by substituting any of the Eqs. (12.2a), (12.2b), or (12.2c) into Eq. (12.1) and remembering that $k = \omega/c$.

The ratio of the square of the amplitude of the transmitted wave compared to that of the incident wave is the quantity which is of most interest, since it tells us the fraction of energy transmitted by the partition. This fraction is called the *transmission coefficient*, $\tau = |A_2|^2/|A_1|^2$ and may be determined using the following two conditions:

1) The acoustic particle velocity normal to the panel surface on each side of the panel must equal the panel velocity v_w at each point.
2) The total pressure acting on the panel equals the mass per unit area times the acceleration $j\omega v_w$ at each point.

The first condition leads to

$$v_w/\cos\theta = p_t/\rho c = p_i/\rho c - p_r/\rho c \tag{12.5}$$

If the panel is very thin, then the pressures p_i, p_r, and p_t near the panel surface are given by putting $x = 0$ in Eqs. (12.2a)–(12.2c) which, when substituted into Eq. (12.5), gives

$$B_1 = A_1 - A_2. \tag{12.6}$$

The second condition gives

$$p_i + p_r - p_t = v_w Z_w. \tag{12.7}$$

Neglecting stiffness K and damping R in the wall, we may put the wall impedance per unit area, $Z_w = j\omega M$, where M is the mass per unit area of the wall. The pressures in Eq. (12.7) again must be evaluated at $x = 0$. Substituting Eqs. (12.2a)–(12.2c) into (12.7), with $x = 0$, and using $v_w = (p_t/\rho c)\cos\theta$ from Eq. (12.5), gives

$$A_1 + B_1 - A_2 = j\omega(M/\rho c)A_2 \cos\theta. \tag{12.8}$$

Substituting Eq. (12.6) into (12.8), and eliminating B_1, gives

$$\frac{A_1}{A_2} = 1 + \frac{j\omega M}{2\rho c/\cos\theta}. \tag{12.9}$$

The ratio of the intensity of the incident and the transmitted waves is

$$1/\tau = \frac{|A_1|^2/\rho c}{|A_2|^2/\rho c} = \frac{|A_1|^2}{|A_2|^2} = 1 + \frac{\omega^2 M^2}{4(\rho c)^2/\cos^2\theta}. \tag{12.10}$$

The quantity, τ in Eq. (12.10) is known as the *sound transmission coefficient*. We define a logarithmic quantity, the *transmission loss* (*TL*), to be

$$TL = 10 \log(1/\tau), \tag{12.11a}$$

$$TL = 10\log\left[1 + \frac{\omega^2 M^2}{4(\rho c)^2/\cos^2\theta}\right] \text{ dB}. \tag{12.11b}$$

Note that *TL*, the transmission loss in decibels, is sometimes known as the *sound reduction index* in Europe and in International Organization for Standardization (ISO) standards (see Section 12.6). The logarithm of $1/\tau$ instead of τ is used in order to obtain positive numbers.

We note that, according to this theory, when a wave approaches the panel at grazing incidence (parallel to its surface), $TL \to 10\log(1) \to 0$ because $\cos\theta \to 0$ and the wave is transmitted through the panel without any attenuation; however, this is an unusual situation in real cases. If a wave approaches the panel in a direction normal to its surface, the transmission loss is maximum. Unless we consider very light panels and very low frequencies, or panels immersed in a fluid medium having a high value of ρc then

$$\frac{\omega^2 M^2}{4(\rho c)^2/\cos^2\theta} \gg 1$$

and

$$TL \approx 10\log\left[\frac{\omega^2 M^2}{4(\rho c)^2/\cos^2\theta}\right] \text{ dB}. \tag{12.12}$$

This result is known as the *mass law*. We note that the *TL* is governed by the mass per unit area, M, of the panel. For a fixed angle of incidence θ, if the frequency, $\omega = 2\pi f$, is kept constant, then each time the mass per unit area of the wall is doubled, the *TL* increases by $10\log(4) = 20(0.301) \approx 6$ dB. For a fixed angle of incidence and fixed mass per unit area M, likewise the *TL* increases by 6 dB for each doubling of frequency (octave) (see Figure 12.2). We have probably all experienced these phenomena in our everyday lives: we know that (i) thick massive walls have a much better *TL* than thin ones, and (ii) we hear the low notes of music transmitted through walls in buildings much better than the high notes.

Example 12.1   Determine the transmission loss for normal incidence at 1000 Hz of a single panel made of aluminum 3 mm-thick and density 2700 kg/m^3.

Solution

The mass per unit area of the panel is $M = 2700 \times 3 \times 10^{-3} = 8.1$ kg/m^2. If we substitute M, $\omega = 2\pi f = 2\pi(1000) = 6283$ rad/s, and $\theta = 0°$ into Eq. (12.12), we obtain the *TL* for normal incidence: $TL = 20\log[(6283 \times 8.1)/(2 \times 1.18 \times 344)] = 35.9$ dB.

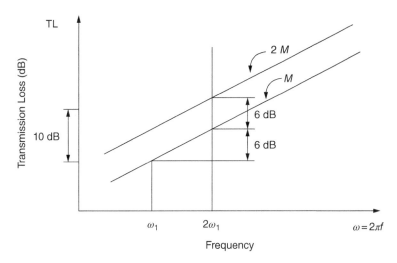

Figure 12.2 Variation of mass law transmission loss of a single panel for sound incident at a given angle θ. (see Eq. (12.12)).

So far in this section, it has been assumed that the panel behaves as a limp mass. It is difficult to include the effects of plate stiffness and damping exactly, but these may be included approximately as follows. We will assume that the panel stiffness and damping can be distributed uniformly over the surface area, so that Z_w in Eq. (12.7) is replaced by $j\omega M + R + K/j\omega$ instead of $j\omega M$. Here, R and K are the damping and stiffness coefficients per unit area. Then instead of Eq. (12.11b) we obtain

$$TL = 10 \log \left[\frac{R_1^2 + (\omega M - K/\omega)^2}{4(\rho c)^2 / \cos^2\theta} \right] \text{dB}, \tag{12.13}$$

where $R_1{}^2 = (R + 2\rho c/\cos \theta)^2$.

Equation (12.13) is plotted in Figure 12.3. At very low frequency, the sound transmission is controlled by the panel stiffness. As the frequency is increased, the panel resonance frequency ω_n is reached, and if the damping R is zero, the panel becomes transparent to sound, and $TL = 0$. As the frequency is increased still further, the TL is dominated by the panel mass (inertia), and of course Eq. (12.13) is equivalent to Eq. (12.12) at high frequency. This simple theory does not predict the coincidence effect predicted by more sophisticated theories (see Section 12.2.3). The effect of higher panel resonances is also not predicted by the theory because the theoretical model is an equivalent single-degree-of-freedom system.

For most building elements, the first panel resonance generally occurs below the practical frequency range of interest, so its effect can be omitted. The TL in the vicinity of this resonance frequency depends strongly upon the panel damping and the characteristics of the incident sound field, both of which are very difficult to estimate. However, in order to improve the attenuation of a panel, the damping should be as large as practical. Also, so far, the attenuation of a panel to sound at only one angle of incidence has been considered. In most practical cases, sound waves strike a panel from many angles simultaneously (random incidence). Figure 12.4 shows the TL of a partition, including the effect of higher order panel resonances and the coincidence effect (see Section 12.2.3). Figure 12.4 is deduced partly from theoretical and experimental considerations.

12.2.2 Random Incidence Transmission Loss

In practice, sound will strike the partition from many angles simultaneously. Thus, some averaging over the angle of incidence of the theoretical results, such as Eq. (12.11) or Eq. (12.12), is necessary in order to predict the partition transmission loss for this case.

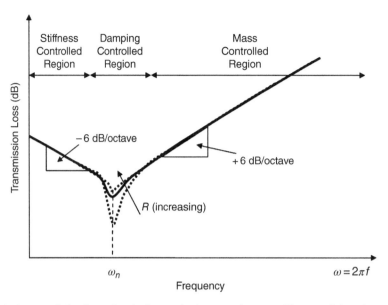

Figure 12.3 Theoretical transmission loss of a single panel when panel mass, stiffness and damping are included. Sound is incident at a given angle θ. (see Eq. (12.13)).

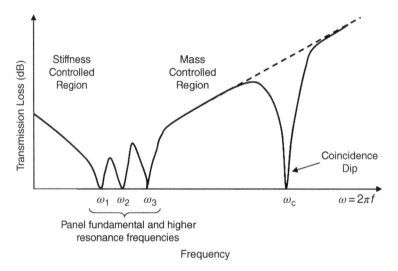

Figure 12.4 Transmission loss of a single panel showing the effects of panel resonances and wave coincidence.

If we consider waves to approach the hemispherical area shown in Figure 12.5 with equal probability, we could choose a limiting angle $\theta' = 60°$ to give an approximate average transmission loss for the waves approaching the surface dS. This is because when $\theta' = 60°$, half of the surface area of the hemisphere is above this angular element dA, and half is below. If we substitute $\theta = 0°$ into Eq. (12.12), we obtain the *TL* for normal incidence:

$$TL_0 \approx 10\log\left[\frac{\omega^2 M^2}{4(\rho c)^2}\right] = 20\log\left[\frac{\omega M}{2\rho c}\right] \text{ dB.} \tag{12.14}$$

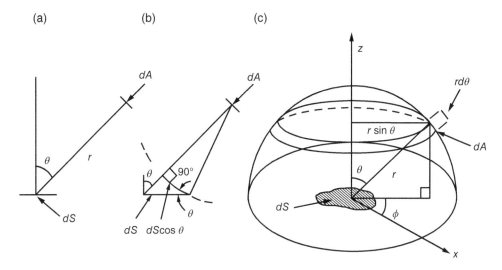

Figure 12.5 Hemispherical area enclosing elemental panel area *dS*.

Then with $\theta = 60°$ in Eq. (12.12), we obtain an estimate for the *TL* for random incidence:

$$TL_{rand} \approx 10 \log \left[\frac{\omega^2 M^2}{4(\rho c)^2 / (1/2)^2} \right] \text{ dB.} \tag{12.15}$$

$$TL_{rand} \approx TL_0 - 6 \text{ dB} \tag{12.16}$$

since $\cos 60° = 1/2$ and $10 \log (1/2)^2 = 6$ dB. This result for TL_{rand} in Eq. (12.15), agrees fairly well with experimental results.

A more rigorous way of estimating *TL* for random incidence sound is to find the average value of the transmission coefficient τ_{rand}, averaged properly over all angles, using Eq. (12.10) and Figure 12.5. Waves approaching the surface of the wall partition at angle θ, only "see" a projected area $dS \cos \theta$. The transmission coefficient depends on angle θ and, thus, in the averaging process must be "weighted" by $\cos \theta$. The average transmission coefficient for random incidence τ_{rand} is thus

$$\tau_{rand} = \frac{\int_\Omega \tau(\theta) \cos \theta \; d\Omega}{\int_\Omega \cos \theta \; d\Omega}, \tag{12.17}$$

where Ω is the solid angle.

Since the differential of the solid angle $d\Omega = \sin \theta \; d\theta \; d\phi$, we may rewrite Eq. (12.17) as

$$\tau_{rand} = \frac{\int\limits_{\theta=0}^{\bar{\theta}} \int\limits_{\phi=0}^{2\pi} \tau(\theta) \cos \theta \sin \theta \; d\theta d\phi}{\int\limits_{\theta=0}^{\bar{\theta}} \int\limits_{\phi=0}^{2\pi} \cos \theta \sin \theta \; d\theta d\phi}. \tag{12.18}$$

For those of us unfamiliar with the use of solid angles, we may care to consider the waves traveling through the small area *dA*. By putting $r = 1$, $dA = \sin\theta \; d\theta \; d\phi$ and after weighting $\tau(\theta)$ by $\cos \theta$ and averaging over all angles, we arrive again at Eq. (12.18). Using Eq. (12.10) we can write

$$\tau(\theta) = 1 / \left(1 + a^2 \cos^2\theta \right), \tag{12.19}$$

where $a = \omega M / 2\rho c$. However, since $d(\sin^2\theta) = 2\sin \theta \cos \theta \; d\theta$ and $d(1 + a^2\cos^2\theta) = -2a^2\cos \theta \sin \theta d\theta$, we may rewrite Eq. (12.18) as

$$\tau_{rand} = \dfrac{-\dfrac{1}{2}\displaystyle\int_{\theta = 0}^{\bar{\theta}} \dfrac{\dfrac{1}{a^2} d\left(1 + a^2 \cos^2\theta\right)}{1 + a^2 \cos^2\theta}}{\dfrac{1}{2}\displaystyle\int_{\theta = 0}^{\bar{\theta}} d\left(\sin^2\theta\right)},$$

$$\tau_{rand} = \frac{\left[-(1/a^2)\ln\left(1 + a^2 \cos^2\theta\right)\right]_{\theta = 0}^{\bar{\theta}}}{\left[\sin^2\theta\right]_{\theta = 0}^{\bar{\theta}}}. \tag{12.20}$$

If we average over all angles $\bar{\theta} = 90°$, $\sin^2\bar{\theta} = 1$, and $\cos^2\bar{\theta} = 0$. Thus,

$$TL_{rand} = 10\log\left(1/\tau_{rand}\right)$$

$$TL_{rand} = 10\log a^2 - 10\log\left[\ln\left(1 + a^2\right)\right] \tag{12.21}$$

and since $a = \omega M/2\rho c$, we see from Eq. (12.14) that $TL_0 = 10\log(a^2)$. Thus,

$$TL_{rand} = TL_0 - 10\log\left[\ln\left(1 + \left(\frac{\omega M}{2\rho c}\right)^2\right)\right] \approx TL_0 - 10\log\left(0.23 TL_0\right). \tag{12.22a}$$

We see as we saw in Eq. (12.16), that the random incidence TL is again predicted to be less than the TL at normal incidence. This is to be expected since the random incidence TL is heavily weighted by waves, for which $\theta > 0°$ and in which $TL < TL_0$.

It is found that Eq. (12.22a) does not agree well with experimental results, and it is normal practice to argue that we should use $\theta = 78°$ instead of $90°$, in order to obtain better agreement between theory and experiment. The result of averaging $\tau(\theta)$ over all angles of incidence from normal ($0°$) to $78°$, using the two Eqs. (12.19) and $TL = 10\log[1/\tau(\theta)]$, is usually known as the *field-incidence*, mass-law transmission loss (TL_{field}). The integration is performed up to $\theta = 78°$ instead of $90°$ to obtain better agreement with experiments. Several reasons are usually quoted to justify this, including (i) the finite size of the partition, (ii) the effect of the test facility and the possible paucity of room modes which would give rise to sound waves near grazing incidence, and (iii) damping effects in the partition for near grazing incidence waves. It should be noted that some authors have used other values of θ, e.g. $80°$ or $81°$ [6]. Figure 12.6 shows a nondimensionalized plot of the mass-law transmission loss of a limp-wall for normal incidence ($\theta = 0°$), several other values of incidence ($45°$, $60°$, and $78°$), field incidence ($\theta = 78°$), and random incidence ($\theta = 90°$).

The transmission loss curves given in Figure 12.6 are only valid in the mass-controlled frequency region of a partition (see Figure 12.4). However, they do illustrate the most important features of the TL of a partition: the fact that the TL increases as the mass per unit area or the frequency of the incident sound are increased. It is noted that a TL somewhat less than that predicted by the mass-law portion of the curve should be expected at the lowest panel resonance in the damping-controlled frequency region. Also, it is seen that the highest TL is for sound approaching at normal incidence. For prediction purposes, it is best to use the field-incidence TL curve if it is expected that the sound approaching the partition comes from many directions instantaneously. This is the normal case in practice. One exception might be, for example, the noise of traffic reaching the windows and walls of the upper stories of high-rise buildings. In this case, the sound would be much nearer to grazing incidence, and a TL lower than the free-field TL would result.

We see from Figure 12.6 that approximately

$$TL_{rand} = TL_0 - 6 \text{ to } 11 \text{ dB}. \tag{12.22b}$$

Often the equation

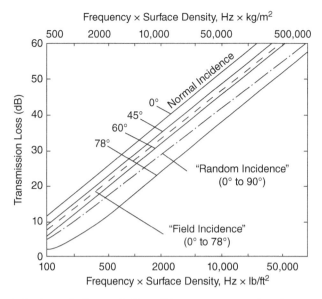

Figure 12.6 Mass law transmission loss of a limp wall. The solid curves correspond to plane waves arriving at given angles of incidence θ. The dashed curve corresponds to "field-incidence mass law," an average over all angles of incidence up to 78°. The dashed-dot curve corresponds to "random-incidence mass law," an average over all angles of incidence 0° to 90°.

$$TL_{field} = 20\log(Mf) - C \tag{12.22c}$$

is used as an empirical fit for Eq. (12.21) or (12.22b). Here, M is the surface density, f is the frequency in Hz, $C = 47$ if the units of ρ_s are kg/m² and $C = 34$ if the units are lb/ft². Equation (12.22b) may be compared with Eq. (12.16). Again it should be emphasized that Eq. (12.22c) does not include the effects of sound transmitted in the frequency region in which coincidence occurs (see Figure 12.4).

Example 12.2 Calculate the TL for normal and field incidence at 500 Hz of a brick wall with a mass per unit area of 415 kg/m².

Solution

We have that $\omega = 2\pi(500) = 3142$ rad/s. From Eq. (12.14) we obtain that for normal incidence $TL_0 = 20\log[3142 \times 415/(2 \times 1.18 \times 344)] = 64.1$ dB.

For field incidence we may use Eq. (12.22c) $TL_{field} = 20\log(Mf) - 47 = 20\log(415 \times 500) - 47 = 59.3$ dB.

Example 12.3 Consider a steel plate of density 7700 kg/m³. At a frequency of 500 Hz, it is desired to have a field-incidence $TL = 35$ dB. Determine the required thickness h of the plate. Consider that 500 Hz is in the mass-controlled frequency region of the plate.

Solution

The required TL is given by Eq. (12.22c): $TL = 20\log(Mf) - 47$. Solving for M we get

$$(TL + 47)/20 = \log(Mf), \text{ then } M = (1/f) \times 10^{(TL + 47)/20}. \text{ Therefore,}$$

$M = (1/500) \times 10^{(35 + 47)/20} = 25.18$ kg/m². Since $M = 7700 \times h = 25.18$, the thickness of the steel panel must be $h = 25.18/7700 \approx 3.3$ mm (1/8 in).

12.2.3 The Coincidence Effect

The phenomenon of coincidence was overviewed in Section 3.16 of this book. A brief explanation of the coincidence effect and its relation with panel *TL* follows.

So far, we have considered only limp partitions with no bending stiffness. Real panels, of course, have bending stiffness; when they are set into motion, bending waves are created (even if a panel vibrates in a vacuum in the absence of sound waves). Unlike the situation of sound waves in a fluid, bending waves on a panel are dispersive. Instead of all traveling at the same speed c independent of frequency, high-frequency waves travel faster than low-frequency waves, i.e. at low frequencies, bending waves on a panel travel slower than the speed of sound in air, c; at high frequency they travel faster. In addition, at low frequency, the wavelength of free-bending waves λ_b is less than the acoustic wavelength λ; at high frequency, the opposite is true. The frequency at which the two speeds and the two wavelengths are equal is called the critical frequency f_c (see Eq. (3.85)). The critical frequency is plotted against thickness in Figure 12.7 for several well-known materials.

For an infinite panel, we have seen that free-bending waves can exist at any frequency. At each particular frequency, if the panel is immersed in air, it will radiate a plane acoustic wave at an angle θ to the normal so that the bending wavelength $\lambda_b = \lambda/\sin\theta$ (see Figure 3.26). This situation can only exist for $f > f_c$, as is seen by studying Figure 3.25, since $\sin\theta < 1$. Theoretically, below the critical frequency, free-bending waves do not radiate any sound. Conversely, if a plane sound wave arrives at an angle θ so that the trace wavelength on the panel $\lambda_T = \lambda/\sin\theta$, and if this wavelength λ_T is equal to the free-bending wavelength at that frequency λ_b, then coincidence is said to occur. The panel response is large, and if the panel damping is zero, the wave is transmitted through the panel without attenuation ($TL = 0$). Panel damping (losses) would create some TL. This coincidence frequency is given by

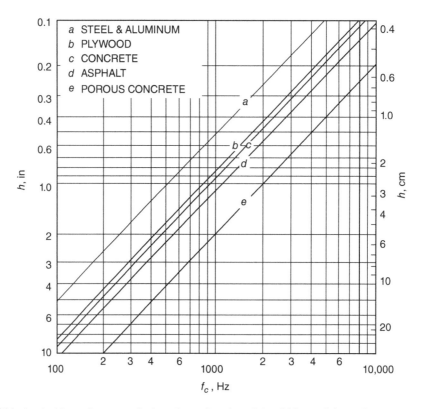

Figure 12.7 Critical coincidence frequency f_c plotted as a function of the thickness h for various construction materials.

$$f_{coinc} = \frac{c^2}{1.8 h c_L \sin^2\theta},\tag{12.23}$$

where h is the thickness, and c_L is the longitudinal wave speed in the panel (see Section 3.16). The critical frequency f_c may be considered to be the lowest possible value of the coincidence frequency and is related to sound waves at grazing incidence ($\theta = 90°$). We notice that because c_l is related to stiffness, the greater the stiffness, the lower the coincidence frequency.

If the panel is excited by a plane sound wave so that the trace wavelength $\lambda_T = \lambda/\sin\theta \neq \lambda_b$ (see Figure 3.26), a forced wave, not a free wave, will be created on the panel. This forced wave will have a trace-wavelength $\lambda/\sin\theta$ and will travel at the trace-wave speed $c/\sin\theta$. We have already calculated the sound transmitted by the panel in the mass-law frequency region assuming forced waves in Section 12.2.1. Cremer first published a theory which accounted for the coincidence effect in sound transmission [11]. However, his theory assumes that the panel is infinite, and only single-leaf partitions are discussed. Real partitions are, of course, finite, and this causes added complications in the theoretical models because we have to take into account bending-wave reflections at the panel boundaries. Careful theoretical study of this case shows that for free vibration of a finite panel only certain discrete frequencies can exist. These frequencies are known as the *resonance* or *natural* frequencies, and their values depend on the spatial boundary conditions (edge constraints) of the panel. In Section 2.4.2 of this book we showed that each natural frequency is associated with a pattern of vibration on the panel, known as a mode shape. These modes of vibration are sometimes known as "standing" waves since they appear to stand still. In reality, each mode or standing wave on a panel is composed of the summation of four traveling waves.

Although Eq. (3.85) (and Eq. (12.23) for $\theta = 90°$) is still valid on a finite panel, the finite size of a panel causes some changes from the previous discussions. Now, below the critical frequency a panel can radiate sound (although inefficiently). Also, coincidence can still occur (above) the critical frequency, when wave matching occurs between acoustic trace wavelengths and free-bending wavelengths. Above the critical frequency, the sound transmission tends to be dominated by resonant motion on the panel and not by forced motion. In this region, panel damping is important and can alter the panel *TL*.

For frequencies greater than the critical frequency, the *TL* can be predicted by the empirical field incidence expression [9, 10]

$$TL = TL_0(f_c) + 10 \times \log(\eta) + 33.22 \times \log(f/f_c) - 5.7,\tag{12.24}$$

where $TL_0(f_c)$ is the transmission loss for normal incidence at the critical frequency and η is the damping loss factor of the panel (see Section 2.5.2 of this book). We see in Eq. (12.24) how the *TL* above the critical frequency is governed by the damping of the panel. If the frequency is kept constant, then each time the damping loss factor of the panel is doubled, the *TL* increases by $10 \log(2) = 10(0.301) \approx 3$ dB. For a fixed damping and fixed mass per unit area M, the *TL* increases by $33.22 \log(2) = 33.22(0.301) \approx 10$ dB for each doubling of frequency (octave) (see Figure 12.4).

Below the critical frequency, trace wave matching cannot occur for free waves on the panel, and the forced transmission dominates. Although this discussion is necessarily simplified, it explains the experimentally observed results for *TL* of panels.

Example 12.4 Calculate the normal-incidence mass law for a 18-mm thick aluminum panel ($c_l = 5420$ m/s and density 2667 kg/m^3) at a frequency of 500 Hz. Also determine the random-incidence and field-incidence mass laws. What is TL_{field} at 2800 Hz when $\eta = 0.01$?

Solution

The mass per unit area of the panel is $M = 2667 \times 18 \times 10^{-3} = 48$ kg/m^2. The normal-incidence mass law is calculated by Eq. (12.14)

$$TL_0 = 20 \log \left[\frac{2\pi f M}{2\rho c} \right] = 20 \log \left[\frac{\pi (500)(48)}{1.18 \times 344} \right] = 45.4 \text{ dB}.$$

Using Eq. (12.22a) the random-incidence is $TL_{rand} = TL_0 - 10\log(0.23 \times TL_0) = 45.4 - 10\log(10.442) = 35.2$ dB. The field-incidence mass law is given by Eq. (12.22c): $TL_{field} = 20\log(Mf) - 47 = 20\log(48 \times 500) - 47 = 40.6$ dB.

The critical frequency of the panel is determined from Eq. (12.23) for $\theta = 90°$,

$$f_c = \frac{c^2}{1.8 h c_l} = \frac{(344)^2}{1.8(18 \times 10^{-3})(5420)} = 674 \text{ Hz}.$$ Since $f = 2800$ Hz is greater than the critical frequency, the field-incidence TL is determined using Eq. (12.24). First,

$$TL_0(f_c) = 20 \log \left[\frac{\pi (674)(48)}{1.18 \times 344} \right] = 48 \text{ dB}. \text{Therefore,}$$

$$TL = 48 + 10 \times \log(0.01) + 33.22 \times \log(2800/674) - 5.7 = 42.8 \text{ dB}.$$

So far we have only discussed the transmission of sound through single-leaf partitions. The transmission of sound through double-leaf partitions is discussed in the following section.

12.3 Sound Transmission for Double and Multiple Panels

In order to achieve high *TL* with a minimum of weight, multiple panels with air spaces between the panels are usually employed. They are used in applications such as aircraft cabin walls and high-rise buildings where weight savings are important. Higher *TL* is possible with a multiple-panel than with a single-panel partition (of the same mass per unit area), particularly if some absorbing material is introduced into the air spaces and if structural "bridges" between the panels which cause short-circuiting are minimized.

12.3.1 Sound Transmission Through Infinite Double Panels

The first theories for multiple panels seem to be due to Constable [12] and Kimball [13] in the 1930s. However, a decade later, because of the need to improve the sound insulation of aircraft, Beranek and Work produced one of the most useful and fundamental theories for sound transmission through multiple partitions in 1949 [14]. The theory is restricted to normal incidence sound waves but can, in principle, be simply extended to cover the case of oblique and random-incidence waves.

The theory rests on the assumption that the particle velocity must be continuous at each boundary as a wave proceeds from left to right. The sound pressure in the air spaces or absorbent layers is given by the solution of the one-dimensional wave equation:

$$p = A \cosh(bz + \phi_b), \tag{12.25}$$

where A is the wave amplitude, z is the distance from a terminal impedance Z_t, b is the propagation constant for the medium, $\phi_b = \coth^{-1}(Z_t/Z_0)$, and Z_0 is the characteristic impedance of the medium (i.e. assumed infinite). The impedance

$$Z = Z_0 \coth(bx + \phi_b). \tag{12.26}$$

In the air space $b = j\omega/c$ and $Z_0 = \rho c$. In the absorbent layers, b and Z_0 will be complex functions of the properties of the material. Beranek and Work then proceeded to determine the impedance seen by a right traveling wave at each interface. From the velocity continuity assumption, the sound pressure ratios across each boundary are deduced to be equal to the relevant impedance ratios.

The theory is found to be very useful and to give reasonable agreement with experiment. Beranek and Work used the flow resistivity (see Section 9.5.3) of various absorbent materials in order to produce "design charts" of panels with various absorbent blankets. The theory does, however, suffer from several shortcomings. First, the theory is only formulated for normal incidence transmission, although it can easily be extended to oblique [15] and random incidence transmission. A more serious shortcoming is the fact that the characteristic impedance is used for the air spaces and absorbent layers and that the "mass law" impedance $j\omega M$ is used for the impervious panels. It is quite easy to use the infinite panel mechanical impedance (see Section 12.2.1) to allow for stiffness and damping. However, if the panels and air spaces are assumed finite, the analysis becomes complicated, although it can still be formulated in principle.

12.3.2 London's Theory

London produced two theoretical analyses for the TL of panels. The first, in 1949, was for single panels [16]; the second, in 1950, was for double panels separated by an air space. London's second theoretical analysis must be acknowledged to be the first for the random incidence of infinite double panels [17]. In both of the analyses, the complete mechanical impedance for the panel (which is assumed to be infinite and very thin) is used (see Section 12.2.1 subsequent to Eq. (12.13)). Unfortunately, however, London included the panel resistance term in an empirical manner and did not relate it to the panel viscous loss factor. It appears that the reason for this (in the case of double panels at least) was that the analysis did not include cavity losses. Thus, London varied his panel resistance term quite arbitrarily with frequency in an attempt to obtain good agreement between theory and experiment. It seems likely that in London's experiment there was also some cavity absorption, which might explain why London found it necessary to vary the panel resistance in this arbitrary manner.

London's analysis for a double panel is formulated by considering an incident and reflected wave at the first panel, a standing wave in the air space (represented by the sum of left and right traveling waves), and a transmitted wave. The theoretical model is shown in Figure 12.8. The sound waves in the three regions can be written:

$$p_1 = P_i e^{j\omega t - jk(x\cos\theta + y\sin\theta)} + P_r e^{j\omega t - jk(-x\cos\theta + y\sin\theta)}, \; -\infty \leq x \leq,$$

$$p_2 = P_+ e^{j\omega t - jk(x\cos\theta + y\sin\theta)} + P_- e^{j\omega t - jk(-x\cos\theta + y\sin\theta)}, 0 \leq x \leq d, \quad (12.27)$$

$$p_3 = P_t e^{j\omega t - jk(x\cos\theta + y\sin\theta)}, d \leq x \leq \infty,$$

where P_i, P_r, P_+, P_- and P_t, are the complex amplitudes of the incident, reflected, right traveling, left traveling, and transmitted waves, respectively. These equations may be solved in a similar way to those in Section 12.2.1 by using the two boundary conditions: (i) the normal particle velocity must be continuous at each wall and (ii) the equation of motion at each wall must be satisfied.

The first boundary condition gives (since for a simple harmonic wave $u = (j/\omega\rho)(\partial p/\partial x)$), at $x = 0$:

$$\left.\frac{\partial p_1}{\partial x}\right|_{x=0} = \left.\frac{\partial p_2}{\partial x}\right|_{x=0},$$

$$P_i - P_r = P_+ - P_- \quad (12.28)$$

at $x = d$:

$$\left.\frac{\partial p_2}{\partial x}\right|_{x=d} = \left.\frac{\partial p_3}{\partial x}\right|_{x=d},$$

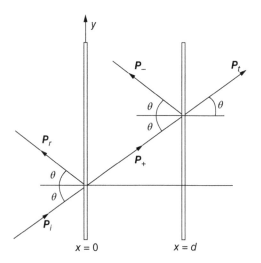

Figure 12.8 Double-panel system with an air gap of width *d*.

$$\boldsymbol{P}_+\, e^{-jkd\cos\theta} - \boldsymbol{P}_-\, e^{jkd\cos\theta} = \boldsymbol{P}_t e^{-jkd\cos\theta}. \tag{12.29}$$

Now the panel velocity must be equal to the particle velocity at each wall; hence,

$$
\begin{aligned}
v_{w1} &= \left.\frac{j}{\omega\rho}\frac{\partial p_1}{\partial x}\right|_{x=0}, \\
v_{w2} &= \left.\frac{j}{\omega\rho}\frac{\partial p_3}{\partial x}\right|_{x=d}
\end{aligned}
\tag{12.30}
$$

where v_{w1} and v_{w2} are the normal velocities of the two walls. Use of Eqs. (12.27) and (12.30) and the equations of motion for the panel (which are similar to Eqs. (12.5) and (12.7)) enable the two following equations to be written:a

$$(\boldsymbol{P}_i + \boldsymbol{P}_r) - (\boldsymbol{P}_+ + \boldsymbol{P}_-) = (Z_w\cos\theta/\rho c)\,(\boldsymbol{P}_i - \boldsymbol{P}_r), \tag{12.31}$$

$$\boldsymbol{P}_+\, e^{-j\beta} + \boldsymbol{P}_-\, e^{+j\beta} - \boldsymbol{P}_t\, e^{-j\beta} = (Z_w\cos\theta/\rho c)\,\boldsymbol{P}_t\, e^{-j\beta}, \tag{12.32}$$

where $\beta = kd\cos\theta$. Using Eqs. (12.28), (12.29), (12.31), and (12.32), the *TL* may be written:

$$TL = 10\log\left(\frac{1}{\tau}\right) = 10\log\left(\frac{|\boldsymbol{P}_i|^2}{|\boldsymbol{P}_t|}\right) = 10\log\left(\left|1 + 2\gamma + \gamma^2\left(1 - e^{-2j\beta}\right)\right|^2\right), \tag{12.33}$$

where $\gamma = Z_w\cos\theta\,/2\rho c$.

If the wall impedance is assumed to be $Z_w = j\omega M$, then Eq. (12.33) becomes

$$TL = 10\log\left[1 + \left(4a^2\cos^2\theta\right)\left(\cos\beta - a\cos\theta\sin\beta\right)^2\right], \tag{12.34}$$

where

$$a = \omega M/(2\rho c). \tag{12.35}$$

It is seen from Eq. (12.34) that a wave will be perfectly transmitted if $\cos\beta = a\cos\theta\sin\beta$, or when

$$\tan\beta = 1/(a\cos\theta). \tag{12.36}$$

If β is small ($d \ll \lambda$), $\tan\beta \approx \beta$, and using Eq. (12.35), the lowest frequency for perfect transmission is given approximately by

$$f_\theta = \frac{1}{2\pi\cos\theta}\left(\frac{2\rho c^2}{Md}\right)^{1/2}, \text{Hz.} \tag{12.37}$$

It is quite easy to show that the frequency predicted by Eq. (12.37) is the same resonance frequency (for the case when $\theta = 0°$) which would be predicted for a mass-spring-mass system with two masses M coupled together with a cavity spring (stiffness) K of $\rho c^2/d$. It is easy to show that the stiffness K of the cavity is $\rho c^2/d$, by moving one panel a distance x and computing the change in pressure which results (and thus the stiffness force) when the other panel is kept still. A mass-spring-mass system has a resonance frequency

$$f = (1/2\pi)\sqrt{2K/M}.$$

Equation (12.36) will of course, predict higher "resonance" frequencies when there is zero *TL* (perfect transmission), since there are additional solutions each time β is increased by approximately π (which results from the repeated nature of the tangent curves and the form of Eq. (12.36)). At high frequency, if $\theta = 0°$, then zero *TL* occurs when $\beta = kd = (2\pi/\lambda)d \approx n\pi$, where n is an integer. That is equivalent to $d \approx n\lambda/2$ or whenever the panel air gap d is equal to an integer number of half-wavelengths. Equation (12.34) is plotted in Figure 12.9 for $\theta = 60°$ [18]. Note that this plot is given in nondimensionalized form. The frequency f is non-dimensionalized in the form f/f_0, where f_0 is the frequency given by Eq. (12.37) with $\theta = 0°$. The other non-dimensional parameter is $2\pi f_0 M/\rho c$. We see in this figure that $\theta = 60°$ and that the fundamental resonance occurs at twice the frequency at which it occurs at normal incidence ($\theta = 0°$). Of more interest, however, is the fact that below the mass-spring-mass resonance, the slope of the curves is somewhat less than 6 dB per octave, while above this frequency the slope is much greater (of the order of 12 or 15 dB per octave, or more).

Obviously, as the angle of incidence is increased, so is the first value of f/f_0 at which zero *TL* occurs. For random-incidence sound, instead of obtaining zero *TL* at one value of f/f_0, we should expect to obtain a "trough" in the *TL* curve appearing below f_0 and continuing into higher frequencies. Unfortunately, Eq. (12.33) cannot be integrated over angle to give a closed-form solution, as we obtained in the case of a single panel in Eq. (12.21). The transmission coefficient τ in Eq. (12.33) must be averaged over angle θ and integrated numerically. London did this in 1948 by hand before the advent of digital computers, and it was extremely laborious. However, now this integration can be done easily by computer, and London's

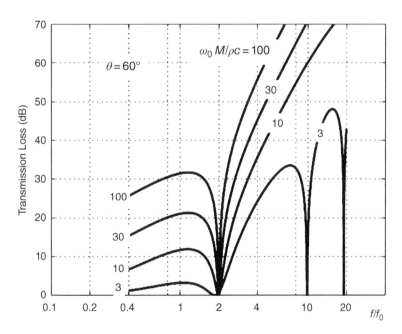

Figure 12.9 Theoretical transmission loss for a double panel predicted by London's theory (see Eq. (12.34)).

theory can be used in designing double partitions. Although useful, London's theory for double partitions has several limitations. It is assumed that the partitions are infinite in extent and thus it does not take into account partition edge effects. Also, cavity absorption is not taken into account.

Example 12.5 In a machinery room of a building, there is a noisy 16-bladed backward-curved centrifugal fan running at 400 rpm. The neighboring room is used as an administrative office. The partition between the two rooms is composed of a double wall made of two plasterboard (density 692 kg/m^3) each having a thickness of 13 mm. The panels are spaced 70 mm apart. Is this a good noise insulation design?

Solution

The mass per unit area of each panel is $M = 692 \times 13 \times 10^{-3} = 9$ kg/m^2. The noise from the fan is predominantly at the blade passing frequency (see Chapter 11 of this book), which is calculated as BPF $= 16(400)/60 = 107$ Hz. The resonance frequency for the double wall is found from Eq. (12.37) for $\theta = 0°$:

$$f_r = \frac{1}{2\pi} \sqrt{\frac{2(1.18)(344)^2}{(0.07)(9)}} = 106 \text{ Hz}.$$

Therefore, it is seen that there is a potential problem since the driving frequency of 107 Hz almost coincides with the resonance frequency of the partition and perfect transmission would be expected. To improve this situation, the resonance frequency must be increased or decreased. A decrease is probably most desirable, and to achieve a sufficient decrease the air gap must be increased to, say, 14 cm and/or the wall mass per unit area increased to 18 kg/m^2. A doubling of either d or M theoretically reduces the resonance frequency by a factor of $1/\sqrt{2}$ or 0.707.

12.3.3 Empirical Approach

In 1978 Sharp [19] presented an empirical approach to estimate the *TL* of a double partition consisting of two solid panels of different mass per unit area with a cavity between them. The approach gives an approximate way of dealing with the low-frequency resonances that often limit the sound insulation of lightweight double-panel walls. Sharp defines three important frequencies: (i) the lowest order acoustic resonance, (ii) the lowest order structural resonance, and (iii) a limiting frequency related to the gap between the panels. The lowest order acoustic resonance is calculated by

$$f_2 = c/2L, \tag{12.38}$$

where *c* is the speed of sound in air and *L* is the longest cavity dimension. The lowest order structural resonance is due to the mass–air–mass resonance; the enclosed air acts as a spring and the panels are the masses. For an air-filled cavity, this frequency can be approximated by

$$f_r = \frac{1}{2\pi} \sqrt{\frac{\rho c^2}{d M_{eff}}}, \tag{12.39}$$

where ρ is the density of air, *d* is the distance between the inner surfaces, M_{eff} is the effective mass per unit area of the two panels given by $M_{eff} = (1/M_1 + 1/M_2)^{-1}$; and M_1, M_2 are the masses per unit area of the two layers. Finally, the limiting frequency f_l is related to the gap width *d* between the panels as

$$f_l = c/2\pi d. \tag{12.40}$$

It is common to add sound-absorbing material to the cavity, which damps cross-cavity resonances which occur at $f_1 = c/2d$ and integer multiples of this frequency. The sound-absorbing material also lowers the frequency at which the mass–air–mass resonance occurs and leads to higher transmission loss values above f_r. As a rough guide, when the wall cavity contains sound-absorbing material, one can multiply the frequency calculated from Eq. (12.39) by 0.7 to estimate the value of f_r [20].

Based on the results presented by Sharp, the following equations can be used to calculate the *TL* when the two panels in a double-leaf partition are completely isolated from each other: [4, 6, 10, 20]

$$TL = \begin{cases} 20 \log\left[(M_1 + M_2)f\right] - 47 & f \leq f_r \\ TL_1 + TL_2 + 20 \log(fd) - 29 & f_r < f < f_l \\ TL_1 + TL_2 + 6 - K & f \geq f_l \end{cases} \tag{12.41}$$

where TL_1, TL_2 are the transmission losses for each leaf of the double wall measured or calculated separately, and $K = 0$ if sound-absorbing material is placed in the air cavity. It is assumed that $K = 10\log(1 + 2/\alpha)$ if the cavity is empty, where α is the average sound absorption coefficient of the panel surfaces. Equation (12.41) indicates that, for frequencies between f_r and f_l, the *TL* increases by 18 dB for each doubling of frequency (octave), although 15 dB/octave is usually obtained in practice. For frequencies above the limiting frequency f_l, the *TL* of a double wall increases by 12 dB for each doubling of frequency when the cavity is filled with a sound-absorbing material.

In summary, in a double-panel wall, high sound insulation can be attained by:

1) Selecting a high mass per unit area of the panels.
2) Avoiding solid connections between the panels: it is found that the use of staggered studs with double panels is particularly effective (see Figure 12.10b). Studs as shown in Figure 12.10a tend to "short-circuit" the two walls throughout the frequency range and the *TL* would be lowered by 5–10 dB for such an arrangement.
3) Selecting a deep cavity between the panels: the gap width of a double-panel system affects the *TL*, particularly at low frequency and the *TL* may be increased slightly by increasing the gap width.
4) Filling the cavity with sound-absorbing material to ensure a low mass–air–mass resonance frequency: a sound-absorbing material, such as a fiberglass blanket, placed in the air cavity, as shown in Figure 12.10c,

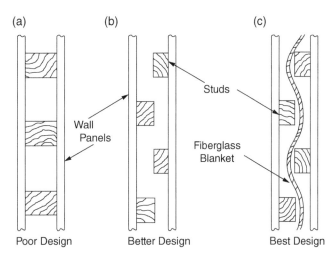

Figure 12.10 Design of double-panel system: (a) straight-through studs; (b) staggered studs; (c) staggered studs with fiberglass blanket.

absorbs some of the sound energy in the cavity and raises the *TL* by 5–10 dB above f_r and increases sound transmission class (STC) values about 6–10 dB [20].

5) Using different materials having different critical frequencies for each face of the wall, so the coincidence dips are less prominent.

Example 12.6   Two glass panels (density 2500 kg/m³, $c_L = 5450$ m/s, and $\eta = 0.002$), each having a thickness of 6 mm are to be used to reduce the sound transmission through an opening. The panels are spaced 75 mm apart. Assume that the surface sound absorption coefficient for the glass is 0.03 for all frequencies. Obtain the expected *TL* of the double wall system at the following frequencies: (a) 63 Hz, (b) 500 Hz, (c) 1000 Hz, and (d) 4000 Hz.

Solution

The mass density of each panel is $M_1 = M_2 = 2500 \times 0.006 = 15$ kg/m², so $M_{eff} = 7.5$ kg/m².

Now, we calculate the mass–air–mass resonance and limiting frequencies for the double wall system, and the critical frequency for each panel as:

$$f_r = \frac{1}{2\pi} \sqrt{\frac{1.18(344)^2}{0.075(7.5)}} = 79.3 \text{ Hz},$$

$$f_l = \frac{344}{2\pi(0.075)} = 730 \text{ Hz},$$

$$f_c = \frac{c^2}{1.8hc_L} = \frac{(344)^2}{1.8(6 \times 10^{-3})(5450)} = 2010 \text{ Hz}.$$

a) *TL* at 63 Hz. Since 63 Hz $< f_r$, $TL = 20 \log [(15 + 15)63] - 47 = 18.5$ dB.
b) *TL* at 500 Hz. Since $f_r < 500$ Hz $< f_l$, the *TL* for one panel is $TL = 20\log(15 \times 500) - 47 = 30.5$ dB. Thus, the *TL* for the double wall is $TL = 30.5 + 30.5 + 20\log(500 \times 0.075) - 29 = 63.5$ dB.
c) *TL* at 1000 Hz. Since 1000 Hz $> f_l$, we obtain

$$TL = 2 \times (20 \log [15 \times 1000] - 47) + 6 - 10 \log [1 + (2/0.03)] = 60.7 \text{ dB}.$$

d) *TL* at 4000 Hz. Since 4000 Hz $> f_l$, but 4000 Hz $> f_c$ we have to use Eq. (12.24) to determine the *TL* of each panel as:

$$TL_0(f_c) = 20 \log \left[\frac{\pi(2010)(15)}{1.18 \times 344}\right] = 47.4 \text{ dB. Therefore,}$$

$TL = 47.4 + 10 \times \log(0.002) + 33.22 \times \log(4000/2010) - 5.7 = 24.6$ dB. Now, the *TL* of the double wall is $TL = 24.6 + 24.6 + 6 - 10\log[1 + (2/0.03)] = 36.9$ dB.

In practice, it is very difficult to construct a double-panel wall where the two panels are not connected in some way, and the transmission loss predicted by Eq. (12.41) is seldom achieved at high frequencies.

Figure 12.11 shows the transmission loss of 16-mm gypsum board constructions for four cases: (i) installed as a single sheet, (ii) installed as two panels screwed together as a single leaf, (iii) installed as two panels with a cavity between them, and (iv) installed as two panels with the cavity filled with fiberglass material. We notice that at f_r there is a dip in the *TL* curve that makes the sound insulation less than that for the two sheets in contact, making it even less than that for the single sheet. The *TL* of the less than double wall becomes equal to that for two panels in contact for frequencies well below f_r. Above this resonance frequency, there are very significant improvements in the transmission loss relative to the curve for the two panels in contact.

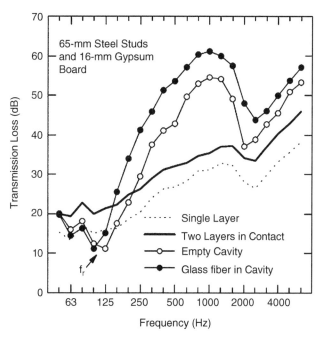

Figure 12.11 Transmission loss of double-leaf and of single-leaf gypsum board walls. The 25 ga. (0.5 mm) steel studs in the double wall are 38 × 65 mm deep. The sound-absorbing material is fiberglass with a thickness of 65 mm and a flow resistivity of 3600 Ns/m² [20].

Example 12.7 A double wall without mechanical connections between panels is made of two 16-mm gypsum boards (density 690 kg/m³, $c_l = 1850$ m/s). The width of the air cavity is 65 mm and it is filled with glass fiber. (a) Determine the TL of this construction at 250 Hz. (b) If the TL of this double wall is greater than 45 dB at 250 Hz, then the partition will fulfill the requirements of sound insulation without increasing the mass per unit area. What is the required gap width d of the double wall?

Solution

a) The mass density of each gypsum board is $M_1 = M_2 = 690 \times 0.016 = 11$ kg/m², so $M_{eff} = 5.5$ kg/m². Now, the relevant frequencies are:

$$f_r = \frac{1}{2\pi} \sqrt{\frac{1.18(344)^2}{0.065(5.5)}} = 100 \text{ Hz},$$

$$f_l = \frac{344}{2\pi(0.065)} = 842 \text{ Hz},$$

$$f_c = \frac{c^2}{1.8 h c_l} = \frac{(344)^2}{1.8(0.016)(1850)} = 2221 \text{ Hz}.$$

Now, since $f_r < 250$ Hz $< f_l$, the TL for one panel is $TL = 20\log(11 \times 250) - 47 = 21.8$ dB. Thus, the TL for the double wall is $TL = 21.8 + 21.8 + 20\log(250 \times 0.065) - 29 = 38.8$ dB

b) We may use the second of Eq. (12.41) to solve the problem but we must check that 250 Hz is between the new f_r and f_l. Thus, $TL = 21.8 + 21.8 + 20\log(250 \times d) - 29 = 45$ dB. Solving for d, we obtain that $d = (1/250)10^{30.4/20} \approx 0.13$ m. Using this gap width, we obtain that $f_r = 70$ Hz and $f_l = 421$ Hz, so $f_r < 250$ Hz $< f_l$. Therefore, a gap width of 13 cm will fulfill the insulation requirements.

12.4 Sound and Vibration Transmission and Structural Response Using Statistical Energy Analysis (SEA)

12.4.1 Introduction

Statistical Energy Analysis is often abbreviated to SEA. SEA was first used in the early 1960s to predict the vibration response of aerospace structures to acoustical environments and to determine the transmission of acoustical and vibrational energy in such structures [21–26]. Ungar, Eichler, Lyon, Maidanik, Smith, Scharton and Manning were some of the main contributors. In 1969, 1970, and 1971, Crocker, Price, and Battacharya published a theoretical approach to predict the sound transmission through finite single and double panels and later panels connected by tie beams [27–30]. The SEA approach has the advantage that (i) sound absorption in the double-panel air gap cavity and (ii) double-panel bridges can be accounted for, theoretically [30].

SEA relies heavily on several results, which have been known in acoustics for many years, such as the modal density and the acoustic energy density of systems. However, one additional result is the assumption that the energy flow between two coupled systems is proportional to the modal energy difference between the two systems [21–26]. In SEA, although the panels and air spaces are assumed to be finite, and modes are assumed to exist in the structures, space–time and frequency averages are made early in the analysis. This enables much simpler solutions to be found than in classical modal analysis theory, which is an alternative approach. For further details about SEA theory, the reader is directed to the books and book chapters by Lyon [21] and De Jong [31], Norton [32], Fahy and Walker [33], Nilsson and Liu [34], Keane and Price [35], Manik [36], and Hambric and Le Bot [37, 38].

SEA has now been applied to a variety of acoustics and vibration problems. More details of SEA theory and its applications to the solution of sound transmission loss problems are given by Crocker and colleagues in journal papers [27–30]. Thus, only a very brief account of the theory – and mainly its applications to sound transmission problems – will be considered in the present Section 12.4.2 of this chapter.

12.4.2 SEA Fundamentals and Assumptions

In order to solve complicated sound/structure response problems, it is necessary to make simplifying assumptions. In SEA these normally involve making averages over time, space and frequency. It is beyond the scope of the discussions in this chapter to go into great depth. Several papers, books and book chapters provide the necessary background [21–26, 31–46]. The NASA CR-160 report by Smith and Lyon is particularly useful [23]. SEA works best at high frequency when there are many modes (in structures and air spaces) resonant in the frequency band under consideration.

When a structural/cavity system is excited by broadband noise, modes of vibration are excited in both the structure and the coupled air cavities. At high frequency, it is found that the shape of the structural panels and of the air cavities is not very important. In most cases the areas of the structural elements and volumes of the cavities are more important than their shapes. Some elementary discussion about the modal behavior of structural panels and air cavities is also included in Sections 3.16 and 3.17 of this book.

a) Vibration of a Simple System

It is assumed that the vibration of extended flat plate structures is almost entirely in bending and it can be modeled as a set of simple resonators. In more complicated built-up structures, however, it is common to include in-plane vibration in SEA models. In steady-state forced vibration, a resonator possesses both kinetic energy $T = \frac{1}{2}Mv^2(t)$ and potential energy $U = \frac{1}{2}Kx^2(t)$ and in steady-state vibration $\langle T \rangle_t = \langle U \rangle_t$, where $\langle \ \rangle_t$ represents a long-time average and is often shortened to $\langle \ \rangle$ (see Eq. (1.7)). In SEA another energy term is of most importance, the dissipated power $W_{diss} = R\langle v^2 \rangle_t$. Here R is the resistance, and $Rv(t)$ is the resistive force $f_R(t)$, so that $W_{diss} = f_R v(t) = 2D$, where D is the Rayleigh dissipation function. For steady state, $\langle W_{diss} \rangle = R_{diss} \langle v^2(t) \rangle$.

b) Energy Loss Factor

We assume small viscous damping for the resonators. Thus, we can write: $\langle W_{diss} \rangle = R\langle v^2 \rangle = \eta \omega_n M\langle v^2 \rangle = \eta \omega_n \langle E \rangle = \beta \langle E \rangle$, where η is the loss factor and $\beta =$ the energy loss factor. For transient vibration we can write: $\langle E \rangle = E_0\, e^{-\omega t}$, where E_0 is the energy at time $t = 0$. Thus $\eta = \langle W_{diss} \rangle /(\omega_n \langle E \rangle)$, where the loss factor η represents the energy dissipated in one cycle of $2\pi/\omega_n$ seconds. In acoustics, we define the reverberation time T_R to be the time in which the energy level in the sound field decays by 60 dB. Likewise in vibration, we write $T_R = 2.2/(\omega_n)$ seconds. Here $f_n = \omega_n/2\pi$ is the natural frequency of the resonator, Hz.

c) Modal Densities of Beams, Plates and Rooms

The modal densities of coupled structures and air spaces are of considerable importance in SEA. For convenience, we normally assume simple supports for the structures and hard walls for the air spaces for mathematical simplicity. At high frequency, the structural and air spatial boundary conditions are found to be of only secondary importance. The modal density can be given in terms of angular frequency ω, rad/s or frequency f, Hz. The modal density represents the number of modes with natural frequencies per radian/second or per hertz, and of course normally depends on frequency.

i) **For a beam:**

The natural frequencies for a simply-supported beam of length l are given by $\omega_m = \kappa c_L\, k^2_m$, where $k^2_m = (m\pi/l)^2$.

The number of modes $N(k)$ below wavenumber k is from Figure 12.12, $N(k) = k/(\pi/l)$, and from the above equations: $N(\omega) = (\omega^{1/2}l)/[\pi(\kappa c_L)^{1/2}]$.

$N(\omega)$ is the total number of modes below frequency ω. The number $N(\omega)$ increases with $\omega^{1/2}$. The modal density $n(\omega)$ may be obtained by differentiating: $n(\omega) = l/[2\pi(\omega \kappa c_L)^{1/2}]$. The modal density of a beam is proportional to its length and decreases with frequency according to $\omega^{-1/2}$.

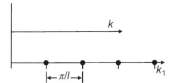

Figure 12.12 Wavenumber diagram for a simply-supported beam.

ii) **For a thin plate:**

The natural frequencies for a simply-supported rectangular plate, of length and width l_1 and l_2, are given by $\omega_m = \kappa c_L\, k_m^2$, where $k_{mn}^2 = k_m^2 + k_n^2 = (m\pi/l_1)^2 + (n\pi/l_2)^2$.

The number of modes below wavenumber k is from Figure 12.13, $N(k) = (\pi\, k^2/4)/(\pi^2/\,(l_1\, l_2)) = k^2 A_p/(4\pi)$, thus $N(\omega) = (\omega A_p)/(4\pi\kappa c_L)$, which increases with ω. The modal density $n(\omega)$ may be

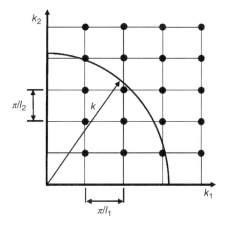

Figure 12.13 Wavenumber diagram for a simply-supported plate.

obtained by differentiation. Thus, the modal density $n(\omega) = A_p/(4\pi\kappa c_L) = \sqrt{3}A_p/(2\pi hc_l)$, which is seen to be a constant and only depends on the plate area A_p. It should be noted, however, that the modal density of thick and sandwich plates depends on the effective wave speed and is more complicated to compute [47].

iii) **For a room:**

The natural frequencies in a rectangular hard-walled room are $\omega_m = ck$, where $k = [(m\pi/l_1)^2 + (n\pi/l_2)^2 + (l\pi/l_3)^2]^{1/2}$, c is the speed of sound in air and l_1, l_2, and l_3 are the room length, width and height.

The number of modes $N(k)$ below wavenumber k (at high frequency ignoring longitudinal and lateral modes) can be obtained from Figure 12.14, $N(k) = (\pi k^3/6)/(\pi^3/(l_1\, l_2\, l_3)) = k^2\, V/(6\pi^2)$ and thus, $N(\omega) = \omega^3 V/(6\pi^3 c^3)$ which increases as ω^3.

Thus, by differentiation the modal density is $n(\omega) = \omega^2 V/(2\pi^2\, c^3)$. The modal density of a room is proportional to its volume and the square of frequency. The mode count $N(\omega)$ and modal density $n(\omega)$ results above are plotted in Figures 12.15 and 12.16.

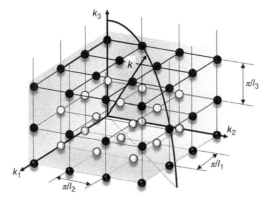

Figure 12.14 Wavenumber diagram for a hard-walled rectangular room.

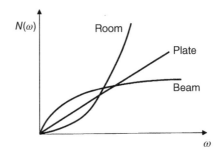

Figure 12.15 Mode number counts for beam, plate and room.

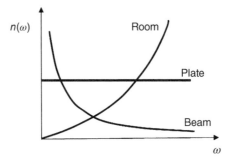

Figure 12.16 Modal densities for beam, plate and room.

In acoustics the modal density $n(\omega)$ is important for several reasons. In structural response and radiation, if $n(\omega)$ is high we can use statistical approaches. If $n(\omega)$ is very low, (only a few modes in the frequency range of interest) we must use the classical approach and work out the response (or radiation) for each mode separately and then sum the modal contributions. In a reverberation room, if there are only a few modes resonant (in say a 1/3 octave band) there will be large fluctuations in panel or air cavity response with the position. This limits the use of SEA to high frequencies.

d) Half Power Points and Bandwidth

Assuming a resonator is excited by a pure tone, we can relate the magnitude of the velocity response V to the magnitude of the force F by its *impedance* $Z = F/V$. Both the force F and velocity V can, in general, be complex and contain both magnitude and phase information. Often, we desire the inverse of the impedance $1/Z = Y$, the so-called *admittance*. At resonance, the admittance becomes real and is called the *conductance* G. The mechanical engineer's critical damping ratio δ (see Eq. (2.16)) is related to the loss factor by $\delta = \eta/2$. The resonator bandwidth $\Delta_{1/2}$ is given by the difference in the half power points $\Delta_{1/2}(\omega) = \eta\omega_n = 2\delta\omega_n$.

If the resonator is excited by a stationary broadband noise source of spectral density $S_f(\omega)$ (see Eq. (1.12)), then a new "effective bandwidth" Δ_e can be defined as $\Delta_e = \pi/2\omega_n\eta = \pi/2R/M = \pi/2\Delta_{1/2}$. See Figure 12.17. Also see Ref. [23].

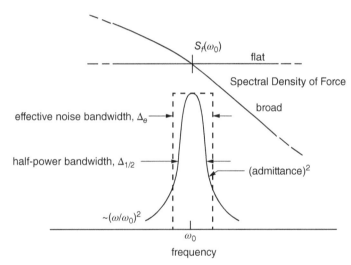

Figure 12.17 Spectral density of random force and square of admittance $|Y|^2$ of a simple resonator [23].

e) Modal Overlap

We can see that for a constant value of η, which is assumed to be independent of frequency, both the half-power points bandwidth $\Delta_{1/2}$ and the effective bandwidth Δ_e of an average mode increase with frequency ω. A frequency is reached where the frequency separation of modes $\Delta\omega_n$ on a two-dimensional structure becomes equal to or less than the half-power bandwidth $\Delta_{1/2} = \eta\omega_n$ of an average mode. Figure 12.18 shows two closely-spaced modes subjected to pure tone excitation. Here we see that the half power bandwidth $\omega_{1/2}$ of mode number ω_1 does not exceed the modal frequency separation, $\Delta\omega$. However, it has become equal to that for mode number ω_2 and by the next mode ω_3, it is expected that the half power bandwidth of the mode of center frequency ω_3 will exceed the frequency separation $\Delta\omega$.

In cases when structures are excited by broadband noise, it is more realistic, however, to use broadband noise excitation rather than pure tone excitation in defining the modal overlap frequency f_{over}. In this case, the modal overlap frequency occurs when the average modal frequency separation $\Delta\omega_n$ equals the effective noise bandwidth Δ_e of a typical resonance. That is when $\Delta\omega_n = \pi/2\eta\omega_n$. See Figure 12.18 in which the magnitude of the admittance squared $|Y|^2$ is plotted against frequency ω. Both the half power bandwidth $\Delta_{1/2}$ and the effective noise bandwidth Δ_e are shown for force excitation of spectral density $S_f(\omega)$ in Figure 12.18.

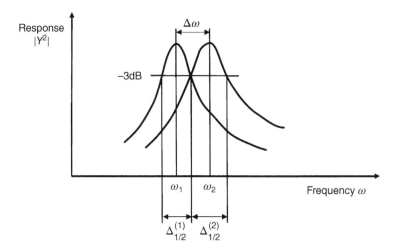

Figure 12.18 Modal overlap factor definition.

It is often stated that SEA works well only if the modal overlap factor is greater than unity. However, this statement is not correct if SEA is viewed in Lyon's original sense as an ensemble theory (R.S. Langley, personal communications, June 21 and 22, 2019). That is the method considers an ensemble of random systems and the predicted energies must be interpreted as ensemble average values. The energy in any one member of the ensemble may differ from the SEA prediction, because the energy has a variance across the ensemble. If the modal overlap is high, then the ensemble variance is low, and every member of the ensemble will have approximately the same energy. But this does not mean that SEA fails when the modal overlap is low – the ensemble average energy can be predicted accurately (R.S. Langley, ibid).

The mean energy is predicted extremely well for very low values of modal overlap (typically as small as 0.5), and the variance can still be predicted well. The variance theory is based on Ref. [49]. So it does not follow that SEA only works for high values of modal overlap. The variance theory can also be applied to coupled FE/SEA models [49] (R.S. Langley, ibid).

Statistical overlap is also important. That is the degree of randomness in the structure for the ensemble theory of SEA to hold requires that there must be sufficient randomness in the system. The statistical overlap is a measure of the rms change in a natural frequency across the ensemble compared to the mean natural frequency spacing (i.e. the "bandwidth" used in the definition of modal overlap is replaced by the rms change in the natural frequency). SEA tends to work when the statistical overlap is greater than unity [49] (R.S. Langley, ibid.). This sets a lower frequency limit on the method, but the modal overlap may be extremely small at this limit [49] (R.S. Langley, ibid.).

Figure 12.19 shows the admittance squared response, $|Y|^2$ of a set of resonators excited in a frequency band W by a force of constant spectral density S_f. As the center frequency ω_c increases, the effective bandwidth of mode number k, $\Delta_k = \frac{1}{2}\pi\eta_k\omega_k$ also increases. In the case of a flat panel, the average mode separation, Δf_{mode}, is independent of frequency. The net result is that, if the panel loss factor η remains constant with frequency, the resonance peaks are distinct and separate at low frequency, but become less and less distinct as the frequency increases above the modal overlap frequency f_{over}. See Example 12.8, part (d). This effect is illustrated in Figure 12.20, where the simplifying assumption is made that the modes have the same frequency separation and all respond with equal energy to the force excitation.

f) Radiation Resistance of a Panel

There are several ways to evaluate the response of a simply-supported panel to noise and the power radiated by the panel when it is excited by noise. We shall assume that either (i) all modes in the frequency band considered have the same mean square velocity (sometimes called equipartition of energy), or (ii) all modes

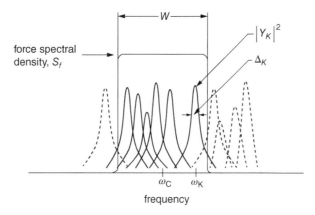

Figure 12.19 A band of force and the resonance curves of the resonators that it drives [23].

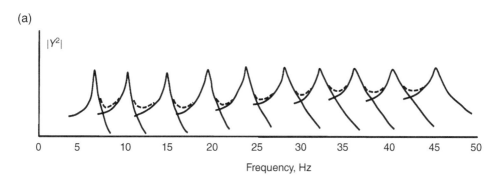

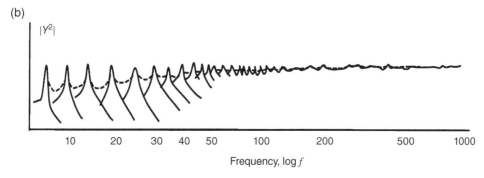

Figure 12.20 (a) Sketch of a multi-resonator system response $|Y|^2$ to a broadband force of constant spectral density S_f. It is assumed that the modes respond with equal energy and have the same constant frequency separation. The effect of the broadening of the effective bandwidth for a constant damping value of η against frequency is seen. Fundamental frequency 6.96 Hz, average modal frequency separation, 4.26 Hz, modal overlap frequency 43.1 Hz. See Example 12.8; (b) Sketch of a multi-resonator system response $|Y|^2$ in Figure 12.20a plotted on a logarithmic frequency scale.

in the band have the same radiation resistance. For modes below the critical frequency, we shall make the first assumption, and for modes above the critical frequency f_c we shall make the second assumption.

The study of finite simply-supported panels has suggested that the major source of radiation from the panel below the critical frequency arises from the interactions of bending waves with edge discontinuities. In particular, those modes of vibration which result in a trace flexural wavelength along the edges, which is greater than the acoustic wavelength, are responsible for most of the radiation from these edges. When the trace flexural wavelength in both directions is greater than the acoustic wavelength, then the corners of the panel alone radiate. Normally these "corner modes," however, radiate much less than the so-called "edge modes" and their contributions can be neglected.

In the analysis presented here, it is assumed at all stages that the panel dimensions are greater than the acoustic wavelength. This simplifies matters in that the interactions between canceled volume velocity elements at the edges and corners of the panel can be treated independently. If the panel dimensions are smaller than the acoustic wavelength, then these simple radiators interact. This case has been considered by Maidanik [39] who, by treating the problem in more depth than is possible here, has derived the equations given in Ref. [39] and in corrected form in Refs. 27, 28. The following simple formulas for radiation to half space for the acoustically slow (A.S.) radiation of a panel below and the acoustically fast (A.F.) radiation above the critical frequency f_c are $\langle R_{rad}^{A.S.} \rangle = (f/f_c)^{1/2}(\rho c \lambda_c P)/\pi^2$, when $f < f_c/2$, and where P is the perimeter of the whole panel. For radiation above f_c, $\langle R_{rad}^{A.S.} \rangle = \rho c A_p (1 - f_c/f)^{-1/2}$. Note that Smith and Lyon give a result for the A.S. radiation in Ref. [23] (Eq. VI.4.2) which is eight times larger and is assumed to be in error [23].

It is of interest to note that the average radiation resistance below the critical frequency is directly proportional to the perimeter P and not to the area A_p of the panel. These results are, of course, only strictly valid for simply-supported panels. It has been shown by Nikiforov that for $f < f_c$ the radiation resistance of a panel with clamped boundaries will give rise to far field mean-square sound pressures twice those for the simply-supported panel.

Example 12.8 The walls between the engine room and a storeroom on a passenger ship are made of 1/16 in (1.6 mm) thick steel. One partition panel wall under consideration measures 3 ft × 4 ft (0.91 m × 1.22 m). Calculate:

a) The approximate critical coincidence frequency;
b) The modal density of the panel;
c) The number of modes in one-third octave bands centered at 10, 100, 1000 and 10000 Hz.
d) The modal bandwidth; assuming the panel damping is 0.5% critical and independent of frequency
e) The lowest natural (resonance) frequency of the panel, and
f) The frequency range in which the panel is acoustically large $l_{min} > \lambda$.

Solution

The solutions are worked in English and SI units. Both systems give the same results. The panel is assumed to have simply supported edges. This is a good engineering assumption to make at high frequency since the boundary conditions have little effect on mode shapes and the panel natural frequencies. This is because most of the vibration takes place away from the boundaries. The working assumes that space-averaging is made over the panel and in the air space contiguous to it.

a) **The critical coincidence frequency**, $f_c = \frac{c^2}{2\pi\kappa c_L}$, where c = speed of sound = 1116 ft/s = 343 m/s and κ is the panel radius of gyration.

For aluminum and steel, the longitudinal wave speed c_L =17000 ft/s = 5280 m/s.

Substituting for c, and $\kappa = h/\sqrt{12}$ for steel and aluminum, where h = panel thickness in inches, gives

$$f_c = 480/h \sim 500/h \sim 500/(1/16) \sim 8000 \text{ Hz}.$$

b) **The modal density**, n_p for a simply-supported panel is given by

$$n_p(\omega) = \frac{\sqrt{3}A_p}{2\pi c_L h} = \frac{\sqrt{3}(3 \times 4)}{2\pi \times 17000 \times 1/192} = 0.037 \text{ modes/rad/s}.$$

Thus $n_p(f) = 2\pi \times 0.037$ modes/Hz. (Remember 1 cycle = 2π rad and 1 Hz = 1 cycle/s = 2π rad/s). Thus, modal density $n_p(f) = 0.235$ modes/Hz.

c) **The frequency bands for one-third octaves** are given by $\Delta f = 23\%\, f_{cent}$, where f_{cent} is the center frequency for the band (see Chapter 1).

Thus, the number of modes n in a band is given by $n = n_p(f) \times 0.23\, f_{cent}$

1) $f_{cent} = 10$ Hz, $n = 0.235 \times 10 \times 0.23 = 0.54$ modes
2) $f_{cent} = 100$ Hz, $n = 0.235 \times 100 \times 0.23 = 5.4$ modes
3) $f_{cent} = 1000$ Hz, $n = 0.235 \times 1000 \times 0.23 = 54$ modes
4) $f_{cent} = 10000$ Hz, $n = 0.235 \times 10000 \times 0.23 = 540$ modes

d) **The Effective Modal Bandwidth**, $b(f) = (\pi/2)\eta_{int}(2\pi f)$. But if the critical damping ratio $\delta = 0.005$, then the internal energy loss factor, $\eta_{int} = 2\delta = 0.01$ (i.e., 1%).

The Effective Modal Bandwidth $b(f) = \pi^2 \times 0.01 \times f$.

The average frequency separation between modes $\Delta f = 1/n_p(f) = 1/0.235 = 4.26$ Hz, which is a constant for a simply-supported plate.

The Modal Overlap is given by $b(f) = \Delta(f)$.

For Modal Overlap, $\pi^2 \times 0.01 \times f_{over} = 1/n_p(f) = 4.26$.

Thus, the Modal Overlap Frequency f_{over} above which mode overlap occurs is

$$\text{Modal overlap frequency } f_{over} = 43.1 \text{ Hz.}$$

e) **The natural frequencies of the panel** are given by $\omega_{mn} = \kappa c_L\, k^2_{mn}$; thus,

$f_{mn} = (1/2\pi)\omega_{mn}$, where k_{mn} is the panel bending wavenumber for the m,n mode, where $k^2_{mn} = k^2_m + k^2_n = (m\pi/l_1)^2 + (n\pi/l_2)^2$.

The lowest natural frequency is given by $f_{1,1}$, where $m = n = 1$.

Thus, $f_{1,1} = (1/2\pi)\left(h/\sqrt{12}\right)(17000)\left[(\pi/3)^2 + (\pi/4)^2\right]$

$$f_{1,1} = (\pi/2)\left(h/\sqrt{12}\right)(17000)[1/9 + 1/16], \text{ where } h = 1/192 \text{ ft (1.6 mm)}$$
$$f_{1,1} = 6.96 \text{ Hz.}$$

Note "engineers rough rule of thumb," $f_{1,1} \approx 1.5\, \Delta f$ (provided $l_1 \approx l_2$)

[Proof: $\Delta f = 1/n_p(f) = c_L h/(\sqrt{3}A_p) = 2c_L\, \kappa/A_p$.

But $f_{1,1} = (1/2\pi)\, \kappa c_L\, \pi^2[1/l_1^2 + 1/l_2^2] = (\pi/2)c_L\kappa\, [l_2/l_1 + l_1/l_2]/A_p$.

Thus, $f_{1,1} \approx 1.5 c_L\kappa\, 2/A_p = 1.5\, \Delta f$ (provided $l_1 \approx l_2$)].

f) **The panel is acoustically large** if the acoustic wavelength, $\lambda < l_{min}$, where l_{min} is the shortest panel length. The frequency limit is given by $c/f < 0.91$ m, or $f > 343/0.91 = 373$ Hz or $c/f < 3$ ft, or $f > 1120/3 = 373$ Hz. Thus, the panel is acoustically large for frequencies > 373 Hz.

Example 12.9   Refer to the panel in Example 12.8. Assume that the partition panel is made of steel (density = 7700 kg/m^3) with internal critical damping ratio $\delta = 0.01$ (1%). Plot 10log radiation efficiency σ_{rad} against log frequency f, where $R_{rad} = \rho c A_p \sigma_{rad}$ is the radiation resistance, A_p is area of panel, ρ = air density, c = speed of sound in air.

Solution

The radiation efficiency σ_{rad} is calculated as follows:
 - **Panel radiation efficiency below coincidence**:

$$\sigma_{rad} = \frac{R_{rad}(2\pi)}{\rho c A_p} = \frac{2}{\pi^2}\left(\frac{f}{f_c}\right)^{1/2} \lambda_c \frac{P}{A_p}, \text{ (for one side of panel),}$$

where λ_c = coincidence wavelength = 343/8000 = 0.042 m, P = panel perimeter = 2(0.91+1.22) = 4.26 m, and A_p = panel area = 0.91×1.22 = 1.11 m^2.

If f/f_c = 1/16 (i.e. f = 500 Hz, assuming f_c = 8000 Hz):

$$\sigma_{rad} = \left(2/\pi^2\right)\,(1/4)\,(0.042) \times 4.26/1.11 = 0.0082.$$

Thus, 10 log σ_{rad} = 10log(0.0082) ≈ −21 dB. This calculation is repeated at f/f_c = 1/8, 1/4, and 1/2 which gives 10log σ_{rad} = −19.5, −18.0, and −16.5 dB, respectively.

- **Panel radiation efficiency above coincidence**:

$$\sigma_{rad} = \frac{1}{\sqrt{1-f_c/f}}.$$

At 16000 Hz, f/f_c = 2, $\sigma_{rad} = \sqrt{2}$. Thus, 10log σ_{rad} = 1.5 dB. The calculation is repeated at 9000 Hz, 10000 Hz and 12000 Hz which gives 10log σ_{rad} = 4.7 dB, 3.5 dB, and 2.3 dB, respectively.

If $f \rightarrow \infty$, $\sigma_{rad} \rightarrow 1$ and 10log σ_{rad} = 0. The results are plotted in Figure 12.21.

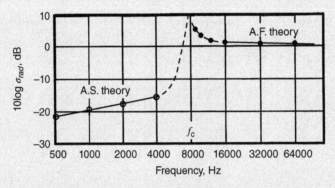

Figure 12.21 Simple plot of radiation efficiency of a simply-supported panel.

Example 12.10 Consider the partition panel in Example 12.8 again. Plot 10log radiation loss factor η_{rad} against log frequency, f.

Plot 10log η_{int} on the same figure. Note $\eta_{rad} = (\rho c/\omega\rho_s)\sigma_{rad}$, where ρ_s is the panel surface density (mass/unit area).

Solution

Radiation loss factor,

$$\eta_{rad} = \frac{R_{rad}(2\pi)}{\omega M} = \frac{\sigma_{rad}\rho c A_p}{\omega\rho_s A_p} = \frac{\sigma_{rad}\rho c}{\omega\rho_s},$$

where ρ_s = panel surface density = 7700 kg/m^3 × 0.0016 m, and ρ = air density = 1.23 kg/m^3. Thus,

$$\eta_{rad} = \frac{\sigma_{rad}(1.23 \times 343)}{7700 \times 0.0016 \times 2\pi f} = 5.5\frac{\sigma_{rad}}{f}.$$

Then, 10log η_{rad} = 7.4 + 10 log σ_{rad} − 10 log f.

Therefore, for f = 500 Hz we obtain 10log η_{rad} = 7.4 −21 −10log(500) = −40.5 dB.

This calculation can be repeated at each frequency for which radiation efficiency data are available.

Now, $\eta_{int} = 2\delta = 2 \times 0.01 = 0.02$ (Remember that $R = \beta M = \omega\eta M = \omega\,2\delta M$, where β is the energy loss factor).

Thus, $10\log \eta_{int} = 10 \log (0.02) = -17$ dB.
The results are plotted in Figure 12.22.

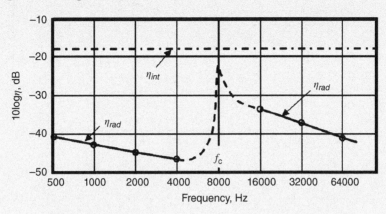

Figure 12.22 Radiation loss factor for a simply-supported panel.

Example 12.11 Vibration energy is fed into the partition panel in Example 12.8 through a point attachment of pipe carrying turbulent flow near one corner producing a force $f_{rms} = 10$ newton in each one-octave band. How much power is injected into the panel by this force?

Assume the input point conductance G is given by $G = \sqrt{3}/\left(4\rho_p c_L h^2\right)$ and that the input power $W_{in} = f_{rms}^2 G$. Plot acceleration levels against $10 \log f$.

Solution

Vibration Energy: Since $f_{rms} = 10$ newton, $W_{in} = (10)^2\, G$,

where G is the assumed point input conductance for the plate $G = \sqrt{3}/\left(4\rho_p c_L h^2\right)$

$[G = \sqrt{3}/\left(4\rho_p c_L h^2\right)$ for an infinite plate].

Note that for steel $\rho_p = 7700$ kg/m^3 and $c_l = 5182$ m/s. Then,

$$G = \frac{\sqrt{3}}{4 \times 7700 \times 5182 \times (0.0016)^2}$$

$$= 4.24 \times 10^{-3} \text{ s/kg. Thus,}$$

$$W_{in} = 100 \times 4.24 \times 10^{-3} = 0.424 \text{ watt.}$$

$$\text{Power level} = 10 \log \left(0.424/10^{-12}\right) = 116.3 \text{ dB.}$$

$$\text{But } W_{in} = \langle v^2 \rangle (R_{tot}) = \omega \eta_{tot} M_p \langle a^2 \rangle / \omega^2.$$

$$\text{Now, } M_p = \rho_s A_p \text{ and } \eta_{tot} = \eta_{int} + \eta_{rad},$$

$$\eta_{tot} = \eta_{int} = 0.02 \text{ (since } \eta_{rad} \ll \eta_{int} \text{ except at coincidence).}$$

$$\langle a^2 \rangle = W_{in}\omega h/\left(\eta_{tot}\, \rho_s\, A_p\right) = 0.424(2\pi\, 1000)/(0.02 \times 7700 \times 0.0016 \times 1.11) = 9750 \text{ m}^2/\text{s}^4,$$

$$\langle a^2 \rangle / g^2 = 9750/(9.81)^2 = 101.3. \text{ Therefore,}$$

$L_a = 10 \log \left(\langle a^2 \rangle / g^2\right) = 20$ dB (in 1000 Hz band). We can easily calculate L_a at different frequencies (see Figure 12.23).

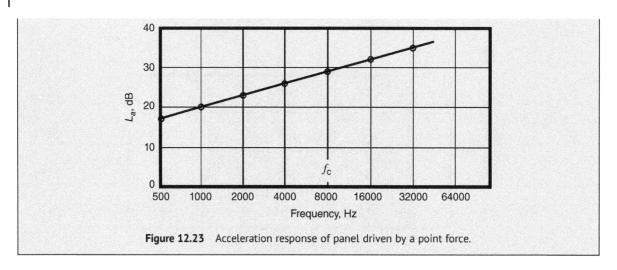

Figure 12.23 Acceleration response of panel driven by a point force.

Example 12.12 In Example 12.8, the storeroom is much too reverberant with a reverberation time of 1 second at all frequencies. Its volume is 85 m³.

Plot the space-average sound pressure level in the cabin against log frequency caused by the force input in part Example 12.11. How much sound absorbing material should be added to the cabin walls, floor and ceiling to reduce the level by first 3 dB and then by 6 dB?

Solution

We can write the following power flow equations using Eq. (12.42).

$$W_{in} = W_{dissP} + W_{rad}, \text{ and}$$
$$W_{rad} = W_{dissR},$$

where W_{dissP} = power dissipated in panel, and W_{dissR} = power dissipated in room.

$$\text{Now, } W_{rad} = \langle v^2 \rangle \, \omega \, M \, \eta_{rad}, \text{ and } W_{in} = \langle v^2 \rangle \, \omega \, M \, (\eta_{rad} + \eta_{int}),$$

$$W_{rad}/W_{in} = \eta_{rad}/(\eta_{rad} + \eta_{int}), \text{ and } W_{rad} = W_{in} \, \eta_{rad}/(\eta_{rad} + \eta_{int}) = W_{in}\eta_{rad}/\eta_{int}, \text{ since } \eta_{int} \gg \eta_{rad}.$$

Also, the equations above in Section (d) and Eq. (12.42) and Figure 12.24b show that all the energy radiated into the room is absorbed in the room, i.e. $W_{rad} = W_{dissR}$.

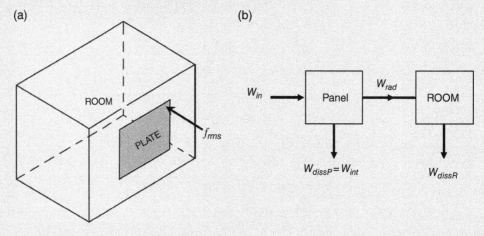

Figure 12.24 (a) Partition panel wall connected to a room; (b) Power flow from panel to room.

Energy density in room is $p^2/\rho c^2$ and total energy in room $= p^2 V/\rho c^2$.

Thus, the rate at which energy is absorbed in room is

$$W_{dissR} = (p^2 V/\rho c^2)\beta_{room},$$

where $\beta_{room} = \omega\eta_{room} = 13.8/T_R$, where T_R is room reverberation time.
Thus, the sound pressure level in room is

$$L_p = 10\log\left(\frac{p^2}{p_{ref}^2}\right) = 10\log\left(\frac{W_{in}}{p_{ref}^2}\frac{\eta_{rad}}{\eta_{rad}+\eta_{int}}\frac{\rho c^2}{V}\frac{T_R}{13.8}\right),$$

$$L_p = 10\log\left(\frac{0.424}{4\times10^{-10}}\frac{1.23\times(343)^2}{85}\frac{1}{13.8}\right) + 10\log\left(\frac{\eta_{rad}}{\eta_{rad}+\eta_{int}}\right)$$

$$L_p = 10\ \log\left\{1.3\times10^{11}\right\} + 10\ \log\left\{\eta_{rad}/(\eta_{rad}+\eta_{int})\right\} = 111\ \text{dB} + 10\ \log\left\{\eta_{rad}/(\eta_{rad}+\eta_{int})\right\}.$$

But we can read $10\log\{\eta_{rad}/(\eta_{rad}+\eta_{int})\}$ from the curves in Example 12.10 (see Figure 12.22) by noting $\eta_{rad} \ll \eta_{int}$ and simply taking the difference between the two curves. Thus, L_p is difference between two curves + 111 dB and the results are plotted in Figure 12.25. Note that if η_{int} were very small, $\eta_{int} \ll \eta_{rad}$ and

$$10\log\left(\frac{\eta_{rad}}{\eta_{rad}+\eta_{int}}\right) = 10\log\left(\frac{\eta_{rad}}{\eta_{rad}}\right) = 10\log(1) = 0.$$

Then, there would be no power loss in the panel, the sound pressure level would be 111 dB in room.

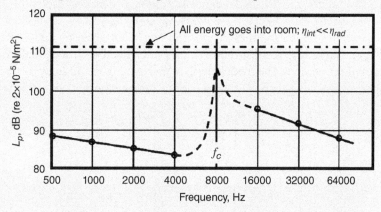

Figure 12.25 Calculated sound pressure level in room.

To reduce the L_p by 3 dB we need to add some absorption area. We first need to calculate the existing absorption area in the room. This is related to the reverberation time T_R by Sabine's formula (see Section 3.14.3 of this book).

Then, $T_R = 0.161V/S\bar{a}$, where V is room volume in m^3 and $S\bar{a} = A$, the absorption area in sabin (m^2).
Now, $V = 85$ m^3, $T_R = 1$ s, and the existing absorption area $A_0 = 0.161\times85/1 = 13.7$ sabins (m^2).
Hence the added absorption A_1 may now be calculated. The reduction in SPL, ΔL_p can be shown to be given by

$$\Delta L_p = 10\log\left(\frac{A_0 + A_1}{A_0}\right),$$

$$3 = 10\ \log\ (1 + A_1/A_0)\ \text{dB}.$$

Hence $1+A_1/A_0$ must be +2 and $A_1 = A_0 = 13.7$ sabins (m^2).

Thus, total absorption at end of modification $= 27.4$ sabins (m^2).

> $$\text{Added absorption} = 13.7 \text{ sabins } (\text{m}^2) = A_1 = S\bar{\alpha}.$$
>
> If we know what $\bar{\alpha}$ is at any frequency, we can calculate the actual absorption area $S\bar{\alpha}$ needed. Analogously, if the desired reduction in sound pressure level is 6 dB, $A_1 = 3A_0 = 41.1$ sabins (m^2), so the added absorption $S\bar{\alpha}$ must be 27.4 sabins (m^2).

12.4.3 Power Flow Between Coupled Systems

It has been shown that the power flow between coupled mode pairs is proportional to the modal energy difference, provided the coupling is weak and linear [40]. Scharton [46] and Ungar [26] have shown that the same result holds even if the coupling is not weak. Newland [42] has extended this result to include nonlinear coupling.

Ungar [26] explains simply how this result can be extended to consider the coupling between sets of modes. The same result can still be assumed to apply, provided either that the modes in each set are assumed to have approximately the same energy or that the coupling factors between the mode pairs are approximately the same.

The power flow W_{12} between systems 1 and 2 can be written as

$$W_{12} = \omega \eta_{12} n_1 \left(\frac{E_1}{n_1} - \frac{E_2}{n_2} \right), \tag{12.42}$$

where two systems are considered (Figure 12.26). E_1 and E_2 are the total energies in systems 1 and 2 in the frequency band $\Delta\omega$ and n_1 and n_2 are the modal densities (number of modes resonant in each system in the frequency band $\Delta\omega$ considered). Thus, W_{12} is the power flow in the frequency band $\Delta\omega$ under consideration. The frequency ω is the center frequency in the band $\Delta\omega$. The result given by Eq. (12.42) may be used for several engineering applications, including airborne sound transmission through wall panels. We will confine ourselves to this application here. The modal energies given in Eq. (12.42) may be those of the air spaces (usually rooms) or the partitions or panels. Some discussion of the modal behavior of a panel is now in order.

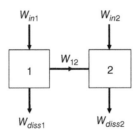

Figure 12.26 Block diagram showing power flow between two resonant systems 1 and 2.

12.4.4 Modal Behavior of Panel

The modes of a panel can be divided into two classes. Modes with natural frequencies above the critical coincidence frequency – and thus having bending-wave speeds greater than the speed of sound in air – are termed AF. Modes with natural frequencies below the critical frequency – and thus having bending-wave speeds less than the speed of sound – are termed AS. Further discussion about the panel modes is included in Section 3.16 of this book.

It can be shown theoretically [21, 24–28, 43] that the AF modes have a high-radiation efficiency, while the AS modes have a low radiation efficiency. The AS modes may further be subdivided into two groups. AS modes, which have bending phase speeds in one edge direction greater than the speed of sound and bending phase speeds in the other edge direction less than the speed of sound, are termed "edge" or "strip" modes. AS modes which have bending phase speeds in both edge directions less than the speed of sound are termed "corner" or "piston" modes.

Corner modes have lower radiation efficiencies than edge modes. The theoretical results for the radiation efficiency and classification of modes can also be given a simple physical explanation. Figure 3.27 in Chapter 3 shows a typical modal pattern in a simply-supported panel or plate. The dotted lines represent panel nodes.

The modal vibration of a finite panel consists of standing waves. Each standing wave may be considered to consist of two pairs of bending waves traveling in opposite directions. Consider a mode which has bending wave phase speeds which are subsonic in directions parallel to both of its pairs of edges. In this case, the fluid will carry pressure waves which will travel faster than the panel bending waves, and the sound pressures created by the quarter-wave cells (as shown in Figure 3.27a) will tend to cancel so that the effective volume source strength approaches zero everywhere except at the corners as shown. If a mode has a bending wave phase speed which is subsonic in a direction parallel to one pair of edges and supersonic in a direction parallel to the other pair, cancelation can only occur in one edge direction and for the mode shown in Figure 3.27b, the quarter-wave cells shown will cancel everywhere except at the *x*-edges. AF modes have bending waves which are supersonic in directions parallel to both pairs of edges. Then the fluid cannot produce pressure waves which will move fast enough to cause any cancelation, and the result is shown in Figure 3.27c.

Since AF modes radiate from the whole surface area of a panel, they are sometimes known as "surface" modes. With surface modes, the panel bending wavelength will always match the acoustic wavelength traced onto the panel surface by sound waves at some particular angle of incidence to the panel; consequently, surface modes have high radiation efficiency. This phenomenon does not happen for AS modes, the acoustic trace wavelength always being greater than the bending wavelength; AS modes have a low radiation efficiency.

At the critical coincidence frequency (when the panel bending wavelength equals the trace wavelength of grazing acoustic waves), the panel vibration amplitude is high [21, 22]. The radiation efficiency which is proportional to the radiation resistance is also high [21, 22]. Thus, at the critical coincidence frequency, the sound transmission is high and is due to modes resonant in a band centered at the frequency. Since the modes are resonant, the transmission can be reduced effectively in this region by increasing the internal damping of the panel.

Well below coincidence, because of their poor coupling with the fluid, the vibration amplitude of resonant modes is low, and the panel radiation efficiency is also low. In this region it is usually found that more sound is transmitted by modes which are not resonant in the frequency band under consideration. Since in such a band these modes are not vibrating at their resonance frequencies, they are little affected by internal damping. The contribution due to the non-resonant modes gives rise to the well-known "mass-law" transmission. Just above the critical coincidence frequency, the panel vibration amplitude and the radiation efficiency are high and the transmission is still resonant. However, as the frequency is increased further, the internal damping usually tends to increase more rapidly than the radiation damping and the non-resonant transmission becomes more important than the resonant transmission.

The relative importance of resonant and non-resonant transmission, of course, depends upon the practical structure under consideration and upon the variation of internal and radiation resistance with frequency. The radiation resistance is normally increased with the addition of stiffeners, which will usually increase resonant transmission. An increase of internal damping which may be achieved in several ways, including the use of screwed or riveted structures or added damping material, will decrease resonant transmission and increase the importance of mass-law transmission.

12.4.5 Use of SEA to Predict Sound Transmission Through Panels or Partitions

Lyon first used SEA to study the sound reduction of a rectangular enclosure with one flexible wall [45]. However he did not compare his results with experiment. White and Powell [50] used SEA to study the transmission of sound through double panels separated by an air space. Unfortunately, White and Powell restricted their theory to resonant transmission and did not allow for absorption in the air space between the panels or for the divergence from the room formula of the air space (cavity) modal density in the low-frequency range (below $f = c/2\ell_2$ or about 1700 Hz in their experiment, where ℓ_2 is the air gap width). It also appears that there is an error in their analysis, since they assume that the acoustic energy flow is proportional to the total energy difference between the systems instead of the modal energy difference.

The discussion which follows is based on work on the SEA sound transmission predictions and experiments on flat panels made by Crocker and his colleagues [27–30].

The transmission of sound through panels is usually measured in the laboratory using a transmission suite consisting of two rooms, as discussed in Section 12.7. The panel or partition under study is mounted between the two rooms. Consider the transmission suite shown in Figure 12.37. This may be considered to consist of three coupled systems, as shown schematically in Figure 12.27. System 1 is the transmission room; 2, the panel; 3, the receiving room. Under steady-state conditions, the power flow balance for the three systems may be written: [21–24]

$$W_{in1} = W_{diss1} + W_{12} + W_{13} \tag{12.43}$$

$$W_{in1} = \omega \eta_1 E_1 + \omega \eta_{12} n_1 \left(\frac{E_1}{n_1} - \frac{E_2}{n_2} \right) + \omega \eta_{13} n_1 \left(\frac{E_1}{n_1} - \frac{E_3}{n_3} \right), \tag{12.44}$$

$$W_{in2} = W_{diss2} - W_{12} + W_{23}, \tag{12.45}$$

$$W_{in2} = \omega \eta_2 E_2 - \omega \eta_{12} n_1 \left(\frac{E_1}{n_1} - \frac{E_2}{n_2} \right) + \omega \eta_{23} n_2 \left(\frac{E_2}{n_2} - \frac{E_3}{n_3} \right), \tag{12.46}$$

$$W_{in3} = W_{diss3} - W_{13} - W_{23}, \tag{12.47}$$

$$W_{in3} = \omega \eta_3 E_3 - \omega \eta_{13} n_1 \left(\frac{E_1}{n_1} - \frac{E_3}{n_3} \right) - \omega \eta_{23} n_2 \left(\frac{E_2}{n_2} - \frac{E_3}{n_3} \right), \tag{12.48}$$

where W_{in1}, W_{in2}, and W_{in3} are the rates of energy flow (in a frequency bandwidth of 1 rad/s, centered on ω) into systems 1, 2, and 3, respectively; W_{diss1}, W_{diss2}, and W_{diss3} are the rates of internal dissipation of energy in systems 1, 2, and 3 (in a 1 rad/s band); E_1, E_2, and E_3 are the total energies of systems 1, 2, and 3 (in a 1 rad/s band), and n_1, n_2, and n_3 are the modal densities of each system (modes per radian per second); η_1, η_2, and η_3 are the internal loss factors of each system; η_{12}, η_{13}, and η_{23} are coupling loss factors between systems 1 and 2, 1 and 3, and 2 and 3, respectively. The formulation so far in Section 12.4.4 is quite general and can be applied to any three coupled systems. Now it will be applied to the particular sound transmission case.

It is assumed that the panel is clamped between the transmission room (volume V_1) and the reception room (volume V_2) of the transmission suite. Reverberant sound is produced in the transmission room by a loudspeaker. In this case, the noise reduction (NR) $NR = 10\log(E_1 V_3 / E_3 V_1)$; consequently, the sound transmission loss produced by the panel may be determined from Eqs. (12.43) to (12.48) with $W_{in2} = 0$ and $W_{in3} = 0$.

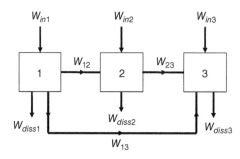

Figure 12.27 Block diagram representing power flows between coupled systems of transmission suite.

Putting $W_{in2} = 0$ in Eq. (12.46) and also using the reciprocity relationship $\eta_{12} n_2 = \eta_{21} n_1$ [31], Eq. (12.49) is obtained:

$$\frac{E_2}{n_2} = \frac{\left(\dfrac{E_1}{n_1}\right)\eta_{21} - \left(\dfrac{E_3}{n_3}\right)\eta_{23}}{\eta_2 + \eta_{21} + \eta_{23}}, \tag{12.49}$$

but $\eta_{21} = \eta_{23} = \eta_{rad}$ (the panel radiation loss factor) and except at low frequency where the present theory does not apply, $E_1/n_1 \gg E_3/n_3$; thus Eq. (12.49) becomes

$$\frac{E_2}{n_2} = \frac{E_1}{n_1}\left(\frac{\eta_{rad}}{\eta_{int} + 2\eta_{rad}}\right), \tag{12.50}$$

where $\eta_{int} = \eta_2$ is the internal damping loss factor of the panel. Putting $W_{in3} = 0$ in Eq. (12.48) yields

$$E_3 = \frac{E_1\eta_{13} + E_2\eta_{23}}{\eta_3 + \eta_{31} + \eta_{32}}. \tag{12.51}$$

The term $E_1\eta_{23}$ represents the mass-law or non-resonant transmission since it occurs without the modes resonant in the frequency band under consideration being excited. The term $E_2\eta_{23}$ represents the resonant transmission. Substituting Eq. (12.50) into Eq. (12.51) gives

$$\frac{E_1}{E_3} = \frac{\eta_3 + (n_1/n_3)\eta_{13} + (n_2/n_3)\eta_{rad}}{\eta_{13} + \eta_{rad}^2(n_2/n_1)/(\eta_{int} + 2\eta_{rad})}. \tag{12.52}$$

Equation (12.52) gives the ratio of total energies in the transmission and receiving rooms. The parameters η_{13}, η_{rad} and η_3 can be evaluated from the following equations:

$$\eta_{rad} = \frac{R_{rad}(2\pi)}{\omega M_p}, \tag{12.53}$$

where M_p is the panel mass, and the panel radiation resistance to half space $R_{rad}(2\pi)$ is given by Maidanik [39]. It should be noted that the expressions given for $R_{rad}(2\pi)$ for $f < f_c$ and $f > f_c$ given in Ref. [26] contain printing errors, but the correct expressions are given in Ref. [22], where, unfortunately, the expression for $f = f_c$ contains a printing error.

The coupling loss factor η_{13} due to non-resonant mass law transmission is obtained from (see Section 12.2.1)

$$10\log\eta_{13} = -TL_{rand} + 10\log\left(\frac{Sc}{4V_1\omega}\right), \tag{12.54}$$

where TL_{rand}, is the random incidence mass-law TL value for the second system (the panel). The value of TL_{rand} given by Eq. (12.21) can be used in Eq. (12.54). Finally,

$$\eta_3 = \frac{2.2}{fT_3}, \tag{12.55}$$

where T_3 is the reverberation time in system 3.

If Eqs. (12.53)–(12.55) are evaluated and a value for η_{int} is chosen, or else measured by experiment, then the $NR = 10\log(E_1V_3/E_3V_1)$ can be found from Eq. (12.53):

$$NR = 10\log\left[\eta_3 + (n_1/n_3)\eta_{13} + (n_2/n_3)\eta_{rad}\right] - 10\log\left[\eta_{13} + \eta_{rad}^2(n_2/n_1)/(\eta_{int} + 2\eta_{rad})\right] - 10\log(V_1/V_3), \tag{12.56}$$

where the room (n_1 and n_3) and panel (n_2) modal densities are (see Section 12.4.2)

$$n_1 = \frac{V_1 \omega^2}{2\pi^2 c^3},$$ (12.57a)

$$n_2 = \frac{\sqrt{3} S}{2\pi h c_L},$$ (12.57b)

$$n_3 = \frac{V_3 \omega^2}{2\pi^2 c^3},$$ (12.57c)

where S, h, c_l are the panel area, thickness, and longitudinal wave speed. The *TL* of the panel is then

$$TL = NR + 10 \log \left[\frac{S c T_3}{24 V_3 \ln(10)} \right],$$ (12.58a)

The response of the panel relative to mass law is easily derived from Eq. (12.50). We first assume that the sound field in the transmission room is reverberant, so that the total energy in a 1 Hz bandwidth $E_1 = S\rho_1 V_1/(\rho c^2)$ and the total panel energy in a 1 Hz bandwidth $E_2 = M_p S_a/\omega^2$. We then also substitute the modal density of the transmission room $n_1(\omega) = V_1 \omega^2/(2\pi^2 c^3)$, the modal density of the panel $n_2(\omega) = \sqrt{3} A_p/(2\pi h c_L)$, and the critical frequency $f_c = \sqrt{3} c^2/(\pi h c_L)$, into Eq. (12.50). Finally, we assume that if the panel were to respond as a limp mass, the response would be $S_{aml}/S_{p\omega} = 1/\rho^2_s$, where $S_{p\omega}$ is the spectral density of the pressure at the panel wall surface. Neglecting panel motion, $S_{p\omega} = 2S_{p1}$, since there is pressure doubling for each wave arriving at the panel surface, although at any instant only half the waves are traveling toward the panel. Thus, $S_{am1}/S_{p1} = 2/\rho^2_s$ and with this final substitution, Eq. (12.58b) is obtained. See Ref. [22].

$$S_a/S_{aml.} = \left[(\eta_{rad}/(\eta_{int} + 2\eta_{rad})) \right] \times \pi^2 f_c \rho_s/2\rho c,$$ (12.58b)

where S_a, $S_{aml.}$ are the spectral densities of the acceleration of the panel given by the present theory and given by mass law theory, respectively; η_{int} is the panel internal loss factor; ρ and ρ_s are the air density and panel surface density.

Figure 12.28 shows a comparison [27, 28] between the measured *TL* of a single wall aluminum panel and that calculated from Eqs. (12.56) and (12.58a). The panel studied was 1/8 in. (3.175 mm) thick and measured 61 in. by 77.5 in. (1.55 m by 1.97 m). The values of η_{int} used in the predictions was 0.005. Measured values of η_{int} varied between 0.005 and 0.01 throughout the frequency range 100–10 000 Hz [27, 28]. The agreement is good with two exceptions. At low frequency $f < 400$ Hz, the disagreement is thought to be caused by room-panel modal coupling effects. Just below the critical coincidence frequency, agreement between theory and experiment is considerably improved if the radiation resistance $R_{rad}(2\pi)$ is assumed to be twice the value for a

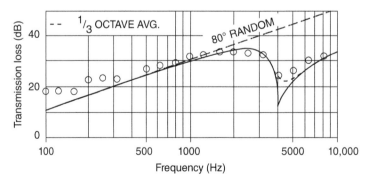

Figure 12.28 Experimental values of transmission loss of 1/8 in. (3.175 mm) thick aluminum panel (○) compared with theoretical prediction (from Eqs. (12.56) and (12.58a) with η_{int} = 0.005); $- \cdot - \cdot$80° random incidence mass law [27, 28].

simply supported panel for $f < f_c$. This is a good assumption to make since the panel edge conditions were intended to be clamped and R_{rad} (2π) for a clamped panel is twice that for a simply supported panel.

Theory for a double-partition system can also be derived [28, 30]. If the double partition consists of two independent panels separated by an air cavity, then the theoretical model shown in Figure 12.27 must be extended to include five oscillators (room-panel-cavity-panel-room). The development is similar to that given in Eqs. (12.43) through (12.48), although now five power flow balance equations are obtained instead of the three in Eqs. (12.43) through (12.48). Complete details of the theory are given in Refs. [21, 23, 26]. Figure 12.29 gives an example of a comparison [28, 30] between the measured *TL* and that predicted by the SEA theory for double panels just discussed. Each panel was 1/8 in. (3.175 mm) thick with the same length and width as the single panel in Figure 12.28. The upper broken curve in Figure 12.29 shows the predicted *TL* assuming non-resonant (or mass-law) transmission through each panel only. This is the result to be expected if each panel became limp (that is, if each panel had no stiffness or damping but maintained the same mass per unit area as the 1/8 in. [3.175 mm] thick aluminum panels).

It is observed, when comparing Figures 12.28 and 12.29, that the double-panel system *TL* curve is steeper than the corresponding one for the single panel. Not only is the double-panel curve everywhere more than 6 dB above the single-panel curve (6 dB increase would be expected for a single panel of twice the mass), but the slope is greater for the double panel. Below the critical coincidence frequency, the *TL* of the double panel increases about 9 dB per doubling of frequency compared with 6 dB for the single panel.

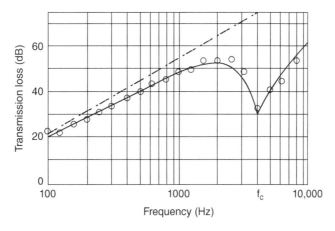

Figure 12.29 Transmission loss for a double aluminum panel, (panel thicknesses = 1/8 in. (3.175 mm), air gap = 2.8 in. (7.11 cm); ○ experiment; —— theoretical prediction including non-resonant and resonant transmission with η_{int} = 0.005; ·—— theoretical prediction for non-resonant transmission only [29].

Figure 12.30 shows measured and predicted results for a double-panel system with panels of the same dimensions as in Figure 12.28, except the panels have different thicknesses (1/4 in [6.350 mm] and 1/8 in. [3.175 mm]). Two dips in the TL are now seen at 2000 and 4000 Hz. The dips are now not as sharp as in Figures 12.28 or 12.29.

The measured acceleration response of the panel (compared with the response predicted by mass law), is given in Figure 12.31. The measured response is compared with the response predicted by Eq. (12.58b) for η_{int} = 0.005. The agreement is good except at low frequency. The low frequency disagreement is again much reduced if R_{rad} (2π) is assumed to be twice that for a simply-supported panel for $f < f_c$. Since all the panels measured had clamped edges, this would seem to be a reasonable assumption. The high frequency disagreement is removed in Figure 12.31, if η_{int} = 0.01 for $f > f_c$. This value is closer to the experimental result for $f > f_c$.

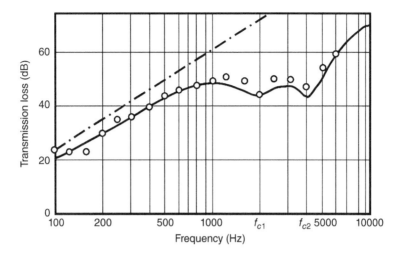

Figure 12.30 Transmission loss for a double aluminum panel system of different thicknesses, (panel thicknesses = 1/4 in. (6.35 mm) and 1/8 in. (3.175 mm), air gap = 2.8 in. 7.11 cm); ○ experiment; —— theoretical prediction including non-resonant and resonant transmission with η_{int} = 0.005; —·— theoretical prediction for non-resonant transmission only [30].

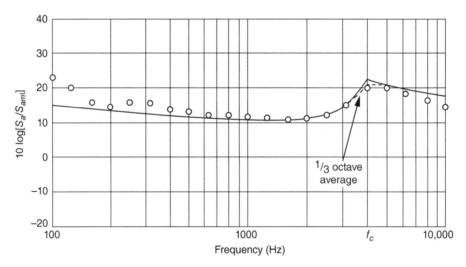

Figure 12.31 Panel acceleration response relative to mass law. ○ experiment; — theoretical prediction including resonant and non-resonant vibration with η_{int} = 0.005 [27, 28].

Figure 12.32 shows experimental and theoretical results for a 1/8 in. (3.18 mm) thick aluminum double-panel system with and without tie beams [30]. The experimental results were obtained both for the case where the panels were connected together by 50 steel coupling ties and for the case where they were independent. The ties were each 2.8 in. (7.11 cm) long, 1 in. (2.54 cm) wide and 0.027 in. (0.69 mm) thick. It is observed that, although the independent double panel is identical to that given in Figure 12.29, the transmission loss is higher in Figure 12.32 above 500 Hz, presumably due to higher internal panel damping. A theoretical prediction for this case shown in Figure 12.32 with η_{int} = 0.005 for f < 800 Hz and η_{int} = 0.02 for f > 800 Hz is seen to give good agreement with the measurements.

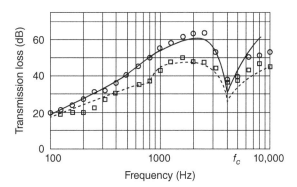

Figure 12.32 Transmission loss for a double aluminum panel with and without tie beams, panel thicknesses = 1/8 in. (6.35 mm), air gap = 2.8 in. (3.75 mm); ○ experiment (without tie beams); —— theoretical prediction for panels without ties; □ experiment with 50 steel tie beams; – – – theoretical prediction for panels with ties; η_{int} = 0.005 for f< 800 Hz, η_{int} = 0.02 for f> 800 Hz in all predictions [30].

12.4.6 Design of Enclosures Using SEA

Airborne noise transmission into enclosures is of interest to designers of operator enclosures in factories, cabs for trucks, cabins for aircraft and farm tractors, ships and even satellite enclosures in spacecraft. In this section of the chapter, such an enclosure is modeled both theoretically and experimentally as a rectangular steel box immersed in a diffuse reverberant sound field [51–55]. The box is structurally isolated from the field and the sound attenuation is determined. The effects of small apertures in the box wall are also examined [55, 56]. SEA has been used for the theoretical analysis and the experimental measurements were performed while the box was being excited by broadband random noise suspended inside a reverberation chamber [55, 56].

The SEA prediction scheme incorporated the presence of small apertures of regular geometry in one panel and acoustical absorbing material inside the box. The attenuation was measured in one-third octave bandwidths whose center frequencies ranged from 63 Hz to 20 kHz. From these results, conclusions can be drawn as to the best ways of maximizing the NR of cabin enclosures [56, 57].

For the purposes of the analysis, the system is separated into eight individual elements, as shown in Figure 12.33a. The box itself is comprised of elements 2 through 7, these being the steel panels which are the sides of the box, while the exterior acoustical space, which is the reverberation room is element 1 and the interior box cavity is element 8.

A schematic power flow diagram is shown in Figure 12.33b. The figure shows each element coupled to every other element and to certain elements is attributed an input power. In reality, panels which are not contiguous are not (directly) coupled and for the purposes of the NR prediction, only element 1 receives any input power. Since only airborne sound transmission is being investigated. The box is assumed to receive no direct mechanical excitation [54].

An eight element system requires an 8×8 matrix equation to model the power flow [54]. At a center radian frequency of ω and for an arbitrary bandwidth $\Delta\omega$, the first power balance equation for power input to element 1 may be written:

$$W_{in1} = W_{diss1} + W_{12} + W_{18} + W_{14} + W_{15} + W_{16} + W_{17} + W_{18}.$$

In terms of the modal energies E/n in that bandwidth:

$$W_{in1} = \omega\eta_1 E_1 + \omega\eta_{12}n_1(E_1/n_1 - E_2/n_2) + \omega\eta_{18}n_1(E_1/n_1 - E_8/n_8)$$
$$+ \omega\eta_{14}n_1(E_1/n_1 - E_4/n_4) + \omega\eta_{15}n_5(E_1/n_1 - E_5/n_5)$$
$$+ \omega\eta_{16}n_6(E_1/n_1 - E_6/n_6) + \omega\eta_{17}n_1(E_1/n_1 - E_7/n_7)$$
$$+ \omega\eta_{18}n_1(E_1/n_1 - E_8/n_8),$$

where $\omega = 2\pi f$.

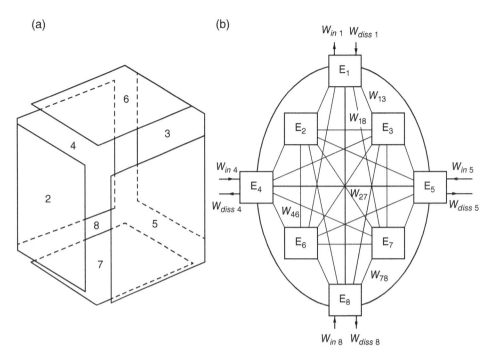

Figure 12.33 (a) The cab model split into its separate elements. (b) Power flow diagram for the cab model [53, 54].

Seven other similar power balance equations may be written for the seven other elements. The entire system of power balance equations over a bandwidth Δf may be solved so that the modal energies in each system may be obtained by a pre-knowledge of the power inputs to each element. In the case of the cab model, W_{in2}, W_{in3}, W_{in4}, ..., and W_{in8} are all assumed equal to zero, while W_{in1} represents the acoustic power flow into the reverberation room from a noise source.

The resonant coupling loss factors are represented in the matrix equation as η_{JI} where I is the panel number and J is the acoustical space number (the loss from the panel to the room/box cavity) and η_{IJ} (the loss from the room/box cavity to the panel).

Included in the non-resonant loss factor is the effect of small holes in the panels. These holes represent direct transmission paths between the outside and the inside of the box, and the loss factor due to these may be calculated and combined with the mass law loss factor.

The power flow equations may be evaluated and solved for a variety of system configurations. The basic calculation is for a completely sealed (no holes) empty box, and thereafter the effect of absorbing material in the box, circular apertures and rectangular apertures may be considered. Each of these situations has been modeled and compared with experimental measurements [54].

The sound pressure level difference outside to inside of the box, sometimes known as NR, is termed *attenuation* in this section of this chapter. The sound pressure level (L_p) in the box is evaluated, and the attenuation may thus be calculated:

$$L_{p_{room}} - L_{p_{box}} = 10 \log \left[\frac{\rho c^2 E_1^m N_1}{V_1 p_{ref}^2} \right] - 10 \log \left[\frac{\rho c^2 E_8^m N_8}{V_8 p_{ref}^2} \right],$$

or

$$\text{Attenuation}(f) = 10 \log \left[\frac{E_1^m N_1 V_8}{E_8^m N_8 V_1} \right],$$

where N_1, V_1, E^m_1, and N_8, V_8, E^m_8 are the mode counts, volumes and modal energies of the exterior and interior air space cavities. The attenuation is thus calculated for each bandwidth at the center frequency f, and is best displayed when plotted against log(f). The experimental values of attenuation, measured in one-third octave bands, are then plotted on the same graph as the theoretical curve for comparison.

Various methods of deriving the sound transmission coefficient of an aperture in a rigid wall have been presented in the literature. For the purposes of this chapter, the transmission coefficient for the aperture, a, derived by Ingerslev and Nielsen [59] and referred to in Mulholland and Parbrook [58] was chosen for its simplicity.

In Figure 12.34a, the predicted and measured results are compared for the case of a sealed box (with no apertures) with $1.2\,\mathrm{m}^2$ absorbing material inside. The shape of the theoretical curve is to be expected; a smooth rise through the low and middle frequencies according to the so-called mass law transmission (see Section 12.2.1) which dominates in this frequency range. As the frequency approaches the critical coincidence frequency, about 8000 Hz, a rapid drop occurs due to the increased vibration response and radiation efficiencies of the panels. Above this frequency, the attenuation increases, since it is in the frequency range where the panel mass and damping control the transmission loss [53].

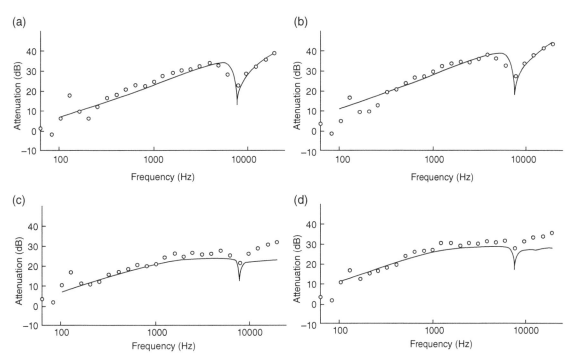

Figure 12.34 (a) Attenuation of a sealed box containing $1.2\,\mathrm{m}^2$ of absorbing material, —— predicted, ○ measured. (b) Attenuation of a sealed box containing $3.5\,\mathrm{m}^2$ of absorbing material, —— predicted, ○ measured. (c) Attenuation of a box with a circular aperture ($a = 0.035$ m) in one panel. Box contains of $1.2\,\mathrm{m}^2$ of absorbing material, —— predicted, ○ measured. (d) Attenuation of a box with a circular aperture ($a = 0.035$ m) in one panel. Box contains $3.5\,\mathrm{m}^2$ of absorbing material, —— predicted, ○ measured [53].

The agreement between experiment and theory is good throughout much of the frequency range. Exceptions occur at low frequencies, below about 200 Hz, and at the critical coincidence frequency. The latter anomaly may be attributed to the fact that the theoretical curve is calculated using frequency bandwidths of 100 Hz, while the experimental values were measured in one-third octave bandwidths.

In Figure 12.34b the predicted and measured results are compared for the case of a sealed box containing 3.5 m^2 absorbing material [53, 54]. All comments pertaining to the previous figure also apply directly to this one. The major difference between Figure 12.34c and d is caused by the increased amount of absorbing material in the box; the attenuation is approximately 4 dB higher throughout the entire frequency range as would be predicted from 10log(3.5/1.2).

Figure 12.34c shows the effect of a circular aperture in one panel and compares the experimental data with predicted values calculated using the value for the aperture transmission coefficient, a, given in Ref. [54]. The aperture has a diameter of 0.035 m and the box contains 1.2 m^2 of absorbing material. Comparison with Figure 12.34c of both the predicted and measured curves, shows how the aperture only affects the attenuation in the middle and high frequency range. Below a frequency of about 500 Hz, the mass law transmission is dominant over the relatively poor transmission of sound by the aperture. Above this frequency, the aperture contributes more and more to the overall transmission as the frequency increases. However, at the critical coincidence frequency the attenuation is the same as that of a sealed box because the efficiently radiating panels dominate in the transmission behavior.

The measured results agree with the theory up to about 1000 Hz. The discrepancies at the lowest frequencies are due to the reasons described previously. It is questionable whether the extreme low frequency theoretical results should be presented, since their meaning is unclear, at least below 163 Hz, which is the lowest frequency at which a mode can exist in the box air cavity. Above 1000 Hz, which corresponds to an aperture ka value of about 0.4, (where a is the aperture radius and k is the wave number) the predicted attenuation is consistently lower than the measured attenuation. As expected, the aperture transmission theory breaks down at high values of ka, and the disagreement becomes more apparent as the frequency increases. The model predicts more efficient transmission by this aperture than is actually observed [53].

Figure 12.34d compares the measured and predicted attenuation of the steel box which has a circular aperture (of diameter 0.035 m) and contains 3.5 m^2 of absorbing material. In the theoretical predictions, the aperture transmission coefficient a given in Ref. [53] was used. In comparing these results with Figure 12.34b the attenuation is found to be increased by about 4 dB at every frequency. This is the same result as obtained for the sealed box (compare Figure 12.34a and b). Comparison with Figure 12.34b again shows how the aperture only affects the attenuation in the middle and high frequency ranges. As in the case of Figure 12.34b the predictions become less accurate at high frequencies [53, 54].

A number of parameters of the enclosure are varied while the others are kept constant. Figures 12.35a–d show the effects of varying panel damping, thickness, internal sound absorption area, and aperture radius in the idealized truck cab enclosure while each of these other variables are kept constant.

Finally, Figure 12.36 shows the difference in experimental results between the attenuation of an idealized cab enclosure with the leaks between panels unsealed and sealed with clay. The attenuation is defined as the difference in space-averaged sound pressure levels outside and inside the enclosure.

12.4.7 Optimization of Enclosure Attenuation

As seen in the earlier analysis, the sound power transmission into the enclosure has a complicated relationship with the different parameters of the system. These effects vary widely with frequency. If the system is subjected to an external diffuse reverberant sound field, it is convenient to take frequency averages of the attenuation over the whole frequency range. For an effective design of the system to minimize the overall attenuation level, all parameters must be considered simultaneously by performing the optimization. The constrained optimization

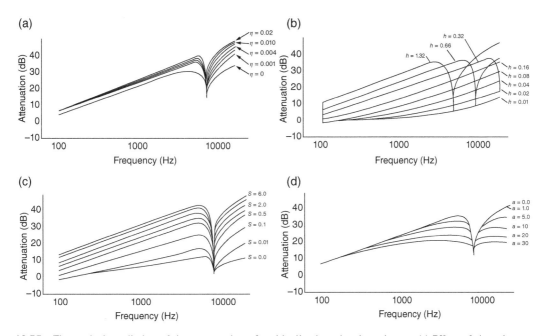

Figure 12.35 Theoretical prediction of the attenuation of an idealized truck cab enclosure. (a) Effect of changing panel damping coefficient (ρ_s = 7700 kg/m³). (b) Effect of changing panel thickness h in mm (S = 1.19 m², ρ_s = 7700 kg/m³). (c) Effect of changing internal sound absorption area, S (in m²). (d) Effect of changing circular aperture radius, a (in mm) [57].

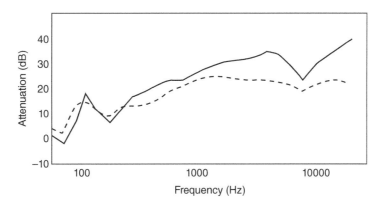

Figure 12.36 Experimentally measured attenuation of the sealed model enclosure. ——— joints sealed: – – – joints unsealed [57].

method was used to consider the bounds on the parameters. The design variables for the optimization problem were set with x_1 as the aperture radius in m, x_2 as the panel thickness in m, x_3 as the absorption area in m² and x_4 as the panel density in kg/m³. The optimization problem can be described as follows:

The initial design variables were taken as $x_1 = 0.01$ m, $x_2 = 0.00159$ m, $x_3 = 0.5$ m², and $x_4 = 7790$ kg/m³. The Constrained Optimization Module of MATLAB version 5.2 was used for the optimization. A typical configuration x_0 as described above was used as an initial estimate for the optimization. This configuration gives after optimization, aperture radius $x_1 = 0.005$ m, panel thickness $x_2 = 0.0011$ m, absorption area $x_3 = 3$ m² and panel density $x_4 = 8000$ kg/m³.

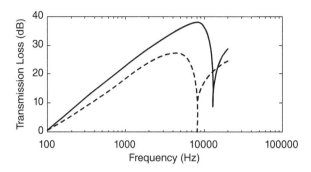

Figure 12.37 Effect of optimization on TL (- - - before optimization, —— after optimization) [57].

The effect of the optimized design on the attenuation is shown in Figure 12.37 [57]. There is a substantial increase in attenuation over the entire frequency range. It shows the importance of considering all the parameters simultaneously for optimization. Further parametric studies of the transmission loss of panels are discussed in a recent paper on single and double-walled plates with absorbing materials [60].

12.4.8 SEA Computer Codes

In the simple calculations presented in Examples 12.8 to 12.12, only hand calculations are needed and these can be performed on the "back of an envelope." This illustrates the beauty and simplicity of SEA. However, for the more complicated sound transmission problems in Sections 12.4.5–12.4.7, computer software codes were written. Such codes allowed more complicated equations to be used, but also for mechanical and acoustical properties to be varied so as to examine their effects on the SEA predictions.

In the 1980s commercial software codes became available such as AutoSEA marketed by Vibro-Acoustic Sciences. This code has been updated as AutoSEA2. Hybrid codes have now been developed, in which SEA is coupled to finite element method (FEM) or BEM. One such code known as VA One was developed after Vibro-Acoustic Sciences was sold to ESI. A new FEM/SEA code, Wave6 is now marketed by Dassault and is used by a wide range of industries. Other computer software codes available for SEA of complex sound and vibration problems are GSSEA-Light marketed by Gothenburg Sound AB and Seam marketed by Altair.

12.5 Transmission Through Composite Walls

Often partitions are composed of several different elements, such as a brick wall having windows and doors. The average transmission loss TL_{av} of this composite structure can be calculated as follows: suppose τ_1 is the transmission coefficient of element of area S_1, etc. and that a diffuse reverberant sound field of intensity $\varepsilon c/4$ (where ε is the energy density in the incident field and c is the speed of sound) strikes the partition, then

$$W_0 = W_1 + W_2 + W_3 + ..., \tag{12.59}$$

where W_0, W_1, W_2, W_3, etc. are the powers transmitted by the overall partition and elements 1, 2, 3, etc. Then, from Eq. (12.59) since $W_0 = (\varepsilon c/4)\tau_0 (S_1 + S_2 + S_3 + ...)$, and $W_1 = (\varepsilon c/4)\tau_1 S_1$, $W_2 = (\varepsilon c/4)\tau_2 S_2$, etc.,

$$\tau_0 = \frac{\tau_1 S_1 + \tau_2 S_2 + \tau_3 S_3 + ...}{S_1 + S_2 + S_3 + ...}, \tag{12.60}$$

and from Eq. (12.11a) the overall TL of the composite wall is

$$TL_{av} = 10 \log (1/\tau_0). \tag{12.61}$$

Example 12.13   Compute the overall partition *TL* at 1000 Hz of a 100 mm thick brick veneer wall measuring 2.3×4.6 m penetrated by a 0.9×1.2 m single-pane window (glass thickness 5 mm). For brick $M_1 = 210 \, \text{kg/m}^2$ and for glass $M_2 = 12.5 \, \text{kg/m}^2$.

Solution

We can estimate the *TL* for each material using the empirical equation (see Section 12.2.2) $TL = 20\log(Mf) - 47$. Thus, for brick $TL_1 = 20\log(210 \times 1000) - 47 = 59$ dB and for glass $TL_2 = 20\log(12.5 \times 1000) - 47 = 35$ dB. Now, for brick $S_1 = 2.3 \times 4.6 - 0.9 \times 1.2 = 9.5 \, \text{m}^2$ and $\tau_1 = 10^{-5.9} = 1.26 \times 10^{-6}$; for glass, $S_2 = 1.08 \, \text{m}^2$ and $\tau_2 = 10^{-3.5} = 3.16 \times 10^{-4}$. Therefore,

$$\tau_0 = \frac{(9.5 \times 1.26 \times 10^{-6}) + (1.08 \times 3.16 \times 10^{-4})}{2.3 \times 4.6} = 3.3 \times 10^{-5};$$ and the overall transmission loss of the partition is $TL_{av} = 10\log(1/3.4 \times 10^{-5}) = 44.8$ dB.

Equations (12.60) and (12.61), after some manipulation, have been presented by different authors as various plots to save us the bother of making the calculations; see, e.g. Figure 12.38. In the previous example $TL_1 - TL_2 = 59 - 35 = 24$. The penetration *k* percent $= (1.08/10.58) \times 100 = 10.2\%$. Interpolating for the value of *k* gives $TL_1 - TL_0 \approx 14$ dB. Thus $TL_0 = 59 - 14 = 45$ dB, which agrees with the numerical calculation in the example.

It is interesting to note in the example just calculated that the sound energy transmission through the window was dominating (even though the window area was only about 10% of the total) since $3.41 \times 10^{-4} \gg 11.97 \times 10^{-6}$, i.e. $\tau_2 S_2 \gg \tau_1 S_1$. This is frequently the case found in practice. Except perhaps where the wall is made of a lightweight material, sound transmission through a window normally gives the situation $\tau_2 S_2 \gg \tau_1 S_1$ and thus Eq. (12.60) becomes

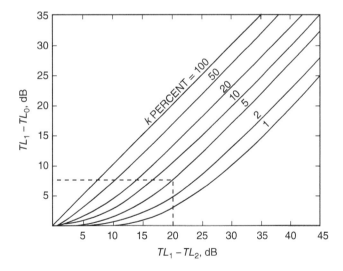

Figure 12.38   Chart for determining the overall sound transmission loss, TL_0, of a wall with a penetration (e.g. door or window) having a transmission loss, TL_2, less than the transmission loss, TL_1, of the basic wall construction. The penetration occupies "*k* percent" of the total wall area. The transmission loss of the combination is shown in relation to that of the wall as $TL_1 - TL_0$.

$$\tau_0 = \frac{\tau_2 S_2}{S_1 + S_2 + \dots},\tag{12.62}$$

and the overall *TL* of the composite partition is

$$TL_0 = TL_2 + 10\log(100/k),\tag{12.63}$$

where k is the ratio of window per total area percent, $k = 100\, S_2/(S_1 + S_2 + \dots) = 100\, S_2/S$ and TL_2 is the *TL* of the window. The upper right part of Figure 12.38 in which the curves are straight lines corresponds to the case where nearly all the energy goes through element S_2, and the curves obey Eq. (12.63). This is easily shown by rearranging Eq. (12.63) in the form:

$$TL_1 - TL_0 = TL_1 - TL_2 - 10\log(100/k),\tag{12.64}$$

which is the result for a straight line of ordinate $TL_1 - TL_0$, plotted against abscissa $TL_1 - TL_2$, where the constant $-10\log(100/k)$ represents the value of $TL_1 - TL_0$, when $TL_1 - TL_2$ is zero.

That this result is correct for the case where the transmission through a window dominates is easily checked from the previous example. Since $k = 10.7\%$, then $10\log(100/k) = 10\log(100/10.7) = 10\log(9.35) = 9.7$ dB; from Eq. (12.63): $TL_0 = 35 + 9.7 = 44.7$ dB. This result agrees with the two previous calculations. The reader may also care to check that Eq. (12.64) is also exactly satisfied by this example, where $10\log(100/k) = 9.7$ dB.

The above example represents a specimen calculation at the frequency of 1000 Hz. This calculation would have to be repeated at each frequency of interest to obtain a curve of *TL*, against frequency. That such calculations for composite walls agree well with experimental measurements on composite walls has been demonstrated in practice [61].

Example 12.14 Calculate the overall transmission loss at 250 Hz of a wall of total area 15 m² constructed from a material that has a transmission loss of 40 dB, if the wall contains a panel of area 5 m², constructed of a material having a transmission loss of 15 dB.

Solution

For the main wall, the transmission coefficient is: $\tau_1 = 1/[10^{(40/10)}] = 0.0001$, while for the panel: $\tau_2 = 1/[10^{(15/10)}] = 0.032$. Therefore, using Eq. (12.60):

$\tau_0 = \dfrac{(0.0001 \times 10) + (0.032 \times 5)}{15} = 0.0107$, and the overall *TL* of the composite wall is $TL_0 = 10\log(1/0.0107) = 19.7$ dB.

Example 12.15 A wall 10×20 m has an initial $TL = 50$ dB. Four windows are added to the wall. The area of each window is 5 m² and its $\tau = 0.01$. What will be the new *TL* of the wall with windows?

Solution

For the wall, the surface is 200 m² and its $\tau_1 = 1/[10^{(50/10)}] = 10^{-5}$. Hence

$\tau_0 = \dfrac{(10^{-5} \times 180) + (0.01 \times 20)}{200} = 0.001$. Thus, the overall transmission loss of the wall with windows is $TL_0 = 10\log(1/0.001) = 30$ dB.

12.6 Effects of Leaks and Flanking Transmission

Leaks in a wall frequently occur in practice and are very serious. Unless great care is taken in construction, leaks will occur and the theoretical TL of a wall will not be achieved. This effect is particularly noticeable at high frequency where the TL of a wall would be expected to be high. In this region, a leak can reduce the TL by over 10 dB. Air leaks can be caused by poorly mortared masonry joints, unsealed holes around pipes, ducts and conduits which penetrate walls, unsealed room partitions, cracks, clearance around doors, partly opened windows, etc. The leak provides an air path for the transmission of sound energy.

It is often assumed that all of the sound energy striking a leak is transmitted (so that $\tau = 1$ for the leak and $TL = 0$). This assumption is approximate only, but gives a starting point for discussion. Under this assumption, Eqs. (12.60) and (12.61) can be used to calculate the TL of the wall including the leak. If the τ for the leak is more accurately known, then this may be incorporated into Eq. (12.60) to obtain better accuracy for TL_0. Under the assumption that $\tau_2 = 1$ for the leak and the sound energy it transmits is much greater than that through the wall, then Eq. (12.63) becomes

$$TL_0 = 10 \log (100/k), \tag{12.65}$$

where k is now the ratio of leak to total wall area, percent.

Example 12.16 The space under a very heavy solid door of $TL = 50$ dB is 1/100 of the total area of the door. If the noise level outside the room is 80 dB, determine the noise level inside the room with the door closed.

Solution

The transmission coefficients of the door and the open space under the door are 10^{-5} and 1, respectively. If the area of the door is S, then

$$TL_{av} = 10 \log \frac{S + 0.01S}{10^{-5} \times S + 1(0.01 \times S)} = 20 \text{ dB. The noise level inside the room with the door closed is}$$

80−20 = 60 dB. Thus, the leak produced by the small space under the door significantly reduces the noise insulation of the heavy door.

It has been shown that assuming $\tau = 1$ ($TL = 0$) for a slit is a very drastic assumption. Gomperts [62] showed that τ can fluctuate from much less to much more than 1. In addition, in the region of resonances, viscous energy losses in the slit must be included in the calculation [63].

Figure 12.39 gives two common examples of an acoustical leak or air path. This air path is much more efficient than the mechanical path through the panel. Only 645 mm^2 (one square inch) of hole can transmit as much energy as a 9.3 m^2 (100 ft^2) wall. Note that a crack 0.25 mm wide × 2.6 m long has an area of 6.5 cm^2. Such a crack, then, might be expected to reduce the TL, on average, of such a wall by 3 dB. If this wall is now made much thicker (say, four times), the effect will be still more serious. The energy transmitted through the crack will remain the same, while the energy transmitted through the wall will be 1/16 of that through the thicker wall or 12 dB less than before. The TL, then, of this thicker wall will be decreased by just over 12 dB by the existence of the crack. This discussion shows why the clearance provided around a door seriously degrades the TL. The area of such gaps is quite large. Grilles and louvers in doors very seriously impair their performance. Likewise, in the construction of industrial buildings, similar care should be taken to avoid leaks where possible. If air must be supplied for cooling purposes, inlet and outlet mufflers or sound traps should be used which have attenuations (transmission losses) equivalent to the walls.

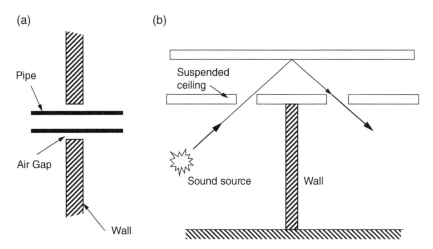

Figure 12.39 Two common construction faults which allow direct air paths to flank and reduce the effectiveness of a wall. (a) Poorly fitting pipe; (b) gap in suspended ceiling.

Plumbing is a very important source of noise in buildings for several reasons including: (i) bends in the pipe and coupling between bending, longitudinal, and other motions of the pipe add vibrational complications, (ii) poorly fitting pipes provide an air path for the transmission of energy, and (iii) the fluid in the pipe can flow in such a way that excites the walls with turbulent pressure fluctuations, in addition to sound waves. It is a common practice to wrap a noisy pipe with an acoustical absorbing material and then to cover this with a thin massive septum (layer) of material such as lead or lead-loaded vinyl. Such an approach is known as wrapping or lagging. The sound radiated from the pipe passes through the absorbing material and is reflected back and forth between the septum and pipe wall. Each time the sound is reflected by the septum some is transmitted through the surrounding space. Thus, a more massive septum should be more effective in reducing the pipe noise to the surroundings, particularly at high frequency, according to the mass law discussed in this chapter. The sound is absorbed during the multiple reflection process between the septum and the pipe wall, and thus thicker acoustical absorbing linings should be more effective.

The layer of acoustical material can consist of fiberglass or open-celled acoustical foam and acts both as an absorbing material and as a vibration isolator to decouple the pipe vibration from that of the septum. The layer of absorbing material can also be used as thermal insulator. In addition, piping must be isolated from structures, such as walls, to avoid structure-borne noise transmission. Resilient pads and vibration-isolation hangers are often very effective in reducing the amount of such vibration being transmitted into the building structure from pipes and ducts. Figure 12.40 shows some examples of noise control measures for pipes in buildings.

We have just discussed "air flanking." It is also common to find that mechanical paths will "flank" energy around a wall and thus reduce its effectiveness (see Figure 12.41). Sound will be transmitted through the wall by the direct path 1. However, energy will also be flanked around the wall by the indirect paths 2, 3, 4, and 5. There will also be indirect paths through the ceiling and floor, as well as the walls. Even more complicated flanking paths are possible, although they have not been shown. The problem is seen to occur at wall-to-wall junctions and wall-to-floor junctions. The solution to this problem is to vibration-isolate the junctions in some way. Several different methods may be used. Breaks can be used in floor slabs at a junction, instead of making them continuous, and these are found to be particularly useful.

Other ways of avoiding flanking transmission in buildings include the use of floating floors and suspended ceilings. See also Chapter 13, Section 13.6.1 and Figure 13.33. Air flanking, through direct air paths at junctions or weak links (i.e. where only a thin wall is presented to a potential air path) at the junction should also be avoided. Sound traps may be used to cut the air paths at such junctions. Caulking should be used wherever

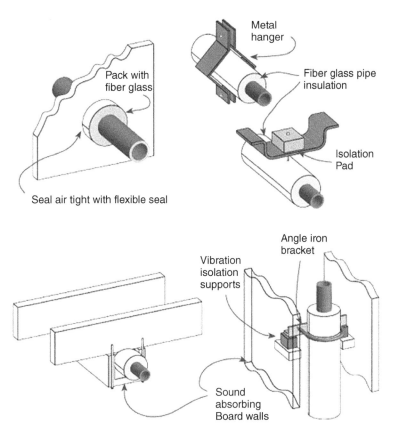

Figure 12.40 Typical noise control measures for piping in buildings. (*Source:* reproduced from Ref. [64] with permission from NAIMA.)

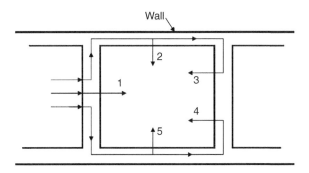

Figure 12.41 Mechanical flanking paths.

cracks may occur. These practical recommendations are of course also valid for the construction of enclosures for machines where similar flanking problems exist (see Section 9.6 of this book). It is possible to estimate the contribution of each flanking transmission to the total insulation provided by a partition [65]. A comprehensive discussion of one of such methods is given in a guide published by the National Research Council in Canada [66].

12.7 Sound Transmission Measurement Techniques

There are several reasons for making sound transmission measurements, including: (i) to make tests under standardized laboratory conditions which should give repeatable comparisons of the performance of different partition structures and materials, (ii) to conduct laboratory tests in development work on partitions to indicate changes in sound *TL* which are caused by small design changes, (iii) to decide from field tests if design goals or commercial bids have been met, and (iv) to determine from field tests if criteria have been met on NR between two locations in a building. There are a number of methods which have been devised for the measurement of *TL*. The most widely used laboratory method consists of the use of a transmission suite. However, there are several other methods which are occasionally used in the laboratory or "in the field." It should be noted that the transmission suite method cannot be used exactly "in the field" since the sound fields are rarely sufficiently diffuse. However, provided the practical difficulties are understood, the transmission suite method can be adapted for field use.

12.7.1 Laboratory Methods of Measuring Transmission Loss

a) Transmission Suite Method

This method consists of the use of two reverberant rooms which are separated by the panel under investigation. Reverberant rooms are used so the sound falling on the panel has equal probability of approaching from any direction (random incidence sound). The rooms are vibration-isolated from each other, by mounting them on vibration isolators, in order to minimize mechanical "flanking" transmission between the rooms. The panel under test is normally attached to one of the rooms if it is a single panel. If it is a double panel, one leaf is normally attached to one room and the other leaf to the other room, unless the additional effect of flanking transmission within the double panel is being considered. A typical transmission suite is shown schematically in Figure 12.42. Procedures for use of a sound transmission suite have been standardized so that results with different partition structures or materials should be comparable even though they were obtained in different laboratories.

The ISO has published ISO 10140-2 [67]. The American Society for Testing Materials (ASTM) has also produced an elaborate standard test method E90-09 [68]. These standards are similar and on agreement on the procedures to be followed in laboratory tests. Unfortunately, in the case of "field" tests, the situation is less satisfactory, since measurement conditions in the field are never ideal, the results are dependent upon the particular installation and building, and the results can be difficult to interpret.

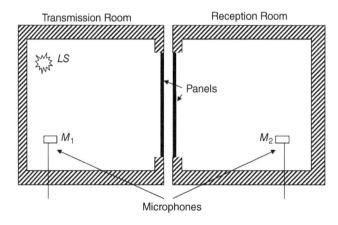

Figure 12.42 Schematic representation of a typical transmission suite.

In the reverberant room or transmission suite method, the walls of the two rooms are made as acoustically "hard" as possible, by keeping absorbing material to a minimum in order to keep the acoustic fields in each room reverberant and diffuse. A steady sound is made in the first room, known as the transmission room, by means of a loudspeaker (LS). The sound pressure level is measured in the first room and also in the second room, known as the reception room, with the two microphones M_1 and M_2. The microphones M_1 and M_2 are moved to randomly selected positions in the two rooms and the measurements repeated. Alternatively, the microphones are continuously moved in each room. Thus, space-average values of the sound pressure levels (L_{p1} and L_{p2}) are determined from one of these methods. This procedure is repeated using sound at several different frequencies. It is normal practice to use one-third octave bands of white noise at each frequency, although sometimes warble tones are used instead.

The noise reduction, NR of the panel is given by

$$NR = L_{p1} - L_{p2}. \tag{12.66}$$

Note that $NR = L_{p1} - L_{p2} = 10 \log \varepsilon_1/\varepsilon_2$ since the energy density ε is $p^2_{rms}/\rho c^2$ in a sound field (see Eq. (3.70)). However, this does not represent exactly the transmission loss TL of the panel since the NR is dependent upon the acoustic absorption in the receiving room. The acoustic absorption in the receiving room can be determined by producing a sound in the room and measuring the reverberation time (T_{R2}) when the sound is cut off (see Section 3.14.3 of this book).

The TL can be deduced from the previously described measurements as follows. It is assumed that there is no flanking transmission and that all the energy transmitted from the transmission room to the receiving room takes place through the panel. It is also assumed that the acoustic absorption is small enough and the measurements made at a frequency, which is high enough for the sound fields in each room to be reverberant and diffuse.

When a steady state is reached, the sound energy densities in transmission room 1 and receiving room 2 are ε_1 and ε_2, and it may be shown [38] that the power flow through the panel $W_{12}|_{\varepsilon_2 = 0}$ is

$$W_{12}|_{\varepsilon_2 = 0} = \tau S \varepsilon_1 c/4, \tag{12.67}$$

where S is the panel area and the transmission coefficient, τ, is the fraction of incident energy transmitted (see, for example, Eq. (12.10)). In the steady state,

$$W_{12}|_{\varepsilon_2 = 0} = W_{diss_2} + W_{21}|_{\varepsilon_1 = 0}, \tag{12.68}$$

where W_{diss_2} is the power dissipated in the receiving room due to absorption and $W_{21}|_{\varepsilon_1 = 0}$ is the power retransmitted back to the transmission room from the receiving room. Thus,

$$\tau S \varepsilon_1 c/4 = A \varepsilon_2 c/4 + \tau S \varepsilon_2 c/4, \tag{12.69}$$

where A is the effective absorption area of the receiving room. Thus,

$$NR = 10 \log (\varepsilon_1/\varepsilon_2) = 10 \log \left[\frac{A + \tau S}{S}\right], \tag{12.70}$$

$$\frac{1}{\tau} = (\varepsilon_1/\varepsilon_2)\left[\frac{S}{A + \tau S}\right] \tag{12.71}$$

and

$$TL = 10 \log (\varepsilon_1/\varepsilon_2) + 10 \log \left[\frac{S}{A + \tau S}\right],$$

$$TL = NR + 10 \log \left[\frac{S}{A + \tau S} \right], \tag{12.72}$$

where *NR* is the noise reduction.

As already described, ε_1 and ε_2 may be determined from measurements of the space-averaged sound pressure levels in rooms 1 and 2 made with the microphones M_1 and M_2. The area of the panel may also be determined. However, the effective absorption of the receiving room must be determined by decay measurements in the receiving room.

This is normally determined at each one-third octave band center frequency. White noise is produced in the receiving room by supplying a one-third octave band of white noise to a loudspeaker. The supply is terminated, and the time for the sound pressure level L_{p2} to decay 60 dB is determined.

The rate of energy decay in the receiving room is given by

$$-V_2 \left(\frac{\partial \varepsilon_2(t)}{\partial t} \right) = (A + \tau S) \varepsilon_2(t) c/4 - \tau S \varepsilon_1(t) c/4. \tag{12.73}$$

However, unless a panel is very lightweight and τ is large, $\tau S \varepsilon_1(t) c/4$ is normally neglected. With this term neglected, Eq. (12.56) is satisfied by the exponential law:

$$\varepsilon_2(t) = \varepsilon_0 e^{-\beta t}, \tag{12.74}$$

where β is the energy decay constant given by

$$\beta = \frac{(A + \tau S)c}{4V_2} \tag{12.75}$$

and ε_0 is the energy density when the acoustical source is terminated.

The reverberation time T_{R2} is given (from Eq. (12.74)) by

$$10^6 = \exp(\beta T_{R2}),$$
$$\beta T_{R2} = \ln(10^6),$$

and, substituting for β from Eq. (12.75),

$$A + \tau S = \frac{24 V_2 \ln(10)}{c T_{R2}}. \tag{12.76}$$

From Eq. (12.72)

$$TL = NR + 10 \log \left[\frac{S c T_{R2}}{24 V_2 \ln(10)} \right]. \tag{12.77}$$

Equation (12.77) is the result which must be used to correct the measured *NR* of a panel (or apparent transmission loss) to obtain the *TL*. Equation (12.77) may also be obtained using SEA (see Eq. (12.58a)). Now, substituting the Sabine equation (see Section 3.14.3 of this book) $T_{R2} = 0.161 \, V/A$ into Eq. (12.77) we obtain

$$TL = NR + 10 \log \left[\frac{S}{A} \right], \tag{12.78}$$

where *S* is the area of the partition common to both source and receiving rooms and *A* is the total sound absorption in the receiving room. Note that *TL* in Eq. (12.78) is known as the *sound reduction index* (*R*) in European and ISO standards.

It should be recognized that Eq. (12.72) is approximate only, since it assumes reverberant conditions in both rooms and does not include the direct field in the receiving room. However, in most practical situations Eqs. (12.72) and (12.78) are sufficiently accurate.

Example 12.17 The sound pressure level on one side of a 3.0×4.5 m wall is measured as 90 dB in the 500 Hz octave band. If the transmission loss of the wall is 45 dB in this band and the absorption in the receiving room is 100 sabins (m^2), what will the sound pressure level be in the receiving room?

Solution

From Eq. (12.78) we get $NR = L_{p1} - L_{p2} = TL - 10 \log (S/A)$, or
$L_{p2} = L_{p1} - TL + 10 \log (S/A) = 90 - 45 + 10 \log (3 \times 4.5/100) = 36.3$ dB. Note that the absorption in the receiving room has reduced the noise level by 8.7 dB more than what is predicted by simply subtracting the TL value from the sound pressure level in the source room.

---

b) Standing wave tube method

If a special apparatus is built, it is possible to measure the normal incidence transmission loss of a material sample directly. The apparatus consists of a set of two tubes of equal internal diameter that can be connected to either end of a test sample holder. A schematic drawing of the apparatus is shown in Figure 12.43.

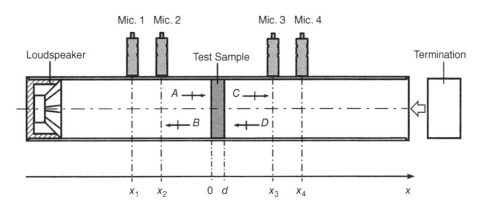

Figure 12.43 Schematic of the apparatus used to measure normal-incidence *TL* of small material samples. (*Source:* Figure courtesy of Bruel & Kjaer.)

This test method is similar to that used to measure the normal-incidence sound absorption coefficient, usually called the two-microphone method [69, 70]. However, for transmission loss, four microphones at two locations on each side of the sample are flush-mounted in the tube wall. The technique was introduced by Song and Bolton [71] and it has been standardized by ASTM [72]. There is no equivalent ISO standard so far. A loudspeaker is placed at one end of the tube and the other end is closed with a minimally reflecting termination. An anechoic termination should ideally be used.

The loudspeaker is fed with a broadband signal (white noise) and this then radiates plane waves down the tube toward the sample. The resulting standing wave pattern is decomposed into forward- and backward-traveling components by measuring the complex sound pressure simultaneously at the four microphone locations, two on either side of the material sample. Their relative amplitude and phase are determined from the measured transfer function between a reference and the four locations by a multichannel digital signal analyzer. Although a large spacing between the microphones improves the accuracy of the measurements, the microphone spacing must be chosen such that it is less than $\lambda/2$. In addition, a careful calibration is required for the measured transfer functions. One of the ways of doing it is the microphone location switching procedure, which prevents the error due to phase mismatch and gain factor between the two sets of measurement channels.

The acoustical transfer matrix is calculated from the acoustic impedance of the traveling waves on either side of the sample. The transmission loss, as well as several other acoustical properties of the material is determined from the transfer matrix. Although this method to determine transmission loss of materials is limited to sound waves normally incident upon the surface of the sample, it provides useful comparison data for small specimens, something that cannot be done in the transmission suite method.

c) Sound intensity probe method

Now that the two-microphone sound intensity method is so fast and accurate when a digital Fast Fourier Transform analyzer is used, a method was conceived to use this technique to considerable advantage in both laboratory and field measurements of *TL*. Although the principles of the method were described more than 40 years ago by Crocker et al. [73], the technique was only standardized at the beginning of the twenty-first century by both ASTM [74] and ISO [75]. The international ISO standard also includes the application of the method for field measurements and laboratory measurements at low frequencies.

The method can be used as an alternative to the transmission suite method for measuring airborne sound insulation of building elements. One important use is when the traditional method fails because of high flanking transmission. The reproducibility of this sound intensity method has been estimated to be equal to or better than that of the transmission suite method. Sound intensity is discussed in more detail in Chapter 8 of this book.

The method can be used in a laboratory setting where the reverberant source room and the specimen mounting conditions satisfy the requirements of the traditional transmission suite method. This method is described in more detail in Chapter 8. The receiving room has no specific acoustical requirements, so in laboratory measurements, only one reverberation room may be used. The acceptability of the receiving room is determined by a set of field indicators that define the quality and accuracy of the sound intensity estimate. Such a procedure reduces the need for two reverberation rooms to one, in sound transmission *TL* measurements. This can result in a considerable savings, since such facilities are very expensive. Specific details of the method can be found in Chapter 8 and in the following standards [74, 75].

The principle of the method is that the incident sound intensity I_i, is calculated from the diffuse field theory assuming $I_i = \langle p^2 \rangle / 4\rho c$, where $\langle p^2 \rangle$ is the space average of the room mean square sound pressure. The panel is mounted directly in a hole in one wall of the room, and the transmitted power W_t, from the panel is measured either by traversing a sound intensity probe close to and all over the panel outside of the room (scanning method) or by placing the probe at a set of fixed points to sample the intensity field normal to the panel surface (discrete point method). Then the *TL* is given directly by $TL = 10 \log (SI_i / W_t)$, where S is the panel surface area, $W_t = SI_t$ and I_t is the space-average of the transmitted intensity.

ISO 15186-2 [76] specifies a sound intensity method for field determination of the sound insulation of walls, floors, doors, windows and small building elements. This method has the advantage that using it, it is possible to determine the energy transmitted directly by the partition and also the energy flanked by walls, floor, and ceiling structures. In addition, the method can be used by transmission suite laboratories, in which the effect of flanking transmission cannot be avoided.

ISO 15186-3 [77] specifies a method to measure *TL* at low frequencies, using sound intensity. The method is intended for the frequency range from 50 to 80 Hz. In this method, the sound pressure level of the source room is measured close to the surface of the test specimen. In addition, the surface opposite the test specimen in the receiving room is highly absorbing and converts the room acoustically into a duct with several propagating cross modes above the lowest cut-on frequency.

12.7.2 Measurements of Transmission Loss in the Field

It is difficult to obtain the idealized conditions of random incidence, diffuse sound fields, and absence of flanking transmission in laboratory situations. In real buildings (field conditions), such conditions are impossible to obtain in practice. However, as discussed before, it is sometimes necessary to determine the *TL* of a partition in the field where flanking transmission is present. Also, it may be necessary to determine the *NR* between two rooms in a building to determine whether a minimum NR requirement has been met to satisfy a building regulation code. Thus, for on-site measurements, it may be necessary to consider the use of one of the methods described as follows.

a) Field Transmission Loss (Apparent Transmission Loss)

If it is required to determine the field transmission loss TL_F of a partition (apparent transmission loss *ATL*) when flanking paths (mechanical and air leaks) are present, then an equation similar to Eq. (12.72) may be used. TL_F is related to the $NR = L_{p1} - L_{p2}$ as follows:

$$TL_F = ATL = NR + 10 \log \left[\frac{S}{A} \right]. \tag{12.79}$$

The symbols are the same as those used in Eq. (12.78). Note that *ATL* is known as the *apparent sound reduction index* (*R'*) in Europe and in ISO standards. In practice, Eq. (12.79) can be used. Since flanking transmission will be present, it is likely that TL_F will be less than the *TL* measured in the laboratory. This will particularly be true if the partition *TL* is high. If the laboratory *TL* is found to be between 35 and 50, then TL_F may be expected to be 1 or 2 dB less because of flanking.

b) Normalized Noise Reduction Difference

Often, for regulation or building code purposes, it is more convenient to measure the NR between a transmission (source) room and a receiving room. However, since the $NR = L_{p1} - L_{p2}$ will be dependent upon the absorption area in the receiving room (caused by furnishings, etc.), it is usual to give a normalized NR, NR_N. There are two ways in which this normalization can be made. The first is to correct the *NR* to give the *NR* which would have been obtained if a reference absorption area A_0 were present in the receiving room:

$$NR_N = L_{p1} - L_{p2} + 10 \log (A_0/A), \tag{12.80}$$

where L_{p1} and L_{p2} are the space-averaged sound pressure levels in the transmission and receiving rooms, A is the measured absorption area in the receiving room, and A_0 is a reference absorption area, usually $A_0 = 10$ sabins (m²) or 108 sabins (ft²). Note that N_{RN} is known as the *normalized level difference* (D_n) in Europe and in ISO standards. The sound pressure levels are spatially sampled in the transmission and receiving rooms using fixed microphone positions, mechanically-scanned microphones, or a manual scanning technique that employs a hand-held microphone or a sound level meter. Note that when A is measured, using the reverberation time method, the quantity $A + \tau S$ is actually measured, but this is of no great consequence, except perhaps at low frequency. Both ASTM [78] and ISO [79] standards define an alternative normalization procedure based on the reverberation time in the receiving room, T_R. In this procedure, the normalized NR (or the *standardized level difference* D_{nT} in ISO standards) becomes

$$NNR = L_{p1} - L_{p2} + 10 \log (T_R/0.5). \tag{12.81}$$

Equation (12.81) results from the fact that in many furnished rooms with moderate absorption, the reverberation time is about 0.5 second, and is almost independent of room volume. Thus, using the Sabine equation: [80] $A_0 \, (\text{m}^2) = 0.32 \, V \, (\text{m}^3)$, and substituting A_0 into Eq. (12.80) gives Eq. (12.81). Note that use of Eq. (12.81) results in a normalized NR value $10 \times \log(0.032 \, V \, [\text{m}^3])$ higher than the value using Eq. (12.80). This is easily seen since

$$NNR - NR_N = 10 \log (T_R/0.5) - 10 \log (10/A)$$
$$= 10 \log (T_R A/0.5 \times 10)$$
$$= 10 \log (0.032V)$$

if the Sabine equation $T_R = 0.161 \, V/A$ is used again. Thus, whenever normalized NRs are determined, the receiving room volume should be noted and reported to facilitate conversion from one normalization procedure to another.

It is known that strong modal responses occur in room volumes less than 25 m^3, which is typically the case of many rooms in dwellings. In these cases, there are often less than five room modes below 100 Hz and consequently the sound field does not approximate to a diffuse field. Therefore, the sound isolation of partitions measured in small room volumes may be under or overestimated at low frequencies. The current ASTM standard [78] states that the transmission and receiving room volumes must be not less than 60 m^3 for measurements down to 100 Hz, 40 m^3 for measurements down to 125 Hz, and 25 m^3 for measurements down to 160 Hz. In order to improve the repeatability of low-frequency TL measurements, the recently published ISO 16283-1 standard [79] describes a procedure for using additional microphone positions to sample the sound pressure near the corners of rooms for frequencies below 100 Hz. Then, the average sound pressure level in the room is obtained from the sound pressure levels measured at the central zone and at the corners of the room using an empirical weighting.

Example 12.18 A wall panel of mass/unit area of 5 kg/m^2 and area 4 m^2 is placed in the "window" between two reverberation rooms of a sound transmission suite. The walls of the receiving room have an area of 100 m^2, and an average absorption coefficient α of 0.05 at all frequencies. (a) Calculate the space-averaged sound pressure level in the receiving room at 1000 Hz, if the sound pressure level in the source (i.e. transmission) room is 90 dB at 1000 Hz. (b) Calculate the space-averaged sound pressure level in the receiving room at 2000 Hz, if the sound pressure level in the source (i.e. transmission) room is 90 dB at 2000 Hz.

Solution

a) We use Eq. (12.78) $NR = L_{p1} - L_{p2} = TL - 10 \log (S/A)$, so

$\quad TL = L_{p1} - L_{p2} + 10 \log (S/A) = 90 - L_{p2} + 10 \log (4/0.05 \times 100)$. Now, using Eq. (12.22c) we obtain:
$TL = 20 \log(Mf) - 47 = 20 \log(5 \times 1000) - 47 = 27$ dB. Therefore

$$TL = 27 = 90 - L_{p2} + 10 \log (4/5); \text{thus } L_{p2} = 62 \text{ dB}.$$

b) By the same process as (a), we get that $L_{p2} = 56$ dB.

12.8 Single-Number Ratings for Partitions

When comparing different wall constructions, it is often convenient to use a single-number rating instead of the complicated transmission loss (or normalized NR) which varies with frequency. Early single-number ratings were obtained by simply averaging the TL (or normalized NR) in the frequency range of interest. Although such a system may be useful in rating walls, it can be misleading because frequency information is lost. This approach is probably satisfactory for brick walls since the critical coincidence frequency is very low (about 80 Hz), and the TL tends to increase continually with frequency in the range of interest. However, with the increasing use of lightweight partitions in buildings, coincidence dips often appear in the frequency range 100–3150 Hz; hence, two partitions with very different TL against frequency curves may have the same average TL. This could be quite serious if the transmission room noise source were intense in the frequency range of the coincidence dip of one of the partitions.

Obviously, knowledge of the noise source spectrum would be necessary in order to specify the acoustical performance required of a partition. However, often the source spectrum is not precisely known and can vary with time and the particular building. For this reason, some researchers have assumed typical transmission room source spectra and criteria for the levels to be tolerated in the receiving room (see Chapter 6 of this book). Another approach which was adopted in the early 1950s in Great Britain and other European countries resulted from several social surveys in Britain, Sweden, Netherlands, and France [81–84]. The British and Swedish studies were conducted on several hundred semidetached houses and apartments. About half of the houses had common 9 in. (230 mm) solid brick party walls, while half had common two-layer concrete party walls separated by an air cavity. The concrete party walls provided higher *TL* at high frequency than the brick walls. The inhabitants of the houses and apartments were questioned about their living conditions and were asked if they felt the walls provided sufficient sound insulation. In short, it appeared that the single brick party wall provided acceptable sound insulation between houses and apartments in most cases. Interestingly, the massive single brick party wall is intended to serve primarily as a fire, rather than an acoustical, barrier.

Since the brick party wall appeared to be acceptable acoustically in most cases, it was adopted in Great Britain in the 1950s as the criterion against which other lightweight structures have to be judged. The approach now used in most countries is to judge the insulation performance of a partition relative to a standard reference curve which is a little higher in the mid-frequency range. The *weighted sound reduction index* (R_w) is the value of the ISO reference curve at 500 Hz when it is shifted vertically so as to ensure that the sum of unfavorable deviations is as large as possible, but not more than 32 dB (for measurements in one-third octave bands) or 10 dB (for one-octave bands). The reference value of the ISO curve is 52. Measurements should comply with ISO 10140-2 [67] or ISO16283-1 [79], although measurements carried out in accordance with the outdated ISO140-1 [85] and ISO140-5 [86] standards are also allowed.

In the U.S., the ISO contour has been adopted by ASTM for use in the STC scheme [87]. Measurements should comply with the standard E90 [68] or E336 [78]. There are some differences between the procedure to calculate R_w and STC values. In the ASTM procedure, measurements of sound attenuation are obtained only in one-third octave bands. The STC value is calculated so that the sum of the deficiencies (the differences between the data points below the contour and the contour value) must be less than or equal to 32 dB. In addition, no deficiency can exceed 8 dB at any one frequency (the 8-dB rule). The STC contour covers the range 125–4000 Hz, while the ISO contour covers the range 100–3150 Hz. For this reason, a partition may have a slightly higher STC rating than the ISO rating because the *TL* of most partitions increases with frequency. However, the numerical values of the two single-number ratings are usually very similar. The 8 dB maximum deficiency limitation was introduced to prevent considerable transmission in a narrow frequency band caused by a narrow-band noise source. Note that sharp *TL* dips are often found in lightweight panels because of coincidence effects. An example of the calculation of a STC rating is illustrated in Figure 12.44 [88]. In Figure 12.44a, ¼-in.

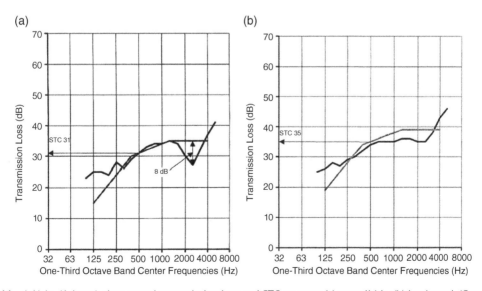

Figure 12.44 A ¼-in. (6.4 mm) glass sound transmission loss and STC contour; (a) monolithic; (b) laminated. (*Source:* adapted from Ref. [88]).

(6.4 mm) monolithic glass is shown to have an STC rating of 31. In this example, the STC contour placement is constrained by the maximum allowed 8 dB deficiency at 2500 Hz. In Figure 12.44b, which shows *TL* data for ¼-in. (6.4 mm) laminated glass, the sharp dip in the *TL* data, characteristic of ¼-in. monolithic glass, is removed by use of the damping interlayer and the ¼-in. laminated glass is shown to have an STC rating of 35, In this case, the STC contour placement is constrained by the maximum 32-dB deficiency requirement.

Figure 12.45 shows the STC contour fitted to the data for a concrete slab and the STC ratings of some common building materials reported by Warnock [20]. Warnock states that rough representative values of STC for block walls [89] can be estimated from STC = 0.51 × block weight (kg) + 38. (The dimensions of the block face are 190 × 390 mm). The estimated values are valid if the wall surfaces are properly sealed and the mortar joints well made.

ISO 717-1 [90] takes into consideration the greater importance of low frequencies in noise sources inside a building and traffic outside a building using weighted summation methods. These methods produce corrections (called *spectrum adaptation terms*) to the R_w ratings by using either an A-weighted pink noise or a specific urban traffic noise spectrum defined in the standard. Note that the overall spectrum levels are normalized to 0 dB for both types of noise sources. Single-number ratings can be combined with one of the spectrum adaptation terms as a sum to characterize the sound insulating properties of building elements or the acoustical performance between rooms inside buildings or from the outside to the inside [91]. In addition, national, special rules have been added in some countries to compensate for shortcomings or difficulties of field test procedures and conditions [92]. ASTM uses a single-number rating called the *outdoor–indoor transmission class* (OITC), calculated in accordance with standard E1332 [93]. This rating is used for comparing the sound insulation performance of building facades and façade elements and its value increases with increasing sound isolation ability. The rating has been devised to quantify the ability of these to reduce the perceived loudness of ground and air transportation noise transmitted into buildings. Like ISO, the procedure to determine the OITC also uses a standard source spectrum and sound transmission loss data. However, the OITC does not involve a contour fitting process.

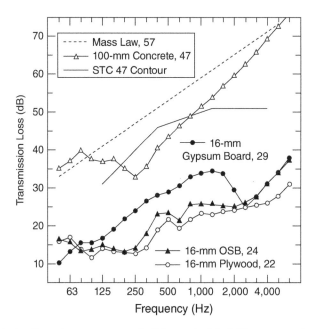

Figure 12.45 Sound transmission through some common building materials: 100-mm concrete (233 kg/m^2), 16-mm plywood (7.5 kg/m^2), and 16-mm oriented strandboard (9 kg/m^2) on joists, and 16-mm gypsum board (11.3 kg/m^2). The dashed line represents mass law predictions for 233 kg/m^2. The numbers next to each curve are STC ratings [20].

The *field sound transmission class* (FSTC) rating is used by the ASTM to assess the sound isolation performance from *TL* measured in the field in accordance with the standard E336 [78]. This rating is similar to *weighted apparent sound reduction index* (R'_w) used in Europe and in the ISO standard [90].

Although single number ratings determined from laboratory and field measurements are intended by standards to be equivalent, practical experience has indicated that field-measured ratings tend to be up to five rating points less than laboratory-measured ratings [88]. Over the years, several other rating or grading schemes have been proposed. Tocci has presented a comprehensive discussion on ratings and descriptors for the built acoustical environment [88].

Example 12.19 A partition of total area of 11 m^2 consists of a double brick wall (STC of 50 and area of 10 m^2) and a 3-mm fixed glazing window (STC of 25 and area of 1 m^2). What is the effective STC of the partition?

Solution

The STC of the composite wall can be approximately calculated using the same procedure described in Section 12.5 with STC treated as *TL*. Then, the effective STC of the combination is $STC_{eff} = 10 \times \log[11/(10 \times 10^{-5} + 1 \times 10^{-2.5})] = 10 \log(11 \times 306.5) = 35$.

Example 12.20 A wall in a recording studio must incorporate a window unit. The wall has an STC = 56. The wall and the window have fractional areas of 0.75 and 0.25 of the total area, respectively. For design purposes the required combined sound transmission class is STC = 50. What are the insulation requirements for the window?

Solution

We write $STC_{eff} = 10 \times \log[1/(0.75 \times 10^{-5.6} + 0.25 \times 10^{-STC/10})] = 50$. Solving for the STC of the window unit:

$$0.25 \times 10^{-STC/10} = 10^{-5} - 0.75 \times 10^{-5.6}, \text{ so}$$

$$STC = -10 \times \log\left[10^{-5} - 0.75 \times 10^{-5.6}\right] - 6 \approx 45.$$

Therefore, the insulation requirements for the window unit are high and a special acoustical window made of either double or triple glazing would be a solution for the recording studio.

12.9 Impact Sound Transmission

Impact noise is other major sound transmission problem in buildings. Although transmission of airborne sound is probably the major problem, the secondary problem of sound transmission from impacts such as footsteps, doors slamming, hammering and other forms of construction, etc. has concerned acousticians for many years. Footsteps are the main source. There are two main problems in the investigation of impact noise: (i) the noise sources are impulsive in nature, and this causes problems with measurement procedures designed to measure steady-state or continuous noises, (ii) the partition under consideration, instead of merely consisting of one passive element in the transmission system in the case of airborne sound, now plays an active role as part of the noise source.

Although it has been shown theoretically that the airborne sound transmission loss (insulation) can be related to the impact insulation the majority of efforts on impact sound transmission seem to have been made in experimental investigations.

12.9.1 Laboratory and Field Measurements of Impact Transmission

As early as 1932, Reiher [94] built an impact machine to simulate footstep noise. This consisted or a single wooden hammer of 280 g which fell 30 mm at time intervals corresponding to footsteps. However, it was unsuccessful because it was difficult to measure the impulsive noise produced, and the level was too low. In 1938, an improved machine was built which consisted of five brass hammers each weighing 500 g which each fell 40 mm producing 10 impacts per second. The problems with Reiher's earlier machine were overcome, and a similar impact machine has since been incorporated in an international standard for laboratory [95] and field [96] measurements of impact sound insulation of floors. The same type of standard tapping machine is used in the corresponding ASTM standards for laboratory [97] and field [98] measurements. Laboratory test methods call for highly diffuse sound fields and the suppression of flanking sound transmission in the receiving room of laboratory.

It has been found that the force spectrum of the standardized tapping machine is dominated by higher frequencies, which is not comparable with real impact noise sources in buildings. International standards for measuring impact sound isolation [96, 99] include a modified tapping machine and a standardized rubber ball as alternative sources to the ISO tapping machine [100]. For many years, tires and rubber balls have been used as low-frequency impact noise sources (between 50 and 630 Hz one-third octave bands) in Japan and Korea to simulate children jumping and running in apartment buildings [101]. The modified tapping machine is intended to make its dynamic source characteristics similar to those of a person walking barefoot and a heavy/soft impact source with dynamic source characteristics similar to those of children jumping.

Basically, in these standards the impact sound source is placed near the center of the floor/ceiling under investigation. With the impact sound source in operation, the space-averaged sound pressure levels are measured in the receiving room directly below the floor specimen. Sound pressure level measurements are conducted in one-third octave bands at several measurement positions either by use of fixed or moving microphones. The ISO standard covers the range 100–5000 Hz, while the ASTM standard covers the range 100–3150 Hz.

The impact sound source is moved to three other locations on the floor, and the one-third octave band level for the four sets of readings, are averaged at each frequency. Finally, the *normalized impact sound pressure level* L_n at each frequency band is obtained from

$$L_n = \langle L_0 \rangle - 10 \log (A_0/A), \tag{12.82}$$

where $\langle L_0 \rangle$ is the impact space-averaged sound pressure level in the receiving room, A_0 is the reference absorption area, 10 sabins (m^2) or 108 sabins (ft^2), and A is the measured absorption area in the receiving room, sabins (m^2). Note that there is no need to normalize the sound pressure level to the surface area of the test element as with the *TL*. Figure 12.46 illustrates the impact sound transmission measurement procedure using a tapping machine.

Example 12.21   A room of dimensions $4 \times 5 \times 7$ m has a reverberation time of 0.85 second. A standard tapping machine is used at four different positions to excite the floor above the room. The average sound pressure levels measured in a third-octave band in the receiving room are 82 dB, 85 dB, 79 dB, and 80 dB. Find the normalized impact sound pressure level in this frequency band.

Solution

The impact space-averaged sound pressure level in the receiving room is

$$\langle L_0 \rangle = 10 \log \left[\frac{10^{82/10} + 10^{85/10} + 10^{79/10} + 10^{80/10}}{4} \right] = 82.1 \text{ dB}.$$

The total absorption area in the receiving room is obtained from the Sabine equation (see Section 3.14.3 of this book): $A = 0.161 \times V/T_R = 0.161 \times (4 \times 5 \times 7)/0.85 = 26.52$ sabins (m^2). Then, replacing the numerical values in Eq. (12.82) we obtain

$$L_n = 82.1 - 10 \times \log(10/26.51) = 86.3 \text{ dB.}$$

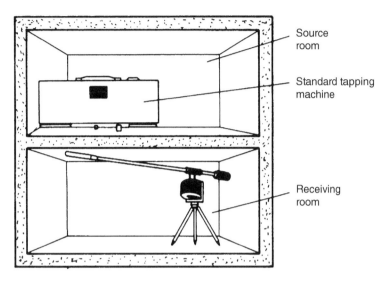

Source room

Standard tapping machine

Receiving room

Figure 12.46 Impact sound transmission measurement procedure.

When measurements are made in the field (real buildings) instead of the laboratory, flanking transmission and different floor sizes alter the results. In addition, the room sound field might not approximate to a reverberant diffuse field. The laboratory results can only be used as an approximate guide to the results expected in practice. Field measurement of impact sound insulation of floors is described in both ISO [96] and ASTM [98] standards. In both standards, the levels are normalized by the reverberation time instead of the absorption area.

For field measurements, the sound pressure levels are normalized with a reverberation time of 0.5 second and the metric is called now *reverberation time normalized impact sound pressure level* (*RTNISPL*) defined as

$$RTNISPL = ISPL - 10\log(T_R/0.5), \tag{12.83}$$

where *ISPL* is the space-averaged impact sound pressure level in the receiving room, and T_R is the measured reverberation time in the receiving room, in seconds. *RTNISPL* is intended for small rooms that can be expected to have a reverberation time of 0.5 second when furnished normally. *RTNISPL* is known as the *standardized impact sound pressure level* (L'_{nT}) in Europe and in ISO standards.

For measuring low frequency (50, 63, and 80 Hz one-third octave bands) impact noise it is needed to take into consideration the effects of small rooms (less than 25 m^3). For this, ISO defines a similar procedure to improve measurement repeatability as in the case of measuring sound transmission loss through a set of room corner measurements. In addition, ISO also defines how operators can measure the sound field using a hand-held microphone or a sound level meter [74].

12.9.2 Rating of Impact Sound Transmission

Just as in the case of rating the airborne isolation of partitions (see Section 12.8), there are several ways of rating the impact sound transmission of floor/ceiling structures. The methods are similar and involve comparing the measured normalized sound pressure levels in the receiving room against a standard contour. Measurements should comply with the standard E492 [97] or E1007 [98] according to the ASTM standard [102]. For the international standard [103], measurements should comply with ISO 10140-3 [95] or ISO16283-2 [96], although measurements carried out in accordance with the outdated standard ISO140-7 [105] are also allowed.

The ISO and ASTM standard contours for one-third octave bands are identical (see Figure 12.47 [20]) although there are some differences between the ISO and ASTM procedures of rating the impact sound transmission. These are the two rating standards most frequently used in building codes.

In the ASTM procedure, the *impact insulation class* IIC, of the floor/ceiling construction is the value of the standard contour at 500 Hz when it has been shifted vertically so as to comply with the following: (1) the sum of positive differences between the measured data and the fitted standard contour for all frequencies is less than or equal to 32 dB, and (2) the greatest of the unfavorable deviations does not exceed 8 dB at any one-third octave frequency band. Note that the IIC rating, like STC and R_w, increases as the impact sound insulation improves. The field impact insulation class FIIC rating is the same as the IIC rating except that it is used to rate the impact sound insulation performance of in-situ floor/ceiling assemblies and is used in conjunction with the ASTM E1007 [98] test method (see Section 12.9.1).

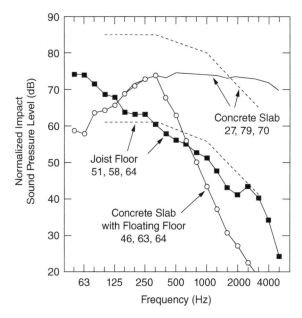

Figure 12.47 Examples of tapping machine levels [20]. The concrete slab is 150 mm thick. The three numbers under each legend are IIC, $L_{n,w}$, and the sum of the energy in the frequency range from 50 to 2500 Hz minus 15 dB. The fitted IIC contour is shown for the bare concrete slab and the joist floor. The joist floor is comprised of 19-mm oriented strandboard, sound-absorbing material, and two layers of 13-mm gypsum board suspended on resilient metal channels. The floating floor is 22-mm parquet on 4-mm cork [20].

The ASTM procedure just described gives a value of IIC roughly comparable in magnitude to the STC of a floor or partition. This is of considerable advantage to an architect who may now specify a floor to have, say, STC = 50 and IIC = 50 for a particular application (see Section 12.11).

In the ISO system, the procedure yields a single number rating termed the *weighted normalized impact sound pressure* $L_{n,w}$. This single-number can be determined from one-third octave (100–3150 Hz) or octave band (125–2000 Hz) measurements. The ISO rating procedure for impact sound insulation is also similar to the ASTM procedure (for one-third octave band measurements), but ISO does not apply the 8-dB rule. Instead, the ISO standard suggests the use of a spectrum adaptation term, C_i, to deal with low-frequency noise typical of that generated below a lightweight joist floor. The rating suggested, $L_{n,w} + C_i$, is just the unweighted sum of the energy in the one-third octave bands from 125 to 2500 Hz minus 15 dB. The 50, 63, and 80 Hz one-third octave band center frequencies may also be included. For additional details on the procedure the reader must consult Annex A of the ISO standard [103].

The relationship between the two ratings is IIC = $110 - L_{n,w}$ if the 8-dB rule has not been invoked [20]. In order to make a distinction between values with and without flanking transmission, primed symbols (e.g. R', L'_{nT} or $L'_{n,w}$) are used in the ISO standard to denote values obtained with flanking transmission [91].

A disadvantage of the use of the weighted normalized impact sound pressure index is that floors/ceilings with higher values of $L_{n,w}$, transmit more impact noise. However, the ranking of structures with airborne sound is opposite, since structures with higher values of averaged *TL*, *NR*, or STC transmit less airborne sound. This opposite ranking is confusing to architects and laymen.

It should be noted that a floor's impact insulation can usually be improved rather easily by installing a soft resilient layer on the surface (such as a carpet, or, better still, a carpet with an underlay). Other, more durable, resilient materials than carpet are available and preferred for industrial or heavy wear. While the use of resilient materials can improve the IIC of a floor/ceiling considerably, they have very little effect on airborne sound transmission loss and STC (or R_w). If impact noise must be reduced even further, consideration should be given to the use of a floating floor construction.

Example 12.22   Consider a room with a ceiling assembly which has a normalized impact sound pressure level $L_n = 51$ dB in the 250 Hz one-octave band. The total absorption in the room is 25 sabins (m²). A standard tapping machine can make impacts on the floor above the room. (a) What impact space-averaged sound pressure level would the standard tapping machine produce in the room in the 250 Hz band? (b) If an impact space-averaged sound pressure level less than 41 dB is required in the room, how much equivalent sound absorption area has to be added to the room?

Solution

From Eq. (12.82) we have that $\langle L_0 \rangle = L_n + 10 \log(A_0/A) = 51 + 10 \log(10/25) = 47$ dB.

a) From Eq. (12.82) we find that $A = A_0 10^{(L_n - \langle L_0 \rangle)/10}$. Therefore, $A = 10 \times 10^{(51-41)/10} = 100$ sabins (m²).
b) Then, an absorption treatment in the room must add 75 sabins (m²) in the 250 Hz band to meet the requirement.

12.10   Measured Airborne and Impact Sound Transmission (Insulation) Data

Laboratories, organizations and manufacturers in several countries have measured, collected, and published data on the measured airborne and impact sound transmission loss data of partitions and floor/ceiling assemblies [105–107]. In the U.S., the National Bureau of Standards (NBS) and NIOSH have published such reports in the past [108, 109].

In 1967 a large report was prepared by the Federal Housing Administration (FHA) [110]. This report is considered one of the most comprehensive compilations of laboratory measurements on interior walls and floor/ceilings. Sound insulation data were presented for 137 wall constructions and 111 ceiling/floor structures. In addition, 345 detailed architectural drawings were included to show the proper construction and installation of wall and floor assemblies required for adequate sound insulation, noise control and privacy in multifamily dwellings. Although the FHA report may be partly outdated, it is still widely used. Another comprehensive report on exterior walls and windows which is still consulted by architects is the one published by Sabine et al. [61] In Canada, the National Research Council has published collections of conventional laboratory test results for wall and floor assemblies evaluated according to ASTM [111–115]. Additional data can be found in some reference books [5, 107, 116, 117].

More recently, manufacturers have published a large amount of *TL* and impact insulation data that have been measured in the laboratory according to the recommendations by ASTM and ISO standards discussed in Sections 12.7.1 and 12.9.1. In addition, some data have been incorporated in insulation prediction software [118, 119]. It has to be noticed that there is scatter in the data measured by different laboratories of similar partitions. The data presented in the reports should be used as an approximate guide to the insulation properties to be expected of different partitions. Representative values of transmission class and impact isolation class for different types of constructions are given in Table 12.1 [120].

12.10.1 Gypsum Board Walls

Gypsum board walls are the most commonly used partitions in homes and apartments in the U.S. and Europe. They normally consist of two layers of gypsum board supported by studs and separated by an air gap. If the air gap is filled with an absorbing batt, then the sound insulation is increased [121, 122]. If the studs supporting the facing and backing walls are rigidly attached in some way, the insulation is decreased. Staggering the facing and backing wall studs and/or mounting the walls to the studs using resilient clips or channels reduces "short-circuiting" of the two walls and increases the insulation.

Differences in construction methods normally account for the differing sound insulation properties and costs of such walls. Unless such walls are properly designed, the cost may increase as additional noise control features are added, with no resulting improvement in sound insulation and STC/R_w rating. There is a considerable scatter in STC/R_w ratings of partitions for the same cost, suggesting inadequate acoustical knowledge is available to partition designers. Table 12.1 gives some examples of the STC/R_w ratings of some typical partitions which are commercially available. For full construction details, readers should consult the original sources or manufacturers' catalogs [61, 64, 105–115].

Figure 12.48 shows four different gypsum double wall constructions and illustrates the different internal designs which will give rise to different STC values. The four partitions contain absorbing material in the cores.

Note that, although the first two single wood-stud walls shown in Figure 12.48 have somewhat similar geometries and constructions, they have different values of STC. The effect of the resilient channel which reduces the direct contact between the studs and the gypsum boards is apparent in the second partition and the STC is significantly increased. The third partition has staggered wood studs which produced the same STC as the second wall. The fourth partition is a double wood stud wall with a particularly thick core which shows the best STC rating and improved *TL* mostly at low frequency.

12.10.2 Masonry Walls

Masonry walls are usually made from concrete block or brick and mortar and designed to carry building loads (in the U.S.). In European countries it is common, in some areas, to build homes completely from brick or

Table 12.1 Representative transmission and impact isolation class data for walls, doors, windows, and floors.

Construction	Single number rating	
	STC/R_w	IIC/L_{nTw}
24-g Metal Studs		
16-mm GB[a] each side of 65-mm studs at 600-mm centers, no absorption	40	
16-mm GB each side of 65-mm studs at 600-mm centers, 40-mm FG[b]	45	
2-×16-mm GB each side of 65-mm studs at 600-mm centers 40-mm FG	55	
2-×16-mm GB each side of 90-mm studs 600 o.c.,[c] no absorption between studs	50	
2×16-mm +1-×16-mm GB on 65-mm studs 600 o.c., 80-mm FG between studs	55	
2×16-mm GB each side of 90-mm studs 600 o.c., 80-mm FG between studs	55	
3×16-mm GB each side of 90-mm studs 600 o.c., 80-mm FG between studs	60	
20-g Metal Studs		
2-×16-mm GB each side of 90-mm studs 600 o.c., 80-mm FG between studs	50	
As above with resilient channels on one side	60	
Wood Studs		
16-mm GB each side of 50-×100-mm studs at 600-mm centers, no absorption	35	
As above but with 80-mm FG and resilient channels on one side	50	
As above but with 2-×16-mm GB on each side of studs and 50-mm FG	60	
Brick		
Single brick	45	
Double brick	50	
Cavity brick	55	
Concrete Block		
100-mm hollow lightweight	45	
As above with 13 mm render both sides	50	
100-mm hollow lightweight with 16-mm GB on resilient channels	55	
150-mm solid concrete with 40-mm FG + 16-mm GB on 50-mm wood furring	60	
Windows		
3 mm, fixed glazing	25	
6 mm, fixed glazing	30	
10 mm, laminated, fixed	35	
Double glazing (4-mm glass, 50-mm airspace, 4-mm glass)	40	
Doors		
Hollow core, no seal/gasket	15	
Solid (35 mm) without seals	20–25	
Solid (35 mm) with seals	25–30	
Concrete slab		
150 mm thick	55	25(85)
150 mm thick with carpet and underlay	55	85(25)
150 mm thick with 50-mm slab on isolation pads	60	<70 (>40)
150 mm thick with wood on furring on pads	60	65(45)

Source: Adapted from Ref. [120].
[a] GB = gypsum board or similar.
[b] FG = fiberglass batt or similar.
[c] On centerline.

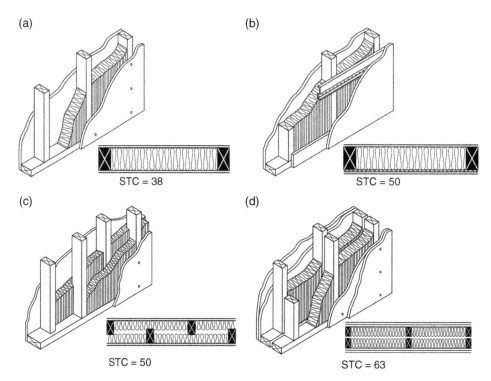

Figure 12.48 Typical wall assemblies using gypsum boards: (a) Single 2 × 4 wood studs, 16″ o.c., single layer ½″ gypsum board each side, one thickness (3½″–4″) fiberglass batt insulation; (b) Single 2 × 4 wood studs, 16″ o.c., with resilient channel, single layer 5/8″ Type X gypsum each side, one thickness (3½″–4″) fiberglass batt insulation; (c) Staggered 2 × 4 wood studs, 16″ o.c., 2 × 6 top and bottom plates, single layer ½″ Type X gypsum board each side, two thicknesses (2½″) fiberglass batt insulation; (d) Double 2 × 4 wood studs, 16″ o.c., double layer ½″ Type X gypsum board each side, two thicknesses (3½″ – 4″) fiberglass batt insulation. (Note: 1″ = 25.4 mm). (*Source:* Reproduced from Ref. [64] with permission from NAIMA.)

concrete blocks. Because of their large mass per unit area, masonry walls normally have high values of *TL* and STC/R_w [115, 124, 125]. Double-masonry walls (with an air gap) have an even higher *TL* than single-masonry walls. Sometimes such double walls are connected together with metal ties which tend to increase structural strength but diminish *TL* and STC/R_w. Some values of airborne sound transmission loss ratings of masonry walls are given in Table 12.1.

12.10.3 Airborne and Impact Insulation of Floors

Two types of floors are commonly used: concrete and wood. Examples of typical constructions are shown in Figure 12.49. Concrete floors are usually quite heavy and provide high *TL* to airborne sound (see Figure 12.49c); however, they are usually poor insulators of impact sound, unless precautions are taken to use a special resilient surface (finish). The airborne and impact insulation of a concrete floor can also be improved if the floor is vibration-isolated in some way, as discussed earlier. Such vibration-isolated floors are termed "floating floors." For further discussion on floating floors, see Chapter 13, Section 13.6.1.

Wood floor-ceiling assemblies are usually much lighter than concrete floors. However, with careful design, despite their low weight, wood floors can have airborne and impact insulation properties almost as good as much heavier concrete floors (compare Figure 12.49a with Figure 12.49c). The impact isolation of a wood floor is usually better than that of a concrete floor when neither has surface finishes. It should be noted that when a resilient surface finish is applied to either a concrete or wood floor, the impact sound insulation is usually improved considerably, although there is little improvement in the airborne sound insulation.

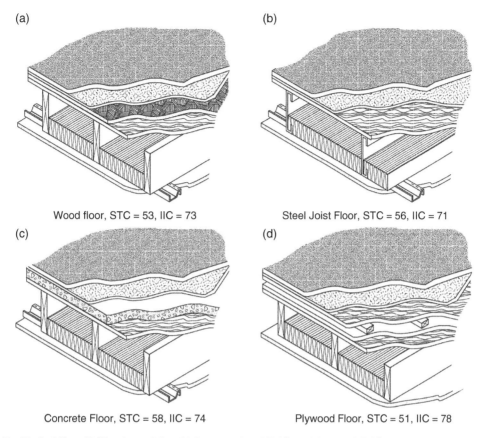

Figure 12.49 Typical Floor/Ceiling Assemblies: (a) Carpet and pad, 3/8″ particle board, 5/8″ plywood subfloor, 2 × 0 joists 16″ o.c., one thickness (3½″ – 4″) fiberglass batt insulation, resilient channel, ½″ Type X gypsum board; (b) Carpet and pad, 3/4″ T&G plywood subwood subfloor, steel joists (7 1/4″, 18 ga.) 24″ o.c., one thickness (3½″ – 4″) fiberglass batt insulation, resilient channel, 5/8″ gypsum board; (c) Carpet and pad, 1½″ lightweight concrete floor, 5/8″ plywood subfloor, 2 × 10 joists 16″ o.c., one thickness (3½″ – 4″) fiberglass batt insulation, resilient channel, ½″ Type X gypsum board; (d) Carpet and pad, 5/8″ plywood floor, 2″ × 3″ furring, ½″ sound deadening board, ½″ plywood subfloor, 2 × 8 wood joists, one thickness (3½″ – 4″) fiberglass batt insulation, 5/8″ Type X gypsum board. (Note: 1″ = 25.4 mm). (*Source:* Reproduced from Ref. [64] with permission from NAIMA.)

For each floor-ceiling assembly, the values of airborne transmission loss and impact insulation ratings are given in Table 12.1 and Figure 12.49. The methods for measuring airborne transmission loss, impact sound pressure levels and the procedures for computing single number ratings are described in Sections 12.8 and 12.9.

a) Floating Floors

Floating floors are very useful in reducing the transmission of airborne sound and structure-borne vibration to spaces below. A second layer is placed above the main structure slab to the floor and separated from it by an air space. The upper layer (floating floor) is supported by and isolated from the structural slab. This is done by pouring the floating concrete floor on plywood panels, which are resiliently mounted on the structural slab. The resilient support might be a precompressed layer of rock wool, fiberglass, foamed plastics, polyurethane, or the like (Figure 12.49a). For further discussion, see Chapter 13, Section 13.6.1.

Instead of a continuous resilient layer, an array of resilient pads may be used and the cavity between the floating slab and the structural floor filled with soft sound-absorbing material (see Figure 12.50b and Figures 13.32 and 13.33). Some thin floating floors are comprised of resilient layers about 4–12 mm thick finished with a layer of hardwood, about 12 mm thick. These can increase IIC values by as much as 20 points.

(a)

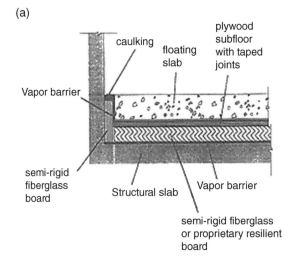

(b)

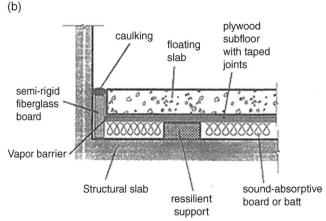

Figure 12.50 A typical floating concrete floor construction of the type commonly used to isolate noise and vibration in buildings. (a) Layer of resilient material used to support a floating slab. (b) Resilient supports are used to support the floating floor slab [125].

The floating floors shown in Figure 12.50 can increase the transmission loss between 125 and 4000 Hz by as much as 20–30 dB. The STC rating can also be increased by a similar amount compared with a standard and structural floor (see Figure 12.51).

A floating floor has a fundamental natural frequency that is determined by the mass per unit area of the floating floor and the dynamic stiffness per unit area of the resilient support. The latter consists of the dynamic stiffness per unit area of the material and that of the enclosed air, which includes the effects of airflow resistivity [20]. A test set-up to determine the dynamic stiffness of an elastic material used under a floating floor has been specified in standard ISO 9052-1 [126, 127]. Typical values of dynamic stiffness per unit area of resilient materials are given by Hopkins [5]. They range from 10 MN/m^3 (rock wool 30-mm thick and 60 kg/m^3) to 28 MN/m^3 (glass wool 13-mm thick and 36 kg/m^3). Thin resilient layers made of nanostructured polymers have been recently shown to exhibit comparable values of dynamic stiffness per unit area [128].

Above the fundamental natural frequency, the improvement in the impact sound transmission of a floating floor varies from between 30 dB per octave for resonantly reacting floors (usually the type used in mechanical rooms) to 40 dB per octave for locally reacting floors. In buildings, if flanking transmission in the walls is not controlled, these improvements may not be achieved.

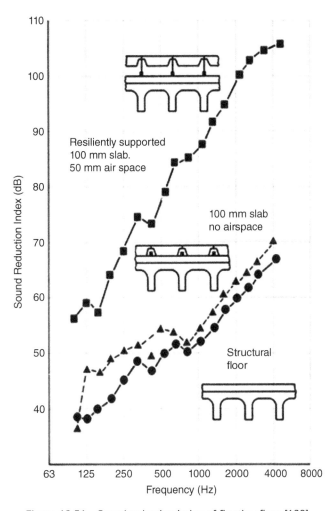

Figure 12.51 Sound reduction index of floating floor [129].

Apart from increasing the airborne transmission loss and the impact noise rating, a floating floor has one more extremely important advantage. It considerably reduces the amount of acoustical and vibrational energy flowing into the mechanical room structural slab and hence into the whole building structure. In modern concrete multistory buildings, energy can be propagated to all parts of the building with very little attenuation and then easily re-radiated. Hence, the attenuation of this energy must be increased and the floating floor meets this requirement extremely well. Of course, the use of floating floors is not confined to mechanical rooms, and they are often employed in other areas of a building. For example, with music practice rooms or with pedestrian malls which may pass over low noise areas, floating floors can be used to considerably reduce impact and airborne noise to such an extent that it is undetectable below.

Some typical floating floor perimeter and dividing wall structural details are shown in Chapter 13. See Figures 13.32 and 13.34.

12.10.4 Doors and Windows

Doors and windows are usually the acoustical "weak links" between rooms in buildings and between interiors of buildings and exterior noise. If a high *TL* is desired for doors and windows, it is essential that a good positive seal is obtained. Inside many residential buildings, some leaks around doors are usually accepted as necessary to provide a path for return air for ventilation systems. However, with external doors and with windows, such leaks must be avoided or poor sound insulation will result (as already discussed in Section 12.6).

Doors are usually made of wood or steel. The sound insulation of a door can be improved if it is made of more than one, at least partly-isolated layers with the air cavities filled with acoustical absorbing material. Extreme care must be taken to provide a positive seal around the edge of the door and the door frame. Many different door seal designs exist; the more commonly used is the gasket type (see Figure 12.52). Unless a good seal is provided and maintained, even the thickest, heaviest door will have a poor *TL*.

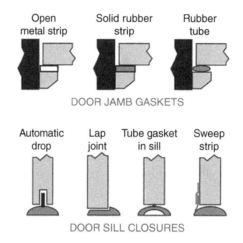

Figure 12.52 Recommended door seal designs. (*Source:* reproduced from Ref. [64] with permission from NAIMA.)

Windows are normally responsible for transmitting most of the acoustical and thermal energy into or out of buildings. However, with careful design, the acoustical and thermal transmission of energy through a window can be considerably reduced. The sound insulation of a window is usually improved if dual glazing (i.e. double windows with an air gap) is used instead of single glazing [130, 131]. A wider air gap between the window panels increases *TL* (particularly at low frequencies) (see Section 12.3). Acoustical absorbing material such as fiberglass packed into the edges of the air gap also improves insulation. Sometimes laminated glass is used to increase the internal damping and reduce the *TL* dip near the critical coincidence frequency.

According to Warnock [20] for single panes with thicknesses of 13 mm or less, STC (or R_w) may be calculated from STC $= 0.61 \times t + 27.9$ for solid panes, or STC $= 0.47 \times t + 31.5$ for laminated panes, where t is the total thickness of the pane including the lamination (mm). Also, use of two different thicknesses of glass in double windows gives different critical frequencies and smooths out the coincidence dip. Examples of outdoor noise insulation provided by windows are shown in Figure 12.53.

Some typical values of STC and R_w ratings are presented for doors and windows in Table 12.1. It should be noted that there is considerable scatter in manufacturers' test results, obviously resulting from different internal design details, care in sealing and measurement procedures and facilities. The values of STC and R_w given may be used as a rough guide of possible values to be expected of different constructions.

12.11 Sound Insulation Requirements

Several criteria have been developed for walls, floors, doors, etc. for use in different buildings. Building codes and standards usually contain requirements for sound insulation, aimed at meeting the demands of different functions of rooms and buildings. Architects and acousticians can use such criteria and select walls and floors with suitable STC/R_w or IIC/$L_{n,w}$ ratings for use in different buildings from information such as that given in

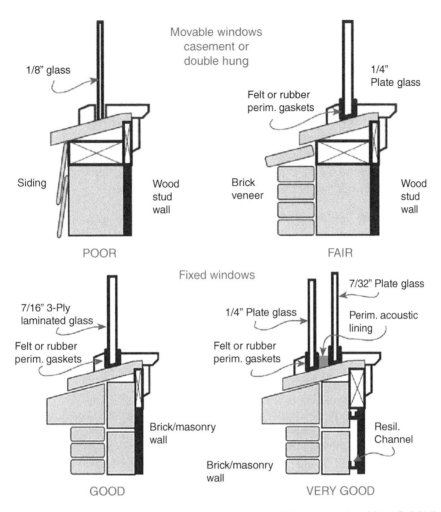

Figure 12.53 Examples of outdoor noise insulation provided by windows. (*Source:* reproduced from Ref. [64] with permission from NAIMA.)

Section 12.10, other handbooks, software, or partition manufacturers' catalogs. Some of the ratings used to assess the acceptability of interior ambient sound in rooms and indoor noise criteria are discussed in detail in Chapter 6 of this book.

Recommended criteria for partitions have been presented in several countries, usually as "building technical codes" and they are discussed in several references [3, 4, 7, 10]. A real challenge for any country is the adoption of simple building standards and of a simple procedure to check whether standards have been met in practice. Some building technical codes include constructional details and recommended wall and floor/ceiling assemblies that are known to satisfy the corresponding building code and have been checked in recognized laboratories.

The International Code Council every three years publishes the International Building Code (IBC). This document is intended to present model building code regulations to protect public health, safety and welfare. Chapter 12 of the 2015 version of the IBC [132] deals with the interior environment and section 1207 defines the minimum standards for noise insulation in buildings. For airborne sound transmission at wall and floor/ ceilings assemblies, the minimum rating is STC = 50 (or FSTC = 45). For impact noise at floor/ceiling assemblies, the minimum rating is IIC = 50 (or FIIC = 45). For more details, the reader should always consult the criteria for a particular application. A discussion on noise control in U.S. building codes is given by Tocci [133]. Rasmussen [92] presented a comprehensive review on the airborne and impact sound insulation between dwellings required by building codes in European countries. It is suggested that readers involved in building design or who are sufficiently interested should obtain the latest versions or these criteria or similar ones recommended by their own countries. An example of one of such criteria is presented in the following.

One of these criteria was recommended by the U.S. FHA and it continues to be referenced as the primary source standard for sound insulation requirements in condominiums, townhouses and other multi-unit residential housing [111]. These recommendations are summarized here. Three grades of construction are specified (Grade II is considered to be the fundamental guide, and constructions which satisfy this criterion should be suitable for most multifamily dwellings in the U.S.):

1) Grade I (luxury rating) applies to quiet suburban and peripheral suburban residential areas where nighttime exterior A-weighted levels do not exceed 35–40 dB. The recommended interior noise environment is NC 20–25. (see Chapter 6 of this book for a definition of NC)
2) Grade II (average rating) is the most important category and applies to urban and suburban residential areas with average noise environments. The nighttime exterior levels are probably about 40–45 dB, and the interior noise environment should not exceed NC 25–30.
3) Grade III (minimum rating) applies to noisy urban areas, and constructions meeting these criteria are regarded as having minimal insulation. Nighttime exterior A-weighted levels might be 55 dB or more, and the interior noise environment should not exceed NC 35.

Note that the IBC establishes only minimum standards of construction, which sometimes mirror the FHA Grade III minima. Minimum ratings are provided for many combinations of adjacent spaces, depending on room use. The partitions chosen for the following applications should have STC and IIC ratings equal to or greater than those listed depending upon the noise environment.

Table 12.2 gives STC values in more detail for partitions separating dwelling units in multifamily buildings. If possible, such buildings should be planned so that partitions separate rooms with similar functions (e.g. kitchen from kitchen, bathroom from bathroom, etc.). Where such layouts are not possible, then the partitions need greater sound insulation properties. See Chapter 13, Section 13.5 for further discussion on space planning and building layouts.

Table 12.3 gives STC and IIC ratings for floor-ceiling assemblies separating dwellings in multifamily buildings. Again, such buildings should be planned, where possible, with floors separating like rooms: bedroom above bedroom, living room above living room, etc.; otherwise undesirable situations will occur, and the insulation properties of the floor-ceiling assemblies must be increased. Dwelling units should not be placed next to mechanical equipment rooms (including furnace rooms, elevator shafts, garages, transformers, emergency power generators, trash shoots, etc.). If this situation is unavoidable, the STC rating of partitions between such mechanical rooms and sensitive areas in dwellings should be: STC $\geq$ 65, STC $\geq$ 62, and STC $\geq$ 58 for Grades I, II, and III: or STC $\geq$ 60, STC $\geq$ 58, and STC $\geq$ 54 for Grades I, II, and III for partitions between mechanical rooms and less sensitive dwelling areas (kitchens, family rooms, etc.).

If possible, dwelling units should not be placed adjacent to business premises (such as restaurants, bars and laundries) in the same buildings. If this situation cannot be avoided, then the partition ratings should exceed STC $\geq$ 60, STC $\geq$ 58, and STC $\geq$ 56 and IIC $\geq$ 65, IIC $\geq$ 63, and IIC $\geq$ 61 for Grades I, II, and III if living areas in dwellings are placed *below* business premises.

Table 12.4 gives suggested criteria for airborne insulation requirements for partitions separating rooms in the same dwelling unit. Again, sensible planning can avoid the use of expensive partitions or the creation of insufficient sound insulation between rooms. See Chapter 13, Section 13.5.

Table 12.2 Criteria for airborne sound insulation of wall partitions between dwelling units.

Partition function between dwellings		Luxury Grade I	Average Grade II	Minimum Grade III
Apt. A	Apt. B	STC	STC	STC
Bedroom	to Bedroom	55	52	48
Living room	to Bedroom	57	54	50
Kitchen	to Bedroom	58	55	52
Bathroom	to Bedroom	59	56	52
Corridor	to Bedroom	55	52	48
Living room	to Living room	55	52	48
Kitchen	to Living room	55	52	48
Bathroom	to Living room	57	54	50
Corridor	to Living room	55	52	48
Kitchen	to Kitchen	52	50	46
Bathroom	to Kitchen	55	52	48
Corridor	to Kitchen	55	52	48
Bathroom	to Bathroom	52	50	46
Corridor	to Bathroom	50	48	46

Table 12.3 Criteria for airborne and impact sound insulation of floor/ceiling assemblies between dwelling units.

Partition function between dwellings		Luxury Grade I		Average Grade II		Minimum Grade III	
Apt. A	Apt. B	STC	IIC	STC	IIC	STC	IIC
Bedroom	above Bedroom	55	55	52	52	48	48
Living Room	above Bedroom	57	60	54	57	50	53
Family Room	above Bedroom	60	65	56	62	52	58
Corridor	above Bedroom	55	65	52	62	48	48
Bedroom	above Living Room	57	55	54	52	50	48
Living Room	above Living Room	55	55	52	52	48	48
Kitchen	above Living Room	55	60	52	57	48	53
Family Room	above Living Room	58	62	54	60	52	56
Corridor	above Living Room	55	60	52	57	48	53
Bedroom	above Kitchen	58	52	55	50	52	46
Living Room	above Kitchen	55	55	52	52	48	48
Kitchen	above Kitchen	52	55	50	52	46	48
Bathroom	above Kitchen	55	55	52	52	48	48
Family Room	above Kitchen	55	60	52	58	48	54
Corridor	above Kitchen	50	55	48	52	46	48
Bedroom	above Family Room	60	50	56	48	52	46
Living Room	above Family Room	58	52	54	50	52	48
Kitchen	above Family Room	55	55	52	52	48	50
Bathroom	above Bathroom	52	52	50	50	48	48
Corridor	above Corridor	50	50	48	48	46	46

Table 12.4 Criteria for airborne sound insulation within a dwelling unit.

Partition function between rooms		Luxury Grade I STC	Average Grade II STC	Minimum Grade III STC
Bedroom	to Bedroom	48	44	40
Living room	to Bedroom	50	46	42
Bathroom	to Bedroom	52	48	45
Kitchen	to Bedroom	52	48	45
Bathroom	to Living room	52	48	45

As mentioned above, walls and floor-ceiling assemblies separating rooms from mechanical equipment rooms within buildings must be considered carefully to meet sound insulation requirements [135, 136]. It is common practice for architects to locate mechanical equipment rooms on upper floors of multistory buildings where they are often supported by lightweight flexible structural slabs and positioned directly over critical areas requiring low noise levels. Since most mechanical rooms in buildings usually contain many pieces of equipment apart from the fans (such as boilers, chillers, and pumps), the noise and vibration levels are accordingly very high, and often the floors need to provide over 50 dB transmission loss at low frequencies in order to achieve acceptable conditions in spaces below. Such a transmission loss cannot be economically obtained at low frequencies by a single layer floor slab.

Since vibration in the audible frequency range can easily be transformed directly into noise or propagated to some other part of the building and then re-radiated as noise, it is of extreme importance to control all vibration in the mechanical room to within tolerable limits. Ideally, the mechanical room should be located well away from critical areas in the building, but when this is not possible, even greater care must be taken with the control of vibration. Although it might seem that the best approach should be to furnish all pieces of equipment with vibration isolators, this is not necessarily so. The first line of attack should be to ensure that each piece of equipment is selected to produce minimum noise and is operated under its specified conditions. For example, one should choose, where possible, rotating equipment in preference to the equivalent reciprocating unit since, in general, the latter type produces much more objectionable noise. Furthermore, it is extremely important to see that each piece of equipment is balanced, both statically and dynamically, to within the recommended tolerances. See Chapter 13 for further discussion on equipment balancing.

All building equipment radiating excessive noise in mechanical rooms should be adequately silenced, if at all possible. For example, some or all of the following noise control techniques may be implemented:

1) Adding acoustical absorption to the walls and ceilings near large centrifugal fans (this may be necessary for hearing conservation and communication)
2) Fitting of silencers at air intakes to forced draft fans which are open to the room
3) Installation of acoustically lined plenum chambers around high power centrifugal fans
4) Installation of resilient mounts and hangers to isolate piping, duct work, wiring, conduit, etc. from the building structure
5) Acoustical enclosures around reciprocating refrigeration machines.

After the sound pressure levels within the mechanical room have been adequately and economically reduced, the next step is to introduce a massive layer between the room and the nearby critical areas. Unfortunately, most of the building mechanical room equipment (especially large centrifugal fans) radiate noise strongly at low frequencies, which causes a problem for the design engineer. Indeed, sound pressure levels in excess of 110 dB in the 63 and 125 Hz octave bands are quite common [132].

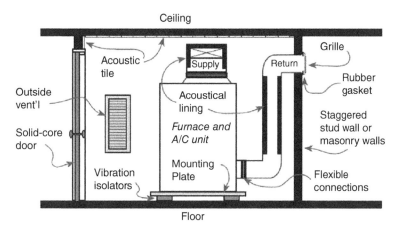

Ceiling

Acoustic tile

Outside vent'l

Solid-core door

Supply

Acoustical lining

Furnace and A/C unit

Mounting Plate

Vibration isolators

Floor

Grille

Return

Rubber gasket

Staggered stud wall or masonry walls

Flexible connections

Figure 12.54 Typical noise and vibration control techniques in a mechanical equipment room in a building. (*Source:* reproduced from Ref. [64] with permission from NAIMA.)

A completely continuous floating floor is not always necessary, and in some cases a floating base beneath certain noisy pieces of mechanical room equipment may be sufficient. When the sound pressure level in a fan plenum chamber far exceeds that from the surrounding mechanical room equipment, then the whole plenum chamber should be mounted on a floating floor. Figure 12.54 shows an example of typical noise control techniques implemented in a mechanical equipment room in a dwelling unit. Noise in HVAC systems is treated in Chapter 13 of this book.

A major concern to the design engineer is to avoid resonances both in the equipment, in its supports, and in the building structure. At resonance, a large vibration amplitude is developed which may be accompanied by excessive radiated noise and stress. The stresses set up during the resonance may ultimately lead to fatigue in equipment or its support, or – even more disastrously – in the building structure (i.e. the floor)! One should, therefore, carefully check whether any equipment is to be operated close to (±25%) any critical speed in the machine, structural resonance in its support, or building structural resonance. Since the floor resonances and machine critical speeds are usually fixed, the operating speeds must be chosen to be well away from both of these. The only concern, then, is the support. In general, support resonances may be avoided by making the support very much less flexible than the flexibility that would result in resonance or by using a separate isolator with much higher flexibility. Such an increase in flexibility automatically means increased static deflection (see Chapter 2).

Example 12.23 In a bank, a mechanical equipment room needs to be located adjacent to a large room which will be used as an open-plan office. The dimensions of the common wall are 3 m × 7 m. The wall is made of concrete ($TL = 50$ dB in the 500 Hz octave band) and it has an access door of 2 m × 1 m ($TL = 25$ dB in the 500 Hz octave band) and a leak under the door 1-cm high. It is estimated that the amount of absorption in the receiving room will be 52 sabins (m^2) at 500 Hz. After a sound-absorbent treatment of the walls of the large room, the absorption is estimated to be 167 sabins (m^2) at 500 Hz. (a) Determine the NR with and without the absorption treatment at 500 Hz octave band. (b) What would be the effect on the NR if the leak is properly sealed?

Solution

(a) The transmission coefficients are: for the concrete: $\tau_1 = 1/[10^{(50/10)}] = 10^{-5}$, for the door: $\tau_2 = 1/[10^{(25/10)}] = 0.00316$; for simplicity we assume that the leak $\tau_3 = 1$. Therefore, from Eq. (12.60):

$$\tau_0 = \frac{(10^{-5} \times (21 - 2 - 0.01)) + (0.00316 \times 2) + (1 \times 0.01)}{21} = 7.86 \times 10^{-4},$$ and the overall TL of the composite wall is $TL_{av} = 10\log(1/7.86 \times 10^{-4}) = 31$ dB.

Now, the NR of the partition is NR = $TL - 10\log(S/A) = 31 - 10\log(21/52) = 35$ dB. Due to the sections with lower TL values (especially the leak) and the hardness of the receiving room, the 50 dB concrete wall results in a NR of only 35 dB. If the absorbent treatment is used, we will have a NR = $31 - 10\log(21/167) = 40$ dB. The NR is increased by 5 dB to 40 dB, which indicates the leak should have been fixed first.

(b) If the leak is plugged with a seal that provides a $TL = 50$ dB, the transmission coefficient of the dividing wall becomes

$$\tau_0 = \frac{(10^{-5} \times (21-2-0.01)) + (0.00316 \times 2) + (10^{-5} \times 0.01)}{21} = 3.1 \times 10^{-4},$$ and the average TL will

be $TL_{av} = 10\log(1/3.1 \times 10^{-4}) = 35$ dB. Consequently, the NR of 35 dB is increased to NR = $TL - 10\log(S/A) = 35 - 10\log(21/52) = 39$ dB, which is almost as much as the sound-absorbing treatment provided. Now, the whole job, sealing the leak under the door and adding sound absorption materials to the room, results in a NR of NR = $TL - 10\log(S/A) = 35 - 10\log(21/167) = 44$ dB, which is 9 dB greater than the 35 dB obtained with the leak and without additional sound absorption. If it is desired to increase the TL even more, it is seen that the door is still the weakest link.

Example 12.24 A common wall between a factory and an office measures 4 m × 10 m, and the office dimensions are 4 m × 10 m × 10 m. The office reverberation times in each one-octave band center frequency are written in row 1 of Table 12.5. The A-weighted sound pressure level in the factory space is 90 dB, the sound field is diffuse, and the spectrum is given in row 2 of Table 12.5. A target level of NC-40 has to be achieved as an acceptable noise environment in a general office (see Chapter 6 of this book). Determine the minimum partition TL required to achieve NC-40 in the office.

Solution

The office volume $V = 400$ m^3 and its total surface area $= (40 \times 4 + 100 \times 2) = 360$ m^2. The absorption area A is calculated from Sabine's formula $A = 0.161 V/T_R$ and written in row 4 of Table 12.5. $10\log(A)$ is calculated in row 5. The values given for NC-40 (see Figure 6.7 in Chapter 6 of this book) are written in row 3 of Table 12.5. Since the common partition area $S = 40$ m^2, $10\log(S) = 16$, and this is written in row 6 of Table 12.5. Since $NR = L_1 - L_2$, then the required TL (Eq. (12.78)) is found from rows (2)–(3) + (6)–(5) = row (7) in Table 12.5.

Table 12.5 Calculation of minimum partition TL required to achieving NC-40 in office of Example 12.19.

	One-Octave band center frequency (Hz)							
	63	125	250	500	1000	2000	4000	8000
1) Reverberation time in office T_R (sec)	1.6	1.4	1.2	1.0	1.0	1.0	1.0	1.0
2) Sound pressure level in factory, L_1 (dB)	79	82	85	85	87	82	75	68
3) Desired office level (NC-40), L_2 (dB)	64	56	50	44	41	39	38	37
4) Calculated absorption area, A sabins (m^2)	40.3	46.0	53.7	64.4	64.4	64.4	64.4	64.4
5) $10\log(A)$	16.1	16.6	17.3	18.1	18.1	18.1	18.1	18.1
6) $10\log(S)$	16	16	16	16	16	16	16	16
7) Required $TL = (2) - (3) + (6) - (5)$	15	26	34	39	44	41	35	29

12.12 Control of Vibration of Buildings Caused by Strong Wind

Structures such as buildings immersed in a moving fluid experience fluid loading forces. These forces are caused by several physical phenomena. The phenomena may be divided into three main categories: (i) steady incoming flow that impinges on the structure, (ii) unsteady incoming flow, and (iii) eddies (vortices) that form in the fluctuating wake flowing past the structure. The manner in which a structure responds to the flow forces is governed by several factors, the most important being: (i) its geometrical shape and dimensions; (ii) its boundary conditions and rigidity (stiffness) and mass distributions, which govern its fundamental and higher natural frequencies; and (iii) its vibration damping [137].

The structures may be divided into two main types: (i) stiff structures that are relatively rigid and that have fundamental bending and torsional frequencies that are above the predominant fluctuation frequencies in the incoming flow and wake eddy formation rates and (ii) flexible structures that have fundamental frequencies much lower than or that coincide with the frequencies of the fluctuating forces in the incoming flow and/ or wakes from any neighboring structures upstream [137].

Dynamic forces experienced by "rigid" structures can be divided into two main types: (i) longitudinal forces in the flow direction mostly caused by unsteady incoming flow and (ii) lateral and torsional forces that are mainly caused by the vortex shedding in the structure's own wake. Flexible structures can experience additional forces caused by motion of the structure itself in response to the flow forces. Motion of the structure, sometimes caused by aeroelastic effects, can alter the flow and result in self-generating feedback mechanisms known such as galloping, flutter, or vortex-generated motion resulting in increased vortex shedding. In addition, aeroelastic effects associated with vortex shedding are also important since, when the vortex shedding frequency coincides with a natural frequency, feedback motion results in which the vortex-generated motion intensifies the vortex shedding.

There is a considerable body of literature describing the vibration of structures caused by fluid flow. Crocker [137] has described the response of structures to fluid flow forces and in particular the response of buildings, towers, and other structures to forces caused by wind. There are many books covering the dynamic response of structures that are also useful in study of this topic [138].

The wind flow speed and direction normally continually change. The wind in most cases is highly turbulent, and its characteristics depend on a number of factors, the most important being: (i) the surface roughness of the terrain (which differs according to whether the surface is water, open terrain, suburban, or urban) and (ii) the height above the ground [139–143].

Wind speeds and profiles are typically referenced with respect to a standard height of 10 m (33 ft) above ground level. See Figure 12.55. It can maintain a fairly steady direction, but it can also change in speed and in direction (known as veering). When an obstruction such as a building is encountered, the flow pattern changes markedly. (see, e.g. Figure 12.56) Buildings introduce distinct individual wind flow patterns. Near large buildings the wind can behave in a very complicated manner due to local flow accelerations, formation of eddies, and flow reversals. For example, winds can reverse direction and head downward in the "upwind roller" that can form in front of a large tall building. The wind can also roll and flow upward behind such a building (see Figure 12.57).

By watching the behavior of vegetation, birds in flight, flags flying, and the like near a building, it is possible to observe such wind flow patterns in real life. The use of scale-model buildings in wind tunnels makes it possible to reveal such wind flows by the use of flow visualization techniques (e.g. smoke). Wool or cotton tufts attached to the model building and ground surfaces are also used sometimes for flow visualization.

Wind velocities vary widely in different parts of the world. For example, in England and Wales, large cities and towns rarely experience wind gusts that exceed 110 km/h (stagnation pressure $p = 560$ Pa) (70 mph). Coastal areas can have gusts of up to 145 km/h (stagnation pressure $p = 970$ Pa) (90 mph). Some areas in the north of England and Scotland can be subjected to gusts that exceed 160 km/h ($p = 1185$ Pa) (100 mph.)

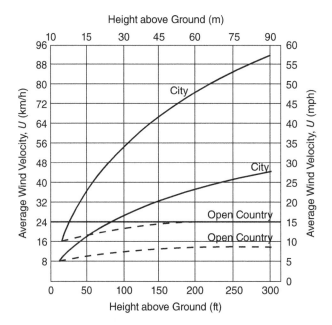

Figure 12.55 Typical wind velocity profiles in city and open country regions with velocities of 8 and 16 km/h referred to a height of 10 m [137].

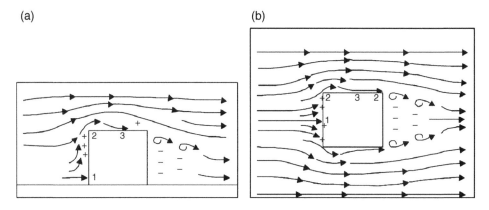

Figure 12.56 Wind flow around a building. (a) section view and (b) plan view. (+) = Positive gauge pressure; (−) = negative gauge pressure. Flow normally separates at the sharp upstream corners of the building. Reattachment in some cases occurs before the wake is formed. (1 = stagnation point, 2 = separation points, and 3 = reattachment points) [137].

12.12.1 Wind Excitation of Buildings

The wind excitation forces on building structures may be divided into two main categories: (i) fluctuating forces in the wind direction on the structures caused by the turbulent pressure fluctuations that exist in the approaching wind flow and (ii) fluctuating forces caused by turbulent eddies shed in the wake of the structure. These eddies (vortices) shed in the structure's wake mostly result in fluctuating forces in the cross-wind direction and to a lesser extent in the torsional direction. Wind excitation forces and the building vibrations they cause may be further subdivided and are now described in more detail in the following sections in this chapter [138–147].

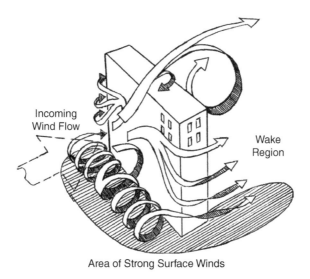

Figure 12.57 Schematic diagram of airflow patterns around a bluff-body or building [137].

As discussed already, the response of a building structure to wind depends not only on the wind characteristics but on the structural parameters as well. Very stiff structures deflect little in the wind, and their response can be treated as simple random forced response problems. However, flexible structures such as road signs, tall bridges, tall buildings, communication towers, and the like can respond appreciably and their movement can change the flow. The interaction of the structural motion and flow is known as an aeroelastic phenomenon. The calculation of the aeroelastic response of building structures is much more complicated than that of a pure forced response problem. Although many attempts have been made to calculate the response of rigid and flexible building structures to wind, many unknowns and variables in the wind flow, speed and direction with time, height above the ground, interactions with other structures, and complicated building geometries make calculations difficult [139–144, 146, 147]. In most cases, scale models need to be tested in wind tunnels to obtain reliable results. Figure 12.58 shows a schematic of wind effects and building excitation effects.

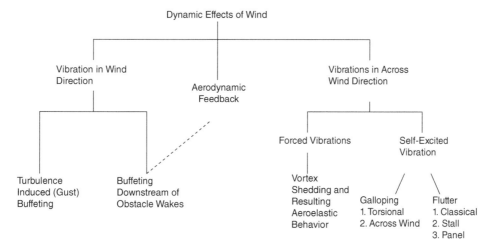

Figure 12.58 Classification of dynamic effects from wind [137].

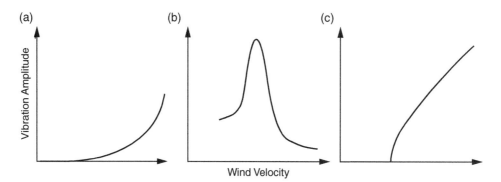

Figure 12.59 Main types of wind-induced oscillations: (a) vibration due to turbulence, (b) vibration due to vortex shedding, and (c) self-excited vibration due to aerodynamic instability [137].

12.12.2 Structural Vibration Response of Buildings and Towers

Figure 12.59 presents a schematic that indicates the vibration amplitude responses of a building structure to the different types of excitation as classified in Figure 12.58.

a) Types of Building Vibrations

As already discussed in Section 12.12.1 of this chapter, fluid flow and wind forces are responsible for a variety of vibratory mechanisms in structures. In the case of inline buffeting, (Figure 12.59a), the vibration amplitudes generated increase nonlinearly and very rapidly with flow speed. This is because motion can be regarded as forced motion, and the turbulent flow forces involved are proportional to flow velocity squared. It is important to know the fundamental natural frequency of a structure if it is in forced motion since below the natural frequency the response is mostly stiffness-controlled, at or near to the natural frequency it is damping-controlled, and above the natural frequency it is mass (or inertia) controlled. In cross-wind vortex shedding motion (see Figure 12.59b), again a knowledge of the structural fundamental natural frequency is needed since by careful building design, it may be possible to modify this frequency and change it so that it only becomes a problem at low wind speed at which excitation forces are minimal.

In cases of aerodynamic instability (galloping or flutter) such instabilities only begin at a critical wind speed (see Figure 12.59c), and again the natural frequencies of the structure are involved in the vibration as described before in this chapter.

b) Natural Frequencies of Building Structures

The natural frequencies of a building can be calculated from knowledge of its stiffness and mass distributions with a numerical program such as the FEM. Figure 12.60 shows a schematic distribution of the mass and stiffness distributions of a 210-m telecommunications tower [144]. The first three natural frequencies and mode shapes of vibration of the tower are shown in Figure 12.60. These are seen to be very low, the first two being less than 1 Hz. Often vibration in the fundamental natural frequency is dominant and most important. Although FEM programs can be used to calculate the fundamental natural frequency, it has been found that the calculation is no more accurate than using an empirical calculation based on the height of the building alone and use of an approximate formula such as Eq. (12.84) [144, 148]

$$f_e = 46/h, \tag{12.84}$$

where h is the building height in metres.

The reason that the fundamental natural frequency is dominated by the building height appears to be that as the buildings grow taller they become more massive, and then stiffer supports must correspondingly be used to support the building's weight. Figure 12.61 shows measured experimental data for building height

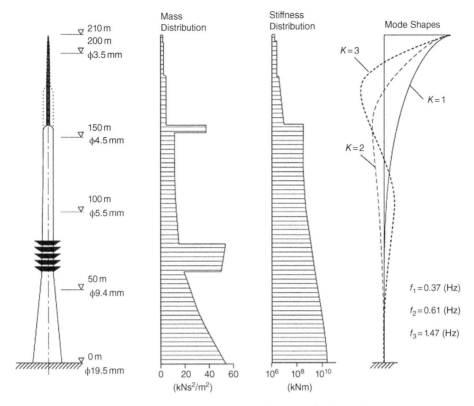

Figure 12.60 Telecommunications tower with mass distribution, stiffness distribution, and natural modes of vibration. (*Source:* From Bachmann et al., Ref. [144], reprinted with permission.)

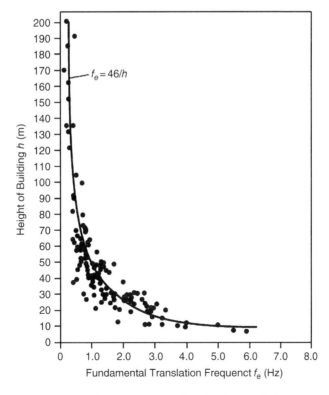

Figure 12.61 Fundamental frequency f_e of tall buildings. (*Source:* From Bachmann et al., Ref. [144]. Reprinted with permission.)

plotted against fundamental natural frequencies of a large number of tall buildings. Equation (12.84) is also plotted in Figure 12.61. It is seen that this equation is a good fit to most of the experimental data. The fundamental natural frequency of a building is of considerable importance, but so are several other building properties. Since there is a continuing trend to construct building structures of increasing height, wind-induced vibration of tall structures is a topic of increasing concern for structural engineers.

The three most important properties of structures that are relevant to wind-induced vibration are: (i) shape, (ii) stiffness (or flexibility), (ii) fundamental natural frequency, and (iv) damping [144].

12.12.3 Methods of Building Structure Vibration Reduction and Control

When structures are excited into motion, the forces opposing the motion are due to inertia, stiffness, and damping. It is impractical to increase the mass of the building. Increasing the stiffness of a building can, in principle, reduce its deflection to wind forces, although often there is a limit to the amount of stiffness increase that can be achieved. However, useful increases in stiffness can be achieved relatively easily in the case of masts, antennas in some cases by the use of guy wires and cables. The dynamics of guyed masts are complicated because tensioning of the guy cables is applied and the vibrations of the guy–mast system tend to be nonlinear in character. The guy cables themselves normally vibrate, resulting in some useful damping [144]. Various types of structural damping are illustrated in Figure 12.62.

For wake buffeting, across-wind vortex-induced vibration, galloping, and flutter, damping forces become dominant in controlling vibration. Passive damping can be applied successfully in such cases. Different passive damping elements have been used in practice, and several new designs have been proposed. The World Trade Center in New York had about 10 000 passive "friction" dampers mounted in the truss structures (see Figure 12.63). Tuned vibration dampers (TMD) are also now used in many tall buildings. The first tall building to have such a device was the 280-m high Citicorp Center in New York City [144]. This employed a 270 000-kg concrete block "floated" on a hydrostatic support system. Springs are attached to the mass and tuned to give the TMD system a natural frequency the same as the fundamental vibration frequency of the building. In addition, hydraulic dampers are applied to dissipate the vibration energy. Stops must be provided to prevent excessive damaging motion of the floating mass during intense storms. A problem is that electrical power is needed to provide hydrostatic pressure for the oil suspension system. If the electrical power fails during a storm, then the TMD system becomes inoperative.

The principle of operation of the tuned mass damper is that as the building vibrates at its natural frequency, the tuned damper will also vibrate at the same frequency, but out of phase and thus applying forces opposing the motion of the building.

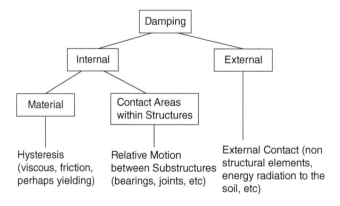

Figure 12.62 Various types of damping [137].

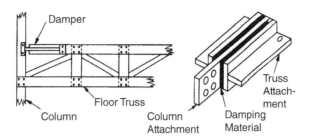

Figure 12.63 Friction dampers used in the load-bearing structure of the now-destroyed World Trade Center towers (New York City) [137].

More recently, tall buildings have used TMDs made of large masses suspended by cables, rather like pendulums. The 500-m building in Taiwan opened in November 2004 has a 500 000-kg mass suspended on four cables near the top of the building between the 86th and 89th floors (see Figure 12.64). As the building vibrates, the mass swings in an opposite direction, and pneumatic dampers are used to dissipate vibration energy. Excessive damaging vibration is prevented by the dampers. The system is mainly designed to reduce wind-induced vibrations by 35%, but it is said to also suppress earthquake vibrations. Since the building is situated only 1 km from a fault line, earthquake damage is of concern. During construction in 2003, a 6.8-scale earthquake was experienced. The building survived the earthquake and the tuned mass system is said to have worked satisfactorily, although a construction crane fell from the 50th floor during the earthquake killing five construction workers. Another example of a tuned-pendulum mass damper is shown in Figure 12.64b. This 3.5-m diameter steel sphere is installed near the top of an 83-m building in Santiago, Chile, and designed to reduce the amplitude of vibration during an earthquake. The 150 000-kg mass is suspended on 12 independent chains with two dampers to dissipate vibration energy.

(a)   (b)

Figure 12.64 Photographs of tuned mass damper (TMD) balls; (a) in Skyscraper Taipei 101 in Taiwan [137]; (b) in CDT building in Chile. (*Source:* photo courtesy of J. Sommerhoff.)

Other types of damping systems have been proposed and, in some cases, implemented. Some such dampers rely on liquid motion inside rigid containers to absorb and dissipate the vibration energy. For example, nutation dampers have been investigated by Modi and Welf [149]. Tuned liquid dampers (TLD) and tuned liquid column dampers (TLCD) have been investigated for use in buildings by Sakai et al. [150] and Xu et al. [151]. These dampers have been used successfully in spacecraft satellites and marine vessels by Amieux and Dureigne [152]. TLD and TLCD systems have some advantages over TMDs, including lower costs, less maintenance, and easier construction and handling [151].

12.12.4 Human Response to Vibration and Acceptability Criteria

In most cases, building vibration caused by wind is insufficient to cause structural damage, except in some cases to the superstructure or cladding (see Figure 12.65). The vibration is, however, often unpleasant and worrying for occupants. The swaying of tall buildings particularly at the tops can be disconcerting for occupants to say the least. Most studies on acceptability of vibration have been for frequencies higher than those normally encountered in tall buildings. Some studies have been conducted with very low frequency vibration using vibration simulators. Other results have been obtained from extrapolations from high frequency to low frequency to arrive at criteria for acceptability of vibration at very low frequencies less than 1 Hz. Most subjects report that they cannot sense vibration with acceleration less than about 0.01 g. If the vibration has an acceleration exceeding about 0.1 g, most subjects find the vibration is becoming intolerable. Table 12.6 lists some human acceptability criteria. Additional criteria for human comfort and annoyance for vibration in buildings are discussed in Chapter 6.

Another source of vibration in buildings is earthquakes. Protection of buildings from seismic-induced vibration is extremely important and this subject has been included in several national and IBCs (e.g. see Chapter 16 of Ref. [110]). Evidently, concerns with building response to earthquakes depend upon the use of the building, the geographical location of the building and the geological characteristics of the soil.

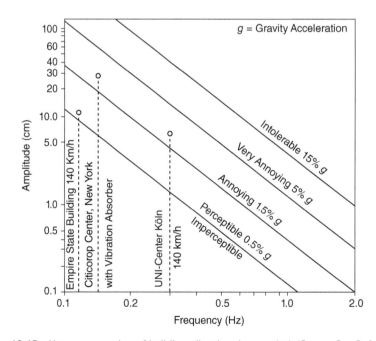

Figure 12.65 Human perception of building vibration due to wind. (*Source:* See Ref. [153].)

Table 12.6 Human acceleration acceptability criteria.

Perception	Acceleration Limits
Imperceptible	$a < 0.005\,g$
Perceptible	$0.005\,g < a < 0.015\,g$
Annoying	$0.015\,g < a < 0.05\,g$
Very annoying	$0.05\,g < a < 0.15\,g$
Intolerable	$a > 0.15\,g$

Source: From Refs. [139, 144, 153].

Building codes to protect buildings from vibrations caused by seismic excitation have been developed in earthquake-prone regions, such as Chile, Greece, and California in the U.S. These codes are mostly aimed to preserve the structural integrity of buildings and prevent their collapse during major earthquakes rather than avoid unpleasantness to occupants. Seismic design of buildings and design approaches of seismic isolators are beyond the scope of this book, but a thorough overview of these approaches is provided in Ref. [154] with support of extensive references.

References

1 Cremer, L., Heckl, M., and Petersson, B.A.T. (2010). *Structure-Borne Sound: Structural Vibrations and Sound Radiation at Audio Frequencies*. Berlin: Springer.

2 Fahy, F.J. and Gardonio, P. (2007). *Sound and Structural Vibration: Radiation, Transmission and Response*, 2e. Oxford: Academic Press.

3 Harris, D.A. (1997). *Noise Control Manual for Residential Buildings*. Boston: Mc Graw-Hill.

4 Hassan, O.A.B. (2009). *Building Acoustics and Vibration: Theory and Practice*. Singapore: World Scientific Publishing Co.

5 Hopkins, C. (2007). *Sound Insulation*. Amsterdam: Elsevier.

6 Beranek, L.L. and Vér, I.S. (1992). *Noise and Vibration Control Engineering*. New York: Wiley.

7 Asselineau, M. (2015). *Building Acoustics*. Boca Raton, FL: CRC Press.

8 Norton, M.P. and Karczub, D.G. (2003). *Fundamentals of Noise and Vibration Analysis for Engineers*, 2e. Cambridge: Cambridge University Press.

9 Beranek, L.L. (1971). *Noise and Vibration Control*. New York: McGraw-Hill.

10 Barron, R.F. (2003). *Industrial Noise Control and Acoustics*. New York: Marcel Dekker.

11 Cremer, L. (1942). *Acoust. Z.* 7: 81. See Reprint and English summary (1997) Theory of the sound attenuation of thin walls with oblique incidence. In: Architectural Acoustics. Benmarch Papers in Acoustics, Vol. 10 (ed. T.D. Northwood), 389. Stroudsburg, PA: Dowden, Hutchinson & Ross.

12 Constable, J.E.R. (1934). Acoustical insulation afforded by double partitions constructed from dissimilar components. *Philos. Mag.* 26 (174): 253–259.

13 Kimball, A.L. (1936). Theory of transmission of plane sound waves through multiple partitions. *J. Acoust. Soc. Am.* 7 (3): 222–224.

14 Beranek, L.L. and Work, G.A. (1949). Sound transmission through multiple structures containing flexible blankets. *J. Acoust. Soc. Am.* 21 (4): 419–428.

15 Utley, W.A. (1967) The transmission of sound through double and triple panels, PhD thesis. University of Liverpool, Liverpool, England.

16 London, A. (1949). Transmission of reverberant sound through single walls, research paper RP1908. *J. Res. Natl. Bur. Stand.* 42: 605–615.

17 London, A. (1950). Transmission of reverberant sound through double walls. *J. Acoust. Soc. Am.* 22 (2): 270–279.

18 Beranek, L.L. (1960). The transmission and radiation of acoustic waves by solid structures. In: *Noise Reduction* (ed. L.L. Beranek) chapter 13. New York: McGraw-Hill.

19 Sharp, B.H. (1978). A study of techniques to increase the sound insulation of building elements, U.S. Dept. of housing and urban development, Washington DC NTIS PB 222 829/4; and prediction methods for the sound transmission of building elements. *Noise Control Eng. J.* 11 (2): 53–63.

20 Warnock, A.C.C. (2007). Sound insulation – airborne and impact. In: *Handbook of Noise and Vibration Control* (ed. M.J. Crocker), 1257–1266. New York: Wiley.

21 Lyon, R.H. (1975). *Statistical Energy Analysis of Dynamical Systems: Theory and Applications*. MIT Press.

22 Lyon, R.H. and Eichler, E. (1964). Random vibration of connected structures. *J. Acoust. Soc. Am.* 36: 1344–1354.

23 Smith, P.W. and Lyon, R.H. (1965) Sound and Structural Vibration. NASA CR-160.

24 Manning, J.E., Lyon, R.H., and Scharton, T.D. (1966) Transmission of Sound and Vibration to a Shroud-Enclosed Spacecraft. Report No. 143. Bolt, Beranek and Newman.

25 Lyon, R.H. (1967) Random Noise and Vibration in Space Vehicles, Monograph SVM-1. Shock and Vibration Information Center. Department of Defence.

26 Ungar, E.E. (1966) Fundamentals of Statistical Energy Analysis of Vibratory Systems, AFFDL-TR-66-52.

27 Crocker, M.J. (1969) The response of structures to acoustic excitation and the transmission of sound and vibration. PhD thesis. Liverpool University, Liverpool.

28 Crocker, M.J. and Price, A.J. (1969). Sound transmission using statistical energy analysis. *J. Sound Vib.* 9 (3): 469–486.

29 Price, A.J. and Crocker, M.J. (1970). Sound transmission through double panels using statistical energy analysis. *J. Acoust. Soc. Am.* 47 (3A): 683–693.

30 Crocker, M.J., Battacharya, M.C., and Price, A.J. (1971). Sound and vibration transmission through panels and tie beams using statistical energy analysis. *J. Eng. Ind. Trans. ASME* 93 (3): 775–782.

31 Lyon, R.H. and De Jong, R.G. (1995). *Theory and Applications of Statistical Energy Analysis*. Newton, MA: Butterworth-Heinemann.

32 Norton, M.P. (1989). *Fundamentals of Noise and Vibration Analyses for Engineers*. Cambridge University Press.

33 Fahy, F. and Walker, J. (2004). *Advanced Applications in Acoustics, Noise and Vibration*. London and New York: Spon Press.

34 Nilsson, C. and Liu, B. (2017). *Vibro-Acoustics*, vol. 3. Heidelberg: Springer.

35 Keane, A.J. and Price, W.G. (1994). *Statistical Energy Analysis. (An Overview with Applications in Structural Dynamics)*. Cambridge University Press.

36 Manik, D.N. (2017). *Vibro-Acoustics: Fundamentals and Applications*. Boca Raton: CRC Press.

37 Hambric, S.A., Sung, S.H., and Nefke, D.J. (2016). *Engineering Vibroacoustic Analysis: Methods and Applications*. Chichester: Wiley.

38 Le Bot, A. (2015). *Foundation of Statistical Energy Analysis in Vibroacoustics*. Oxford University Press.

39 Maidanik, G. (1962). Response of ribbed panels to reverberant acoustic fields. *J. Acoust. Soc. Am.* 34 (6): 809–826.

40 Lyon, R.H. and Maidanik, G. (1962). Power flow between linearly coupled oscillators. *J. Acoust. Soc. Am.* 34 (5): 623–639.

41 Lyon, R.H. and Scharton, T.D. (1965). Vibrational energy transmission in a three-element structure. *J. Acoust. Soc. Am.* 38 (2): 253–261.

42 Newland, D.E. (1965). Energy sharing in the random vibration of nonlinearly coupled modes. *J. Inst. Math. Appl.* 1 (3): 199–207.

43 Lyon, R.H. (1966–67) Program for Advanced Study – Aerospace Noise and Vibration. Los Angeles, Chapter 9, Typed Lecture Notes from Bolt, Beranek and Newman, Inc.

44 Newland, D.E. (1968). Power flow between a class of coupled oscillators. *J. Acoust. Soc. Am.* 43 (3): 553–559.

45 Lyon, R.H. (1963). Noise reduction of rectangular enclosures with one flexible wall. *J. Acoust. Soc. Am.* 35 (11): 1791–1797.

46 Scharton, T.D. (1965) Random vibration of coupled oscillators and coupled structures. Doctoral thesis. Massachusetts Institute of Technology.

47 Zhou, R. and Crocker, M.J. (2010). Sound transmission loss of foam-filled honeycomb sandwich panels using statistical energy analysis and theoretical and measured dynamic properties. *J. Sound Vib.* 329 (6): 673–686.

48 Langley, R.S. and Cotoni, V. (2004). Response variance prediction in the statistical energy analysis of built up systems. *J. Acoust. Soc. Am.* 115 (2): 706–718.

49 Langley, R.S. and Cotoni, V. (2004). Response variance prediction for uncertain vibro-acoustic systems using a hybrid deterministic-statistical method. *J. Acoust. Soc. Am.* 122 (6): 3445–3463.

50 White, P.H. and Powell, A. (1965). Transmission of random sound and vibration through a rectangular double wall. *J. Acoust. Soc. Am.* 40 (4): 821–832.

51 Nakayama, K., Prediction of the Sound Pressure Level Inside a Vibrating Steel Box Using SEA, M. S. Dissertation, Purdue University, August 1979.

52 Cole, V., Prediction of the Noise Reduction of a Rectangular Steel Box Using Statistical Energy Analysis, MSME Thesis, Purdue University, December 1980.

53 Cole, V., Crocker, M.J., and Raju, P.K. (1983). Theoretical and experimental studies of the noise reduction of an idealized cabin enclosure. *Noise Control Eng. J.* 20 (3): 122–132.

54 Cole, V., Crocker, M.J. and Cheng, M.K. (1984) The use of statistical energy analysis to predict the noise reduction of idealized cabin enclosures. SECTAM XII Proceeding, Auburn University, 63–68.

55 Crocker, M.J. (1994) A systems approach to the transmission of sound and vibration through structures. Noise-Con'94 Proceedings, 525–540.

56 Patil, A.R., Crocker, M.J. (1999) Sound transmission onto an enclosure with an aperture using statistical energy analysis. Inter-Noise 99 Proceedings, 199: 529–534.

57 Crocker, M.J., Patil, A.R. and Arenas, J.P. (2002) Theoretical and experimental studies of the acoustical design of vehicle cabs – a review of truck noise sources and cab design using statistical energy analysis. IUTAM Symposium on designing for quietness, Bangalore, India, Kluwer Academic Publishers, Dordrecht, 47–65.

58 Mulholland, K.A. and Parbrook, H.D. (1967). Transmission of sound through apertures of negligible thickness. *J. Sound Vib.* 5 (3): 499–508.

59 Ingerslev, F. and Nielsen, A.K. (1944) On the transmission of sound through small apertures. Ingeniorvidenska-beliga Skrifter, 5–31.

60 Oliazadeh, P., Farshidianfar and Crocker, M.J. (2019). Study of sound transmission through single- and double-walled plates with absorbing material: Experimental and analytical investigation. *Appl. Acoust.* 145: 7–24.

61 Sabine, H.J., Lacher, M.B., Flynn, D.R., and Quindry, T.L. (1975) Acoustical and thermal performance of exterior residential walls, doors, and windows. Building Science Series 77, National Bureau of Standards.

62 Gomperts, M.C. (1964). The sound insulation of circular and slit-shaped apertures. *Acustica* 14 (1): 1–16.

63 Gomperts, M.C. (1965). The influence of viscosity on sound transmission through small circular apertures in walls of finite thickness. *Acustica* 15 (4): 191–198.

64 North American Insulation Manufacturers Association (NAIMA), *Sound Control for Commercial and Residential Buildings*, PUB # BI405, 12/97.

65 Gerretsen, E. (1979). Calculation of sound transmission between dwellings by partitions and flanking structures. *Appl. Acoust.* 12 (6): 413–433.

66 Zeitler, B., Quirt, D., Hoeller, C. et al. (2016) Guide to Calculating Airborne Sound Transmission in Buildings, 2. National Research Council Canada (NRCC).

67 ISO 10140-2 (2010) *Acoustics – Laboratory measurement of sound insulation of building elements – Part 2: Measurement of airborne sound insulation* Geneva: International Standards Organization.

68 ASTM E90-09 (2009) Standard Test Method for Laboratory Measurement of Airborne Sound Transmission Loss of Building Partitions and Elements.

69 ASTM E1050-98 (1998) Standard Test Method for Impedance and Absorption of Acoustical Materials Using a Tube, Two Microphones, and a Digital Frequency Analysis System.

70 ISO 10534-2 (1998) *Acoustics – Determination of sound absorption coefficient and impedance in impedance tubes – Part 2: Transfer-function method*. Geneva: International Standards Organization.

71 Song, B.H. and Bolton, J.S. (2000). A transfer-matrix approach for estimating the characteristic impedance and wave numbers of limp and rigid porous materials. *J. Acoust. Soc. Am.* 107: 1131–1152.

72 ASTM E2611-09 (2009) Standard Test Method for Measurement of Normal Incidence Sound Transmission of Acoustical Materials Based on the Transfer Matrix Method.

73 Crocker, M.J., Raju, P.K., and Forssen, B. (1981). Measurement of transmission loss of panels by direct determination of acoustic intensity. *Noise Control Eng.* 17 (1): 6–11.

74 ASTM E2249-02 (2002) Standard Test Method for Laboratory Measurement of Airborne Transmission Loss of Building Partitions and Elements Using Sound Intensity (reapproved in 2016).

75 ISO 15186-1 (2000) *Acoustics – Measurement of sound insulation in buildings and of building elements using sound intensity – Part 1: Laboratory measurements*. Geneva: International Standards Organization.

76 ISO 15186-2 (2003) *Acoustics – Measurement of sound insulation in buildings and of building elements using sound intensity – Part 2: Field measurements*. Geneva: International Standards Organization.

77 ISO 15186-3 (2002) *Acoustics – Measurement of sound insulation in buildings and of building elements using sound intensity – Part 3: Laboratory measurements at low frequencies*. Geneva: International Standards Organization.

78 ASTM E336-16a (2016) Standard Test Method for Measurement of Airborne Sound Attenuation between Rooms in Buildings.

79 ISO 16283-1:2014 (2014) *Acoustics – Field measurement of sound insulation in buildings and of building elements – Part 1: Airborne sound insulation*. Geneva: International Standards Organization.

80 Crocker, M.J. and Price, A.J. (1975). *Noise and Noise Control*, vol. I. Cleveland, OH: CRC Press; M.J. Crocker and F.M. Kessler (1982) *Noise and Noise Control*. Vol. II, Boca Raton, FL: CRC Press.

81 Parkin, P.H., Purkis, H.J., and Scholes, W.E. (1960) *Field measurements of sound insulation between dwellings*, National Building Studies, Research Paper No. 33, her Majesty's Stationary Office, London.

82 Brandt, O. and Dalen, I. (1952). *Is the Sound Insulation in Our Dwellings Sufficient?* Sweden: Byggmastaren.

83 Bitter, C. and van Wieren, P. (1955) *Sound nuisance and sound insulation in blocks of dwellings*, Report No. 24, Research Institute for Public Health Engineering, T.N.O., The Hague, Netherlands.

84 Josse, M. (1969) *Etude sociologioque de la satisfaction des occupants de locaux conformes aux regles qui sont supposes garantir un comfort acoustique sufficant*. Paris: Centre Scientifique et Technique du Batiment.

85 ISO 140-1 (1998) *Acoustics – Measurement of Sound Insulation in Buildings and of Building Elements: Part 1: Requirements for Laboratory Test Facilities with Suppressed Flanking Transmission*. Geneva: International Standards Organization.

86 ISO 140-5 (1998) *Acoustics – Measurement of Sound Insulation in Buildings and of Building Elements – Field Measurement of Airborne Sound Insulation of Façade Elements and Facades*. Geneva: International Standards Organization.

87 ASTM E413-16 (2016) Classification for Rating Sound Insulation.

88 Tocci, G.C. (2007). Ratings and descriptors for the built acoustical environment. In: *Handbook of Noise and Vibration Control* (ed. M.J. Crocker), 1267–1282. New York: Wiley.

89 Warnock, A.C.C. (1998) Controlling Sound Transmission through Concrete Block Walls, Construction Technology Update 13, IRC, NRCC.

90 ISO 717-1 (2013) *Acoustics – Rating of sound insulation in buildings and of building elements – Part 1: Airborne sound insulation*. Geneva: International Standards Organization.

91 Metzen, H.A. (2007). ISO ratings and descriptors for the built acoustical environment. In: *Handbook of Noise and Vibration Control* (ed. M.J. Crocker), 1283–1296. New York: Wiley.

92 Rasmussen, B. (2007). Sound insulation of residential housing – building codes and classification schemes in Europe. In: *Handbook of Noise and Vibration Control* (ed. M.J. Crocker), 1354–1366. New York: Wiley.

93 ASTM E1332-16, Standard Classification for Rating Outdoor-Indoor Sound Attenuation, 2016.

94 Reiher, H. (1932). Uber den Schallschutz durch Baukonstruktionstelle. *Beith. Ges. Ing.* 2 (11): 2.

95 ISO 10140-3 (2010) *Acoustics – Laboratory measurement of sound insulation of building elements – Part 3: Measurement of impact sound insulation.* Geneva: International Standards Organization.

96 ISO 16283-2 (2018) *Acoustics – Field measurement of sound insulation in buildings and of building elements – Part 2: Impact sound insulation.* Geneva: International Standards Organization.

97 ASTM E492-09 (2009) Standard Test Method for Laboratory Measurement of Impact Sound Transmission Through Floor-Ceiling Assemblies Using the Tapping Machine (reapproved 2016).

98 ASTM E1007-16 (2016) Standard Test Method for Field Measurement of Tapping Machine Impact Sound Transmission through Floor-Ceiling Assemblies and Associated Support Structures.

99 ISO 10140-5 (2010) *Acoustics – Laboratory measurement of sound insulation of building elements – Part 5: Requirements for test facilities and equipment.* Geneva: International Standards Organization.

100 Tachibana, H., Tanaka, H., Yasuoka, M., and Kimura, S. (1998) Development of new heavy and soft impact source for the assessment of floor impact sound insulation in buildings, in Proceedings of Internoise 98, Christchurch, New Zealand.

101 Jeong, J.H. (2015) Evaluation method of rubber ball impact sound, in Proceedings of EuroNoise, 1901–1904.

102 ASTM E989-06 (2006) Standard Classification for Determination of Impact Insulation Class (IIC) (reapproved 2012).

103 ISO 717-2 (2013) *Acoustics – Rating of sound insulation in buildings and of building elements – Part 2: Impact sound insulation.* Geneva: International Standards Organization.

104 ISO 140-7 (1998) *Acoustics – Measurement of Sound Insulation in Buildings and of Building Elements – Part 7: Field Measurements of Impact Sound Insulation of Building Elements.* Geneva: International Standards Organization.

105 Hedeen, R.A. (1980) NIOSH Compendium of Materials for Noise Control, DHEW Publication No. 80-116. U.S. Department of Health, Education and Welfare, 4676 Columbia Parkway, Cincinnati, OH. 45226.

106 Dupree, R.B. (1981) Catalog of STC and IIC Ratings for Wall and Floor/Ceiling Assemblies, Office of Noise Control, California Department of Health Services, Berkeley, CA.

107 The American Institute of Architects (2007). *Architectural Graphic Standards*, 11e. Hoboken, NJ: Wiley.

108 Berendt, R.D. and Winzer, G.E. (1964) Sound Insulation of Wall, Floor, and Door Constructions. U.S. Department of Commerce, National Bureau of Standards, Monograph 77 November.

109 Compendium of Materials for Noise Control, HEW Publication No. 75-165 (1975) NIOSH, Illinois Institute of Technology.

110 Berendt, R.D., Winzer, G.E., and Burroughs, C.B. (1967) *A guide to airborne, impact, and structure borne noise-control in multifamily dwellings.* FT/TS-24, prepared for the Federal Housing Administration, U.S. Dept. of Housing and Urban Development, Washington.

111 Warnock, A.C.C. and Birta, J.A. (1998) IR-761 Gypsum Board Walls: Transmission Loss Data. NRCC.

112 Nightingale, T.R.T., Halliwell, R.E., Quirt, J.D., and Birta, J.A. (2002) IR-832 Sound Insulation of Load Bearing Shear Resistant Wood and Steel Stud Walls, NRCC.

113 Warnock, A.C.C. and Birta, J.A. (2000) IR-811 Detailed Report for Consortium on Fire Resistance and Sound Insulation of Floors: Sound Transmission and Impact Insulation Data in 1/3 Octave Bands. NRCC.

114 Warnock, A.C.C. (2005) RR-169 Summary Report for Consortium on Fire Resistance and Sound Insulation of Floors: Sound Transmission and Impact Insulation Data. NRCC.

115 Warnock, A.C.C. (1990) IR-586 Sound Transmission Loss Measurements through 190 mm and 140 mm Blocks with Added Drywall and Through Cavity Block Walls. NRCC.

116 Harris, C.M. (1998). *The Handbook of Acoustical Measurements and Noise Control*, 3e. Melville, NY: Acoustical Society of America.

117 Bies, D.A. and Hansen, C.H. (2009). *Engineering Noise Control: Theory and Practice*, 4e. Abingdon: Spon Press.

118 INSUL, Sound insulation prediction software. Auckland, New Zealand: Marshall Day Acoustics.

119 Software application soundPATHS. www.nrc-cnrc.gc.ca/eng/solutions/advisory/soundpaths/index.html (accessed 20 February 2020).

120 Field, C. and Fricke, F. (2007). Noise in commercial and public buildings and offices – prediction and control. In: *Handbook of Noise and Vibration Control* (ed. M.J. Crocker), 1367–1374. New York: Wiley.

121 Loney, W. (1971). Effect of cavity absorption on the sound transmission loss of steel stud gypsum wallboard partitions. *J. Acoust. Soc. Am.* 49 (2): 385–390.

122 Novak, R.A. (1992). Sound insulation of lightweight double walls. *Appl. Acoust.* 37: 281–303.

123 Warnock, A.C.C. (1991). Sound transmission through concrete blocks with attached drywall. *J. Acoust. Soc. Am.* 90: 1454–1463.

124 Warnock, A.C.C. (1992). Sound transmission through two kinds of porous concrete blocks with attached drywall. *J. Acoust. Soc. Am.* 92: 1452–1460.

125 Harris, C.M. (ed.) (1994). *Noise Control in Buildings*. McGraw Hill.

126 ISO 9052-1 (1989) *Acoustics – Determination of dynamic stiffness – Part 1: Materials used under floating floors in dwellings*. Geneva: International Standards Organization.

127 Kim, C., Hong, Y.-K., and Lee, J.-Y. (2017). Long-term dynamic stiffness of resilient materials in floating floor systems. *Constr. Build. Mater.* 133: 27–38.

128 Arenas, J.P., Castaño, J.L., Troncoso, L., and Auad, M.L. (2019). Thermoplastic polyurethane/laponite nanocomposite for reducing impact sound in a floating floor. *Appl. Acoust.* 155: 401–406.

129 Fry, A. (ed.) (1988). *Noise Control in Building Services*. Oxford: Pergamon Press.

130 Quirt, J.D. (1982). Sound transmission through windows. I. Single and double glazing. *J. Acoust. Soc. Am.* 72 (3): 834–844.

131 Quirt, J.D. (1983). Sound transmission through windows. II. Double and triple glazing. *J. Acoust. Soc. Am.* 74 (2): 534–542.

132 2015 International Building Code, Chapter 1207 Sound Transmission, International Code Council, Washington, DC, 2015.

133 Tocci, G.C. (2007). Noise control in U.S. building codes. In: *Handbook of Noise and Vibration Control* (ed. M.J. Crocker), 1348–1353. New York: Wiley.

134 Federal Housing Administration (1974) TS-24 Guide to Airborne, Impact and Structure Borne Noise Control in Multifamily Dwellings. Washington, DC: U.S. Department of Housing and Urban Development (HUD).

135 Keith, R.H. (2007). Noise control for mechanical and ventilation systems. In: *Handbook of Noise and Vibration Control* (ed. M.J. Crocker), 1328–1347. New York: Wiley.

136 Noise and Vibration Control for Mechanical Equipment, Army Manual TM 5-805-4 (1983) Washington, DC: Headquarters Department of Army, The Air Force and Navy.

137 Crocker, M.J. (2007). Vibration response of structures to fluid flow and wind. In: *Handbook of Noise and Vibration Control* (ed. M.J. Crocker), 1375–1392. New York: Wiley.

138 Paz, M. and Leigh, W. (2004). *Structural Dynamics – Theory and Computation*, 2, 5e. Springer.

139 Simiu, E. and Scanlan, R. (1996). *Wind Effects on Structures*, 3e. New York: Wiley.

140 Liu, H. (1991). *Wind Engineering – A Handbook for Structural Engineers*. Englewood Cliffs, NJ: Prentice Hall.

141 Blevins, R.D. (1990). *Flow-Induced Vibration*. New York: Van Nostrand Reinhold.

142 Sachs, P. (1977). *Wind Forces in Engineering*, 2e. Oxford, UK: Pergamon.

143 Melaragno, M.G. (1982). *Wind in Architectural and Environmental Design*. New York: Van Nostrand Reinhold.

144 Bachmann, H., Ammar, W.J., Deisch, F., and Eisenmann, J. (1995). *Vibration Problems in Structures – Practical Guidelines*. Basel: Birkhäuser.

145 Blevins, R.D. (1995). Part I: Vibration of structures, induced by fluid flow. In: *Shock and Vibration Handbook*, 5e (eds. C.M. Harris and A.G. Piersol), 29.1–29.20. New York: McGraw-Hill.

146 Davenport, A.G. and Novak, M. (1995). Part II: Vibration of structures, induced by wind. In: *Shock and Vibration Handbook*, 4e (ed. C.M. Harris). New York: McGraw-Hill.

147 Kijewski, T., Hann, F., and Kareem, A. (2002). *Wind-induced vibrations*. In: *Encyclopedia of Vibration*, vol. III (eds. S. Braun, D.J. Ewins and S.S. Rao), 1578–1587. London: Academic.

148 Jeary, A.P. and Ellis, B.R. (1983). On predicting the response of tall buildings to wind excitation. *J. Wind Eng. Ind. Aerodyn.* 13: 173–182.

149 Modi, V. and Welt, F. (1988). Damping of wind induced oscillations though liquid sloshing. *J. Wind Eng. Ind. Aerodyn.* 30: 85–94.

150 Sakai, F., Takaeda, S., and Tamaki, T. (1989) Tuned liquid column dampers – new type devise for suppression of building vibrations. Proceedings of the International Conference on High Rise Buildings, Vol. 2, Nanjing, China, 926–931.

151 Xu, Y.L., Samali, B., and Kwok, K.C.S. (1990) Effect of passive damping devices on wind-induced response of tall buildings. *National Conference Publication – Institution of Engineers, Australia*, No. 90, pt 9, *Vibration and Noise-Measurement Prediction and Control*, 127–132. Barton, ACT, Australia: The Institution of Engineers, Australia.

152 Amieux, J. and Dureigne, M. (1972). Analytical design of optimal nutation damper. *J. Spacecraft* 9 (12): 934–936.

153 Chang, F.K. (1967). Wind and movement of tall buildings. *J. Struct. Div. ASCE* 37 (8): 70–72.

154 Kappos, A.J. and Sextos, A.G. (2007). Protection of buildings from earthquake-induced vibration. In: *Handbook of Noise and Vibration Control* (ed. M.J. Crocker), 1393–1403. New York: Wiley.

13

Design of Air-conditioning Systems for Noise and Vibration Control

13.1   Introduction

There is widespread use of heating ventilation and air-conditioning (HVAC) systems in a variety of small and large buildings. The main consideration facing the designer of such systems is to ensure that the plant provides the required amount of heated or cooled air throughout the building and does not create objectionable noise or vibration either in the areas served by the system or in adjacent areas. Because air must be supplied (and in many cases extracted) by a fan of some kind (either axial, centrifugal, or mixed-flow), it is inevitable that some noise and vibration are generated. It is becoming common practice to use systems with very high airflow velocities which introduce additional problems due to noise generated by the turbulent airflow, which is created. However, despite the many difficulties that may be encountered, the noise produced by the system can be controlled if the system is correctly sized and care is taken to ensure that all elements of the system are properly installed. Air-conditioning also makes it possible to use sealed windows, thus giving good sound isolation from most of the outdoor noise. This can be important with buildings situated close to airports, railroads, and highways.

Once a completed HVAC system has been installed in a building, it is often very difficult and expensive to correct noise and vibration problems. Thus, great care should be taken at the design stage of a system to select all of the equipment items carefully and to minimize all possible sources of noise and vibration. A successful design can be obtained only by careful cooperation between the architect, ventilation engineer, and acoustical consultant.

The primary considerations in selecting the mechanical equipment necessary for cooling, heating, and ventilating a building are related to (i) satisfying its intended use, and (ii) providing acceptable sound and vibration conditions in occupied spaces in the building. In critical cases such as conference rooms, auditoria, bedrooms in homes and hotels and lightweight buildings, the sound and vibration produced by the equipment must be minimized. In order to meet these mechanical and acoustical requirements, it is important to use the source-path-receiver concept discussed in Chapter 9 and throughout this book. The major chapter topics include

- Noise level criteria
- General features of HVAC systems
- Errors in system installations
- Fans and fan noise
- Space planning
- Mechanical room noise and vibration
- Ductwork and duct attenuation
- Sound attenuators (plenums and silencers)

Engineering Acoustics: Noise and Vibration Control, First Edition. Malcolm J. Crocker and Jorge P. Arenas.
© 2021 John Wiley & Sons Ltd. Published 2021 by John Wiley & Sons Ltd.

- Sound generation by flow
- Air terminal devices (grilles and diffusers)
- Duct breakout and breakin noise
- Sound radiation from mixing boxes and plenum walls
- Prediction of room sound pressure level from sound power level

This chapter concentrates on the acoustical aspects of HVAC systems, although these are intimately related to each part of the mechanical systems installed. Useful reviews on HVAC noise and vibration control can be found as chapters in several books [1–17]. The ASHRAE Handbooks on HVAC applications, updated periodically, contain valuable chapters on noise and vibration control [18–20].

13.2 Interior Noise Level Design Criteria

When specifying interior noise level objectives for different areas of a building, many factors have to be considered. In general, the ventilation system noise should be low enough so as not to interfere with speech or other communicative sounds. In determining the acceptability of the air-conditioning system noise, it is necessary not only to aim for a particular noise level or loudness but also to consider its relationship with any other noise which exists in the area. An attempt to achieve a good balance between the two is important. In particular, other sounds which may be important in deciding upon a particular noise level include:

1) Necessary sounds, such as speech, music, audible warning signals, etc. Here the HVAC system noise level should be kept low enough to ensure it does not mask such sounds.
2) Sounds which should not be heard, such as speech from adjacent areas. In this case, the ventilation system noise can be used to mask the unwanted sounds.
3) General ambient noise, both from outside and within the building. In this case, the ventilation system noise created should be kept just below the general ambient noise level.

Beranek was one of the first to suggest suitable noise criteria [21, 22]. Table 6.4 in Chapter 6 shows acceptable ranges of background noise levels produced by air-conditioning systems in various types of buildings, evaluated with the NCB curve approach suggested by Beranek [23, 24].

The lower NCB value is usually used only in situations where the general ambient level is very low and where good noise reduction techniques are employed. The upper end of the scale is used in buildings with a relatively high ambient noise level caused by other sources. It should be noted that these ranges are based on the assumption that the ventilation system noise is steady and has a broadband frequency spectrum. If this is not the case, then more stringent limits should be specified (i.e. for impulsive or pure tone sounds). The NCB curves can be used to evaluate the acceptability of the air-conditioning noise (see Figure 6.10 in Chapter 6) [23, 24].

13.3 General Features of a Ventilation System

In many countries, HVAC systems have come into widespread use in residential and commercial buildings (small houses, apartments, shops, stores, warehouses, hotels, and large office buildings, etc.). The energy sources for the systems include gas, oil, and electricity. In residential housing the heating or cooling medium can be air, water, steam, or refrigerant requiring ducting, piping or free delivery systems. After delivery, terminal devices are needed such as diffusers, registers, grilles, radiators, radiant panels, and fan-coil units. The only exceptions are hydroponic systems, which usually consist of underfloor heating elements. In some dry climates, humidification is needed. In humid climates, it is necessary to dehumidify the air, which is normally accomplished when the air is passed over an evaporating coil.

13.3.1 HVAC Systems in Residential Homes

In residential homes, the HVAC systems can be of the gas furnace type, split-system air-conditioner type (see Figure 13.1) or the split-system heat pump type (see Figure 13.2) [20]. However, manufactured homes now amount to about 10% of new single-family homes built in the USA. They are constructed in factories rather than on-site and normally have HVAC systems that are factory installed. A typical unit is shown in Figure 13.3 [20].

In the first residential type, air from air-conditioned spaces enters the equipment via a return air duct, usually through an air filter (Figure 13.1). The air-circulation blower is integrated with the furnace which provides heat during the winter. In the summer, when cooling is required, cooling occurs and the moisture is removed when the air passes over an evaporator coil. The refrigerant medium passes through tubes to an outdoor unit where heat is rejected by the condensing equipment.

In the second type (Figure 13.2), air from the air-conditioned space travels to the HVAC equipment through a return air duct. The circulating blower is an integral part of the heat pump system, which supplies heated air after the air passes through the indoor coil during the heating season. When cooling is required, heat and moisture are removed when the air passes through the evaporator coil. Refrigerant lines are used to connect the indoor coil to the outdoor condensing unit. Condensate from the indoor coil is removed via a drain and a trap.

The HVAC systems in manufactured homes (Figure 13.3) normally consist of forced air downflow units feeding main supply ducts that are built into the subfloor supplying floor registers situated throughout the homes [20]. A small number of homes in the far south and southwest of the USA use upflow units, which feed overhead units in the attic spaces [20]. Noise control measures may be needed in the units, which incorporate large forced air systems installed close to bedrooms. Such units may be installed in closets or alcoves usually in a hallway.

13.3.2 HVAC Systems in Large Buildings

The remainder of this chapter will deal mostly with HVAC systems used in large buildings. Although the principles of operation of the systems used in large buildings are similar to those used in residential homes, (Figures 13.1–13.3), the systems used are of a much larger physical scale and complexity. This is because the heating and/or cooling air requirements will normally be different in various rooms, zones and occupied and unoccupied areas of a building. Provision must be made for the different requirements throughout the building. It is sometimes even necessary to provide heat to one side of a building while it is in the shade and cooling to the other side of the building while it is in the direct sun. All air-conditioning and ventilation (HVAC) systems have to supply air to many areas of large buildings, and the major source of noise and vibration is usually associated with the supply fan or fans.

There are two main types of HVAC systems that are used in large multistory buildings such as hotels, office blocks, shopping malls, etc. Each type of system has advantages and disadvantages [12].

1) A centrally located system normally has the supply fan (or fans) situated in a location, which is distant from the areas to be served by long runs of ductwork. The longer is this ductwork and the greater volume it occupies in the building, the smaller is the remaining useful space [12].
2) A distributed system comprises self-contained "package" units, each comprised of a fan, heating/cooling coils, humidifier/dehumidifier, dampers, etc. Such systems are normally located as a separate item or items on each floor of a multistory building.

Figure 13.4 shows the main paths of noise and vibration for a typical centrally located HVAC system installed in a multistory building. Whenever possible, centrally located and distributed systems should be located in between toilets, storage rooms, stairs, and elevators to reduce noise and vibration reaching occupied building spaces [12]. Space planning is discussed further in Section 13.5.

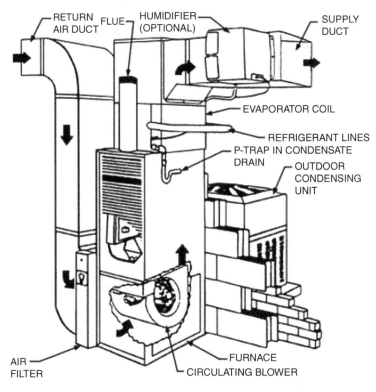

Figure 13.1 Typical residential installation of heating, cooling, humidifying, and air filtering system [20]. © ASHRAE, www.ashrae.org. 2015 ASHRAE Handbook-Guide & Data Book, Chap 1 Fig 1.

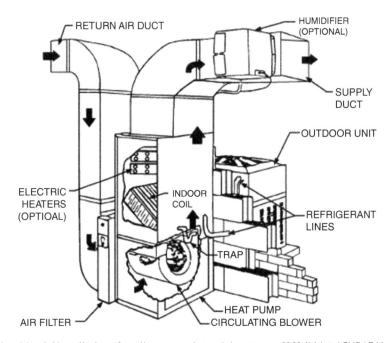

Figure 13.2 Typical residential installation of a split-system air-to-air heat pump [20]. ibid © ASHRAE Handbook, Chap 1, Fig 2.

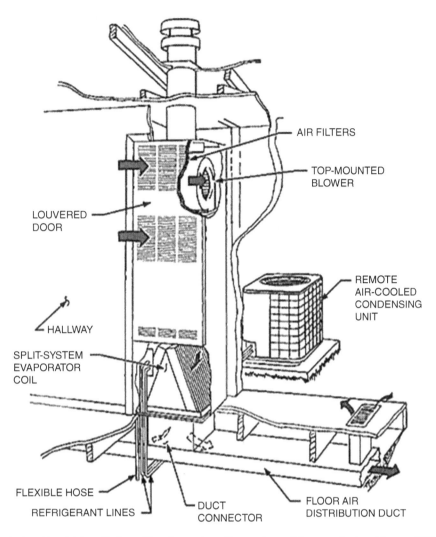

Figure 13.3 Typical residential installation of heating and cooling equipment for manufactured homes [20]. ibid © ASHRAE Handbook, Chap 1, Fig 4.

Initially in a large air-conditioning system, air from the fan is sent through a fan plenum chamber to attenuate some of the fan noise and, in some cases, to allow the air to be heated or chilled. This plenum may have several outlets of various cross-sectional areas each leading from a main supply duct to the rest of the building. As it passes through the ductwork, the air may encounter many radius bends, elbows, and right angled take-off junctions. These provide some attenuation of the noise traveling down the duct from the plenum. The attenuation depends upon the geometry of the bends and whether or not the duct is acoustically lined. In relatively high-velocity flow systems, however, more noise is often generated by the turbulent flow itself at the bends and take-offs than is attenuated by them.

Finally, before the air is sent into a room, it often passes through a mixing box or terminal box to attenuate generated noise and to be mixed with cold or hot air. Although such boxes provide some attenuation, they also radiate noise into the space around them, which may happen to be close to or above critical areas of the building. From here the air is sent into the room via outlet grille(s). At this point, end reflection effects may cause a

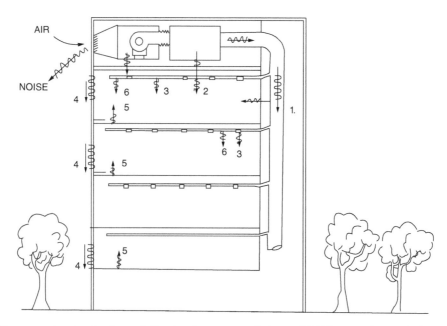

Figure 13.4 Sources and paths of noise and vibration from a centrally located HVAC system in a typical multistory building: (1) fan noise in duct, (2) noise transmitted through mechanical room floor, (3) grille noise, (4) structure-borne noise, (5) structure "re-radiated" noise, (6) duct mixing box radiated noise.

great deal of low-frequency attenuation to be achieved; but some high-frequency noise may also be generated by the flow of air through the grille. The whole process is, therefore, one of continual sound attenuation (by the plenum, ductwork, bends, etc.) and noise generation both by the flow of air through the system (at bends, take-offs, and air terminal devices) and by the radiation of noise from the vibration of the fan/blower. See Figure 13.5a.

There are three main types of air terminal devices: grilles, registers, and diffusers (GRDs). All allow air to pass through them from the supply side to the occupied space (see Figure 13.5b). The grille is the simplest design and normally only incorporates deflectors, which may be fixed or adjusted to supply the air in a certain direction. See Figure 13.5b(1). A register is similar to a grille but in addition a dumper is incorporated, which may be used to restrict the airflow. See Figure 13.5b(2). Grilles and registers normally have inlets (necks) and outlets (at the face side) of similar cross-section areas and may be used with the return instead of the supply air. A diffuser can also be used for the air supply. However, here the air goes through a 90° turn. The inlet (neck) area here is smaller than the exit area (or face.) See Figure 13.5b(3).

13.3.3 Correct and Incorrect Installation of HVAC Systems

The mechanical equipment room itself is normally a major source of noise and vibration problems, especially since it seems to be becoming common practice to locate fan rooms on the upper floors in a building where they are usually supported by lightweight flexible structural slabs. In fact, these are very often situated directly over critical executive office spaces, conference rooms, apartment units or other areas that require especially low noise levels. It is very important that great care is taken to install the complete air-conditioning system properly using the system approach to minimize sources and paths of noise and vibration.

The A-weighted sound pressure level within a fan plenum chamber may be as high as 115–120 dB; hence, the noise level within the mechanical equipment room itself – which is made up of noise radiated from the fan, plenum, and ductwork – is often very high. High transmission-loss floors and suspended ceilings are thus

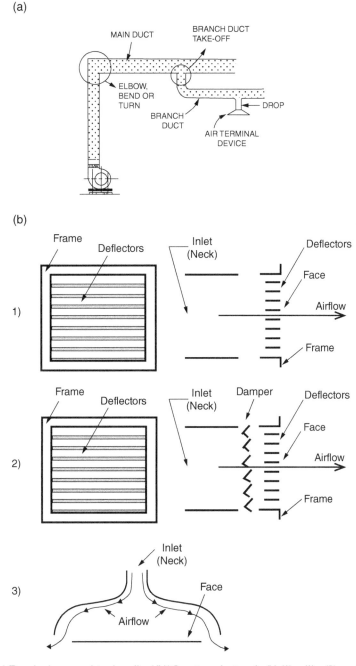

Figure 13.5   (a) Terminology used to describe HVAC system ductwork. (b) (1) grille, (2) register, (3) diffuser.

usually required to help isolate the mechanical room from the rest of the building. Some vibration from the fan, plenum chamber, and other auxiliary mechanical room equipment can often be easily transmitted into the mechanical room floor slab. Because modern buildings are almost entirely built out of concrete with low internal damping, especially good vibration isolation techniques are required in the mechanical room to prevent energy being transmitted to and ultimately radiated from various other parts of the building.

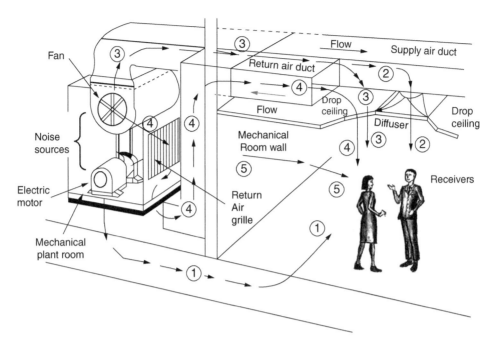

Figure 13.6 Typical paths in HVAC systems [20]. 1: Structure-borne path through floor; 2: Airborne path through supply air system; 3: Duct breakout from supply air duct; 4: Airborne path through return air system; 5: Airborne path through mechanical equipment room wall.

There are many possible airborne and structure-borne paths between the source (usually a fan) and the receiver (normally the building occupants). (See Figure 13.6.) These paths should be suppressed as much as possible.

It is particularly important that once the air-conditioning system components have been correctly sized and chosen that they be carefully installed. Figure 13.7 shows an air-handling unit (AHU) which has been installed poorly resulting in a very noisy system [25]. Twelve faults can be seen. Most of them are system-related and caused by improper installation.

The airflow through the AHU and its ductwork system shown in Figure 13.7 results in turbulence and noise. The small mechanical room results in poor airflow, noise, and rumble in adjacent rooms; the vibration isolators are inadequate and incorrectly located on a flexible floor without adequate structured support directly under the unit.

Figure 13.8 shows the same AHU installed after care is taken to reduce most of the problems in the installation in Figure 13.7 [25]. The improved ductwork system results in better airflow, reduced pressure drop and thus greater efficiency and reduced fan horsepower requirements. The larger mechanical room has better clearance between the unit and walls reducing noise transmission and rumble in adjacent rooms. The improved vibration isolators with the stiffened floor and a structural beam directly under the unit reduce vibration transmission throughout the entire building structure.

13.3.4 Sources of Noise and Causes of Complaints in HVAC Systems

Figure 13.9 shows the noise sources and frequency ranges of sources most likely to cause complaints. At low frequency, the most common complaints concern throb, rumble, and roar and are caused by turbulence often created by improper layout of components in HVAC systems. In the mid- and high-frequency ranges, the complaints concern hiss and result from poorly designed or situated grilles and diffuser systems. Occupant

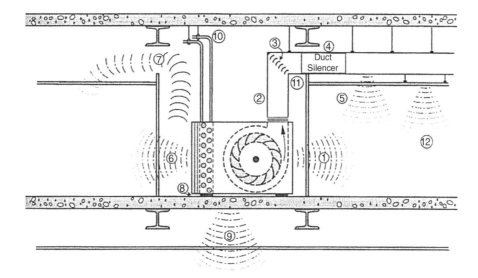

Figure 13.7 Example of an air-handling unit room with numerous acoustical and vibration problems [25]. 1. AHU panel vibration "couples" to the lightweight, flexible gypsum wall just a few centimetres away. This coupling lets low-frequency noise pass easily through the wall, 2. The counterclockwise rotation of the fan's discharge airstream is forced to change its spin direction at the downstream elbow. The turbulence generated at the change can produce unstable flow with a very high, fluctuating pressure drop, thereby resulting in fan instability that is heard as rumble, 3. Problem 2 is aggravated if the elbow's turning vanes do not have long trailing edges to straighten the airflow and control the turbulence, 4. The duct silencer is too close to the elbow. This compounds the turbulence problem, 5. Rectangular ductwork and duct silencers do not control the rumble produced by the turbulent airflow, 6. The AHU's air inlet is too close to the wall. This causes two acoustical problems: unstable fan operation leading to surge and rumble, and direct exposure of the inlet noise to the mechanical room wall, 7. The lack of a duct silencer in a mechanical room return air opening allows fan noise to travel into the ceiling cavity, then through the lightweight acoustical ceiling into the occupied space, 8. The unit is resting on thin cork/neoprene isolation pads that are too stiff to adequately isolate the fan vibration, 9. The poorly isolated unit is resting on a relatively flexible floor slab without sufficient structural support. This arrangement allows unit vibration to enter the slab, 10. The chilled water piping is rigidly attached to the slab above, thereby letting unit vibration; enter the slab, 11. Duct wall vibration in the duct silencer (or any other part of the trunk duct system) touching the drywall partition can cause the partition to act as a sounding board and radiate low-frequency noise into the occupied space, 12. Suspending ceiling from supply duct causes ceiling to be a sound radiator.
© ASHRAE www.ashrae.org, M. E. Schaffer, Practical Guide for HVAC, 2nd Ed., 2011, RP 526 Fig A.

complaints can occur, however, even in well-designed HVAC systems, since some people are more susceptible to noise than others.

Figure 13.10 shows the frequency ranges in which different mechanical sources are normally dominant. Low-frequency sources include fan instability and periodic ingestion of turbulent flow, variable air volume (VAV) unit noise is important at mid-frequency and diffusers and grilles are some of the main causes at high frequency.

Figure 13.11 shows the spectrum of a typical HVAC system and the contributions made by the fan, VAV valve and diffuser. Fan noise is particularly important in the low and mid-frequency range, VAV noise at mid-frequency and diffuser noise at high frequency [20].

All the above-mentioned sources and paths will now be dealt with in detail to show how a heating, ventilating, and air-conditioning system can be designed and evaluated acoustically.

13.4 Fan Noise

It should be noted that the selection of a fan is based primarily not on its acoustical characteristics but rather on its ability to move a required amount of air against the installed static pressure created by the downstream system elements. The first cost of the fan, its size, and maintenance costs are of primary concern [12]. Once

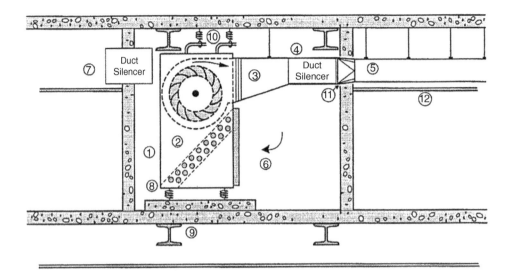

Figure 13.8 The AHU with a greatly improved installation [25]. 1. Keeping a minimum 0.6 m (2 ft) clearance reduces coupling between AHU and wall. Masonry wall provides excellent low frequency sound isolation, 2. Use of a horizontal discharge AHU eliminates the need for a turbulence-producing airflow, 3. Gradual transition at AHU outlet minimizes turbulence, 4. Duct silencer is far enough away from AHU outlet to avoid excessive regenerated noise and turbulence, 5. Circular ductwork controls the transmission of low-frequency noise and rumble into the occupied space, 6. The large clearance at the AHU inlet keeps the unit away from the wall and avoids excessive inlet turbulence, 7. The return air duct silencer controls AHU noise via the return air path, 8. The unit is resting on high-deflection, steel spring vibration isolators, 9. The floor assembly supporting the unit has a housekeeping pad and at least one major beam under the unit. Additional stiffness and mass help to control the transmission of unit vibration into the slab, 10. The chilled water pipes are suspended by vibration isolation hangers, 11. The supply trunk duct does not touch the wall. A 1 cm (0.5 in.) gap surrounding the duct is filled with a non-hardening sealant, 12. Ceiling not suspended from supply duct. ibid © ASHRAE Schaffer Guide.

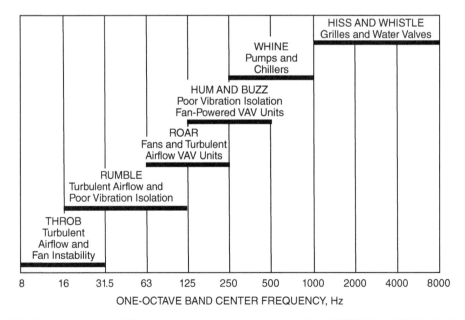

Figure 13.9 Frequency ranges of likely sources of sound-related complaints [25]. ibid © ASHRAE Schaffer Guide.

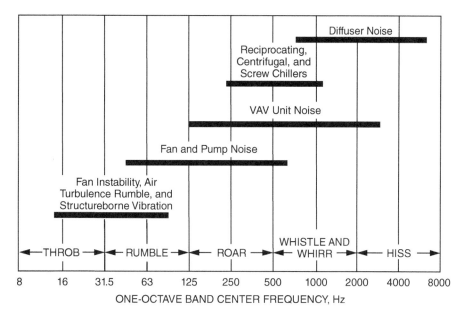

Figure 13.10 Frequencies at which different types of mechanical equipment generally control sound spectra [25]. ibid © ASHRAE Schaffer Guide.

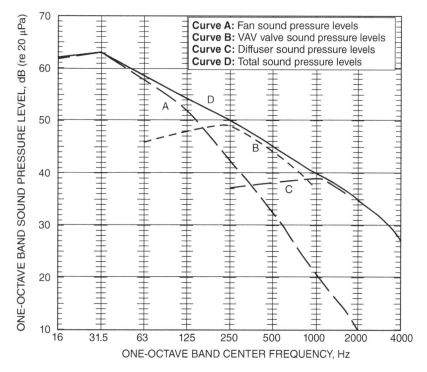

Figure 13.11 Illustration of a typical HVAC sound spectrum for occupied spaces [20]. ibid © ASHRAE Handbook, Chap 1, Fig 2.

the operating requirements have been decided, the fan noise characteristics will automatically become known. It is usually not practical to try to choose a quieter fan since it is unlikely that it can meet the operating requirements already determined.

13.4.1 Types of Fans Used in HVAC Systems

As already discussed in Chapter 11, there are two main types of fans: axial and centrifugal. There is also a third type, called a mixed-flow fan, which combines elements of the axial and centrifugal types. See Figures 13.12 and 13.13 in the present chapter. Figure 11.7 in Chapter 11 presents more details about centrifugal and axial fans concerning their use and characteristics. A typical fan sound power spectrum consists of a broadband portion with superimposed discrete pure tone peaks at the blade passing frequency and at integer multiples of this frequency, i.e. harmonics. The relative contribution of the broadband and discrete frequency components depends upon the type and geometry of the fan. For example, many of the centrifugal fans used in air-conditioning systems produce a mainly broadband frequency spectrum at low tip speed.

From a noise control perspective, the pure tones produced by a fan are very important. In many HVAC systems, in which fan noise is a problem, it is often found that pure tone components are the main cause of complaints. The ear is particularly sensitive to pure tones and can detect them in the general broadband noise making them irritating and annoying. For this reason, it is important to determine the octave band in which the fundamental pure tone (known as the blade passing frequency [BPF]) occurs. Care must be taken to make sure that noise control measures in this frequency band are adequate [12].

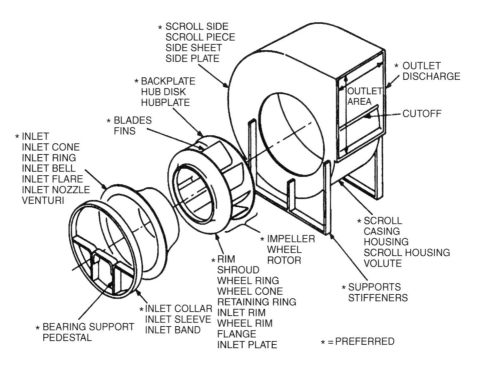

Figure 13.12 Components of a centrifugal fan.

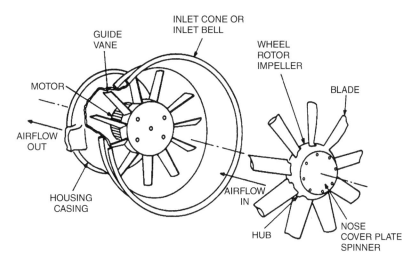

Figure 13.13 Exploded view of a typical axial-flow fan.

13.4.2 Blade passing Frequency (BPF)

The pure tones are primarily produced by the interaction of the rotating fan blades with the almost stationary air and by the interaction of the turbulent airflow with the stators. Each point in the surface through which the blade passes will receive a number of impulses per second equal to the number of fan blades times the number of fan revolutions per second. Thus, the fundamental blade passing frequency f_B, is given by

$$f_B = NR/60 \, \text{Hz}, \tag{13.1}$$

where N = number of fan blades, and R = number of revolutions per minute (rpm). Since the periodic impulses created by the fan are not wholly sinusoidal and never completely identical, also because the inflow to the fan is not completely steady and for other reasons, tones at integer, multiples, n, of the BPF also occur (i.e. nf_B).

Example 13.1 An eight-bladed axial fan is operated at 1800 rpm. Calculate the fundamental BPF and the first three harmonics using Eq. (13.1).

Solution

$N = 8$ blades and $R = 1800$ rpm. The fundamental BPF is $f_B = 8 \times 1800/60 = 240 \, \text{Hz}$.
The first three harmonics, for $n = 2, 3, 4$, are $nf_B = 480 \, \text{Hz}, 720 \, \text{Hz}, 960 \, \text{Hz}$.

The broadband noise is generated by sources which are random in time and may be dipole or quadrupole in nature. The noise can be generated either by lift fluctuations on the fan blade or by turbulent flow in the wake of the blade. In short, broadband noise is generated primarily by vortex shedding, by the interaction of the fan blades with turbulence, and by fan blade flow separation. The analytical prediction of this type of aerodynamic fan noise is beyond the scope of this book. However, reasonably accurate estimates can be obtained relatively simply. This is explained in Chapter 11 and later in this chapter. Lauchle has an extensive chapter dealing with fan noise in Ref. [26].

Since the primary purpose of a fan is to move a given quantity of air against a given pressure difference as efficiently as possible, it might seem that acoustical considerations are of secondary importance in the selection of a fan. However, it is found that when a fan is operated at peak aerodynamic efficiency, it also produces least noise. In other words, the same factors which increase a fan's efficiency usually also tend to reduce the sound power it produces. A decrease in efficiency of only a few percent can result in a 3 dB change (doubling) in the sound power emitted. If an undersized fan were selected, it would not operate at its peak efficiency, and thus it would produce more noise.

The apparent cause of the higher noise at the lower efficiency appears to be the higher airflow velocity through the fan. On the other hand, an oversized fan (again not operating at its peak efficiency) is not only uneconomical but also noisier than one of the optimum size. The separation of airflow over the blades causes this effect.

In modern ventilation systems, the two main types of fans, centrifugal and vane-axial, are employed in approximately equal numbers. Figure 13.12 shows an exploded view of a typical centrifugal fan. Figure 13.13 shows a similar view of a typical axial-flow fan.

Figure 11.7 in Chapter 11 and Figure 13.14 shows the main types of centrifugal fans. Of these three types, forward-curved centrifugal fans are the most compact and operate at the lowest speed when delivering a given volume flow rate and pressure. However, air is delivered into the scroll at a higher speed than the blade tip speed resulting in considerable turbulence and noise generation in the scroll-impeller blade region. Since the fan can operate at a lower rotational speed than other centrifugal fans in delivering the required flow rate, the mechanical and bearing noise and wear, are, however, less than other centrifugal fans [26].

Centrifugal fans with backward-curved blades release air into the scroll with a lower speed than the tip speed resulting in less noise. However, in order to deliver the air volume flow rate and pressure required, this type of fan must operate at a rotational speed of up to twice that of the forward-curved blade type resulting in higher mechanical noise and vibration, wear and maintenance costs. Fans with backward-curved aerofoil blades are the most efficient of these three types, but are the most expensive. They are often used when variable volume flow rates and pressures are needed [26].

Centrifugal fans with radial blades are rarely used in air-conditioning systems. Their main use is in industrial applications, often when transporting particles such as coal dust, wood chippings, and plastic waste. Fan blades for such applications are normally made stronger and thicker.

Figure 13.15 shows typical octave band sound power spectra for vane-axial, centrifugal, and mixed-flow fans. (The latter are special fans incorporating features of axial and centrifugal designs.) The manufacturer's electrical power rating given of 37.3 kW for each of these fans is the same. They also develop the same fluid power 23.0 kW when working at their maximum efficiency.

The sound power spectrum of each of these three fans is different. The spectrum for the vane-axial type is fairly flat, while the spectrum for a centrifugal fan decreases with frequency at a rate of about 4–6 dB per octave. The centrifugal fan usually produces a little more sound power at low frequency but much less at high frequency than the equivalent axial type. The blade passing frequency sound is more pronounced in the vane-axial fan spectrum and relatively little low-frequency noise is generated. This is characteristic of axial fans, where there is usually a strong peak at the blade passing frequency, and there are quite often peaks

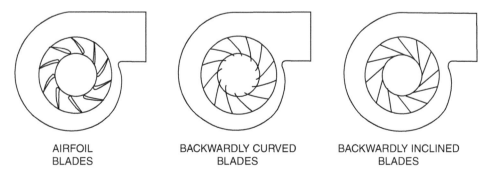

AIRFOIL
BLADES

BACKWARDLY CURVED
BLADES

BACKWARDLY INCLINED
BLADES

Figure 13.14 The three principal types of high-efficiency centrifugal fans.

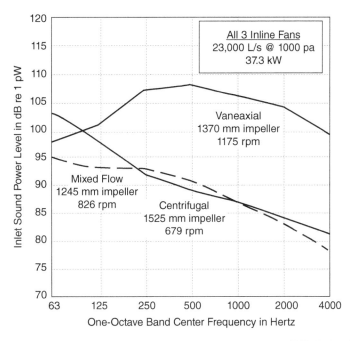

Figure 13.15 Inline fan sound power level comparison [25]. ibid © ASHRAE Schaffer Guide.

evident at higher harmonics of the BPF. It should be noted that the fans shown in Figure 13.15 have the flow (inlet and exhaust) aligned with the duct work. If duct bends or obstructions located upstream or downstream of the fan are situated close to the fan, the pure tone BPF sound power will increase. The pure tone components have the added effect of making the noise very objectionable. This is particularly noticeable with fans which have less than 15 blades.

13.4.3 Fan Efficiency

There is a common incorrect belief that a large fan operating at a low speed is less noisy than a small fan running at a higher speed. This is incorrect. The correct solution is to have the fan operating at its peak efficiency for the required airflow volume and pressure, and this will automatically produce the quietest fan. The ventilation engineer should always attempt to design the system so that the lowest possible pressure will be required, since the generated sound power rapidly increases with the pressure regardless of the type of fan used. Figure 13.16 shows the A-weighted inlet sound power level, L_{WA}, for a typical plenum fan. Also, shown are the fan curve, volume flow rate litre/second, and static pressure produced. The maximum efficiency for this fan is in a small operating range in the flow region of 17000 litre/second and pressure of 1500 Pa.

If a fan is not operated near its designed maximum efficiency, its sound power level will increase. The increase in each octave band can be calculated as follows:

Calculate the fan's static efficiency:

$$\text{Static efficiency} = (\text{volume flow rate} \times \text{static pressure})/(k \times \text{mechanical input power}) \qquad (13.2)$$

If the volume flow rate is given in litre/second, and the static pressure is given in pascals, then the constant $k = 1$ and the mechanical input power developed by the fan shaft is in kilowatts. If the volume flow rate is given in ft^3/min, and the pressure is given in inches of water, $k = 6354$ and the input power is given in brake horsepower [9, 27]. For vane-axial fans, the pressure in Eq. (13.2) should be the static pressure plus the

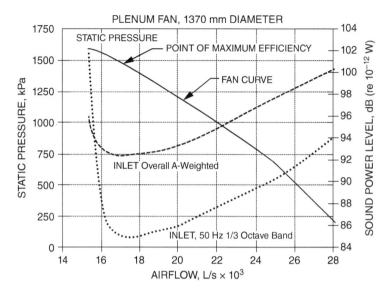

Figure 13.16 A-weighted sound power level test data for a typical plenum fan, for different operating points (static pressures and volume flow rates.) [20] Note that A-weighted sound power level and 50 Hz 1/3 octave band sound power level rise as operating point moves away from maximum efficiency point. ibid © ASHRAE Handbook, Chap 48, Fig 8.

Table 13.1 Fan efficiency adjustment, i.e. the number of decibels by which the sound power level of a fan should be increased because of its operation at other than peak efficiency (these values are for different types of fans) [12].

Airfoil centrifugal and vane-axial fan		Backward-curved centrifugal fan		Forward-curved centrifugal fan	
Efficiency, %	Increase, dB	Efficiency, %	Increase, dB	Efficiency, %	Increase, dB
80 to 72	0	75 to 67	0	65 to 58	0
71 to 68	3	66 to 64	3	57 to 55	3
67 to 60	6	63 to 56	6	54 to 49	6
59 to 52	9	55 to 49	9	48 to 42	9
51 to 44	12	48 to 41	12	41 to 36	12

dynamic pressure created in the moving airstream [9]. Table 13.1 gives an estimate of the increase in sound power level expected for fans not operated at their designed maximum efficiency.

As already discussed, fans should be operated as close to their points of maximum efficiency as possible. This ensures the fan will require minimum power and in general minimum noise will be produced. See Figure 13.16. As the operating point is shifted to the right and down along the fan curve, the fan produces greater airflow but achieves a lower static pressure. Also the inlet A-weighted sound power level, L_{WA}, increases. Fans produce noise not only at the exhaust but at the inlet as well. In the USA, manufacturers usually provide the inlet sound power level of fans in octave bands with center frequencies from 63 to 8000 Hz. Normally the inlet sound power values given are measured under the ideal laboratory conditions specified in AMCA Bulletin 300 using a reverberation room.

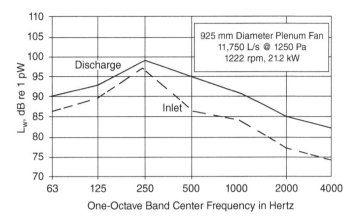

Figure 13.17   Inlet and discharge octave band L_w values for a 925 mm plenum fan [25]. ibid © ASHRAE Schaffer Guide.

13.4.4   Sound Power and Frequency Content of Fans

Some manufacturers now provide both inlet and exhaust sound power level data. As discussed, these levels will increase if obstructions, abrupt duct bends or cross-section changes are situated in inlet or exhaust locations close to the fan. Figure 13.17 shows the inlet and discharge octave band sound power levels for a typical plenum fan. In many cases, the discharge sound power levels can be as much as 5–10 dB above the inlet sound power levels (which are those normally supplied by manufacturers).

Example 13.2   Consider the fan in Figure 13.15. Assume that the electric motor driving the fan converts 75% of the 40 kW electrical power into shaft power. Also assume that the fan only converts 85% of the shaft power into fluid flow power. Define the fan efficiency as fluid flow power divided by shaft power. The type of fan in Figure 13.16 is unknown. Use the fan curve for the plenum fan of diameter 1370 mm in Figure 13.16. With these various assumptions, calculate the fan efficiency, fluid flow power and sound power level increase of the fan for representative pressure/airflow conditions in Figure 13.16 by using Table 13.1.

Solution

Use the following relationships:

$$\text{litre} = 0.001 \text{ m}^3, \text{Pa} = \text{N/m}^2, \text{watt} = \text{J/s} = \text{N m/s}$$

Fluid flow power = $Q \times P$. For fan in Figure 13.15, electrical power = 40 kW. Assume shaft power = $0.75 \times 40 = 30$ kW. Assume fluid flow power at maximum efficiency = $0.85 \times 30 = 25.5$ kW. Note fluid flow power is given by $Q \times P$ in Figure 13.16.

$$\text{Fluid Flow power at maximum efficiency} = (17\,000\,\text{l/s})\,(1500\,\text{kPa})$$

$$= (17\,\text{m}^3/\text{s})\,(1500\,\text{N/m}^2) = 25.5 \times 10^3\,\text{N m/s}$$

$$= 25.5\,\text{kW}$$

This result agrees with the initial assumption that at maximum efficiency the fan converts 85% of shaft power into fluid flow power. The values of fluid flow power for other fan curve conditions are found from Figure 13.16. The efficiency and sound power level increases are estimated from Table 13.1 and given in Table 13.2. We note that the level increases predicted here for the lower static pressures of 750 kPa and 500 kPa are higher than in Figure 13.16, which are, however, A-weighted.

Table 13.2 Estimated sound power level increases for different fan pressures and volume flows.

Static Pressure kPa	Air Flow L/s × 10³	Fluid Flow Power Q × P kW (Figure 13.16)	Efficiency (Q × P/30)100%	Estimated Sound Power Level Increase dB (Table 13.1)
1500	17.0	25.35	25.4/30 = 85%	0
1250	19.9	25.35	25.4/30 = 85%	0
1000	22.5	22.5	22.5/30 = 75%	3 dB
750	24.5	18.4	18.4/30 = 61%	6 dB
500	26.0	13.0	13.0/30 = 43%	12 dB

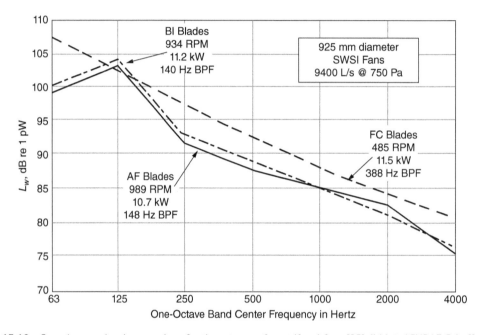

Figure 13.18 Sound power level comparison for three types of centrifugal fans [25]. ibid © ASHRAE Schaffer Guide.

Figure 13.18 shows a comparison between the unweighted sound power levels of three types of centrifugal fans operating at 9400 L/s and 750 Pa total pressure. The fans all have similar electrical power ratings of between 10.7 kW and 11.5 kW. The forward-curved fan (FC) type does not show as strong a BPF peak at 125 Hz, but the greater level at 63 Hz is more difficult to suppress at this low frequency, at which sound-absorbing and barrier materials are less effective. Figure 13.19 gives inlet sound power levels for three types of axial-flow fans with electrical power ratings of 2.2 kW.

13.4.5 Sound Power Levels of Fans and Predictions

As already discussed, although, all of the different axial fan types shown in Figure 13.19 have different overall sound power levels and frequency spectra, they all produce their lowest noise levels when operated in the region of peak efficiency on their performance curves. See Figure 13.16. The cast-aluminum blade type is the

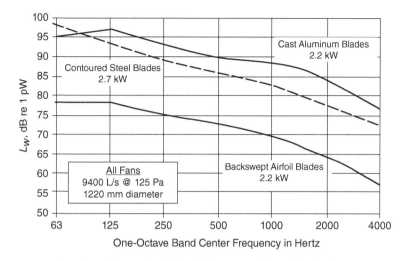

Figure 13.19 Inlet one-octave band sound power levels L_w of three types of propeller fans [25]. ibid © ASHRAE Schaffer Guide.

least expensive, but it is the noisiest. The contoured steel blade is quieter but costs more. The backward swept airfoil blade fan is the quietest type but the most expensive.

It is often necessary for an engineer to be able to estimate fan sound power levels at the design stage within ±5 dB, if manufacturers' data should not be available. Consequently, several prediction schemes involving empirical equations have been developed and used with some success for sound power predictions. However, when using such empirical prediction schemes, one should ensure that the scheme chosen is valid for the particular fan and its operating conditions. Some of the fan sound power prediction schemes are now reviewed briefly.

One of the earliest fan sound power prediction schemes was produced by Beranek et al. in 1955 [21]. They assumed an acoustical efficiency of 10^{-6}. This efficiency is defined to be sound power divided by source mechanical power. By measuring the sound power spectra of 14 different centrifugal fans, they determined an empirical result for the overall fan sound power level. They found that the octave band levels sloped downwards at about 5 dB per octave. The first one-octave band centered at 20 Hz was normally found to be also downward and given by 1 dB less than that predicted by the overall level. Beranek's result gives better accuracy for centrifugal fans developing large static pressures.

In 1967, Groff gave a relation for the sound power level generated for a variety of centrifugal fans [26]. In 1972, Graham proposed a simplified fan sound power prediction scheme partly based on the earlier methods discussed [5, 6]. This scheme was adopted by ASHRAE in the 1973 Guide and Data Book and was widely used. Graham later discussed this scheme further, giving several examples, and also converted it into SI units [10]. This scheme assumes that the fan is well designed and operating near its maximum efficiency. In order to estimate fan sound power, fans are divided into the main types shown in Figure 11.7 in Chapter 11.

Most manufacturers of centrifugal fans now provide measured sound power levels for their products along with detailed information on their operating conditions for both FC, backward inclined (BI), and aerofoil fan blades. One should note carefully whether these are quoted in decibels (re 10^{-12} or 10^{-13} W). Manufacturers' data are preferred to estimated levels since predictions can be in serious error. The sound power levels in Table 11.2 in Chapter 11 are given in decibels with stated reference quantities.

13.4.6 Prediction of Fan Sound Power Level

Although Graham's method [3–6] no longer appears in the latest ASHRAE Handbook [20], it is still widely used. Table 11.2 in Chapter 11 presents a method based on the work by Graham and Hoover [10]. In this

Table 13.3 Fan sound power frequency spectrum corrections in dB [8].

Fan type	One-octave band center frequency (Hz)						
	63	125	250	500	1000	2000	4000
Forward-curved centrifugal	−2	−7	−12	−17	−22	−27	−32
Backward-curved centrifugal	−7	−8	−7	−12	−17	−22	−27
Axial-flow	−5	−5	−6	−7	−8	−10	−13
Mixed-flow	−12	−11	−10	−10	−13	−17	−22

approach, specific sound power levels are given for different types of fans based on the reference quantities $W_{ref} = 10^{-12}$ W, volume flow $Q_{ref} = 1$ m³/s, and $\Delta P_{ref} = 1$ kPa. For fans operating at other values of Q and ΔP, the correction can be made by adding $10 \log (Q) + 20 \log (\Delta P)$ to the specific sound power levels. A pure tone correction is also made to the octave band in which the BPF occurs by adding the last increment in Table 11.2 to the level in that band.

Fry gives a very similar prediction scheme [8]. In this, the sound power level is given by

$$L_w = 10 \log (Q) + 20 \log (P) + 40 \text{ dB}, (\text{re } 10^{-12} \text{ W}). \tag{13.3}$$

It should be noted, however, that in Fry's method, $Q_{ref} = 1$ m³/s, and $P_{ref} = 1$ Pa, not 1 kPa.

Fry provides Table 13.3 for frequency spectrum corrections as follows. Both Eq. (13.3) and Table 13.3 are derived from the earlier work of Beranek and Allen.

Example 13.3 Calculate the sound power level in one-octave bands for the FC and BI fans in Figure 13.18. Each fan operates with a volume flow of 9400 L/s and pressure of 750 Pa. The FC fan has a BPF of 388 Hz and the BI fan has a BPF of 148 Hz. Use both the Graham and Hoover (Chapter 11, Table 11.2) and the Fry method [8]. Note 1 m³ = 1000 l.

Solution

First for the FC fan with the Graham and Hoover method, we use the values for the forward-curved centrifugal fan in Table 11.2 to obtain the octave band sound power levels. See the second row in Table 13.4.

Table 13.4 Graham and Hoover calculation of the one-octave band sound power of FC and BI fans.

One-Octave Band Center Frequency, Hz	63	125	250	500	1000	2000	4000	8000
Forward-curved (FC) Fan	98	98	88	81	81	76	71	66
Flow and Pressure Correction	+7.2	+7.2	+7.2	+7.2	+7.2	+7.2	+7.2	—
Total (FC) Fan, L_w, dB	105	105	95	88 + 2	88	83	78	—
Backward Inclined (BI) Fan $d > 0.75$ m	85	85	84	79	75	68	64	62
Flow and Pressure Correction	+7.2	+7.2	+7.2	+7.2	+7.2	+7.2	+7.2	+7.2
Total (BI) Fan, L_w, dB	95	95 + 3	96	86	82	75	71	69
Backward Inclined (BI) Fan $d < 0.75$ m	90	90	88	84	79	73	69	64
Flow and Pressure Correction	+7.2	+7.2	+7.2	+7.2	+7.2	+7.2	+7.2	+7.2
Total (BI) Fan, L_w, dB	97	97 + 3	95	91	86	80	76	71

Table 13.5 Fry calculation of the sound power levels, dB, of FC and BI fans.

Frequency, Hz	63	125	250	500	1000	2000	4000	8000
Forward-curved (FC) Fan	107	107	107	107	107	107	107	107
Spectrum Correction, dB (Table 13.3)	−2	−7	−12	−17	−22	−27	−32	—
Total FC Fan L_w, dB	105	100	95	90	85	80	75	—
Backward Inclined (BI) Fan	107	107	107	107	107	107	107	107
Spectrum Correction, dB (Table 13.3)	−7	−8	−7	−12	−17	−22	−27	—
Total BI Fan, L_w, dB	100	99 + 3	100	95	90	85	80	—

Then we make the flow and pressure correction log $Q + 20 \log \Delta P = 10 \log (9.4) + 20 \log (0.750) = 7.2$ dB and put this into row 3. Finally, the total FC fan values of L_w are obtained in row 4 of the table. A similar approach is used to obtain the prediction for the BI fan using the values for the backward-curved centrifugal fan in Table 11.2. The total BI fan values are given in row 7.

Table 13.5 shows the results for the FC and BI fans using the Fry method [8]. First the one-octave band sound power levels are calculated using Eq. (13.3) and the flow Q and pressure P values in Figure 13.18. See rows 2 and 5 in Table 13.5. Then the spectrum corrections for the FC and BI fans are found from Table 13.2. See rows 3 and 6. The final L_w values are calculated after the corrections are made and are given in rows 4 and 7.

The sound power level predictions made in Tables 13.4 and 13.5 are shown in Figures 13.20 and 13.21 and compared with the manufacturer's sound power level data for backwardly inclined and forward-curved centrifugal fans. The Graham and Hoover, and the Fry prediction methods both give predictions for the FC fan in Figure 13.21 within ±5 dB except in the low frequency range (< 250 Hz) and high frequency range (>1000 Hz) for the centrifugal fan in Figure 13.20. Both the Graham and Hoover, and the Fry methods make predictions within ±5 dB for the FC centrifugal fan in Figure 13.21. The Fry method overpredicts the sound power level for the BI fan levels in Figure 13.20 in the mid frequency range by a few decibels more than 5 dB. The Graham and Hoover method predicts the level well, provided it is assumed incorrectly that the fan diameter d < 0.75 m. But assuming correctly that d > 0.75 m makes it underpredict the level by almost 10 dB for frequencies >2000 Hz.

13.4.7 Importance of Proper Installation of Centrifugal Fans

The noise generated by centrifugal fans is strongly dependent on the installation. Figure 13.22 shows the worst installation at the right to good in the middle and best at the left. With installations shown as bad or fair in Figure 13.22, duct rumble is likely to occur. While with installations shown as good, very good or best, rumble is less likely to occur. There are several ways to reduce the occurrence of duct rumble. One method is to change the speeds of the motor, fan belt or fan. Another method involves mass loading the duct to change the wall resonance frequencies excited by the flow. Noise reductions between 5 and 10 dB in the 31.5 and 63 Hz frequency bands have been recorded by this approach [20]. Complete enclosure of the duct wall with absorbing material placed between the enclosure wall and duct wall can reduce rumble and also breakout noise and break-in noise, provided the enclosing system is decoupled from the duct wall [20].

It is essential that great care is taken to ensure that an HVAC system is installed properly and that faults do not occur resulting in an unsatisfactory system. In the case of projects with large buildings, it is best that architects, engineers, and acoustical consultants should all be involved to ensure a satisfactory outcome. It is often found that architects and mechanical engineers have sized the HVAC items of equipment and specified noise and vibration control elements such as duct liners, silencers, and vibration isolators correctly,

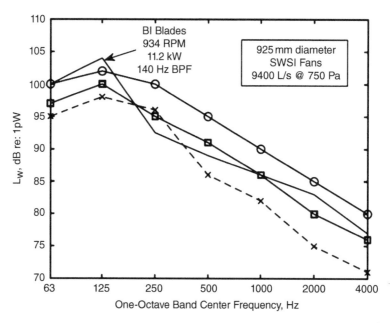

Figure 13.20 Sound power level comparison between two prediction methods for backward inclined (BI) type of centrifugal fan. Graham and Hoover Method (*d* > 0.75 m) ×- -×- -×; Graham and Hoover Method (*d* < 0.75) □-□-□; Fry Method ○-○-○.

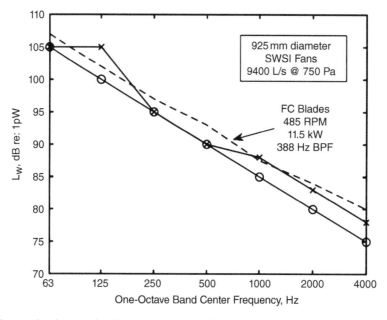

Figure 13.21 Sound power level comparison between two prediction methods for forward- blade (FC) type of centrifugal fan. Graham and Hoover Method ×- -×- -×; Fry Method ○-○-○.

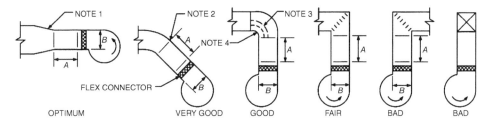

Figure 13.22 Guidelines for centrifugal fan installations [20]. Notes: 1. Slopes of 1 in 7 preferred. Slopes of 1 in 4 permitted below 10 m/s, 2. Dimension *A* should be at least 1.5 times *B*, where *B* is the largest discharge duct dimension, 3. Rugged turning vanes should extend full radius of elbow, 4. Minimum 150 mm radius required. ibid © ASHRAE Handbook, Chap 28, Fig 25.

Table 13.6 Poor intake and discharge condition corrections [8].

(a) Abrupt Entry

Frequency, Hz	63	125	250	500	1000	2000	4000
Correction, dB	+2	+5	+7	+5	+5	+5	+5

(b) Upstream interference for example trimming vane, idling impeller, radiused bend, acute transformations, expanders.

Frequency, Hz	63	125	250	500	1000	2000	4000
Correction, dB	+6	+6	+6	+6	+6	+6	+6

(c) Flexible connectors (misaligned or concave)

Frequency, Hz	63	125	250	500	1000	2000	4000
Correction, dB	+6	+6	—	—	—	—	—

(d) Form of running (motor upstream of impeller)

Frequency, Hz	63	125	250	500	1000	2000	4000
Correction, dB	+6	+7	+3	+8	—	—	+2

but they have been incorrectly installed resulting in complaints. Table 13.6 shows the increase in noise level expected from some poor fan inlet and discharge conditions.

Exhaust fans are needed in bathrooms and sometimes in conference rooms and theaters in which many people are present at the same time. In bathrooms, exhaust fans are often poorly installed with short ducts, obstacles located before or after the fan and with sharp bends after it. If the fan is not vibration-isolated, vibration and duct noise can be transmitted throughout the whole building. Figure 13.23 shows an exhaust fan with necessary vibration isolation correctly undertaken and a flexible duct connection provided. In addition, exhaust fans may be needed in mechanical rooms where substantial heat build-up can occur. Figure 13.24 shows examples of noisy and quiet installations of exhaust fans in a conference room ceiling set-up.

13.4.8 Terminal Units (CAV, VAV, and Fan-Powered VAV Boxes)

In recent years, different types of terminal units have come into widespread use. Constant air volume systems (CAV) are designed to supply a constant airflow volume to a room at a variable temperature. In contrast, VAV systems vary the airflow at a constant temperature. The advantages of VAV systems include more precise temperature control, better dehumidification, reduced system wear, lower fan energy consumption and less

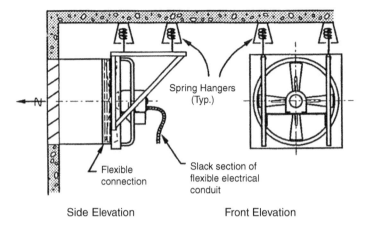

Figure 13.23 Vibration isolation suspension for propeller fans. Note: Position hangers on line of center of gravity of fan unit. Supplemental sections of steel angle or channel may be secured to fan mounting frame, as required, for support. ibid © ASHRAE Schaffer Guide.

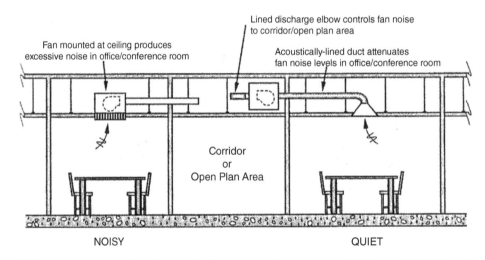

Figure 13.24 Noisy and quiet installation of ceiling-mounted exhaust fans [25]. ibid © ASHRAE Schaffer Guide

noise. Figure 13.25 shows a variable volume flow system (VAV) at the right top and a constant flow system (CAV) at the right bottom. Figure 13.26 shows guidelines for the installation of VAV systems.

Control of the VAV system's fan is very important. Without proper control, the system and its ductwork can be damaged through over pressurization. In the cooling mode, once the required temperature is reached, the VAV "box" valve partially or completely closes. As the temperature rises, the box valve reopens to reduce the temperature. The fan must be designed to provide a constant pressure regardless of the VAV box setting. As the box closes, the fan must slow down to reduce the airflow volume but as the box reopens it must speed up again to increase the volume flow and maintain the constant static pressure. Figure 13.26 shows guidelines for the installation of VAV units. Flexible connectors and a lined sheet metal plenum are advisable. The unit should be located as far away from the drop ceiling as possible. Figure 13.27 shows curves which can be used for guidance in selecting terminal VAV systems. If VAV fans are required to produce large pressures at low flow rates, surge, increased noise and even damage can occur.

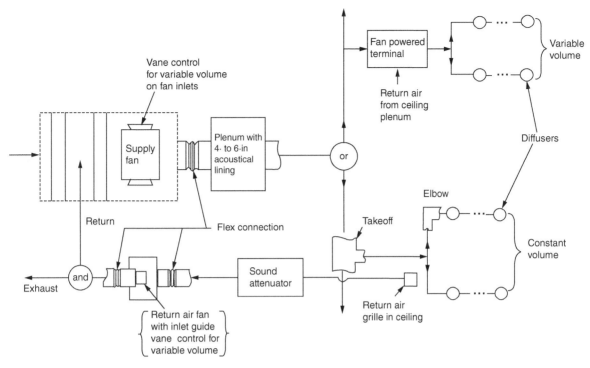

Figure 13.25 Line diagram illustrating the major components of an HVAC system, related to the generation and control of noise [9].

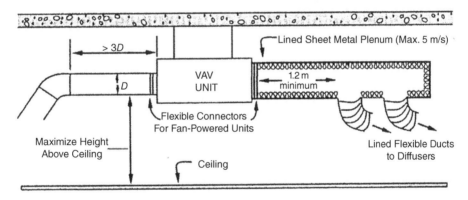

Figure 13.26 Guideline for VAV unit installation [25]. Note: Parallel or side-pocket fan-powered units often require an up or side-looking lined sheet metal inlet elbow to control fan noise. ibid © ASHRAE Schaffer Guide.

13.5 Space Planning

The main acoustical goal should be to locate the major noise sources, mechanical equipment rooms (housing the AHU), cooling towers, chiller rooms and roof top and other package units as far as possible from noise-sensitive areas in the building. In large buildings it is often necessary to locate the mechanical room in the core of a building. Figure 13.28 shows several possible building layouts, in which stairs, toilets, elevators, and storage rooms are used to separate the mechanical room from noise-sensitive areas of the building. In the layout at the top, three walls of the mechanical room are potentially connected to occupied spaces and these walls need to be made more massive. In the two middle plans, the number of connected walls is reduced to two, and in the bottom plan, no walls are connected. In the middle plans, only one of the walls needs to be made more massive if the other wall is the exterior wall of the building.

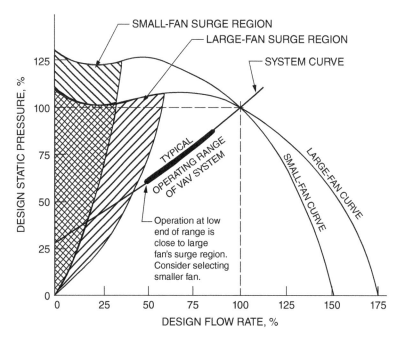

Figure 13.27 This figure shows a set of curves giving guidance for the selection of VAV systems [20]. ibid © ASHRAE Handbook, Chap 48, Fig 25.

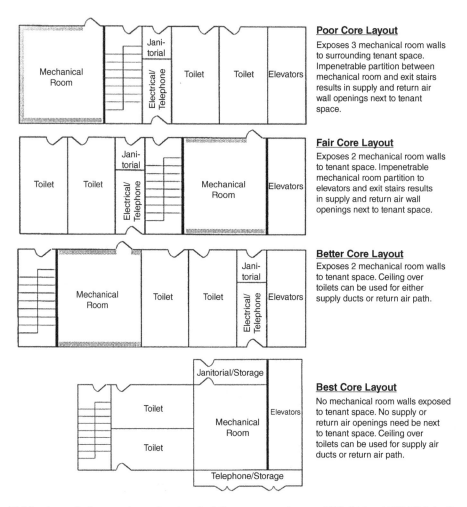

Poor Core Layout

Exposes 3 mechanical room walls to surrounding tenant space. Impenetrable partition between mechanical room and exit stairs results in supply and return air wall openings next to tenant space.

Fair Core Layout

Exposes 2 mechanical room walls to tenant space. Impenetrable mechanical room partition to elevators and exit stairs results in supply and return air wall openings next to tenant space.

Better Core Layout

Exposes 2 mechanical room walls to tenant space. Ceiling over toilets can be used for either supply ducts or return air path.

Best Core Layout

No mechanical room walls exposed to tenant space. No supply or return air openings need be next to tenant space. Ceiling over toilets can be used for supply air ducts or return air path.

Figure 13.28 Acoustical comparison of various building core area layouts [25]. ibid © ASHRAE Schaffer Guide.

13.6 Mechanical Room Noise and Vibration Control

If the mechanical room is located on the ground floor of a building, transmission of noise and vibration through the floor is normally not a problem. However, care still needs to be taken to prevent noise and vibration propagating into adjacent spaces. Figure 13.29 shows a ground floor mechanical room where the intake air flow passes through a sound-absorbent section, which prevents direct line of sight of the fan at the intake. The airflow into the fan does not make an abrupt change of direction at the fan entry. The fan unit is mounted on neoprene vibration isolators so that structure-borne noise propagation throughout the building is reduced.

It has become common for architects to locate mechanical equipment rooms on the upper floors of multistory buildings, where they are often supported by lightweight flexible structural slabs and sometimes even positioned directly over critical areas requiring low noise levels. If considerable care is taken with the rooftop installation, the mechanical equipment room noise can be suppressed considerably.

As a result, however, some architects have now become even more daring and sometimes locate ventilation equipment rooms over auditoriums and other extremely critical occupied spaces. Since most mechanical rooms usually also contain many items of equipment in addition to fans (such as boilers, chillers, and pumps), the noise and vibration levels are accordingly very high, and often the floors need to provide over 50 dB transmission loss at low frequencies in order to achieve acceptable conditions below. Such a transmission loss cannot be economically obtained at low frequencies by a single layer floor slab [27].

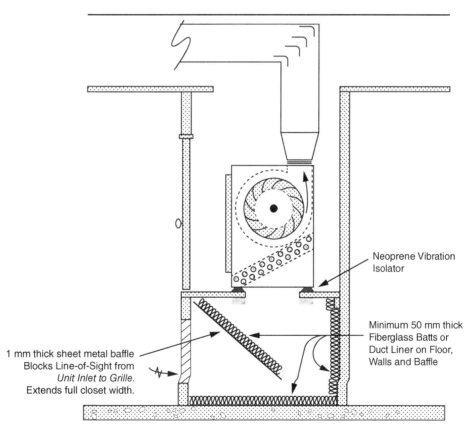

Figure 13.29 Mechanical room on ground floor of building. ibid © ASHRAE Schaffer Guide.

In any noise control problem, one should begin remedial action by attempting to reduce the strength of the source. In this respect, mechanical rooms are not exceptions. All equipment radiating excessive sound power in mechanical rooms should be adequately silenced, if at all possible. For example, some or all of the following noise control techniques may be implemented [27]:

1) Adding acoustical absorption to the walls and ceilings near large centrifugal fans (this may also be necessary for hearing conservation and communications between maintenance staff)
2) Fitting of silencers at air intakes to forced draft fans which are open to the room
3) Installation of acoustically lined plenum chambers around high power centrifugal fans
4) Installation of resilient mounts and hangars to isolate piping, ductwork, wiring, conduit, etc. from the building structure
5) Use of acoustical enclosures around reciprocating refrigeration machines.

To cope with these problems, mechanical and acoustical engineers have developed techniques to reduce and control both the noise and vibration emanating from such rooms. Figures 13.30 and 13.31 show noisy and quiet mechanical room rooftop installations. In the poorly designed HVAC system in Figure 13.30, the plenum liner is too thin to absorb low-frequency noise, bends after the fun exhaust cause turbulence, low-frequency noise and rumble and the fan vibration isolation is absent.

In the HVAC system in Figure 13.31, the sound-absorbing plenum lining is increased to 100 mm. The plenum and the refrigeration system are mounted on a separate floor from which it is also vibration-isolated. Cylindrical ducts rather than rectangular ones are used to reduce rumble and throbbing.

13.6.1 Use of Floating Floors

After the sound pressure levels within the mechanical room have been adequately and economically reduced, the next step is to introduce a massive layer between the room and the nearby critical areas. The primary purpose of this massive layer, normally called a floating floor, is to reduce airborne transmission

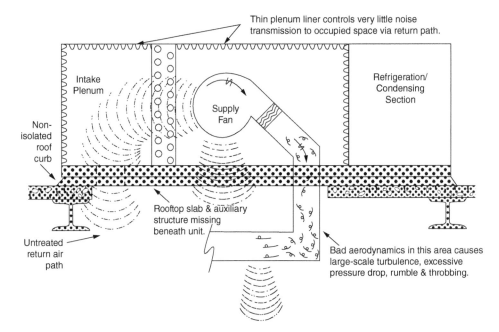

Figure 13.30 Very noisy rooftop unit installation [25]. ibid © ASHRAE Schaffer Guide.

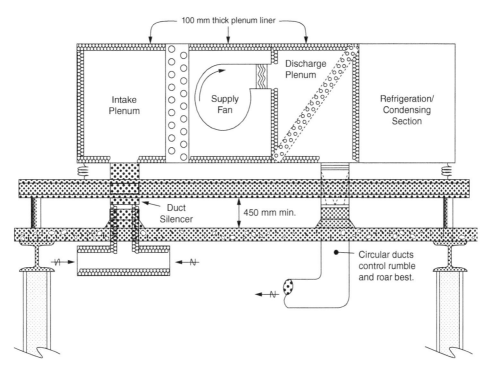

Figure 13.31 Very quiet rooftop installation [25]. ibid © ASHRAE Schaffer Guide.

through the floor rather than reduce vibration transmission [28, 29]. Although lightweight walls are used effectively as dividing partitions between two areas where speech privacy is required, these provide only little transmission loss at low frequencies.

Unfortunately, most of the mechanical room equipment (especially large centrifugal fans) radiate strongly at low frequencies, which causes a problem for the design engineer. Indeed, sound pressure levels in excess of 110 dB in the 63 and 125 Hz one-octave bands are quite common. Since doubling the mass of the floor slab results in an increase in its transmission loss of only about 5 dB, in order to attain the required transmission loss, the floor design would have to become both economically and structurally intolerable. This is exactly the same situation as discussed for dividing walls in Chapter 12 in which it was shown that a pair of double walls separated by an air space was very much more effective than an equivalent double-weight single wall. In this case, a second layer is added to the floor, which is separated from the structural slab by an air space. The upper layer, which is generally known as a "floating floor," is supported by and isolated from the structural slab.

The floating floor is normally constructed by pouring the floating concrete floor on plywood panels which are resiliently mounted on the structural slab by small vibration isolation pads (see Figure 13.32) [20, 27]. The floating floor slab usually has a minimum thickness of 100 mm with a separation of 25–100 mm between the floating and structural slabs. The natural frequency of the floating floor system depends on the stiffness of the vibration isolation pads and the stiffness of the airspace between the two slabs. With a 50 mm airspace and a 100 mm thick floating floor, the natural frequency is normally about 18 Hz.

The vibration isolation pads are from 5 to 7.5 cm (2 to 3 in.) thick, spaced approximately 0.3 m (1 ft) apart, and made of molded glass fiber. They are coated with flexible, moisture-impervious elastomeric membranes, usually made of neoprene, as shown in Figure 13.33. It is essential that the pads do not break down under load, even over a long period of time, and that they are resistant to attack by oil and water, which may leak through the floating floor slab. Once the floating floor and the mechanical room equipment have been installed, it would be extremely costly, if not impossible, to take up the floor and renew the isolation pads.

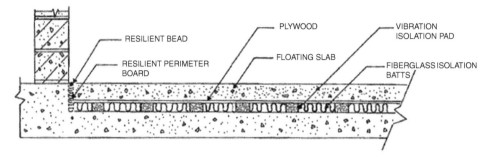

Figure 13.32 A typical floating concrete floor construction of the type commonly used to isolate noise and vibration from mechanical rooms [27].

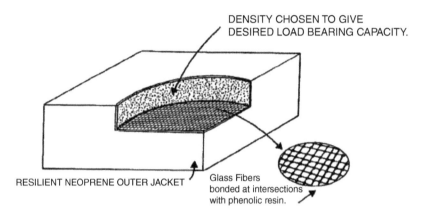

Figure 13.33 Floating floor vibration isolators-molded precompressed glass fibers [27].

Under load, these pads have the unique property of being able to maintain a constant natural frequency of approximately 14–16 Hz throughout their entire operating range, whether lightly or heavily loaded. Hence they can perform well, even though the exact floor load distribution may not have been accurately determined. This low-frequency region (14–16 Hz) lies below the lower human audible limit and below the natural frequency of most mechanical room equipment. One should be careful to check that the natural frequency of the isolators does not correspond to the lowest natural frequency of the structural slab. This would cause resonance and not only result in zero or negative isolation, but could result in the fracture of the structural slab itself [27].

The resonance (natural) frequency, f_n, of a standard concrete floor slab may be determined from

$$f_n = \frac{1}{2\pi}\sqrt{\frac{S_f}{\rho_f}}, \tag{13.4}$$

where S_f = dynamic stiffness per unit area in N/m^3 and ρ_f = mass per unit area in kg/m^2. A useful "rule of thumb" estimate for the natural frequency of such a concrete slab may be obtained from

$$f_n = 55/\sqrt{\ell}, \tag{13.5}$$

where ℓ is floor span in metres, or

$$f_n = 100/\sqrt{\ell}, \tag{13.6}$$

where ℓ is the floor span in feet.

Example 13.4   Estimate the fundamental natural frequency of concrete floor span in feet and metres with (a) 9.09 m (30 ft) span, and (b) 18.18 m (60 ft) span.

Solution

a) From Eq. (13.5), $f_n = 55/\sqrt{9.09} = 55/3.014 = 18.25$ Hz
   From Eq. (13.6), $f_n = 100/\sqrt{30} = 100/5.48 = 18.25$ Hz
b) From Eq. (13.5), $f_n = 55/\sqrt{18.18} = 55/4.26 = 12.91$ Hz
   From Eq. (13.6), $f_n = 100/\sqrt{60} = 100/7.746 = 12.91$ Hz.

The improvement ΔL_n in the impact sound transmission of a floating floor varies from between 30 dB per octave for resonantly reacting floors (usually the type used in mechanical rooms) to 40 dB per octave for locally reacting floors. See also discussion on impact sound transmission through floors in Chapter 12.

Apart from increasing the airborne transmission loss and the impact noise rating, a floating floor has one more extremely important advantage. It considerably reduces the amount of acoustical and vibrational energy flowing into the mechanical room structural slab and hence into the whole building structure. In modern concrete multistory buildings, vibration energy can be propagated to all parts of the building with very little attenuation and then can be easily re-radiated as sound. Hence, the attenuation of this vibration energy must be increased and the floating floor meets this requirement extremely well. Of course, the use of floating floors is not confined to mechanical rooms, and they are often employed in other areas of a building. For example, with music practice rooms or with pedestrian malls, which may pass over low noise areas, floating floors can be used to considerably reduce impact and airborne noise to such an extent that it is undetectable below [27].

Some typical mechanical room floating floor perimeter and dividing wall structural details are shown in Figure 13.34. The waterproof membrane under the plywood on which the concrete is poured prevents water and oil from reaching the isolation pads and prevents the concrete from forming bridging paths between the floating and structural floor slabs. A completely continuous floating floor is not always necessary, and in some cases a floating base beneath certain noisy pieces of mechanical room equipment may be sufficient. When the sound pressure level in a fan plenum chamber far exceeds that from the surrounding mechanical room equipment, then the whole plenum chamber should be mounted on a floating floor.

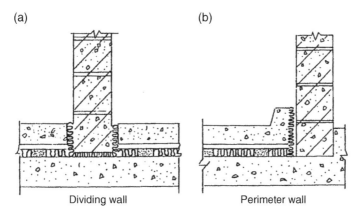

(a)   (b)

Dividing wall   Perimeter wall

Figure 13.34   Constructional details of floating floors at the base of dividing and perimeter walls [27].

13.6.2 Vibration Control of Equipment

Since vibration in the audible frequency range can easily be transformed directly into noise, or propagated to some other part of the building and then radiated as noise, it is of extreme importance to control all vibration in the mechanical room to within tolerable limits. Ideally, the mechanical room should be located well away from critical areas in the building, but when this is not possible, even greater care must be taken with the control of vibration. Although it might seem that the best approach should be to furnish all pieces of equipment with vibration isolators, this is not necessarily so. The first line of attack should be to ensure that each piece of equipment is selected to produce minimum noise and is operated under its specified conditions. For example, one should choose, where possible, rotating equipment units in preference to the equivalent reciprocating ones since, in general, the latter type produces much more objectionable noise. Furthermore, it is extremely important to see that each piece of equipment is balanced, both statically and dynamically, to within the recommended tolerances [27].

A major concern to the design engineer is to avoid resonances both in the equipment, in its supports, and in the building structure. At resonance, a large vibrational amplitude is developed, which may be accompanied by excessive radiated noise and stress. The stresses set up during the resonance may ultimately lead to fatigue in equipment or its support, or – even more disastrously – in the building structure (i.e. the floor)! One should, therefore, carefully check whether any equipment is to be operated close to (within ±25%) any critical speed in the machine, structural resonance in its support, or building structural resonance. Since the floor resonances and machine critical speeds are almost fixed, the operating speeds must be chosen to be well away from both of these. The only concern, then, is the support. In general, support resonances may be avoided by making the support very much less flexible than the flexibility that would result in resonance or by using a separate isolator with much higher flexibility. Such an increase in flexibility automatically leads to increased static deflection.

13.6.3 Selection of Vibration Isolators

The theory of vibration isolation presented in Chapter 2 shows that in order to reduce the magnitude of the force transmitted by an isolator to its support, one has to choose its natural frequency, f_n, so that the ratio of the forcing frequency to the natural frequency, (f/f_n) is much greater than $\sqrt{2}$. This means that for most air-conditioning and ventilating equipment operating in a general range of 500 to 1800 rpm (i.e. 8–30 Hz), efficient vibration isolators should have a natural frequency well below this range. Such a low natural frequency requires a large static deflection, which in such cases may be of the order of 0.5–5 cm (0.25–2 in.).

In describing practical vibration isolators, the following parameters are often used:

$$\text{Isolation efficiency} = 100\left[1 - \frac{1}{(f/f_n)^2 - 1}\right], \text{for } (f/f_n) \gg 2 \tag{13.7}$$

and

$$f_n = 4.99\sqrt{1/d}, \tag{13.8}$$

for d = static deflection in centimetres, or

$$f_n = 3.13\sqrt{1/d}, \tag{13.9}$$

for d = static deflection in inches.

The natural frequency in cycles per minute would clearly be $f_n \times 60$.

Figure 13.35 shows the relationship between f, f_n, d, and efficiency for a linear one-degree-of-freedom mass-spring-isolator system.

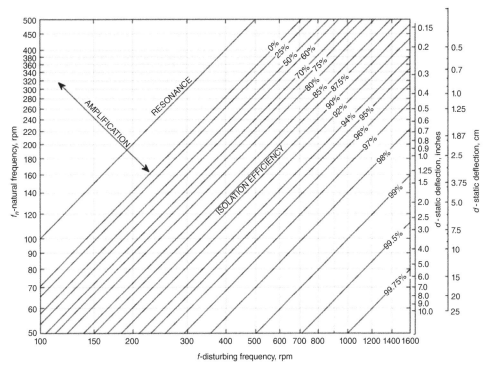

Figure 13.35 Relationship between isolation efficiency, disturbing frequency, natural frequency, and static deflection.

Example 13.5 Various items of equipment each of 500 kg mass are operated at 600, 1200 and 1800 rpm. Determine the natural frequency f_n and isolation efficiency achieved at each of the three operating conditions for isolators which produce static deflections of 0.5 cm (0.2 in.), 1.0 cm (0.5 in.), 2.5 cm (1.0 in.) and 5 cm (2.0 in.).

Solution

The natural frequency is unchanged by the operating rpm of each machine. The natural frequency and isolation efficiency are determined for each operating condition using Eq. (13.7) and Eq. (13.8), and are given in Table 13.7.

Table 13.7 Calculation of natural frequency f_n and static deflection needed to produce the isolation efficiency, η, for each rpm.

Machine rpm	600 rpm	1200 rpm	1800 rpm
Frequency, f	10 Hz	20 Hz	30 Hz
d	0.5 cm	0.5 cm	0.5 cm
f_n	7.07 Hz	7.07 Hz	7.07 Hz
f/f_n	1.414	2.83	4.24
$(f/f_n)^2 - 1$	1.0	7.0	17
η %	0%	86%	94%
d	1.0 cm	1.0 cm	1.0 cm
f_n	5.0 Hz	5.0 Hz	5.0 Hz
f/f_n	2.0	4.0	6.0

(Continued)

Table 13.7 (Continued)

Machine rpm	600 rpm	1200 rpm	1800 rpm
Frequency, f	10 Hz	20 Hz	30 Hz
$(f/f_n)^2 - 1$	3.0	15.0	35.0
η %	67%	93%	97%
d	2.5 cm	2.5 cm	2.5 cm
f_n	3.16 Hz	3.16 Hz	3.16 Hz
f/f_n	3.16	6.33	9.49
$(f/f_n)^2 - 1$	9.0	39.0	89.0
η %	89%	97%	99%
d	5.0 cm	5.0 cm	5.0 cm
f_n	2.24 Hz	2.24 Hz	2.24 Hz
f/f_n	4.46	8.92	13.4
$(f/f_n)^2 - 1$	19	79	179
η %	95%	99%	99%

Example 13.6 Spring isolators normally have small damping but are durable with well-defined performance. Figure 13.35 provides the performance of isolators, in which damping is neglected. Calculate the properties of a spring isolation system to achieve 95% isolation efficiency using Figure 13.35 for a disturbing frequency of (a) 600 rpm and (b) 1200 rpm.

Solution

a) For 600 rpm (10 Hz), Figure 13.35 gives a static deflection $d = 5.0$ cm (2 in.) for an isolation efficiency of 95% and a natural frequency f_n of 130 rpm (2.17 Hz).
b) For 1200 rpm (20 Hz), $d = 1.3$ cm (0.5 in.) and $f_n = 260$ rpm (4.3 Hz).

Although there are no exact criteria for the minimum isolation efficiency, one should attempt to choose an isolator providing the maximum possible isolation efficiency for practical conditions of static deflection, stability, and vibration amplitude. Generally, values in excess of 90% are adequate.

The nature of the exciting force should first be determined to see whether it is periodic or impulsive. If it is impulsive, then a different approach is required. However, in mechanical rooms, the exciting forces are almost certain to be random or periodic in nature and may arise from a variety of sources including rotor imbalance or wear, pressure fluctuations, and turbulence in fan housings or in combustion processes. Unbalanced forces vary as the square of the speed and so can easily become intolerable.

Equipment isolators with natural frequencies close to the floor resonance frequency should be strictly avoided, and hence some estimate of the floor resonance frequency should be made at the outset of any isolator selection. As the floor span increases, its deflection also increases; hence, isolators used on large span floors require high static deflections so that their natural frequencies do not coincide with the floor resonances. Table 13.8 shows recommended isolator static deflections for a variety of mechanical room equipment operating at various speeds on several concrete floors with spans ranging from 6 to 18 m (20 to 50 ft). Isolators should be selected not only to have adequate vibration isolation efficiency but also to compensate

Table 13.8 Recommended minimum static deflections for vibration isolators used with a variant of mechanical room equipment.

Type of equipment		Noncritical locations						Critical location			
		Basement below grade		20-ft (6.1 m) floor span		30-ft (9.15 m) floor span		40-ft (12.2 m) floor span		50-ft (15.2 m) floor span	
		mm	in.	mm	in.	mm	in.	mm	in.	mm	in.
Centrifugal fans and high-pressure package units											
Floor-mounted units	Up to 20 hp										
	175 to 300 rpm	8.9	0.35	63.5	2.5	63.5	2.5	88.9	3.5	127	5.0
	301 to 500 rpm	8.9	0.35	44.45	1.75	44.45	1.75	63.5	2.5	88.9	3.5
	Above 500 rpm	8.9	0.35	25.4	1.0	25.4	1.0	44.45	1.75	63.5	2.5
	Above 20 hp										
	175 to 300 rpm	8.9	0.35	63.5	2.5	88.9	3.5	127	5.0	139.7	5.5
	301 to 500 rpm	8.9	0.35	44.45	1.75	63.5	2.5	88.9	3.5	127	5.0
	Above 500 rpm	8.9	0.35	25.4	1.0	44.45	1.75	63.5	2.5	88.9	3.5
Vent sets and low-pressure package units											
Suspended units	Up to 5 hp	19.05	0.75	25.4	1.0	25.4	1.0	25.4	1.0	25.4	1.0
	Above 5 hp										
	175 to 500 rpm	31.75	1.25	31.75	1.25	31.75	1.25	44.45	1.75	63.5	2.5
	Above 500 rpm	25.4	1.0	25.4	1.0	25.4	1.0	31.75	1.25	43.2	1.7
Floor-mounted units	Up to 5 hp	8.9	0.35	25.4	1.0	25.4	1.0	25.4	1.0	25.4	1.0
	Above 5 hp										
	175 to 500 rpm	8.9	0.35	44.45	1.75	44.45	1.75	44.45	1.75	63.5	2.5
	Above 500 rpm	8.9	0.35	25.4	1.0	25.4	1.0	38.1	1.5	43.2	1.7
Pumps											
Close coupled	Up to 5 hp	8.9	0.35	8.9	0.35	25.4	1.0	25.4	1.0	25.4	1.0
	Above 5 hp	19.05	0.75	25.4	1.0	38.1	1.5	63.5	2.5	63.5	2.5
Base coupled	—	—	—	—	—	—	—	—	—	—	—
Refrigeration machines and boilers											
Reciprocating air or refrigeration compressors	500 to 750 rpm	25.4	1.0	38.1	1.5	63.5	2.5	69.9	2.75	88.9	3.5
	Above 750 rpm	25.4	1.0	25.4	1.0	38.1	1.5	63.5	2.5	68.6	2.7
Reciprocating chillers or heat pumps	500 to 750 rpm	25.4	1.0	38.1	1.5	63.5	2.5	69.9	2.75	88.9	3.5
	Above 750 rpm	25.4	1.0	25.4	1.0	38.1	1.5	63.5	2.5	68.6	2.7
Package boiler units	—	6.35	0.25	6.35	0.25	25.4	1.0	44.45	1.75	68.6	2.7

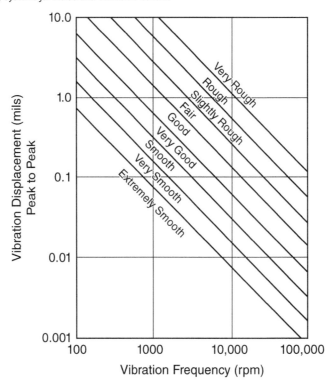

Figure 13.36 General machinery vibration amplitude severity chart (1 mil = 0.001 in. = 0.0254 mm).

for floor flexibility [20]. Longer floor spans are more flexible and more prone to vibration. Table 13.8 shows how isolators can be selected with consideration of equipment operating speeds and power as well as floor span. By specifying isolator deflections rather than isolation efficiency, the design engineer can compensate for floor stiffness by selecting isolators with greater deflection than the supporting floor [20].

Typical vibration amplitudes and vibration severity curves are shown in Figure 13.36 for various steady machine speeds (in rpm). However, while the machine is being started or stopped, much larger amplitudes may occur as the machine speed (and hence the related forcing frequency) momentarily passes through and coincides with the isolator system's natural frequency. For this reason, the isolator should have sufficient internal damping to keep the displacement within an acceptable limit at such resonances. It is also necessary to ensure that all connections (piping, wiring, and ducts) to the isolated equipment allow the machine to move at least 0.3 cm (0.1 in.) and that they have much less stiffness than the equipment isolators [24].

The vibration displacement of a machine mounted on isolators of adequate flexibility is independent of the isolator's static deflection. Furthermore, the displacement can be significantly reduced by adding a massive concrete inertia base to the equipment. Inertia blocks are usually required for fans giving static pressures in excess of 9.0 cm (3.5 in.) of water or having a mechanical power greater than 40 HP and for base-mounted pumps of over 10 HP. The base should have a weight at least 1.5–3 times the weight of the fan or compressor mounted on it.

The addition of an inertia base also lowers the center of gravity of the isolated equipment and in doing so helps to decouple the horizontal and rotational modes of vibration. The extra specification that each isolator shall have equal horizontal and vertical stiffnesses (iso-stiff) also helps to decouple these modes. If they were well coupled, the result would be a rocking motion (rotation about the center of gravity) which, apart from being objectionable for the satisfactory performance of the equipment, might very well lead to failure. In order to achieve complete decoupling, the following conditions must be met:

1) The center of gravity of the base of the equipment must be on the same horizontal plane as the isolators. This can be achieved only by using a T-shaped inertia block.
2) Each isolator should be iso-stiff.
3) The spacing between the isolators must be twice the radius of gyration of the system.

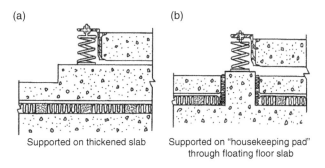

(a)   (b)

Supported on thickened slab   Supported on "housekeeping pad" through floating floor slab

Figure 13.37 Isolation and support of concrete inertia bases from a mechanical room floor [27].

When mode coupling is considered, it is generally found that isolators have at least one natural frequency (either horizontal or rotational) greater than the vertical natural frequency [27].

Consequently, in order to avoid resonance and ensure low transmissibility, all these natural frequencies should fall below the limit imposed by the forcing frequency. This results in an even greater static deflection being required than if mode coupling were not considered. The greater static deflection is provided by softer isolators. The consequent reduction in vertical and horizontal stiffness, hence, further increases the possibility of isolator instability. This possibility can be reduced by decoupling the isolator modes as described above. Figure 13.37 shows the mounting of some commonly used vibration-isolated inertia bases on a floating floor, as might be found in a ventilation mechanical room in a modern building.

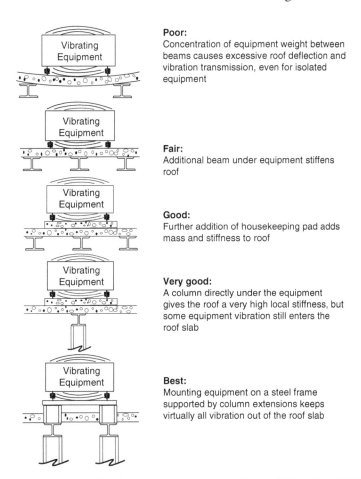

Poor:
Concentration of equipment weight between beams causes excessive roof deflection and vibration transmission, even for isolated equipment

Fair:
Additional beam under equipment stiffens roof

Good:
Further addition of housekeeping pad adds mass and stiffness to roof

Very good:
A column directly under the equipment gives the roof a very high local stiffness, but some equipment vibration still enters the roof slab

Best:
Mounting equipment on a steel frame supported by column extensions keeps virtually all vibration out of the roof slab

Figure 13.38 Structural support for vibration control of rooftop equipment [20]. ibid © ASHRAE Schaffer Guide.

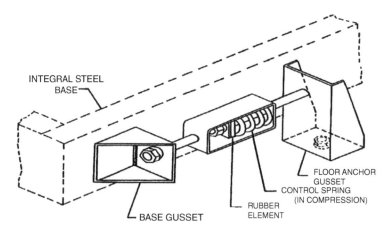

Figure 13.39 A typical method used to provide horizontal restraint for a vibration-isolated inertia base [27].

Rooftop mounted mechanical systems require special attention to avoid structural transmission of vibration and noise. Figure 13.38 shows how a roof should be stiffened and a housekeeping pad (inertia base) added to reduce machine vibration for roof-mounted equipment. Columns placed immediately below the equipment can minimize vibration transmitted through the building and in particular to noise-sensitive occupied spaces below.

In cases where large static or dynamic horizontal forces may be present (i.e. in the case of a fan producing a water pressure of 15 cm (6 in.), some restriction must be placed on the horizontal movement of the inertia base. This is done with horizontal vibration isolators positioned so as to be in compression under load. When extreme movement has to be prevented, concrete or steel snubbers are often incorporated. Control of horizontal motion is of particular importance for equipment mounted in moving vehicles since sudden impacts could easily knock the equipment off its vibration mounts, with disastrous results! A commonly used method for controlling and restraining horizontal motion is shown in Figure 13.39.

In selecting a particular resilient material for use in a vibration isolator, many factors have to be considered. The foremost requirement is being able to achieve the desired static deflection. Other factors of considerable importance include cost, mechanical, and chemical stability, internal damping, linearity, size, and shape. Internal damping is necessary to reduce the vibration amplitude at resonance (i.e. during start-up or shutdown), but it results in less isolation in the operating region. It is partly for this reason that steel spring isolators, which have very little internal damping, are usually equipped with or rested on a rubber or neoprene base. This has the added advantage of reducing the effect of standing wave resonances set up in the spring at frequencies where the spring height is an integral number of half-wavelengths. These resonances often occur in the mid-speech frequency range and appear as successive peaks in the isolator transmissibility curve. The heights of the resonance peaks are found to decrease with increasing forcing frequency, and for the resonances to disappear entirely at high frequencies (Figure 13.40). The addition of a rubber pad to a spring isolator thus increases its damping and noise isolation characteristics.

Another important consideration in the selection of an isolator material is the ease and accuracy with which its natural frequency and transmissibility can be calculated. In this respect, steel springs are superior since both the natural frequency and transmissibility can be easily and accurately predicted. On the other hand, it is fairly difficult to perform an equivalent calculation for rubber or neoprene. This is mainly due to the fact that the dynamic modulus of rubber is greater than its static modulus, although this depends upon the hardness of the material. Nevertheless, rubber and neoprene have several special advantages over steel springs and other resilient materials which make them particularly useful for many applications. The advantages and disadvantages of several commonly used resilient materials employed in vibration isolators are compared in Table 13.9 [27].

In general, steel springs should be used when static deflections in excess of 1.25 cm (0.5 in.) are required; while rubber isolators should be used for static deflections of less than 0.5 cm (0.2 in.).

When selecting a spring isolator, it is very important to check that the spring will not be overstressed or unstable when installed under the desired load. Such a system is assumed stable when the following condition is met:

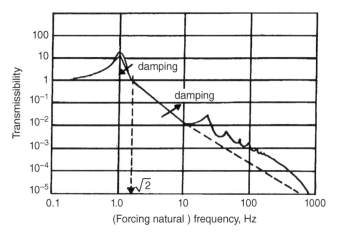

Figure 13.40 Typical standing wave resonance as observed in a spring isolator.

Table 13.9 Comparison of material used for vibration isolators [27].

Material	Advantages	Disadvantages
Rubber	1) Easily molded and bonded to metal for easy attachment to equipment 2) Relatively good internal damping 3) Good for shock isolation because of its great energy storage capacity 4) Can be used in compression and shear; combinations of these result in a wide range of characteristics to suit many needs 5) Can be designed to have a constant natural frequency independent of load once in working region.	1) Cannot withstand strain for long periods or else creep will set in; usually limits of 10% in compression and 35% in shear are applicable 2) Cannot withstand very long or high temperatures 3) Cannot be used in tension or else will fail 4) Extremely difficult to predict characteristics 5) It is compressible thus requiring room to bulge outwards under compression 6) Useful only for static deflections less than 0.5 cm (0.2 in.).
Steel springs	1) Almost unlimited life, even under large stress, without drift or creep 2) Easily designed to give required characteristics of natural frequency and stiffness (in both directions) 3) Can be loaded in tension without failure, although almost always used in compression 4) Large static deflections can be obtained, hence used mainly for low-frequency isolation.	1) Low internal damping resulting in large amplitude at resonance 2) Possibility of instability, especially if static deflection is large or lateral stiffness too small 3) Usually requires steel housing 4) Internal resonances may be set up in spring.
Cork	1) Useful under concrete foundations; since it can cover the entire area, it simplifies the pouring of concrete 2) It is compressible and hence does not need room to bulge, unlike rubber 3) High internal damping.	1) Very limited flexibility, hence rarely used to isolate primary vibrations.

$$\frac{K_x}{K_y} \geq 1.2\frac{d}{h}, \tag{13.10}$$

where K_x and K_y = stiffnesses of the spring in the horizontal and vertical directions, respectively; d = static deflection (vertical); and h = working height.

For the previously stated reasons, one usually requires a spring to be iso-stiff, and in order for such a spring to be stable, the static deflection should be less than 80% of the working height. A static deflection of 0.3–0.6 of the working height should be the design objective. In order to prevent dynamic spring fatigue, helical springs should be precompressed and should not be used in circumstances where the

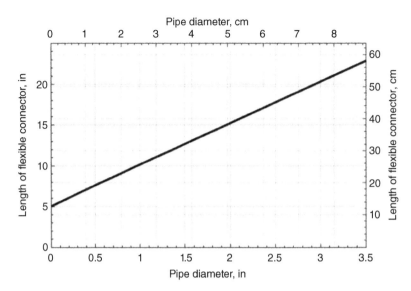

Figure 13.41 Length of flexible pipe connector required to give adequate vibration isolation [24].

shear stress would exceed 620 000 kPa (90 000 psi) at full static deflection for hot coiled springs of wire diameter greater than 0.5 cm (0.2 in.) [30].

13.6.4 Vibration Isolation of Ducts, Pipes, and Wiring

Because a machine mounted on vibration isolators is subjected to large amplitude motions, all connections to the machine from piping, wiring, and ductwork should be flexible enough so as not to restrict the movement of the machine or allow significant amounts of noise and vibration to pass through the connections into the building structure. In order to achieve these requirements, commercially available flexible connectors are commonly used. In many cases, however, the fluid in a pipe or duct short-circuits such a flexible connector. For example, this is particularly true for metal types, which do not reduce significantly fluid-borne vibration from pumps. Similarly, flexible duct connectors are usually ineffective in providing good isolation of air pulsations from fans. The length of the connector required for adequate isolation may be obtained from Figure 13.41.

Example 13.7 Find the recommended length of flexible pipe connector needed to provide adequate vibration isolation for pipes of diameter (a) 2.5 cm (1 in.) and (b) 5.0 cm (2 in.), and (c) 7.5 cm (3 in.)

Solution

From Figure 13.41, the flexible pipe connector lengths are: (a) 25 cm (10 in.), (b) 37.5 cm (15 in.), and (c) 50 cm (20 in.).

If ducts and pipes pass through the walls of a mechanical room, great care should be taken to prevent leaks that would allow transmission of airborne noise. Sealing material should not be too rigid or vibration transmission can also occur. Figure 13.42 shows the penetration of pipes and ducts through walls and floors and how noise and vibration transmission can be minimized.

Resilient vibration-isolation hangers are often very effective in reducing the amount of such vibration being transmitted into the building structure from pipes and ducts. Hangers used close to a piece of isolated equipment (i.e. the first few hangers) should have a static deflection equal to or greater than that of the main machine isolators, in order not to restrict the machine movement. The remaining hangers should be selected to have

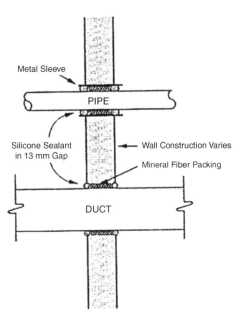

Figure 13.42 Duct and pipe penetrations through walls. Note: Support pipes and ducts on both sides of the wall without permitting contact with the wall or its framing. ibid © ASHRAE Schaffer Guide.

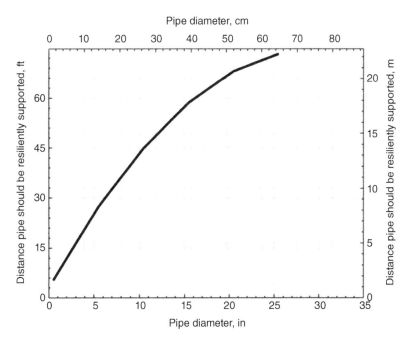

Figure 13.43 All piping connections to a vibrating source should be resiliently supported for a distance as shown above (or to the nearest flexible connector) [24].

either 2.5 cm (1 in.) static deflection or one-half the static deflection of the isolated equipment, whichever is the lesser. These hangers should be spaced at least 15 pipe diameters apart and should be continued for a distance, as shown in Figure 13.43. Spring hangers should also be able to take a 15° angular misalignment without binding. All wiring connections to vibration-isolated units should be provided with flexible wire with a 360° loop.

Example 13.8   Pipes are connected to equipment, which has undesirable vibration particularly during start up and shut down. Determine the distance from the equipment that should be resiliently supported for pipe diameters: (a) 2.5 cm (1 in.) and (b) 5.0 cm (2 in.), and (c) 7.5 cm (3 in.).

Solution

From Figure 13.43, the pipe lengths that should be resiliently supported are: (a) 2.5 m (7.5 ft); (b) 4 m (12 ft), and (c) 5 m (16.5 ft).

13.7   Sound Attenuation in Ventilation Systems

Many of the elements which make up a typical ventilation system are capable of providing some degree of sound attenuation. It is, therefore, very important to be able to estimate this attenuation in order to ensure acceptable acoustical conditions in all the areas to be served by the ventilation unit. This implies that there should be neither too little attenuation, since an excessively noisy condition would result, nor too much attenuation, since this could result in extremely low background noise levels which might significantly affect privacy in the area [27].

The elements and physical phenomena which should be considered in evaluating the sound attenuation characteristics of a complete ventilation system are as follows:

1) Fan plenum chamber(s)
2) Ducts
3) Specially constructed sound attenuator(s) (silencer(s))
4) Openings of ducts into large spaces which cause end reflections
5) Changes in duct cross-sectional area
6) Elbows and take-offs.

The natural attenuation given by these elements may be substantially increased when necessary by relatively simple means.

13.7.1   Use of Fiberglass in Plenum Chambers, Mufflers, and HVAC Ducts

Fiberglass materials are commonly used in HVAC systems. Such materials provide the most convenient and cost-effective way to control noise in such systems. In the 1990s there was concern that fiberglass fibers may be carcinogen and that such materials used in ductwork may promote microbial growth. At that time the use of fiberglass in some institutional, educational, and medical establishments was severely limited or completely banned. However, in 1995 the International Agency for Research on Cancer (IARC) performed extensive research on carcinogenicity of fiberglass and found inadequate evidence to link its use to cancer in humans. Current evidence suggests that both moisture and dirt are required for microbial growth and care should be taken to remove both of these with filters and evaporator coils. The complete elimination of fiberglass duct liners in HVAC systems would require longer duct runs, larger fans and plenums, etc. which would result in much less satisfactory and more expensive systems.

13.7.2   Attenuation of Plenum Chambers

Plenum chambers are normally inserted into the HVAC system at the outlet section of a fan before the air distribution system of a building. They have two major purposes: (i) to smooth out the turbulent flow leaving the fan and (ii) to absorb some of the fan noise. Acoustically a plenum chamber may be simply defined as a large enclosed

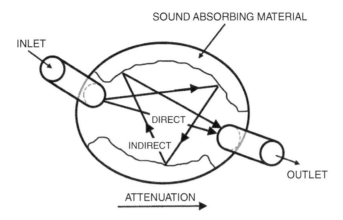

Figure 13.44 Schematic representation of an acoustical plenum chamber.

space containing sound-absorbing material as shown schematically in Figure 13.44. In the design of plenum chambers, it is normal to assume that the typical dimension d, of the chamber is much greater than the acoustical wavelength, i.e. $d \gg \lambda$, so that the sound field inside can be considered to have direct and reverberant components. This assumption, of course, fails at low frequency. There must be at least one inlet and one outlet in order that the flow and the acoustical path is continuous. Of the sound energy which enters the chamber, part will be radiated directly to the outlet and part will enter the outlet only after the successive reflections from the chamber walls. The remainder will be dissipated by any absorbing material present in the space.

Assuming the sound field in the chamber is reverberant, and that $d \gg \lambda$, the amount of energy reaching the outlet by a direct path is directly proportional to the area of the outlet and falls off with the square of the distance between the inlet and outlet. It is also reduced by a directional effect if the outlet is not exactly opposite the inlet and will disappear altogether if the outlet and inlet are on the same plane. The amount of energy reaching the outlet from the reverberant field is also proportional to the outlet area and to the energy set up in the reverberant field within the chamber. This is dependent upon the total amount of acoustical absorption present in the space.

Acoustically lined plenum chambers are very frequently used in large air-conditioning or ventilation systems where considerable attenuation is required to reduce the sound power produced by a fan. Although other methods of obtaining such attenuation are available from lined ductwork or commercial silencers, these are often expensive or insufficient on their own. Furthermore, unlike a plenum chamber, these are not capable of providing such high attenuation at low frequencies. A typical plenum chamber is shown diagrammatically in Figure 13.45 [31].

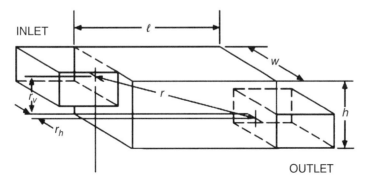

Figure 13.45 Schematic of end-in/end-out plenum.

The attenuation of a plenum chamber is approximately given by:

$$TL = -10\log\left\{S_{out}\left(\frac{Q\cos\theta}{4\pi r^2} + \frac{1-\overline{\alpha}_A}{S\overline{\alpha}_A}\right)\right\}, \tag{13.11}$$

where (see Figure 13.45): S_{out} = area of plenum outlet, m²; S = total inside surface area of the plenum, m²; r = distance between centers of inlet and outlet of plenum, m; Q = directivity factor (assumed to be 4); $\overline{\alpha}_A$ = average absorption coefficient of lining; θ = angle between vector r to long axis l of duct (see Eq. (13.13)).

The average absorption coefficient $\overline{\alpha}_A$ of the plenum lining is given by

$$\overline{\alpha}_A = \frac{S_1\alpha_1 + S_2\alpha_2}{S_1 + S_2}, \tag{13.12}$$

where α_1 = sound absorption coefficient of unlined inside surfaces of plenum; S_1 = surface area of unlined inside surfaces of plenum, m²; α_2 = sound absorption coefficient of lined inside surfaces of plenum; S_2 = surface area of lined surfaces inside plenum, m². If 100% of the inside surfaces of the plenum chamber are lined with a sound-absorbing material, $\overline{\alpha}_A = \alpha_2$. Table 9.3 gives the values of sound absorption coefficients for selected common plenum materials.

The value for $\cos\theta$ is obtained from

$$\cos\theta = \frac{\ell}{r} = \frac{l}{\sqrt{\ell^2 + r_v^2 + r_h^2}}, \tag{13.13}$$

where (see Figure 13.45): ℓ = length of plenum, m; r_v = vertical offset between axes of plenum inlet and outlet, m; r_h = horizontal offset between axes of plenum inlet and outlet, m.

Equation (13.11) assumes the plenum behaves as if it were a large enclosure. Thus, Eq. (13.11) is valid only for the case where the wavelength of sound is small compared to the characteristic dimensions of the plenum. For frequencies less than the "cut-on" frequency f_{co} higher modes cannot exist and only plane wave propagation is possible in the duct. Then the results predicted by Eq. (13.11) are usually not valid. Plane wave propagation in a duct exists at frequencies below f_{co} as follows:

$$f_{co} = c/2a \tag{13.14}$$

or

$$f_{co} = 0.586\,c/d, \tag{13.15}$$

where f_{co} = cut-on frequency, Hz; c = speed of sound in air, m/s; a = larger cross-sectional dimension of rectangular duct, m; d = diameter of circular duct, m.

The cut-on frequency f_{co} is the frequency above which cross modes and spinning modes (in circular section ducts) can exist in addition to plane waves. At frequencies below f_{co}, the actual attenuation normally is less than the values given by Eq. (13.11) by 5–10 dB. Eq. (13.11) usually applies best at frequencies of 1000 Hz and higher.

Example 13.9 A small plenum chamber is 2.0 m high, 1.0 m wide, and 2.0 m long. The shape of the plenum is similar to that shown in Figure 13.45. The inlet and outlet are each 0.6 m wide by 0.5 m high. The horizontal distance between centers of the plenum inlet and outlet is 0.35 m. The vertical distance is 1.45 m. The plenum is lined with 25 mm thick 64 kg/m³ density fiberglass insulation board. The inside surfaces of the plenum are lined with the fiberglass insulation. Determine the transmission loss associated with this plenum. See Table 9.3 for the values of the absorption coefficients.

Solution

The areas of the inlet section, outlet section, and plenum cross-section are:

$S_{in} = S_{out} = 0.6 \times 0.5 = 0.3 \text{ m}^2$, and $S_{pl} = 2.0 \times 1.0 = 2.0 \text{ m}^2$.

$\ell = 2.0 \text{ m}$, $r_v = 1.45 \text{ m}$, and $r_h = 0.35 \text{ m}$. The values of r and $\cos \theta$ are

$r = \sqrt{2^2 + 1.45^2 + 0.35^2} = 2.495 \text{ m}$, and $\cos \theta = 2.0/2.495 = 0.802$. The total inside surface area of the plenum is $S = 2 (2 \times 1) + 2 (2 \times 1) + 2 (2 \times 2) = 16 \text{ m}^2$. The total area of lined surfaces inside plenum is: $16 - 2(0.3) = 15.4 \text{ m}^2$. The value of f_{co} is:

$f_{co} = 343/(2 \times 0.6) = 285 \text{ Hz}$.

The results using Eq. (13.11) are tabulated below in Table 13.10.

As discussed before, the predicted attenuation values (TL) for 125 and 250 Hz may be too large, since these frequencies are less than the cut-on frequency, f_{co} of 285 Hz. The ASHRAE Handbook [20] gives a somewhat complicated empirical method to give better low frequency attenuation predictions below f_{co}.

Table 13.10 Predicted attenuation in Example 13.9.

	One-Octave Band Center Frequency, Hz					
	125	250	500	1000	2000	4000
α	0.07	0.23	0.48	0.83	0.88	0.80
$\bar{\alpha}_A$	0.10	0.26	0.50	0.84	0.88	0.81
$1 - \bar{\alpha}_A$	0.90	0.74	0.50	0.16	0.12	0.19
$S\bar{\alpha}_A$	1.62	3.99	7.69	12.88	13.62	12.44
$A = 10^3 \cos \Theta / \pi r^2$	41.0	41.0	41.0	41.0	41.0	41.0
$B = 10^3 (1 - \bar{\alpha}_A)/S\bar{\alpha}_A$	554.2	185.9	65.1	12.7	8.5	15.5
$C = S_{out}(A + B)$	0.18	0.07	0.03	0.02	0.01	0.02
TL, dB	7.5	11.7	15.0	17.9	18.3	17.7

At high frequencies, the above equation may yield values for the plenum attenuation which are sufficiently close to measured values for purposes of chamber design. However, at low frequencies, where the wavelength approaches or exceeds any of the chamber dimensions, the attenuation calculated from the above equation may be 5–10 dB less than that actually obtained with such a plenum chamber. In order to obtain good low frequency attenuation, the following requirements should be met:

1) The distance between the inlet and outlet should be made as large as possible.
2) The exits should be positioned so as to be well out of the line of sight of the inlet.
3) The acoustical lining on the plenum chamber walls should have good low-frequency absorption properties. This means that the lining should be between 5 and 10 cm (2 and 4 in.) thick. However, one should check carefully to see whether the extra cost involved – for example, in doubling the lining thickness – is justified by the extra attenuation obtained.

The chamber lining has a dual function in that it provides both sound absorption and thermal insulation. Although a thickness of 1.25–5 cm (0.5–2 in.) is normally adequate for the thermal insulation, a greater thickness of material is usually required to give adequate low-frequency sound absorption and, hence, plenum attenuation.

It is normal to include the necessary heating and cooling coils within or adjacent to the plenum chamber, and usually these coils do not produce any significant noise. However, their auxiliary equipment (such as pumps) can produce objectionable noise and vibration and should therefore be properly isolated. With large systems, it is usual to mount the fan on a separate vibration-isolated inertia block which, in turn, is isolated from a floating floor on which the whole plenum chamber rests. In smaller installations, the fan may be mounted directly on isolators on the floating floor. See the previous discussion concerning Figures 13.31, 13.34, and 13.38.

Many variations of the simple plenum chamber have been proposed. These have usually resulted from placing two or three single plenum chambers in series. Each additional chamber usually gives an extra 10 dB attenuation at high frequencies but very little improvement at low frequencies. It is for this reason that most of the plenum chambers found in practical HVAC systems are of the single-chamber type. Other variations include the placement of partial-height solid barriers within the chamber to reduce the direct sound pressure level at the outlet. However, in high-velocity systems, these can result in turbulence-generated noise produced at the edges of the barriers, which tends to nullify the attenuation provided by the chamber.

Many examples related to plenum chambers are to be found in architectural acoustics. A common example is found when the space above a drop ceiling is used as a common volume for return airflow from several rooms causing cross talk. Another example, is when a corridor is used in this way to reduce "cross talk" between nearby rooms with open doors. Here the corridor provides both acoustic absorption and a break in the direct path of sound propagation between the two doors. Ceiling tiles and carpet in the corridor can reduce the cross talk between open doors of bedrooms in a hotel or offices in a building. Cross talk can also occur between rooms through the supply or return air ducts linking two adjacent rooms. See Figure 13.46. This can be reduced by lining the supply or the return air duct with sound-absorbing materials.

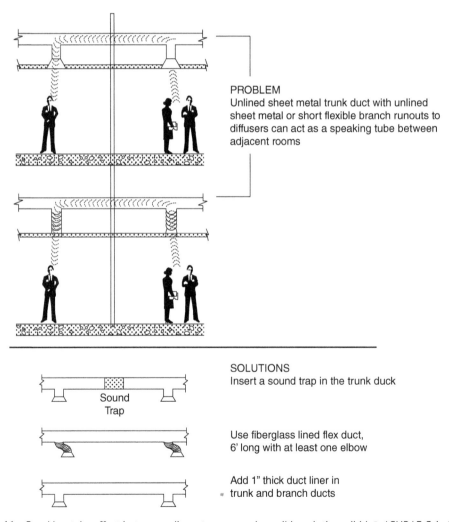

PROBLEM
Unlined sheet metal trunk duct with unlined sheet metal or short flexible branch runouts to diffusers can act as a speaking tube between adjacent rooms

SOLUTIONS
Insert a sound trap in the trunk duck

Use fiberglass lined flex duct, 6' long with at least one elbow

Add 1" thick duct liner in trunk and branch ducts

Figure 13.46 Speaking tube effect between adjacent rooms and possible solutions. ibid © ASHRAE Schaffer Guide.

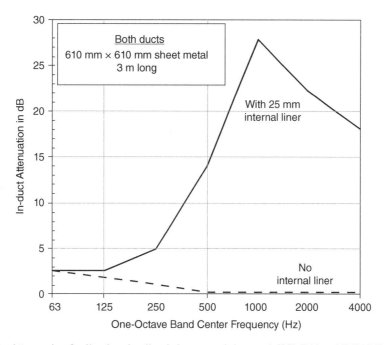

Figure 13.47 Attenuation for lined and unlined sheet metal ductwork [25]. ibid © ASHRAE Schaffer Guide.

13.7.3 Duct Attenuation

As acoustic energy is propagated through a duct, some of the energy is lost by transmission through the duct walls, and some is lost by internal vibration energy dissipation within the walls. Above the fundamental frequency of mechanical resonance of the duct walls, more acoustic energy is transmitted through the walls at low than at high frequencies; consequently, more sound attenuation is observed in unlined ducts at low than at high frequencies. See the broken line in Figure 13.47. However, if the duct walls are made very stiff, then they radiate much less low-frequency sound. This is the case for circular ducts where the increased mechanical stiffness results in lower low-frequency attenuation characteristics than for a corresponding rectangular duct (see Tables 13.11–13.13.)

Fry states that "when calculating the attenuation of rectangular ducts, it is necessary to take both sidewall duct dimensions into account." The attenuation for each dimension should be calculated separately and then added. Fry also suggests that upper limits for the predicted attenuations should be set in the octave bands as follow: 40 dB at 63 Hz, 30 dB at 125 Hz, 35 dB at 250 Hz, 40 dB at 500 Hz and 45 dB at 1000 Hz. Thus, large rectangular ducts should be chosen when good low frequency attenuation properties are required. Conversely, when ductwork passes directly over critical acoustical areas, circular ducts should be chosen (in preference to rectangular) in order to reduce the radiated low-frequency noise. This effect can be explained by the higher (breakout) transmission loss (TL) of circular ducts than rectangular ducts in the 63 and 125 Hz one-octave bands (see Figure 13.48). Flat oval ducts have a greater breakout TL at these frequencies than rectangular ducts, but a smaller TL than circular ones.

Table 13.11 In-duct attenuation in unlined straight ducts [8].

| | | **Attenuation in dB/metre run** | | | | | | |

Straight Duct Circular/oval or Rigid Walled (unlined)	x (mm)	One-Octave Band Center Frequency (Hz)						
		63	125	250	500	1 k	2 k	4 k
	75–200	0.07	0.10	0.10	0.16	0.33	0.33	0.33
	200–400	0.07	0.10	0.10	0.16	0.23	0.23	0.23
	400–800	0.07	0.07	0.07	0.10	0.16	0.16	0.16
	800–1500	0.03	0.03	0.03	0.07	0.07	0.07	0.07

Straight Duct rectangular (unlined)	x (mm) side dimension	One-Octave Band Center Frequency (Hz)						
		63	125	250	500	1 k	2 k	4 k
	75–200	0.16	0.33	0.49	0.33	0.33	0.33	0.33
	200–400	0.49	0.66	0.49	0.33	0.23	0.23	0.23
	400–800	0.82	0.66	0.33	0.16	0.16	0.16	0.16
	800–1500	0.66	0.33	0.16	0.10	0.07	0.07	0.07

Table 13.12 In-duct attenuation within externally lagged straight ducts [8].

| | | **Attenuation in dB/metre run** | | | | | | |

Circular/oval	x (mm)	One-Octave Band Center Frequency (Hz)						
		63	125	250	500	1 k	2 k	4 k
	75–200	0.14	0.20	0.20	0.32	0.33	0.33	0.33
	200–400	0.14	0.20	0.20	0.32	0.23	0.23	0.23
	400–800	0.14	0.14	0.14	0.20	0.16	0.16	0.16
	800–1500	0.06	0.06	0.06	0.14	0.07	0.07	0.07

Rectangular	x (mm) side dimension	One-Octave Band Center Frequency (Hz)						
		63	125	250	500	1 k	2 k	4 k
	75–200	0.33	0.66	1.00	0.66	0.33	0.33	0.33
	200–400	1.00	1.32	1.00	0.66	0.23	0.23	0.23
	400–800	1.64	1.32	0.66	0.32	0.16	0.16	0.16
	800–1500	1.32	0.66	0.32	0.20	0.07	0.07	0.07

Table 13.13 Sound attenuation dB/m of unlined sheet metal ducts at one-octave band center frequencies.

	63 Hz	125 Hz	250 Hz	> 250 Hz
Rectangular Duct Sizes				
150 × 150 mm	0.98	0.66	0.33	0.33
305 × 305 mm	1.15	0.66	0.33	0.20
610 × 610 mm	0.82	0.66	0.33	0.10
1220 × 1220 mm	0.49	0.33	0.23	0.07
Round Duct Sizes				
≤ 150 mm	0.10	0.10	0.16	0.16
180–380 mm	0.10	0.10	0.10	0.16
380–760 mm	0.07	0.07	0.07	0.10
760–1500 mm	0.03	0.03	0.03	0.07

Source: based on ASHRAE Ref. [20] and other sources.

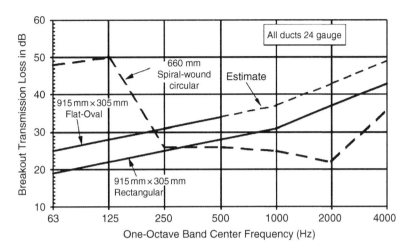

Figure 13.48 Breakout transmission loss for three types of sheet metal ductwork [25]. If duct is externally insulated, double attenuation values. ibid © ASHRAE Schaffer Guide.

Example 13.10 Determine the attenuation of a 610 mm × 610 mm cross-section of unlined duct of length 10 m. Use all available methods of prediction including Table 13.11, Figure 13.47 and Table 13.13.

Solution

Note that values given in the tables are dB/m, while in Figure 13.47, they are attenuation, dB, for a 3 m long duct. The 10 m values will be assumed to be 10 times those given in the tables, while the 10 m values obtained from Figure 13.47 are assumed to be given by $10/3 = 3.33$ times the values in the figure. The results of the calculations are given in Table 13.14.

Table 13.14 Predicted attenuation, dB, of a 610 mm × 610 mm unlined, 10 m long duct, in Example 13.10.

	One-Octave Band Center Frequency, Hz						
Source	63	125	250	500	1000	2000	4000
Table 13.11	8.2	6.6	3.3	1.6	1.6	1.6	1.6
Figure 13.47	2.5 × 3.33 = 8.33	2.0 × 3.33 = 6.7	1.0 × 3.33 = 3.33	0.5 × 3.33 = 1.7	0.5 × 3.33 = 1.7	0.5 × 3.33 = 1.7	0.5 × 3.33 = 1.7
Table 13.13	8.2	6.6	3.3	1.0	1.0	1.0	1.0

If the walls of a duct are partially or wholly internally lined with a sound-absorbing material, a significant increase in the resulting attenuation is observed (Figures 13.47 and 13.49). This increase is caused by the absorption of sound upon successive reflections within the duct and also by the extra damping given to the walls. If the duct is also externally lagged (usually for thermal reasons), approximately twice the attenuation values shown in Figure 13.47 are obtained at 125 and 250 Hz. For large lengths of ductwork, it is not possible to obtain more than approximately 30 dB attenuation because of flanking energy transmitted along the walls of the duct.

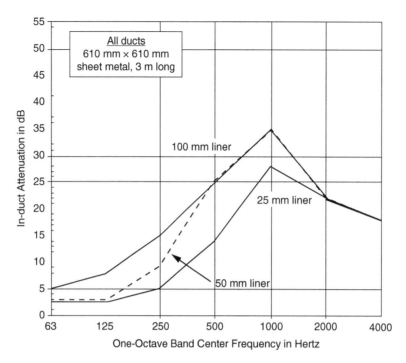

Figure 13.49 In-duct attenuation for various duct liner thicknesses [25]. ibid © ASHRAE Schaffer Guide.

Example 13.11   Determine the attenuation of a 610 mm × 610 mm cross-section internally lined duct of length 6 m. Use all available data for lining thickness of 25, 50, and 100 mm.

Solution

Figure 13.49 gives attenuation for a 3 m long duct. We will convert attenuations given to dB/m and then to attenuation for 6 m. See the results in Table 13.15.

Table 13.15   Predicted attenuation in decibels of a 610 mm × 610 mm lined, 6 m long duct, in Example 13.11.

Lining Thickness, mm	Source	One-Octave Band Center Frequency, Hz						
		63	125	250	500	1000	2000	4000
25	Figure 13.49	2.5 × 2 = 5	2.5 × 2 = 5	5 × 2 = 10	14 × 2 = 28	28 × 2 = 56	22 × 2 = 44	18 × 2 = 36
	Table 17 ASHRAE [20]	0.7 × 6 = 4.2	0.7 × 6 = 4.2	1.6 × 6 = 9.6	4.6 × 6 = 27.6	9.2 × 6 = 55.2	7.2 × 6 = 43.2	5.9 × 6 = 35.4
50	Figure 13.49	2.5 × 2 = 5	3.6 × 2 = 7	10 × 2 = 20	25 × 2 = 50	35 × 2 = 70	22 × 2 = 44	18 × 2 = 36
	Table 18 ASHRAE [20]	—	1.0 × 6 = 6	3.0 × 6 = 18	8.2 × 6 = 49.2	11.5 × 6 = 69	7.2 × 6 = 43.2	5.9 × 6 = 35.4
100	Figure 13.49	5 × 2 = 10	7.5 × 2 = 15	15 × 2 = 30	25 × 2 = 50	35 × 2 = 70	22 × 2 = 44	18 × 2 = 36

Example 13.12 Determine the attenuation of a 610 mm × 610 mm cross-section externally lagged duct of length 10 m. Use all available data.

Solution

Table 13.13 gives attenuation values in dB/m for unlined rectangular ducts. The values given will be doubled. Table 13.12 gives values for rectangular externally lagged ducts with side wall dimensions in the range 400–800 mm, which will be selected. The results of the calculations are given in Table 13.16.

Table 13.16 Calculated attenuation values in decibels for 10 m long duct in Example 13.12.

Source	Duct Length	One-Octave Band Center Frequency, Hz						
		63	125	250	500	1000	2000	4000
Table 13.13	1 m length	$0.82 \times 2 = 1.64$	$0.66 \times 2 = 1.32$	$0.33 \times 2 = 0.66$	$0.1 \times 2 = 0.2$	0.2	0.2	0.2
	10 m length	16.4	13.2	6.6	2.0	2.0	2.0	2.0
Table 13.12	1 m length	1.64	1.32	0.66	0.22	0.16	0.16	0.16
	10 m length	16.4	13.2	6.6	2.2	1.6	1.6	1.6

13.7.4 Sound Attenuators (Silencers)

When the natural sound-attenuation characteristics of a complete ventilation system are inadequate, some additional attenuation has to be introduced into the system. This is often accomplished by the installation of a specially constructed sound attenuator or silencer. Such a device may simply be a lined duct incorporating a splitter system as described above, designed to give the required additional attenuation. However, because air flows over the lining on the walls and splitters, there is a corresponding drop in static air pressure across the device. This pressure drop is dependent upon the airflow velocity and the percentage open area, as shown in Figure 13.50. More discussion concerning pressure drop in duct systems can be found in Ref. [32].

The following expression is useful for estimating the pressure drop ΔP across a typical splitter-type attenuator

$$20 \log (\Delta P) = 20 \log V - 10 \log S - 74 \text{ dB,} \qquad (13.16a)$$

where ΔP = pressure drop in inches of water; V = air flow velocity, ft/min; and S = open area, ft^2, or

$$20 \log (\Delta P) = 20 \log V - 10 \log S + 9.5 \text{ dB,} \qquad (13.16b)$$

where ΔP = pressure drop in pascals; V = air flow velocity, m/s; and S = open area, m^2.

In mechanical systems where large static pressure drops cannot be tolerated, alternative silencer designs are utilized. These silencers are usually circular in cross-section and contain not only acoustical lining on their inside walls but also a lined center body which provides additional sound attenuation and also streamlines the airflow through the unit. This reduces both the pressure drop across the silencer and the generation of noise by the flow of air through it.

The construction of common splitter cylinders is shown in Figure 13.51 including a cylindrical silencer. See also discussion in Chapter 10 concerning splitter silencers. The attenuation of a typical cylindrical silencer is

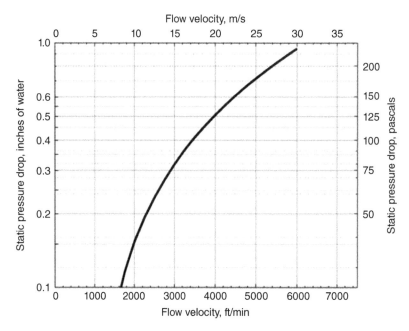

Figure 13.50 Variation of static pressure drop across a typical splitter-type sound attenuator for increasing air flow velocities with open area 30 cm^2 (1 ft^2) [27].

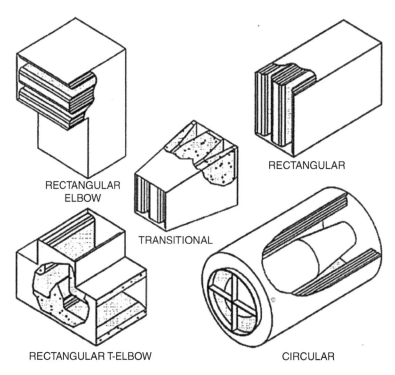

Figure 13.51 Construction of typical commercial silencers including a circular center body silencer [25]. ibid © ASHRAE Schaffer Guide.

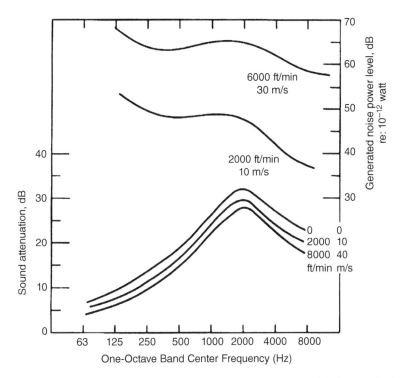

Figure 13.52 Attenuation and noise generation characteristics of typical cylindrical center body silencer [27].

shown in Figure 13.52. The attenuation usually decreases somewhat as the flow velocity is increased. Of more concern is the considerable increase in flow noise with increased flow velocity.

Care must be taken as to where to locate such silencers. Improper placement can result in severe degradation of performance so that noise paths bypass the silencer. See Figure 13.53. The best location for the silencer is in the wall of the mechanical room shown in Figure 13.53. A good location is for the silencer to be just inside the room. The two bottom locations give poor results, since noise can bypass the silencer in both cases.

The performance of silencers depends on their length, thickness of absorbing material and geometries. Figure 13.54 shows typical examples of available silencers with the lowest performance and best performance. Also shown is the attenuation of a 3 m long duct lined with 25 mm absorbing material. It is seen that as much as 30 dB attenuation can be obtained above 500 Hz, but below this frequency the attenuation is poor.

Figure 13.55 compares the attenuation achieved by typical 1.5 m long silencers. Coating the absorbing material with film protection and use of honeycomb spacers can reduce the effectiveness at low and medium frequencies, but increase this at high frequency. A typical reactive silencer can have a similar performance at low frequency but a poorer one at high frequency. See also Chapter 9.

13.7.5 Branches and Power Splits

Whenever a given volume of airflow is split up into two or more paths, there is an associated division of sound power between these paths. In such a case, the power loss is determined either from the ratio of the areas of the duct before and after the split or to the ratio of the air volume flow rates before and after the split. For example, if a duct with air having a volume flow rate Q_1 and of cross-sectional area A_1 branches into several other ducts of which one has a cross-sectional area A_2 and flow rate Q_2, then the power loss (i.e. attenuation) from the main duct to this branch is

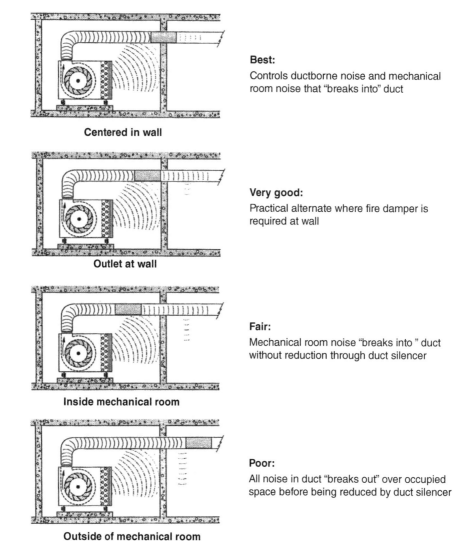

Best:
Controls ductborne noise and mechanical room noise that "breaks into" duct

Centered in wall

Very good:
Practical alternate where fire damper is required at wall

Outlet at wall

Fair:
Mechanical room noise "breaks into " duct without reduction through duct silencer

Inside mechanical room

Poor:
All noise in duct "breaks out" over occupied space before being reduced by duct silencer

Outside of mechanical room

Figure 13.53 Duct silencer placement near a mechanical room wall [25]. ibid © ASHRAE Schaffer Guide.

$$\Delta L_W = 10 \log \left(A_1/A_2 \right), \text{dB} \qquad (13.17)$$

or

$$\Delta L_W = 10 \log \left(Q_1/Q_2 \right), \text{dB} \qquad (13.18)$$

This is equal to the ratio of the air volume flow rates before and after the division only for a constant velocity system. In such a system, the power split losses can be determined from the percentage of air reaching a given outlet grille. Equations (13.17) and (13.18) are strictly only valid for constant velocity systems. If the sum of the branch areas A_2 is not equal to A_1, then there is a reflection.

13.7.6 Attenuation Due to End Reflection

When a duct discharges air into a large acoustical space (i.e. a room or plenum chamber), there is a sudden change in both the cross-sectional area and acoustic impedance. This results in some of the sound being reflected. Such a loss (attenuation) is known as an "end reflection" and can result in a significant quantity

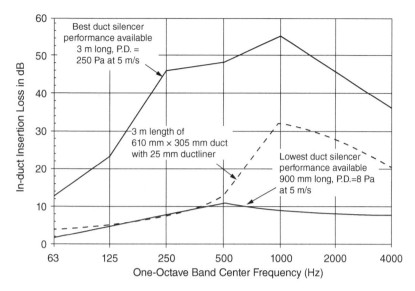

Figure 13.54 In-duct attenuation of duct silencers and lined ductwork [25]. ibid © ASHRAE Schaffer Guide.

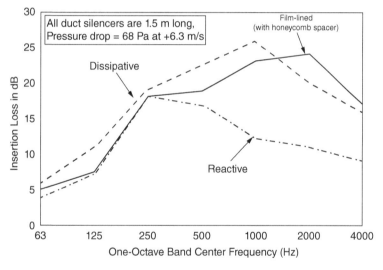

Figure 13.55 Comparative insertion loss of dissipative, film-lined, and reactive duct silencers [25]. ibid © ASHRAE Schaffer Guide.

of low-frequency attenuation being introduced into a system. Figure 13.56 shows some typical end reflection losses of a duct opening into a room. The end reflection is frequency-dependent (see Figure 13.57). This effect plays an important role in the attenuation obtained at the entrance to a plenum chamber.

Example 13.13 Determine the end reflection for 150 mm × 150 mm, 305 mm × 305 mm, and 610 mm × 610 mm ducts flush-mounted with a wall. Use Figures 13.56 and 13.57.

Solution

Use of the two figures gives slightly different results. The end reflection is seen to be greatest at low frequency where the acoustic wavelength is much greater than the duct sidewall dimension(s). The end reflection effect is also greater if the duct exit is located in free space or at wall-floor or wall-ceiling

Table 13.17 End reflections, dB, obtained from Figures 13.56 and 13.57.

Duct dimensions, mm	One-octave band center frequency, Hz						
	63 Hz	125 Hz	250 Hz	500 Hz	1000 Hz	2000 Hz	4000 Hz
150 × 150 Figure 13.56	16	12	8	3	1	0	0
150 × 150 Figure 13.57	18	15	10	4	1	0	0
305 × 305 Figure 13.56	12	7	3	1	0	0	0
305 × 305 Figure 13.57	15	10	3	1	0	0	0
610 × 610 Figure 13.56	7	3	1	0	0	0	0
610 × 610 Figure 13.57	10	4	1	0	0	0	0

junctions. The reflection increases further at low frequency if the duct exit is located near to a room corner (see Table 3.2 in Chapter 3). Typical end reflections are shown in Table 13.17.

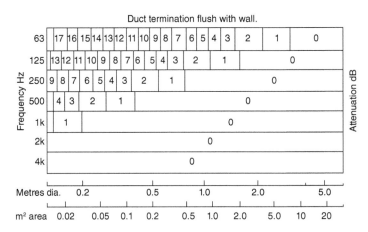

Figure 13.56 Attenuation (dB) at duct termination due to end reflection loss. Dimensions or areas shown, are "gross, overall" figures and ignore the presence of grilles, louvers, etc., which at low frequencies do not significantly affect end reflection losses [8].

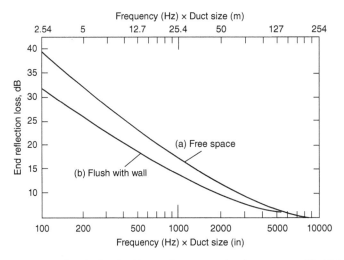

Figure 13.57 This shows the expected "end reflection" losses for rectangular ducts mounted in (a) free space and (b) flush with a wall [27].

13.7.7   Attenuation by Mitre Bends

Whenever the airflow in a ventilation system is caused to change its direction, some attenuation is obtained because some of the incident sound is reflected hack toward the source. Noise is thus reflected and attenuated at mitre bends in ductwork, provided the duct cross dimensions are large compared to a wavelength. This attenuation can be further improved by adding a sound-absorbing lining to the duct walls before and/or after the bend. See Figures 13.58 and 13.59.

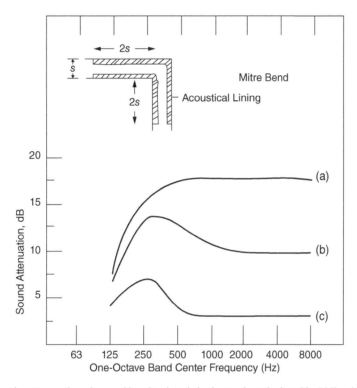

Figure 13.58   This shows the attenuation observed in mitre bends in ducts when the bend is: (a) lined before and after bend, (b) lined after bend, and (c) unlined [27].

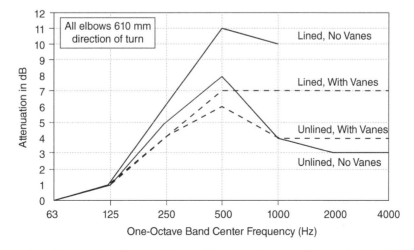

Figure 13.59   Attenuation of rectangular elbows with and without turning vanes (lined and unlined) [25]. ibid © ASHRAE Schaffer Guide.

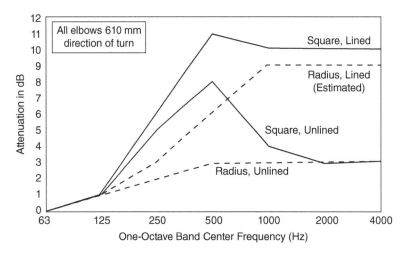

Figure 13.60 Attenuation of rectangular and radius elbows (lined and unlined) [25]. ibid © ASHRAE Schaffer Guide.

The effect is also shown in Figure 13.60. Large radius bends or bends with large radius turning vanes provide very little attenuation. However, such bends do not produce any significant amounts of flow-generated noise. This is discussed in the next section of the chapter. Tables 13.18 and 13.19 show that the useful high frequency attenuation achieved with mitre bends is significantly reduced if they are fitted with large turning vanes instead of vanes of small width. If turning vanes are used, they should have a width less than 1/8 of the wavelength of sound at which the greatest attenuation is needed. For frequencies of 1000 Hz and lower, the vanes should be no more than 50 mm wide [8].

13.8 Sound Generation in Mechanical Systems

In high-velocity systems, the flow of air through the various elements (such as ducts, elbows, take-offs, mixing boxes, and grilles) can produce high noise levels. This type of flow-generated noise is often neglected by system designers who are later surprised to find that some areas of a building are much noisier than they had estimated from fan sound power and system attenuation data alone. This generated noise is mainly caused by air turbulence and vortex shedding. Indeed, it is quite common to find a condition in which the generated noise level in part of the system (i.e. an elbow) is far in excess of the naturally attenuated sound from the fan. Since the number of high-velocity ventilation and air-conditioning systems in use continues to increase, it is essential to be able to obtain a reasonably accurate estimate of the sound power level of such generated noise at common points within the system. Of these, the most important points are found to be elbows, branch take-offs, and grilles (see Figure 13.5).

13.8.1 Elbow Noise

In order to transport air from the fan plenum chamber system to all areas of a building, it is necessary to incorporate bends and take-offs in the supply ductwork (see Figure 13.5). These may be smooth radius or sharp-angled bends. The latter types are of extreme concern because of the turbulence and noise they

Table 13.18 Attenuation (dB) of unlined mitre bends. Attenuation is given for bends with the following geometry. (a) Mitre bend or mitre with small chord turning vanes. (b) Mitre bend with long chord turning vanes or radiused bends [8].

(a)

Duct width	One-Octave band center frequency							
D mm	63	125	250	500	1 k	2 k	4 k	Hz
75–100	—	—	—	—	1	7	7	dB
100–150	—	—	—	—	5	8	4	
150–200	—	—	—	1	7	7	4	
200–250	—	—	—	5	8	4	3	
250–300	—	—	1	7	7	4	3	
300–400	—	—	2	8	5	3	3	
400–500	—	—	5	8	4	3	3	
500–600	—	—	6	8	4	3	3	
600–700	—	1	7	7	4	3	3	
700–800	—	2	8	5	3	3	3	
800–900	—	3	8	5	3	3	3	
900–1000	—	5	8	4	3	3	3	
1000–1100	1	6	8	4	3	3	3	
1100–1200	1	7	7	4	3	3	3	
1200–1300	1	7	7	4	3	3	3	
1300–1400	2	8	7	3	3	3	3	
1400–1500	2	8	6	3	3	3	3	
1500–1600	3	8	5	3	3	3	3	
1600–1800	5	8	4	3	3	3	3	
1800–2000	6	8	4	3	3	3	3	

(b)

Duct width/diameter	One-Octave band center frequency							
D mm	63	125	250	500	1 k	2 k	4 k	Hz
150–250	—	—	—	—	1	2	3	dB
250–500	—	—	—	1	2	3	3	
500–1000	—	—	1	2	3	3	3	
1000–2000	—	1	2	3	3	3	3	

generate. Although a great deal of research has been done on jet noise, there has been relatively little work on noise generation by turbulent flow in ducts. This is because duct noise is very much more complex than jet noise since it is affected by many parameters, of which some are unknown or unavailable. For example, the noise spectrum generated is very dependent upon the coupling between the turbulent airflow and the duct walls, and the flow is also greatly influenced by the duct geometry and wall conditions. A detailed analytical study would require information about the turbulent flow structure in the duct and this is usually not available to the system design engineer. Various workers have attempted to

Table 13.19 Attenuation in decibels of lined mitre bends [8].

Lining thickness = $\dfrac{D}{10}$

Lining to extend distance 2D or greater

Duct width	One-Octave band center frequency							
D mm	63	125	250	500	1 k	2 k	4 k	Hz
75–100	—	—	—	—	2	13	18	dB
100–150	—	—	—	1	7	16	18	
150–200	—	—	—	4	13	18	18	
200–250	—	—	1	7	16	18	16	
250–300	—	—	2	11	18	18	17	
300–400	—	—	4	14	18	18	17	
400–500	—	1	5	16	18	16	17	
500–600	—	1	8	17	18	16	17	
600–700	—	2	13	18	18	17	18	
700–800	—	3	14	18	17	16	18	
800–900	—	4	15	18	18	17	18	
900–1000	—	5	16	18	17	17	18	
1000–1100	1	7	17	18	16	17	18	
1100–1200	1	8	17	18	16	17	18	
1200–1300	1	10	17	18	16	18	18	
1300–1400	2	11	18	18	16	18	18	
1400–1500	2	12	18	18	16	18	18	
1500–1600	3	14	18	18	17	18	18	
1600–1800	4	15	18	18	17	18	18	
1800–2000	5	16	18	17	17	18	18	

observe, through experiments, empirical relationships between the noise level generated and several physical variables such as airflow velocity and duct geometry.

The generated noise level spectrum is generally found to be [27]:

1) Proportional to a power of the flow velocity between the fifth and seventh
2) Practically independent (±1 dB) of the angle of the bend
3) Practically independent of the aspect ratio of the duct (i.e. width/breadth) except at very low frequencies (i.e. 63 Hz) where the sound power generated may increase by 10 dB for a change of aspect ratio from 1 : 1 to 1 : 4
4) Strongly proportional to the area of the duct elbow at low frequencies but only slightly at high frequencies.

The general features concerning noise levels and frequency spectra are presented in Figure 13.61. In order to reduce turbulence at elbows and take-offs, turning vanes are often placed in the airflow. These are found to reduce the very low and high frequency noise levels. The reduction can be as much as 3–5 dB above about 2000 Hz and 5–10 dB or more below 250 Hz. These reductions can be important in reducing rumble at low frequency and hiss at high frequency. See Figure 13.61. The turning vanes do not provide any reduction in noise in the mid-frequency range, 500–2000 Hz, however.

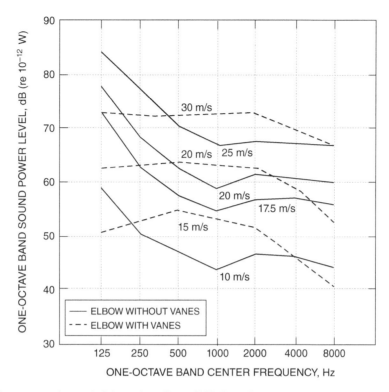

Figure 13.61 Velocity-generated sound of duct mitre elbows [20]. Note: Comparison of one-octave band sound power levels produced by airflow through 200 × 200 mm rectangular elbow with and without seven circular arc turning vanes.

Some detailed empirical methods of computing flow-generated noise levels in elbows with and without turning vanes have been proposed, and these should be consulted when accuracies of within ±5 dB are required. Price and Crocker give an empirical equation in Reference [27] for use with turning vanes in rectangular duct elbows. Both the equation and measured results use the foot-pound system.

13.8.2 Take-off Noise

Aerodynamically generated noise resulting from turbulence at abrupt round or square take-offs can be of particular importance when small lengths of duct are taken off the main supply duct from the fan and fed directly to rooms. In such cases, the generated noise cannot be attenuated by the duct system, and often excessive noise is radiated from the room grille.

It is usually found that a round edge tee produces significantly higher sound power levels than a square edge tee for low velocity flow, but less sound is generated at high velocities. The round edge tee also has less velocity dependence than the sharp edge branch. Once again, an exact analysis of such a problem is extremely complex and empirical relationships have been proposed [27]. These depend upon the velocity at the branch and the main upstream ducts. The following empirical relations have been found to give good estimates of tee-branch generated noise:

1) Square edge branch tee

$$L_W = 23 \log (V_2 V_3) + 10 \log (A_3/A_2) - 15 \log (f) - 40, \quad \text{dB re}: 10^{-12} \, \text{W} \qquad (13.19\text{a})$$

where V_2, A_2 = velocity in and area of downstream main duct in ft/min and ft^2, and V_3, A_3 = velocity in and area of tee branch in ft/min and ft^2, or

$$L_W = 23 \log (V_2 V_3) + 10 \log (A_3/A_2) - 15 \log (f) + 66, \quad \text{dB re} : 10^{-12}\,\text{W} \qquad (13.19\text{b})$$

where $V_2, A_2 =$ velocity in and area of downstream main duct in m/s and m^2, and $V_3, A_3 =$ velocity in and area of tee branch in m/s and m^2.

2) Round edge branch tee

$$L_W = 23 \log (V_2) + 15 \log (V_3) - 15 \log (f) + 10 \log (A_3/A_2) - 12, \quad \text{dB re} : 10^{-12}\,\text{W} \qquad (13.20\text{a})$$

or

$$L_W = 23 \log (V_2) + 15 \log (V_3) - 15 \log (f) + 10 \log (A_3/A_2) + 75.5, \quad \text{dB re} : 10^{-12}\,\text{W} \qquad (13.20\text{b})$$

if V_2 and V_3 are in m/s.

Note the lower dependence on V_3, the tee branch velocity in Eq. (13.20). It can be seen from the above discussion and from Eqs. (13.19) and (13.20) that the noise level generated is strongly dependent upon velocity and that it increases by some 10–20 dB for each doubling of both the downstream and tee branch flow velocities. Therefore, it might seem logical at first glance to use large size ducts wherever possible in order to reduce high flow velocities and corresponding generated noise. However, the extra cost involved in using large ducts far exceeds the cost of using small high-velocity ducts in conjunction with special sound attenuation devices.

13.8.3 Grille Noise

Reduction of grille noise is very important in the successful acoustical design of a mechanical system in the mid- to high-frequency range, but it is of little or no importance at low frequencies where fan noise is the most objectionable source. It is found that certain types of grilles produce more noise than others for the same conditions of static pressure drop and airflow velocity. For example, grilles producing a wide angled spread of airflow tend to produce some 10–15 dB more sound power than equivalent narrow spread units. Grilles consisting of a perforated plate give a small amount of spread and are usually the quietest type (Figure 13.62).

Although a comprehensive analysis of the mechanism is difficult, it is useful to note that at low frequencies the generated sound power levels are dependent upon the volume flow of air passing through the grille, the grille shape, and its open area. At high frequencies, the sound power is proportional to the grille area, the

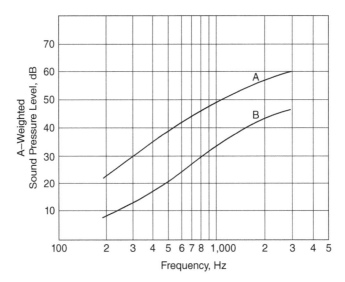

Figure 13.62 A-weighted sound pressure levels measured 1.5 m (5 ft) from surface of two types of grille: (A) wide angle spread; (B) narrow angle spread [27].

pressure drop across it, and the velocity of the air leaving it. In the ease of a widespread grille, it should be remembered that the velocity of the air leaving certain parts may be well in excess of the main duct airflow velocity.

An approximate value for the generated sound power level, L_W, produced by a narrow spread grille can be obtained from:

$$L_W = 10 \log S + 20 \log P + 20 \log f - 16, \quad \text{dB re } 10^{-12}\,\text{W} \tag{13.21a}$$

where S = grille area, ft^2; P = static pressure drop, inches of water; and f is the one-octave band center frequency, or

$$L_W = 10 \log S + 20 \log P + 20 \log f - 54, \quad \text{dB re } 10^{-12}\,\text{W} \tag{13.21b}$$

where S = grille area, m^2 and P = static pressure drop, pascals.

The location of a grille in a room also affects the effective sound power emitted from it. For example, a grille placed at the junction of a wall and ceiling (but not a corner) will produce 3 dB more sound power than if the grille were in the center of the ceiling. This can be compared to the mirror image effect in optics. Similarly, a grille placed in a corner of a room at the ceiling will effectively produce 6 dB more sound power than if it is at the center of the ceiling. It is, therefore, desirable to place grilles at the center of a wall or ceiling, or at least near the center of the edge at which the ceiling and wall join.

If there is more than one grille in a room, then the total sound power fed into the space is found by adding the individual sound power levels logarithmically using $L_{WT} = 10 \log \Sigma \, (10^{L_w/10})$. For example, if there were n grilles, each contributing a sound power level L_W, then the net sound power would be $(L_W + 10 \log n)$ dB.

Figure 13.63a shows the effect of proper and improper airflow conditions into the grille. Figure 13.63b shows that misalignment of a flexible connector can increase turbulence and noise by as much as 12–15 dB. The importance of installing an equalizing grid before the grille is seen in Figure 13.63a. Excessive bends present in the flexible duct connector before the grille can increase noise levels by 12–15 dB or more.

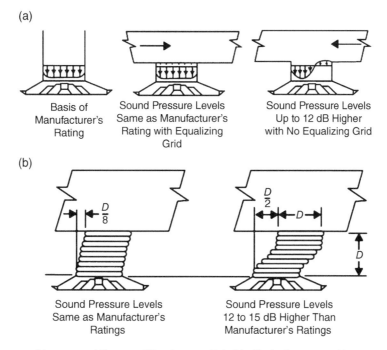

Figure 13.63 (a) Proper and improper airflow condition to an outlet; (b) effect of proper and improper alignment of flexible duct connector [20]. ibid © ASHRAE Handbook, Chap 1, Fig 2.

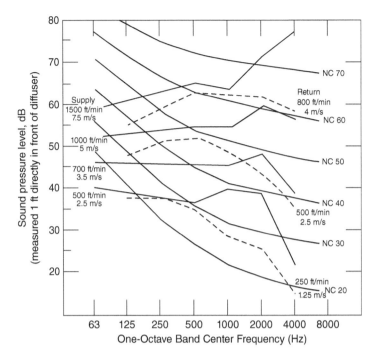

Figure 13.64 Typical generated sound pressure levels vs. flow velocity for sidewall diffusers: — supply, - - - - return. These are for 0.1 m² (1.0 ft²) open area [27]. Measurements made at 0.3 m (1 ft) in front of diffusers.

13.8.4 Diffuser Noise

This is a very similar in character to grille noise but usually tends to be at a slightly higher level for the same air velocity and open area. It is furthermore characterized by a broad peak in the frequency spectrum in the range 1000–2500 Hz. The position of this peak tends to move toward higher frequencies as the neck airflow velocity is increased. This phenomenon and the strong velocity dependence are shown in Figure 13.64.

Once again, making an exact analysis is extremely complicated. However, a good indication of supply diffuser generated sound pressure levels, within ±5 dB, may be obtained from the use of Figure 13.65.

13.8.5 Damper Noise

In order to achieve a correctly aerodynamically balanced ventilation system, the flow of air at various outlets has to be regulated after the system has been installed. To achieve design objectives for air volume flow rates, site adjustments are usually made by the regulation of system dampers. Very often it becomes necessary to significantly close the opposed blades on these dampers, thus altering the effective open area. This decrease in open area results in an increase in airflow velocity which causes a sharp rise (10–15 dB) in the generated sound power level. This increase can be calculated from the static pressure drop across the unit:

$$\text{Increase(dB)} = 20 \log\left(P_c/P_0\right), \tag{13.22}$$

where P_c = static pressure drop, damper partially closed; and P_0 = static pressure drop, damper fully open. The effect of partially closing the damper on a supply grille is shown in Figures 13.66 and 13.67. This suggests strongly that, if possible, dampers should be placed away from the terminal grille (Figures 13.67 and 13.68) so that there is enough space to insert a sound attenuator after the damper, if this should be required. See Figure 13.69.

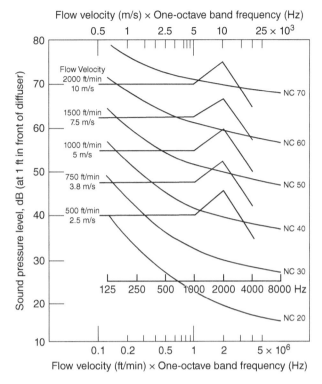

Figure 13.65 This figure can be used to estimate generated sound pressure levels for a supply sidewall diffuser to ±5 dB, measured 0.3 m (1 ft) directly in front of the diffuser [27].

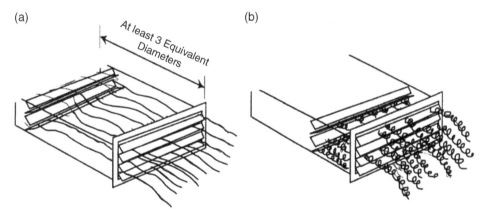

Figure 13.66 The effect of installing a damper behind a grille; (a) Without damper or with damper at least three equivalent duct diameters upstream. Manufacturer's grille noise data are valid for design purposes; (b) A damper within three equivalent duct diameters of a grille increases the noise – up to 5 dB if the damper is wide open, as much as 15–40 dB if the damper is half closed.

13.9 Radiated Noise

Although one is often more concerned with air noise emanating from grilles and diffusers in an occupied area, careful consideration should also be given to noise which may be radiated directly from the outer casings of various system elements conveying air to or from the grille. For example, although ducts and mixing

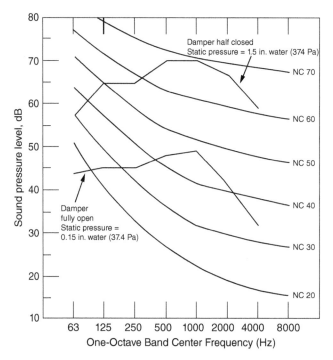

Figure 13.67 The effect of partially closing a damper on a 50 cm × 50 cm (20 in. × 20 in.) exhaust grille operating at 5 m/s (1000 ft/min) on the generated sound pressure level measured 0.3 m (1 ft) directly in front of the grille [27].

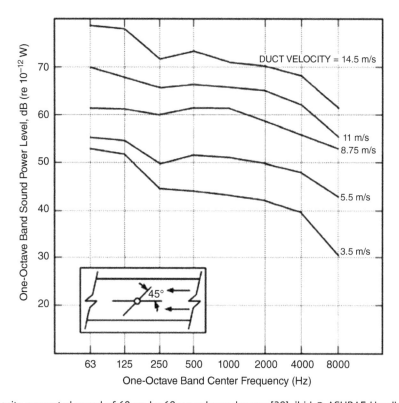

Figure 13.68 Velocity-generated sound of 60 cm by 60 cm volume damper [20]. ibid © ASHRAE Handbook, Chap 1, Fig 2.

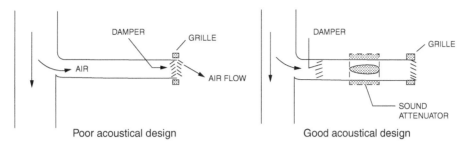

Figure 13.69 If possible, dampers should be placed well back behind outlet grille so as to leave room for a sound attenuator if required [27].

boxes are often situated in a ceiling plenum directly over critical acoustical areas, the possibility that radiated noise from these elements could give rise to a serious acoustical problem is usually not adequately considered. The system designer should, therefore, be aware of this possibility and be able to estimate the noise radiated from outer casings of these elements.

The main sources of radiated noise are ducts, mixing boxes, and often fan plenum chamber panels. These will now be dealt with separately.

13.9.1 Duct-Radiated Noise

The noise radiated directly from the walls of a duct as a result of the flow of air inside can result in excessively noisy adjacent areas. This is usually caused by the poor low-frequency transmission loss characteristics of duct walls which results in a rumble sound in the surrounding areas. An exact analysis of the low-frequency transmission loss properties of rectangular and circular ducts is beyond the scope of this hook. However, it is sufficient to say that the mass law is obeyed at frequencies above that at which the cross-sectional duct dimensions equals the acoustic wavelength, up to an octave below the critical frequency, f_c, of the duct walls (usually 10–15 kHz). At low frequencies, when the duct dimensions are much smaller than the acoustic wavelength, the transmission loss is controlled by the stiffness of the duct walls. It is found that at frequencies well below the first acoustic cross mode in the duct, the transmission loss increases with frequency at a rate of 3 dB per octave rather than 5–6 dB per octave, as predicted by the mass law. The following empirical relationship has been found to be useful in predicting the effective transmission loss of a rectangular duct at low and high frequencies:

$$L_{WD} - L_{WR} = \begin{cases} 20\log\left(M\right) + 10\log\left(f\right) - 10\log\left(S_W/S\right), & \mathrm{dB}(f < 550/d) \\ 20\log\left(M\right) + 20\log\left(f\right) - 10\log\left(S_W/S\right), & \mathrm{dB}(f > 550/d) \end{cases}, \tag{13.23a}$$

where L_{WD}, L_{WR} = sound power levels within the duct, and radiated from its walls, respectively; and M = weight of duct wall (lb/ft²); S_W = total duct wall area, for length considered (ft²); S = cross-sectional area (ft²); d = largest duct cross-sectional dimension (ft); and f = frequency (Hz), or

$$L_{WD} - L_{WR} = \begin{cases} 20\log\left(M\right) + 10\log\left(f\right) - 10\log\left(S_W/S\right) - 13.8, & \mathrm{dB}(f < 168/d) \\ 20\log\left(M\right) + 20\log\left(f\right) - 10\log\left(S_W/S\right) - 13.8, & \mathrm{dB}(f > 168/d) \end{cases}, \tag{13.23b}$$

where M = weight of duct wall (kg/m²); S_W = total duct wall area, for length considered (m²); S = cross-sectional area (m²); and d = largest duct cross-sectional dimension (m).

Once the radiated sound power level has been calculated, the sound pressure level in the space around the duct can be calculated from

$$L_p = L_{WR} - 10\log\left(R\right) + 16.5, \quad \mathrm{dB} \tag{13.24}$$

where R = room constant for the absorption in the surrounding space, sabins. The approximations used in the derivation of the above equation break down under certain conditions at very low frequencies or when very large lengths of duct are considered. Clearly the radiated sound power can never exceed the power within the duct, and if Eq. (13.24) should ever predict such a situation, then a power loss of 3 dB may be assumed as a fair estimate (see Eq. (13.25))

$$L_{WD} - L_{WR} = 3 \text{ dB}. \tag{13.25}$$

Circular ducts have a greater stiffness than equivalent area rectangular ducts and are therefore found to radiate much less (5–10 dB) low-frequency noise. For this reason, circular ducts should always be used in preference to rectangular ducts in areas where low-frequency (rumble) radiated noise is likely to be a problem. It is found that lagging the duct externally (e.g. with 5 kg/m^2 (1 lb/ft^2) lead also tends to reduce the radiated noise level by increasing the effective weight of the duct walls and by adding damping to them.

Although the designer is usually more concerned with attempting to reduce duct-radiated noise, it should be pointed out that such a mechanism can sometimes be used advantageously. For example, the noise in supply ducts situated directly after a fan plenum chamber is usually "rich" in low-frequency noise. These components may not be sufficiently attenuated by the time that the outlet grille is reached; therefore, some additional low frequency attenuation is required. In normal circumstances, the designer would have several alternatives for providing this additional low-frequency attenuation, but these would usually involve the use of commercial silencers. However, often this low-frequency attenuation can be achieved very inexpensively by allowing the low-frequency noise to be transmitted out through the duct walls of a long run of thin-walled rectangular duct out into noncritical areas of the building such as a storeroom. Alternatively, the noise may be allowed to pass out into a ceiling plenum lined with sound-absorbing material and situated over an acoustically noncritical area.

13.9.2 Sound Breakout and Breakin From Ducts

Breakout sound is caused by the fan or airflow noise inside a duct. This noise can be transmitted through the duct walls and then radiated into the surrounding space (Figure 13.70). Breakout noise should be adequately attenuated before the duct runs over or through an occupied space. Sound that is transmitted into a duct from the surrounding area is called breakin (Figure 13.71). The main factors affecting breakout and breakin sound transmission are the transmission loss of the duct, the total exposed surface area of the duct, and the presence of acoustical duct lining. The breakout sound power transmitted is given by [20]

$$L_{W(out)} = L_{W(in)} + 10 \log (S/A) - TL_{out}, \tag{13.26}$$

where $L_{W(out)}$ = sound power level of sound radiated from the outside surface of duct walls, dB; $L_{W(in)}$ = sound power level of sound inside duct, dB; S = surface area of outside sound-radiating surface of duct, m₂; A = cross-section area of inside of duct, m^2; TL_{out} = normalized duct breakout transmission loss, dB.

Equation (13.26) is a simplified expression that assumes that the sound power level inside the duct does not decrease with distance over the length of the duct. For very long ducts (when $S \gg A$), the radiated sound power level. Equation (13.26) predicts that $L_{W(out)}$ could become greater than the sound power level inside the duct, which would violate the conservation of energy principle. A more accurate expression for breakout sound is presented in the ASHRAE Handbook which allows for attenuation of the sound along the duct section. Values of TL_{out} for rectangular ducts, round ducts and for flat oval ducts are given in the ASHRAE Handbook [20].

Equations for S and A for rectangular ducts are

$$S = 2L(a + b), \tag{13.27}$$

$$A = ab, \tag{13.28}$$

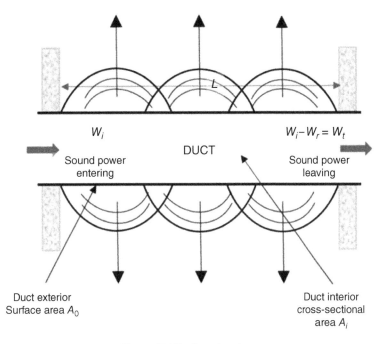

Figure 13.70   Duct breakout.

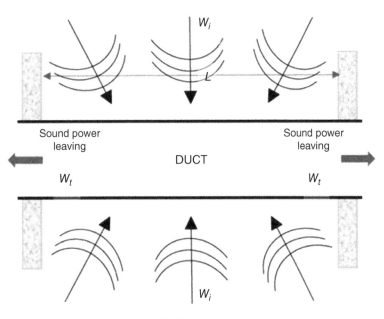

Figure 13.71   Duct breakin.

where a = larger duct cross-section dimension, m; b = smaller duct cross-section dimension, m; L = length of duct sound-radiating surface, m.

Equations for S and A for round ducts are

$$S = L\pi d, \tag{13.29}$$

$$A = \pi d^2/4, \tag{13.30}$$

where d = duct diameter, m; L = length of duct sound-radiating surface, m.

For flat oval ducts

$$S = L[2(a - b) + \pi b], \tag{13.31}$$

$$A = b(a - b) + \pi b^2/4, \tag{13.32}$$

where a = length of major axis, m; b = length of minor axis, m; L = length of duct sound-radiating surface, m.

Equation (13.26) assumes no decrease in the internal sound power level with distance along the length of the duct. Thus, it is valid only for relatively short lengths of unlined duct. The ASHRAE Handbook deals with long ducts or ducts that have internal acoustical lining [20]. In most rooms where the listener is close to the duct, an estimate of the breakout sound pressure level can be obtained from

$$L_p = L_{W(out)} - 10 \log(\pi r L), \tag{13.33}$$

where L_p = sound pressure level at a specified point in the space, dB; $L_{W(out)}$ = sound power level of sound radiated from the outside surface of duct walls, dB, given by Eq.(13.26); r = distance between duct and position for which L_p is calculated, m; L = length of the duct sound-radiating surface, m.

Note that Eq. (13.33) gives the sound pressure radiated from a duct that is in a wide-open ceiling plenum space. If the duct is in a restricted space between floor slab and ceiling, the level may be up to 6 dB higher.

Example 13.14   A 610 mm × 610 mm × 7.5 m long rectangular supply duct is constructed of 0.75 mm sheet metal. Given the following sound power levels produced for a 0.75 m diameter backward-curved centrifugal fan in the duct, what are the breakout sound pressure levels 2.5 m from the surface of the duct?

Solution

Using Eqs. (13.26) and (13.33), the sound pressure level in the room is calculated and given in Table 13.20. It is observed that the low frequency levels below 500 Hz are quite high. The A-weighted level which is only about 50 dB is probably acceptable for many spaces in a building and in particular store rooms.

Table 13.20   Sound pressure level predictions for breakout noise in Example 13.14.

	One-Octave Band Center Frequency, Hz								
	63	125	250	500	1000	2000	4000	8000	BPF
Sound Power Level, dB	85	85	84	79	75	68	64	62	3
$-TL_{out}$	−20	−23	−26	−29	−32	−37	−43	−45	
$10\log S/A$	+18	+18	+18	+18	+18	+18	+18	+18	
$L_{W(out)}$	83	80	80	68	65	49	39	35	
$-10\log(\pi r L)$	−18	−18	−18	−18	−18	−18	−18	−18	
Sound Pressure Level, dB	65	62	62	50	47 + 2 BPF = 49	31	21	17	

13.9.3 Mixing Box Radiated Noise

Noise radiating from the outer metal case of mixing boxes into a ceiling plenum space can also result in a significant acoustical problem. Indeed, high-velocity mixing boxes should always be placed as far away as possible from critical acoustical areas and positioned over corridors or other high noise level regions of a building. The careful selection and positioning of mixing boxes are, therefore, of primary importance for good acoustical conditions, and manufacturers' figures for measured radiated sound power levels should be consulted whenever possible. Some typical radiated sound power levels are shown in Figure 13.72 for dual-inlet mixing boxes for various airflow volumes for a 375 Pa (1.5 in. water) static pressure drop. For other static pressure drops, a correction of $20 \log (P/375)$ dB for SI units (or $20 \log (P/1.5)$ dB where P is inches water) should be added to these sound power levels.

Since the radiated noise level is dependent upon the pressure drop across the box, it is advisable to choose a box with the lowest possible pressure differential and to arrange to take up the extra pressure elsewhere – for example, at an elbow, take-off, or attenuator. Several manufacturers now provide mixing boxes with lead-lined aluminum casings which can reduce the radiated sound power by 5–10 dB at low frequencies and 15–20 dB at high frequencies. The effect is shown in Figure 13.73.

A reduction in the radiated noise may also be obtained by:

1) Bolting the unit hard against a structural slab. This allows the box to dissipate energy into the slab and also reduces the total radiating area. Such a procedure can reduce the radiated noise by as much as 3 dB throughout the frequency spectrum.
2) Making the inlet connections to the box acoustically tight but flexible. If possible, the inlet connection should be covered with plaster or some other similar material to reduce the low-frequency noise transmitted through the connector.
3) Fitting the entire box within a metal airtight shroud. This can often reduce the radiated noise level by 10–15 dB. For example, an unshrouded box operating at 25.3 m^3/min, 425 L/s (900 ft^3/min) and 750 Pa (3 in. water) pressure drop may produce a noise spectrum corresponding to Noise Criterion NC45; however, when this box is enclosed in a separate shroud, this level may be reduced to Noise Criterion NC30.

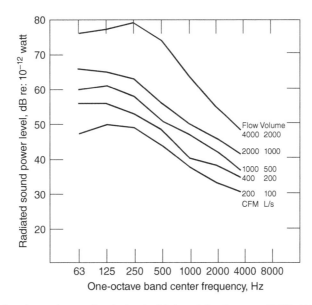

Figure 13.72 Typical radiated sound power levels for dual-inlet mixing boxes at 374 Pa (1.5 in. water) static pressure drop vs. airflow volume [27].

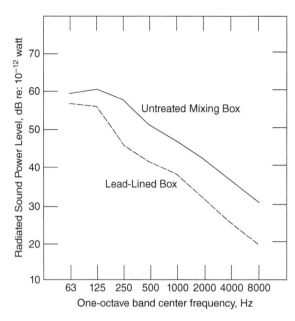

Figure 13.73 Effect on radiated sound power level of applying a layer of 1/64 in. thick lead on inside of a dual-inlet mixing box operating at 470 L/s, 0.47 m³/s, (1000 cfm) and 250 Pa (1.0 in. water) static pressure drop — ordinary untreated box; - - - - lead-lined box [27].

13.9.4 Radiation From Fan Plenum Walls

It is often necessary to estimate the reverberant sound pressure level within a fan room in order to determine if sound transmission through the fan room walls or floor may create a problem in adjacent areas. This involves an estimation of the sound transmission loss of the fan plenum chamber. Fortunately, the material used in such constructions is usually 14–18-gauge galvanized steel which has a relatively high critical frequency, in the range 8–25 kHz. This means that the "mass law" gives a fairly good estimate of its transmission loss characteristics in the frequency range of interest: – for example, 200–8000 Hz. However, one should be particularly careful to avoid mechanical resonances in the walls of the fan plenum. These may be excited by the rotation of the fan rotor and can result in excessive low-frequency noise in the surrounding space, from which it may be readily transmitted to other adjacent areas. Such resonances can usually be avoided by providing adequate mechanical vibration isolation between the fan and its surrounding plenum. However, if such a resonance phenomenon should be observed, then the problem is best solved by either stiffening the resonant panel or changing its dimensions, or by slightly altering the fan rotational speed.

13.9.5 Overall Sound Pressure Level Prediction

Once the fan system is chosen for a building for HVAC purposes, it is useful to predict the sound pressure level in different rooms and other occupied and non-occupied spaces in the building. Commercial software is available to make the necessary calculations and to examine different ways to attenuate the sound. Examples of such software include the Trane Acoustic Program – Tap ™ version 2.3, the Vibro-Acoustics® V-A Select Release 5.0, and Acousticcalc (www.acousticcalc.com). Ryherd and Wang compare results from the first two softwares with a spreadsheet they developed in Ref. [14]. The calculations presented in Example 13.15 are based on the approach of Fry in Ref. [8]. If such software is not available, simple preliminary calculations such as in Example 13.15 may be useful. In the following example, assume that the direct, reverberant and total sound pressure levels can be calculated using Eqs. 13.34–13.36.

$$L_{p\,(total)} = L_{p\,(dir)} + L_{p\,(rev)} \quad \text{dB}, \tag{13.34}$$

$$L_{p\,(dir)} = L_{W\,(per\,grille)} + 10\,\log\,(QF) - 10\,\log\,(2\pi r^2) \quad \text{dB}, \tag{13.35}$$

$$L_{p\,(rev)} = L_{W\,(to\,room)} + 10\,\log\,T - 10\,\log\,V + 14 \quad \text{dB}, \tag{13.36}$$

where $L_{p\,(total)}$ = total sound pressure level, dB; $L_{p\,(dir)}$ = direct sound pressure level, dB; $L_{p\,(rev)}$ = reverberant sound pressure level, dB; $L_{W\,(per\,grille)}$ = sound power level from each grille, dB; $L_{W\,(to\,room)}$ = sound power level entering rood, dB; QF = directivity factors; r = distance from source to receiver, m; T = reverberation time, s; and V = volume, m³.

Example 13.15 Design a HVAC system for an office block as shown in Figure 13.74. Use an axial fan with volume flow 2 m³/s and pressure of 40 N/m² to supply air to staff offices, the reception room and the store room. The primary duct work is 8 m long with cross-section 915 × 610 mm and the secondary ducts are 615 × 305 mm and 610 × 610 mm. The total sound pressure level $L_{p\,(total)}$ will be calculated at the critical distance from a grille by logarithmically adding $L_{p\,(dir)}$ to $L_{p\,(rev)}$. A distance of 0.8 m (2 ft) is assumed to be this distance. Allow for the mitre bend shown in the mechanical room and for the take-offs to the directors' and other staff offices using Eq. (13.17). Assume the diffusers are flush-mounted in the ceiling away from the walls. The air flow to each diffuser is 0.25 m³/s.

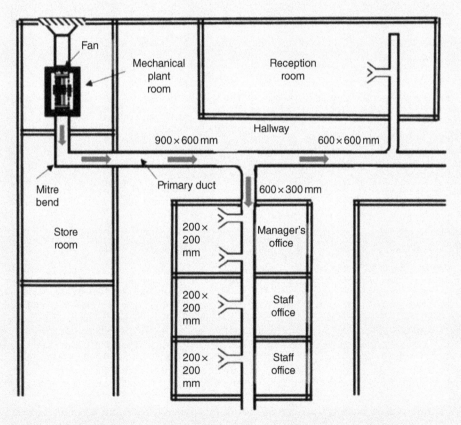

Figure 13.74 HVAC systems for an office block. Dimensions are in mm.

Solution

Using the Fry method, the in-duct sound power level of the fan is calculated from Eq. (13.3). The reverberation time T and grille directivity factor will depend on frequency. For simplicity, we will assume $T = 0.5$ second and $Q = 1$ independent of frequency. Then,

$$L_W = 10 \log (2) + 20 \log (40) + 40 \, dB = 3 + 32 + 40 = 75 \, dB.$$

Assume that 50% of the sound power is radiated equally in each direction to the fan inlet and exhaust. Thus, the supply in duct sound power is $75 - 3 = 72$ dB. The steps in the calculation are shown as lines 1 through 17 of Table 13.21.

Table 13.21 Sound pressure level, dB, predicted in Example 13.15.

Item		One-Octave Band Center Frequency (Hz)							
		63	125	250	500	1000	2000	4000	8000
1	Fan sound power level	72	72	72	72	72	72	72	72
2	Fan correction	−5	−5	−6	−7	−8	−10	−13	−15
3	Corrected unit L_W	67	67	66	65	64	62	59	57
4	10 m 915 × 610 mm duct	−15	−10	−5	−3	−2	−3	−2	−2
5	5 m 610 × 305 mm duct	−7	−7	−4	−2	−2	−2	−2	−2
6	Mitre bend	0	−5	−8	−4	−3	−3	−3	−3
7	End reflection	−14	−10	−5	−2	0	0	0	0
8	L_W Leaving system	31	35	44	54	57	55	52	50
9	% L_w to room = 20%	−8	−8	−8	−8	−8	−8	−8	−8
10	Volume and reverb time correction	−4	−4	−4	−4	−4	−4	−4	−4
11	L_p reverberant	19	23	32	42	45	43	40	38
12	L_W to outlet 20%	−8	−8	−8	−8	−8	−8	−8	−8
13	Distance to listener	−4	−4	−4	−4	−4	−4	−4	−4
14	L_p direct	19	23	32	42	45	43	40	38
15	L_p total	22	26	35	45	48	46	43	41
16	NCB 30 Criterion	55	46	40	35	32	29	25	22
17	Additional attenuation needed	0	0	0	10	16	17	18	19

Item: (1) Volume flow and pressure substituted in Eq. (13.3); (2) Fan correction from Table 13.3; (3) Corrected sound power level, L_W; (4) Duct attenuation from Table 13.11 by combining each duct-wall side attenuation calculated separately and combining; (5) Same procedure as (4) above; (6) Mitre bend without turning vanes or absorbing material. No attenuation assumed for secondary duct take-offs (Table 13.18); (7) End reflection Figures 13.56 and 13.57 for duct exit flush in center of ceiling; (8) Sound power level L_W leaving system wall; (9) L_W leaving two grilles; (10) Reverb Time corrections; (11) Reverberant L_p; (12) L_W to grilles; (13) Listener distance from grille 0.8 m (2.5 ft); (14) L_p direct; (15) Total $L_{p \, (reverb)} + L_{p \, (direct)}$; (16) NCB 30 Criterion; (17) additional attenuation needed to meet NCB 30.

Example 13.15 shows that to meet the NCB 30 criterion requires substantial additional attenuation in the frequency range 500–8000 Hz. Various approaches could be used to achieve the attenuation required. (a) A centrifugal fan could be chosen instead of an axial-flow one since low frequency is not of concern and centrifugal fans generate less high-frequency noise. See Figure 13.15. However, the axial fan probably cannot be changed at this stage since it was chosen for HVAC reasons, not noise. (b) A lined mitre bend could be chosen (Figure 13.59 and Table 13.19) since the attenuation above 500 Hz is just sufficient. (c) Lined ducts (Figure 13.49) could be considered, since with sufficient lining thickness (25 or 50 mm) and 5 or 10 m length, adequate attenuation could be achieved. (d) A plenum chamber located either in the mechanical room, or soon after (Example 13.9 and the results of Table 13.10) would also be a good choice to achieve the necessary attenuation. (e) Mufflers could also be considered since they can provide more than adequate attenuation at this frequency (see Figure 13.54 and Chapter 10).

References

1 Leonard, R.W. (1957). Heating and ventilating system noise. In: *Handbook of Noise Control* (ed. C.M. Harris), 27.1–27.17. New York: McGraw-Hill.

2 Allen, C.H. (1960). Noise control in ventilation systems. In: *Noise Reduction* (ed. L.L. Beranek), 541–570. New York: McGraw-Hill.

3 Graham, J.B. (1972). How to estimate fan noise. *Sound Vib.* 5: 24–27.

4 Graham, J.B. (1975). Noise of fans and blowers. In: *Reduction of Machinery Noise* (ed. M.J. Crocker), 305. Lafayette, IN: Purdue University.

5 Graham, J.B. and Faulkner, L.L. (1976). Fan and flow system noise. In: *Handbook of Industrial Noise Control* (ed. L.L. Faulkner), 368–438. New York: Industrial Press Inc.

6 Graham, J.B. (1979). Fans and blowers. In: *Handbook of Noise Control* (ed. C.M. Harris). New York: McGraw-Hill.

7 Kingsbury, H.F. (1979). Heating, ventilating, and air-conditioning systems. In: *Handbook of Noise Control* (ed. C.M. Harris), 28.1–28.13. New York: McGraw-Hill.

8 Anon (1988). Ductborne noise-transmission. In: *Noise Control in Building Services* (ed. A. Fry). Oxford: Pergamon Press.

9 Hoover, R.M. and Blazier, W.E. (1991). Noise control in heating, ventilating, and air-conditioning systems. In: *Handbook of Acoustical Measurements and Noise Control* (ed. C.M. Harris), 42.1–42.31. New York: McGraw-Hill.

10 Graham, J.B. and Hoover, R.M. (1991). Fan noise. In: *Handbook of Acoustical Measurements and Noise Control* (ed. C.M. Harris), 41.1–41.22. New York: McGraw-Hill.

11 Blazier, W.E. Jr. (1991). Noise control criteria for heating, ventilating, and air-conditioning systems. In: *Handbook of Acoustical Measurements and Noise Control* (ed. C.M. Harris), 43.1–43.18. New York: McGraw-Hill.

12 Hoover, R.M. and Blazier, W.E. (1994). Chapter 7, part 1: noise control in heating, ventilating, and air-conditioning systems. In: *Noise Control in Buildings* (ed. C.M. Harris), 7.1–7.62. New York: McGraw-Hill.

13 Hoover, R.M. and Keith, R.H. (1997). Noise control for mechanical and ventilation systems. In: *Encyclopedia of Acoustics* (ed. M.J. Crocker), 1219–1241. New York: Wiley.

14 Ryherd, S.R. (2005) Acoustical prediction methods for heating, ventilating, and air-conditioning (HVAC) systems. Proceedings of NOISE-CON 2005.

15 Fry, A.T. and Sturz, D.H. (2006). Noise control in heating, ventilating, and air-conditioning systems. In: *Noise and Vibration Control Engineering* (eds. I.L. Ver and L.L. Beranek), 685–719. New York: Wiley.

16 Kingsbury, H.F. (2007). Noise sources and propagation in ducted air distribution systems. In: *Handbook of Noise and Vibration Control* (ed. M.J. Crocker), 1316–1322. New York: Wiley.

17 Keith, R.H. (2007). Noise control for mechanical and ventilation systems. In: *Handbook of Noise and Vibration Control* (ed. M.J. Crocker), 1328–1347. New York: Wiley.

18 ASHRAE Guide and Data Book (1970) American Society of Heating, Refrigeration, and Air-Conditioning Engineers. New York, chap 33.

19 ASHRAE Guide and Data Book (1973) American Society of Heating, Refrigeration, and Air-Conditioning Engineers. New York, chap 35.

20 ASHRAE Guide and Data Book (2015) American Society of Heating, Refrigeration, and Air-Conditioning Engineers. New York, chap. 48.

21 Beranek, L.L. (1957). Revised criteria for noise in buildings. *Noise Control* 3 (1): 19–27.

22 Beranek, L.L., Blazier, W.E., and Figwer, J.J. (1971). Preferred noise criterion (PNC) curves and their application to rooms. *J. Acoust. Soc. Am.* 50 (5): 1223–1228.

23 Beranek, L.L. (1989). Balanced noise criterion (NCB) curves. *J. Acoust. Soc. Am.* 86 (2): 650–654.

24 (1958) *Heating, Piping and Air-Conditioning*, Table 1 January, 211.

25 Schaffer, M.E. (2011). *A Practical Guide to Noise and Vibration Control for HVAC Systems*, 2e. Atlanta, GA: American Society of Heating, Refrigerating and Air-Conditioning Engineers.

26 Lauchle, G.L. (2007). Centrifugal and axial fan noise prediction and control. In: *Handbook of Noise and Vibration Control* (ed. M.J. Crocker), 868–884. New York: Wiley.

27 Crocker, M.J. and Kessler, F.M. (1982). *Noise and Noise Control*, vol. II. Boca Raton, FL: CRC Press.

28 Ver, I.L. (1971). Impact noise isolation of composite floors. *J. Acoust. Soc. Am.* 50 (4 Pt 1): 1043–1050.

29 Cremer, L. (1952). Theorie des Kolpfschalles bei Decken mit Schwimmenden Estrich. *Acustica* 2 (4): 167–178.

30 (1962) *Manual on Design and Applications of Helical and Spiral Springs*. Warrendale, PA: Society of Automotive Engineers.

31 Wells, R.J. (1958). Acoustical plenum chambers. *Noise Control* 4 (4): 9–15.

32 Bies, D.A. and Hansen, C.H. (2009). *Engineering Noise Control – Theory and Practice*, 4e. London: Spon Press.

14

Surface Transportation Noise and Vibration Sources and Control

14.1 Introduction

The number of vehicles on the road continues to increase worldwide. Road traffic noise is really a greater problem than aircraft noise in most countries since it affects many more people. The noise of railroad and rapid transit vehicles is also a problem for people living near to rail lines. Noise and vibration sources in road and rail vehicles and aircraft affect not only the occupants but nontravelers as well. Major sources consist of (i) those related to the power plants and (ii) those non-power-plant sources generated by the vehicle motion. Most rail and rapid transit vehicles have power plant noise and vibration sources that are similar to road vehicles. With rail and rapid transit vehicles, however, tires are mostly replaced with metal wheels, and the wheel–rail interaction becomes a major source of noise and vibration. Brake, gearbox, and transmission noise and vibration are additional problems in road and rail vehicles. Interior noise level criteria depend upon the vehicle type and the occupied space considered. For passengers, speech communication and privacy are the main concerns, while for truck, railroad vehicle, and ship engine room operators and crew, hearing protection and speech communication remain important issues.

14.2 Automobile and Truck Noise Sources and Control

In cars, trucks, and busses, major power plant noise sources include gasoline and diesel engines, cooling fans, gearboxes and transmissions, and inlet and exhaust systems. Other major sources include tire/road interaction noise and vibration and aerodynamic noise caused by flow over the vehicles [1]. (See Figure 14.1) Although vehicle noise and vibration have been reduced over the years, traffic noise remains a problem because of the continuing increase in the numbers of vehicles. In addition, most evidence suggests that the exterior noise of most new cars, except in first gear, is dominated in normal operation by *rolling noise* (defined here as tire/road interaction noise together with aerodynamic noise), which becomes increasingly important at high speed and exceeds *power train noise* (defined here as engine, air inlet, exhaust, cooling system, and transmission). See Table 14.1 and the detailed review of vehicle noise by Nelson. This review concludes that rolling noise has a negligible effect on the noise produced by heavy vehicles at low speed, but at speeds above 20 km/h for cars and 80 km/h for heavy vehicles, rolling noise contributes significantly to the overall noise level [2]. At speeds above 60 km/h for cars, rolling noise becomes the dominant noise source [2].

Figure 14.2 shows that there is little difference between the exterior noise generated by (i) a modern car at steady speeds (in the top three gears) and (ii) the car operating in a coast-by condition at the same steady speeds without the power plant in operation. This suggests that tire noise together with aerodynamic noise

Engineering Acoustics: Noise and Vibration Control, First Edition. Malcolm J. Crocker and Jorge P. Arenas.
© 2021 John Wiley & Sons Ltd. Published 2021 by John Wiley & Sons Ltd.

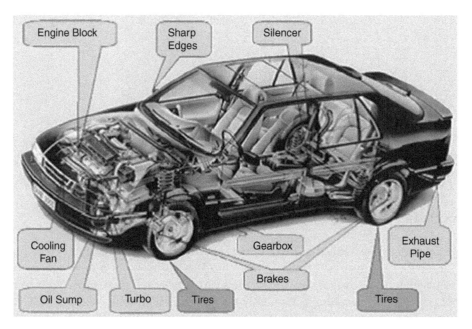

Figure 14.1 Location of sources of power plant, tire, and wind noise on an automobile [1].

Table 14.1 Comparison of rolling and power train A-weighted sound pressure levels.

Road Speed (km/h)	Vehicle Class	Rolling Noise (dB)	Power Train Noise (dB)	Total Noise (dB)
20	Heavy[a]	61	78	78
	Light	58	64	65
80	Heavy[a]	79	85	86
	Light	76	74	78

[a] Heavy vehicles are defined as having an unloaded mass of greater than 1525 kg.
Source: From Nelson [2].

are the dominant sources for most normal operations of modern cars at steady highway speeds. Trucks are dominated by power train noise at low speeds, but at higher speeds above about 80 km/h, exterior truck noise is mostly dominated by tire/road noise and aerodynamic noise. Figure 14.3 shows that above 70–80 km/h there is little difference between the exterior noise of the truck whether it is accelerating, cruising at a steady speed, or coasting by with the power plant turned off [1]. Again this suggests that above 70–80 km/h this truck's exterior noise is dominated by tire/road noise and aerodynamic noise.

At high speeds above about 130 km/h, vehicle noise starts to become dominated by aerodynamic flow noise [1]. Due to different noise standards for vehicles in different countries and regions, and the condition of the vehicles, the speed at which tire noise starts to dominate may be higher than indicated above. The data here are for modern European vehicles in new condition. For example, in the United States the truck power plants are generally noisier than in Europe. One interesting approach to predict vehicle pass-by noise involves the use of the reciprocity technique [3].

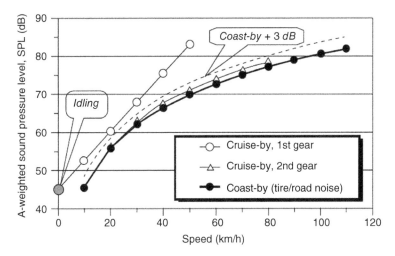

Figure 14.2 A-weighted vehicle sound pressure level, SPL, at constant speeds (cruise-by), as well as tire/road noise (coast-by) for a Volvo S40 (2000) car in new condition. Cruise-by includes power unit and tire/road noise. Cruise-by levels for gears 3–5 are very close to the coast-by curve and are therefore not shown. (*Source:* From Tyre/Road Noise Reference Book [1].)

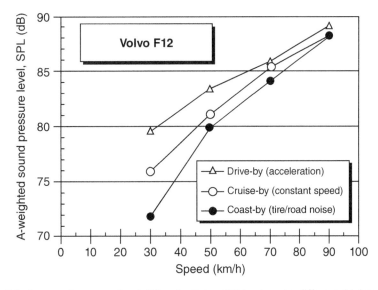

Figure 14.3 Exterior sound pressure level, SPL, of a Volvo F12 truck under different driving conditions [1].

14.2.1 Power Plant Noise and Its Control

Although hybrid vehicles (HVs), which partially use relatively quiet electric motors, are increasing in use, the internal combustion engine (ICE) remains a major source of noise in transportation and industry. ICE intake and exhaust noise can be effectively silenced [4]. The noise radiated by vibrating engine surfaces, however, is more difficult to control. In gasoline engines a fuel–air mixture is compressed to about one-eighth to one-tenth of its original volume and ignited by a spark. In diesel engines air is compressed to about one-sixteenth to one-twentieth of its original volume, liquid fuel is injected in the form of a spray, then spontaneous ignition and combustion occurs. Because the rate of cylinder pressure rise is initially more abrupt with a diesel engine than with a gasoline engine, diesel engines tend to be noisier than gasoline engines. The noise of internal combustion engine (ICE) diesel engines has consequently received the most attention from both manufacturers and researchers. The noise of engines can be divided into two main parts: combustion noise and mechanical noise. The combustion noise is caused mostly by the rapid pressure rise created by ignition, and the mechanical noise is caused by a number of mechanisms with perhaps piston slap being one of the most important, particularly in diesel engines [5].

The noise radiated from the engine structure has been found to be almost independent of load, although it is dependent on cylinder volume and even more dependent on engine speed. Measurements of engine noise over a wide range of cylinder capacities have suggested that the A-weighted sound pressure level (SPL) of engine noise increases by about 17 dB for a 10-fold increase in cylinder capacity [6]. A-weighted engine noise levels have been found to increase at an even greater rate with speed than with capacity (at least at twice the rate) with about 35 dB for a 10-fold increase in speed. Engine noise can be reduced by attention to details of construction. In particular, stiffer engine structures have been shown to reduce radiated noise. Partial add-on shields and complete enclosures have been demonstrated to reduce the A-weighted noise level of a diesel engine of the order of 3–10 dB.

Although engine noise may be separated into two main parts – combustion noise and mechanical noise – there is some interaction between the two noise sources. The mechanical noise may be considered to be the noise produced by an engine that is motored without the burning of fuel. Piston slap occurs as the piston travels up toward top dead center and is one of the mechanical sources that results in engine structural vibration and radiated noise. But piston slap is not strictly an independent mechanical process since the process is affected by the extra forces on the piston generated by the combustion process. The opening and closing of the inlet and exhaust valves, the forces on the bearings caused by the system rotation, and the out of balance of the engine system are other mechanical vibration sources that result in noise. The mechanical forces are repeated each time the crankshaft rotates, and, if the engine is multicylinder, then the number of force repetitions per revolution is multiplied by the number of cylinders. Theoretically, this behavior gives rise to forces at a discrete frequency, f, which is related to R, the number of engine revolutions/minute (rpm), and N, the number of cylinders:

$$f = NR/60 \text{ Hz}. \tag{14.1}$$

Since the mechanical forces are not purely sinusoidal in nature, harmonic distortion occurs. Thus, mechanical forces occur at integer multiples of f given by the frequencies $f_n = nf$, where n is an integer, 1, 2, 3, 4,.... Assuming that the engine behaves as a linear system, these mechanical forces result in forced vibration and mechanical noise at these discrete frequencies. Combustion noise is likewise partly periodic in nature, and this part is related to the engine rpm because it occurs each time a cylinder fires. This periodic combustion noise frequency, f_p, is different for a two-stroke than for a four-stroke engine and is, of course, related to the number of cylinders, N, multiplied by the number of firing strokes each makes per revolution, m. Some of the low-frequency combustion noise is periodic and coherent from cylinder to cylinder. Some of the combustion noise is not periodic because it is caused by the unsteady burning of the fuel–air mixture. This burning is not exactly the same from cycle to cycle of the engine revolution, and so combustion noise, particularly at the higher frequencies, is random in nature.

Example 14.1   Determine the frequencies of mechanical forces in a four-cylinder engine operated at 2000 rpm.

Solution

First we determine the fundamental frequency from Eq. (14.1) as

$$f = 4 \times 2000/60 = 133.3 \text{ Hz}$$

Then, the mechanical forces occur at $n \times f = n \times 133.3$ with $n = 1,2,3,...$, i.e.

$$f = 133.3 \text{ Hz}, 266.6 \text{ Hz}, 400 \text{ Hz, etc.}$$

As mentioned above, engine noise varies with engine size, speed, and combustion system. Usually, noise is predicted from these parameters using empirical relationships. Although it is probable that changes in emissions regulations have caused a new forcing function, gear train rattle, to become the dominant noise forcing function in many heavy-duty diesel engines. A study [7] showed that the average A-weighted SPL at 1 m from the engine, for an engine running at full load is given by

for naturally aspirated (NA) direct injection (DI) diesels:

$$L_P = 30 \log (N) + 50 \log (B) - 106, \tag{14.2}$$

for turbocharged diesels:

$$L_P = 25 \log (N) + 50 \log (B) - 86, \tag{14.3}$$

for indirect injection (IDI) diesels:

$$L_P = 36 \log (N) + 50 \log (B) - 133, \tag{14.4}$$

for gasoline:

$$L_P = 50 \log (N) + 30 \log (B) + 40 \log (S) - 223.5, \tag{14.5}$$

where N is the speed in rpm, B is the bore in mm, and S is the stroke in mm. NA refers to natural aspiration, DI to direct injection, and IDI to indirect injection. The bore is the diameter of the circular chambers cut into the cylinder block and the stroke is the distance the piston travels.

Example 14.2 Consider a Chevrolet 153-cubic-inch (2.5 L) gasoline four-cylinder engine. The bore and stroke of each cylinder are 3.875 and 3.25 in., respectively. Determine the average A-weighted SPL at 1 m from the engine running at full load at 3000 rpm.

Solution

We must use Eq. (14.5) for a gasoline engine. First we transform inches into mm. Then we have a bore of 3.875 in. = 98.4 mm and a stroke of 3.25 in. = 82.6 mm. Substituting the values into Eq. (14.5) we obtain the A-weighted SPL at 1 m:

$$L_p = 50 \times \log (3000) + 30 \times \log (98.4) + 40 \times \log (82.6) - 223.5$$
$$= 173.9 + 59.8 + 76.7 - 223.5 = 86.9 \, dB$$

Larger engines tend to have a lower maximum speed than small engines, and inherently loud diesel engines have a lower maximum speed than quieter gasoline engines. The net result is that many engines have similar noise levels at maximum speed and load, although noise levels compared at a given speed can vary by up to 30 dB [5]. Research continues on understanding engine noise sources and how the noise energy is transmitted to the exterior and interior of vehicles [8, 9].

Mechanical noise reduction (NR) in an engine has been mostly aimed at piston slap and valve train noise, although NRs of accessories, reduction of structural response to force inputs, and use of noise shields and enclosures are often considered.

A number of options are available to reduce radiated noise from the exterior surfaces of engines. They include stiffening of exterior surfaces, reducing the stiffness of surfaces, adding damping treatments, or isolating the connection between the structure and covers [5]. However, these options have to be carefully designed. Figure 14.4 shows the radiated sound power level by a heavy-duty diesel engine

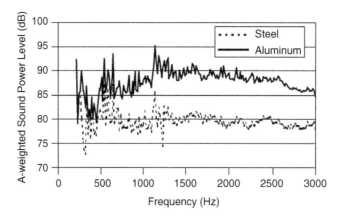

Figure 14.4 Comparison of A-weighted radiated sound power level for a stiff cast-aluminum oil pan and a relatively flexible stamped steel pan on the same heavy-duty diesel engine [5].

for two different oil pans. It is observed that, although stiffer, the aluminum pan is about 10 dB louder than the flexible stamped oil pan at many frequencies. This difference is due to the higher radiation efficiency of the stiffer aluminum oil pan. The stamped steel pan has a 7-dB lower overall A-weighted sound power level than the much stiffer cast-aluminum pan [5].

In recent years, the number of electric vehicles continues to increase worldwide. Even though electric cars are mainly intended to reduce CO_2 emissions, there is also a possibility that they can help to reduce environmental noise, since electric vehicles in general are found to be very quiet at low speed. This is particularly important in congested streets where low traffic flow is prevalent and tire/road noise becomes less important than power plant noise. However, there is some concern about the effects of the low noise emission from electric vehicles on traffic safety. Some studies have discussed the importance of implementing artificial sound warning signals in electric vehicles, in particular to protect blind or visually impaired pedestrians [10].

A Japanese study measured the noise emissions from two electric vehicles, one hybrid vehicle (HV), and two ICE vehicles [11]. The measurements also included noise artificially added to the electric and HVs following the Japanese guidelines that recommend sounds which simulate the sound of ICE vehicles. The results are shown in Table 14.2 when the cars were driven at 10 and 20 km/h. The results show that for 10 km/h, there is a 6–9 dB difference between the noise of ICE and electric vehicles without the artificially added sound. At 20 km/h the difference is 5 dB between EV-2 and the ICE cars, and for EV-1 there is no

Table 14.2 A-weighted sound pressure levels from pass-by measurements of two electric cars (EV), one hybrid vehicle (HV), and two ICE cars. The microphone was placed 2 m from the center of the track and 1.2 m above the ground.

Vehicle	10 km/h	20 km/h
EV-1	50 dB	62 dB
EV-1 with artificial sound	55 dB	62 dB
EV-2	47 dB	57 dB
EV-2 with artificial sound	56 dB	62 dB
HV-1	50 dB	60 dB
HV-1 with artificial sound	54 dB	60 dB
ICE-1	56 dB	62 dB
ICE-2	58 dB	62 dB

Source: Data from Ref. [11].

apparent difference. With the artificially added noise there is no difference in SPL for either of the electric cars and the ICE cars, although a difference of 2 dB between the ICE cars and the hybrid car is observed.

A recent Danish report has presented a literature survey on the noise from electric vehicles [12]. The report concluded that electric cars are quieter than ICE vehicles only at low speeds. At speeds between 25 and 50 km/h, no difference is found between ICE and electric car noise. Several studies in the frequency domain have reported that the noise of electric vehicles can have some peaks at middle frequencies, which may be perceived as annoying [12].

14.2.2 Intake and Exhaust Noise and Muffler Design

With each intake stroke, ICE engine noise is generated by an unsteady airflow produced by the volume of air that each cylinder draws in. Ducted sources are found in many different mechanical systems. Common ducted-source systems include engines and mufflers (also known as silencers), fans, and air-moving devices (including flow ducts and fluid machines and associated piping). Silencers (also known as mufflers) are used as well on some other machines including compressors, pumps, and air-conditioning systems. In these systems, the source is the active component and the load is the path, which consists of elements such as mufflers, ducts, and end terminations. The acoustical performance of the system depends on the source–load interactions. Models based on electrical analogies have been found useful in predicting the acoustical performance of systems. Various methods exist for determining the internal impedance of ducted sources. The acoustical performance of a system with a muffler as a path element is usually best described in terms of the muffler insertion loss (IL) and the sound pressure radiated from the outlet of the system [4, 13].

There are two main types of mufflers that are fitted to the intake (inlet) and exhaust (outlet) pipes of machinery ductwork. Reactive mufflers function by reflecting sound back to the source and also to some extent by interacting with the source and thereby modifying the source's sound generation. Absorptive silencers on the other hand reduce the sound waves by the use of sound-absorbing material packed into the silencer. The exhaust and intake noise of ICEs is so intense that they need to be "muffled" or "silenced." ICE pressure pulsations are very intense and nonlinear effects should be included. Exhaust gas is very hot and flows rapidly through the exhaust system. The gas stream has a temperature gradient along the exhaust system, and the gas pressure pulsations are of sufficiently high amplitude that they may be regarded almost as shock waves. Some of these conditions violate normal acoustical assumptions. Modeling of an engine exhaust system in the time domain has been attempted by some researchers to account for the nonlinear effects. But such approaches have proved to be challenging. Most modeling techniques have used the transmission matrix approach in the frequency domain and have been found to be sufficiently effective.

The acoustical performance of a ducted-source system depends on the impedances of the source and load and the four-pole parameters of the path. This complete description of the system performance is termed *insertion loss (IL)*. The IL is the difference between the SPLs measured at the same reference point (from the termination) without and with the path element, such as a muffler, in place. (See Figure 14.5a.)

Another useful description of the path element is given by the *transmission loss (TL)*. The TL is the logarithmic ratio of the incident to transmitted sound powers $(S_i I_i)/(S_t I_t)$. (See Figure 14.5b.)

The *noise reduction (NR)* is another descriptor used to measure the effect of the path element and is given by the difference in the measured SPLs upstream and downstream of the path element (such as the muffler), respectively. (See Figure 14.5c.)

The *insertion loss (IL)* is the most useful description for the user since it gives the net performance of the path element (muffler) and includes the interaction of the source and termination impedances with the muffler element. It can be shown that the IL depends on the source impedance. It is easier to measure IL than to predict it because the characteristics of most sources are not known.

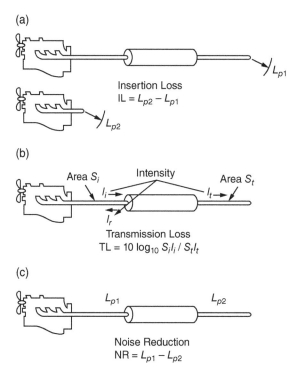

Figure 14.5 Definitions of muffler performance.

The TL is easier to predict than to measure. The TL is defined so that it depends only on the path geometry and not on the source and termination impedances.

The TL is a very useful quantity for the acoustical design of a muffler system path geometry. However, it is difficult to measure since it requires two transducers to separate the incident and transmitted intensities. The description of the system performance in terms of the NR requires knowledge of both the path element and of the termination.

Figure 14.5 shows various acoustical performance descriptors used for ducted-source systems. It should be noted that similar system terminology to that described here and shown in Figure 14.5 is commonly used for the acoustical performance of air-conditioning system components, machinery enclosures, and partition walls in buildings. The terminology used for the acoustical performance of barriers is also similar, although the additional descriptor "attenuation" is also introduced for barriers. Note that with barriers the IL descriptor can have a slightly different meaning. (See Chapter 9 in this book.)

Work on muffler design and modeling continues with approaches varying from completely experimental to mostly theoretical [14–18]. The acoustical design of reactive and absorptive mufflers and silencers is discussed in more detail in Chapter 10.

14.2.3 Tire/Road Noise Sources and Control

In most developed and developing countries, vehicle traffic is the main contributor to community noise. Aircraft noise is a lesser problem since it mostly affects small areas of urban communities that are located near to major airports. The substantial increase in road vehicle traffic in Europe, North America, Japan, and other countries suggests that road vehicle traffic will continue to be the dominant source of community noise into the foreseeable future. Legislative pressures brought to bear on vehicle manufacturers, particularly in Europe and Japan, have resulted in lower power unit noise output of modern vehicles. Tire/road interaction noise, caused by the interaction of rolling tires with the road surface, has become the predominant noise source on new passenger cars when operated over a wide range of constant speeds, with the exception of the first gear. When operated at motorway speeds, the situation is similar for heavy trucks and again tire road noise is found to be dominant.

A European study has suggested that, for normal traffic flows and vehicle mixes on urban roads, about 60% of traffic noise sound power output is due to tire/road interaction noise [19]. On motorways at high speed, this increases to about 80%. Both exterior power plant noise (the noise from the engine, gearbox/transmission, and exhaust system) and tire/road noise are strongly speed dependent. Measurements show that exterior tire/road noise increases logarithmically with speed (about 10 dB for each doubling of speed). Since an increase of 10 dB represents an approximate doubling of subjective loudness, the tire/road noise of a vehicle traveling at 40 km/h sounds about twice as loud as one at 20 km/h; and at 80 km/h, the noise will sound about four times as loud. Studies have shown that there are many possible mechanisms responsible for tire/road interaction noise generation. Although there is general agreement on the mechanisms, there is still some disagreement on their relative importance. The noise generation mechanisms may be grouped into two main types: (i) vibrational (impact and adhesion) and (ii) aerodynamic (air displacement).

There are five main methods of measuring tire noise. These methods include (i) close proximity CPX (trailer), (ii) cruise-by, (iii) statistical pass-by, (iv) drum, and (v) sound intensity. Figure 14.6 shows tire noise being measured by a CPX trailer built at Auburn University [20].

Figure 14.7 shows A-weighted SPLs measured with the Auburn CPX trailer. Note that the A-weighted spectrum peaks between 800 and 1000 Hz, and the peak increases slightly in magnitude and frequency as the vehicle speed is increased.

Knowledge of tire/road aerodynamic and vibration noise generation mechanisms suggests several approaches, which if used properly can be used to help suppress tire/road noise [19]. Perhaps the most hopeful, lower cost, approach to suppress tire/road noise is the use of porous road surface mixes [20]. Porous roads, with up to 20–30% or more of air void volume, are being used increasingly in many countries. They can provide A-weighted SPL reductions of up to 5–7 dB, drain rain water, and reduce splash-up behind vehicles as well. If the porous road can be designed to have maximum absorption at a frequency between 800 and 1000 Hz, it can be most effective in reducing tire/road interaction noise. Reduction of tire/road noise is also very important for reducing the environmental noise in urban areas (see Chapter 16 of this book). Tire/road

Figure 14.6 View of finished CPX trailer built at Auburn University [20].

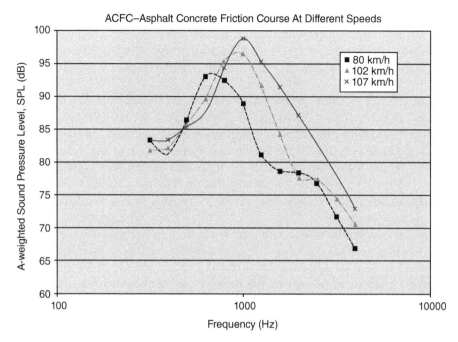

Figure 14.7 A-weighted one-third octave band sound pressure level tire/road noise measurements made with a CPX trailer built at Auburn University.

interaction noise generation and measurement are discussed in detail in Ref. [19]. Much research continues to be conducted into understanding the origins of tire/road noise [21–26].

14.2.4 Aerodynamic Noise Sources on Vehicles

The interaction of the flow around a vehicle with the vehicle body structure gives rise to sound generation and noise problems both inside and outside the vehicle. Turbulent boundary layer fluctuations on the vehicle exterior can result in sound generation. The pressure fluctuations also cause structural vibration, which in turn results in sound radiated both to the exterior and to the vehicle interior. Abrupt changes in the vehicle geometry result in regions of separated flow that considerably increase the turbulent boundary layer fluctuations. Poorly designed or leaking door seals result in aspiration (venting) of the seals, which allows direct communication of the turbulent boundary layer pressure fluctuations with the vehicle interior. Appendages on a vehicle, such as external rear-view mirrors and radio antennas also create additional turbulence and noise. The body structure vibrations are also increased in intensity by the separated flow regions. Although turbulent flow around vehicles is the main cause of aerodynamic noise, it should be noted that even laminar flow can indirectly induce noise. For example, the flow pressure regions created by laminar flow can distort body panels, such as the hood (bonnet), and incite vibration. Figure 14.8 shows various locations on a vehicle body where flow separation regions exist creating fluctuating pressures. Multiple and extensive separated regions may be present between rear window and trunk lid (notch-back), and at the base of fast-back or hatch-back vehicles [27].

At vehicle speeds above about 130 km/h the vehicle flow-generated noise exceeds the tire noise and increases with speed to the sixth power. Because of the complicated turbulent and separated flow interactions with the vehicle body, it is difficult to predict accurately the aerodynamic sound generated by a vehicle and its radiation to the vehicle interior and exterior. Vehicle designers often have to resort to empiricism and/or full-scale vehicle tests in wind tunnels for measurements of vehicle interior and exterior aerodynamic flow

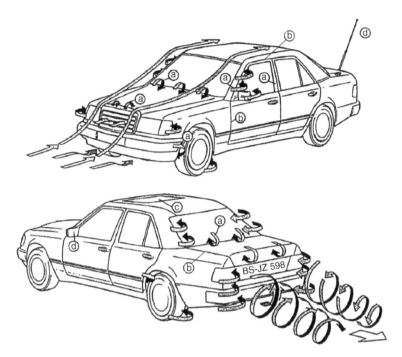

Figure 14.8 Some locations of sound sources on an automobile body. Type (a) locations are, for example, hood front edge, windshield base, windshield/front door transition (A-pillar), around recessed window panes, wheel wells, and the body underside. Type (b) locations are at structure gaps such as around doors, rain gutters, hood/windshield base, trunk lid, and the like. Type (c) locations are around doors, windows, trunk lid, sunroof, and the like. Type (d) locations are at external rear-view mirrors, windshield wipers, radio antenna, sunroof, popup head lamps, and the like. Type (e) locations are in the ducting and venting circuit of high-volume air-conditioning equipment [27].

generation and interior noise predictions. Statistical energy analysis and computational fluid dynamics have also been utilized to predict interior wind noise in vehicles [28].

14.2.5 Gearbox Noise and Vibration

Transmissions and gearbox systems are used in cars, trucks, and busses to transmit the mechanical power produced by the engine to the wheels. Similar transmission systems are used in propeller aircraft to transmit power to the propeller(s) from the engine(s) or turbine(s). Transmission gearboxes are also used in some railroad systems and ships. Some modern high-speed rail vehicles, however, use motive power systems (electric motors) mounted directly onto the axles of each rail vehicle, resulting in quieter operation and reduced noise problems.

The gearbox can be the source of vibration and radiated noise and should be suitably soft mounted to the vehicle structure, wherever possible. Shaft misalignment problems must be avoided, however, with the mounting system chosen. The principal components of a gearbox are comprised of gear trains, bearings, and transmission shafts.

Unless substantial bearing wear and/or damage have occurred, gear meshing noise and vibration are normally the predominant sources in a gearbox. The vibration and noise produced depend upon gear contact ratios, gear profiles, manufacturing tolerances, load and speed, and gear meshing frequencies. Different gear surface profiles and gear types produce different levels of noise and vibration. In general, smaller gear tolerances result in smoother gear operation but require increased manufacturing costs.

Gearboxes are often fitted with enclosures to reduce noise radiation, since the use of a low-cost gearbox combined with an enclosure may be less expensive than the use of a high-performance gear system and gearbox without an enclosure. Gearbox enclosures, however, can result in reduced accessibility and additional maintenance difficulties. A better approach, where possible, is to try to utilize a lower noise and vibration gear system so that a gearbox enclosure is unnecessary. Reference [29] deals with gearbox noise and vibration and Ref. [30] specifically addresses gear noise and vibration prediction and control methods.

14.2.6 Brake Noise Prediction and Control

Brake noise has been recognized as a problem since the mid-1930s. Research was initially conducted on drum brakes, but recent work has concentrated on disk brakes since they are now widely used on cars and trucks. The disk is bolted to the wheel and axle and thus rotates at the same speed as the wheel. The brake caliper does not rotate and is fixed to the vehicle. Hydraulic oil pressure forces the brake pads onto the disk, thus applying braking forces that reduce the speed of the vehicle.

From a dynamics point of view a braking system can be represented as two dynamic systems connected by a friction interface [31]. The normal force between the two systems results from the hydraulic pressure and is related to the friction force. The combination of the friction interface and the dynamic systems makes it difficult to understand and reduce brake noise. Brake noise usually involves a dynamic instability of the braking system.

There are three overlapping "stability" parameters of (i) friction, (ii) pressure, and (iii) temperature. If the brake operates in the unstable area, changes in the parameters have little or no effect, and the brake is likely to generate noise. Such conditions may be caused by excessively low or high temperatures, which cause changes in the characteristics of the friction material. The "stable" area may be regarded to represent a well-designed brake in which the system only moves into the "unstable" region when extreme changes in the system parameters occur. Brake noise and vibration phenomena can be placed into three main categories: (i) judder, (ii) groan, and (iii) squeal.

Judder occurs at a frequency less than about 10 Hz and is related to the wheel rotation rpm or a multiple of it. It is a *forced* vibration caused by nonuniformity of the disk, and the vibration is of such a low frequency that it is normally sensed rather than heard. There are two types of judder: cold and hot judder. Cold judder is commonly caused by the brake pad rubbing on the disk during periods when the brakes are not applied. Hot judder is associated with braking at high speeds or excessive braking when large amounts of heat can be generated causing transient thermal deformations of the disk.

Groan occurs at a frequency of about 100 Hz. It usually happens at low speed and is the most common unstable brake vibration phenomenon. It is particularly noticeable in cars and/or heavy trucks coming to a stop or moving along slowly and then gently braking. It is thought to be caused by the stick–slip behavior of the brake pads on the disk surface and because the friction coefficient varies with brake pad velocity.

Squeal normally occurs above 1 kHz. Brake squeal is an unstable vibration caused by a geometric instability. It can be divided into two main categories: (i) low-frequency squeal (1 kHZ to 4 kHz), diametrical nodal spacing in the disk, and (ii) high-frequency squeal (> 4 kHz). See Ref. [31] for a more detailed discussion of brake noise. References [32–37] describe recent research on brake noise.

14.3 Interior Road Vehicle Cabin Noise

14.3.1 Automobiles and Trucks

The interior noise in the occupied spaces of automobiles, busses, and trucks is mainly caused by the engine, exhaust, transmission, power train, tire/road interaction, and wind/structure interaction. With automobiles, trucks, and busses, structure-borne noise usually tends to be dominant below about 400–500 Hz, while airborne noise from tire/road interaction and airflow/structure interaction (wind noise) tends to dominate in the mid- and high-frequency ranges. Reference [38] describes the relative contributions from different noise and vibration sources and paths to the vehicle interior. Interior noise levels depend also upon vehicle operating conditions and in particular on speed. During acceleration, engine noise tends to dominate; while during cruise conditions, above about 80 km/h, tire noise becomes the major contributor, and at higher speeds above about 120 km/h, wind noise becomes dominant. Improper sealing around vehicle doors, windows, dashboards, windshields, and the like causes excessive external turbulent pressure fluctuations from airflow over the vehicle (wind noise) to be transmitted to the vehicle interior. This is known as aspiration [38]. Some luxury vehicles have multiple door seals to accomplish effective sealing to reduce the transmission of these external turbulent pressure fluctuations to the occupied interior.

To reduce airborne and structure-borne noise reaching the vehicle's occupants, various well-known techniques are often used, including (i) increasing structural damping, (ii) improving the TL of body panels and windows, (iii) increasing the use of sound-absorbing materials in the engine, passenger, and luggage compartments, and (iv) vibration isolation of mechanical components. Hirabayashi et al. have described how all four of these techniques are used the in the automotive industry to reduce noise and vibration in vehicles [39]. Rao has provided a useful review of the use of vibration damping materials in automobiles and commercial aircraft [40]. Polce et al. have also presented a detailed study on improvements in cabin noise obtained by using damping treatments [41]. Add-on constrained layer, spray-on, and integral damping materials are often used. Table 14.3 lists various locations on a typical automobile, such as illustrated in Figure 14.9, where these are often used.

Damping materials are generally more effective at reducing structure-borne vibration (particularly panel resonances) rather than at reducing the airborne sound transmitted through panels. Of the damping approaches, spray-on damping material is easiest to apply, but constrained damping layers comprised of

Table 14.3 Automotive applications [39].

Engines and Power Trains	Body Structures	Brakes and Accessories
Oil pans	Dash panels	Brake insulators
Valve covers	Door panels	Backing plates
Engine covers	Floor panels	Brake covers
Push rod covers	Wheelhouses	Steering brackets
Transmission covers	Cargo bays	Door latches
Timing belt covers	Roof panels	Window motors
Transfer case covers	Upper cowl	Exhaust shields

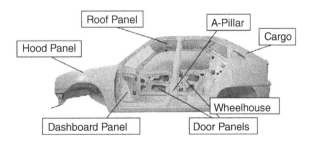

Figure 14.9 Locations in a typical automobile where damping treatments are often applied [40].

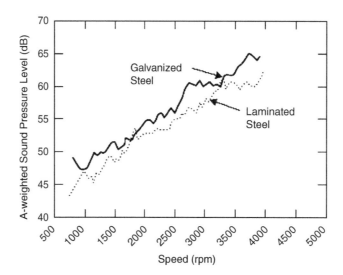

Figure 14.10 A-weighted sound pressure levels obtained during stationary engine run-up tests for two different oil pans with increasing engine rpm [40].

a viscoelastic layer constrained by a thin surface layer normally provide higher damping for the same weight or less weight through shearing action in the viscoelastic layer. The oil pan (or sump) of an engine can be responsible for radiating 50% of the sound power of an engine [40, 42]. Figure 14.10 shows the A-weighted interior SPL in a vehicle at the driver's ear position with: (i) a regular galvanized steel oil pan and (ii) with the oil pan replaced by a high damping laminated steel oil pan [40]. The A-weighted SPLs were obtained during a stationary engine run-up test with increasing engine revolutions per minute (rpm). Figure 14.10 shows that reductions in the A-weighted interior noise level of about 5 dB were obtained with the laminated steel oil pan.

Water-based spray-on damping materials are also used in automobile manufacture. Their advantage is that they can be sprayed robotically on areas such as floor panels and other locations that are hard to reach (such as wheel housings) [40]. Thicknesses of between 1 and 3 mm are normally used. The disadvantage of the use of such materials is that they require costly spray and robotic equipment [40]. It is believed that the use of body and floor panel damping is effective mostly at reducing structure-borne interior automobile noise in the 100- to 500-Hz range. Although increased damping normally has disappointing results in increasing sound insulation, spray-on damping materials do also add *mass*, which can reduce airborne sound transmission through areas such as floor panels.

The use of laminated glass to reduce automobile interior noise continues to increase [43]. Figure 14.11 shows the effect of using laminated glass in reducing wind noise and tire/road noise [38]. The reduction in airborne noise by the use of the laminated glass is probably caused by the material impedance mismatch, which results in sound reflection at each interface layer; while the reduction in structure-borne noise is likely caused mostly by the increase in vibration damping produced by use of the laminated glass. At some frequencies, the damping loss factor for the laminated glass is twice that of the standard tempered glass. The increased damping can reduce sound transmission in the coincidence frequency region.

The use of laminated vibration-damped steel (LVDS) is now being studied for its capability in reducing airborne and structure-borne noise reaching the passenger compartment from the engine and power train components. See an example of an LDVS dash panel in Figure 14.12. The advantage of LVDS is that no add-on damping treatment is needed. Improvements in sound quality and speech interference level through the use of LVDS must be considered in light of cost and other factors.

In the automotive industry, materials used to enclose a noise source (such as the engine) or the passengers (the cabin enclosure) are usually termed *barrier* materials. Such barrier materials are required to reduce

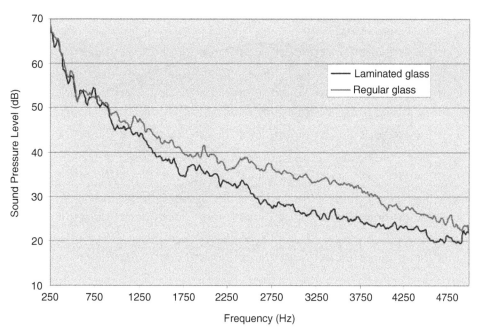

Figure 14.11 Wind noise at driver's ear location and speed of 180 km/h in vehicle interior when laminated side windows are used [38].

Figure 14.12 Dashboard panel made of laminated vibration damped steel [40].

airborne sound reaching the cabin from noise sources, including the engine, fan, exhaust system, tires, and wind. Below the coincidence frequency region, the *TL* of such materials is mostly dominated by the mass/unit area (m) of a partition (see Chapter 12 of this book). For single layer partitions this means that *TL* increases by 6 dB for each doubling of frequency or by 6 dB at a given frequency if m is doubled. Multilayer partitions, particularly with intervening air gaps between layers, can achieve *TL* results much better than mass law would predict, and the *TL* for such panels can be more like 12 dB/octave rather than 6 dB/octave. Great care must be taken to avoid air leaks for high values of *TL* to be achieved in practice. Figure 14.13 shows locations where barrier materials are often applied in an automobile.

Once airborne and structure-borne sound has penetrated into the cabin, the sound can be absorbed effectively by the use of sound-absorbing material (see Chapter 9 of this book). Generally, thicker sound-absorbing

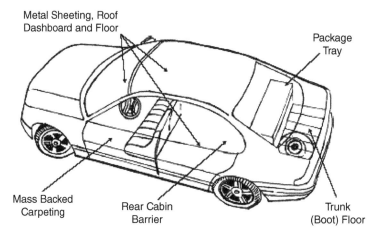

Figure 14.13 Typical locations in an automobile where "barrier" materials are utilized.

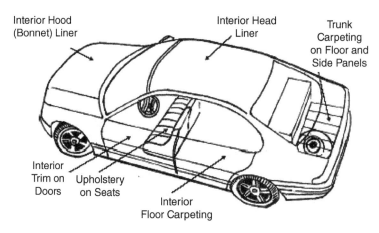

Figure 14.14 Typical locations in an automobile where sound-absorbing materials are utilized.

materials are better than thinner ones, and a material works best if its thickness approaches one quarter wavelength of the sound. This thickness can only be approached at quite high frequency in an automobile. Figure 14.14 shows locations where sound-absorbing materials are often applied. In particular, sound absorption materials based on acoustical textiles are widely used in vehicles to reduce interior noise and improve the sensation of ride comfort for the vehicle's occupants. See Ref. [44] for a detailed discussion on the use of acoustical textiles in the automotive industry.

A variety of theoretical methods, such as statistical energy analysis, finite element analysis (FEA), boundary element analysis, and computer-aided engineering methods are used to predict sound and vibration energy transmission from exterior sources to the vehicle interior as described in many publications [45–70]. Because of complicated structural geometries, analytical approaches must often be supplemented with laboratory noise testing in anechoic rooms, wind tunnels, and actual tests with real full-scale vehicles on dedicated test tracks [38]. With automobiles and trucks, it is insufficient to simply reduce noise reaching the passenger compartment. Passengers expect different types of vehicles such as passenger cars, sports cars, sport utility vehicles, luxury cars, and light trucks to have distinctive sounds. In some cases, manufacturers' brand names are often associated with a particular vehicle sound, and it is important not just to reduce interior noise but also to consider the quality of the sound [71].

14.3.2 Off-Road Vehicles

Off-road vehicles are used for a variety of tasks including moving earth, road, railway and airport construction, and use on building sites and in agriculture. The vehicles are provided either with wheels or tracks for propulsion. Cabs are used to protect the operator from rollover and other operational hazards, from the weather, and from noise and vibration. One of the main concerns with these vehicles is to provide a safe acoustical environment for the operator inside the cab. To achieve this, the time-averaged A-weighted equivalent SPL inside the operator cab should be no higher than about 75 dB. Since noise levels vary with time as vehicles undertake various activities such as moving earth or grading, noise measurement methods to record the SPLs generated have been standardized internationally, which take account of this operational variability. A-weighted internal cab noise measurements are normally made with a single microphone in the presence of the operator. The microphone is oriented horizontally at the operator head height and pointed forward or in the direction in which the operator normally looks.

Interior noise levels depend upon the strength of the noise and vibration sources and paths including the power plant, exhaust, transmission, hydraulic systems, mounting systems, and wheel and/or track ground surface interactions. It is important to ensure that the sources are vibration-isolated from the vehicle structure and that proper care is taken to achieve satisfactory operator sound isolation by providing a suitable TL (sound reduction index) for the cab enclosure. Air leaks should be minimized if the cab noise level can only be achieved with windows closed. Structural damping and interior sound-absorbing material, if properly used in the design stage, can help achieve the necessary interior cab noise level design goals.

Figure 14.15 Operator cabin of an agricultural harvester machine [72].

Analytical models, which are often employed in the design of off-road vehicle cabs, include statistical energy analysis, FEA and boundary element methods (BEM) [72, 73]. The track undercarriage can become a dominant noise source for tracked vehicles. Experimental confirmation that cabin noise goals have been achieved is necessary during prototype vehicle development and testing. The SPL depends on vehicle operations including moving earth and other working cycles. Reference [74] discusses the interior noise and vibration of off-road vehicles and methods to reduce the interior noise.

A-weighted SPLs in off-road vehicle cabs have been measured to be as high as 107 dB [75]. By proper passive noise control approaches, these levels can be reduced considerably [74]. Approaches vary from mostly experimental to almost completely theoretical [72, 73, 76]. Both structure-borne and airborne noise and vibration paths to the cab interior must be considered [72]. A quite sophisticated study of the noise transmitted into the operator cab of an agricultural machine using both scale models and theoretical FEA and BEM has been reported [72]. Figure 14.15 shows the operator cabin studied by Desmet et al. [72]. The scale model was used to obtain experimental results. Both FEA and BEM meshes were used for the cabin structure and acoustical space. A comparison between the measured and predicted sound *IL* of the model cabin is shown in Figure 14.16.

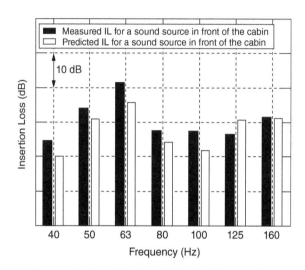

Figure 14.16 Measured and predicted agricultural cabin sound insertion loss [72].

14.4 Railroad and Rapid Transit Vehicle Noise and Vibration Sources

14.4.1 Wheel–Rail Interaction Noise

Noise produced by wheel–rail interaction continues to be of concern in railway operations. Many studies have been conducted on wheel–rail interaction noise. Most of the studies have involved various measurement approaches [77–85]. Main wheel–rail sources include (i) rolling noise, which is caused by small-scale vertical profile irregularities (roughness) of wheel and rail, (ii) impact noise caused by discrete discontinuities of the profile such as wheel flats, rail joints, or welds, and (iii) squeal noise that occurs in curves. In each case, the noise is produced by vibrations of the wheels and track. The dynamic properties of the wheel and track have an affect on the sound radiation. Control measures for rolling noise include reduced surface roughness, wheel shape optimization and added wheel passive damping treatments, increased rail support stiffness, or use of local wheel–rail shielding. The use of trackside noise barriers is becoming common for railways [86]. For squeal noise, mitigation measures include friction control by lubrication or friction modifiers. Reference [87] discusses causes of wheel–rail noise and methods for its control.

A train running on straight unjointed track produces rolling noise. This is a broadband, random noise radiated by wheel and track vibration over the range of about 100–5000 Hz. The overall radiated SPL increases at about 9 dB per doubling of train speed. This represents almost a doubling of subjective loudness for a doubling of speed. Rolling noise is induced by small vertical profile irregularities of the wheel and rail running surfaces. This is often referred to as roughness, although the wavelength range is between about 5 and 250 mm, which is greater than the range normally considered for microroughness. Wheel and rail roughness may be considered incoherent and their noise spectra simply added. The roughness causes a relative displacement between the wheel and rail and makes the wheel and rail vibrate and radiate noise [87].

14.4.2 Interior Rail Vehicle Cabin Noise

The main concern in the design of railroad passenger cars and rapid transit system (RTS) vehicles is the provision of a comfortable noise environment for the passengers. A balance must be achieved between acoustical privacy and speech interference. Passengers want the SPL to be low enough so that they can carry on conversations and use cell (mobile) telephones easily. On the other hand, they do not want the level to be so low that their conversations can be overheard by other passengers nearby. The noise environment in railroad car and RTS vehicles is caused by a variety of sources that depend mostly upon the power plant setting and vehicle speed.

Figure 14.17 shows the various noise and vibration sources that exist within railroad cars and RTS vehicles [88]. At low speed, interior noise is caused mainly by the power plant and air-conditioning systems. Wind

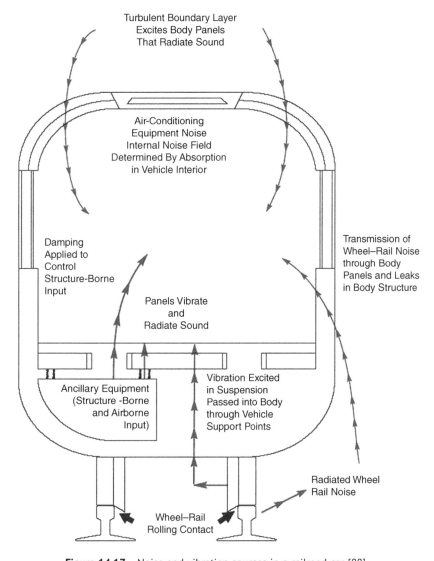

Figure 14.17 Noise and vibration sources in a railroad car [88].

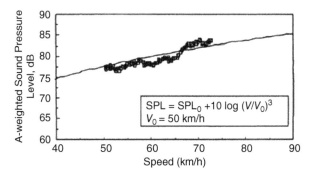

Figure 14.18 Interior sound pressure level as a function of train speed, V [80].

noise is normally unimportant for slow or medium-speed RTS vehicles, but it can become the dominant noise source with railroad systems when they are operated at very high speeds. The balance between acoustical privacy and speech interference changes with speed and it is difficult to optimize this balance without the use of artificial masking noise during low speed or stationary operations of the vehicles.

Various power plants are used to propel the railroad and RTS vehicles. In some high-speed railroad vehicle designs several sets of traction motors are used to drive each passenger car, instead of just the locomotive. In some cases, individual electric motors are employed to drive the vehicle wheels directly. Wheel–rail noise becomes important at medium and high speeds and depends upon rail roughness and wheel wear [89–91].

Soft rubber wheel suspensions have been used on some RTSs (e.g. Mexico City) in an attempt to reduce noise. Rubber or elastomeric damping blocks have also been built into the wheel systems of some other passenger railcars and crew locomotives [80]. If damping blocks and damping rims are implemented, they should be properly tested to demonstrate that they can be used safely in service.

Figure 14.18 shows the A-weighted interior noise level at moderate speeds for one particular railcar [80]. It is observed that the A-weighted SPL increases at a rate of 9–10 dB for each doubling of speed in the range of measurements shown. Subjectively, this amounts to approximately a doubling of linear loudness (in sones) for a doubling of train speed. It should be emphasized, however, that this result may be somewhat different for other railcar designs.

Example 14.3 Estimate the change in A-weighted interior SPL in a railcar when the speed is increased from 50 to 85 km/h.

Solution

Considering the empirical formula in Figure 14.18, we obtain the SPL at 50 km/h as: SPL (at 50 km/h) = $SPL_0 + 10 \times \log (50/50)^3 = SPL_0$. Now, the SPL at 85 km/h is

SPL (at 85 km/h) = $SPL_0 + 10 \times \log (85/50)^3 = SPL_0 + 6.9$ dB. Therefore, the change in SPL is ΔSPL = SPL (at 85 km/h) – SPL (at 50 km/h) = 6.9 dB. Thus, the A-weighted SPL is increased by 7 dB, approximately.

Figure 14.19 shows acceleration levels measured on the railcar floor with conventional solid wheels and low-noise wheels. Figure 14.20 shows a comparison of the interior A-weighted SPL of the same railcar as in Figure 14.18 with conventional wheels and with the low noise wheels. Figure 14.20 demonstrates that in this particular railcar, at the speeds shown, the wheel/rail interaction noise is very important. This study shows that there is not complete correspondence between the reduction in the floor acceleration level and the reduction in interior SPL, achieved by the use of low noise wheels. This indicates the presence of other noise sources, such as power plant, wind noise, and air-conditioning noise.

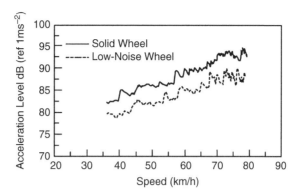

Figure 14.19 Comparison of floor vibration levels for solid wheel and low-noise wheel [80].

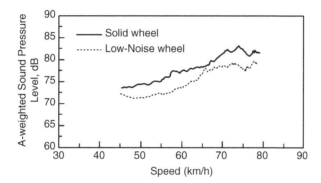

Figure 14.20 Comparison of interior noise levels for solid wheel and low-noise wheel [80].

It is normal practice to set target noise level goals for passenger comfort and to specify maximum power plant and air-conditioning system source sound powers from equipment manufacturers and suppliers in order to achieve the noise targets [88]. Target A-weighted equivalent interior SPLs of between 65 and 70 dB are commonly chosen. To achieve such target noise levels, the vehicle structure must also be carefully designed from the start so that the noise and vibration paths sufficiently attenuate the power plant, wheel/ rail, wind, and air-conditioning noise reaching the passenger compartments.

The contribution to the internal noise caused by wheel/rail interaction can be predicted from knowledge of the roughness of the rails. It is important for the rail roughness levels to be defined at the beginning of the acoustical design process. Both airborne and structure-borne paths must be considered. Cabin wall airborne sound transmission properties, structural damping, and interior sound absorption all affect the transmission of sound and vibration and the resulting SPL in the cabin interior. Accurate prediction of the interior cabin noise can only be made with a knowledge of all of the external and internal cabin noise and vibration sources and paths.

There is extensive literature on noise and vibration sources and paths in rail vehicles [77, 81, 84, 85, 92–96]. Squeal noise is a problem experienced on curved tracks [92, 94]. Reference [88] discusses noise sources and noise level targets in high-speed railroad cars. Reference [97] describes target noise levels in RTS cars and the acoustical design processes commonly used to achieve them.

14.5 Noise And Vibration Control in Ships

Many of the same noise problems exist in passenger ships as in surface transportation vehicles and aircraft. Airborne and structure-borne paths from noise and vibration sources can be of similar and sometimes of equal concern during different ship operations. The main sources include the power plant machinery and the propulsion units including screws and propellers. Figure 14.21 shows the noise transmitted over airborne, structure-borne, and secondary structure-borne paths in a ship model [98]. The latter results from airborne noise impinging on the structure, which then transmits the noise along the structural path.

Noise levels can be intense in the engine room compartments of ships, and consideration needs to be given to providing low enough SPLs so that speech communication is possible and audible alarm signals can be heard. Crew comfort must also be considered in crew rest areas so that they can recuperate from the high noise levels experienced in engine rooms and other high-noise regions of ships.

Noise levels in rest areas must be low enough so that the effects of noise on sleep and overall crew performance are minimized. Figure 14.22 shows that the acoustical environment on cruise ships often receives most complaints [99]. This study also showed that the most irritating noises for cruise passengers fall into three categories: (i) squeaking, clattering, cracking, creaking noises, (ii) noise of engines, and (iii) ventilation and whistles [99].

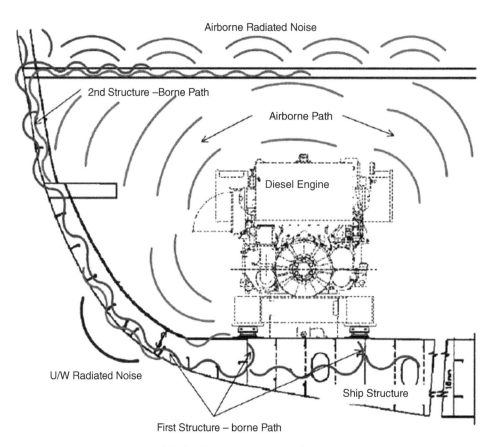

Figure 14.21 Source/path acoustical model [98].

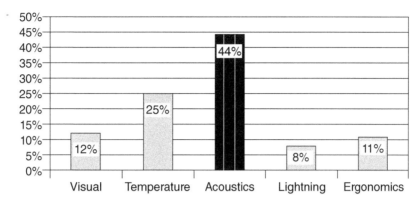

Figure 14.22 Percentage of passengers choosing various criteria needing improvement on a cruise ship [99].

Some A-weighted noise level criteria for crew rest areas are as low as 45 dB, while others permit levels as high as 60 dB in such areas. Noise levels in passenger spaces in ships are generally low enough that speech communication and sleep interference are not issues; providing passenger comfort is the predominant concern in this case. Acoustical prediction methods and the design of ships to achieve effective noise and vibration control are discussed in Ref. [98]. Various studies have been made on the noise in ships [99–101]. Studies have been made using sound intensity to identify noise sources in ships. In one shipboard study the sound intensity field was measured close to a double cabin window to try to identify the source of ship cabin noise problems [100].

The turbine and engine room is usually the noisiest place on a ship, and noise levels can be reduced using composite multilayer structures similar to the layers used in a car hoodliner. In addition, honeycomb panels made of raw fibers are used in ship hulls, bulkheads, fixed walls, and other interior structures of ships to reduce weight and avoid corrosion. Corrosion-resistance composites are also being used internally in ships as space dividers and doors, replacing metal to reduce weight [44].

Example 14.4 An engine room of dimensions 7 m × 10 m with a height of 4 m in a cargo ship has an average SPL in the reverberant field of 95 dB at 500 Hz. Suppose the average absorption coefficient is $\bar{a} = 0.02$ at 500 Hz. It is required to reduce the sound pressure to 85 dB at that frequency by placing a sound absorption material with a sound absorption coefficient of $\bar{a} = 0.6$ at 500 Hz to completely cover the ship's engine room overhead. Does this noise control work?

Solution

The NR can be calculated using Eqs. (3.81) and (3.82). The engine room surface area = $2(70) + 2(28) + 2(40) = 276 \, \text{m}^2$, therefore the room constant $R_1 = 276(0.02)/0.98 = 5.6$ sabins (m²). The new average absorption coefficient is $\bar{a}_2 = [206(0.02) + 70(0.6)]/276 = 0.17$, and the new room constant is $276(0.17)/0.83 = 56.53$ sabins (m²). Thus, the predicted noise level reduction is $10 \times \log(56.53/5.6) = 10$ dB. Therefore, the average SPL in the reverberant field of the engine room is reduced to 85 dB.

In crew and passenger's rest areas, in particular in luxury cruise ships, high noise levels are expected to adversely impact sleep, comfort, and speech intelligibility. Sound absorption is usually provided by carpeting cabins in passenger ships. Furnishing acoustical textiles used in ships comply with strict fire retardant standards [44]. Floating floors have also been used for noise and vibration reduction in ship cabins. In some cases where cabins are located above extremely noisy rooms such as engine or auxiliary machinery rooms, use of floating floors may be the only alternative for reducing the noise levels in the cabin. Commonly, the floating floor consists of an upper panel and mineral wool or recycled textile, which is in turn laid on the steel deck plate. Mathematical models to predict IL of floating floors used in ship cabins have been reported by some authors [102, 103].

References

1 Sandberg, U. and Ejsmont, J.A. (2002) *Tyre/Road Noise Reference Book*. Kisa, Sweden: Informex. www.informex.info (accessed 20 February 2020).

2 Nelson, P. (1992). Controlling vehicle noise – a general review. *Acoust. Bull.* 1 (5): 33–57.

3 Maruyama, S., Aoki, J., and Furuyama, M. (1997). Source: application of a reciprocity technique for measurement of acoustic transfer functions to the prediction of road vehicle pass-by noise. *JSAE Rev.* 18 (3): 277–282.

4 Bodén, H. and Glav, R. (2007). Exhaust and intake noise and acoustical design of mufflers and silencers. In: *Handbook of Noise and Vibration Control* (ed. M.J. Crocker), 1034–1054. New York: Wiley.

5 Reinhart, T.E. (2007). Internal combustion engine noise prediction and control – diesel and gasoline engines. In: *Handbook of Noise and Vibration Control* (ed. M.J. Crocker), 1024–1033. New York: Wiley.

6 Beranek, L.L. and Ver, I.L. (eds.) (1988). *Noise and Vibration Control Engineering*. New York: Wiley.

7 Anderton, D. (2004) *Basic origin of automotive engine noise*. Lecture E2 of Engine Noise and Vibration Control Course. University of Southampton.

8 Steel, J.A. (1998). Study of engine noise transmission using statistical energy analysis. *Proc. Inst. Mech. Eng. Part D J. Automobile Eng.* 212 (3): 205–213.

9 Lee, J.-H., Hwang, S.-H., Lim, J.-S., et al. (1998) A new knock-detection method using cylinder pressure, block vibration and sound pressure signals from a SI engine. SAE International Spring Fuels and Lubricants Meeting and Exposition. Warrendale, PA: Society of Automotive Engineers.

10 Glaeser, K.-P., Marx, T., and Schmidt, E. (2012) Sound detection of electric vehicles by blind or visually impaired. Proceeding of Internoise. New York.

11 Sakamoto, I., Houzu, H., Tanaka, T. et al. (2012) Report on basic research for standardization of measures for quiet vehicles in Japan (Intermit report). Proceedings of Internoise, New York.

12 Marbjerg, G. (2013) Noise from electric vehicles – a literature survey, Competitive Electric Town Transport COMPETT. Danish Road Directorate.

13 Prasad, M.G. and Crocker, M.J. (1998). Acoustic modeling (ducted-source systems). In: *Handbook of Acoustics* (ed. M.J. Crocker), 169–174. New York: Wiley.

14 Lee, I.J., Selamet, A., and Huff, N.T. (2003) Acoustic characteristics of coupled dissipative and reactive silencers. SAE Noise and Vibration Conference and Exposition. Warrendale, PA: Society of Automotive Engineers.

15 Steel, J.A., Fraser, G., and Sendall, P. (2000). A study of exhaust noise in a motor vehicle using statistical energy analysis. *Proc. Inst. Mech. Eng. Part D J. Automobile Eng.* 214 (D1): 75–83.

16 Gerges, S.N.Y., Jordan, R., Thieme, F.A. et al. (2005). Muffler modeling by transfer matrix method and experimental verification. *J. Braz. Soc. Mech. Sci. Eng.* 27 (2): 132–140.

17 Goossens, S., Osawa, T., and Iwama, A. (1999) Quantification of intake system noise using an experimental source-transfer-receiver model. SAE Noise and Vibration Conference and Exposition. Warrendale, PA: Society of Automotive Engineers.

18 Møller, N. and Gade, S. (2002) Operational modal analysis on a passenger car exhaust system. Congresso 2002 SAE Brasil – 11th International Mobility Technology Congress and Exhibition. Warrendale, PA: Society of Automotive Engineers.

19 Sandberg, U. and Ejsmont, J.A. (2007). Tire/road noise – generation, measurement, and abatement. In: *Handbook of Noise and Vibration Control* (ed. M.J. Crocker), 1054–1071. New York: Wiley.

20 Crocker, M.J., Li, Z., and Arenas, J.P. (2005). Measurements of tyre/road noise and of acoustical properties of porous road surfaces. *Int. J. Acoust. Vib.* 10 (2): 52–60.

21 Kim, G.J., Holland, K.R., and Lalor, N. (1997). Identification of the airborne component of tyre-induced vehicle interior noise. *Appl. Acoust.* 51 (2): 141–156.

22 Perisse, J. (2002). A study of radial vibrations of a rolling tyre for tyre-road noise characterization. *Mech. Syst. Signal Proc.* 16 (6): 1043–1058.

23 Boulahbal, D., Britton, J.D., Muthukrishnan, M. et al. (2005) High frequency tire vibration for SEA model partitioning. Noise and Vibration Conference and Exhibition. Warrendale, PA: Society of Automotive Engineers.

24 Donavan, P.R. and Rymer, B. (2003) Assessment of highway pavements for tire/road noise generation. SAE Noise and Vibration Conference and Exposition. Warrendale, PA: Society of Automotive Engineers.

25 Constant, M., Leyssens, J., Penne, F., and Freymann, R. (2001) Tire and car contribution and interaction to low frequency interior noise. SAE Noise and Vibration Conference and Exposition. Warrendale, PA: Society of Automotive Engineers.

26 Saemann, E.-U., Ropers, C., Morkholt, J., and Omrani, A. (2003) Identification of tire vibrations. SAE Noise and Vibration Conference and Exposition. Warrendale, PA: Society of Automotive Engineers.

27 Ahmed, S.R. (2007). Aerodynamic sound sources in vehicles – prediction and control. In: *Handbook of Noise and Vibration Control* (ed. M.J. Crocker), 1072–1085. New York: Wiley.

28 Bremner, P.G. and Zhu, M. (2003) Recent progress using SEA and CFD to predict interior wind noise. SAE Noise and Vibration Conference and Exposition. Warrendale, PA: Society of Automotive Engineers.

29 Tuma, J. (2007). Transmission and gearbox noise and vibration prediction and control. In: *Handbook of Noise and Vibration Control* (ed. M.J. Crocker), 1086–1095. New York: Wiley.

30 Houser, D.R. (2007). Gear noise and vibration prediction and control methods. In: *Handbook of Noise and Vibration Control* (ed. M.J. Crocker), 847–856. New York: Wiley.

31 Brennan, M.J. and Shin, K. (2007). Brake noise prediction and control. In: *Handbook of Noise and Vibration Control* (ed. M.J. Crocker), 1133–1137. New York: Wiley.

32 Cunefare, K.A. and Rye, R. (2001) Investigation of disc brake squeal via sound intensity and laser vibrometry. SAE Noise and Vibration Conference and Exposition. Warrendale, PA: Society of Automotive Engineers.

33 Oberst, S. and Lai, J.C.S. (2015). Nonlinear transient and chaotic interactions in disc brake squeal. *J. Sound Vib.* 342: 272–289.

34 Lee, H. and Singh, R. (2003) Sound radiation from a disk brake rotor using a semi-analytical method. SAE Noise and Vibration Conference and Exposition. Warrendale, PA: Society of Automotive Engineers.

35 Gesch, E., Tan, M., and Riedel, C. (2005) Brake squeal suppression through structural design modifications. Noise and Vibration Conference and Exhibition. Warrendale, PA: Society of Automotive Engineers.

36 Oberst, S. and Lai, J.C.S. (2015). Squeal noise in simple numerical brake models. *J. Sound Vib.* 352: 129–141.

37 Bettella, M., Harrison, M.F., and Sharp, R.S. (2002). Investigation of automotive creep groan noise with a distributed-source excitation technique. *J. Sound Vib.* 255 (3): 531–547.

38 Bernhard, R.J., Moeller, M., and Young, S. (2007). Automobile, bus, and truck interior noise and vibration prediction and control. In: *Handbook of Noise and Vibration Control* (ed. M.J. Crocker), 1159–1169. New York: Wiley.

39 Hirabayashi, T., McCaa, D.J., Rebandt, R.G. et al. (1999). Automotive noise and vibration control treatments. *Sound Vib.* 33 (4): 22–32.

40 Rao, M.D. (2003). Recent applications of viscoelastic damping for noise control in automobiles and commercial airplanes. *J. Sound Vib.* 262: 457–474.

41 Polce III, C.T., Saha, P., Groening, J.A., and Schappert, R. A data analysis approach to understand the value of a damping treatment for vehicle interior sound. SAE Noise and Vibration. Warrendale, PA: Society of Automotive Engineers.

42 Reinhart, T.E. and Crocker, M.J. (1982). Source identification on a diesel engine using acoustic intensity measurements. *Noise Control Eng.* 18 (3): 84–92.

43 Esposito, R.A. and Freeman, G.E. (2002) *Glazing for Vehicle Interior Noise Reduction*. International Body Engineering Conference and Exhibition. Warrendale, PA: Society of Automotive Engineers.

44 Arenas, J.P. (2016). Applications of acoustic textiles in automotive/transportation. In: *Acoustic Textiles* (eds. R. Nayak and R. Padhye), 143–163. Singapore: Springer.

45 Misaji, K., Tada, H., Yamashita, T. et al. (2003) *Testing Uniqueness of a Hybrid SEA Modeling Solution for a Passenger Car*. SAE Noise and Vibration Conference and Exposition. Warrendale, PA: Society of Automotive Engineers.

46 Fredö, C.R. (2005) A modification of the SEA equations: a proposal of how to model damped car body systems with SEA. Noise and Vibration Conference and Exhibition. Warrendale, PA: Society of Automotive Engineers.

47 Nishikawa, K., Misaji, K., Yamazaki, T., and Kamata, M. (2003) SEA model building of automotive vehicle body in white using experiment and FEM. SAE Noise and Vibration Conference and Exposition. Warrendale, PA: Society of Automotive Engineers.

48 Sol, A. and Van Herpe, F. (2001) Numerical prediction of a whole car vibro-acoustic behavior at low frequencies. SAE Noise and Vibration Conference and Exposition. Warrendale, PA: Society of Automotive Engineers.

49 Zhang, Q., Wang, D., Parrett, A., and Wang, C. (2003) SEA in vehicle development Part II: Consistent SEA modeling for vehicle noise analysis. SAE Noise and Vibration Conference and Exposition. Warrendale, PA: Society of Automotive Engineers.

50 Misaji, K., Tada, H., Yamashita, T. et al. (2003) Development of a hybrid SEA modeling scheme for a passenger car. SAE 2003 World Congress. Warrendale, PA: Society of Automotive Engineers.

51 Moore, J.A., Powell, R., and Bharj, T. (1999) Development of condensed SEA models of passenger vehicles. SAE Noise and Vibration Conference and Exposition. Warrendale, PA: Society of Automotive Engineers.

52 Cotoni, V., Gardner, B., Shorter, P., and Lane, S. (2005) Demonstration of hybrid FE-SEA analysis of structure-borne noise in the mid frequency range. Noise and Vibration Conference and Exhibition. Warrendale, PA: Society of Automotive Engineers.

53 Zhang, W., Wang, A., and Vlahopoulos, N. (2001) Validation of the EFEA method through correlation with conventional FEA and SEA results. SAE Noise and Vibration Conference and Exposition. Warrendale, PA: Society of Automotive Engineers.

54 Aubert, A.C., Long, J.T., Powell, R.E., and Moeller, M.J. (2003) SEA for design: a case study. SAE Noise and Vibration Conference and Exposition. Warrendale, PA: Society of Automotive Engineers.

55 Yan, H.H. and Parrett, A. (2003) A FEA based procedure to perform statistical energy analysis. SAE Noise and Vibration Conference and Exposition. Warrendale, PA: Society of Automotive Engineers.

56 Galasso, A., Montuori, G., De Rosa, S., and Franco, F. (2003) Vibroacoustic analyses of car components by experimental and numerical tools. SAE Noise and Vibration Conference and Exposition. Warrendale, PA: Society of Automotive Engineers.

57 Huang, L., Krishnan, R., Connelly, T., and Knittel, J.D. (2003) Development of a luxury vehicle acoustic package using SEA full vehicle model. SAE Noise and Vibration Conference and Exposition. Warrendale, PA: Society of Automotive Engineers.

58 Parrett, A., Zhang, Q., Wang, C., and He, H. (2003) SEA in vehicle development part I: balancing of path contribution for multiple operating conditions. SAE Noise and Vibration Conference and Exposition. Warrendale, PA: Society of Automotive Engineers.

59 Wang, D., Goetchius, G.M., and Onsay, T. (1999) *Validation of a SEA Model for a Minivan: Use of Ideal Air and Structure-Borne Sources*. SAE Noise and Vibration Conference and Exposition. Warrendale, PA: Society of Automotive Engineers.

60 Kozukue, W. and Hagiwara, I. (1996). Development of sound pressure level integral sensitivity and its application to vehicle interior noise reduction. *Eng. Comput. (Swansea, Wales)* 13 (5): 91–107.

61 Alt, N.W., Wiehagen, N., and Schlitzer, M.W. (2001) *Interior Noise Simulation for Improved Vehicle Sound*. SAE Noise and Vibration Conference and Exposition. Warrendale, PA: Society of Automotive Engineers.

62 Eisele, G., Wolff, K., Alt, N., and Hüser, M. (2005) *Application of Vehicle Interior Noise Simulation (VINS) for NVH Analysis of a Passenger Car*, Noise and Vibration Conference and Exhibition. Warrendale, PA: Society of Automotive Engineers.

63 Alt, N.W., Nehl, J., Heuer, S., and Schlitzer, M.W. (2003) Prediction of combustion process induced vehicle interior noise. SAE Noise and Vibration Conference and Exposition. Warrendale, PA: Society of Automotive Engineers.

64 Nunes, R.F., Botteon, A., Jamaguiva, J. et al. (2002) Cabin interior noise acoustic assessment under the influence of air intake system: simulation and experimental investigation. Congresso 2002 SAE Brazil, 11th International Mobility Technology Congress and Exhibition. Warrendale, PA: Society of Automotive Engineers.

65 Patil, A.R. and Crocker, M.J. (2000) *Prediction and Optimization of Radiated Sound Power and Radiation Efficiency of Vibrating Structures Using FEM*. SAE 2000 World Congress. Warrendale, PA: Society of Automotive Engineers.

66 Bocksch, R., Schneider, G., Moore, J.A., and Ver, I. (1999) Empirical noise model for power train noise in a passenger vehicle. SAE Noise and Vibration Conference and Exposition. Warrendale, PA: Society of Automotive Engineers.

67 Mealman, M., Unglenieks, R., and Naghshineh, K. (2001) A method to determine the power input associated with rain excitation for SEA models. SAE Noise and Vibration Conference and Exposition. Warrendale, PA: Society of Automotive Engineers.

68 Guedes Sampaio, R. and Goncalves, P.J.P. (2003) Investigation of sub-system contribution to a pickup truck boom noise using a hybrid method based on noise path analysis to simulate interior noise. 12th SAE Brazil Congress and Exposition. Warrendale, PA: Society of Automotive Engineers.

69 Schroeder, L.E. (2001) Feasibility of using acoustic room models and measured sound power to estimate vehicle interior noise. SAE Noise and Vibration Conference and Exposition. Warrendale, PA: Society of Automotive Engineers.

70 Kim, M.-G., Jo, J.-S., Sohn, J.-H., and Yoo, W.-S. (2003) Reduction of road noise by the investigation of contributions of vehicle components. SAE Noise and Vibration Conference and Exposition. Warrendale, PA: Society of Automotive Engineers.

71 Crocker, M.J. (2007). Psychoacoustics and product sound quality. In: *Handbook of Noise and Vibration Control* (ed. M.J. Crocker), 805–828. New York: Wiley.

72 Desmet, W., Pluymers, B., and Sas, P. (2003). Vibro-acoustic analysis procedures for the evaluation of the sound insulation characteristics of agricultural machinery cabins. *J. Sound Vib.* 266 (3): 407–441.

73 Tsujiuchi, N., Koizumi, T., Kubomoto, I., and Ishida, E. (1999) Structural optimization of tractor frame for noise and vibration reduction. SAE International Off-Highway and Powerplant Congress and Exposition. Warrendale, PA: Society of Automotive Engineers.

74 Ivanov, N. and Copley, D. (2007). Noise and vibration in off-road vehicle interiors – prediction and control. In: *Handbook of Noise and Vibration Control* (ed. M.J. Crocker), 1186–1196. New York: Wiley.

75 Legris, M. and Pouline, P. (1998). Noise exposure profile among heavy equipment operators, associated laborers, and crane operators. *AIHA J.* 59: 774–778.

76 Wang, S. and Bernhard, R. (1999) Energy Finite Element Method (EFEM) and Statistical Energy Analysis (SEA) of a heavy equipment cab. SAE Noise and Vibration Conference and Exposition. Warrendale, PA: Society of Automotive Engineers.

77 Thompson, D.J. and Remington, P.J. (2000). The effects of transverse profile on the excitation of wheel/rail noise. *J. Sound Vib.* 231 (3): 537–548.

78 Dine, C. and Fodiman, P. (2000). New experimental methods for an improved characterisation of the noise emission levels of railway systems. *J. Sound Vib.* 231 (3): 631–638.

79 Frid, A. (2000). Quick and practical experimental method for separating wheel and track contributions to rolling noise. *J. Sound Vib.* 231 (3): 619–629.

80 Koo, D.H., Kim, J.C., Yoo, W.H., and Park, T.W. (2002). An experimental study of the effect of low-noise wheels in reducing noise and vibration. *Transport. Res. Part D Transport Environ.* 7 (6): 429–439.

81 Dittrich, M.G. and Janssens, M.H.A. (2000). Improved measurement methods for railway rolling noise. *J. Sound Vib.* 231 (3): 595–609.

82 Sakamoto, H., Hirakawa, K., and Toya, Y. (1996) Sound and vibration of railroad wheel. *Proceedings of the 1996 ASME/IEEE Joint Railroad Conference* (Cat. No. 96CH35947), 75–81.

83 de Beer, F.G. and Verheij, J.W. (2000). Experimental determination of pass-by noise contributions from the bogies and superstructure of a freight wagon. *J. Sound Vib.* 231 (3): 639–652.

84 Kalivoda, M., Kudrna, M., and Presle, G. (2003). Application of MetaRail railway noise measurement methodology: comparison of three track systems. *J. Sound Vib.* 267 (3): 701–707.

85 Hardy, A.E.J. (2000). Measurement and assessment of noise within passenger trains. *J. Sound Vib.* 231 (3): 819–829.

86 Arenas, J.P. (2007). Use of barriers. In: *Handbook of Noise and Vibration Control* (ed. M.J. Crocker), 714–724. New York: Wiley.

87 Thompson, D.J. (2007). Wheel–rail interaction noise prediction and its control. In: *Handbook of Noise and Vibration Control* (ed. M.J. Crocker), 1138–1146. New York: Wiley.

88 Frommer, G.H. (2007). Noise management of railcar interior noise. In: *Handbook of Noise and Vibration Control* (ed. M.J. Crocker), 1170–1177. New York: Wiley.

89 Verheijen, E. (2006). A survey on roughness measurements. *J. Sound Vib.* 293 (3–5): 784–794.

90 Diehl, R.J. and Holm, P. (2006). Roughness measurements – have the necessities changed? *J. Sound Vib.* 293 (3–5): 777–783.

91 Hardy, A.E.J., Jones, R.R.K., and Turner, S. (2006). The influence of real-world rail head roughness on railway noise prediction. *J. Sound Vib.* 293 (3–5): 965–974.

92 De Beer, F.G., Janssens, M.H.A., and Kooijman, P.P. (2003). Squeal noise of rail-bound vehicles influenced by lateral contact position. *J. Sound Vib.* 267 (3): 497–507.

93 Gautier, P.-E. (2000). A review of railway noise research and results since the 5th IWRN in Voss (Norway). *J. Sound Vib.* 231 (3): 477–489.

94 Vincent, N., Koch, J.R., Chollet, H., and Guerder, J.Y. (2006). Curve squeal of urban rolling stock – part 1: state of the art and field measurements. *J. Sound Vib.* 293 (3–5): 691–700.

95 Mellet, C., Létourneaux, F., Poisson, F., and Talotte, C. (2006). High speed train noise emission: latest investigation of the aerodynamic/rolling noise contribution. *J. Sound Vib.* 293 (3–5): 535–546.

96 Nagakura, K. (2006). Localization of aerodynamic noise sources of Shinkansen trains. *J. Sound Vib.* 293 (3–5): 547–556.

97 Thrane, H.W. (2007). Interior noise in railway vehicles – prediction and control. In: *Handbook of Noise and Vibration Control* (ed. M.J. Crocker), 1178–1185. New York: Wiley.

98 Fischer, R. and Collier, R.D. (2007). Noise prediction and prevention on ships. In: *Handbook of Noise and Vibration Control* (ed. M.J. Crocker), 1216–1231. New York: Wiley.

99 Goujard, B., Sakout, A., and Valeau, V. (2005). Acoustic comfort on board ships: an evaluation based on a questionnaire. *Appl. Acoust.* 66 (9): 1063–1073.

100 Weyna, S. (1995). The application of sound intensity technique in research on noise abatement in ships. *Appl. Acoust.* 44 (4): 341–351.

101 Zheng, H., Liu, G.R., Tao, J.S., and Lam, K.Y. (2001). FEM/BEM analysis of diesel piston-slap induced ship Hull vibration and underwater noise. *Appl. Acoust.* 62 (4): 341–358.

102 Cha, S.-I. and Chun, H.-H. (2008). Insertion loss prediction of floating floors used in ship cabins. *Appl. Acoust.* 69 (10): 913–917.

103 Nilsson, A. and Liu, B. (2015). *Vibro–Acoustics*. Vol. 1, 2 ed., 347–353. Heidelberg: Springer.

15

Aircraft and Airport Transportation Noise Sources and Control

15.1 Introduction

The number of aircraft used for civilian and military air transportation has increased steadily during the last decades. Thus, aircraft noise has been a problem for people living in the vicinity of airports for many years. In the case of aircraft and helicopters, similar power plant and motion-related noise sources as in surface transportation vehicles exist. Some small aircraft are powered by internal combustion engines. Nowadays many general aviation and all medium-size airliners and helicopters are powered by turboprop power plants. Aircraft propeller and helicopter rotors are major noise sources that are difficult to control. All large civilian aircraft are now powered by jet engines in which the high-speed (HS) exhaust and turbomachinery are significant noise sources. Although the latest passenger jets with their bypass turbofan engines are significantly quieter than the first generation of jet airliners, which used pure turbojet engines, the noise of passenger jet aircraft remains a serious problem, particularly near airports. The noisiest pure jet passenger aircraft have been or will soon be retired in most countries. Airport noise, however, is likely to remain a difficult problem since in many countries the frequency of aircraft operations continues to increase and because of the public opposition to noise voiced by some citizens living near airports. This opposition has prevented runway extensions to some airports and the development of some new airports entirely because of environmental concerns.

15.2 Jet Engine Noise Sources and Control

The introduction of commercial passenger jet aircraft in the 1950s brought increasing complaints from people living near airports. Not only were the jet engines noisier than corresponding piston engines on airliners, since they were more powerful, but the noise was more disturbing because it had a higher frequency content than piston engines. This was most evident during the 1950s and 1960s when pure turbojet engines were in wide use in civilian jet airliners. Pure jet engines are still in use on some supersonic aircraft and in particular on HS military aircraft.

A pure turbojet engine takes in air through the inlet, adds fuel, which is then burnt, resulting in an expanded gas flow and an accelerated very high speed exhaust jet flow. The whole compression, combustion, and expansion process results in the thrust produced by the engine. The kinetic energy in the exhaust is non-recoverable and can be considered as lost energy. Since the 1970s, turbofan (or bypass) jet engines have come into increasing use on commercial passenger airliners. Turbofan engines have a large compressor fan at the front of the engine, almost like a ducted propeller. A large fraction of the air, after passing through the fan stage, bypasses the rest of the engine and then is mixed with the HS jet exhaust before leaving the engine tail pipe. This results in an engine that has a much lower exhaust velocity than for the case of a pure jet engine.

Engineering Acoustics: Noise and Vibration Control, First Edition. Malcolm J. Crocker and Jorge P. Arenas.
© 2021 John Wiley & Sons Ltd. Published 2021 by John Wiley & Sons Ltd.

The thrust can still be maintained if a larger amount of air is processed through the turbofan engine than with the pure turbojet engine (in which no air is bypassed around the combustion chamber and turbine engine components). The efficiency of the turbofan engine is greater than that of a pure turbojet engine, however, since less kinetic energy is lost in the exhaust jet flow.

A simplified calculation clearly shows the advantage of a turbofan engine over a turbojet engine. For instance, if a turbofan engine processes twice as much air as a turbojet engine, but only accelerates the air half as much, the kinetic energy lost will be reduced by half (1–2 × ¼), while the engine thrust is maintained. If the engine processes four times as much air, but accelerates the air only one quarter as much, the kinetic energy lost will be reduced by three quarters (1–4 × 1/16) and the engine thrust is still maintained. In each successive case described, the engine will become more efficient as the lost kinetic energy is reduced. Fortunately, as originally shown by Lighthill, there is an even more dramatic reduction in the sound power produced since in a jet exhaust flow the sound power produced is proportional to the exhaust velocity to the eighth power [1–3]. A halving in exhaust velocity then, theoretically, can give a reduction of 2^8 in exhaust sound power and mean square sound pressure, and in sound power level and sound pressure level of about 80 × log(2) or about 24 dB. But, of course, if the mass flow is twice as much, then the reduction in sound power and in the mean square pressure is about 2^7, or a reduction in sound power level and sound pressure level of about 21 dB.

The major noise sources for a modern turbofan engine are discussed in detail in Ref. [4]. Reference [5] is devoted to a review of the theory of aerodynamic noise and in particular jet noise generation. The major noise sources for a modern turbofan engine are shown in Figure 15.1 [4]. The relative sound pressure levels generated by each engine component depend on the engine mechanical design and power setting. During take-off, the fan and jet noise are both important sources with the exhaust noise usually dominant. During landing approach, the fan noise usually dominates since the engine power setting and thus jet exhaust velocity are reduced. Noise from other components such as the compressor, combustion chamber and turbine is generally less than that from the fan and the jet. The noise radiated from the inlet includes contributions from both the fan and compressor but is primarily dominated by the fan. Downstream radiated noise is dominated by the fan and jet, but there can also be significant contributions from the combustor and turbine, whose contributions are very much dependent on each engine design.

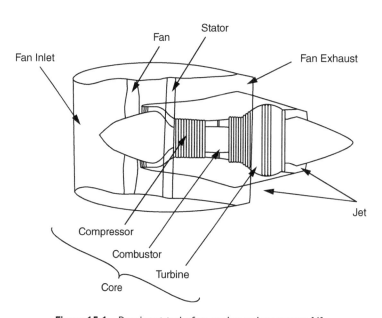

Figure 15.1 Dominant turbofan engine noise sources [4].

Reliable jet engine noise prediction methods are difficult to develop since they depend on accurate prediction of the unsteady flow field in and around the engine. Methods for reducing jet engine noise include modifying the unsteady flow field, redirecting the sound generated, absorbing the sound using acoustical treatments, and/or combinations of all three [4].

15.3 Propeller and Rotor Noise Sources and Control

As described in Ref. [6], propellers are used on small general aviation aircraft as well as small to medium-sized passenger airliners (See Figure 15.2) [7]. In small general aviation aircraft, propellers operate with a fixed-blade pitch. In larger general aviation and commuter aircraft, they operate with adjustable pitch to improve aircraft take-off and flight performance. Smaller aircraft have two-blade propellers while larger aircraft have three or more blades. The propeller operation gives rise to blade thrust and drag forces.

For a single propeller, tones are generated, which are harmonics of the blade passing frequency. This phenomenon occurs even for the case of tones generated by blade–turbulence interaction, which is caused by turbulent eddies in the airflow approaching the propeller. The BPF is the product of the shaft frequency and the blade number.

Theoretically, the BPF is a discrete frequency, f_{BPF}, which is related to the number of blades, N, and the engine rpm, R, and is given by $f_{BPF} = NR/60$ Hz. Since the thrust and drag forces are not purely sinusoidal in nature, harmonic distortion occurs as in the case of the noise generated by fans, diesel engines, pumps, compressors, and the like. Thus, blade passing harmonic tones occur at frequencies f_{BPn} given by $f_{BPn} = nNR/60$ Hz, where n is an integer, 1, 2, 3, 4... Figure 15.3 shows a typical spectrum of the cabin noise level in a turboprop aircraft during cruise flight. To avoid unpleasant acoustical beating, aircraft are equipped with a unit to synchronize the rotational speed of propellers [8].

Prediction of propeller noise is complicated. Accurate noise predictions require methods that include the influence of the flow field in which the propeller operates. Predictions may be made both in the time domain and the frequency domain. Noise reduction approaches are normally based both on experimental tests and theoretical predictions [6].

The spectrum of the propeller noise has both discrete and continuous components. The discrete frequency components are called tones. The continuous component of the spectrum is called broadband noise. There are three main kinds of propeller noise sources. These are (i) thickness (monopole-like), (ii) loading (dipole-like), and (iii) nonlinear (quadrupole-like) noise sources. All three of these source types can be steady or unsteady in nature. The loading dipole axis exists along the local normal to the propeller plane surface. For the first few harmonics of a low-speed propeller, the simple Gutin formula [9] gives the noise level in terms of the net thrust and torque. Unsteady sources can be further classified as periodic, aperiodic, or random. Steady and periodic sources produce tonal noise while random sources produce broadband noise.

Quadrupole sources are normally only important when the flow over the propeller airfoil is transonic or supersonic. Quadrupole sources are not important noise generators for conventional propellers, but they can be important in the generation of the noise of highly loaded HS propellers such as propfans. See Refs. [5] and [10] for more detailed discussions on aerodynamic noise and nonlinear acoustics.

15.4 Helicopter and Rotor Noise

The generation of helicopter rotor noise is very complicated [11]. Sources of helicopter noise include (i) main rotor, (ii) tail rotor, (iii) the engines, and (iv) the drive train components. The dominant noise contributors are the main rotor and the tail rotor. Engine noise is normally less important, although for large helicopters

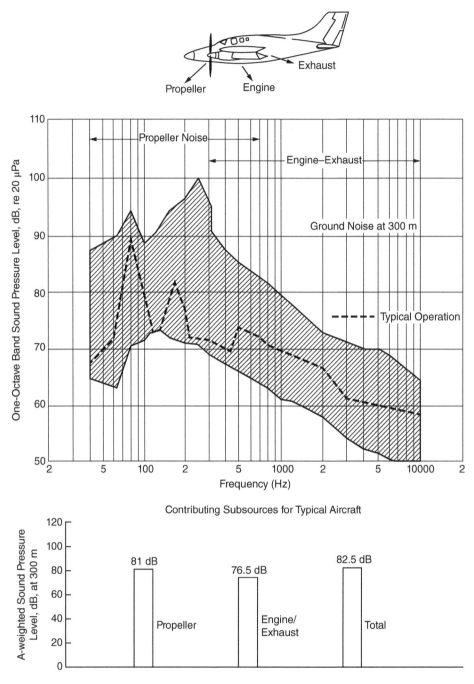

Figure 15.2 Noise levels and spectra of general aviation aircraft [7].

engine noise can be dominant at take-off. Rotor noise including main rotor and tail rotor noise can be classified as (i) discrete-frequency rotational noise, (ii) broadband noise, and/or (iii) impulsive noise (also of discrete-frequency character). The main rotor head kinematics are shown in Figure 15.4 [11].

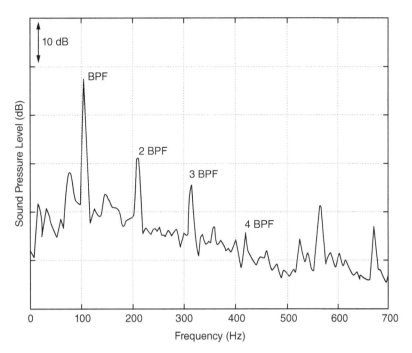

Figure 15.3 Sound pressure level spectrum for typical cabin noise in a turboprop aircraft [8].

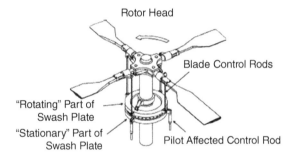

Figure 15.4 Main rotor head kinematics [11].

Helicopter rotor noise is comprised of *thickness noise* (the noise generated from the periodic volume displacement of the rotating blades) and *loading noise* (caused by the rotating lift and drag forces). Thickness noise is more important in the low-frequency range of the rotor–noise spectrum (at the BPF and first few harmonics). It also contains mid- and high-frequency components since it is of an impulsive nature. At low rotational speeds and for low blade loading, the thickness noise line spectrum can be exceeded by the *broadband noise* components. Broadband noise is a result of turbulent inflow conditions, blade/wake interferences, and blade self-noise ("airframe noise").

Of great importance are *impulsive-type noise sources*, resulting in the familiar "bang, bang, bang" sound. There are two main kinds: HS impulsive noise and blade–vortex interaction (BVI) impulsive noise. Tail rotor noise has similar characteristics to the main rotor noise. The flow around the tail rotor is the sum of the interacting flows generated by the wakes of the main rotor, the fuselage, the rotor hub, as well as the engine exhaust and empennage flows in addition to its own wake. For most helicopters, the tail rotor noise dominates at moderate speed straight flight conditions and during climb. Practical *rotor noise reduction measures* include passive reduction of high-speed impulsive noise, reduction of tail rotor noise, and active reduction of BVI [11]. Lowson has provided an in-depth review on helicopter noise [12].

15.5 Aircraft Cabin Noise and Vibration and Its Control

15.5.1 Passive Noise and Vibration Control

Low interior cabin noise is important for passenger and crew comfort [13]. High cabin noise levels experienced in passenger jet aircraft in the 1960s and 1970s have now been considerably reduced by the use of turbofan engines instead of noisy pure jet engines. Obtaining satisfactory cabin noise environments, which satisfy both speech interference and speech privacy criteria, remains a difficult problem since there is a large variety of noise sources that become dominant during different aircraft operating conditions. As is the case with surface transportation vehicles, with aircraft there are many different noise sources that contribute to the acoustical environments inside aircraft cabins.

The dominant cabin noise sources are mainly those exterior to the aircraft cabin and include power plant noise and vibration and turbulent boundary level excitation. The relative strength of the sources depends upon the aircraft operating conditions and flight speed. Internal cabin noise sources include air-conditioning systems, hydraulic systems, and electrical and mechanical equipment. These sources are mainly of concern with aircraft during ground operations before take-off. Power plant noise tends to be dominant at low flight speeds during take-off and landing, while the noise generated by turbulent boundary layer excitation of the cabin walls is dominant at higher speeds, after take-off and during subsequent climb, during landing descent, and in cruise conditions.

In the front of the cabin, turbulent boundary layer noise tends to peak at high frequency. While, in the middle to rear of the cabin, the boundary layer noise peaks at a much lower frequency. This is because the turbulent eddies in the thick boundary layer at the rear of the aircraft are larger than those in the thinner boundary layer at the front of the cabin. Jet noise is likewise more pronounced at the rear of the aircraft cabin for wing-mounted engines since jet noise is predominantly radiated downstream [4, 14]. For passenger aircraft with rear-mounted jet engines, the engine noise is mainly experienced at the rear part of the cabin. In cruise conditions, the noise from the engines is mostly structure-borne and is caused by small out-of-balance forces created by minor aircraft engine manufacturing imperfections.

To reduce interior cabin noise caused by engine noise and boundary layer noise, it has been normal practice to make use of lightweight damping materials. In the case of propeller aircraft, passive dampers are sometimes used, which are tuned to the blade passing frequency and/or the second and third multiples and that are attached to the fuselage interior skin panels.

In the case of jet passenger aircraft, constrained layer dampers are normally used and are placed at the center of skin pockets between the stringers and ring frames. Figure 15.5 shows a typical "stacked" damper system, and Figure 15.6 shows the skin pockets of a passenger jet aircraft where they are usually located. In most designs, a layer of viscoelastic damping material is constrained by a thin metal constraining layer of metal or Kevlar. A spacer is located between the base structure (airplane fuselage skin) to move the viscoelastic damping material as far as possible away from the neutral axis of the fuselage skin. The spacer is normally slotted to reduce its weight and minimize its bending stiffness. Ideally, the spacer should also have high

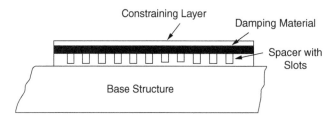

Figure 15.5 Stacked constrained layer damper system [15].

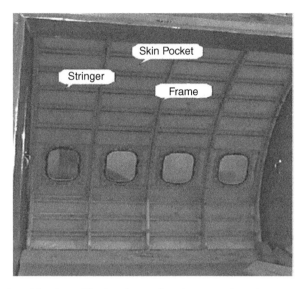

Figure 15.6 Interior of fuselage skin showing pockets between the stringers and ring frames [15].

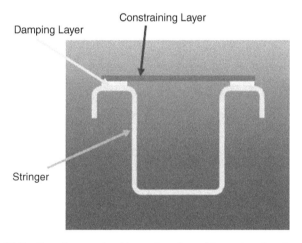

Figure 15.7 Use of constrained layer damping with an aircraft stringer [15].

shear stiffness. By this means, the shear distortion in the viscoelastic damping layer is magnified as the aircraft skin flexes in bending, and the damping effectiveness of the stacked damper is increased. Since the aircraft fuselage skin gets very cold at cruising altitude and the stacked damper is in close contact with the aircraft skin, special viscoelastic material must be used that has a maximum damping loss factor at the skin temperature during normal cruise conditions.

Viscous constrained damping layers are often used on aircraft stringers. Figure 15.6 shows the location of stringers on an aircraft fuselage skin. Figure 15.7 shows how the damping material is constrained between an aircraft stringer and the fuselage skin, which in this case acts as the constraining layer. Damping layers are also often applied to ring frames. Since the ring frames are in contact with the cabin air, they are not as cold as the fuselage skin and the viscoelastic material used is normally selected to have a maximum damping loss factor closer to the normal cabin air temperature.

On some parts of the aircraft cabin, separated flow and shock waves can occur near abrupt changes in the airplane geometry. For example, such geometric changes usually occur near the cockpit and can cause intense noise to be experienced nearby in the cabin. Cabin noise in propeller-driven aircraft poses a special problem, which is somewhat different from the interior noise of passenger jet aircraft. The interior noise from the power plants of propeller-driven aircraft is dominant over the entire flight regime, not just during take-off and landing. Propeller noise occurs at the fundamental BPF and the first few integer multiples, and for wing-mounted engines this noise can be intense if the propellers pass close to the passenger cabin fuselage. The propeller noise, being predominantly low frequency, is difficult to control using passive methods. There is obviously a limit to the amount of mass which can be added to reduce airborne sound transmission. Sound-absorbing materials are not very effective either at low frequency. The cabin noise levels can be reduced to some extent by the use of damping materials, which will also improve speech intelligibility. However, vibration transmitted from the engines to the airframe and thus resulting in cabin noise is also a matter of concern [16, 17]. Research on aircraft noise reduction techniques continues [18–24].

15.5.2 Active Noise and Vibration Control

Although passive noise source and path control methods have been improved considerably in recent years, in some cases aircraft cabin noise environments are still unsatisfactory. This is particularly true in the case of propeller-powered aircraft, and active control methods have been successfully employed to reduce the low-frequency cabin noise at the fundamental blade passing frequency and the first few multiples.

Active noise control systems are particularly successful in the low-frequency region in which the cabin sidewall sound transmission loss is poor and at which the blade passing frequency occurs. The active control is achieved by introducing multiple secondary sources in the cabin and the use of active headsets or "silent"

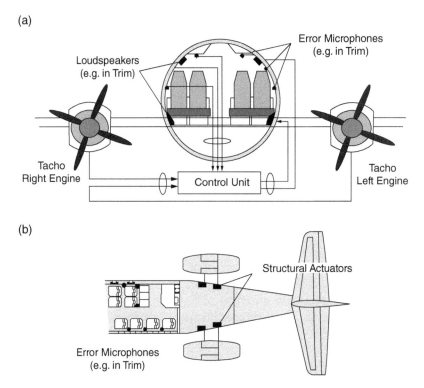

Figure 15.8 Active control of aircraft interior noise: (a) active noise control system in a turboprop aircraft, and (b) active structural/acoustical control in a jet aircraft [8].

seats. In the case of silent seats, small secondary sources are located in the top of the passenger seats in order to create a zone of "silence" near to the passenger's head. Structure-borne noise is also reduced on some aircraft by locating active 180° out-of-phase vibration sources near to aircraft engine mounts. Figure 15.8a shows an illustration of an active noise control system synchronized with both engines. In situations where the aim is to reduce the sound radiation from the fuselage, the loudspeakers are replaced with structural actuators controlling the fuselage vibrations as shown in Figure 15.8b [25–27].

Attenuation of broadband cabin noise is a more difficult task but can be undertaken, provided sufficient numbers of secondary 180° out-of-phase noise sources are located inside the cabin. Since jet power plant noise has been successfully reduced in recent years, and in any case is minimal during cruise conditions, broadband turbulent boundary layer noise is becoming recognized as an increasing problem during cruise conditions. Active noise control of aircraft cabins is discussed in Ref. [8]. Active control of the structure close to the engine can also be undertaken to reduce structure-borne noise reaching the passenger cabin [28]. See Ref. [29] for further discussion on active vibration control, and Ref. [30] for further discussion on active noise control. Research on aircraft active noise and vibration control continues [31–36].

15.6 Airport Noise Control

One of the major environmental noise problems throughout the world is the noise associated with civil aircraft operations nearby airports. In the US, the Federal Aviation Administration (FAA) recognizes that airport noise issues can be highly technical and complex. The European Union (EU) has also recognized the problem and noise emission limitations from civil aircraft have been in force in European legislation since 1980. Currently Directive 2002/30/EC [37] of the EU provides guidance on "the establishment of rules and procedures with regard to the introduction of noise-related operating restrictions at community airports." Noise metrics or indicators must be used to analyze aircraft operations noise and its effect on noise-sensitive land use. In the US, the primary metric for airport assessment is the A-weighted day–night average sound pressure level, DNL, in which a penalty of 10 dB is added to noise made at night. The Community Noise Equivalent Level, CNEL, metric is sometimes used to assess aircraft noise exposure in communities surrounding airports located in California. The EU uses the day–evening–night sound level, DENL, and penalties of 5 dB are applied to noise in the evening and 10 dB at night (see Chapter 6 in this book for a discussion on these descriptors). However, both the EU and the US legislation recognized that some cases may benefit from the use of supplementary noise indicators, in particular for measuring single-events shorter than 24 hours.

Most airports in the world have noise abatement programs. Approaches to mitigate perceived noise problems associated with civil aircraft operations in airports include specific noise control approaches to reduce noise at the source and specific airport noise control measures.

15.6.1 Noise Control at the Source

Reducing the noise at the aircraft source requires that for all aircraft, specific noise testing or certification procedures and associated limits must be defined. In addition, louder aircraft must be phased out of operation over time [38].

In 1969 the FAA published a regulation establishing noise certification standards for most airplane types, generally requiring newly designed and manufactured aircraft to be significantly quieter than older aircraft. In this Federal Aviation Regulation (FAR) Part 36, noise limits were defined in terms of the EPNL descriptor (see Chapter 6). In an amendment of 1977 [39], FAR Part 36 established three stages (called Chapters by the International Civil Aviation Organization, ICAO [40]) of aircraft noise levels for subsonic large transport aircraft and subsonic turbojets. Stage 1 aircraft were those that did not meet current noise standards and hence must be modified or replaced according to a previously established schedule of 1976. Stage 2 aircraft

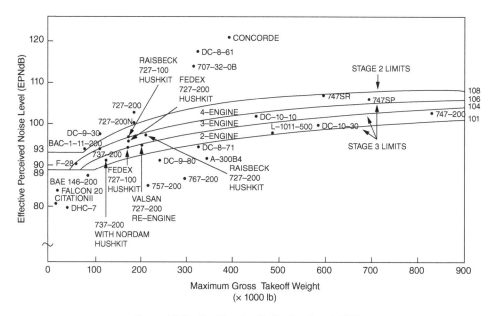

Figure 15.9 Certification limits for aircraft [38].

met the current standards, while stage 3 aircraft were able to meet the more rigorous noise standards for the next generation of jet transports prescribed by the rule. For helicopters, two different stages exist: stage 1 and stage 2. As with civil turbojet aircraft, stage 2 is quieter than stage 1. The FAA has been working to adopt the latest international standards for helicopters, which are to be called stage 3 and to be quieter than stage 2.

It is recognized that FAR Part 36 has produced major improvements in the noise radiated from civil aircraft. In fact, the noise output of the larger and more powerful jet engines has not increased with the increased mechanical power of the jet engines themselves. Although air traffic has been on the increase, aircraft noise exposure in communities around airports has been on the decrease as quieter aircraft become more prevalent. Figure 15.9 presents the limits that define FAA's stage 1, stage 2, and stage 3 aircraft [38].

Currently, a new certification limit for jet aircraft has been adopted by the FAA, called stage 4 (Chapter 4 internationally) and has been applied since 2013 and has resulted in quieter aircraft. The international community has been looking to approve a more stringent aircraft noise standard which the FAA will call Stage 5, which will be effective for new type designs submitted after December 31, 2017 and/or December 31, 2020, depending on the aircraft weight.

15.6.2 Airport-specific Noise Control Measures

The FAA has developed a number of programs aimed to increase the knowledge about noise impacts, advice on solutions to reduce those impacts, and to educate the public on the noise-related issues. One of these programs is called FAR Part 150 [41, 42] which is a comprehensive FAA program of airport planning for noise compatibility that works as a compliment to Part 36. Eldred [43] has summarized the principal actions that can be taken to abate noise and its perceived problems (see Table 15.1).

When developing noise reduction at a specific airport, one of the first factors to consider is the location of its flight tracks with respect to land use. Thus, flight tracks that use noise compatible land (water, industrial, commercial, and agriculture) must be preferred. In this sense, preferential runways with least noise impact must be defined and further extension or building of new runways should consider the best use of compatible land by improving airport layout [43].

Table 15.1 Possible airport actions to abate noise [43].

Flight tracks	Get aircraft away from people
Preferential runways	Increase use of runways with least impact
Restrict noisy aircraft	Minimize operations in day or night
Noise abatement flight procedures	Require use of noise abatement throttle and flap management procedures for take-off and/or approach
Airport layout	Extend or build new runways and taxiways to make best use of compatible land and water
Shielding barriers	Shield people from noise of ground operations
Soundproof	Soundproof schools, homes, and churches
Land use control	Assure compatible land use through acquisition of property or other rights
Monitor and model	Monitor airport noise and flight tracks to provide data for the public and for evaluating proposed alternatives
Communication	Listen to complaints and suggestions; develop and institutionalize continuing effective dialog and information transfer among all concerned parties.

Restriction on the use of noisy aircraft is another possible action to abate noise in airports. Such restrictions should be employed only after consideration of other alternatives and after thorough consultation with the affected parties to avoid uneven economic consequences. Some of the forms that such restrictions might take are: (i) based on cumulative impact, (ii) based upon certificated noise levels, and (iii) based upon estimated single-event noise levels. A restriction placed upon all or certain classes of aircraft by time of day appears to have considerable potential to reduce airport noise impact. Curfews during nighttime when most people are resting and are most sensitive to noise intrusions are common in modern airports. However, curfews have economic impacts upon airport users, upon those providing airport-related services, and upon the community as a whole. Therefore, curfews should only be considered after careful consideration of other alternatives [42].

Landing fees based on noise have also been suggested as a strategy for noise mitigation. The strategy encourages the use of quieter aircraft while producing additional revenue to offset airport noise-induced expenses. Some airports have implemented reverse strategies, such as rewarding air carriers who reduce noise generated by their aircraft by providing a discount or a reduction in landing fees.

Another noise mitigation strategy is the use of noise abatement take-off and landing procedures. There are a number of alternatives within this strategy that include runway selection, take-off and landing profiles and power settings, and approach or departure paths. Take-off and landing profiles and their attendant throttle and aircraft configuration techniques can be adjusted so as to offer relief to nearby noise-sensitive areas.

Soundproofing and noise barriers are noise reduction techniques that can be implemented in airport noise problems. Ground-level noise sources on an airport include run-up and maintenance areas, taxiways and freight warehouse areas. Because the noise is generated on the ground, its impact is usually confined to those areas immediately adjacent to the source. An effective method of mitigating this type of noise impact is use of sound barriers or berms (see Chapter 9 in this book). New hangar or terminal structures placed strategically in the airport may also shield adjacent neighborhoods. Movement of run-up and maintenance operations to an area of the airport away from the community might also do the work.

Acquisition of full or partial interest in compatible land may be the only way the airport can be assured of long-term protection, although purchase of sufficient land area to totally contain the significant noise impacts of an airport is usually impractical [42].

The implementation of continuous airport monitoring systems is essential to provide important input to the process of refining airport noise contours and supply data for public discussion. In brief, any FAA approved noise monitoring system should have the following minimum capabilities: (i) Provide continuous measurement of A-weighted noise levels at each site, (ii) Provide hourly L_{eq} data, (iii) Provide daily L_{dn} data, and (iv) Provide single-event maximum A-weighted sound pressure level data. Monitoring of data can be used to develop a statistical data base of noise levels for each aircraft type category and to the further examination of alternatives that might provide future mitigation of noise.

Airport noise and mitigation measures are discussed in detail in Refs. [38] and [43]. Research on airport noise assessment and airport-specific noise control continues [44–59].

Example 15.1 Consider an aircraft noise event that has a duration of five seconds (with a maximum A-weighted sound pressure level of 65 dB) according to Table 15.2. Determine the sound exposure level (SEL) of the noise event.

Table 15.2 Data for an aircraft noise event of five seconds.

Seconds	A-weighted noise level, dB
1	60
2	63
3	65
4	63
5	60

Solution

The SEL is calculated according to the definitions discussed in Chapter 6, Section 6.9. Then, the A-weighted

$$SEL = 10 \times \log\left(10^{60/10} + 10^{63/10} + 10^{65/10} + 10^{63/10} + 10^{60/10}\right) = 69.6\,dB.$$

Example 15.2 Suppose there are 10 aircraft noise events in a period of 24 hours at a small airport, according to the schedule shown in Table 15.3.

a) Determine the day-night equivalent level, DNL.

b) Consider now a curfew that restricts all nighttime aircraft operations for the airport. Determine the change in DNL.

Table 15.3 A-weighted aircraft noise level events in an airport.

Hour	SEL, dB
12:00 a.m.	86.1
01:00 a.m.	
02:00 a.m.	
03:00 a.m.	
04:00 a.m.	
05:00 a.m.	90.0
06:00 a.m.	86.1
07:00 a.m.	

Table 15.3 (Continued)

Hour	SEL, dB
08:00 a.m.	93.6
09:00 a.m.	
10:00 a.m.	82.6
11:00 a.m.	
12:00 p.m.	90.3
01:00 p.m.	
02:00 p.m.	
03:00 p.m.	
04:00 p.m.	
05:00 p.m.	94.8
06:00 p.m.	
07:00 p.m.	
08:00 p.m.	
09:00 p.m.	86.1
10:00 p.m.	85.2
11:00 p.m.	89.5

Solution

a) Five A-weighted noise level events occur during daytime (from 07:00 a.m. to 10:00 p.m.) and five A-weighted noise level events during nighttime (from 10:00 p.m. to 07:00 a.m.). We determine the day-night equivalent level for 24 hours (86 400 seconds) applying the 10-dB penalty to the five night-time aircraft noise events:

$$L_{DN} = 10\log\left[\left(10^{9.36} + 10^{8.26} + 10^{9.03} + 10^{9.48} + 10^{8.61} + 10^{9.52} + 10^{9.95} + 10^{9.61} + 10^{10} + 10^{9.61}\right)/86400\right]$$
$$= 56.4 \text{ dB.}$$

b) The new A-weighted DNL is calculated now without considering the aircraft events affected by the curfew between 10:00 p.m. and 07:00 a.m.:

$$L_{DN} = 10\log\left[\left(10^{9.36} + 10^{8.26} + 10^{9.03} + 10^{9.48} + 10^{8.61}\right)/86400\right]$$
$$= 49 \text{ dB.}$$

Therefore, the curfew reduces the DNL by 7.4 dB.

References

1 Lighthill, M.J. (1952). On sound generated aerodynamically. I. General theory. *Proc. Roy. Soc. A* 211: 564–587.
2 Lighthill, M.J. (1954). On sound generated aerodynamically. II. Turbulence as a source of sound. *Proc. Roy. Soc. A* 222: 1–32.

3 Powell, A. (1997). Aerodynamic and jet noise. In: *The Encyclopedia of Acoustics*, vol. 1, 301–311. New York: Wiley.

4 Huff, D.L. and Envia, E. (2007). Jet engine noise generation, prediction, and control. In: *Handbook of Noise and Vibration Control* (ed. M.J. Crocker), 1096–1108. New York: Wiley.

5 Morris, P.J. and Lilley, G.M. (2007). Aerodynamic noise: theory and applications. In: *Handbook of Noise and Vibration Control* (ed. M.J. Crocker), 128–158. New York: Wiley.

6 Metzger, F.B. and Farassat, F. (2007). Aircraft propeller noise – sources, prediction, and control. In: *Handbook of Noise and Vibration Control* (ed. M.J. Crocker), 1109–1119. New York: Wiley.

7 Wyle Laboratories (1971) Transportation Noise and Noise from Equipment Powered by Internal Combustion Engines. EPA Report No. NTID 300.13.

8 Johansson, S., Hakansson, L., and Claesson, I. (2007). Aircraft cabin noise and vibration prediction and active control. In: *Handbook of Noise and Vibration Control* (ed. M.J. Crocker), 1207–1215. New York: Wiley.

9 Nelson, P. (1992). Controlling vehicle noise – a general review. *Acoust. Bull.* 1 (5): 33–57.

10 Rudenko, O.V. and Crocker, M.J. (2007). Nonlinear acoustics. In: *Handbook of Noise and Vibration Control* (ed. M.J. Crocker), 159–167. New York: Wiley.

11 Heller, H.H. and Yin, J. (2007). Helicopter rotor noise: generation, prediction, and control. In: *Handbook of Noise and Vibration Control* (ed. M.J. Crocker), 1120–1132. New York: Wiley.

12 Lowson, M.V. (1992). Progress towards quieter civil helicopters. *Aeronaut. J.* 96 (956): 209–223.

13 Wilby, J.F. (2007). Aircraft cabin noise and vibration prediction and passive control. In: *Handbook of Noise and Vibration Control* (ed. M.J. Crocker), 1197–1206. New York: Wiley.

14 Crocker, M.J. (2007). Introduction to transportation noise and vibration sources. In: *Handbook of Noise and Vibration Control* (ed. M.J. Crocker), 1013–1023. New York: Wiley.

15 Rao, M.D. (2003). Recent applications of viscoelastic damping for noise control in automobiles and commercial airplanes. *J. Sound Vib.* 262 (3): 457–474.

16 DePriest, J. (2000) Aircraft Engine Attachment and Vibration Control, SAE General Aviation Technology Conference and Exposition. Warrendale, PA: Society of Automotive Engineers.

17 Choi Chow, L., Lempereur, P., and Mau, K. (1999). Aircraft airframe noise and installation effects – research studies. *Air Space Eur.* 1 (3): 72–75.

18 Neise, W. and Enghardt, L. (2003). Technology approach to aero engine noise reduction. *Aerospace Sc. Tech.* 7 (5): 352–363.

19 James, M.M. and Gee, K.L. (2010). Aircraft jet plume source noise measurement system. *Sound Vib.* 44 (8): 14–17.

20 da Rocha, J., Suleman, A., and Lau, F. (2011). Flow-induced noise and vibration in aircraft cylindrical cabins: closed-form analytical model validation. *J. Vib. Acoust.-Trans. ASME* 133 (5): 051013.

21 Li, Y., Wang, X., and Zhang, D. (2013). Control strategies for aircraft airframe noise reduction. *Chinese J. Aeron.* 26 (2): 249–260.

22 Hileman, J.I., Spakovszky, Z.S., Drela, M. et al. (2010). Airframe design for silent fuel-efficient aircraft. *J. Aircr.* 47 (3): 956–969.

23 Munday, D., Cuppoletti, D., Perrino, M. et al. (2013). Supersonic turbojet noise reduction. *Int. J. Aeroacoust.* 12 (3): 215–243.

24 Huff, D.L. (2013). NASA Glenn's contributions to aircraft engine noise research. *J. Aerosp. Eng.* 26 (2): 218–250.

25 Fuller, C.R., Elliott, S.J., and Nelson, P.A. (1996). *Active Control of Vibration*. London: Academic.

26 Ross, C.F. (1998) A comparison of techniques for the active control of noise and vibration in aircraft. Proc. of ISMA23, 831–835.

27 UltraQuiet Noise and Vibration Systems (2011) www.ultraquiet.com (accessed 6 January 2011).

28 MacMartin, D.G. (1996). Collocated structural control for reduction of aircraft cabin noise. *J. Sound Vib.* 190 (1): 105–119.

29 Fuller, C. (2007). Active vibration control. In: *Handbook of Noise and Vibration Control* (ed. M.J. Crocker), 770–784. New York: Wiley.

30 Elliott, S.J. (2007). Active noise control. In: *Handbook of Noise and Vibration Control* (ed. M.J. Crocker), 761–769. New York: Wiley.

31 Huang, Y.-M. and Tseng, H.-C. (2008). Active piezoelectric dynamic absorbers on vibration and noise reductions of the fuselage. *J. Mech.* 24 (1): 69–77.

32 Kochan, K., Sachau, D., and Breitbach, H. (2011). Robust active noise control in the loadmaster area of a military transport aircraft. *J. Acoust. Soc. Am.* 129 (5): 3011–3019.

33 Testa, C., Bernardini, G., and Gennaretti, M. (2011). Aircraft cabin tonal noise alleviation through fuselage skin embedded piezoelectric actuators. *J. Vib. Acoust.- Trans. ASME* 133 (5): 051009.

34 Griffin, S., Weston, A., and Anderson, J. (2013). Adaptive noise cancellation system for low frequency transmission of sound in open fan aircraft. *Shock. Vib.* 20 (5): 989–1000.

35 Haase, T., Algermissen, S., Unruh, O., and Misol, M. (2014). Experiments on active control of counter-rotating open rotor interior noise. *Acta Acust. United Acust.* 100 (3): 448–457.

36 Bernardini, G., Testa, C., and Gennaretti, M. (2016). Tiltrotor cabin noise control through smart actuators. *J. Vib. Control.* 22 (1): 3–17.

37 Directive 2002/30/EC (2002) European Parliament and Council Directive, 26 March, OJ L 85, 28.3.2002, 40.

38 Miller, N.P., Reindel, E.M., and Horonjeff, R.D. (2007). Aircraft and airport noise prediction and control. In: *Handbook of Noise and Vibration Control* (ed. M.J. Crocker), 1479–1489. New York: Wiley.

39 Federal Aviation Regulations Part 36 (1977) Noise Standards: Aircraft Type and Airworthiness Certification. U.S. Department of Transportation, Federal Aviation Administration, Amendment 7 February.

40 ICAO Assembly Resolution A33-7 (2001) Consolidated Statement of Continuing ICAO Policies and Practices Related to Environmental Protection: Appendix C, Policies and Programs Based on a "Balanced Approach" to Aircraft Noise Management.

41 FAA (1988) Airport noise compatibility planning. 14 CFR Part 150, as amended, March.

42 FAA Advisory Circular 150/5020-1 (1983) Noise Control and Compatibility Planning for Airports. U.S. Department of Transportation, Federal Aviation Administration, August 5.

43 Eldred, K.M.K. (1998). Airport noise. In: *Handbook of Acoustics* (ed. M.J. Crocker), 883–896. New York: Wiley.

44 Khardi, S. (2011). Development of innovative optimized flight paths of aircraft takeoffs reducing noise and fuel consumption. *Acta Acust. United Acust.* 97 (1): 148–154.

45 Hogenhuis, R.H., Hebly, S.J., and Visser, H.G. (2011). Optimization of area navigation noise abatement approach trajectories. *Proc. Inst. Mech. Eng. G- J. Aerospace Eng.* 225 (G5): 513–521.

46 Khardi, S. and Abdallah, L. (2012). Optimization approaches of aircraft flight path reducing noise: comparison of modeling methods. *Appl. Acoust.* 73 (4): 291–301.

47 Netjasov, F. (2012). Contemporary measures for noise reduction in airport surroundings. *Appl. Acoust.* 73 (10): 1076–1085.

48 Vogiatzis, K. (2012). An assessment of airport environmental noise action plans with some financial aspects: the case of Athens international "Eleftherios Venizelos". *Int. J. Acoust. Vib.* 17 (4): 181–190.

49 Goldschagg, P.L. (2013). Using supplemental aircraft noise information to assist airport neighbours understand aircraft noise. *Transp. Res. D: Transp. Environ.* 21: 14–18.

50 Asensio, C., Recuero, M., and Pavón, I. (2014). Citizens' perception of the efficacy of airport noise insulation programmes in Spain. *Appl. Acoust.* 84 (10): 107–115.

51 Ozkurt, N. (2014). Current assessment and future projections of noise pollution at Ankara Esenboga airport, Turkey. *Transp. Res. D: Transp. Environ.* 32: 120–128.

52 Lu, C. (2014). Combining a theoretical approach and practical considerations for establishing aircraft noise charge schemes. *Appl. Acoust.* 84 (10): 17–24.

53 Vogiatzis, K. (2014). Assessment of environmental noise due to aircraft operation at the Athens international airport according to the 2002/49/EC directive and the new Greek national legislation. *Appl. Acoust.* 84 (10): 37–46.

54 Xie, H., Li, H., and Kang, J. (2014). The characteristics and control strategies of aircraft noise in China. *Appl. Acoust.* 84 (10): 47–57.

55 Iglesias-Merchan, C., Diaz-Balteiro, L., and Solino, M. (2015). Transportation planning and quiet natural areas preservation: aircraft overflights noise assessment in a National Park. *Transp. Res. D: Transp. Environ.* 41: 1–12.

56 Park, S.G. and Clarke, J.P. (2015). Optimal control based vertical trajectory determination for continuous descent arrival procedures. *J. Aircr.* 52 (5): 1469–1480.

57 Ganic, E.M., Netjasov, F., and Babic, O. (2015). Analysis of noise abatement measures on European airports. *Appl. Acoust.* 92 (5): 115–123.

58 Postorino, M.N. and Mantecchini, L. (2016). A systematic approach to assess the effectiveness of airport noise mitigation strategies. *J. Air Transp. Manag.* 50: 71–82.

59 Lawton, R.N. and Fujiwara, D. (2016). Living with aircraft noise: airport proximity, aviation noise and subjective wellbeing in England. *Transp. Res. D: Transp. Environ.* 42: 104–118.

16

Community Noise and Vibration Sources

16.1 Introduction

The main sources of urban community noise are (i) road traffic, that is, trucks, cars, and motorcycles, (ii) aircraft/airport noise, (iii) railroads, (iv) construction noise, (v) noise from light and heavy industry, and (vi) noise from recreation activities. Road traffic noise is the most important of these and is discussed in this chapter in some detail. Aircraft/airport noise, although often cited as a major problem, is generally considered to be of less importance because, although it can be intense, it is mostly concentrated around major airports. However, from another point of view, aircraft noise is of extreme importance since strong public resistance to airport expansion in many countries is driven by aircraft noise complaints around airports. Railroad noise is a problem for residential strips situated along major rail lines, but it is much less pervasive than road traffic noise. Construction noise is often a problem too since in large cites there are normally some new building projects being undertaken; in addition, noise from the construction of new highways is a concern where heavy equipment contributes to the noise problem.

16.2 Assessment of Community Noise Annoyance

Several noise indicators and rating measures are in use. The equivalent sound pressure level L_{eq} (see Section 6.8 of this book) is used in many countries for the assessment of road traffic noise, although the statistical 10% level L_{10} (see Section 6.11 of this book) is used in Australia, Hong Kong, and the United Kingdom for target values and insulation regulations for new roads and planning values for new residential areas. The A-weighted day–night average sound pressure level, DNL is currently the main descriptor of community noise in the United States (See Section 6.10 of this book). With DNL a penalty of 10 dB is added to noise made at night. The European Union (EU) has specified the use of the day–evening–night sound pressure level, DENL. Penalties of 5 dB are applied to noise in the evening and 10 dB at night. Chapter 6 discusses some of the descriptors that have been used. Reference [1] describes the current practice of evaluating community noise problems in Europe, the United States, and several other countries. Coelho has written a review of community noise in which the different types of community noise ordinances are described [2].

In the United States, the day–night average sound pressure level (DNL) has been in widespread use to assess community noise annoyance since its adoption by several federal agencies in the mid-1970s. If the DNL of the community noise is below 70 dB, there is little danger of hearing loss. However, many people become highly annoyed (HA), and there can be significant sleep disturbance in the community. As discussed in Section 6.15.2 of this book, by evaluating the results of 11 clustering social surveys on noise annoyance in

Engineering Acoustics: Noise and Vibration Control, First Edition. Malcolm J. Crocker and Jorge P. Arenas.
© 2021 John Wiley & Sons Ltd. Published 2021 by John Wiley & Sons Ltd.

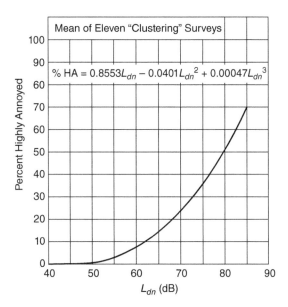

Figure 16.1 Synthesis of all the clustering survey results. The mean of the "clustering surveys" data, shown here, is proposed as the best currently available estimate of public annoyance due to transportation noise of all kinds. It may also be applicable to community noise of other kinds [3].

several countries, Schultz [3] produced in 1978 a curve relating the average annoyance response percentage of people HA with the day–night level (see Figure 16.1). The Schultz curve has been widely used since then, although in recent years considerable doubt has been expressed whether it can be applied with equal confidence to different sources of community noise, including road traffic, railroad, and aircraft.

As shown in Figure 6.16, later studies sponsored by the U.S. Air Force in the early 1990s have indicated that people have a moderate, but a different, reaction to aircraft, traffic, and railway noise at the same day–night average sound pressure level [4]. One possible explanation may be that aircraft noise near a busy airport has a greater variation in level in time and has a different frequency content (generally higher) than road or rail traffic. Another explanation may be the inadequacy of the day–night sound pressure level measure itself, which does not make any allowance for level or, even more importantly, for loudness.

In a more recent study, Miedema and Vos [5] came to a similar conclusion as Finegold et al. [4], although their results were slightly different in magnitude. At high values of DNL between 60 and 70 dB, the ranking of road traffic and railway noise differ a little from the earlier results of Finegold et al. [4] Their conclusions may be summarized as follows. Below a DNL of 40–45 dB, virtually no one is HA. As the DNL increases, so does the percentage of the population who are HA (%HA). The rate of increase in the %HA is greater for aircraft noise than road traffic noise, which in turn has a greater rate of increase than railway noise [5]. Some of the results of this study are given in Figure 16.2, which shows curves fitted to the data points by a least squares regression procedure. Since the 95% confidence limits for the three curves do not overlap, for the higher levels of DNL it can be concluded that the percentage of people that are HA, %HA, depends on the mode of transportation causing the noise [5].

Equations fitting the data [5], assuming zero annoyance at an A-weighted sound pressure level of 42 dB, are

$$\text{Aircraft}: \quad \%\text{HA} = 0.20 \times (\text{DNL} - 42) + 0.0561 \times (\text{DNL} - 42)^2, \tag{16.1}$$

$$\text{Road traffic}: \quad \%\text{HA} = 0.24 \times (\text{DNL} - 42) + 0.0277 \times (\text{DNL} - 42)^2, \tag{16.2}$$

$$\text{Rail}: \quad \%\text{HA} = 0.28 \times (\text{DNL} - 42) + 0.0085 \times (\text{DNL} - 42)^2. \tag{16.3}$$

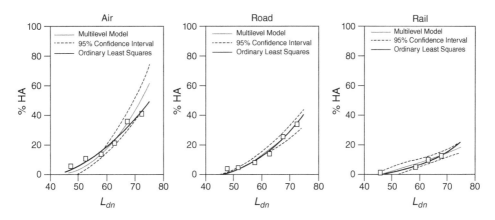

Figure 16.2 Percentage highly annoyed persons (%HA) as a function of DNL. Two synthesis curves per mode of transportation, and the data points are shown. For the curves obtained with multilevel analysis, the 95% confidence intervals are shown [5].

Example 16.1 Suppose there are five transportation noise events in a period of 24 hours near a residential community. Three events occurred during daytime and two events during nighttime. A-weighted sound exposure levels of each event at the residential community are 96, 93, and 91 dB during daytime and 90 and 92 dB during nighttime. Determine the percentage of HA persons, if all the events correspond to (a) airplanes, (b) vehicles, and (c) trains.

Solution

We determine the A-weighted day-night equivalent level for 24 hours (86 400 seconds) applying the 10-dB penalty to the two nighttime events (see Chapter 6):

$$\text{DNL} = 10 \log \left[\left(10^{9.6} + 10^{9.3} + 10^{9.1} + 10^{(90+10)/10} + 10^{(92+10)/10} \right) / 86400 \right],$$

$$= 10 \log \left[\left(10^{9.6} + 10^{9.3} + 10^{9.1} + 10^{10} + 10^{10.2} \right) / 86400 \right] = 55.8 \text{ dB}.$$

Now, replacing this level in Eqs. (16.1)–(16.3), we obtain

a) Aircraft: %HA $= 0.20 \times (55.8 - 42) + 0.0561 \times (55.8 - 42)^2 = 13.4\%$
b) Road traffic: %HA $= 0.24 \times (55.8 - 42) + 0.0277 \times (55.8 - 42)^2 = 8.6\%$
c) Rail: %HA $= 0.28 \times (55.8 - 42) + 0.0085 \times (55.8 - 42)^2 = 5.5\%$

Therefore, 13.4, 8.6, and 5.5% of the residents would be HA if the noise is produced by aircraft, road traffic, and rail, respectively.

More recently Fidel and Silvati have also found that aircraft noise is more annoying than noise from other forms of transportation at the same values of DNL [6]. In this study, the authors find 14% HA by aircraft noise at DNL of 55 dB and 5% HA at DNL of 50 dB, so that they estimate that the airport attitudinal survey data, when grouped in 5-dB wide "bins," will yield 12% HA at about 54 dB [6]. As a result of these and other more recent studies, many standards that continue to use A-weighted metrics such as L_{eq}, DNL, and DNEL have applied various penalties for aircraft noise in using the original Schultz curve (Figure 16.1). The International Organization for Standardization (ISO) standard [7] applies a 3- to 6-dB penalty, while the American National Standards Institute (ANSI) standard uses a 5-dB penalty [8]. The ANSI penalty is phased in between

Table 16.1 Penalties for aircraft noise applied by ANSI to original Schultz curve.

Measured Value of DNL	New value of DNL
50	50
55	55
56	57
58	61
60	65
65	70

Source: Personal communication with P.D. Schomer, September 6, 2006.

55 and 60 dB as shown in Table 16.1 (personal communication with P.D. Schomer, September 6, 2006). Schomer et al. have questioned the continuing use of metrics such as DNL, which are based on A-weighting [9]. It is well known that the A-weighting filter is independent of the sound pressure level, while the apparent loudness of the sound is not. Since annoyance is obviously related strongly to loudness, then use of metrics based on A-weighting are likewise fraught with problems (see discussion in Chapter 4 of this book).

Schomer et al. [9] have suggested instead that consideration should be given to the use of a loudness level-weighted sound exposure level (LLSEL) and loudness-level-weighted equivalent level (LL-LEQ). They suggest that LLSEL and LL-LEQ can be used to assess the annoyance of environmental noise. They conclude from their annoyance studies that, compared with A-weighting, loudness-level weighting better orders and assesses transportation noise sources, and with the addition of a 12-dB adjustment, loudness level weighting better orders and assesses highly impulsive sounds. Thus, they state that significant improvements can be made to the measurement and assessment of environmental noise without resorting to the large number of adjustments that are required when assessing sound using just the A-weighting [9]. Implementation of LLSEL and LL-LEQ capabilities on type 1, hand-held one-third octave band sound level meters would also be inexpensive [9].

Some community noise ordinances and test codes or standards, such as those used to evaluate building site noise in Germany and the United Kingdom, require the intrusive noise to be compared with the ambient noise. If the noise is more than 10 dB above the ambient, it is deemed excessive, while if it is only 5 dB above, it may be tolerated. Fields claims such a procedure is not supported by large amounts of noise data and surveys [10]. According to Fields, the results of 70 000 evaluations of 51 noise sources by over 45 000 residents show that there is no evidence to support the long-held assumption that the reactions of residents in a community to an intrusive noise is reduced when there are other environmental noises present [10]. In fact, Fields asserts that, provided residents are able to ascertain a logical estimate of their long-term exposure to the actual noise levels of the intrusive events, the presence or absence of ambient noise does not affect their judgment of the amount of noise annoyance produced by the intrusive noise events [10]. This is an important finding since it suggests that noise limits should be given as an absolute number of decibels, rather than an amount exceeding the ambient noise. Of course, if DNL or the new European DNEL is used as the noise limit, different limits should be set for aircraft, road traffic, and railway noise.

16.3 Community Noise and Vibration Sources and Control

16.3.1 Traffic Noise Sources

There are several reasons for the emergence of traffic noise as the main source of community noise annoyance in most developed countries. The power–weight ratio of trucks and cars has been constantly increased to permit higher payloads and more speed and acceleration; the resulting higher power engines are usually noisier than the earlier lower power ones. The number of vehicles has increased dramatically in most

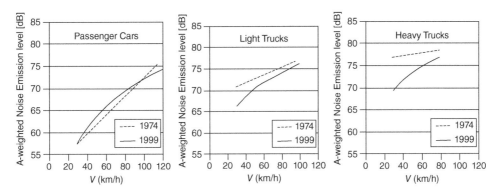

Figure 16.3 A-weighted sound pressure levels of cars and light and heavy trucks measured at different speeds, V, in 1974 and 1999 [12].

countries over the last 20 or 30 years. This, combined with the movement of people from country to city and the natural increase in urban population, has exposed more and more people to more and more traffic noise. Reference [11] discusses road traffic noise in more detail.

Although aircraft noise near airports is more intense than road traffic noise, the large number of vehicles in use and the fact that traffic noise is created in close proximity to residential housing ensures that it is a greater problem than aircraft and airport noise in most countries. The numbers of vehicles in use continues to increase. Studies in North America and Europe suggest that the external noise of many vehicles has not been significantly reduced in recent years. Figure 16.3 shows the cruise-by noise levels of cars, light trucks, and heavy trucks over the period of 1974–1999.

The main sources on vehicles include power plant (and power train) noise, tire–road interaction noise, and wind noise. The noise of cars is dominated by tire noise, except under accelerating conditions at low speed, during which power plant noise exceeds tire noise. With medium and heavy trucks, however, power plant noise is not exceeded normally until about 40 or 50 km/h is reached. At very high speeds wind noise on cars and trucks can become a major source, but below about 130 km/h it normally does not exceed tire noise (See Chapter 14 of this book).

The sound pressure levels generated by heavy vehicles exceed those of most cars by about 10–15 dB (See Figure 16.4). Trucks are vastly outnumbered by cars, but since they are normally in service for much longer periods than cars each day and they are so much noisier, they are a very important contributor to the overall road traffic noise problem. Noise levels near highways depend upon traffic flow rates and the mix of light and heavy vehicles with cars. Traffic noise tends to increase during mornings and evenings as people travel back and forth to work and other activities. Traffic noise is normally at a minimum during nighttime hours between about 1:00 a.m. and 4:00 a.m.

There are two main methods of evaluating vehicle noise. The first consists of measuring the pass-by noise of a vehicle at 7.5 or 15 m from the road centerline. The maximum A-weighted sound pressure level is recorded for single vehicles under controlled conditions normally on a special test track [11]. Figure 16.4 is an example of the noise of individual vehicles traveling at constant speed, normally known as the "cruise-by" condition. For traffic noise prediction schemes, statistical pass-by measurements of randomly occurring vehicles are made near selected highways. The levels measured in the statistical pass-by approach are dependent on the mix of light and heavy vehicles with cars and are also dependent on the type of road surface.

The second type of test, normally used for regulatory purposes, consists of a full-throttle acceleration test performed on a vehicle, which approaches the measurement zone AA in Figure 16.5 at a controlled speed. The measurement is again made at 7.5 or 15 m from the road centerline. The maximum A-weighted sound pressure level measured between lines A–A and B–B is recorded (See Figure 16.5).

Traffic noise prediction schemes normally include statistical information on the numbers of vehicles, the vehicle mix, the noise characteristics of each vehicle, the road surfaces, and the shielding effects of residential buildings on the propagation of the sound to the prediction points [11].

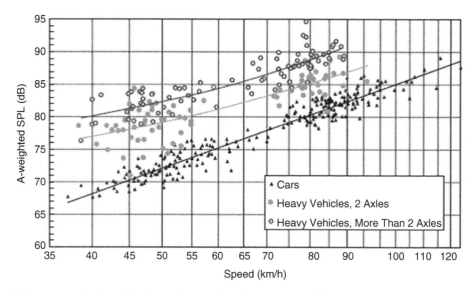

Figure 16.4 Cruise-by exterior A-weighted sound pressure levels measured at 7.5 m for cars, heavy vehicles with two axles, and vehicles with more than two axles [13].

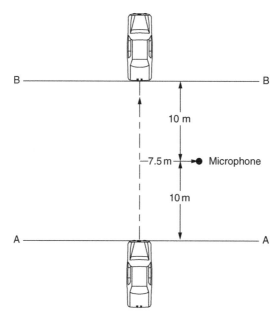

Figure 16.5 Measurement positions used for cruise-by or acceleration noise tests.

Community noise surveys and the creation of noise maps (see Section 16.6.3) are used in communities as a basis for checking results of some of these traffic noise prediction schemes. Also, traffic noise prediction schemes and community noise surveys of the one octave, one-third octave, and/or A-weighted sound pressure level generated by traffic are useful in deciding on noise abatement strategies. The most common traffic noise abatement strategy employed is the construction of roadside barriers, although porous sound-absorbent road surfaces are coming into use in some countries, as discussed in Refs. [14, 15]. See Ref.

[16] for a general discussion on barrier performance and Ref. [17] for a discussion on their use to control road and rail noise in the community.

Although well-established prediction schemes are available to predict environmental noise from road traffic in communities, it has been demonstrated that specific prediction schemes can be established, which are more reliable for individual cities during workday hours. Of course, such specific prediction schemes cannot be transferred to other cities [18]. There are many other studies of road traffic noise and its prediction in the community [19–25].

16.3.2 Rail System Noise Sources

Railway noise is generally less of a problem than road traffic noise and aircraft/airport noise. This is because the numbers of rail vehicles are much smaller than road vehicles, and railroad noise in general only adversely affects smaller regions of most cities. Rail system noise is, however, a major problem for communities situated near railroad routes [26–37]. The major sources on railway and rapid transit systems are (i) power plant noise, (ii) wheel–rail interaction noise, and (iii) aerodynamic noise. The main railway and rapid transit system power plants in use include (i) electric motors, (ii) diesel engines, and (iii) combined diesel–electric systems. See Ref. [38] for a detailed discussion on rail system noise sources and methods for their control.

Diesel power plants can be very noisy if the noise is not properly suppressed. Wheel–rail noise depends upon wheel–rail roughness and train speed. Wheel–rail roughness can be increased by the use of cast-iron brake systems. Disk brakes, which are coming into increasingly widespread use, have been found to reduce wheel–rail wear, roughness, and thus noise. Aerodynamic noise, although a problem inside rail vehicles at very high speeds, has not normally been found to be a major community noise problem, even at the very high speeds of 300 km/h [38].

Community noise prediction schemes for railway noise must include data on the power plant and wheel–rail noise of the rail vehicles, the number of railcars in operation and the rail vehicle speeds. In addition, the prediction schemes must account for the attenuation caused by air and ground surface absorption and by the distance to the observation points. The screening caused by obstacles, including buildings, railway cuttings, embankments, and purpose-built noise barriers must also be included in the schemes. There is some evidence to suggest that railway noise causes less sleep disturbance than road traffic noise at the same noise level [32–37]. In the United Kingdom, an A-weighted sound pressure level differential in favor of railway noise has been used in the development of railway noise legislation, using the equivalent A-weighted sound pressure level and existing road traffic noise legislation as the base. Railway noise and its effects continue to obtain the attention of many researchers and authorities [32–37].

16.3.3 Ground-Borne Vibration Transmission from Road and Rail Systems

Vibration generated by road and rail vehicles, some industrial enterprises, and building sites is transmitted through the ground and into buildings nearby. Vibration at frequencies up to 200–250 Hz can be transmitted at distances as far as about 200 m from roads or railway lines. Vibration at higher frequencies tends to be attenuated more rapidly with distance. The vibration caused in the buildings results in floor and wall vibrations, the movement of household or office objects, the rattling of doors and windows, and indirectly as re-radiated noise. Vibration is annoying to people at frequencies up to 50 or 100 Hz because various body organs resonate at low frequencies. For example, the stomach and other internal organs resonate in the region of 8–10 Hz, and the eyes and head resonate at frequencies of about 20–40 Hz. The chest wall cavity resonates in the range of 50–100 Hz. Chapter 5 of this book discusses some of these phenomena in more detail. Damage to buildings from vibration is unusual, although there are some cases where construction of new highways or railway lines has not been allowed because of the fear that the vibration they would cause could damage ancient historical buildings. Vibration of the ground or building foundations is normally measured in the vertical direction with velocity or acceleration transducers such as accelerometers. The quantities usually recorded consist of the maximum velocity or acceleration levels. The levels are normally recorded in one-third octave frequency bands, and each band is weighted according to human response to vibration. Reference [39] describes the procedures for measurement and prediction of these quantities. Chapter 5 of this book discusses human response to vibration and suggests suitable vibration limits.

The sources of ground vibration from road vehicles include passage of the vehicle wheels over road irregularities including bumps and holes. With rail vehicles, the source mechanisms are related to the travel of the wheels over the rail, which causes periodic and random forces. The periodic forces are created by the passage of the wheels over the spatially periodic supports of the rails and any discontinuities located at rail joints. The broadband random vibration is caused by unevenness or roughness in the rail and wheel contours. At high speed, vibration can also be caused when the vehicle speed exceeds either the Rayleigh surface wave speed in the ground or the bending-wave speed in the rails [39].

Vibration is transmitted through the ground by various wave mechanisms. The wave motion is quite complicated and consists of three main types: dilatational or pressure waves, equivolume or shear waves, and free surface or Rayleigh waves [39]. The main methods of protecting buildings from ground vibration include reduction of vibration at the source, such as better road and rail maintenance, the use of softer suspension systems for road and rail vehicles, resiliently mounted and better maintained rail tracks, and grinding of the wheels and rails to reduce roughness. Other methods to protect buildings include base isolation of the buildings themselves, as is described in Ref. [40].

Models exist for predicting the propagation of ground vibration from road and rail traffic [41, 42]. The models are mainly different because of the different input force mechanisms. With road vehicles, for the purpose of generating the wheel–road interface forces, the road itself can be considered to be rigid while the vehicle and its suspension are assumed to generate the dynamic forces in response to the road roughness. With rail vehicles, however, the excitation is different and is related to the unsprung mass of the wheel and axle combined with the mass of part of the rail system. Existing models range from mostly empirical to completely theoretical. Two-dimensional theoretical models are simpler but do not include the complete effects of fully three-dimensional models. Finite element (FEM) and coupled finite element/boundary element (FEM–BEM) models have been used and are becoming available in commercial software packages. Finite difference methods are also in use and have some advantages over FEM–BEM models since the computational code is simpler. Using those models to calculate absolute vibration levels [43] requires significant modeling details. Greatest accuracy is achieved when making predictions of insertion loss, even with relatively simple models [39, 40].

16.3.4 Aircraft and Airport Noise Prediction and Control

Air travel is projected to continue increasing in the foreseeable future. These increases will require the expansion of existing major airports and the creation of new airports. Airport expansion provokes public resistance because of the annoyance, speech interference, and sleep interference caused by aircraft noise in nearby residential districts. Aircraft noise is a much more localized problem than surface transportation noise since it is significant primarily around major airports. Most of the noise energy is produced by the operations of scheduled airliners, the contribution of the large numbers of general aviation aircraft being relatively small [44].

The introduction of early pure jet passenger aircraft in the late 1950s brought much higher noise levels during both take-off and landing than the piston engine airliners that they replaced. The majority of piston engine propeller-driven airliners have now been phased out of service, although twin-engine propjet aircraft continue to be used on some low-density, short-range routes. Jet aircraft operate at a much higher cruising speeds than do propeller types, and the aerodynamic configuration necessary for them to achieve these speeds results in higher take-off and landing speeds. Required runway length is normally greater with jet than propeller aircraft, partly as a consequence of these higher speeds, and partly because jet engine thrust is reduced when an aircraft is stationary or nearly so. This naturally brings the airport noise closer to residential communities. The generally large size and inertia of long-range passenger jet aircraft, and their greater throttle response times compared with those of piston engine propeller-driven aircraft, require them to use long approach paths, resulting in more extensive low-level flight over surrounding neighborhoods. Moreover, jet aircraft normally use considerable amounts of power on approach to counteract the drag of their high-lift devices. The noise produced by jet powered aircraft is primarily from their engines. The aerodynamic noise produced by the passage of the aircraft through the air (termed airframe noise) is still normally insignificant in comparison.

In the early 1960s, the first fanjet engines (also known as turbofans or bypass jets) entered service. Although developed mainly to improve fuel economy, the fanjet engines were quieter than the first pure jet (or turbojet) engines of the same thrust. A fanjet engine can be regarded as essentially a ducted propjet engine, in which the propeller (called the fan) is ducted to improve efficiency. All long-range airliners and almost all medium-range airliners are now powered by fanjet engines. Such fanjet engines mostly produce broadband noise from their high-speed jet exhausts during take-off, although with lower power conditions during landing approach, the tonal whine produced by the compressor stages usually becomes prominent. At a distance

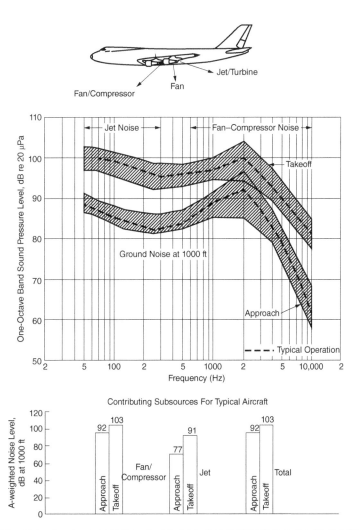

Figure 16.6 Noise levels and spectra of wide-body fanjet aircraft (e.g. the Boeing 747) [46].

of 3 km, the A-weighted sound pressure level during a jet aircraft take-off is of the order of 60–65 dB and is sufficient to interfere with speech [44]. Current wide-body fanjets (e.g. Boeing 717, 737, 767, 777, and 787 and Airbus A-300/310, A-318, A-320, A-330/340, A-350, and A-380) have considerable noise control technology built into their engines and are much quieter than the early pure jet and fanjet aircraft.

Some airliners have been particularly designed to have quiet operational characteristics. The BAe 146, first introduced in the 1980s, was an early, very quiet, four-engine jet airliner, which has special high-lift devices giving it short take-off and landing capabilities that make it able to operate from very short runways at small downtown airports. It is able to operate from downtown airports such as those at Monchengladbach, Aspen, and at the London City Airport (a converted dock). In May 2006 landing and take-off tests at the London City Airport of the newer Airbus A-318 has shown it can also use this downtown airport. Following evaluations by Airbus, London City Airport, and the UK airworthiness authorities, in March 2006, these authorities granted the A-318 a steep landing approach certification that enabled the airport compatibility tests to take place. With its very low noise characteristics, the A-318 makes it possible for it to use downtown airports such as the one in downtown London. The modern narrow-body Bombardier's C Series aircraft has very quiet operational characteristics that increase airport utilization. This airplane is a single-aisle short-haul jet, equipped with turbofan engines, and made of advanced composite materials. C Series aircraft's community noise level is below the stage 4 limit defined by the Federal Aviation Authority (see Chapter 15), making it ideal for downtown airport operations.

As discussed in more detail in Ref. [44], propeller-driven airliners are used on some medium-range and most short-range routes, and such aircraft are almost exclusively powered by gas turbine engines. Such turboprop aircraft, as they are commonly known, mainly produce tonal noise instead of the broadband jet noise. The noise of propellers is quite directional and is mostly radiated in the propeller plane. Helicopters are less commonly used than jet and turboprop aircraft, but they can be quite noisy and produce A-weighted noise levels of about 60 dB at 1 km. Like propeller aircraft, the helicopter blade noise is very directional and is created both by the main rotor and the high-speed tail rotor.

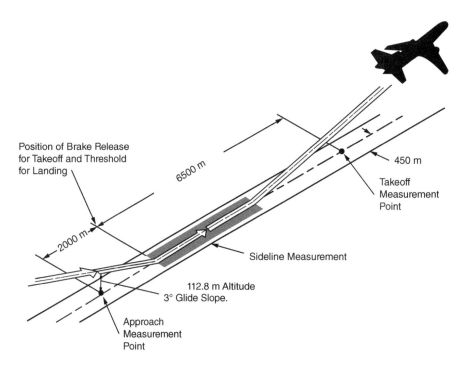

Figure 16.7 Measurement locations for certification testing of aircraft to FAR Part 36 noise standards.

Aircraft noise is evaluated for two main reasons: (i) for certification purposes and (ii) to monitor the noise around airports. For certification, the noise of individual aircraft is measured using the effective perceived noise level (EPNL). For monitoring noise at airports it is normal to use a measure that accounts for many aircraft movements and the time of day in which the noise is produced. In the EU a composite measure known as day–evening–night level (DENL or L_{den}) is used [44]. The DENL includes components for daytime, evening, and nighttime hours. During the evening, a 5-dB penalty is applied and at night a 10-dB penalty.

Figure 16.6 shows the noise levels and spectra of an early fanjet aircraft both for take-off and approach. Figure 16.7 shows the measurement locations used for certification of aircraft according to FAR Part 36 noise standards [45]. Research on evaluating aircraft and airport noise and community annoyance continues [47–58].

16.3.5 Off-road Vehicle and Construction Equipment Exterior Noise Prediction and Control

Off-road vehicles and heavy construction equipment are used for roadway and railway construction, earth moving and excavation, laying of pipes and cables, and construction of new buildings. They are responsible for high levels of environmental noise and cause annoyance, speech interference, and sleep disturbance in residential areas. In addition, there is the possibility they can cause hearing damage to people operating this equipment or people working in close proximity. A-weighted sound pressure levels can be as high as 75– 90 dB at distances of 15 m. Sound pressure levels at distances of several hundred metres can be of the order of 60–70 dB, which is above acceptable community noise limits. The main noise sources on off-road vehicles and construction equipment consist of the engine, exhaust, intake, cooling fans, and the tracked wheels. Mobile compressors are also significant noise contributors and are often used in conjunction with impulsive noise generators, such as pneumatic drills, cutters, and vibrating roller equipment.

In Europe, off-road vehicles and heavy construction equipment are required to have product labels giving their sound power outputs. Noise emission limitations are set by EC Directives. In the United States, such equipment noise output is governed by community noise ordinances or state regulations. Mobile compressors are one exception, since their noise output is regulated by the U.S. Environmental Protection Agency (EPA). Other countries do not seem to have uniform approaches to regulate the noise of off-road vehicle and construction equipment. In Europe, the sound power output of such equipment must be provided by display of a label [59]. The sound power is normally determined through measurements of the sound pressure level at discrete microphone locations on a hemispherical surface as described in ISO standards. Alternatively, the sound power can be determined with the use of sound intensity equipment (see Chapter 8).

Off-road vehicles and heavy construction equipment can be divided into three main categories [59]: (i) wheeled vehicles, including excavators, loaders, and graders, with which the engine is the dominant noise source, (ii) tracked vehicles, in which the track noise is comparable to the engine noise, and (iii) vibration and impact-generating tools, including pneumatic drills and vibrating rollers, in which the noise is generated by the tool itself and no engine is involved. The exterior noise can be reduced by a combination of passive means, including enclosures, vibration isolators, use of vibration damping materials, sound-absorbing materials, baffles, and barriers. Enclosures can be very effective, particularly if absorbing materials are used inside and they are properly sealed. Unfortunately, this is not always possible because of the heat build-up created by most items of machinery, in particular the engine. The use of inlet and exhaust mufflers is essential on such equipment [59]. Sound-absorbing material is sometimes placed inside the mufflers, although the material can lose its effectiveness because of contamination with moisture and carbon particles. Normally, for such reasons, reactive mufflers are preferred. Reference [59] gives an example of noise reduction on a mobile compressor. Other examples of mobile compressor noise control are given in Ref. [60]. Compressor noise is discussed in detail in Ref. [61]. Efforts continue on reducing interior and exterior noise and vibration of off-road vehicles [62–64].

16.3.6 Industrial and Commercial Noise in the Community

The annoyance produced by industrial and commercial noise is very similar to that produced by road traffic noise when the long-term energy- sound pressure levels are the same [65–68]. However, in most countries industrial noise and road traffic noise are treated differently [69]. As explained in Ref. [2], it is common now for a city to have different district plan noise rules. Normally, industrial enterprises are located in special zones in the city. It is desirable for the industrial zones to be separated from zones for residential housing. The separation in distance must not be too great; otherwise people will be inconvenienced by having to travel a long way to work. But the industrial and residential zones must not be situated too close to each other either. The noise created in residential areas from the sources in the industrial zones is normally predicted on the basis of the sound power outputs from the different noise source components in the industrial zone [69].

Industrial and commercial noise sources can be divided into two main categories [69]. The first category includes steady noise sources that have little variation in level and character during the day and night. The second category includes intermittent noise resulting from different industrial production cycles and caused by vehicles coming and going to the industrial and commercial areas. Steady noise containing pure-tone and impulsive components is generally found to be less acceptable in residential areas compared with steady noise of the same long-term equivalent sound pressure level. This is particularly true when steady noise contains pure-tone components below 90 Hz [69]. Impulsive noise is also known to be more annoying than steady noise of the same equivalent sound pressure level.

Reference [69] describes a procedure for predicting the sound pressure levels in residential areas, which is based on knowledge of the sound power output per unit area of the industrial operation. Included in the procedure are adjustments for wind effects on the noise propagation. Such sound pressure level predictions are valuable in deciding whether new industrial operations should be permitted near residential communities and/or residential developments should be allowed near existing industry. It is important to monitor the noise levels produced by the industrial and commercial operations and to make comparisons with those that are predicted. Also, in order to avoid community complaints, it is important to inform the public in the residential areas of such noise monitoring and also to provide details about all activities being undertaken to reduce noise reaching the residential areas from the industrial and commercial zones [70–80].

16.3.7 Construction and Building Site Noise

Noise created on building sites in a city can interfere with speech, sleep, and other human activities [66–68, 70, 71]. The noise created comes from a variety of machinery and mechanical equipment and includes demolition of buildings, construction of new buildings, laying of pipelines, sewers, and cables, and construction of new roadways and railways. Some noise is impulsive in character, such as caused by pile driving. Other noise is more continuous in nature from sources such as compressors and heavy earth-moving equipment. Noise levels on building sites can be predicted from knowledge of the sound power output of the different items of machinery. Reference [81] explains how predictions of the noise at different locations near a building site can be made. Normally the contributions from the different sources are added on an energy basis. In some countries the predicted or measured sound pressure level from all the sources is compared with the ambient noise level when none of the sources is acting. There are few standards or test codes for the prediction and control of noise from building sites. In 2006, the only such test codes were those in Germany and the United Kingdom. These test codes provide tools for the calculation of the daytime and nighttime rating levels. These are then compared to standard ambient-noise values for decisions on the acceptability of the building site noise. Transient noise peaks on the building site can also be considered in these assessments.

Local authorities sometimes are tolerant of intense noise on a building site in order to speed up construction for economic or political purposes or to limit disruptions to road and rail traffic and the operations of public utilities. If the A-weighted sound pressure levels caused by the sources are no more than 5–10 dB greater than the ambient levels, they may be allowed by some authorities. But levels that are greater than 10 dB above the ambient are normally determined to be excessive and require regulation and/or control.

Using city or national ordinances, some local authorities impose penalties for impulsive noise and/or noise containing prominent pure-tone components. Other local authorities restrict noise in certain city zones and others only during nighttime. In some cases noise control programs are initiated when repeated and/or frequent complaints are received. If noise control measures are found to be necessary, these should start with the noisiest sources and incorporate the normal source–path–receiver concepts as explained in Chapter 9 of this book and Refs. [60, 81]. As discussed in Refs. [59, 81], the EU Directive 2000/14/EC requires construction equipment machinery used on building sites to be subject to noise labeling. This EU directive requires the manufacturer to state the guaranteed sound power level in the operating instructions and on the machine itself with a label. The determination of the sound power level is usually carried out in accordance with ISO 3744. Some of these construction machines are subject to sound power level limits.

16.4 Environmental Impact Assessment

Environmental noise impact assessments (sometimes known also as environmental noise impact statements) are used to balance the negative noise impact of a proposed development versus the benefits that the development, such as a new highway industrial development area or recreation facility, could bring to a nearby community [1, 82–86]. Existing community noise exposure guidelines are consulted when preparing the assessments. These are based on the premise that there is a level of noise that is acceptable to the majority of the community. If this level is likely to be exceeded by a proposed development, appropriate mitigation measures need to be selected with due consideration of cost and technical feasibility. The mitigated noise impact is assessed for acceptability to the potentially affected community as well as the benefit of the development to the community as a whole. Sometimes the terms community noise and environmental noise are used interchangeably.

An environmental impact assessment can be formally defined as a procedure for considering all the environmental consequences of a decision to endorse legislation, putting into practice policies and plans, or to initiate infrastructure projects. An environmental impact statement corresponds to the final step of an environmental assessment exercise where the conclusions of the assessment are published in a communicable form to the concerned developers, authorities, and the general public [87].

The assessment should report the analysis of the impact of the proposed development on both the natural and social environment. It includes assessment of long- and short-term effects on the physical environment, such as air, water, and noise pollution, as well as effects on employment, living standards, local services, and aesthetics [88].

In general, there is typically a two- to five-year decision-making process required before any major project can be built. The authors of an environmental impact assessment usually represent many areas of expertise and possibly will include biologists, sociologists, economists, and engineers [87].

Since their origin 50 years ago, environmental impact assessments have become widely accepted tools in environmental management for both planning and decision-making. Environmental impact assessments have been adopted in several countries with different degrees of enthusiasm, where they have evolved to varying levels of sophistication [88]. Since 1969, the National Environmental Policy Act (NEPA) has put into practice the environmental impact assessment procedures in the US. Other industrialized countries have also implemented procedures. For example, Canada adopted the legislation in 1973 while Australia approved it in 1974. The Netherlands and Japan approved the legislation in 1981 and 1984, respectively. In 1985, after nearly a decade of deliberation, the European Community adopted a directive making environmental impact assessment mandatory for certain categories of projects [88]. The Resource Management Act (RMA), passed in 1991, is New Zealand's main legislation for environmental assessment. In 1974, Colombia became the first Latin American country to establish these procedures. In addition, environmental impact assessment procedures have been endorsed by law in many other countries. These procedures are normally conducted by local and government agencies [88, 89].

Public acceptance of an environmental impact assessment procedure is clearly supported by community participation [1]. A community involvement process warrants that residents, businesses, and others have an opportunity to participate. It is well known that, in some cases, litigation arises from environmental groups who want to block a project or from parties who feel that the assessment exaggerates the threat to the environment to the detriment of economic interests [87]. Therefore, a collaborative planning process with the community does not have to begin only after claims and conflicts occur. In addition, initiating a collaborative planning process does not require extraordinary resources or leadership at the very highest levels of government. This is particularly important during scoping to incorporate new ideas from the community that will serve as the basis for alternative development, screening, and environmental evaluation [87].

16.5 Environmental Noise and Vibration Attenuation

16.5.1 Attenuation Provided by Barriers, Earth Berms, Buildings, and Vegetation

Noise barriers are being used increasingly to protect residential communities from road traffic, rail and rapid transit noise. Chapter 9 of this book and Ref. [16] describes empirical formulas that can be used to predict barrier performance. Barriers are of limited use to protect residential areas from aircraft and airport noise, and construction site noise, with the possible exception of their use to offer protection from the noise caused by the testing of aircraft engines during ground run-up. Likewise, the mobility and elevation of noise sources of construction equipment used on building sites and highway construction sites often make barriers of limited use. Urban barriers must also be designed to be acceptable aesthetically. The formulas for barrier performance given in Ref. [16] are mostly based on idealized theoretical considerations or experimental studies conducted in the laboratory. Reference [17] describes some of the practical considerations in the use of urban barriers.

The attenuation of a barrier is normally defined in two main ways. The first involves the barrier *attenuation*, which is defined as the difference between the sound pressure levels measured (or predicted) at a location and the sound pressure level at the same location under free-field conditions. The second definition involves the reduction in sound pressure level (known as *insertion loss*) at the receiving location achieved by the *insertion* of the barrier. The attenuation provided by a barrier is a function of frequency. More exactly it can be related to the difference between the two path lengths from the source to the receiver divided by the wavelength: (i) over the barrier and (ii) straight though the barrier. This quantity is known as the Fresnel number (See Eq. (9.49)), which is also used in optics [16].

The Fresnel number can also be related to the effective nondimensionalized height parameter of the barrier (defined as the ratio of barrier height perpendicular to the incident sound divided by its wavelength). The effective height of the barrier increases with wavelength and so does the barrier attenuation. This means that a barrier of fixed height is more effective in attenuating high-frequency sound, and stronger shadows are created for high-frequency sound than low-frequency sound. A similarity can be observed with the behavior of light since an obstacle produces a stronger shadow for short-wavelength (high-frequency) violet light than long-wavelength (low-frequency) red light [16].

Theoretically, noise barriers can be shown to provide the same attenuation if placed at the same distance from the source or from the receiver. In practice, however, as common with other passive noise control measures, their placement nearer to the source is usually more effective. This is because receivers, such as the upper regions of high-rise buildings, can extend in height above a barrier placed near to them, and thus sound is not blocked from reaching them so they are not protected [90]. Also, single or multiple-road and rail vehicles are normally located close to the ground, and barriers placed close to them block the sound better, which would otherwise be traveling to the multiple elevated receivers.

It is particularly important to ensure that barriers do not have holes or leaks that can degrade their performance and that they are constructed from material with an adequate transmission loss (sound reduction index) to sufficiently attenuate the sound penetrating them and reaching the receivers directly in that manner. Urban barriers are sometimes made to absorb sound on the side facing the source, so as to reduce the sound reflected back to the source. Care must then be made to ensure that the sound-absorbing material can survive the local environmental conditions and not become degraded too rapidly. Reference [17] describes the use of sound-absorbing materials with urban barriers. Ref. [16] presents formulas for predicting the attenuation of barriers, while Ref. [17] gives formulas and nomograms for predicting the attenuation of barriers used in urban situations. These formulas are now incorporated in commercial software. The sound attenuation of barriers with complicated shapes such as cantilever, parabolic barriers, or earth berms can now be predicted with boundary element method (BEM) approaches.

Unfortunately, when barriers are used in the field, the atmospheric effects of turbulence or wind and/or temperature gradients above the barriers normally degrade their attenuation and/or insertion loss performance. Reference [17] describes factors that must be considered when barriers are used in practical situations to reduce road and rail traffic noise in the community. Wind is probably the main cause of the degradation of the acoustical performance of barriers. Wind has been found to have two main effects. First, the turbulence in the wind causes the sound waves to be scattered so that some of the sound energy propagates into the shadow zone behind the barrier. Second, wind gradients, which exhibit increasing wind speed with height above a barrier (with wind blowing in the same direction as the sound propagation), can bend the sound downward into the so-called shadow region of the barrier, thereby decreasing its attenuation. Temperature gradients, in which the temperature increases with height (called temperature inversions), can have a similar effect in bending sound downward into the shadow zone behind a barrier. In practice, the attenuation of an urban

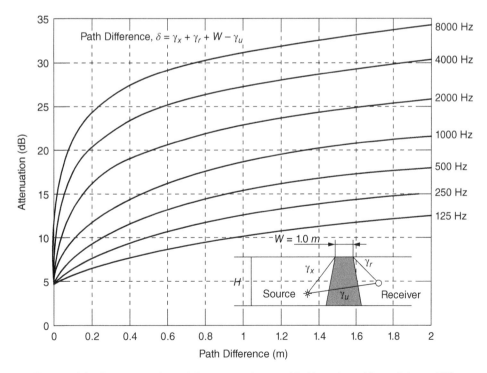

Figure 16.8 Predicted values of the attenuation provided by a 1 m wide earth berm [17].

barrier does not reach its theoretical value because of such environmental effects, and in real-use urban barriers normally have an upper attenuation limit of about 20 dB, unless they are of double construction, in which case the upper limit is about 25 dB [17].

An example of a noise barrier is an earth mound. An earth mound (berm) is a noise barrier constructed of soil, stone, rock, or rubble, often landscaped, running along a highway to protect adjacent land users from noise pollution. The sound attenuation provided by a 1 m wide earth berm as a function of the path difference (see Chapter 9) is shown in Figure 16.8. There is a cost advantage in using earth mounds since they can often be constructed using surplus materials at project sites, provided there is sufficient land area available for their construction. In general, earth mounds represent the lowest cost alternative to construct a noise barrier. An earth mound is an obvious solution to reduce visual impact because it can be made to fit in with the landscape more naturally than any vertical structure, especially as it can support planting which greatly improves its appearance in most rural contexts. In other words, the soft natural outline of an earth mound, in conjunction with planting, is likely to be more attractive to both local residents and to road users [87]. When plants are selected for use in conjunction with a barrier, they should generally be of hardy evergreen species (native plantings are preferable), which require a low level of maintenance. Concerning the acoustical performance of earth mounds, some studies have indicated that earth mounds may provide more sound attenuation than vertical walls of the same height, although experimental and theoretical assessments have yielded mixed results [91].

There is a prevalent popular belief that plantings, shrubs, or rows of big trees produce a significant reduction in road traffic noise. However, noise reductions are irrelevant unless the vegetation is very dense and wide. Although some authors have suggested that vegetation produces beneficial effects in improving public perception of the noise due to visual and psychological relief, other studies have shown the opposite [92]. In addition, care must be exercised with placing plants on barriers, since the scattering they cause can actually reduce a barrier's attenuation. Further discussion on the noise attenuation provided by trees can be found in Ref. [17].

16.5.2 Base Isolation of Buildings for Control of Ground-Borne Vibration

Road and rail traffic and some industrial and/or road building operations cause ground vibration of a fairly continuous nature. Railways are usually the sources of most intense ground vibration since they often carry vehicles with relatively heavy loads at high speeds, which results in significant rapidly created forces transmitted to the ground. Although it is possible to reduce the vibration at the source in some cases, there is some limit to the reduction that can be achieved economically.

When new buildings, particularly those of a sensitive nature, such as hospitals, auditoriums, and concert halls are to be constructed near to existing highways and railway lines, suppression of the building vibration and base isolation of the buildings themselves should be considered [40]. Some simple, relatively low-cost measures are available for reducing building vibration, including (i) increasing the damping in the structure, (ii) stiffening certain regions of the building to move natural frequencies away from forcing frequencies thereby avoiding resonances, and (iii) the installation of floating floors in sensitive parts of the buildings.

In cases where the ground vibration is severe, such as where a building must be constructed near to a railway, it may be necessary to vibration-isolate the complete building from the ground by means of base isolation [40]. This may be essential only for sensitive buildings. The amount of vibration that is acceptable in a building will depend upon its use and other factors such as the duration and nature of the vibration. The principles of base isolation are similar to those used to protect buildings against earthquakes; however, the lower level of vibration experienced from road, rail, and some industrial sources compared with earthquakes, makes the design criteria somewhat different for each case studied. Each building design will be different and building use, acceptable vibration limits, and other criteria will determine the design chosen. Typical isolation frequencies are in the range of 5–15 Hz.

There are two main types of isolators normally used for base isolation of buildings. These consist of (i) laminated rubber isolators and (ii) helical steel spring isolators. The rubber isolators can either be made from natural rubber or from synthetic rubber. Steel spring isolators usually combine several helical springs in one unit. Rubber isolators normally have higher inherent damping than spring isolators. However, the spring isolators sometimes include additional damping elements to suppress internal coil resonances of the springs at high frequencies, and to limit vibration during the rapid vibration onset or vibration decay caused by passing trains or vehicles. Steel springs are expensive but can be manufactured to have precise stiffness values and a long life. Rubber isolators are usually less expensive but have the drawback that they can be subject to degradation more rapidly than steel springs unless they are protected against any possible hostile environmental conditions, which could cause the degradation. If adequate protection is provided, however, rubber isolators can be made to have sufficient life in terms of both degradation and creep performance.

The design of vibration isolators is described in Refs. [40, 60, 93]. Their performance in buildings is normally measured in terms of the insertion loss they provide, since this quantity describes the benefit obtained from the use of the isolators [39]. Single-degree-of-freedom models are useful to give some approximate indication of the isolator performance, although building vibration is very complicated, and more sophisticated vibration models must be used for better vibration predictions. With continuous excitation, which is almost steady-state, such as is caused by passing trains, the vibration problem can be treated in the frequency domain. In some cases, where the excitation is more impulsive in nature, such as is caused by some industrial applications, then a time-domain modeling approach is more convenient. More complicated building base models employ FEM methods. Various software programs are available commercially to create such sophisticated building vibration models. In some cases, simple two-dimensional numerical models are used to reduce computational demands. Three-dimensional approaches are more suitable to obtain more accurate building models [40]. Studies on base isolations of buildings continue [94–100].

16.5.3 Noise Control Using Porous Road Surfaces

Paved roads have been in existence for at least 2000 years since Roman times. Obviously such roads created considerable noise by the impacting of wheels on the uneven road surface. In South America, the first allusion to community noise dates back to the end of eighteenth century, when a vice royal decree recommended the use of leather strips on carriage wheels. This resulted from the numerous complaints of inhabitants of the Spanish colonial cities having cobblestone roads [101]. Modern road surfaces are normally composed of asphalt or concrete.

One way to reduce tire/road noise is by the use of porous road pavement surfaces. Such surfaces have the advantage that they not only reduce the tire/road noise at the point of its generation, but they also attenuate it (and vehicle power plant noise) by absorption of sound as it propagates to nearby residential areas. Such surfaces have the further advantage that they drain water well and reduce the splash up behind vehicles during heavy rainfalls. So far there has been greater use of porous road surfaces in Europe, although there is now increasing interest in their use in the USA [13, 102, 103].

The sound absorption of porous road pavement surfaces is affected by several geometrical and other parameters of the road pavements. These include: (i) the *thickness d* of the porous layer, (ii) the *air voids* (V_a) or "porosity," (iii) the *airflow resistance per unit length*, R, (iv) the *tortuosity*, q, and (v) the *coarseness of the aggregate mix* (use of small or large aggregate, etc.). For most common dense asphalt mixes, V_a is about 5%, while for new porous mixes, the air void content V_a varies from about 15–30%. The airflow resistance R is the resistance experienced by air when it passes through open pores in the pavement. The tortuosity or "structural form factor" as it is sometimes known is a measure of the shape of the air void passages (whether they are almost straight or twisted and winding and slowly or rapidly change cross-section area) and the effect this has on the pavement sound absorption properties (see Chapter 9).

Hamet et al. [104] and others have shown that porous surfaces exhibit one or more regions of high sound absorption in the frequency range of most interest (200–2000 Hz). These high sound absorption regions often peak and can have sound absorption coefficients of almost unity. The thickness of the porous surface has a large effect on the sharpness of the peaks. Generally the thicker the porous surface, the lower is the peak frequency. With thicker porous surfaces, the peaks generally also become broader and the peak absorption is somewhat reduced. The airflow resistance and tortuosity have an influence on these effects too. Hamet et al. measured the absorption of various thickness surfaces from thicknesses of 50–400 mm. They found that their porous surface of 50 mm thickness has a sharp absorption peak of almost unity at about 900 Hz. While their 100 mm porous surface has at first a smoother peak at about 450 Hz and a second sharper peak at about 1350 Hz. Their 150 mm thick surface has peaks at about 300, 900, and 1500 Hz. It is observed that the first peak frequency is proportional to the thickness. Also, it is observed, where the second and third peaks can be seen in the measurements of Hamet et al., that these higher peak frequencies are at almost exactly twice and three times the frequency of the first peak. But the frequencies are only a little more than one half what would be expected from a simple one-quarter, three-quarter, and five-quarter wavelength matching with the thickness. This is presumably because the tortuosity of the air passages makes the distance from the surface to the dense pavement below effectively almost twice as great as the thickness itself.

Von Meier et al. [105, 106] have made theoretical studies of the effect of air void content and flow resistance on the sound absorption coefficient of porous surfaces. They found that both the air void content and flow resistance have a strong effect on the value of the absorption coefficient of the peaks of a 40 mm thick porous surface with a tortuosity value of 5. The air void content leads to higher values of the sound absorption coefficient at both of the peaks predicted for such surfaces; while higher values of air flow resistance also initially lead to higher values of the absorption coefficient at the peaks. After a certain value of air flow resistance per unit length R is reached, however, the peak absorption values start to decline. Additional discussion on porous road surfaces can be found in Ref. [13].

16.6 City Planning for Noise and Vibration Reduction and Soundscape Concepts

16.6.1 Community Noise Ordinances

The major sources of noise in the community are caused by surface transportation, aircraft/airports, industry, and construction [72–80]. As discussed in this part of the book, the main contribution to community noise, in most countries, is caused by road and rail traffic. The annoyance caused by noise depends upon the level, duration, time of day, frequency content, and other factors. Road traffic noise depends upon the road surface, road inclination, traffic density, and mix, and speed of automobiles and light and heavy trucks. Railroad noise depends upon vehicle speed load, wheel and rail roughness, and other factors. Some of these sources can be controlled locally by regulation, such as with speed limits. Some factors are outside local control.

Community noise regulations and noise ordinances are of two main types: qualitative and quantitative. Qualitative types of noise ordinances prohibit excessive noise during certain hours of the day or night and/or prohibit some noisy activities during certain hours and in certain defined noise zones. The difficulty with such ordinances is that they are vague in nature and difficult to enforce. Quantitative noise ordinances, however, restrict noise generated and/or received. The levels allowed at the boundary where they are produced or the boundary where they are received are normally given as A-weighted sound pressure levels. These quantitative ordinances need trained staff for enforcement who can use calibrated measurement instruments to determine the sound pressure levels accurately. Since noise levels fluctuate in the community and vary during day, evening, and night periods, average levels are often used such as the equivalent sound pressure level. The A-weighted equivalent sound pressure level came into use in the United States in the 1970s, and recommendations have been published by the World Health Organization (WHO) for day and nighttime to protect people from becoming moderately or seriously annoyed. In the United States the day–night average sound pressure level is also widely used. In Europe the day–evening–night level is now increasingly used.

Some noise ordinances specify different noise zones that are defined for the activities performed in the zones. Figure 16.9 shows an example of such ordinance. Zones can be specified for heavy industry, light industry,

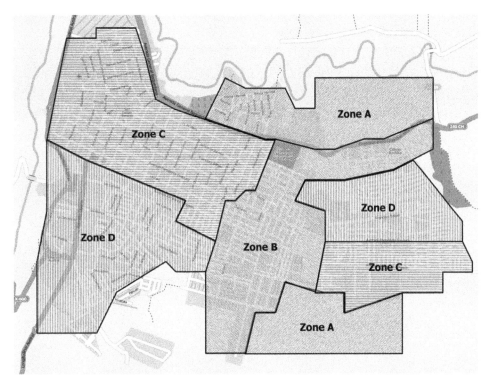

Figure 16.9 Definition of four zones according to a city plan and the corresponding use of land. Zone A corresponds to heavy industry (noisiest zone), Zone B is for commercial and light industry, Zone C is a commercial area, and Zone D is residential (quietest zone).

residential, school, hospital, and other uses. Normally different noise zones are created with steps of 5 dB from noisiest to quietest zone with restrictions on the maximum noise permitted in each zone. For instance, some activities such as construction may be prohibited in some zones. Obviously, enforcement is essential if the ordinance is to have any use at all. Fines and other financial penalties can be used for enforcement. In the United States, the EPA Office of Noise Abatement and Control set limits for noise output of some sources such as heavy trucks and compressors on building sites and was active in producing a model community noise ordinance standard. Since 1980 centralized planning by the EPA Office of Noise Abatement and Control has largely been disbanded and taken up by different government agencies and local, citywide, or state authorities. In Europe a different situation exists. European directives in recent years have required strategic noise maps to be created for major transportation facilities such as airports and railway stations and for large cities. Action plans are required to manage noise problems and recommend noise reduction if necessary. The goal is to harmonize noise criteria and noise zoning and limits among the different European member states. The experience in other countries such as Australia, Canada, China, and Japan is described in detail in Ref. [2]. Further work and discussion on community noise evaluation and ordinances and regulations continues [72–80].

Example 16.2   Consider that a quantitative noise ordinance restricts the noise generated at the city plan of Figure 16.9 according to Table 16.2.

The day time is defined as from 7 a.m. until 10 p.m. and night time as from 10 p.m. to 7 a.m. An industry located in zone A increases its capacity by adding more machines to a production line. The energy-equivalent hourly A-weighted sound pressure levels are measured at a point in zone A using an integrating sound level meter, and the results are presented in Table 16.3. Assess the community noise according to the city noise ordinance.

Table 16.2 Example of a community noise ordinance.

Zone code	Description	Limits in A-weighted L_{eq}, dB	
		Day time	Night time
A	Heavy industry	75	70
B	Commercial and light industry	65	55
C	Commercial	60	50
D	Residential	55	45

Table 16.3 Data used in Example 16.2.

Number	Time	A-weighted sound pressure level, dB
1	07:00	72
2	08:00	70
3	09:00	69
4	10:00	71
5	11:00	73
6	12:00	70
7	13:00	76
8	14:00	77
9	15:00	75
10	16:00	73
11	17:00	71
12	18:00	72
13	19:00	80
14	20:00	73
15	21:00	71
16	22:00	72
17	23:00	75
18	00:00	75
19	01:00	72
20	02:00	73
21	03:00	78
22	04:00	79
23	05:00	72
24	06:00	70

Solution

The A-weighted daytime average sound pressure level is calculated considering measurements 1–15 giving

$$L_d = 10 \times \log\left[\left(10^{7.2} + 10^7 + 10^{6.9}... + 10^{7.1}\right)/15\right] = 74\,\text{dB}.$$

The A-weighted nighttime average sound pressure level is calculated considering measurements 16–24 resulting in

$$L_n = 10 \times \log\left[\left(10^{7.2} + 10^{7.5} + 10^{7.5}... + 10^{7}\right)/9\right] = 75 \, \text{dB}.$$

Therefore, the industry complies with the noise limit set for zone A by the noise ordinance during the daytime but it does not act in accordance during night time.

16.6.2 Recommendations for Urban Projects

Modern urban legislation should consider all of the environmental impacts caused by new projects and how to limit the noise in the city, particularly in residential areas. For this, it is necessary to have on hand the regulations corresponding to each community that establish the types of areas (in terms of land-use) and their corresponding maximum permissible limits. As a rule, proper general and specific plans should always bring together economic and industrial development with quality of life in cities [107]. A general plan contains a set of goals for the future growth and development of the city. Instead, the specific plan is essential to carry out the goals of the general plan. According to the experience in several cities the following proposals for consideration in urban projects should be considered:

1) plan the city specifying a set of zones according to their use, which is also suitable for noise issues,
2) locate all community noise sources (airports, industrial areas, railways, highways, etc.) far away from the indoor and outdoor sensitive areas (residences, schools, parks, hospitals, etc.),
3) make preference for underground instead of surface modes of transportation, and take measures to ensure that the inhabitants use public rather than individual transportation,
4) plan the location of traffic lanes (especially heavy road traffic) far from residences and take steps to ensure free traffic flow,
5) eliminate uneven surface roads and use porous road pavement surfaces,
6) use screening attenuation (sound barriers, buildings, berms, etc.) between noisy roads and residences,
7) enforce that the noise levels produced during the construction of the public and private infrastructure are below the noise limits,
8) develop regulations for noise control at the facades, and
9) conduct public awareness campaigns on the effects of noise.

16.6.3 Strategic Noise Maps

A noise map is a cartographic representation of the noise level distribution in a determined area and period of time. Strategic noise maps are important tools that can provide relevant information for local and global action plans. In urban administration and planning, noise mapping is a very useful tool for generating information about environmental impacts and enabling the visualization of noise pollution in the urban landscape. Many cities all over the world have produced noise maps for different modeling areas, several noise sources and with varied outcomes. Research on noise maps is a subject of great interest in European countries [108] where the use of geographic information systems (GIS) has provided relevant information for noise maps and specific maps have been developed for airport noise [109, 110]. Figure 16.10 presents a typical contour set of equal noise levels near the runways of an airport [44]. Since 2002, European cities are required to have a strategic noise map to meet the Environmental Noise Directives and Regulations [111, 112]. In this scenario, noise maps are tools that have several potential applications to help with: identifying noise levels within cities, identifying areas with greater exposure, and assessing future scenarios. They also serve as a

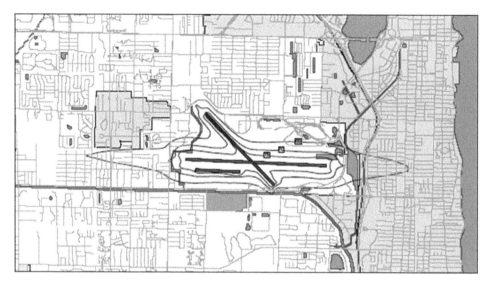

Figure 16.10 Typical noise contours for an airport [44].

basis for adopting actions to reduce noise levels through structural and nonstructural measures such as use of urban planning tools [113]. In addition, noise maps are an important tool for environmental education and awareness.

Currently, computer-based calculation methods have become adopted into general practice in the creation of noise maps. Regardless of their advantages and shortcomings with respect to sampling methods based on in-situ measurements, maps are still needed to verify the results obtained [112]. The computer software needs information on cartographic data and noise sources, among other inputs to perform the modeling process. Of course, the cost of producing a noise map for a large city could become very high and recent studies have been aimed to produce cost-effective noise maps through use of simplified methodologies [114, 115].

During the period 2009–2012 the European Commission developed CNOSSOS-EU (Common Noise Assessment Methods in Europe), the common and harmonized methodological framework for strategic noise mapping under the European Directive 2002/49/EC for the assessment and management of environmental noise [111, 116]. The main goal was to have the common noise assessment methodology operational in 2017. However, due to a number of challenging issues that had taken too long to overcome [116, 117] CNOS-SOS-EU will probably only be implemented and operational for a future round of strategic noise mapping.

16.6.4 Soundscapes

For the last 30 years, the concept of soundscapes has been studied in a multi-disciplinary approach and nowadays is an important issue to consider in environmental acoustics. The term soundscape was coined by Schafer [118] as the study of the effects of the acoustical environment on the physical responses or behavioral characteristics of creatures living within it. More recently, in 2014 the ISO standardized the definitions of an acoustical environment as the sound at the receiver from all sound sources as modified by the environment. The same standard defines a soundscape as the acoustical environment as perceived or experienced and/or understood by a person or people [119].

Therefore, a soundscape is composed of the natural acoustical environment (animal vocalizations, weather, natural elements) and environmental sounds created by humans (music, ordinary human activities, and sounds originated from industry). Thus, a person may consider noise pollution to consist of any disruption of their acoustical environment.

It is well known that acoustical environments may have negative or positive effects on human health and wellbeing. Quiet areas, such as natural environments make a positive contribution to public health since they enhance physiological recovery from stress, induce positive emotions and reduce mental exhaustion. People relate to their acoustical environments on an emotional level by interpreting the sensory information they receive. Thus, soundscape characteristics define a particular environment as a desirable or undesirable place to occupy, as uniquely judged by each individual [120]. Environments having acoustical stressors (i.e. noise) tend to induce negative emotions, and motivate an avoidance response, while environments free from such stressors (i.e. parks, green spaces, and natural or wilderness areas) may induce positive emotions and motivate an acceptable response [120]. In addition, certain acoustical environments can be linked by people to particular events in their lives and eventually they like to keep these memories by preserving the associated sounds, even though these sounds can induce negative emotions in others.

Listening is one of the psychological functions through which people perceive the world. Evaluating the effects of sounds on people is primarily a subjective issue rather than one merely based on objective parameters [121]. To explore people's perceptions of a soundscape, the comprehension between the acoustical stimulus and opinions of inhabitants has been investigated qualitatively by means of the grounded theory approach [122].

Therefore, identification of all the factors related to the psychological needs of the urban soundscape is extremely important in creating a comfortable acoustical environment and to help in identifying the existing urban areas and exploring the developmental trend of soundscapes [123]. Many other studies on soundscapes in both urban and rural environments exist and further research on the subject continues [120–132].

References

1 Burgess, M.A. and Finegold, L.S. (2007). Environmental noise impact assessment. In: *Handbook of Noise and Vibration Control* (ed. M.J. Crocker), 1501–1508. New York: Wiley.

2 Bento Coelho, J.L. (2007). Community noise ordinances. In: *Handbook of Noise and Vibration Control* (ed. M.J. Crocker), 1525–1532. New York: Wiley.

3 Schultz, T.J. (1978). Synthesis of social surveys on noise annoyance. *J. Acoust. Soc. Am.* 64 (2): 377–405.

4 Finegold, L.S., Harris, S.C., and von Gierke, H.E. (1994). Community annoyance and sleep disturbance: updated criteria for assessing the impacts of general transportation noise on people. *Noise Control Eng. J.* 42 (1): 25–30.

5 Miedema, H.M.E. and Vos, H. (1998). Exposure-response relationships for transportation noise. *J. Acoust. Soc. Am.* 104 (6): 3432–3445.

6 Fidell, S. and Silvati, L. (2004). Parsimonious alternative to regression analysis for characterizing prevalence rates of aircraft noise annoyance. *Noise Control Eng. J.* 52 (2): 56–68.

7 ISO 1996-1 (2016) *Acoustics – Description, Measurement and Assessment of Environmental Noise – Part 1*: Basic Quantities and Assessment Procedures.Geneva: International Standards Organization.

8 ANSI S12.9 Part 4 (2005) American National Standard, Quantities and Procedures for Description and Measurement of Environmental Sound – Part 4: Noise Assessment and Prediction of Long-Term Community Response. Melville, NY: Acoustical Society of America (reviewed in 2015).

9 Schomer, P.D., Suzuki, Y., and Saito, F. (2001). Evaluation of loudness-level weightings for assessing the annoyance of environmental noise. *J. Acoust. Soc. Am.* 110 (5): 2390–2397.

10 Fields, J.M. (1998). Reactions to environmental noise in an ambient noise context in residential areas. *J. Acoust. Soc. Am.* 104 (4): 2245–2260.

11 Donovan, P.R. and Schumacher, R. (2007). Exterior noise of vehicles-traffic noise prediction and control. In: *Handbook of Noise and Vibration Control* (ed. M.J. Crocker), 1427–1437. New York: Wiley.

12 de Graff, D.F. (2001) A speed and acceleration limit in the noise type approval of vehicles will enable silent cars to reveal their silence. Proceedings of Inter-Noise 2001. The Hague, The Netherlands.

13 Sandberg, U. and Ejsmont, J.A. (2002). *Tyre/Road Noise Reference Book*. Kisa, Sweden: Informex.

14 Crocker, M.J. (2007). Introduction to transportation noise and vibration sources. In: *Handbook of Noise and Vibration Control* (ed. M.J. Crocker), 1013–1023. New York: Wiley.

15 Sandberg, U. and Ejsmont, J.A. (2007). Tire/road noise-generation, measurement, and abatement. In: *Handbook of Noise and Vibration Control* (ed. M.J. Crocker), 1054–1071. New York: Wiley.

16 Arenas, J.P. (2007). Use of barriers. In: *Handbook of Noise and Vibration Control* (ed. M.J. Crocker), 714–724. New York: Wiley.

17 Horoshenkov, K., Lam, Y.W., and Attenborough, K. (2007). Noise attenuation provided by road and rail barriers, earth berms, buildings, and vegetation. In: *Handbook of Noise and Vibration Control* (ed. M.J. Crocker), 1446–1457. New York: Wiley.

18 Ausejo, M. and Recuero, M. (2006) Study and proposal of the prediction equation for traffic noise during workable days in Buenos Aires Downtown. Proceedings of the Thirteenth International Congress on Sound and Vibration. Vienna, Austria.

19 Steele, C. (2001). A critical review of some traffic noise prediction models. *Appl. Acoust.* 62 (3): 271–287.

20 Watts, G.R., Chandler, -Wilde, S.N., and Morgan, P.A. (1999). The combined effects of porous asphalt surfacing and barriers on traffic noise. *Appl. Acoust.* 58 (3): 351–377.

21 Gündogdu, Ö., Gökdag, M., and Yüksel, F. (2005). A traffic noise prediction method based on vehicle composition using genetic algorithms. *Appl. Acoust.* 66 (7): 799–809.

22 De Coensel, B., De Muer, T., Yperman, I., and Botteldooren, D. (2005). The influence of traffic flow dynamics on urban soundscapes. *Appl. Acoust.* 66 (2): 175–194.

23 Thorsson, P.J., Ögren, M., and Kropp, W. (2004). Noise levels on the shielded side in cities using a flat city model. *Appl. Acoust.* 65 (4): 313–323.

24 Cho, D.S., Kim, J.H., Choi, T.M. et al. (2004). Highway traffic noise prediction using method fully compliant with ISO 9613: comparison with measurements. *Appl. Acoust.* 65 (9): 883–892.

25 Tansatcha, M., Pamanikabud, P., Brown, A.L., and Affum, J.K. (2005). Motorway noise modelling based on perpendicular propagation analysis of traffic noise. *Appl. Acoust.* 66 (10): 1135–1150.

26 Öhrström, E. (1997). Effects of exposure to railway noise – a comparison between areas with and without vibration. *J. Sound Vib.* 205 (4): 555–560.

27 Öhrström, E. and Skanberg, A.-B. (1996). A field survey on effects of exposure to noise and vibration from railway traffic, part I: annoyance and activity disturbance effects. *J. Sound Vib.* 193 (1): 39–47.

28 Van Leeuwen, H.J.A. (2000). Railway noise prediction models: a comparison. *J. Sound Vib.* 231 (3): 975–987.

29 Makarewicz, R. and Yoshida, M. (1996). Railroad noise in an open space. *Appl. Acoust.* 49 (4): 291–306.

30 Heng, C.C. (1997). Propagation of train noise in housing estates. *Appl. Acoust.* 51 (1): 1–12.

31 Xiaoan, G. (2006). Railway environmental noise control in China. *J. Sound Vib.* 293 (3–5): 1078–1085.

32 Griefahn, B., Marks, A., and Robens, S. (2006). Noise emitted from road, rail and air traffic and their effects on sleep. *J. Sound Vib.* 295: 129–140.

33 Yano, T., Yamashita, T., and Izumi, K. (1997). Comparison of community annoyance from railway noise evaluated by different category scales. *J. Sound Vib.* 205 (4): 505–511.

34 Gore, C. (1996). Railway noise: principles for an EU policy – the CER view. *J. Sound Vib.* 193 (1): 397–401.

35 Lim, C., Kim, J., Hong, J., and Lee, S. (2006). The relationship between railway noise and community annoyance in Korea. *J. Acoust. Soc. Am.* 120 (4): 2037–2042.

36 Schulte-Werning, B., Beier, M., Degen, K.G., and Stiebel, D. (2006). Research on noise and vibration reduction at DB to improve the environmental friendliness of railway traffic. *J. Sound Vib.* 293 (3–5): 1058–1069.

37 Oertli, J. (2006). Developing noise control strategies for entire railway networks. *J. Sound Vib.* 293 (3–5): 1086–1090.

38 Hemsworth, B. (2007). Rail system environmental noise prediction, assessment, and control. In: *Handbook of Noise and Vibration Control* (ed. M.J. Crocker), 1438–1445. New York: Wiley.

39 Hunt, H.E.M. and Hussein, M.F.M. (2007). Ground-borne vibration transmission from road and rail systems: prediction and control. In: *Handbook of Noise and Vibration Control* (ed. M.J. Crocker), 1458–1469. New York: Wiley.

40 Talbot, J.P. (2007). Base isolation of buildings for control of ground-borne vibration. In: *Handbook of Noise and Vibration Control* (ed. M.J. Crocker), 1470–1478. New York: Wiley.

41 Hunt, H.E.M. (1996). Modelling of rail vehicles and track for calculation of ground-vibration transmission into buildings. *J. Sound Vib.* 193 (1): 185–194.

42 Kirzhner, F., Rosenhouse, G., and Zimmels, Y. (2006). Attenuation of noise and vibration caused by underground trains, using soil replacement. *Tunnel. Underground Space Tech.* 21 (5): 561–567.

43 Nagy, A.B., Fiala, P., Marki, F. et al. (2006). Prediction of interior noise in buildings generated by underground rail traffic. *J. Sound Vib.* 293 (3–5): 680–690.

44 Miller, N.P., Reindel, E.M., and Horonjeff, R.D. (2007). Aircraft and airport noise prediction and control. In: *Handbook of Noise and Vibration Control* (ed. M.J. Crocker), 1479–1489. New York: Wiley.

45 Crocker, M.J. (2007). Introduction to community noise and vibration prediction and control. In: *Handbook of Noise and Vibration Control* (ed. M.J. Crocker), 1413–1426. New York: Wiley.

46 Wyle Laboratories (1971) Transportation Noise and Noise from Equipment Powered by Internal Combustion Engines. EPA Report No. NTID 300.13.

47 Miller, N.P., Reindel, E.M., Senzig, D.A., and Horonjeff, R.D. (1998) Low-frequency noise from aircraft start of takeoff. Proceedings of Inter-Noise 1998, Christchurch, NZ.

48 Raman, G. and McLaughlin, D.K. (1999). Highlights of aeroacoustics research in the U.S.–1998. *J. Sound Vib.* 228 (3): 589–610.

49 Clemente, J., Gaja, E., Clemente, G., and Reig, A. (2005). Sensitivity of the FAA integrated noise model to input parameters. *Appl. Acoust.* 66 (3): 263–276.

50 Hsu, C.-I. and Lin, P.-H. (2005). Performance assessment for airport noise charge policies and airline network adjustment response. *Transport. Res. Part D: Transport and Environ.* 10 (4): 281–304.

51 Vogt, J. (2005). The relative impact of aircraft noise and number in a full-factorial laboratory design. *J. Sound Vib.* 282 (3–5): 1085–1100.

52 Kruppa, P. (2000). Research on reduction of aircraft noise, supported by the European Union. *Air Space Eur.* 2 (3): 18–19.

53 Carlsson, F., Lampi, E., and Martinsson, P. (2004). The marginal values of noise disturbance from air traffic: does the time of the day matter? *Transport. Res. Part D: Transport Environ.* 9 (5): 373–385.

54 McMillen, D.P. (2004). Airport expansions and property values: the case of Chicago O'Hare airport. *J. Urban Econ.* 55 (3): 627–640.

55 Ollerhead, J. and Sharp, B. (2001). MAGENTA – assessments of future aircraft noise policy options. *Air Space Eur.* 3 (3–4): 247–249.

56 Zaporozhets, O.I. and Tokarev, V.I. (1998). Predicted flight procedures for minimum noise impact. *Appl. Acoust.* 55 (2): 129–143.

57 Rylander, R. and Björkman, M. (1997). Annoyance by aircraft noise around small airports. *J. Sound Vib.* 205 (4): 533–537.

58 Aasvang, G.M. and Engdahl, B. (2004). Subjective responses to aircraft noise in an outdoor recreational setting: a combined field and laboratory study. *J. Sound Vib.* 276 (3–5): 981–996.

59 Drozdova, L., Ivanov, N., and Kurtsev, G.H. (2007). Off-road vehicle and construction equipment exterior noise prediction and control. In: *Handbook of Noise and Vibration Control* (ed. M.J. Crocker), 1490–1500. New York: Wiley.

60 Crocker, M.J. (2007). Introduction to principles of noise and vibration control. In: *Handbook of Noise and Vibration Control* (ed. M.J. Crocker), 649–667. New York: Wiley.

61 Crocker, M.J. (2007). Noise of compressors. In: *Handbook of Noise and Vibration Control* (ed. M.J. Crocker), 910–934. New York: Wiley.

62 Kennes, P., Anthonis, J., Clijmans, L., and Ramon, H. (1999). Construction of a portable test rig to perform experimental modal analysis on mobile agricultural machinery. *J. Sound Vib.* 228 (2): 421–441.

63 Hostens, I. and Ramon, H. (2003). Descriptive analysis of combine cabin vibrations and their effect on the human body. *J. Sound Vib.* 266 (3): 453–464.

64 Deprez, K., Moshou, D., and Ramon, H. (2005). Comfort improvement of a nonlinear suspension using global optimization and *in situ* measurements. *J. Sound Vib.* 284 (3–5): 1003–1014.

65 Osada, Y., Yoshida, T., Yoshida, K. et al. (1997). Path analysis of the community response to road traffic noise. *J. Sound Vib.* 205 (4): 493–498.

66 Gottlob, D. (1995). Regulations for community noise. *Noise/News Int.* 3: 223–236.

67 von Gierke, H.E. and Eldred, K.M.K. (1993). Effects of noise on people. *Noise/News Int.* 1 (2): 67–89.

68 Shaw, E.A.G. (1996). Noise environments outdoors and the effects of community noise exposure. *Noise Control Eng. J.* 44 (3): 109–120.

69 Kuehner, D. (2007). Industrial and commercial noise in the community. In: *Handbook of Noise and Vibration Control* (ed. M.J. Crocker), 1509–1515. New York: Wiley.

70 State of NSW and the Environment Protection Authority (2013). *Noise Guide for Local Government.* Sydney: Environment Protection Authority.

71 European Union (1996) Future Noise Policy. European Commission Green Paper. Brussels: European Union.

72 Schomer, P.D. (2000). Loudness-level weighting for environmental noise assessment. *Acta Acust* 86 (1): 49–61.

73 Suzuki, Y., Kono, S., and Sone, T. (1988). An experimental consideration of the evaluation of environmental noise with tonal components. *J. Sound Vib.* 127 (3): 475–484.

74 Sato, T., Yano, T., Morihara, T., and Masden, K. (2004). Relationships between rating scales, question stem wording, and community responses to railway noise. *J. Sound Vib.* 277 (3): 609–616.

75 Hume, K., Gregg, M., Thomas, C., and Terranova, D. (2003). Complaints caused by aircraft operations: an assessment of annoyance by noise level and time of day. *J. Air Transp. Manag.* 9 (3): 153–160.

76 King, R.P. and Davis, J.R. (2003). Community noise: health effects and management. *Int. J. Hyg. Environ. Health* 206 (2): 123–131.

77 García, A., Faus, L.J., and García, A.M. (1993). The community response to aircraft noise around six Spanish airports. *J. Sound Vib.* 164 (1): 45–52.

78 Knoepfel, I.H. (1996). A framework for environmental impact assessment of long-distance energy transport systems. *Energy* 21 (7–8): 693–702.

79 Omiya, M., Kuno, K., Mishina, Y. et al. (1997). Comparison of community noise ratings by $L50$ and Laeq. *J. Sound Vib.* 205 (4): 545–554.

80 Visser, H.G. (2005). Generic and site-specific criteria in the optimization of noise abatement trajectories. *Transport. Res. Part D: Transport Environ.* 10 (5): 405–419.

81 Trautmann, U. (2007). Building site noise. In: *Handbook of Noise and Vibration Control* (ed. M.J. Crocker), 1516–1524. New York: Wiley.

82 Arana, M. and García, A. (1998). A social survey on the effects of environmental noise on the residents of Pamplona, Spain. *Appl. Acoust.* 53 (4): 245–253.

83 Kurra, S., Morimoto, M., and Maekawa, Z.I. (1999). Transportation noise annoyance – a simulated-environment study for road, railway and aircraft noises, part 1: overall annoyance. *J. Sound Vib.* 220 (2): 251–278.

84 Kurra, S., Morimoto, M., and Maekawa, Z.I. (1999). Transportation noise annoyance – a simulated-environment study for road, railway and aircraft noises, part 2: activity disturbance and combined results. *J. Sound Vib.* 220 (2): 279–295.

85 Fields, J.M., De Jong, R., Brown, A.L. et al. (1997). Guidelines for reporting Core information from community noise reaction surveys. *J. Sound Vib.* 206 (5): 685–695.

86 Minoura, K. and Hiramatsu, K. (1997). On the significance of an intensive survey in relation to community response to noise. *J. Sound Vib.* 205 (4): 461–465.

87 Arenas, J.P. (2008). Potential problems with environmental sound barriers when used in mitigating surface transportation noise. *Sci. Total Environ.* 405: 173–179.

88 Wathern, P. (2004). *Environmental Impact Assessment: Theory and Practice.* London: Routledge.

89 Horstmann, K., Klennert, K., and Phantumvanit, D. (1985) Environmental Impact Assessment (EIA) for development. Proceedings of a Joint DSE/UNEP International Seminar, DSE/ZEL, 9–12 April 1984. Feldafing, Germany.

90 Godinho, L., Antonio, J., and Tadeu, A. (2002). The scattering of 3D sound sources by rigid barriers in the vicinity of tall buildings. *Eng. Anal. Boundary Elements* 26 (9): 781–787.

91 Arenas, J.P. (2007) Sound attenuation provided by earth mounds used for reducing traffic noise from highways. Proceedings Inter-Noise. Istanbul, Turkey.

92 Watts, G., Chinn, L., and Godfrey, N. (1999). The effects of vegetation on the perception of traffic noise. *Appl. Acoust.* 56: 39–56.

93 Ungar, E.E. (2007). Use of vibration isolation. In: *Handbook of Noise and Vibration Control* (ed. M.J. Crocker), 725–733. New York: Wiley.

94 Loh, C.-H. and Chao, C.-H. (1996). Effectiveness of active tuned mass damper and seismic isolation on vibration control of multi-storey building. *J. Sound Vib.* 193 (4): 773–792.

95 Kareem, A. (1997). Modelling of base-isolated buildings with passive dampers under winds. *J. Wind Eng. Ind. Aerodyn.* 72: 323–333.

96 Vulcano, A. (1998). Comparative study of the earthquake and wind dynamic responses of base-isolated buildings. *J. Wind Eng. Ind. Aerodyn.* 74–76: 751–764.

97 Matsagar, V.A. and Jangid, R.S. (2003). Seismic response of base-isolated structures during impact with adjacent structures. *Eng. Struct.* 25 (10): 1311–1323.

98 Calió, I., Marletta, M., and Vinciprova, F. (2003). Seismic response of multi-storey buildings base-isolated by friction devices with restoring properties. *Comput. Struct.* 81 (28–29): 2589–2599.

99 Baratta, A. and Corbi, I. (2004). Optimal Design of Base-Isolators in multi-storey buildings. *Comput. Struct.* 82 (23–26): 2199–2209.

100 Matsagar, V.A. and Jangid, R.S. (2004). Influence of isolator characteristics on the response of base-isolated structures. *Eng. Struct.* 26 (12): 1735–1749.

101 Arenas, J.P. (2005). Acoustics is making waves in South America. *Int. J. Acoust. Vib.* 10 (1): 2.

102 Hanson, D.I., James, R.S., and NeSmith, C. (2004) Tire/Pavement Noise Study. Report NCAT 04–02. Auburn University, AL: National Center for Asphalt Technology.

103 National Academy of Engineering of the National Academy of Sciences (2010). *Technology for a Quieter America.* Washington, DC: Committee on Technology for a Quieter America, The National Academy Press.

104 Hamet, J.-F., Deffayet, C., and Palla, M.-A. (1990) Air pumping phenomena in road cavities. Proceedings of Int. Tyre/Road Noise Conf., Gothenburg, STU information no. 794–1990, NUTEK, Stockholm.

105 von Meier, A. (1988) Acoustically porous road surfaces, recent experiences and new developments. Proceedings Inter-Noise 88, Avignon, France.

106 von Meier, A., van Blockland, G.J., and Heerkens, J.C.P. (1990) Noise of optimized road surfaces and further improvements by tyre choice. Proceedings INTROC 90, Gothenburg, Sweden.

107 Ayuntamiento de Madrid (1991). *The noise in the city: Management and control (in Spanish).* Madrid: Spanish Acoustical Society.

108 Vogiatzis, K. and Remy, N. (2014). From environmental noise abatement to soundscape creation through strategic noise mapping in medium urban agglomerations in South Europe. *Sci. Total Environ.* 482–483: 420–431.

109 Vogiatzis, K. (2012). An assessment of airport environmental noise action plans with some financial aspects: the case of Athens international "Eleftherios Venizelos". *Int. J. Acoust. Vib.* 17: 181–190.

110 Vogiatzis, K. (2014). Assessment of environmental noise due to aircraft operation at the Athens international airport according to the 2002/49/EC directive and the new Greek national legislation. *Appl. Acoust.* 84: 37–46.

111 Directive 2002/49/EC (2002) Directive of The European Parliament and of the Council of 25 June 2002 Relating to the Assessment and Management of Environmental Noise; 2002/49/EC. European Commission.

112 Assessment of exposure to noise (2007) Good practice guide for strategic noise mapping and the production of associated data on noise exposure, version 2. European Commission Working Group, August.

113 Suárez, E. and Barros, J.L. (2014). Traffic noise mapping of the city of Santiago de Chile. *Sci. Total Environ.* 466–467: 539–546.

114 Rey Gozalo, G., Barrigón Morillas, J.M., and Gómez Escobar, V. (2013). Urban streets functionality as a tool for urban pollution management. *Sci. Total Environ.* 461–462: 453–461.

115 Bastián-Monarca, N.A., Suárez, E., and Arenas, J.P. (2016). Assessment of methods for simplified traffic noise mapping of small cities: casework of the city of Valdivia, Chile. *Sci. Total Environ.* 550: 439–448.

116 Kephalopoulos, S., Paviotti, M., Anfosso-Lèdèe, F. et al. (2014). Advances in the development of common noise assessment methods in Europe: the CNOSSOS-EU framework for strategic environmental noise mapping. *Sci. Total Environ.* 482–483: 400–410.

117 Bento Coelho, J.L., Vogiatzis, K., and Licitra, G. (2011) The CNOSSOS-EU initiative: a framework for road, railway, aircraft and industrial noise modeling for strategic noise mapping in EU member states. Proceedings of 18th International Congress on Sound and Vibration, Rio de Janeiro.

118 Schafer, R.M. (1977). *The Tuning of the World*. New York: A. A. Knopf.

119 ISO 12913-1 (2014) *Acoustics – Soundscape – Part 1: Definition and conceptual framework*. Geneva: International Standards Organization.

120 Medvedev, O., Shepherd, D., and Hautus, M.J. (2015). The restorative potential of soundscapes: a physiological investigation. *Appl. Acoust.* 96 (9): 20–26.

121 Kang, J. (2007). *Urban Sound Environment*. London: Taylor & Francis.

122 Fiebig, A. and Schulte-Fortkamp, B. (2004). Soundscapes and their influence on inhabitants – new findings with the help of a grounded theory approach. *J. Acoust. Soc. Am.* 115 (5): 2496.

123 Liu, F. and Kang, J. (2016). A grounded theory approach to the subjective understanding of urban soundscape in Sheffield. *Cities* 50 (2): 28–39.

124 Schulte-Fortkamp, B. and Dubois, D. (eds.) (2006). Recent advances in soundscape research. *Acta Acust. United Acust.* 92 (6): 857–1076.

125 Brown, A.L., Kang, J., and Gjestland, T. (2011). Towards standardization in soundscape preference assessment. *Appl. Acoust.* 72 (6): 387–392.

126 Davies, W.J., Adams, M.D., Bruce, N.S. et al. (2013). Perception of soundscapes: an interdisciplinary approach. *Appl. Acoust.* 74 (2): 224–231.

127 Bruce, N.S. and Davies, W.J. (2014). The effects of expectation on the perception of soundscapes. *Appl. Acoust.* 85 (11): 1–11.

128 Hong, J.Y. and Jeon, J.Y. (2015). Influence of urban contexts on soundscape perceptions: a structural equation modeling approach. *Landsc. Urban Plan.* 141: 78–87.

129 Watts, G., Khan, A., and Pheasant, R. (2016). Influence of soundscape and interior design on anxiety and perceived tranquility of patients in a healthcare setting. *Appl. Acoust.* 104 (3): 135–141.

130 Kogan, P., Turra, B., Arenas, J.P., and Hinalaf, M. (2017). A comprehensive methodology for the multidimensional and synchronic data collecting in soundscape. *Sc. Total Environ.* 580: 1068–1077.

131 ISO/TS 12913-2 (2018) *Acoustics – Soundscape – Part 2: Data collection and reporting requirements*. Geneva: International Standards Organization.

132 ISO/TS 12913-3 (2019) *Acoustics – Soundscape – Part 3: Data analysis*. Geneva: International Standards Organization.

Glossary

References

1) ISO Standards

ISO 11201: *Acoustics – Noise Emitted by Machinery and Equipment – Measurement of Emission Sound Pressure Levels at a Work Station and at Other Specified Positions Engineering Method in an Essentially Free Field over a Reflecting Plane*

ISO 11688: *Acoustics – Recommended Practice for the Design of Low-Noise Machinery and Equipment*

ISO 140: *Acoustics – Measurement of Sound Insulation in Buildings and of Building Elements*

ISO 1996: *Acoustics – Description, Measurement and Assessment of Environmental Noise*

ISO 1999: *Acoustics – Determination of Occupational Noise Exposure and Estimation of Noise-Induced Hearing Impairment*

ISO 226: *Acoustics – Preferred Frequencies*

ISO 354: *Acoustics – Measurement of Sound Absorption in a Reverberation Room*

ISO 3741: *Acoustics – Determination of Sound Power Levels of Noise Sources Using Sound Pressure – Precision Methods for Reverberation Rooms*

ISO 3744: *Acoustics – Determination of Sound Power Levels of Noise Sources Using Sound Pressure – Engineering Method in an Essentially Free Field over a Reflecting Plane*

ISO 3745: *Acoustics – Determination of Sound Power Levels of Noise Sources Using Sound Pressure – Precision Methods for Anechoic and Hemi-anechoic Rooms*

ISO 3751 : 1988

ISO 389: *Acoustics – Reference Zero for the Calibration of Audiometric Equipment*

ISO 4869: *Acoustics – Hearing Protectors*

ISO 6081-1986 (replaced by ISO 11201-1995)

ISO 7235: *Acoustics – Laboratory Measurement Procedures for Ducted Silencers and Air-Terminal Units – Insertion Loss*, Flow Noise and Total Pressure Loss

ISO 7849: *Acoustics – Estimation of Airborne Noise Emitted by Machinery Using Vibration Measurement*

ISO 80000-8: *Quantities and Units – Part 8: Acoustics*

ISO 9053: *Acoustics – Materials for Acoustical Applications – Determination of Airflow Resistance*

ISO 9614: *Acoustics – Determination of Sound Power Levels of Noise Sources Using Sound Intensity*

2) IEC Standards

IEC 61043: *Electroacoustics – Instruments for the Measurement of Sound Intensity – Measurements with Pairs of Pressure Sensing Microphones*

IEC 61260: *Electroacoustics – Octave-Band and Fractional-Octave-Band Filters*

IEC 61672-1: *Electroacoustics – Sound Level Meters – Part 1: Specifications*

Engineering Acoustics: Noise and Vibration Control, First Edition. Malcolm J. Crocker and Jorge P. Arenas.
© 2021 John Wiley & Sons Ltd. Published 2021 by John Wiley & Sons Ltd.

3) ANSI Standards

   ANSI S1.1-1994: *Acoustical Terminology*

   ANSI S1.11: *Specification for Octave-Band and Fractional-Octave-Band Analog and Digital Filters*

   ANSI S1.42: *Design Response of Weighting Networks for Acoustical Measurements*

   ANSI S12.12: *Engineering Method for the Determination of Sound Power Levels of Noise Sources using Sound Intensity*

4) ASTM Standards

   ASTM E336–97 *Standard Test Method for Measurement of Airborne Sound Insulation in Buildings*

   ASTM C384–04 *Standard Test Method for Impedance and Absorption of Acoustical Materials by Impedance Tube Method*

   ASTM C423–02 *Standard Test Method for Sound Absorption and Sound Absorption Coefficients by the Reverberation Room Method*

   ASTM C634–02: *Standard Terminology Relating to Environmental Acoustics*

   ASTM E413–87: *Classification for Rating Sound Insulation*

   ASTM E492–90: *Standard Test Method for Laboratory Measurement of Impact Sound Transmission through Floor-Ceiling Assemblies Using the Tapping Machine*

   ASTM E596–96 *Standard Test Method for Laboratory Measurement of Noise Reduction of Sound-Isolating Enclosures*

   ASTM E989–89: *Standard Classification for Determination of Impact Insulation Class (IIC)*

5) Morfey, C.L. (2001). *Dictionary of Acoustics.* London: Academic.

6) U.S Department of Labor, Occupational Safety and Health Administration, 1910.95 Appendix I – Definitions (http://www.osha.gov/pls/oshaweb/owadisp.show_document&p_table=STANDARDSp_id=9744?).

7) U.S Department of Health and Human Services, 98–126, Occupational Noise Exposure (http://www.cdc.gov/niosh/pdfs/98-126.pdf).

8) Lapedes, D.N., Editor-in-Chief, (1978). *McGraw-Hill Dictionary of Physics and Mathematics.* New York: McGraw-Hill.

9) Crystal, D. (1991). *A Dictionary of Linguistics and Phonetics*, 3e. Oxford, UK: Blackwell.

10) Titze, I.R. (2000). *Principles of Voice Production*, 2nd printing. Denver, CO: National Center for Voice and Speech.

11) O'Shaughnessy, D. (2000). *Speech Communications: Human and Machine*, 2e. New York: IEEE Press.

12) Vibration Institute, *Vibration Terminology* (https://www.vi-institute.org/vibration-terminology-project/).

13) Brüel & Kjær's definitions. Briüel & Kjær, Naerum, Denmark.

Glossary

A-weighted sound exposure	*See* Sound exposure.
Absorption	Conversion of sound energy into another form of energy, usually heat, when passing through an acoustical medium or at a boundary.
Absorption area	*See* Equivalent sound absorption area.
Absorption factor, absorption coefficient, absorbance	Ratio of the dissipated and transmitted sound power to the incident sound power. In a specified frequency band, the absorption factor is a measure of the absorptive property of a material or surface. Also known as *(sound) absorption coefficient.*

Unit: none. Symbol: α. *See* Dissipation factor, Transmission factor, Reflection factor. *Note*: (1) The absorption factor equals the dissipation factor plus the transmission factor, $\alpha = \delta + \tau$, where δ is the dissipation factor, and τ is the transmission factor. (2) The summation of the dissipation, transmission, and reflection factors is unity, $\delta + \tau + r = 1$, where r is the reflection factor.

Accelerance	A form of the frequency response function, which is the complex ratio of the acceleration to the excitation force. Units: (metre per square second) per newton [$(\text{m/s}^2)/\text{N} = \text{kg}^{-1}$]. *See* Compliance, Mobility, Dynamic Mass. *Note:* The alternative term *inertance* is not recommended because it has a conflicting interpretation. Sometimes inertance is referred to as the reciprocal of accelerance.
Acceleration	A vector quantity equal to the rate of change of velocity. Units: metre per square second (m/s^2). *See* Displacement, Velocity, Jerk.
Accelerometer	A transducer whose electrical output is proportional to acceleration in a specified frequency range.
Acoustical holography	An inspection method using the phase interference between sound waves from an object and a reference signal to obtain an image of reflectors in the object under test.
Acoustical trauma	Ear injury caused by a sudden and intense sound pressure that causes a degree of permanent or temporary hearing loss.
Acoustic impedance	At a specified surface, the complex ratio of the average sound pressure over that surface to the sound volume flow rate through that surface. Units: pascal per (cubic metre per unit time) [$\text{Pa}/(\text{m}^3/\text{s})$]. Symbol: Z_a. *See* Characteristic impedance, Mechanical impedance, Specific acoustic impedance. **Acoustic reactance** Imaginary part of acoustic impedance. Symbol: R_a.
Acoustic resistance	Real part of acoustic impedance. Symbol: X_a.
Acoustic reactance	*See* Acoustic impedance.
Acoustic resistance	*See* Acoustic impedance.
Acoustics	Science of the production, control, transmission, reception, and effects of sound and of the phenomenon of hearing.
Action level	The cumulative work shift noise dose at which hearing conservation program is mandated by the Occupational Safety and Health Administration (OSHA). An 8-hour time-weighted average of 85 dB measured with A-weighting and slow response, or equivalently, a dose of 50%. *See* Hearing conservation program, Noise dose.
Active control	Of sound and vibration, the use of secondary sources of excitation to cancel, or reduce, the response of a system to given prime sources; also to suppress self-excitation oscillations of a system that is unstable.

Active sound control   *See* Active control.

Active sound field   A sound field in which the particle velocity is in phase with the sound pressure. All acoustic energy is transmitted; none is stored. A plane wave propagating in free field is an example of a purely active sound field.

Active vibration control   *See* Active control.

Admittance   The reciprocal of an impedance. *See* Acoustic impedance, Characteristic impedance, Mechanical impedance, Mobility.

Airborne sound   Sound that propagates though air. *See* Liquid-borne sound, Structure-borne sound.

Airflow resistance   A quantity defined by $R = \Delta p/q_V$, where Δp is the air pressure difference, in pascals, across the test specimen; q_V is the volumetric airflow rate, in cubic metres per second, passing through the test specimen. Units: pascal second per cubic metre $(\mathrm{Pa\cdot s/m^3})$. Symbol: R. *See* Specific airflow resistance.

Airflow resistivity   If the material is considered as being homogenous, a quantity defined by $r = R_s/d$, where R_s is the specific airflow resistance, in pascal seconds per metre, of the test specimen; d is the thickness, in metres, of the test specimen in the direction of flow. Units: pascal second per square metre $(\mathrm{Pa\cdot s/m^2})$. Symbol: r. *See* Specific airflow resistance.

Aliasing error   An error in digital signal processing in which high-frequency signal energy is folded into lower frequency components due to inadequate sampling frequency that is less than twice the maximum frequency in the signal. *See* Sampling theorem, Antialiasing filter.

Alignment   The fact of being in line or bringing into line. For rotating machinery, alignment refers to the axial continuity of machine components, such as shaft and bearings, during operation.

Ambient noise   (1) Sound present in a given situation at a given time, in absence of that from sources of immediate concern, usually being composed of sound from many sources near and far. *See* Background noise. (2) In underwater acoustics, the naturally occurring acoustical environment in the ocean, caused by wave breaking, marine life, and other sources. (3) In sonar detection, the noise from all unwanted sources of sound, apart from those directly associated with the sonar equipment and the platform on which it is mounted.

Amplitude distribution   A method of representing the amplitude (or level) variation of noise by indicating the percentage of time that the noise level occurs in a series of amplitude intervals.

Analog-to-digital conversion (ADC)   The process of sampling an analog signal to convert it to a series of quantized numbers, a series that is a digital representation of the same signal. The sampling frequency must be at least twice as high as the highest frequency present in the signal to prevent aliasing errors. *See* Aliasing error, Nyquist frequency, Sampling theorem.

Anechoic room
A test room whose surfaces absorb essentially all of the incident sound energy over the frequency range of interest, thereby approximating free-field conditions over a measurement surface. *See* Semianechoic room, Free field.

Angular frequency
Frequency of a periodic quantity in units of radians per unit time (rad/s). Symbol: ω. *See* Frequency.

Angular repetency
See Wavenumber.

Annoyance
A person's internal response to a noise. Is quantifiable (1) psychologically by rating or (2) technically by a physical noise descriptor, for example, the equivalent continuous A-weighted sound pressure level $L_{Aeq,T}$. For a given person, usually the correlation coefficient between descriptor and related ratings does not exceed 0.5 due to the influence of other factors in determining annoyance. *See* Noise, Rating, Equivalent continuous A-weighted sound pressure.

Antialiasing filter
An analog low-pass filter applied to continuous analog signals prior to analog-to-digital conversion, to eliminate frequency components higher than one-half the sampling frequency (Nyquist frequency). *See* Aliasing error, Sampling theorem, Nyquist frequency.

Antiresonance
A condition in forced vibration whereby a specific point has zero amplitude at a specific frequency of vibration. For a certain frequency response function (FRF), an antiresonance may correspond to a local minimum in the magnitude. For a point FRF, there must be an antiresonance between two resonances. However, different FRFs have different antiresonances. Some FRFs have no antiresonance. *See* Resonance.

Apparent sound reduction index
See Sound reduction index.

Argand diagram
A plot of a complex function in which the x and y coordinates denote the real and imaginary parts of the function. *See* Frequency response function, Nyquist diagram.

Articulation index (AI)
A number that specifies the output intelligibility of a speech transmission channel, based on the signal-to-noise ratio at the listener. The articulation index takes values between zero (completely unintelligible) and unity (completely intelligible). *See* Intelligibility, Signal-to-noise ratio, Privacy index.

Articulator
Any of the moveable parts in and around the vocal tract that can affect its shape, and therefore, the speech sounds generated. The tongue is the most important articulator; lips, teeth, and velum are others. *See* Vocal tract, Articulatory.

Articulatory
An articulatory model or synthesizer uses parameters related to properties of the articulators: for example, position of tongue tip or tongue dorsum, degree of lip rounding, or length of the vocal tract. *See* Articulator.

Attenuation coefficient Coefficient α in the spatial attenuation factor $e^{-\alpha x}$, which describes the amplitude reduction of a single-frequency progressive wave with distance x, in the propagation direction. Unit: reciprocal metre (m^{-1}). Symbol: α. *Note:* the quantity $1/\alpha$ is called the *attenuation length*, and $m = 2\alpha$ is called the *power attenuation coefficient*.

Audibility threshold The lowest sound pressure level, at a specified frequency, at which persons with normal hearing can hear. Sometimes also known as *hearing threshold*. *See* Hearing threshold.

Audio frequency Frequency of a sound in the range normally audible to humans. Unit: hertz (Hz). Audio frequencies range roughly from 20 Hz to 20 kHz.

Autocorrelation function A statistical measure of the extent to which one part of a signal is related to another part, offset by a given time, of the same signal. The Fourier transform of the auto-correlation function gives the autospectrum. *See* Cross-correlation function, Power spectrum.

Autospectrum Also known as *autospectral function, power spectrum*. *See* Power spectrum, Cross spectrum.

A-weighted sound power level *See* Sound power level.

A-weighted sound pressure level *See* Sound pressure level.

Background noise (1) Total of all interference from unwanted sound, independent of the sound source being studied. *See* Ambient noise. (2) Undesired signals that are always present in electrical or other systems, independent of whether or not the desired signal is present. Background noise may include contributions from airborne sound, structure-borne vibration, and electrical noise in instrumentation. *Note* According to ANSI S1.1–1994, ambient noise detected, measured, and recorded with the signal is part of the background noise.

Balancing The procedure of adjusting the mass distribution of a rotor to reduce the transverse vibration of the rotor.

Band pressure level *See* Sound pressure level.

Bandwidth (1) The difference between the upper and lower frequency limits of a frequency band. (2) Of a signal, the range of frequencies between the upper and lower frequency limits within which most of the signal energy is contained. (3) Of a linear system or transducer, the range of frequencies over which the system is designed to operate. (4) Of a bandpass filter, the difference in frequency between the upper and lower cutoff frequencies. *See* Critical bandwidth, Effective noise bandwidth, Half-power bandwidth.

Bearing

A device that supports, guides, and reduces the friction of motion between fixed and moving machine parts. There are primarily two types, rolling-element bearings and fluid-film bearings. A rolling-element bearing consists of an inner race, an outer race, balls or rollers, and a cage to maintain the proper separation of the rolling elements. A fluid-film bearing supports the shaft on a thin film of oil or on a thin film of pressurized air.

Bearing misalignment

Misalignment that results when the bearings supporting a shaft are not aligned with each other. The bearings may not be mounted in parallel planes, cocked relative to the shaft, or distorted due to foundation settling or thermal growth.

Beat

A phenomenon of slow amplitude modulation produced by interference between two simple harmonic signals, when the frequency difference is a small fraction of either frequency.

Beat frequency

The frequency of the amplitude modulation, equal to the difference in frequency between the two original signals.

Bel (B)

Unit of level in terms of logarithm to base 10, abbreviated as B. *See* Level, Decibel.

Bode diagram, Bode plot

(1) A form of presentation of a complex function, such as a frequency response function, in two plots. One plot shows the magnitude of the complex function against frequency, plotted in decibels. The other plot shows the phase of the complex function against frequency. Both plots use the same logarithmic frequency scale on the horizontal axis. *See* Frequency response function, Argand diagram, Nyquist diagram. (2) Specifically for rotating machines, a pair of two plots displaying the magnitude and phase of a particular order component of a vibration vector. Both plots are in Cartesian format and use the same horizontal axis, which represents rotational speed or perturbation frequency. *See* Order analysis.

Boundary element method

A numerical approach for solving a boundary integral equation in which the domain of integral is divided into subdomains referred to as elements. An advantage of the boundary element method is the reduction of dimensionality. The wave equation in a three-dimensional region can be rewritten as boundary integral equation on a two-dimensional surface. Unlike the finite element method, the boundary element method is based on the Green's function, which satisfies the governing differential equation exactly over the domain and solves the unknown nodal values that approximately satisfy the boundary conditions. *See* Finite element method.

Bulk modulus

Of a fluid or solid, the slope of the curve of pressure plotted against the logarithm of the density, with either adiabatic or isothermal conditions imposed.

Center frequency

The arithmetic center of a constant bandwith filter or the geometric center (midpoint on a logarithmic scale) of a constant percentage filter. *See* Constant bandwidth filter, Constant percentage filter.

Cepstrum   For a real signal, a nonlinear signal processing technique that involves the inverse Fourier transform of the logarithm of the power spectrum or the Fourier transform of the signal. Since both the Fourier transform and the inverse Fourier transform are complex processes, the cepstrum is complex if the phase information is preserved. The complex cepstrum has the corresponding inverse complex cepstrum. However, if the input of the inverse Fourier transform is real, for example, the power spectrum, or the magnitude of the Fourier transform of the signal, the cepstrum is real. Exact definitions vary across the literature.

Characteristic equation   (1) Any equation that has a solution, subject to specified boundary conditions, only when a parameter occurring in it has certain values. (2) Specifically, the equation $A\mathbf{u} = \lambda\mathbf{u}$, which can have a solution only when the parameter λ has certain values, where A can be a square matrix that multiplies the vector $\mathbf{u}$, or a linear differential or integral operator that operates on the function $\mathbf{u}$, or in general, any linear operator operating on the vector $\mathbf{u}$ in a finite or infinite dimensional vector space. Also known as *eigenvalue equation*. (3) An equation that sets the characteristic polynomial of a given linear transformation on a finite dimensional vector space, or of its matrix representation, equal to zero. *See* Eigenvalue, Eigenvector.

Characteristic impedance   Of a medium, at a point and for a plane progressive wave, the magnitude of the ratio of the sound pressure to the component of the sound particle velocity, in the direction of the wave propagation. Characteristic impedance is defined by $Z_c = \rho c$, where ρ is the density, in kilograms per cubic metre; c is the (phase) speed of sound, in metres per second. Units: pascal per metre per second [Pa/(m/s)]. Symbol: Z_c. *See* Acoustic impedance, Specific acoustic impedance, Mechanical impedance.

Charge amplifier   An amplifier with a low input impedance whose output voltage is proportional to the output charge from a piezoelectric transducer.

Close-proximity (CPX) method   A measurement method used for measuring noise properties of tires and/or road surfaces (pavements), utilizing either a trailer with a test tire or a regular vehicle such as an automobile, SUV, or van on which one of its tires constitutes the test tire. Generally, at least two microphones are placed close to the test tire. In case of a trailer, the test tire is often covered by an enclosure that protects the tire and the microphones from noise from other vehicles and from air turbulence noise. Attempts to standardize this procedure and to specify a couple of reference tires for the CPX method are made within the ISO as part of the ISO 11819 series of standards. *See* Tire/road noise, Statistical pass-by method.

Coherence function   Of two time signals, a measure of the degree to which the two signals are linearly related at any given frequency. The coherence function takes values between zero and unity.

Coincidence effect   For a panel or partition, a great reduction of the sound reduction index as a function of frequency when the trace wave speed (speed at which the incident wave propagates along the panel) and the speed of the flexural wave in the panel are the same. *See* Sound reduction index.

Compliance	(1) The reciprocal of stiffness. Sometimes also known as *flexibility*. (2) A form of the frequency response function: the complex ratio of the displacement to the excitation force. Also known as *dynamic compliance*, or receptance. *See* Accelerance, Mobility.
Condensation	In acoustics, the increase in the density of a material under stress, divided by the original density. *See* Density.
Conductance	Real part of an admittance. *See* Admittance, Susceptance.
Conductive hearing loss	*See* Hearing loss.
Constant bandwidth filter	A filter that has a fixed-frequency bandwidth, regardless of center frequency.
Constant percentage filter	A filter whose bandwidth is a fixed percentage of center frequency, for example, octave filter.
Correlation coefficient	(1) Of two real signals, $x(t)$ and $y(t)$, that are joint stationary with respect to time, the value of the covariance function at zero time delay $C_{xy}(0)$, normalized by the standard deviations of the two signals, σ_x and σ_y:

$$\rho_{xy} = \frac{C_{xy}(0)}{\sigma_x \sigma_y}.$$

	See Cross-correlation coefficient. (2) Of two random variables X and Y, the normalized covariance, defined by

$$\rho_{XY} = \frac{\text{cov}(X, Y)}{\sigma_X \sigma_Y}.$$

Critical bandwidth	The widest frequency band within which the perceived loudness of a band of continuously distributed random noise of constant bandwidth sound pressure level is independent of its bandwidth.
Critical frequency, critical coincidence frequency	Of a panel or partition, the lowest frequency at which the coincidence effect occurs. If the panel is orthotropic or anisotropic, the critical frequency may vary with direction along the panel. Unit: hertz (Hz). Symbol: f_c. *See* Coincidence effect.
Critical speed	Rotational speed of a rotating system that corresponds to a resonance frequency of that same system.
Cross-correlation coefficient	Of two real signals that are joint stationary with respect to time, the normalized covariance function, defined by

$$\rho_{xy} = \frac{C_{xy}(\tau)}{\sigma_x \sigma_y}$$

where $C_{xy}(\tau)$ is the covariance function of the two signals $x(t)$ and $y(t)$, evaluated at time delay τ, and σ_x and σ_y are the respective standard deviations. *See* Correlation coefficient.

Cross-correlation function	The statistical correlation between two different signals as a function of relative time between the signals. The Fourier transform of the cross-correlation function gives the cross spectrum. *See* Autocorrelation function, Cross spectrum.
Cross spectrum, cross-power spectrum	A complex spectrum containing the product of the power spectra of two correlated signals. The Fourier transform of the cross-correlation function. If the two time signals are $x(t)$ and $y(t)$, then the cross spectrum is $G_{xy}(j\omega) = X^*(j\omega)Y(j\omega)$, where $*$ denotes complex conjugate. The cross spectrum is conjugate even. *See* Cross-correlation function, Power spectrum.
Cumulative distribution	A method of representing time-varying noise by indicating the percentage of time that the noise level is present above (or below) a series of amplitude levels.
Damped natural frequency	Frequency of free vibration of a damped linear system. Unit: hertz (Hz). *See* Natural frequency.
Damping	(1) Any means of dissipating vibration energy within a vibrating system to reduce the amplitude of movement, or to shorten the decay time of free vibration. (2) Of an enclosure, removal of echoes and reverberation by the use of sound-absorbing materials. Also known as *sound proofing*.
Coulomb damping	Dissipation of energy that occurs when the motion of a vibrating system is resisted by a force whose magnitude is constant and independent of displacement and velocity, and whose direction is opposite to the direction of the velocity of the system. Also known as *dry friction damping*.
Critical damping	The minimum viscous damping that will allow a displaced system to return to its original position without oscillation.
Damping ratio	For a system with viscous damping, the ratio of viscous damping to critical damping.
Hysteretic damping, hysteresis damping	Damping of a vibrating system due to energy dissipation through mechanical hysteresis. The area of the hysteresis loop is a measure of damping. Also sometimes called *structural damping*.
Structural damping	An equivalent term for *hysteretic damping*.
Viscous damping	Dissipation of energy that occurs when the motion of a vibrating system is resisted by a force that has a magnitude proportional to the magnitude of the velocity of the system and direction opposite to the direction of the velocity of the system.
Day average sound pressure level	Time-average sound pressure level between 07:00 and 22:00 hours (15 hours). Unit: decibel (dB). Symbol: L_d. *See* Time-average sound pressure level, Day–night average sound pressure level.

Day–night average sound pressure level

The 24-hour time-average sound pressure level for a given day, after addition of 10 dB to night levels from between 22:00 and 07:00 hours. Unit: decibel (dB). Symbol: L_{dn}. *See* Time-average sound pressure level, Day average sound pressure level.

Decade

A unit of logarithmic frequency interval from f_1 to f_2, defined as $\log_{10}(f_2/f_1)$. 1 decade = 3.322 octaves. *See* Octave.

Decibel (dB)

The unit of measure for powerlike quantities in acoustics, noise, and vibration to denote level and level difference where the base of the logarithm is $10^{1/10}$. Abbreviated as dB. One tenth of a Bel. One decibel is equal to the ratio $10^{1/10} = 1.26$. Multiplying powerlike quantities (such as sound power or mean square sound pressure) by the factor $10^{n/10}$ increases the level by n decibels.

Decibel scale

A linear numbering scale used to define a logarithmic amplitude scale, in units of decibels, thereby compressing a wide range of amplitude values to a small set of numbers.

Degrees of freedom

The minimum number of independent coordinates required to define completely the positions of all parts of a mechanical system at any instant of time.

Density

In continuum mechanics, the mass per unit volume of a material. Also known as *mass density*. Units: kilograms per cubic metre (kg/m^3).

Diffraction

A phenomenon by which the propagation direction and intensity of a sound wave are changed when it passes by an obstacle or through an aperture if the wavelength of the sound wave is the same size or greater than the size of the obstacle or aperture.

Diffuse field

An idealized sound field in which the sound pressure level is the same everywhere and the flow of energy is equally probable in all directions.

Diffuse sound

Sound that is completely random in phase; sound that appears to arrive uniformly from all directions.

Dipole

An idealized sound source consisting of two closely spaced out-of-phase monopoles. *See* Monopole, Quadrupole.

Directivity factor

(1) Of a sound source, the ratio of the far-field mean-square pressure at a given frequency and radius, in a specified direction from the source, to the average mean square pressure over a sphere of the same radius, centered on the source. (2) The ratio of the mean-square pressure (or intensity) on the axis of a transducer, at a certain distance, to the mean-square pressure (or intensity) that a monopole source radiating the same power would produce at that point.

Directivity index	A measure of the extent to which a source radiates sound predominantly in one direction. The directivity index, DI_i, in the direction from a source to microphone position i, is defined as $DI_i = L_{pi}^* + \overline{L_p^*}$, where L_{pi}^* is the sound pressure level at microphone position i, corrected for background noise, $\overline{L_p^*}$ is the sound pressure level averaged over the measurement surface, corrected for background noise. Unit: decibel (dB). Symbol: DI.

Discrete Fourier transform (DFT)

A version of the Fourier transform applicable to a finite number of discrete samples. For a discrete sequence $x[n]$, its discrete Fourier transform is

$$X[k] = \sum_{n=0}^{N} x[n] e^{-j(2\pi kn/N)}.$$

$X[k]$ is also discrete and has the same number of samples as does $x[n]$. The original sequence can be recovered from the inverse DFT:

$$x[n] = \frac{1}{N} \sum_{k=0}^{N} X[k] e^{j(2\pi kn/N)}$$

See Fourier transform, Fast Fourier transform.

Displacement

A vector quantity that specifies the change of position of a body, usually measured from the rest or equilibrium position. Unit: metre (m). *See* Velocity, Acceleration, Jerk.

Dissipation factor, dissipance

Ratio of the dissipated sound power to the incident sound power. Unit: none. Symbol: δ. *See* Absorption factor, Reflection factor, Transmission factor.

Doppler effect

Phenomenon evidenced by the shift in the observed frequency from the source frequency caused by relative motion between the source and the observer. The Doppler effect is described quantitatively by

$$f_r = f_s \frac{1 + v_r/c}{1 - v_s/c}$$

where f_r is the received frequency, in Hz; f_s is the source frequency, in Hz; v_r is the component of velocity relative to the medium of the receiving point toward source, in metres per second; v_s is the component of velocity relative to the medium of the source toward the receiving point, in metres per second; c is (phase) speed of sound in a stationary medium, in metres per second.

Dose

See Noise dose.

Dosimeter(s)

Noise dosimeters measure and store sound pressure levels (SPL) and, by integrating these measurements over time, provide a cumulative noise exposure reading for a given period. Dosimeters may also provide a time history of sound exposure that is useful in determining contributions to the cumulative noise exposure.

Duct silencer	*See* Muffler.
Dynamic capability index	Given by $L_d = \delta_{pI_0} - K$, where δ_{pI_0} is the pressure-residual intensity index, K is selected according to the grade of accuracy required (see the table below). Unit: decibel (dB). Symbol: L_d. *See* Pressure-residual intensity index.

Grade of Accuracy	Bias Error Factor (dB)
Precision (grade 1)	10
Engineering (grade 2)	10
Survey (grade 3)	7

Dynamic mass	Ratio of applied force amplitude to resulting acceleration amplitude during simple harmonic motion. Reciprocal of accelerance magnitude. *See* Accelerance.
Dynamic modulus	Ratio of stress to strain under vibratory conditions. Generally a complex quantity.
Dynamic range	(1) Of a transducer, the ratio of the highest to the lowest input quantities within the linear range of the transducer, expressed in decibels. Unit: decibel (dB). (2) Of a spectrum analyzer, the maximum ratio of two signals simultaneously present at the input that can be measured to a specified accuracy, expressed in decibels. Unit: decibel (dB). (3) Of an acoustical variable, the ratio of the maximum and minimum mean-square values. Unit: decibel (dB).
Dynamic stiffness	Of a point-excited mechanical system, the complex ratio of applied force to displacement. It is the reciprocal of dynamic compliance. Units: newton per metre (N/m). *See* Compliance.
Eccentricity	Distance of the center of gravity of a revolving body from the axis of rotation.
Effective noise bandwidth	Bandwidth of an ideal filter that would pass the same amount of power from a white noise source as the filter described. Used to define bandwidth of one-third octave and octave filters.
Effective perceived noise level	A complex rating used to certify aircraft types for flyover noise. Includes corrections for pure tones and for duration of the noise. If $L(t)$ denotes the tone-corrected perceived noise level as a function of time, the effective perceived noise (EPN) level is defined as

$$L_{\text{EPN}} = 10 \log_{10} \left[\frac{1}{t_{\text{ref}}} \int 10^{L/10} dt \right]$$

The reference time used for normalization is $t_{\text{ref}} = 10$ s. Unit: dB re $(20 \text{ μPa})^2 \cdot 10$ s. Symbol: L_{EPN}. *See* Noise Exposure forecast.

Eigenfrequency	An alternative term for *natural frequency*. *See*: Natural frequency, Eigenvalue.

Eigenvalue

(1) Scalar λ such that $T(v) = \lambda v$, where T is a linear operator on a vector space, and v is an eigenvector. Specifically, of a square matrix, any scalar λ such that the determinant of $\mathbf{A} - \lambda \mathbf{I}$ vanishes, where $\mathbf{A}$ is the matrix concerned, and $\mathbf{I}$ is the identity matrix of the same size. Also known as *characteristic number, characteristic root, characteristic value. See* Eigenvector. (2) Of an acoustical system, usually refers to one of the values κ_n^2 ($n = 1, 2, 3...$) for which the equation describing the acoustic pressure at a specified frequency, e.g., the Helmholtz equation $(\nabla^2 + \kappa^2)p = 0$, has a solution matching the boundary conditions. (3) Root of a characteristic equation. Specifically, in modal analysis, the eigenvalue of the characteristic equation of a dynamic system is related to the square of one of the natural frequencies of the system. *See* Characteristic equation, Modal analysis.

Eigenvector

(1) Nonzero vector v whose direction is not changed by a given linear transformation T; that is, $T(v) = \lambda v$ for some scalar λ. Also known as *characteristic vector. See* Eigenvalue. (2) Specifically, in modal analysis an eigenvector of the characteristic equation of a dynamic system refers to a mode shape of the system.

Equivalent continuous A-weighted sound pressure level

Value of the A-weighted sound pressure level of a continuous, steady sound that, within a specified time interval T, has the same mean-square sound pressure as a sound under consideration whose level varies with time. It is given by

$$L_{Aeq,T} = 10 \log_{10} \left[\frac{1}{t_2 - t_1} \int_{t_1}^{t_2} \frac{p_A^2(t)}{p_0^2} \, dt \right]$$

where $p_A(t)$ is the instantaneous A-weighted sound pressure of the sound signal, p_0 is the reference sound pressure (20μPa). Unit: decibel (dB). Symbol: $L_{Aeq,T}$. *See* Equivalent continuous sound pressure level, Sound pressure.

Equivalent continuous sound pressure level

Value of sound pressure level of a continuous steady sound that, within a measurement time interval T, has the same mean-square sound pressure as a sound under consideration which varies with time, given by

$$L_{peq,T} = 10 \log_{10} \left[\frac{1}{t_2 - t_1} \int_{t_1}^{t_2} \frac{p^2(t)}{p_0^2} \, dt \right]$$

where t_1 and t_2 are the starting and ending times for the integral, $p(t)$ is the instantaneous sound pressure, p_0 is the reference sound pressure (20μPa). Also known as *time-averaged sound pressure level*. Unit: decibel (dB). Symbol: $L_{peq,T}$. If the equivalent continuous sound pressure level is A-weighted, then the symbol is $L_{pAeq,T}$, which can be abbreviated as L_{pA}. In general the subscripts "*eq*" and "*T*" can be omitted. *See* Sound pressure level, Equivalent continuous A-weighted sound pressure level.

Equivalent sound absorption area

(1) Of a room, the hypothetical area of a totally absorbing surface without diffraction effects that, if it were the only absorbing element in the room, would give the same reverberation time as the room under consideration. Units: square metre (m^2). Symbol: A_1, for the empty reverberation room; A_2, for the reverberation room containing

a test specimen. *See* Sabin, Eyring absorption coefficient. (2) Of a test specimen, the difference between the equivalent sound absorption area of the reverberation room with and without the test specimen. Units: square metre (m^2). Symbol: *A*.

Equivalent threshold level

For monaural listening, at a specified frequency, for a specified type of transducer, and for a stated force of application of the transducer to the human head, the vibration level, or sound pressure level set up by that transducer in a specified coupler or artificial ear when the transducer is activated by that voltage that, with the transducer applied to the ear concerned, would correspond with the hearing threshold. Unit: decibel (dB). *See* Hearing threshold (1).

Ergodic process

A stationary time-dependent stochastic process whose time average is equal to the ensemble average.

Excitation

An external force or motion applied to a system that causes the system to respond in some way. **Force excitation** The excitation force is independent of the properties of the excited structures; an example of this is the effect of a light and flexible source on a relatively stiff and heavy structure. **Velocity excitation** The excitation velocity is independent of the properties of the excited structures; an example of this is a light and flexible structure excited by a relatively massive source.

Eyring absorption coefficient

Sound absorption coefficient attributed to a surface according to the Eyring reverberation time equation

$$T = \frac{24V \ln 10}{-cS \ln (1 - \overline{\alpha})}$$

where *V* is the volume of the room, in cubic metres; *c* is the (phase) speed of sound in air, in metres per second; *S* is the total internal surface area of the room, in square metres. Unit: none. Symbol: $\overline{\alpha}$.

Far field

Portion of the radiation field of a sound source in which the sound pressure level decreases by 3 dB for each doubling of the area of the measurement surface. This is equivalent to a decrease of 6 dB for each doubling of the distance from a point source. In the far field, the sound waves can be considered locally planar, and the mean-square pressure is proportional to the total sound power radiated by the source.

Fast Fourier transform (FFT)

A rapid numerical technique for computing the discrete Fourier transform. *See* Discrete Fourier transform, Fourier transform.

Fast Hilbert transform

A rapid numerical technique for computing the Hilbert transform. The Fourier transform of the Hilbert transform of $x(t)$ is $H(f) = -j\text{sgn}(f)X(f)$, where $\text{sgn}(f)$ is a sign function, and $X(f)$ is the Fourier transform of $x(t)$. So the fast Hilbert transform can be realized using the fast Fourier transform. *See* Hilbert transform, fast Fourier transform.

Finite element method	A numerical approach for obtaining approximate solutions to variational problems. The domain of interest is divided into subdomains referred to as elements. The finite element method uses interpolation functions that satisfy the essential boundary conditions and solves the unknown nodal values that approximately satisfy the governing differential equation over the domain. *See* Boundary element method.
Flanking transmission	Sound that travels between a source and a receiving room along paths other than through the partition dividing the two rooms.
Flexibility	*See* Compliance.
Force	Vector quantity that causes a massive body to accelerate, equal to the time rate of change of momentum of the body. *See* Newton.
Force excitation	*See* Excitation.
Forced vibration, forced oscillation	Response of a system caused by external excitation. *See* Free vibration.
Formant	(1) Noun. A resonance of the vocal tract. The frequencies of the formants are determined in large part by the vocal tract shape; changes in the shape change the frequencies and so result in different speech sounds. *See* Vocal tract. (2) Adjective. A formant synthesizer uses parameters related to the formant frequencies and bandwidths of different speech sounds.
Fourier transform	A mathematical operation for decomposing a time function $x(t)$ into its frequency components (amplitude and phase) given by

$$X(f) = \int_{-\infty}^{\infty} x(t)e^{-j2\pi ft}dt$$

The inverse Fourier transform is

$$x(t) = \int_{-\infty}^{\infty} X(f)e^{j2\pi ft}df$$

See Discrete Fourier transform, Spectrum, Fast Fourier transform, Power spectrum.

Free field	A sound field in a homogeneous, isotropic medium free of boundaries. In practice, it is a field in which reflections at the boundaries are negligible in the frequency range of interest. See Free field over a reflecting plane, Diffuse field.
Free field over a reflecting plane	A sound field in a homogeneous, isotropic medium in the half-space above an infinite, rigid plane surface on which the source is located. See Free field.

Free vibration, free oscillation

A phenomenon that occurs in a mechanical system when it vibrates in the absence of external excitation. Free vibration consists of natural modes of the system, each vibrating at a natural frequency. See Forced vibration, Natural frequency, Normal mode of vibration.

Frequency

The number of cycles of a periodic phenomenon per unit time interval. It is the reciprocal of the period. Unit: hertz (Hz). Symbol: f. See Angular frequency, Period, Hz.

Frequency response function (FRF)

Of a linear time-invariant system, the complex ratio of the Fourier transforms of the output signal to the input signal. Mathematically, the frequency response function is the Fourier transform of the impulse response function. By extension, the frequency response function can be used for any two linearly related signals. See Impulse response function, Transfer function.

Frequency weighting

Modification of the spectrum of an acoustical signal by means of an analog or digital filter having one of the standardized response characteristics known as A, B, C, etc., defined in IEC 61672–1. The A-weighting filter is the one most commonly used. *See* Weighting network.

Fundamental frequency

The lowest natural frequency of a dynamic system. *See* Natural frequency.

Ground effect

In outdoor sound propagation, the influence of the sound reflected from a surface on horizontal propagation of sound traveling directly from source to receiver. Over distances of 25 m and above, this can be significant at audio frequencies, particularly when reflection at near-grazing angles is involved.

Group speed of sound

Travel speed of the energy of a sound wave through a medium, given by $c_g = d\omega/dk$, where ω is the angular frequency, in radians per second; k is the wavenumber, in reciprocal metre. Units: metre per unit time (m/s). Symbol: c_g. *See* Phase speed of sound.

Half-power bandwidth

(1) Of a resonance response curve in which the squared gain factor of a linear system is plotted against frequency, the frequency separation between the 3-dB down points on the two sides of the resonance peak, at which the curve is at half of its peak value. (2) Of a signal with a peaked power spectrum, the frequency separation between the points on two sides of the spectral peak where the spectrum level is 3 dB less than the peak value. *See* Bandwidth.

Hand-transmitted vibration

Vibration transmitted to the hand, often from powered hand tools.

Harmonic

A frequency component whose frequency is an integer multiple of the fundamental frequency of a periodic quantity to which it is related.

Hearing conservation program   A system to identify noise-exposed workers, and to monitor their exposure and audiometric function.

Hearing loss   Increase in the threshold of audibility due to disease, injury, age, or exposure to intense noise. *See* Presbyacusis.

    Conductive hearing loss   Hearing loss caused either by blockage of the external ear or by disease or damage in the middle ear, so that the signal amplitude reaching the inner ear is reduced.

    Noise-induced hearing loss (NIHL)   Cumulative hearing loss associated with repeated exposure to noise.

    Sensorineural hearing loss   Hearing loss due to a lesion or disorder of the inner ear or of the auditory nervous system.

    Nonoccupational hearing loss   Hearing loss that is caused by exposures outside of the occupational environment.

Hearing threshold   (1) For a given listener and specific signal, the minimum sound pressure level that is capable of evoking an auditory sensation in a specified function of trials. Sound reaching the ears from other sources is assumed to be negligible. *See* Audibility threshold, Masking. (2) The level of a sound at which, under specified conditions, a person gives a predetermined percentage of correct detection responses on repeated trials.

Hearing threshold level (HTL)   For a specified signal, amount in decibels by which the hearing threshold for a listener, for either one or two ears, exceeds a specified reference equivalent threshold level. Unit: decibel (dB). *See* Equivalent threshold level, Hearing threshold (1).

Helmholtz resonator   Hollow, rigid-walled cavity filled with gas or liquid and having a small opening called the neck. Its fundamental frequency f_0 is approximated by

$$f_0 = \frac{c}{2\pi} \sqrt{\frac{A}{LV}}$$

where c is the (phase) speed of sound, in metres per second, A and L are the cross-sectional area, in square metres and effective length, in metres, of the neck, and V is the cavity volume, in cubic metres. *See* Muffler.

Hertz (Hz)   A unit of frequency measurement, representing cycles per second. *See* Frequency.

Hilbert transform   A mathematical transform that shifts the phase of each frequency component of the instantaneous spectrum by 90° without affecting the magnitude. For a time signal $x(t)$, its Hilbert transform is defined as

$$h(t) = H\{x(t)\} = \frac{1}{\pi} \int\limits_{-\infty}^{+\infty} \frac{x(\tau)}{t - \tau} d\tau$$

See Fast Hilbert transform, Fourier transform.

Ideal filter

A filter having a rectangularly shaped characteristic: unity amplitude transfer within its passband and zero transfer outside its passband.

Impact

Excitation of a structure with a force pulse, for example, by using an impact hammer.

Impact insulation class (IIC)

A single number representing 110 dB minus the normalized impact sound index. *See* Normalized impact sound index.

Impact sound pressure level

In architectural acoustics, average sound pressure level in a specified frequency band in the receiving room under a test floor being excited by the standardized impact sound source specified in ASTM E492–90. Unit: decibel (dB). *See* Normalized impact sound index, Impact insulation class.

Normalized impact sound pressure level

For a specified frequency band, average sound pressure level in decibels due to the standardized impact sound source, plus 10 times the logarithm to the base 10 of the ratio of the Sabine absorption in the receiving room to the reference Sabine absorption of 10 metric sabins.

Impedance tube

A uniform rigid-walled tube with a sound generation device, for example, loud-speaker, at one end to excite plane sound waves. When such a tube is terminated by a sample of material, the acoustic impedance (or the absorption factor) of the material can be determined using a traversing microphone or two fixed microphones. Also known as *standing-wave tube* or *Kundt's tube*.

Impulse response function

Of a linear time-invariant system, the function that gives the system output when the input is an impulse function (Dirac's delta function) at time $t = 0$. Its Fourier transform is the frequency response function of the system. For any other type of input, the system output is determined by the convolution between the input and the impulse response function. *See* Frequency response function.

Infrasound

Sound at frequencies below the audible range, generally below about 20 Hz.

Insertion loss

(1) Of a silencer or other sound reduction element, in a specified frequency band, the decrease in sound pressure level, measured at the location of the receiver, when the sound reduction element is inserted in the transmission path between the source and the receiver. Unit: decibel (dB). (2) Of a silencer or other sound reduction element, in a specified frequency band, the decrease in sound power level, measured when the sound reduction element is inserted in the transmission path between the source and the receiver. Unit: decibel (dB).

Instantaneous sound intensity

Instantaneous rate of flow of sound energy per unit area in the direction of the local instantaneous acoustic particle velocity. This is a vector quantity that is equal to the product of the instantaneous sound pressure at a point and the associated particle velocity: $\mathbf{I}(t) = p(t) \cdot \mathbf{u}(t)$, where $p(t)$ is the instantaneous sound pressure at a point, and $\mathbf{u}(t)$ is the associated particle velocity. Units: watt per square metre (W/m^2). Symbol: $\mathbf{I}(t)$. *See* Sound intensity. *Note*: The sound intensity is generally complex.

The real part is the propagating part of the sound field (sometimes called the active part). The imaginary part is the nonpropagating part of the sound field (sometimes called the reactive part).

Integrator
A device the output of which is the integrated input. For noise and vibration applications, the integrator is generally an electrical device, used to convert a signal proportional to a vibratory acceleration to one that is proportional to velocity or displacement, or to convert a signal proportional to a vibratory velocity to one that is proportional to displacement.

Intelligibility, speech intelligibility
Of speech in a particular listening environment, a qualitative term that describes the ability of the acoustical environment to transmit speech intelligibly, usually expressed relative to perfect listening conditions. Intelligibility can be quantified for a particular speech sample by asking listeners to record their interpretation of what they hear and processing the data to obtain the percentage syllable articulation. *See* Percentage syllable articulation, Articulation index.

Intensity
See Instantaneous sound intensity, Sound intensity.

Isolation
Vibration isolation
Resistance to the transmission of sound or vibration by materials and structures. Reduction, usually attained by the use of a resilient coupling, in the vibration of a system in response to mechanical excitation.

Jerk
A vector quantity that specifies time rate of change of acceleration. Units: metre per second-cubed (m/s^3). *See* Displacement, Velocity, Acceleration.

Leakage
See Spectral leakage error.

Level
Logarithm of the ratio of a powerlike quantity to a reference quantity of the same kind. The base of the logarithm, the reference quantity, and the kind of level shall be specified.

Level difference
In architectural acoustics, in a specified frequency band, the difference in the space and time-average sound pressure levels produced in two rooms by one or more sound sources in one of them: $D = L_1 - L_2$, where L_1 is the average sound pressure level in the source room, L_2 is the average sound pressure level in the receiving room. Also known as *noise reduction*. Unit: decibel (dB). Symbol: D or NR. *See* Sound reduction index.

Standardized level difference
The level difference corresponding to a reference value of the reverberation time in the receiving room, given by $D_{nT} = D + 10\log_{10}(T/T_0)$, where D is the level difference, T is the reverberation time in the receiving room, and T_0 is the reference reverberation time. Unit: decibel (dB). Symbol: D_{nT}. *Note*: For dwellings, $T_0 = 0.5$ s. The standardizing of the level difference corresponding to the reverberation time in the receiving room of $T_0 = 0.5$ s is equivalent to standardizing the level difference with respect to an equivalent absorption area of $A_0 = 0.32\,V$, where A_0 is the equivalent absorption area, in square metres, and V is the volume of the receiving room, in cubic metres.

Linearity	A characteristic that satisfies the superposition property. (1) Of a system, if two separate inputs x_1 and x_2 produce respective outputs y_1 and y_2, then the combined input $x_1 + x_2$ produces an output $y_1 + y_2$; if input x produces output y, then input ax (where a is a constant) produces output ay. (2) Of an operator L, it satisfies $L(x_1 + x_2) = L(x_1) + L(x_2)$ and $L(ax) = aL(x)$, where x_1 and x_2 are variables on which the operator L acts, and a is a constant. (3) Of an equation, two solutions can be added, or a solution can be multiplied by a constant, the result is also a solution. *See* Nonlinearity.
Liquid-borne sound	Sound that propagates through a liquid. *See* Airborne sound, Structure-borne sound.
Loss factor	A measure of the damping capability of a system. For viscous damping, the loss factor is twice the damping ratio. The loss factor may appear in different forms, for example, internal loss factor, radiation loss factor, or coupling loss factor. *See* Damping.
Loudness	The attribute of auditory sensation in terms of which sounds may be ordered on a scale extending from soft to loud. *See* Sone, Rating, Magnitude scaling.
Loudness level	Of a given sound, the sound pressure level of a reference sound, consisting of a sinusoidal plane progressive wave of frequency 1000 Hz coming from directly in front of the listener, which is judged by otologically normal person to be equally loud to the given sound. Unit: phon. *See* Phon.
Magnitude scaling	A method to quantify psychological variables. Typically, a fixed physical stimulus is provided that evokes a perceptual strength to be taken as an internal standard (unit). The subject is requested to assess the ratio of the perceptual strength of an unknown stimulus with respect to that internal standard. *See* Rating, Loudness, Sone.
Masking	(1) The process by which the threshold of audibility of one sound is raised by the presence of another (masking) sound. (2) The amount by which the hearing threshold level is so raised, expressed in decibels.
Mechanical impedance	(1) At a surface, the ratio of the total force on the surface to the component of the average sound particle velocity at the surface in the direction of the force. Called driving point impedance if force and velocity are measured at the same point and in the same direction, otherwise called transfer impedance. Units: newton per metre per second [N/(m/s)]. Symbol: Z_m. (2) A form of the inverse frequency response function, which is the complex ratio of the excitation force to the velocity response. *See* Acoustic impedance, Characteristic impedance, Specific acoustic impedance, Mobility.
Microphone	Transducer that converts a sensed sound pressure signal to an electric output signal, with minimal distortion over its designed frequency range.

Mobility   (1) An alternative term for *mechanical admittance*. (2) The ratio of the following complex quantities:

> Mechanical mobility Z_M = velocity across/force through
>
> Rotational mobility Z_R = angular velocity across/torque through
>
> Acoustic mobility Z_A = volume velocity across/sound pressure through

See Admittance, Accelerance, Compliance, Mechanical impedance, Acoustic impedance.

Modal analysis   Process of determining the mode shapes and associated parameters, natural frequencies, and damping.

Modal mass   In modal analysis, of a particular mode of a dynamic system, an element in the generalized mass matrix. For mass-normalized mode shapes, the modal mass is unity.

Modal stiffness   In modal analysis, of a particular mode of a dynamic system, an element in the generalized stiffness matrix. For mass-normalized mode shapes, the modal stiffness is equal to the eigenvalue of the characteristic equation of the system. *See* Eigenvalue.

Mode shape   A pattern of vibration exhibited by a structure at a natural frequency. Generally, described as a vector of values, defining the relative displacement amplitudes and phases of each point on the structure at a specified natural frequency. *See* Eigenvector.

Modal density   For a given system, the average number of modal resonances per unit interval of frequency. Unit: reciprocal hertz (1/Hz).

Mode of vibration   Characteristic pattern assumed by a system undergoing vibration in which the motion of every particle is simple harmonic with the same frequency. *See* Normal mode of vibration.

Mode shape   *See* Modal analysis.

Monopole   An idealized sound source that is concentrated at a single point in space. A monopole can be represented by a pulsating sphere producing spherical wave fronts. *See* Dipole, Quadrupole.

Muffler, duct silencer Device designed to reduce the level of sound transmitted along a duct system.

Natural frequency   Frequency of free vibration of a system. If the system is damped, then the natural frequency is the damped natural frequency. Unit: hertz (Hz). *See* Damped natural frequency, Eigenvalue, Free vibration.

Near field   The portion of the radiation field of a sound source that lies between the source and the far field. *See* Far field.

Neper (Np)	A unit of level of a field quantity in terms of logarithm on the Napierian base $e \approx$ 2.7183, and unit of level of powerlike quantity when the base of the logarithm is e^2.
Newton (N)	A unit of force. The force of 1 N accelerates a 1 kg mass at $1 \, \text{m/s}^2$. *See* Force, Acceleration.
Night average sound pressure level	Time-average sound pressure level between 22:00 and 07:00 hours (9 hours). Unit: decibel (dB). *See* Day average sound pressure level, Day–night sound pressure level, Equivalent continuous sound pressure level.
Node	(1) A point, curve, or surface, on a vibrating structure or in a fluid volume that remains stationary. *See* Standing wave. (2) A grid point in the discrete model of a structure.
Noise	(1) Undesired, unpleasant sound. *See* Periodic noise, Random noise, Tonal noise, Pink noise, White noise. (2) Erratic, unwarranted disturbance, for example, electrical noise.
Noise dose	(1) According to the definition given by Occupational Safety and Health Administration (OSHA), the ratio, expressed as a percentage of (1) the time integral, over a stated time or event, of the 0.6 power of the measured "S" (slow) exponential time-averaged, squared A-weighted sound pressure and (2) the product of the criterion duration (8 hours) and the 0.6 power of the squared sound pressure corresponding to the criterion sound pressure level (90 dB). (2) According to the definition given by the National Institute for Occupational Safety and Health (NIOSH), the percentage of actual exposure relative to the amount of allowable exposure, and for which 100% and above represent exposures that are hazardous. The noise dose is calculated using: $D = \sum_{i=1}^{n} C_i / T_i \times 100\%$, where C_i is the total time of exposure at a specified noise level, and T_i is the exposure time at which noise for this level becomes hazardous.
Noise emission level	A-weighted sound pressure level measured at a specified distance and direction from a noise source, in an open environment, above a specified type of surface. Generally follows the recommendation of a national or industry standard.
Noise exposure forecast (NEF)	A complex criterion for predicting future noise impact of airports. The computation considers effective perceived noise level of each type of aircraft, flight profile, number of flights, time of day, etc. Generally used in plots of NEF contours for zoning control around airports. *See* Effective perceived noise level.
Noise exposure level	The level given by $L_{Aeq,T_e} + 10 \log_{10}(T_e/T_0)$, where T_e is the effective duration of the working day, in hours; T_0 is the reference duration (=8 hours); L_{Aeq,T_e} is the A-weighted equivalent sound pressure level during the time interval T_e. If T_e does not exceed 8 hours, $L_{EX,8h}$ is numerically equal to $L_{Aeq,8h}$. Unit: decibel (dB). Symbol: $L_{EX,8h}$. *See* Equivalent continuous A-weighted sound pressure level, Sound exposure. *Note*: The noise exposure level $L_{EX,8h}$ may be calculated from

the A-weighted sound exposure E_{A,T_e}, in pascal-squared second (Pa²·s), using the following formula:

$$L_{EX,8h} = 10 \log_{10} \frac{E_{A,T_e}}{1.15 \times 10^{-5}}$$

where the reference sound exposure is 1.15×10^{-5} Pa²·s.

Noise immission level(NIL)
A measure of cumulative A-weighted sound exposure during a person's lifetime, based on a modification of the noise exposure level concept. The reference sound exposure is chosen in such a way that after working for N years in an environment with a constant A-weighted sound pressure level L_A, a person who works a typical 1740 hours per year will have a cumulative noise immission level given by NIL = L_A + $10\log_{10}N$, assuming that person's leisure time exposure and previously cumulated work exposure are negligible. The equation implies a reference sound exposure of $E_0 = 2.5 \times 10^{-3}$ Pa²·s, i.e., (1740/8) times the reference sound exposure used for noise exposure level. Unit: decibel (dB). Symbol: NIL. *See* Noise exposure level, Sound exposure level.

Noise reduction coefficient (NRC)
Arithmetic average of the sound absorption coefficients of a material at 250, 500, 1000, and 2000 Hz.

Noise reduction (NR)
An alternative term for *level difference*. *See* Level difference, Sound reduction index.

Noise reduction rating (NRR)
A single-number rating that indicates the noise reduction capabilities of a hearing protector. Unit: decibel (dB).

Noise-induced hearing loss (NIHL)
See Hearing loss.

Nonlinearity
Characteristic that does not satisfy linearity. *See* Linearity.

Normal mode of vibration
Mode of free vibration of an undamped system. *See* Eigenvector.

Normal sound intensity
Component of the sound intensity in the direction normal to a measurement surface defined by the unit normal vector $\mathbf{n}$: $I_n = \mathbf{I} \cdot \mathbf{n}$. Units: watt per square metre (W/m²). Symbol: I_n. *See* Sound intensity.

Normal sound intensity level
Logarithmic measure of the unsigned value of the normal sound intensity $|I_n|$, given by: $L_{I_n} = 10 \log_{10} |I_n|/I_0$, where I_0 is the reference sound intensity and is 10^{-12} W/m². When I_n is negative, the level is expressed as (−) XX dB, except when used in the evaluation of residual intensity index. Unit: decibel (dB). Symbol: L_{I_n}. *See* Normal sound intensity, Sound intensity.

Normalized impact sound index (NISI)
A single-number rating of normalized impact sound pressure level for 16 successive one-third octave bands from 100 to 3150 Hz inclusive. The calculation procedure is specified in ASTM E989–89. Unit: decibel (dB). *See* Impact sound pressure level, Impact insulation class.

Normalized impact sound pressure level	*See* Impact sound pressure level.
Noy	A linear unit of noisiness to quantify the annoyance potential of complex sound. *See* Perceived noisiness, Sone.
Nyquist diagram	A plot of the real part versus the imaginary part of the frequency response function.
Nyquist plot	For a single-degree-of-freedom system, the Nyquist plot is a circle. *See* Argand diagram, Bode diagram, Frequency response function.
Nyquist frequency	One-half the sampling frequency in analog-to-digital conversion, which is the theoretical maximum frequency that can be correctly sampled. *See* Sampling theorem, Aliasing error.
Octave	A unit of logarithmic frequency interval from f_1 to f_2, defined as $\log_2(f_2/f_1)$. 1 octave = 0.301 decade. *See* Decade.
One-octave filter	A filter whose upper and lower passband limits are in the ratio of 2 and centered at one of the preferred frequencies given in ISO 266. Should meet the attenuation characteristic of IEC 61260 and ANSI S1.11–1986.
One-octave band	A frequency band whose upper and lower frequency limits are in the ratio of 2. *See* One-third octave band.
One-third octave band	A frequency band whose upper and lower frequency limits are in the ratio of $2^{1/3}$.
One-third octave filter	A filter whose upper and lower passband limits are in the ratio of $2^{1/3}$ and centered at one of the preferred frequencies given in ISO 266. Should meet the attenuation characteristics of IEC 61260 and ANSI S1.11–1986.
Order analysis	A form of frequency analysis, used with rotating machines where the amplitudes of signal frequency components are plotted as a function of multiples of the rotational speed. The frequency component with frequency of n times the rotational speed is called the nth order, generally symboled as nX.
Ototoxic chemical	Chemical that causes damage specifically to the cochlea, the auditory nerve, or vestibular system, which impairs the ability to hear. For example, organic solvents, lead, and mercury are ototoxic.
Particle velocity	Velocity of a medium particle about it rest positions, usually due to a sound wave. Units: metre per second (m/s).
Pascal (Pa)	A unit of pressure corresponding to a force of 1 N acting uniformly upon an area of 1 square metre. $1\,\text{Pa} = 1\,\text{N/m}^2$.
Passband	Of a filter, the frequency range over an the frequency response function has a magnitude close to unity.

Perceived noise level (PNL) — A frequency-weighted sound pressure level obtained by a stated procedure that combines the sound pressure levels in the 24 one-third octave bands with midband frequencies from 50 Hz to 10 kHz. Unit: decibel (dB). Symbol: L_{PN}. *See* Perceived noisiness.

Perceived noisiness, noisiness — Prescribed function of sound pressure levels in the 24 one-third octave bands with nominal midband frequencies from 50 Hz to 10 kHz that is used in the calculation of the perceived noise level. Unit: noy. *See* Noy.

Percentage syllable articulation — In a given speech sample presented to a listener, the percentage of syllables that are heard correctly. *See* Intelligibility, Articulation index.

Percentile level — A-weighted sound pressure level obtained by using time-weighting "F" (see IEC 61672–1) that is exceeded for $N\%$ of the time interval considered. Unit: decibel (dB). Symbol: $L_{AN,T}$, for example, $L_{A95,1h}$ is the A-weighted level exceeded for 95% of one hour.

Period — The smallest interval (of time or distance) over which an oscillation repeats itself.

Periodic noise — A noise event that is periodically repeated. Typical sources of periodic noise are gear wheels and piston machines.

Periodic vibration — Oscillatory motion whose amplitude pattern is repeated after fixed increments of time.

Permissible exposure level (PEL) — Regulatory limit of sound exposure. The OSHA (Occupational Safety and Health Administration) PEL is a noise dose of 1.0 based on 8-hour A-weighted sound exposure level at 90 dB with a 5 dB exchange rate. European PEL is generally 8-hour A-weighted sound exposure level at 85 dB with a 3-dB exchange rate.

Phase speed of sound — Travel speed of the phase of a sound wave through a medium, given by $c = \omega/k$, where ω is the angular frequency, in radians per second; k is the wavenumber, in reciprocal metre. Units: metre per second (m/s). Symbol: c. *See* Group speed of sound.

Phon — Unit of loudness level of a sound. It is numerically equal to the sound pressure level of a 1-kHz free progressive wave that is judged by reliable listeners to be as loud as the unknown sound. *See* Loudness level.

Phone — Acoustic realization of a phoneme. Phones can represent the variation in pronunciation of a phoneme, due to context, dialect, speaking style, or rate, that does not cause a change in meaning. *See* Phoneme.

Phoneme — The minimal sound unit in a language that can change meaning of a word. Each language has its own set of phonemes, which vary both in number of phonemes, from 15 to over 100, and in the sounds included. *See* Phone.

Pink noise	A broadband random noise whose power spectrum is inversely proportional to frequency (-3 dB per octave or -10 dB per decade), thus giving it a constant power spectrum per octave or one-third octave band. *See* Random noise, White noise.
Power spectrum	Magnitude square of frequency spectrum. Also known as *autospectrum*. The power spectrum is real and even. *See* Autocorrelation function, Power spectrum density, Cross spectrum.
Power spectrum density, power spectral density	Limit of the power of a signal (displacement, velocity, acceleration, etc.) divided by bandwidth, as the bandwidth approaches zero. *See* Power spectrum.
Power spectrum level	The level of the power in a band 1 Hz wide referred to a given reference power.
Power unit noise	Unwanted sound generated by the propulsion system of a road vehicle, except the tires; such sources including, e.g., engine, air intake, exhaust silencer, transmission, and fan. Other terms sometimes (but incorrectly and inconsistently) used for this include power train noise and drive train noise. Also known as *propulsion noise*. *See* Tire/road noise.
Preferred frequencies	A set of standardized octave and one-third octave center frequencies defined by ISO 266, DIN 45401, and ANSI S1.6–1967.
Presbyacusis, presbycusis	Progressive hearing impairment with age, in the absence of other identifiable causes. *See* Hearing loss.
Pressure-residual intensity index	Of a sound intensity measurement device, the difference between the indicated sound pressure level L_p and the indicated sound intensity L_I when the device is placed and oriented in a sound field such that the sound intensity is zero: $\delta_{pI_0} = L_p - L_I$. Details for determining the pressure-residual intensity index are given in IEC 61043. Unit: decibel (dB). Symbol: δ_{pI_0}. *See* Dynamic capability index.
Privacy index, speech privacy index (PI)	A number that is a measure of speech privacy or lack of speech intelligibility. The privacy index is calculated from the articulation index (AI): $PI = (1 - AI) \times 100\%$. *See* Articulation index.
Quadrupole	An idealized sound source consisting of four closely spaced monopoles with adjacent monopoles out-of-phase.
Quality factor, Q factor	Of a lightly damped linear system, a measure of the sharpness of a peak in the frequency response, which is equal to the reciprocal of the loss factor of the corresponding vibration mode. *See* Loss factor, Half-power bandwidth.
Radiation efficiency, radiation factor	The ratio of the sound power radiated by a vibrating surface, with a given time-mean-square velocity, to the sound power, which would be emitted as a plane wave by the same vibrating surface with the same vibration velocity. The radiation factor is given by the following equation:

$$\sigma = \frac{P_S}{\rho c S_S \overline{v^2}}$$

where P_S is the airborne sound power emitted by the vibrating surface, ρc is the characteristic impedance of air, S_S is the area of the vibrating surface, and $\overline{v^2}$ is the squared rms value of the vibratory velocity averaged over the area S_S. Unit: none. Symbol: σ. *See* Sound power, Characteristic impedance, Vibratory velocity.

Radiation impedance   A generic term for the impedance presented to a vibrating surface by the adjacent acoustical medium. Units: newton per metre per second [N/(m/s)]. Symbol: Z_r. Also known as *fluid loading impedance*. Its real part is the *radiation resistance* from which the emitted sound power is obtained by multiplying it by the mean-square velocity of the body. Symbol: R_r. The imaginary part is called the *radiation reactance*. Symbol: X_r. *See* Acoustic impedance, Mechanical impedance.

Random noise   Stationary noise whose instantaneous amplitude cannot be specified at any given instant of time. Instantaneous amplitude can only be defined statistically by an amplitude distribution function. *See* Pink noise, White noise.

Random vibration   Vibration whose instantaneous amplitude cannot be specified at any given instant of time. The instantaneous amplitude can only be defined statistically by a probability distribution function that gives the fraction of the total time that the amplitude lies within specified amplitude intervals. Random vibration contains no periodic or quasi-periodic constituents. Pseudo, periodic, and burst random are special forms.

Rating   A method to quantify psychological variables, such as annoyance, loudness, and others. Typically, the subject is requested to choose one out of several locations on a scale labeled, for instance, from 0 to 10, whereby the selected location indicates the strength of the internal response. *See* Annoyance, Loudness.

Rayleigh's criterion   In thermoacoustics, a necessary condition for the onset of instability, in an irrotational flow at low Mach number that contains a compact heat source of time-varying output $Q(t)$. It states that oscillations can become unstable when the heat input is in phase with the local sound pressure $p(t)$, at any given frequency. A more precise statement applicable to nonlinear oscillation is that instability requires $\oint p(t)Q(t)dt > 0$ (integral over one cycle).

Reactance   The imaginary part of an impedance. *See* Acoustic impedance, Mechanical impedance, Resistance.

Reactive sound field   Sound field in which the particle velocity is 90° out of phase with the pressure. An ideal standing wave is an example of this type of field, where there is no net flow of energy and constitutes the imaginary part of a complex sound field. *See* Standing wave.

Receptance   An alternative term for *dynamic compliance*. *See* Compliance.

Reflection coefficient	The ratio of the reflected sound pressure amplitude to the pressure amplitude of the sound wave incident on the reflecting object. Unit: none. Symbol: r_a.
Reflection factor, reflectance	The ratio of the reflected sound power to the incident sound power. Unit: none. Symbol: r. *See* Absorption factor, Dissipation factor, Transmission factor.
Refraction	A phenomenon by which the propagation direction of a sound wave is changed when a wavefront passes from one region into another region of different phase speed of sound.
Repetency	*See* Wavenumber.
Resistance	The real part of an impedance. *See* Acoustic impedance, Mechanical impedance, Radiation impedace, Reactance.
Resolution	The smallest change or amount a measurement system can detect.
Resonance	Conditions of peak vibratory response where a small change in excitation frequency causes a decrease in system response.
Resonance frequency	Frequency at which resonance exists. Unit: Hz.
Response	Motion or other output resulting from an excitation, under specified conditions. *See* Excitation.
Reverberant sound field	Portion of the sound field in the test room over which the influence of sound received directly from the source is negligible.
Reverberation	Persistence of sound in an enclosure after a sound source has stopped.
Reverberation room	A room with low absorption and long reverberation time, designed to make the sound field therein as diffuse as possible.
Reverberation time	Of an enclosure, for a given frequency or frequency band, the time required for the sound pressure level in an initially steady sound field to decrease by 60 dB after the sound source has stopped. Unit: second (s).
Room constant	A quantity used to describe the capability of sound absorption of an enclosure, determined by $R = S\bar{\alpha}/(1 - \bar{\alpha})$, where S is the total internal surface area of the enclosure, in square metres; $\bar{\alpha}$ is the average sound absorption coefficient of the enclosure. Units: square metre (m^2). Symbol: R.
Root mean square (rms)	The square root of the arithmetic average of a set of squared instantaneous values.

Sabin, metric sabin	A measure of sound absorption of a surface. One metric sabin is equivalent to 1 square metre of perfectly absorptive surface. *See* Absorption, Equivalent sound absorption area.
Sampling theorem	Theorem that states that if a continuous time signal is to be completely described, the sampling frequency must be at least twice the highest frequency present in the original signal. Also known as *Nyquist theorem, Shannon sampling theorem.*
Scaling	*See* Magnitude scaling, Rating.
Semianechoic field	*See* Free field over a reflecting plane.
Semianechoic room	A test room with a hard, reflecting floor whose other surfaces absorb essentially all the incident sound energy over the frequency range of interest, thereby affording free-field conditions above a reflecting plane. *See* Anechoic room, Free field over a reflecting plane.
Sensitivity	(1) Of a linear transducer, the quotient of a specified quantity describing the output signal by another specified quantity describing the corresponding input signal, at a given frequency. (2) Of a data acquisition device or spectrum analyzer, a measure of the device's ability to display minimum level signals. (3) Of a person, with respect to a noise, the extent of being annoyed.
Sensorineural hearing loss	*See* Hearing loss.
Shock	Rapid transient transmission of mechanical energy.
Shock spectrum	Maximum acceleration experienced by a single-degree-of-freedom system as a function of its own natural frequency in response to an applied shock.
Signal-to-noise ratio (SNR)	In a signal consisting of a desired component and an uncorrelated noise component, the ratio of the desired-component power to the noise power. For a signal $x(t)$, if $x(t) = s(t) + n(t)$, where $s(t)$ is the desired signal, and $n(t)$ is noise, then the signal-to-noise ratio is defined as $\text{SNR} = 10\log_{10}\langle s^2 \rangle / \langle n^2 \rangle$, where $\langle \rangle$ indicates a time average.
Significant threshold shift	Shift in hearing threshold, outside the range of audiometric testing variability (± 5 dB), that warrants followup action to prevent further hearing loss. The National Institute of Occupational Safety and Health (NIOSH), defines significant threshold shift as an increase in the hearing threshold level of 15 dB or more at any frequency (500, 1000, 2000, 3000, 4000, or 6000 Hz) in either ear that is confirmed for the same ear and frequency by a second test within 30 days of the first test. *See* Hearing threshold level.
Silencer	Any passive device used to limit noise emission.

Simple harmonic motion	Periodic motion whose displacement varies as a sinusoidal function of time.
Single-event sound pressure level	Time-integrated sound pressure level of an isolated single sound event of specified duration T (or specified measurement time T) normalized to $T_0 = 1$ s. It is given by the formula:

$$L_{p,1s} = 10 \log_{10} \left[\frac{1}{T_0} \int_0^T \frac{p^2(t)}{p_0^2} \, dt \right]$$

$$= L_{peq,T} + 10 \log_{10} \left(\frac{T}{T_0} \right) dB,$$

where $p(t)$ is the instantaneous sound pressure, p_0 is the reference sound pressure, and $L_{peq,T}$ is the equivalent continuous sound pressure level. Unit: decibel (dB). Symbol: $L_{p,1s}$.

Sone	A linear unit of loudness. One sone is the loudness of a pure tone presented frontally as a plane wave of 1000 Hz and a sound pressure level of 40 dB, referenced to 20 μPa. See Loudness, Magnitude scaling.
Sound	Energy that is transmitted by pressure waves in air or other materials and is the objective cause of the sensation of hearing. Commonly called noise if it is unwanted.
Sound absorption coefficient	See Absorption factor.
Sound energy density	Mean sound energy in a given volume of a medium divided by that volume. If the energy density varies with time, the mean shall be taken over an interval during which the sound may be considered statistically stationary. Units: joule per cubic metre (J/m^3).
Sound energy, acoustic energy	Total energy in a given volume of a medium minus the energy that would exist in that same volume with no sound wave present. Unit: joule (J).
Sound exposure	Time integral of squared, instantaneous sound pressure over a specified interval of time, given by $E = \int_{t_1}^{t_2} p^2(t) \, dt$, where $p(t)$ is the instantaneous sound pressure, t_1 and t_2 are the starting and ending times for the integral. If the instantaneous sound pressure is frequency weighted, the frequency weighting should be indicated. Units: pascal-squared second $(Pa^2 s)$. Symbol: E. *See* Sound pressure.
A-weighted sound exposure	Exposure given by $E_{A,T} = \int_{t_1}^{t_2} p_A^2(t) \, dt$, where $p_A(t)$ is the instantaneous A-weighted sound pressure of the sound signal integrated over a time period T starting at t_1 and ending at t_2. *See* Frequency weighting.

Sound exposure level (SEL) Measure of the sound exposure in decibels, defined as $L_E = 10\log_{10}(E/E_0)$ dB, where E is sound exposure, and the reference value $E_0 = 400\,\mu\text{Pa}^2\,\text{s}$. Unit: decibel (dB). Symbol: L_E. *See* Sound exposure, Single-event sound pressure level.

Sound intensity Time-averaged value of the instantaneous sound intensity $\mathbf{I}(t)$ in a temporally stationary sound field:

$$\mathbf{I} = \frac{1}{t_1 - t_2} \int_{t_1}^{t_2} \mathbf{I}(t)\ dt$$

where t_1 and t_2 are the starting and ending times for the integral. Units: watt per square metre (W/m^2). Symbol: $\mathbf{I}$. *See* Instantaneous sound intensity. *Note:* Sound intensity is generally complex. The symbol J is often used for complex sound intensity. The symbol I is used for active sound intensity, which is the real part of the complex sound intensity.

Sound intensity level A measure of the sound intensity in decibels, defined as $L_I = 10\log_{10}(I/I_0)$, where I is the active sound intensity, and the reference value $I_0 = 10^{-12}\,\text{W/m}^2 = 1\,\text{pW/m}^2$. Unit: decibel (dB).

Sound level *See* Sound pressure level.

Sound level meter Electronic instrument for measuring the sound pressure level of sound in accordance with an accepted national or international standard. *See* Sound pressure level.

Sound power Power emitted, transferred, or received as sound. Unit: watt (W). *See* Sound intensity.

Sound power level Ten times the logarithm to the base 10 of the ratio of a given sound power to the reference sound power, given by $L_W = 10\log_{10}(P/P_0)$, where P is the rms value of sound power in watts, and the reference sound power P_0 is 1 pW ($=10^{-12}$ W). Unit: decibel (dB). Symbol: L_W. The weighting network or the width of the frequency band used should be indicated. If the sound power level is A-weighted, then the symbol is L_{WA}. *See* Frequency weighting.

Sound pressure Dynamic variation in atmospheric pressure. Difference between the instantaneous pressure and the static pressure at a point. Unit: pascal (Pa). Symbol: p.

    A-weighted sound pressure The root-mean-square sound pressure determined by use of frequency weighting network A (see IEC 61672–1). Symbol: p_A.

Sound pressure level (SPL) Ten times the logarithm to the base 10 of the ratio of the time-mean-square sound pressure to the square of the reference sound pressure, given by $L_p = 10\log_{10}(p^2/p_0^2)$, where p is the rms value (unless otherwise stated) of sound pressure in pascals, and the reference sound pressure p_0 is 20 μPa ($=20 \times 10^{-6}\,\text{N/m}^2$) for measurements in air. Unit: decibel (dB). Symbol: L_p.

If p denotes a band-limited, frequency or time-weighted rms value, the frequency band used or the weighting shall be indicated. Frequency and time weightings are specified in IEC 61672–1.

A-weighted sound pressure level

Sound pressure level of A-weighted sound pressure, given by $L_{pA} = 10 \log_{10}(p_A^2/p_0^2)$, where p_A is the A-weighted sound pressure, and p_0 is the reference sound pressure. Symbol: L_{pA}. *See* Sound pressure.

Band pressure level

The sound pressure level in a particular frequency band.

Sound reduction index

Of a partition, in a specified frequency band, 10 times the logarithm to the base 10 of the reciprocal of the sound transmission coefficient, given by $R = 10\log_{10}(1/\tau) = 10\log_{10}(W_1/W_2)$, where τ is the sound transmission coefficient, W_1 is the sound power incident on the partition under test, and W_2 is the sound power transmitted through the specimen. In practice, the sound reduction index is evaluated from

$$R = L_1 - L_2 + 10 \log_{10} \frac{S}{A} = D + 10 \log_{10} \frac{S}{A}$$

where L_1 and L_2 are the average sound pressure levels in the source and receiving rooms, S is the area of the test specimen, A is the equivalent sound absorption area in the receiving room, and D is the level difference. Also known as *sound transmission loss*. Unit: decibel (dB). Symbol: R, or TL. *See* Sound transmission coefficient, Level difference, Equivalent sound absorption area, Coincidence effect.

Apparent sound reduction index

Ten times the logarithm to the base 10 of the ratio of the sound power W_1, which is incident on the partition under test to the total sound power transmitted into the receiving room if, in addition to the sound power W_2 transmitted through the specimen, the sound power W_3 transmitted by flanking elements or by other components, is significant:

$$R' = 10 \log_{10} \left(\frac{W_1}{W_2 + W_3} \right)$$

Unit: decibel (dB). Symbol: R'. *See* Flanking transmission.

Sound source

Anything that emits acoustic energy into the adjacent medium.

Sound transmission class (STC)

A single-number rating for describing sound transmission loss of a wall or partition. Unit: decibel (dB). The standardized method of determining sound transmission class is provided in ASTM E413–87. *See* Sound reduction index.

Sound transmission loss

See Sound reduction index.

Sound volume velocity

Surface integral of the normal component of the sound particle velocity over an area through which the sound propagates. Also known as *sound volume flow rate*. Units: cubic metre per second (m³/s). Symbol: q or q_v.

Sound energy flux   Time rate of flow of sound energy through a specified area. Unit: watt (W).

Specific acoustic impedance   Complex ratio of sound pressure to particle velocity at a point in an acoustical medium. Units: pascal per metre per second [Pa/(m/s)], or rayls (1 rayl = $1 \, N \cdot s/m^3$). *See* Characteristic impedance, Acoustic impedance, Mechanical impedance.

Specific airflow resistance   A quantity defined by $R_s = RA$, where R is the airflow resistance, in pascal seconds per cubic metre, of the test specimen, and A is the cross-sectional area, in square metres, of the test specimen perpendicular to the direction of flow. Units: pascal second per metre (Pa·s/m). Symbol: R_s. *See* Airflow resistance.

Spectral leakage error   In digital spectral analysis, an error that the signal energy concentrated at a particular frequency spreads to other frequencies. This phenomenon results from truncating the signal in the time domain. The leakage error can be minimized by applying a proper window to the signal in the time domain. *See* Window.

Spectrum   Description of a signal resolved into frequency components, in terms of magnitude, and sometimes as well as phase, such as power spectrum, one-third octave spectrum. *See* Fourier transform, Power spectrum, Power spectrum density, Cross spectrum.

Speech quality   Degree to which speech sounds normal, without regard to its intelligibility. Measurement is subjective and involves asking listeners about different aspects of speech such as naturalness, amount and type of distortion, amount and type of background noise.

Standard threshold shift   Increase in the average hearing threshold of 10 dB or more at 2000, 3000, and 4000 Hz in either ear. *See* Hearing loss, Hearing threshold.

Standardized level difference   *See* Level difference.

Standing wave   Periodic wave motion having a fixed amplitude distribution in space, as the result of superposition of progressive waves of the same frequency and kind. Characterized by the existence of maxima and minima amplitudes that are fixed in space.

Static pressure   Pressure that would exist in the absence of sound waves.

Statistical pass-by (SPB) method   A measurement method used for measuring noise properties of road surfaces (pavements), utilizing a roadside microphone (7.5 m from the center of the road lane being measured) and speed measurement equipment. Vehicles passing by in the traffic are measured and classified according to standard types, provided no other vehicles influence the measurement. The measured values are treated statistically, by vehicle type; being plotted as noise level versus speed. Either the regression curve is determined or the noise level is read at one or a few reference speeds. The method is standardized as ISO 11819-1. *See* Tire/road noise, Close-proximity method.

Stiffness

The ratio of the change in force to the corresponding change in displacement of an elastic element, both in specified direction.

Structure-borne sound

Sound that propagates through a solid structure. *See* Airborne sound, Liquid-borne sound.

Subharmonic

A frequency component whose frequency is an integer fraction of the fundamental frequency of a periodic quantity to which it is related. *See* Harmonic.

Susceptance

The imaginary part of an admittance. *See* Admittance, Conductance.

Swept sine

A test signal consisting of a sine wave whose frequency is changing according to a certain pattern, usually a linear or logarithmic progression of frequency as a function of time, or an exponential sweep.

Thermoacoustical excitation

Excitation of sound wave by periodic heat release fluctuations of a reacting flow (flame). A necessary condition for thermoacoustical excitation is given by the Rayleigh criterion. *See* Rayleigh criterion.

Time-averaged sound pressure level

An alternative term for *equivalent continuous sound pressure level*. *See* Equivalent continuous sound pressure level.

Time-weighted average (TWA)

The averaging of different exposure levels during an exposure period. For noise, given an A-weighted 85-dB sound exposure level limit and a 3-dB exchange rate, the TWA is calculated using: $\text{TWA} = 10\log_{10}(D/100) + 85$, where D is the noise dose. *See* Noise dose.

Tire/road noise

Unwanted sound generated by the interaction between a rolling tire and the surface on which it is rolling. Also known as *tire/pavement noise*. *See* Close-proximity method, Statistical pass-by method.

Tonal noise

Noise dominated by one or several distinguishable frequency components (tones).

Transducer

A device designed to convert an input signal of a given kind into an output signal of another kind, usually electrical.

Transfer function

Of a linear time-invariant system, the ratio of the Fourier or Laplace transform of an output signal to the same transform of the input signal. *See* Frequency response function.

Transmissibility

The ratio of the response amplitude of a system in steady-state forced vibration to the excitation amplitude. The input and output are required to be of the same type, for example, force, displacement, velocity, or acceleration.

Transmission factor, transmission coefficient, transmittance

The ratio of the transmitted sound power to the incident sound power. Unit: none. Symbol: τ. *See* Absorption factor, Dissipation factor, Reflection factor.

Transmission loss

(1) Reduction in magnitude of some characteristic of a signal between two stated points in a transmission system, such as a silencer. The characteristic is often some kind of level, such as power level or voltage level. Transmission loss is usually in units of decibels. It is imperative that the characteristic concerned (such as sound pressure level) be clearly identified because in all transmission systems more than one characteristic is propagated. (2) An equivalent term for *sound transmission loss*. *See* Sound reduction index. (3) In underwater acoustics, between specified source and receiver locations, the amount by which the sound pressure level at the receiver lies below the source level. Also known as *propagation loss*.

Turbulence

A fluid mechanical phenomenon that causes fluctuation in the local sound speed relevant to sound generation in turbo machines (pumps, compressors, fans, and turbines), pumping and air-conditioning systems, or propagation from jets and through the atmosphere.

Ultrasound

Sound at frequencies above the audible range, i.e., above about 20 kHz.

Velocity

A vector quantity that specifies time rate of change of displacement. Units: metre per second (m/s). *See* Displacement, Acceleration, Jerk, Particle velocity, Vibratory velocity.

Velocity excitation

See Excitation.

Vibration

(1) Oscillation of a parameter that defines the motion of a mechanical system. Vibration may be broadly classified as transient or steady state, with further subdivision into either deterministic or random vibration. (2) The science and technology of vibration. *See* Forced vibration, Free vibration.

Vibration absorber

A passive subsystem attached to a vibrating machine or structure in order to reduce its vibration amplitude over a specified frequency range. At frequencies close to its own resonance, the vibration absorber works by applying a large local mechanical impedance to the main structure. Also known as *vibration neutralizer, tuned damper*.

Vibration isolator

A resilient support that reduces vibration transmissibility. *See* Isolation, Transmissibility.

Vibration meter

An instrument for measuring oscillatory displacement, velocity, or acceleration.

Vibration severity

A criterion for predicting the hazard related to specific machine vibration levels.

Vibratory velocity level, vibration velocity level

Velocity level given by the following formula $L_v = 10 \log_{10}\left(v^2/v_0^2\right)$, where v is the rms value of the vibratory velocity within the frequency band of interest, v_0 is the reference velocity and is equal to 5×10^{-8} m/s (as specified in ISO 7849) or 10^{-9} m/s (as specified in ISO 1683). Unit: decibel (dB). Symbol: L_v. *See* Vibratory velocity.

Vibratory velocity, vibration velocity

Component of the velocity of the vibrating surface in the direction normal to the surface. The root-mean-square value of the vibratory velocity is designated by the symbol v. *See* Vibratory velocity level.

Viscosity

Of a fluid, in a wide range of fluids the viscous stress is linearly related to the rate of strain; such fluids are called *newtonian*. The constant of proportionality relating fluid stress and rate of strain is called the viscosity. Units: pascal seconds (Pa·s).

Vocal folds, vocal cords

Paired muscular folds of tissue layers inside the larynx that can vibrate to produce sound.

Vocal tract

Air passage from the vocal folds in the larynx to the lips and nostrils. It can be sub-divided into the pharynx, from larynx to velum, the oral tract, from velum to lips, and the nasal tract, from above the velum through the nasal passages to the nostrils. Its shape is the main factor affecting the acoustical characteristics of speech sounds. *See* Vocal folds.

Voicing, voiced, voiceless, unvoiced, devoiced

Voicing is one of the three qualities by which speech sounds are classified; a sound with voicing is called voiced, which means that the vocal folds are vibrating and produce a quasi-periodic excitation of the vocal tract resonances. A phoneme that is normally voiced but is produced without voicing, or in which the voicing ceases, is devoiced. A phoneme that is intended not to be voiced is voiceless or unvoiced. *See* Vocal folds, Phoneme.

Voltage preamplifier

A preamplifier that produces an output voltage proportional to the input voltage from a piezoelectric transducer. Input voltage depends upon cable capacitance.

Volume velocity

(1) *See* Sound volume velocity. (2) For speech, a measure of flow rate in the absence of sound, as through a duct, including through the vocal tract. Units: cubic metre per second (m³/s).

Wavefront

(1) For a progressive wave in space, the continuous surface that is a locus of points having the same phase at a given instant. (2) For a surface wave, the continuous line that is a locus of points having the same phase at a given instant.

Wavelength

Distance in the direction of propagation of a sinusoidal wave between two successive points where at a given instant of time the phase differs by 2π. Equals the ratio of the phase speed of sound in the medium to the fundamental frequency.

Wavenumber

At a specified frequency, 2π divided by wavelength, or angular frequency divided by the phase speed of sound: $k = 2\pi/\lambda = \omega/c$, where λ is wavelength, in metres; ω is angular frequency, in radians per second; c is the phase speed of sound, in metres per second. Unit: reciprocal metre (1/m). Symbol: k. *Notes:* (1) The ISO standards prefer to use the term *angular repetency* and *repetency*. A remark in ISO 8000 says that in English the names repetency and angular repetency should be used instead of wavenumber and angular wavenumber, respectively, since these quantities are not numbers. (2) Angular repetency is defined as the same as wavenumber. (3) Repetency: at a specified frequency, the reciprocal of wavelength: $\sigma = 1/\lambda$, where λ is wavelength. Unit: reciprocal metre (1/m). Symbol: σ.

Weighting

(1) *See* Frequency weighting. (2) *See* Window. (3) Exponential or linear time weighting defined in IEC 61672–1.

Weighting network

Electronic filter in a sound level meter that approximates under defined conditions the frequency response of the human ear. The A-weighting network is most commonly used. *See* Frequency weighting.

White noise

A noise the power spectrum of which is essentially independent of frequency. *See* Pink noise.

Whole-body vibration

Vibration of the human body as a result of standing on a vibrating floor or sitting on a vibrating seat. Often encountered near heavy machinery and on construction equipment, trucks, and buses.

Window

In signal processing, a weighting function with finite length applied to a signal. Usually applied in the time domain, as a multiplying function applied to the time signal. For spectral analysis, Hanning, Hamming, triangle, Blackman, flat top, Kaiser windows are commonly used. Force and exponential windows are special for impact testing.

Index

Note: A g following a page number indicates a citation to an entry in the Glossary on that page.

Engineering Acoustics: Noise and Vibration Control, First Edition. Malcolm J. Crocker and Jorge P. Arenas.
© 2021 John Wiley & Sons Ltd. Published 2021 by John Wiley & Sons Ltd.